T0344859

Transdisciplinary Engineering Design Process

# Transdisciplinary Engineering Design Process

*Atila Ertas*

*Registered Office*
John Wiley & Sons, Inc., 111 River Street, Hoboken, NJ 07030, USA

*Editorial Office*
The Atrium, Southern Gate, Chichester, West Sussex, PO19 8SQ, UK

For details of our global editorial offices, customer services, and more information about Wiley products visit us at www.wiley.com.

Wiley also publishes its books in a variety of electronic formats and by print-on-demand. Some content that appears in standard print versions of this book may not be available in other formats.

*Library of Congress Cataloging-in-Publication Data*
Names: Ertas, Atila, 1944- author.
Title: Transdisciplinary engineering design process/by Atila Ertas.
Description: First edition. | Hoboken, NJ : John Wiley & Sons, 2018. | Includes index. |
Identifiers: LCCN 2018002610 (print) | LCCN 2018012070 (ebook) | ISBN 9781119474777 (pdf) | ISBN 9781119474661 (epub) | ISBN 9781119474753 (cloth)
Subjects: LCSH: Engineering design. | Multidisciplinary design optimization.
Classification: LCC TA174 (ebook) | LCC TA174 .E7824 2018 (print) | DDC 620/.0042—dc23
LC record available at https://lccn.loc.gov/2018002610

Cover image: © DDCoral/Shutterstock
Cover design: Wiley

Set in 10.5/13pt WarnockPro by SPi Global, Chennai, India

C10004035_092118

*This book is dedicated to Professors Jesse C. Jones and Timothy T. Maxwell for their endless support.*

# Table of Contents

# About the Author

Dr. Atila Ertas, Professor of Mechanical Engineering at Texas Tech University, received his masters and PhD from Texas A&M University. He had 12 years of industrial experience prior to pursuing graduate studies. He has been the driving force behind the conception and the development of the transdisciplinary model for education and research. His pioneering efforts in transdisciplinary research and education have been recognized internationally by several awards from the Society for Design and Process Science (SDPS). He is the director of the transdisciplinary PhD program at Texas Tech University. He is the creater and the founding president of the non-profit organization The Academy for Transdisciplinary Learning and Advanced Studies (TheATLAS). He is a Senior Research Fellow of the ICC Institute at the University of Texas Austin, a Fellow of ASME, a Fellow of SDPS, and a Founding Fellow of the Luminary Research Institute in Taiwan. He is also an honorary member of the International Center for Transdisciplinary Research (CIRET), France. He has earned a national and international reputation in engineering design. He is the author of a number of books and the editor/coeditor of more than 35 conference proceedings. His contributions to teaching and research have been recognized by numerous honors and awards. These include: President's Excellence in Teaching; Pi Tau Sigma Best Professor Award; Pi Tau Sigma Outstanding Teaching Award; Halliburton Award in recognition of outstanding achievement and professionalism in education and research; College of Engineering Outstanding Researcher Award; George T. and Gladys Hanger Abell Faculty Award for overall excellence in teaching and research; and President's Academic Achievement Award. He has published over 150 scientific papers that cover many engineering technical fields. He has been principal investigator or co-princial investigator on over 40 funded research projects. Under his supervision more than 190 MS and PhD graduate students have received degrees.

# Preface

During the last decade, the number of complex problems facing engineers has exploded, and the technical knowledge and understanding in science and engineering required to attack these problems is rapidly evolving. The world is becoming increasingly interconnected as new opportunities and highly complex problems connect the world in ways we are only beginning to understand. When we don't solve these problems correctly and in a timely manner, they rapidly become crises. Problems, such as energy shortages, pollution, transportation, the environment, natural disasters, health, hunger and the global water crisis, threaten the very existence of the world as we know it today. Recently, fluctuating fuel prices and environmental concerns have sent car manufacturers in search of new, zero polluting, fuel efficient engines. None of these complex problems can be understood from the sole perspective of a traditional discipline. The last two decades of designing large–scale engineering systems have taught us that neither mono–disciplinary nor inter– or multi–disciplinary approaches provide an environment that promotes the collaboration and synthesis necessary to extend beyond existing disciplinary boundaries and produce truly creative and innovative solutions to large–scale, complex problems.

Large–scale, complex problems include not only the design of engineering systems with numerous components and subsystems which interact in multiple and intricate ways; they also involve the design, redesign and interaction of social, political, managerial, commercial, biological, medical, etc., systems. Further, these systems are likely to be dynamic and adaptive in nature. Solutions to such large–scale, complex problems require many activities which cross discipline boundaries.

One of the widely agreed to characteristics of transdisciplinary research is that it is performed with the intent of solving problems that are complex and multidimensional, particularly those related to sustainability in a human environment. Transdisciplinary research tends to focus on collaborations that transcend specific disciplines to define new knowledge.

The anticipated results of transdisciplinary research and education are: emphasis on teamwork, the bringing together of investigators from differing disciplines, and sharing of methodologies to generate fresh, invigorating ideas that expand the boundaries of problem solutions. The Transdisciplinary approach develops people with the desire to seek collaboration outside the bounds of their professional experience in order to explore different ideas. A truly transdisciplinary research and educational system is needed to address large–scale, complex problems and to educate the researchers and designers of the future.

Transdisciplinary education involves students from many areas of knowledge crossing disciplinary boundaries such as economics, modeling and simulation, optimization, reliability, statistical decisions, ethics, and project management, all of which are included in this book. Hence, students can understand issues from a broad point-of-view in order to synthesize potential solutions.

Over the past decade, awareness and understanding of the complexity of the environmental impact on human activity is growing. Issues of environmental change are of increasing concern for both developed and developing nations of the world. Design and processes are central to the concept of transdisciplinary education. Social, political and cultural aspects of problems and issues must be recognized if workable and economically feasible solutions are to be developed. Students will emerge from this transdisciplinary education program with a broad perspective of the world and its problems, including a wide range of tools that will equip them to address such problems and apply them to socially relevant issues. It is a program that will teach students to integrate and manage knowledge in technical, social and scientific areas that require the collaboration of engineers, planners, physicists, biologists, psychologists, sociologists, economists, and other specialists. Transdisciplinary methodologies and tools covered in this book can be applied in a wide variety of disciplines including economics, business, management, operations research, engineering, chemistry, genetics, and the social and behavioral sciences.

It is a pleasure to make grateful acknowledgment of the many valuable suggestions which have been contributed by Professor Jesse C. Jones. Two chapters, (Chapters 10 and 11), and a majority of the problems which are included were used without change from "The Engineering Design Process," co-authored by Ertas and Jones. The first five chapters of this book are about the Transdisciplinary education, the remaining chapters are devoted to fundamental engineering knowledge adapted from the earlier book, with a significant amount of new material, example problems and case studies.

In conclusion, the author takes this opportunity to express his thanks to Ms. Lauren Newmyer, Mr. Utku Gulbulak, Dr. Adam Stroud, Dr. Turgut B. Baturalp and Dr. Bugra H. Ertas for their help in the preparation of this book.

**Atila Ertas**

# 1

# Systemic Thinking and Complex Problem Solving

We live in a highly complex, technological world – and it's not entirely obvious what's right and what is wrong in any given situation, unless you can parse the situation, deconstruct it. People just don't have the insight to be able to do that very effectively.

**Christopher Langan**

## 1.1  Introduction

The world's population continues to increase rapidly, which causes technology to develop at a geometric pace. Modern communication systems offer each of us overwhelming mountains of information, much of which is disorganized, not relevant, redundant, or inaccurate, and thus may well provide more confusion than clarity. We are faced with the necessity to wrestle with and solve many large-scale problems if we are to maintain sources of clean water, clean air, food, energy, adequate medical services, political stability, and a civilized social structure. Improving the condition of our world will prove even more difficult.

The area of study known as complexity is a very popular area of research. Complexity arises from the nature of large interconnected systems and is escalated by the background, personal characteristics, and perspectives of the individuals working on the design teams. It is important for designers to understand complexity and how it affects the understanding and projection of system behavior. It is also important to manage complexity so that it does not overwhelm the design effort and prevent the development of effective solutions. This chapter presents an overview of complexity, discusses how complexity can increase almost without bound(s), and suggests ways to control the impact of complexity on design process.

## 1.2  What Is Complexity?

During the last two decades of designing large-scale engineering systems it has been demonstrated that mono-, inter-, and multi-disciplinary approaches do not provide an environment that promotes the collaboration and synthesis necessary for extending disciplinary boundaries and producing innovative solutions to large-scale, complex problems. Such problems include the design of engineering systems with numerous components and subsystems which interact in multiple and intricate ways with social, political, managerial, commercial, biological, and medical systems. Furthermore, these systems are likely to be dynamic and adaptive in nature. Solutions to such complex problems require activities that cut across traditional disciplinary boundaries; this is what we call transdisciplinary research and education.[1]

Complexity is difficult to understand due to the variety of competing proposed solutions and explanations for what constitutes complexity. Many researchers have proposed that complexity can described by size, entropy, information content, thermodynamic and information required to construct, computational capacity, statistical complexity, as well as others.[2]

Size was proposed as a level of complexity based on the presumption that larger things are inherently more complex. Information content refers to the length of computer program required to define a message or pattern.[2] Many of the proposed definitions of complexity are based on identifying a quantifiable parameter for a system or problem; however, proposals to date have not provided an agreed-upon definition.[2]

A distinction must be drawn between complex and complicated systems because this association is a source of confusion. Complicated systems and complex systems may both have multiple individual interactive components; however, in complicated systems the behavior is well understood, while complex systems lack this clear understanding.[3]

An additional problem in defining complexity comes about due to the association of randomness as an indicator of complexity. This is likely due to the synonymous use of complexity with unpredictability, which is a characteristic of random systems.[4]

Although it is possible for complex systems to produce random outputs, this perception of randomness is often relative to individual observers and their knowledge base. The presumption of order or randomness cannot be definitely or certainly demonstrated; therefore, this is not an effective measure of complexity.[4]

To add further difficulty in providing a physical definition of complexity, Pierce argued, in one of the earlier texts addressing complexity, against it being a quantifiable parameter. He states: "Complexity is that sensation experienced in the human mind when, in observing or considering a system, frustration arises from lack of comprehension of what is being explored."[5] In this paradigm, complexity is dependent on the individual examiner of a system, not the system itself. Warfield presents seven necessary conditions for a situation to be complex:[6]

1) a human presence;

---

1 Ertas, A., "Understanding of transdiscipline and transdisciplinary process," *Transdisciplinary Journal of Engineering & Science*, 1, 48–64, 2010.
2 Mitchell, M., *Complexity: A Guided Tour*, Oxford University Press, Oxford, 2011.
3 Suh, N.P., "A theory of complexity, periodicity and the design axioms," *Research in Engineering Design*, 11, 116–131, 1999.
4 Biggiero, L., "Sources of complexity in human systems," *Nonlinear Dynamics, Psychology, and Life Sciences*, 5(1), 3–19, 2001.
5 Peirce, C.S., "How to make our ideas clear," *Popular Science*, pp. 286–302, 1878.
6 Warfield, J.N., "Structural thinking: Organizing complexity through disciplined activity," *Systems Research*, 13, 47–67, 1996.

2)  a generic purpose associated with the human presence;
3)  an inquiry into the system by the human presence;
4)  a human purpose related infrastructure to allow inquiry;
5)  a system related environment;
6)  a sensing mechanism to measure inquiry; and
7)  cognition of the human presence.

In Warfield's analysis, the human observer is involved with every requirement of complexity. Complexity has been described as a degree of ignorance. Objects are more or less complex depending on our ignorance or lack of information about it, our ability to make distinctions and perceptions about it, and our ability to infer information from it. Encrypted messages highlight the concept of observer dependent complexity. Encrypted messages are commonly broadcast during wartime and are received by all parties within the broadcast range. Observers with the correct cipher, or knowledge related to the information, can interpret the message. Those without knowledge about the data may see the same information as without meaning and, in fact, random. This is underlined by the ideas presented by Gell-Mann, proposing that systems are hard to predict not because they are random, but because their regularities cannot easily be described, or are unknown.[7]

The perception driven definition of complexity is considered incomplete by many. Axelrod and Cohen argue that the source of complexity is fundamental to the system and cannot be eliminated.[8] They go on to describe the structure of complex systems as being composed of artifacts and agents. The agents are the interacting entities with some level of functioning behavior. Agents have memory and capability, and can formulate strategies and interact with agents. The artifacts are unanimated objects, manipulated by agents and properties of the agents, such as location and capabilities. In this description, the agent's use of selective intervention is a source of complexity as it is very difficult to develop predictions for the system.[8] This concept of complexity originating from a system was expanded with the identification of three sources of system complexity: environmental influence, initial conditions of the system, and the system structure itself.[9] Stability was also studied by Simon by examining the chaotic nature of complex systems from minor changes in environment or initial condition.[10]

The two distinct aspects of complexity were characterized by Warfield as cognitive complexity and situational complexity. Situational complexity is complexity inherent to the system being examined. Cognitive complexity refers to the complexity associated with interpretation by the observer.[11] This duality will be examined in greater detail later in this chapter.

Time has also been introduced as having an impact on complexity. The identification of unstable responses of the system related to influencing parameters indicates that the system is not stable in time.[10] Works by Suh identify time, and specifically time periods, as a key parameter

---

7  Gell-Mann, M., "What is complexity?," *Complexity*, 1(1), 16–19, 1995.
8  Axelrod, R., and Cohen, M., *Harnessing Complexity*, Free Press, New York, 1999.
9  Wolfram, S., *A New Kind of Science*, Wolfram Media, Champaign, IL, 2002.
10  Simon, H.A., *The Sciences of the Artificial*, 3rd edn, MIT Press, Cambridge, MA, 1999.
11  Warfield, J.N., *A Science of Generic Design: Managing Complexity through Systems Design*, Iowa State University Press, Ames, 1994.

for prediction.[12] It is argued that the lack of ability to predict system behavior is due to an excessively long period of prediction.[12] This indicates that a less daunting task is to predict the state of a system incrementally, in near time, rather than at the end state, such as many years from now. It is argued that the introduction of incremental or periodic analysis will reduce complexity and improve the ability to predict system behavior.[12]

Suh provides one of the most quantifiable definitions of complexity by focusing on functional requirements (FRs) of a system rather than the physical components.[12] In this way, the relationship of system output relative to a desired output range defines the system's complexity.[12] Figure 1.1 describes the concept of complexity represented by a system output in the form of a distribution as it falls compared to a desired range (design range). The overlap range between the design range and system range shown in Figure 1.1 is the conformance or predictability of success. If there is no overlap area between two ranges, then there is a finite probability that the system will fail in time. The use of a small range relative to the distribution will constitute higher complexity in the image.

The predictability in achieving a desired outcome with a specified system becomes the measure of complexity. This effectively defines complexity, but does so with a relative quantity, instead of a physical measure.[12]

This is illustrated with the example of cutting a rod to a length. The length accuracy (design range) may be ±10 micrometers. If the tool can always cut the rod with an accuracy of ±10 micrometers, then the FR is satisfied and we assume that the complexity is zero. However, to perform this with a chop saw would be complicated due to its inaccuracy. Thus, cutting the rod within the design range (±10 micrometers) may not be possible. In this case, we assume that the task is complex. The same task with a laser capable of extremely accurate cutting is much less complex. The relative overlap of system capability relative to desired outcome gives a quantifiable measure of complexity provided that the interrelations between entities are known and fixed.[12] This is ineffective in circumstances where the response of agents varies significantly and rapidly, as is common in human systems.

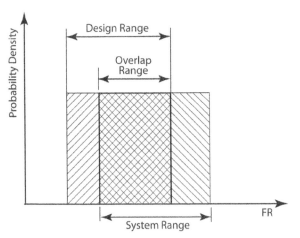

**Figure 1.1** Uniform probability density function of an FR.

12 Suh, N.P., *Complexity: Theory and Application*, Oxford University Press, New York, 2005.

It is clear that the definition of complexity itself is not simply defined. To carry the discussion forward we will focus on the common denominators of complex systems instead of focusing on formal definitions. Examining complexity in the context of known complex systems proves very insightful, and provides a basis for a useful working knowledge. Although there is much ambiguity in the formal definition of complexity as a concept, the attributes of a complex system are relatively well known. A complex system is loosely defined as "a system in which large networks of components with no central control nor any simple rules of operation give rise to complex collective behavior, sophisticated information processing, and adaptation via learning or evolution."[2] This behavior is unexpected considering the collective ability of the individual pieces. These systems should exhibit complex collective behavior, information processing, and apparent adaptation.[2] Simon specifies the four characteristics of complex systems as follows:[10]

1) Complex systems are frequently hierarchical.
2) The structure of complex systems emerges through evolutionary processes, and hieratic systems will evolve much more rapidly than non-hierarchic systems.
3) Hierarchically organized complex systems may be decomposed into sub-systems for analysis of their behavior.
4) Because of their hierarchical nature, complex systems can frequently be described, or represented, in terms of a relatively simple set of symbols.

We can conclude from these four statements that complex systems, at least ones that are likely to be of interest to us, are not random, but rather, have structure. This statement agrees with Axelrod and Cohen's comments mentioned earlier. Further, this structure is based on the interactions between the various parts (components and sub-systems) of the system. Simon also indicates that complex systems, because of their hierarchical nature, evolve and change rapidly with time. Axelrod and Cohen explain the adaptation of systems due to the intervention of agents (agents are individuals, organizations, political entities, computer software routines, etc.). Simon's third aspect of complex systems indicates a likelihood that they can be decomposed for study and analysis. This characteristic, which provides hope for dealing with complex systems, may not be easily realized. The interconnections and interactions of the parts of a complex system will not typically be obvious without significant study. Finally, Simon notes that complex systems can often be represented by seemingly simple arrangements of symbols which follow from the hierarchical structure of the system.

Warfield expresses his view of complex systems a bit differently:[11]

> For better or worse, our society has accepted the idea of large and complex systems. If we are going to have them, it behooves us to learn how to manage them....
>
> One of the primary motivations comes from recognizing that society today involves large sociotechnical systems whose performance is far from ideal. It is clear that many of these large systems have taken their present forms primarily through evolutionary change that did not involve any systematic overview design, but may have involved some systematic design of parts. Other systems are said to have been designed, but still fail in ways that produce disasters.

Warfield defines three classes of systems:

1) (class A) systems found in the physical world;
2) (class B) systems based on intellectual technology or artificial intelligence;
3) (class C) systems synergistically composed of class A and class B systems.

Class B and C systems often do not have readily available metrics to measure performance. Relative metrics must be generated to evaluate and conceptualize these systems.[11]

Maxwell, Ertas, and Tanik draw the parallels between class C systems and Warfield's definition of a socio-technical system.[13] These types of systems are increasingly commonplace and must be understood to be effective in modern society. This will require a transdisciplinary process to design and manage the complexity associated with these systems.[13] Complex systems are hierarchical, with many agents and subsystems that interact, often in non-trivial and varying ways that make them difficult to understand and predict. They can be completely decomposed within their hierarchic structure; high levels of redundancy make them relatively simple to represent symbolically. They adapt and evolve over time, and can contain physical and artificial entities within them. Finally, the understanding and management of complex systems is a transdisciplinary methodology. Herb Simon stated:[14]

> Today, complexity is a word that is much in fashion. We have learned very well that many of the systems that we are trying to deal with in our contemporary science and engineering are very complex indeed. They are so complex that it is not obvious that the powerful tricks and procedures that served us for four centuries or more in the development of modern science and engineering will enable us to understand and deal with them. We are learning that we need a science of complex systems, and we are beginning to construct it.
>
> **Herb Simon**

In *The Sciences of the Artificial*, Simon further states:[10]

> The proper study of mankind is the science of design, not only as the professional component of a technical education but as a core discipline for every liberally educated person
>
> **Herb Simon**

Our success in solving complex problems through design will depend largely on our ability to manage the complexity associated with these problems.

---

13 Maxwell, T.T., Ertas, A., and Tanik, M.M., "Harnessing complexity in design," *SDPS Transactions: Journal of Integrated Design and Process Science*, 6(3), 63–74, 2002.

14 Simon, A.H., Keynote Speech, 2000 Integrated Design and Process Technology (IDPT) Conference, Dallas.

## 1.3   Source of Complexity

Considerable work has been done to identify and classify the sources of complexity; the majority of work has been done in human systems. Human systems are affected by several sources of complexity which fall under three categories: logical, gnosiological, and computational.

Systems impacted by logical complexity are not predictable at all. Logical complex systems are observable and comprehendible by an observer, but behavioral relationships cannot be generated to represent them. This classification of complex systems has two subclasses: logical and relational. Logical complexity applies to systems that are irreducible, problems within the system isolated. Order cannot be recognized in system data, and each theory is unavoidably temporary. The relational complex human system occurs when the modification of a human in a system is effective due to the presence of other actors in the system. This interactive modification of behavior, also known as the "Hawthorn Effect," represents a shifting between individual and group behavior patterns. Individuals within groups become less objective in their observed behavior.[4] The cases of logical complexity indicate that the observer is unable to develop any relationships between components in a system and, to them, it truly appears random.

Gnosiological systems are only predictable through an infinite computational capacity. Four subcategories exist for this type of complex human system: pure gnosiological, evolutionary, semiotic, and semantic. The pure gnosiological complex system exists when there is an irreducible subjectivity in perceiving the environment or system. The viewer is unable to draw distinctions from objects in the system, and without distinction no information can be obtained. In an evolutionary complex system, the characteristic of human systems to face continuous change is the source of complexity. The ability of humans to learn drives continuous change in the reaction of individuals to certain conditions. This affects the ability to predict behavior greatly. The second type of gnosiological complex system is the semiotic complex system. This source of complexity arises due to an inability to derive meaning from system behavior due to ambiguity. The lack of self-evident facts in individual behavior prevents the building of computationally predictive models. The semantic complex human system arises when ambiguity arises due to differences in language and culture of individuals in a system. Ambiguity in expression, interpretation, and translation of information provides a source of unpredictability in a system.[4] This correlates with Warfield's description of design group dynamics, including knowledge, point of view, values, and objectives influencing the complexity of a system.[11]

Systems of the computational category are only predictable through transcomputational ability.[4] The third class of sources of complexity can be reduced into three subcategories: pure computational complexity, chaotic complexity, and self-organizational complexity. Pure computational complexity is a source arising from the inability to perform computer predictive models at a rate equal to event occurrence. Suh would argue that this is actually an instance of complicatedness instead of complexity.[3] Chaotic complexity is a source arising from an inability to perfectly model behavior of objects in a system, particularly with large quantities of individual members. Self-organizational complexity is the source of complexity resulting from the unexpected and unexplainable emergence of behavior from an otherwise chaotic system. This category aims to classify the inability to model system behavior.

Of these three categories of complexity, the first two categories essentially describe human limitations: the inability to recognize a pattern or relation, and the ambiguity of interpretation. The third category describes the difficulty in describing a physical system, or situational complexity as described by Warfield.[11]

## 1.4 Two Aspects of Complexity

It can be seen from the proposed definitions of complexity that a divergence is occurring. While some define complexity with physical metrics, others suggest it is dependent on the observer.[2,4,5] It has been proposed that there exist two distinct classifications of complexity: *situational* and *cognitive* complexity, also referred to as *real* and *imaginary* complexity.[11,12]

Warfield defines situational complexity as the complexity associated with a system being analyzed, and does not account for the observer's ability to perceive the systems' behavior.[11] Suh, who takes a relative perspective on complexity, reinforces this by defining real complexity as a measure of uncertainty associated with achieving a task.[12] Many of the other proposed physical descriptions of complexity, including information content, free energy, statistical complexity, computational requirements, level or hierarchies, and size, would align as instances of situational complexity as they are measures independent of the observer.[2]

Cognitive complexity, as Warfield describes it, arises from the aspects in the system that makes interpretation difficult.[11] Imaginary complexity, as Suh calls it, is not a real complexity but is cognitive and arises from an observer's lack of familiarity or ability to understand the system. Imaginary complexity can also exist even in the absence of real complexity.[12] This complexity aligns with early descriptions of complexity by Pierce, in which complexity is not quantifiable but, instead, persists as a sensation of frustration from the inability to interpret a system.[5] Biggiero illustrated cognitive complexity with his observations that objects are more or less complex depending on the observer's ignorance of them.[4]

**Example 1.1**

Discuss the complex system of the stock market.

**Solution**

The stock market consists of traders behaving in their best interest, which produces an emergent systemic behavior.[2] The collection of interacting components represents the situational complexity in the system. Now imagine an observer standing on the floor of the New York Stock Exchange evaluating the performance of the market. The observer's ability to assess the market performance comprises the cognitive complexity. Now let us take away the observer's (and only the observer's) access to the various indexes and displays of stock prices. The observer has lost much of his ability to perceive the stock market. He must rely on observing the behavior of the other brokers and listening to interactions occurring in their immediate vicinity. To the observer, the system has become more complex because of a diminished ability to perceive and evaluate the system. This is an example of an increase in cognitive complexity without any change in the situational complexity of the system.

Both situational and cognitive complexity must be accounted for when defining complexity. Every system with any number of components will have some level of situational complexity, however small it may be. There also must exist some level of cognitive complexity, due to the finite cognitive abilities of humans.[10] Human cognitive abilities must always be considered as the human is involved with all aspects of complexity, according to Warfield.[6]

## 1.5  Complexity and Societal Problems

Societal problems are real-life problems that are highly complex because of their dynamic character and impact on society.[15] They are highly transdisciplinary, with social, cultural, economical, political, and emotional issues intertwined with technology. They cannot be easily solved by experiment, and the implementation of a correction to a problem changes the social system in a complex way.[16] Analyzing a complex societal problem requires knowledge from diverse domains: bringing together non-academic experts and academic researchers from different disciplines to share concepts, methodologies, processes, and tools. Cronin stated:[17]

> There is a need for transdisciplinary research (TR) when knowledge about a societally relevant problem field is uncertain, when the concrete nature of problems is disputed, and when there is a great deal at stake for those concerned by problems and involved in dealing with them. TR deals with problem fields in such a way that it can: a) grasp the complexity of problems, b) take into account the diversity of life world and scientific perceptions of problems, c) link abstract and case specific knowledge and d) constitute knowledge and practices that promote what is conceived to be the common good.

### 1.5.1  Causes of Societal Problems

Many actors are connected with societal problems that have legal, environmental, political, technical, safety, transportation, and economic aspects. Even though the real reason for complex

---

15 DeTombe, D.J., "Compram, a method for handling complex societal problems," *European Journal of Operational Research*, 128(2), 266–281, 2001.
16 Warfield, N.J., *Societal Systems*, Intersystems Publications, Salinas, CA, USA, 1989.
17 Cronin, K., "Transdisciplinary research (TDR) and sustainability," Environmental Science and Research (ESR) Ltd., 2008.

societal problems is not always clear, many are caused by humans. In particular, some of the fundamental problems are political and economic, and rooted in human nature. Human involvement is apparent in complex societal problems such as antisocial behavior, sexually transmitted diseases, drug and alcohol abuse, wars, crime, and others. The main cause of social problems is unemployment, poverty, economic deprivation, urban problems, inflation, hunger, water crises, and diseases which are also social problems themselves. To solve such problems, important interactions between problem elements need to be considered. Failure to do so leads to unexpected and often undesirable solutions.

### 1.5.2   Process for Societal Problem Solving

As shown in Figure 1.2, societal problem-solving processes include problem orientation and problem-solving style.[18] Positive problem orientation is a constructive problem-solving process by which a team attempts to identify or discover effective solutions using a transdisciplinary approach to produce positive solution outcomes. Negative problem orientation is a

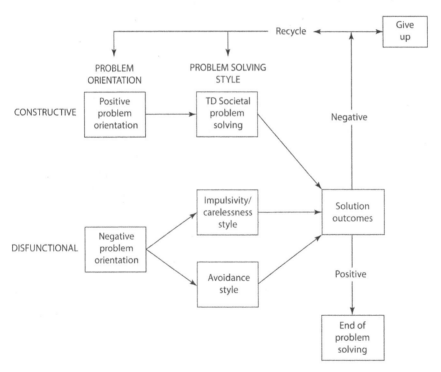

**Figure 1.2** Societal problem-solving process. (Adapted from D'Zurilla, T.J., Nezu, A.M., and Maydeu-Olivares, A., *Social Problem-Solving Inventory-Revised (SPSI-R): Technical Manual*, Multi-Health Systems, North Tonawanda, NY, 2002. Used with permission.)

---

18 D'Zurilla, T.J., Nezu, A.M., and Maydeu-Olivares, A., *Social Problem-Solving Inventory-Revised (SPSI-R): Technical Manual*, Multi-Health Systems, North Tonawanda, NY, 2002.

dysfunctional problem-solving process which contributes to an impulsivity/carelessness or avoidance style; both are likely to produce negative problem solution outcomes.[19]

When the problem solution results are negative or unacceptable, two cases are possible:

1) give up and stop working on the problem solution, or
2) recycle or return to the problem-solving process by changing the problem-solving goals for more realistic outcomes.

For example, consider a magnitude 8 earthquake – it can be tremendously destructive with collapsing buildings and loss of life, but the destruction is often compounded by mudslides, fires, floods, sicknesses or tsunamis. Assume that such a devastating earthquake happened in a poor country with terrible diplomatic relations with other countries in the same region. Perhaps the first attempted goal would be to solve compounding problems individually. Pride is the main reason they may not want to ask for help. The reality is that these problems cannot be solved in a short time and the initial outcome would likely be negative. The following two cases are possible:

1) What can a poor country do to solve the many issues that this tragic event causes? Government mismanagement and lack of resources may delay the cleanup process and the result will not be a solution for the societal problems. In other words, give up and end with disaster.
2) Finding that the problems are not solved in a timely manner could be disastrous; their goals would have to be changed and plans should be developed to ask for help from other countries to minimize further damage and undesirable consequences. The solution is to return to the problem-solving process, to find a better way forward with collaborative effort with the help of other countries.

### 1.5.2.1 Transdisciplinary Societal Problem Solving

As mentioned earlier, for the positive problem orientation process, a transdisciplinary (TD) approach can be used. A schematic representation of the TD process is shown in Figure 1.3. In this figure, the process starts by understanding the problem (issue or situation), identifying the TD team members from diverse disciplines, and selecting or developing an appropriate methodology to solve the problem in hand. Chain links shown in Figure 1.3 represent the required interactions among the three elements of triangle. The three connections (interrelationships) are usually more difficult to deal with than the three elements accomplished separately (problem, team, and methodology).

***Problem.*** As shown in Figure 1.3, the societal problem in hand can be divided conceptually into three groups: social problems, policy problems, and organizational problems. A social problem is an issue that influences a significant number of individuals within a society – social problems are closely related to the well-being of people such as obesity, poverty, domestic violence, homelessness, hunger, and healthcare problems.

---

19 D'Zurilla, T.J., Nezu, A.M., and Maydeu-Olivares, A., "Social problem solving: theory and assessment," in E.C. Chang, T.J. D'Zurilla, and L.J. Sanna (eds), *Social Problem Solving: Theory, Research, and Training*, American Psychological Association, Washington, DC.

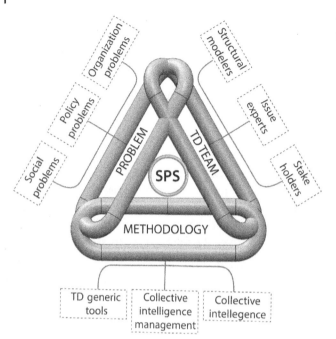

**Figure 1.3** Transdisciplinary societal problem-solving process.

Some of the technical policy problems can be very complex and highly resistant to solution; these are *wicked problems*. Examples of such problems are: housing and urban renewal policies, sustainable communities, and large infrastructure projects.

In the case of complex organizational problems, there are no simple and clear solutions. Issues facing corporate, government, and non-profit organizations are extremely complex. Some of the examples of complex organizational problems are: global marketing, starting new joint ventures, achieving goals such as strategic planning, working more efficiently, addressing conflicts, and promoting diversity.

*Transdisciplinary team.* As shown in Figure 1.3, the transdisciplinary team is divided into three groups:

1) Stakeholders – the people who have a stake in the problem being considered (have skin in the game).
2) Issue experts – the people who have specialized knowledge that is related to a problem under consideration.
3) Structural modelers – the people are concerned with describing "things" in a system and how these things are related to each other.

These three groups from diverse disciplines can be brought together in different combinations of people to collaborate together in solving complex unstructured problems. In some cases, issue experts may not be knowledgeable of the stakeholder's concern and interest and it may be necessary to involve stakeholders even in structuring the problem. The structural modeler establishes the reliability of the structural approach to the problem, hence arranging the basis for further development of content along the lines of the structure.

#### 1.5.2.2  Methodology

Figure 1.3 indicates that methodology includes transdisciplinary generic tools, collective intelligence management (CIM) and development of collective intelligence. Use of TD tools and creating collective intelligence are a necessary part of a methodology for solving societal problems. The process is also involved in dealing with academic and non-academic experts for conducting CIM in developing possible solutions for societal problems.

## 1.6  Understanding and Managing Complexity

The understanding and management of a complex intervention can be a lengthy process. First, it is essential to understand the complexity of a situation in order to manage it. Complexity will either be managed or it will overwhelm the people or society. The understanding of complexity and the principal paths to the management of complexity shown in Figure 1.4 will be discussed in this section.

### 1.6.1  Understanding Complexity

Although there are many definitions and measures of complexity in the literature, the concept has proven to be very difficult to understand. The following factors can be considered for definitions and measures associated with complexity:[20]

1) Numeric size of basic elements in a system. Although a larger size corresponds to a higher degree of complexity, numeric size in itself is not adequate to define complexity in its whole.
2) Variety of elements. Even though disorder alone cannot adequately define complexity, many researchers have used the degree of disorder or entropy in information theories as the measure for variety or complexity. Drożdż et al. stated that complexity lies somewhere between order and disorder.[21]
3) Relation between elements. Relations (interactions, etc.) among the components of a system contribute to complexity. Individual components of a system are held together through the relations of its internal structure. As an example, a chess pattern can be of great complexity to a player because the player counts on the relations between the elements, not just the number and the variety of the elements.[20]
4) Observer dependency. As shown in Figure 1.5 (similar figures were used by Grassberger in experiments),[22] the variety of the images changes as the disorder of image pixels and the number of pixels increases from left to right. However, the middle image was selected by the observers as the most complex one. This experiment shows that complexity depends on how observers process information.

---

20  Xing, J., "Measures of information complexity and the implications for automation design," Office of Aerospace Medicine, Washington DC, 2004.
21  Drożdż, S., Kwapień, J., Speth, J., and Wójcik, M., Identifying complexity by means of matrices," *Physica A*, 314, 355–361, 2002.
22  Grassberger, P., "Information and complexity measures in dynamical systems," in H. Atmanspacher and H. Scheingraber (eds), *Information Dynamics*, Plenum Press, New York, 1991, pp. 15–33.

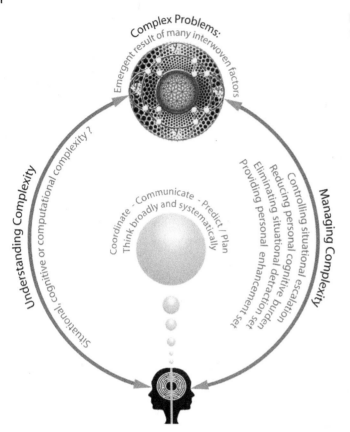

**Figure 1.4** Understanding and managing complexity.

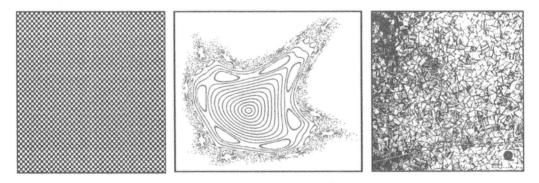

**Figure 1.5** Human perspective of complexity. (Adapted from Grassberger, P., "Information and complexity measures in dynamical systems," in H. Atmanspacher and H. Scheingraber (eds), *Information Dynamics*, Plenum Press, New York, 1991.)

5) Task dependency. The complexity also depends on the task requirements a person is involved with. For example, if the task is to count ants in a colony, then the complexity of the ants does not vary with the number of the ants and the variation in the shape or color of the ants. If the task is the investigation of the social life of an ant colony, then the complexity assessment will change with the behavior of the ants.

Thus, we may conclude from Simon's words and from our own deliberate consideration of ongoing efforts to wrestle with complex problems that complexity is a major obstacle, and that design and process are the keys to solutions of large-scale problems. Our success in solving complex problems through design will depend largely on our ability to manage the complexity associated with these problems. In summary, the degree of complexity is a combination of three factors: numeric size, variety, and relation. All three factors must be considered within the constraints of task requirements.[20]

### 1.6.1.1   Twenty Laws of Complexity

Miller developed 20 laws of complexity in order to quantify and evaluate complexity. These laws aim to identify commonalities that can be useful in understanding the aspects of complex systems.[23]

1) *Law of triadic compatibility.* This law states that the human mind's ability to recall and conceptually explore seven conceptual components simultaneously limits the size of system that the human brain can process to three components and the four possible combinations (interactions) of them. Systems of a larger size than this are difficult for the human brain to develop an understanding of component relationships.

2) *Law of requisite parsimony.* This law also represents a limitation of human capacity with respect to the flow of information. A human being can only collect and process information at a certain rate. A flow of information exceeding a person's ability to internalize it will result in overburdened thinking and the reasoning produced will be flawed and untrustworthy. In this way, systems generating huge quantities of data may appear complex simply due to the inability to process information into behavioral predictions.

3) *Law of structural underconceptualization.* An example of the law of structural underconceptualization is where system behavior is being developed and the relationships and components of a system are incomplete. The lack of a complete mapping of the system presents challenges to defining behavior and increases complexity.

4) *Law of organizational linguistics.* According to this law, individuals within organizations and disciplines operate in a linguistic domain to allow communication of ideas. These varying layers of these organizations and disciplines develop working linguistics to promote better communication within the level but hinder communication to different layers. Consider how scientists of different disciplines (e.g. physics, chemistry, biology, medicine) would describe an event like an X-ray of a human body and their ability to convey their ideas to members from different disciplines.

---

23  Warfield, J.A., "Twenty laws of complexity: Science applicable in organizations," *Systems Research and Behavioral Science*, 16, 3–40, 1999. Used with permission.

5) *Law of vertical incoherence of organizations.* In organizations there exist repeating patterns of behavior that can be categorized and subcategorized to describe behavior and certain problem resolutions within the structure. This represents a means of simplifying behavioral descriptions by building generalizations.

6) *Law of validation.* The validity of knowledge within a discipline or organization requires a high degree of consensus within the specific community. Divergence from this consensus requires an extensive burden of proof and may represent a barrier to developing new concepts and understanding.

7) *Law of diverse beliefs.* When dealing with complex issues, often teams of experts are assembled of differing background in order to encompass the conceptual scope of the problem. The use of diverse teams utilizing different linguistic domains produces an inability to cooperatively approach a problem from a unified point of view. This provides further inability to understand behavior and increases complexity.

8) *Law of gradation.* This law acknowledges that all conceptual bodies of knowledge can be graded into varying layers and subcategories. The applicability of each level of the science is situation dependent. It is uncommon for every grade of a discipline to be utilized to evaluate a problem. The division of knowledge into subcategories allows specialization into specific tasks by individuals. It is no surprise that academic disciplines are hierarchical in nature, as this is a common structure for humans and allows the discipline to evolve quickly.[10]

9) *Law of universal priors.* The prerequisites for science and human understanding are humans possessing common language, reasoning of relationships, and an archival representation of information. This seems very basic, but can explain difficulties experienced when attempting to learn entirely new systems or concepts. This law is particularly important in a global economy involving participants from a broad spectrum of nations and cultures.

10) *Law of inherent conflict.* Similar to the law of diverse beliefs, groups assembled of individuals of different backgrounds and linguistic domains will disagree on the relative importance of different factors affecting a complex issue. This law also has significant impact in collaborative global design efforts. This was experienced in the development of the Boeing 787 Dreamliner when designs were modified by different groups, resulting in a structural failure.[24]

11) *Law of limits.* For all activities and complex problems, there exist limits that define the relationships and performance. The understanding of these limits, their relationship to other components, and how they may change is key to designing for complexity. The development of the 787 Dreamliner provides an example of limits contributing to complexity. The limits of the titanium supply chain were not known, and severe shortages emerged in the system as production began.[20]

12) *Law of requisite saliency.* When a designer is developing design targets for a system, it is seldom the case that each factor affecting the target is of equal weight. With each system there is an inherent need to develop a fundamental understanding of the relative importance of factors in a system to define performance.

---

24 Norris, G. and Wagner, M., *Boeing 787 Dreamliner*, Zenith Press, Minneapolis, MN, 2009.

13) *Law of success and failure.* When designing solutions to complex problems, there exist prerequisites that the design group must possess in order to achieve success, including leadership, financial support, component availability, design environment, designer participation, documentation support, and design process. Without each component, the successful design of solutions could be jeopardized.

14) *Law of uncorrelated extremes.* During the learning process of a design group consisting of members from various backgrounds, it was found that the relative importance of the factors affecting the problem evolved away from the individuals' disciplines toward a more comprehensive understanding. In this way the initial perceptions of individuals could be described as a collection of uncorrelated extremes prior to the learning process.

15) *Law of induced groupthink.* Groups attempting to address complex issues under a time constraint will have a tendency to exhibit a behavior known as "groupthink." Groupthink is a phenomenon that occurs with collaborative group decisions. The results of the group can represent courses of action that have not been well thought out by any individual group members. The collective decision can also be in stark contrast to the views of individual members.[11]

16) *Law of requisite variety.* A designer attempting to address the behavior of a complex issue has numerous specifications or design variables to produce the desired outcomes.

17) *Law of forced substitutions.* In design groups experiencing difficulties of conceptualization and internal conflict, it is common to see personnel substitutions in an attempt to induce results. This may compromise the ability of the group to fully define complex issues and develop a predictive control of the complex issue. The change in the group could also be the source of additional complexity.

18) *Law of precluded resolution.* The absence of interpretive modeling to describe the structural patterns representing the complex issue will result in the failure to properly address the issue itself. Without a clear methodology the inherent nature of any diverse group or organization is to focus on persuading the group of their perceived requirements of the individual members. This results in a failure to address the actual issue set out by the group or organization.

19) *Law of triadic necessity and sufficiency.* Complex relationships exhibit a sort of modularity consisting of three relational components. All complex problems can be reduced to a combination of triadic modules. In order to effectively approach a complex problem of any magnitude, the triadic modules to represent the system.

20) *Law of small displays.* The natural tendency of individuals is to accommodate problems to the size of media most familiar, including paper and computer monitors. This is in contrast to sizing the display to the size of the problem at hand. Too limited a display will create difficulty in conceptualizing large and complex problems.

### 1.6.1.2 Relationships between Components: Block Diagram

A block diagram can be used to visualize relationships between components in a system. The system diagram provides a visual mapping of the relationships between components. Lines should connect each element with all the other elements it can affect or be affected by.

Consider a basic, simplified fuel oil system powering a two-stroke diesel engine, as shown in Figure 1.6. The system has three sub-systems: a fuel heating sub-system, a fuel supply sub-system, and a fuel injection sub-system. Fuel oil must be heated to a certain temperature to provide the correct viscosity for combustion. Booster oil pumps are used to pump the fuel oil through heaters to the engine-driven fuel pumps. The fuel pumps will discharge high-pressure fuel to their respective injectors. The energy source for heating fuel oil can be waste heat, electricity, or steam.

For this example, assume that the heating source is steam and produced by a boiler shown in Figure 1.7(a). The block diagram shown in Figure 1.7(b) can be used to visualize the relationships between components in the fuel heating

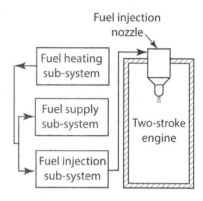

**Figure 1.6** Fuel oil system to power two-stroke engine.

sub-system. However, if a block diagram is used to visualize the relationships between the components in the entire fuel oil system as shown in Figure 1.6, the diagram becomes cluttered and difficult to interpret.

## 1.7 Managing Complexity

### 1.7.1 Processes to Manage Complexity

We have discussed the meaning of complexity, complexity escalation, and the limitations of human beings in dealing with complexity. Now we would like to suggest a process for

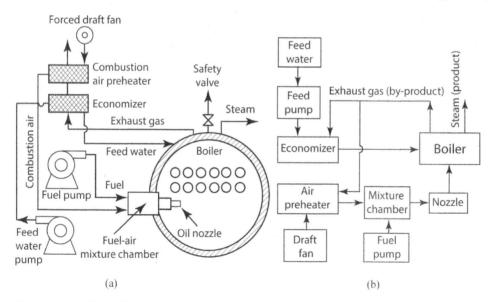

**Figure 1.7** (a) Physical boiler sub-system; (b) boiler sub-system block diagram.

harnessing or managing the complexity associated with a large-scale design project. A management methodology will be proposed for a real-world example of a design team addressing a complex system. The example involves the design of a fuel cell powered vehicle by a design team at Texas Tech University (TTU). The methodology will be compared with the applicable literature to produce a method to manage complexity.[25]

The development of the TTU fuel cell vehicle, the TTU entry in the 2002 FutureTruck competition sponsored by the US Department of Energy and the Ford Motor Company, was initiated with a new 2002 Ford Explorer, an 80 kW fuel cell stack, and high-pressure tanks to store hydrogen. The development of this vehicle is truly a transdisciplinary design exercise including mechanical, electrical, chemical, control components, and sub-systems to be developed and integrated. Vehicle emissions, energy efficiency, and consumer acceptability are the important parameters to be considered.

As the quotes from Simon mentioned previously indicate, complex systems are frequently hierarchical in nature, and thus can be divided into sub-systems. Both Simon and Wolfram indicate that complex hierarchical systems can be represented in terms of simple symbols or rules. Thus, what needs to be done is to develop a simple representation of the system, and to divide the system into its individual sub-systems. If necessary, the sub-systems can also be divided along similar lines until a manageable hierarchy of systems and sub-systems is achieved. This process is not that easy to accomplish.

First, any large-scale design project will require a team effort – probably a large team. Teams, especially large teams, do not just happen. A manager cannot merely identify a number of persons and form an effective team, even if the appointed team members represent all of the knowledge areas related to the project. It takes time for team members to develop confidence and trust in each other. Confidence and trust are built on interactions and relationships. The team members must learn to communicate effectively among themselves. Face-to-face interactions are best, but there is no reason why internet or other electronic communication and interfacing cannot provide a healthy team environment. Confidence, trust, and a good working relationship can be attained with effort, frequent communication, and time.

Large teams should be organized to complement the hierarchical nature of the system to be designed. That does not imply a hard vertical structure; rather, a combination of a vertical and horizontal structures is needed. The horizontal aspect of the structure provides for easy and frequent communication among sub-teams working on the details of various sub-systems that must interact. In hierarchical systems not all sub-systems directly interact with each other, thus not every sub-team will need to communicate directly or frequently with every other sub-team. The vertical aspect of the team structure provides a unifying effect on the overall team by defining the relationships of the various sub-systems and related sub-groups, and it provides the capability to continually reorganize the sub-teams as required during the design process.

Frequent and regular meetings of the team members are very important. Members of sub-teams should communicate at least daily about the details of the sub-systems on which they are working. Sub-groups working on interacting sub-systems should meet at least weekly to

---

25  This example is taken from Maxwell, T.T., Ertas, A., and Tanik, M.M., "Harnessing complexity in design," *SDPS Transactions: Journal of Integrated Design and Process Science*, 6(3), 63–74, 2002.

ensure that the respective sub-systems are compatible. The entire team should meet frequently to ensure that everyone involved with the project has an understanding of how the overall project is progressing and is able to keep his or her part in perspective. Meetings, especially of the larger groups, do not require that everyone literally meet face to face at one location. Internet or televised meetings can be very effective, particularly if two-way communication is available.

The fuel cell vehicle development team was comprised of undergraduate students, graduate students, technicians, and faculty advisors – about 10 graduate students, about 50 undergraduate students, two technicians, and two faculty advisors. The students were divided into several sub-groups. Each sub-group was assigned to a specific portion or sub-system of the vehicle; however, everyone was expected to be familiar with the vehicle in general, and members of one sub-group frequently worked with other sub-groups to accomplish specific tasks. Regular weekly meetings of the entire team were held. During the weekly meetings, each sub-group would summarize accomplishments, problems, decisions made during the last week and anticipated accomplishments, etc., for the upcoming week. Thus, all team members were kept aware of what other sub-groups were doing and what decisions that might affect their activities were being made by other sub-groups.

It is very important for all team members to understand the basic objective of the project and how their part fits into the whole. An iterative process that must begin immediately and continue throughout the entire project involves defining the project objective(s), requirements, constraints, etc. This iterative process should include the team members, customers, and stakeholders. The vertical component of the team organization should allow for the upper sub-teams to address primarily the overall aspects of the project and the basic relationships of the major sub-systems, while the lower sub-teams address the more detailed aspects of various sub-systems and the relationships among sub-systems. This process continues because the full understanding of a large-scale, complex project can only be accomplished as an iterative process as more of the underlying aspects become visible to the team and because the project requirements, constraints, and resources will very likely change with time. To assist members of the student team with the objective of the vehicle design project and to provide as much background information as possible, a library was set up on the internet. This library contained all information collected on fuel cells, hybrid electric vehicles, and previous TTU vehicle projects. Team members could add any information to the library that they developed or identified as being relevant.

After the team has been organized, has significantly completed its background reviews, and has initiated the process to define the objective, requirements, and constraints of the project, it must then contemplate decomposing the project into sub-systems and components. To begin this process it is useful to utilize one or more creativity sessions with some or all of the team members to identify aspects of the system to be designed, sub-systems, components, potential concerns, etc. Various creativity models can be used, for instance brainstorming, or any of the methods described by De Bono.[26] Table 1.1 shows a list generated for the fuel cell powered SUV.

---

26 De Bono, E., *Serious Creativity*, HarperCollins, New York, 1992.

**Table 1.1** Typical components, sub-systems, etc., required for fuel cell vehicle.

| | | |
|---|---|---|
| Fuel cell stack | Air humidifier | Sensor monitoring |
| Traction motor(s) | Humidity sensors | De-ionized water storage |
| 12 volt components | Pressure sensors | Radiator |
| High-voltage battery pack | Temperature sensors | Cooling water pump |
| Hydrogen humidifier | Vehicle controller | Compressor cooling |
| 12 volt battery | Stack preload pressure | Humidification water heater |
| Water injectors | Air exhaust | Air plumbing |
| Cell voltage monitoring system | Hydrogen storage tanks | Power steering |
| Hydrogen sensors | Hydrogen plumbing | Power brakes |
| Hydrogen exhaust | Air filtering | Motor controllers |
| Battery monitoring | Compressor | Stack cooling |
| Humidifier drains | Hydrogen flow control | Air conditioning/heating |
| Instrumentation for driver | Energy management | Fuel efficiency |
| Vehicle parking control | Gearbox(es) | Reactant metering |
| Reactant pressure control | Packaging | Gear reducer(s) |
| High-voltage wiring | Low-voltage wiring | Sensor/actuator wiring |
| Fuel cell control | DI water filter(s) | Reactant humidification control |

One session will not likely produce an exhaustive list. Once an initial list is developed, each item should be critiqued, duplicates should be combined, and extraneous items should be deleted. For small and medium size projects, this list can be maintained on large sheets of paper and the entire team can simultaneously participate in the development process. For large projects, a computerized list may be necessary and creativity sessions may be limited to smaller sub-sets of the team; however, all team members should participate.

Decomposing the overall system begins with developing the list of components, sub-systems, concerns, etc., as described above and continues with the organizing of the list into a hierarchical structure. The topics on the list should be correlated by relationships. The more encompassing sub-systems should be identified and then the topics within each upper sub-system must be correlated and organized similarly.

Table 1.2 provides a possibility for the first tier of sub-systems related to the development of the fuel cell SUV. The six sub-systems listed in Table 1.2 indicate the primary systems to be required for vehicle operation. Of course other variations could have been tried. For example, the high- and low-voltage systems could be lumped together as the electrical system, or the control system could have been split and included partly in the fuel cell system and partly in the power train. The first arrangement may often not be best. In fact it may be necessary to organize the next level, or levels, of sub-systems under each of the blocks in the first tier. Similarly, each tier of the structure may change after possible scenarios for the previous tier are considered.

**Table 1.2** Possible tier 1 sub-systems for fuel cell vehicle.

| | | |
|---|---|---|
| 1 | Power train | Drive motors and associated components |
| 2 | Fuel cell system | Stack, compressor, humidifiers, sensors, etc. |
| 3 | High-voltage system | 300 V battery pack, mounting, interface to fuel cell, etc. |
| 4 | Low-voltage system | 12 V electrical system to run vehicle lights, accessories, etc. |
| 5 | Control system | Fuel cell monitoring and control, vehicle control |
| 6 | Accessories | Power brakes, power steering, air conditioning, etc. |

Interactions between sub-systems, components, etc., should be noted on the hierarchical structure. Indeed, there will likely be several interactions or linkages that horizontally connect sub-systems on different vertical legs of the structure. Such cross-linkages are extremely important to identify and include. The organizational process, which will be iterative at each step, must be repeated until all levels of sub-systems are identified. We emphasize that there are many potential hierarchical structures that could be developed for a system; a part of the iterative procedure is to consider several possible structures in the process of developing the best structure. A schematic diagram of a typical hierarchical structure for a system is shown Figure 1.8. Note the gray lines which indicate an interaction between sub-groups in different vertical parts of the structure. The mindmapping process as described by Wycoff[27] and others is a good model for a tool to help with the decomposition and organization of project. The graphical representation inherent in the mindmapping process clearly shows both the

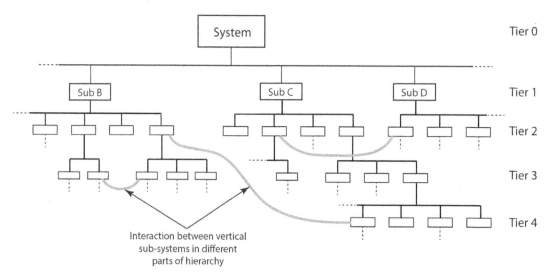

**Figure 1.8** Possible hierarchical structure for fuel cell SUV development.

---

27 Wycoff, J., *Mindmapping*, Berkley Books, New York, 1991.

hierarchical structure and the interactions among sub-systems and components. Traditional mindmapping works well for small and medium size projects where the entire map can be displayed on a single large sheet of paper. At least one software package, Inspiration from Inspiration Software, Inc., which supports mindmapping is widely available.

The organizational process is iterative in nature. First, it is likely that additional items will be added to the initial list as the process progresses. Additional items will be recognized as part of the organizational process because the very efforts that are needed to organize the list will help clarify understanding of the system and provide additional insight. Secondly, as the project develops, additional information will be discovered and additional sub-systems, components, etc., will be identified. Further, additional and more involved interactions or linkages between various sub-systems will be realized. The organizational structure being developed must be expanded to include these new components and interactions. Thus, the task of organizing and fully understanding the system being designed is not completed until the design process is completed.

Alongside the decomposition process there should be a team member organizational process. The team for a large project is likely to be large. Thus, the structure and communication within the team and sub-teams are critical. The project organizational structure should clearly indicate the major areas of work and the important interactions among sub-systems, hence members of the design team should be grouped into sub-teams similarly and the project organization should be reviewed frequently to ensure that no aspects of the overall design are neglected.

In any endeavor that requires effective coordination of the effort of many people, leadership and management are critical issues. The leader must provide vision, inspiration, and resources for the team and the leader must remove obstacles which limit the efforts of the team. The vision for the project, major resources, etc., are provided by the official team leader, and team members must provide leadership with respect to specific efforts of the team or sub-teams. Clearly, team management is also important, but management is not leadership.

The working environment for the design team is important. Appropriate space and communication capabilities must be available. In earlier times, space meant a situation room with ample table space, means to display information on a scale providing easy access, and secretaries or computer terminal operators to record activities. Today, team members may be located remotely all over the world. Work space probably refers to computer input software, display, and storage, while communication is related to interactive computer sessions, e-mail, online conferences, other machine interfaces, and perhaps even telephone discussions. What is important is convenient communication among team members with easy transmission of data to each other and the capability to display data and ideas as structured images.

In summary, management of complexity requires an overt, integrated mode of operation so as to:

- Develop a strong and well-organized team.
- Recognize that large-scale systems are complex and significant effort must be directed toward reducing this complexity.
- Recognize that large-scale systems typically have a hierarchical structure.

- Decompose the system to be designed in terms of its hierarchical structure.
- Identify interactions among sub-systems and components.
- Continually update the system structure to incorporate new information and changes in requirements, resources, etc.
- Identify and provide tools needed by the design team, especially for communication, manipulation and display of data and information.

### 1.7.2   Escalation of Complexity

As if complexity were not enough, the level of complexity associated with a system can and will escalate with time. Interactions between sub-systems increase the complexity of the system. One of the distinctions Suh made in the types of complexity relates the complexity to time. Time-independent complex systems do not experience change in their level of complexity with time. Time-dependent complex systems result when changing component relationships degrade the ability to predict performance. Most systems, if studied for a long period, will have some time dependency. In mechanical systems, this could be the result of component wear. A vehicle viewed over the period of one day may appear to have no time dependence, but monitored over a period of 20 years would certainly be time dependent.[3]

One of Simon's defined characteristics of complex systems is that they evolve over time.[10] Biggiero establishes evolutionary complexity as originating from two sources: increasing entropy within a system and the progress of advancing technology in biological and human systems.[4]

The time dependence of complexity represents a possible source of complexity escalation. As the interaction behavior between components changes over time, the ability to predict the system behavior is less robust, and the system becomes less stable. Suh argues that the introduction of functional periodicity is essential to mitigate the escalation of complexity. Functional periodicity can be viewed as an interruption resetting period, to reestablish the current state and relationships of components, thus preventing unstable behavior. This also shortens the period of prediction for the system to one period and makes prediction easier and more reliable.

There are also time-dependent complexity issues with the activity surrounding the management of complex systems. This evaluation includes the observer(s) as part of the system. When discussing systems of significant complexity which one individual would be incapable of fully perceiving, a multidisciplinary team is assembled. As new perspectives are introduced to a group, new facets, components, and relationships may be revealed. This will change the known complexity of the system being evaluated. The introduction of group members also introduces new components and relationships in the form of group dynamics. This also increases the complexity of the system and the team. The presence of a number of values, policies, practices, terminology, self-interest, and organizations aligned with the members imposes new relationships and behavior to the system. The valuation and prediction of a system can be diminished due to reduced communication, trust, and consensus of original system relationships from group dynamics.[11]

Consider the complex system of a traveling ant colony. This biological system necessitates higher-level emergent behavior with higher cognitive capacity than would be required for individual agents or ants.[2] One example of an emergent behavior is the tendency to find the most efficient paths of travel through the environment.[28]

Consider a hypothetical group evaluating the complex system of an ant colony. If an entomologist were added to the group, he would share the fact that ants use deposited pheromones to indicate the path that they have taken. The shortest path will have the most travel due to the shortest arrival and return time; this will mean the highest pheromone levels. This is how ants find the best path.[15] The ignorance of the group evaluating the ant colony would be reduced due to this previously unknown knowledge of the relationships of the ants. On top of this, there are new components and relationships introduced into the system, including pheromone glands, scent receptors, and behavioral responses to different pheromone conditions that must now be accounted for in the analysis.

There can be other escalations of the complexity of a problem. The nature of the problem may change due to adaptation of the system or because the team's understanding of the system changes. The design target may be redefined for any of many reasons, or the resources available to the project may change, usually for the worse. Incompatible organizational values or objectives may provide escalation. Typically education and training concentrate on much simpler situations and ignore all important complexity escalations.

### 1.7.3  Human Limitations

The cognitive aspect of complexity comprises an inseparable component of a system's complexity.[11] This is expanded by Peirce's argument that (cognitive) complexity exists as the sense of frustration due to a lack of ability to understand a problem, and directly ties human understanding to complexity.[5] The understanding of human limitations is vitally important to the understanding and management of complexity.

The human mind has fixed limitations on cognitive ability, just as all physical things have limitations. Miller demonstrated that a person is capable of mentally working with between five and nine concepts or ideas simultaneously.[29]

Wolfram proved in his experiments that computer models based on a small number of simple rules will generate extremely complex results.[9] This creates a severe mismatch, as many complex systems have large numbers of interacting components. The model of a system created of agents and artifacts describes components with elaborate and intelligent behaviors, further increasing difficulties to the human mind.[8]

As discussed earlier, complex problems by definition have many parts, thus it is not possible for a person to consider the interactions of hundreds of system parts simultaneously. Table 1.3 indicates some of the typical measures of the span of immediate recall. It is interesting to note that most people can recall seven-digit phone numbers, but are unlikely to remember 16-digit credit card numbers.

---

28  Dorigo, M., "Ant colony system: A cooperative learning approach to the traveling salesman problem," *IEEE Transactions on Evolutionary Computation*, 1(1), 53–66, 1997.

29  Miller, G., "The magical number seven, plus or minus two: Some limits on our capacity for processing information," *Psychological Review*, 63(2), 81–97, 1956.

**Table 1.3** Some measures of the span of immediate recall.[11,26]

| Stimulus set | Variables | Limit |
| --- | --- | --- |
| Single words | # syllables/word | 5–7 |
| Phrases | # words/phrase | 2–4 |
| Digits | # digits | 8 |
| Digits | Age of subjects | 2–8, increasing with age* |

*Can be increased with persistent practice.

Wolfram indicates that even computer programs based on very simple rules can generate very complex results. He presents results of many types of simple programs to support his conclusions. However, it seems significant that his small programs began to generate complex results when only a small number of parameters were included – usually between four and eight parameters. In line with Wolfram's theory that simple program experiments can be related to natural phenomena, perhaps there is a relationship between Miller's seven plus or minus two concept and Wolfram's simple program theory.

Beyond the mere cognitive limitation, humans also have behaviors that limit their ability to address complex systems. Biggiero describes the interactive or Hawthorne effect in organizational science, as the modification of human behavior by observation of it.[4]

This implies that all complex systems involving people may behave differently when not being observed. Predictions based on observed behaviors would be invalid in instances where observation was not taking place. Additionally, humans exhibit an ambiguity in the interpretation of systems. Differences in knowledge, subjectivity, beliefs, culture, and language result in considerable differences in interpretation of data from a complex system.[4] Together this contributes to a lack of interpretation of complex systems and poses a major limitation to understanding complex systems.

## 1.8 Complex Systems, Hierarchies, and Graphical Representations

According to Simon's definition of "complex systems," they have a large number of elements that have many interactions.[10] In studying complex issues related to systems or other human endeavor, it is necessary to create system hierarchies, and then to synthesize and analyze them. In other words, the complex system should be divided into a number of simpler sub-systems (such that they are easier to grasp and comprehend), and their solutions then coordinated to solve the original complex system problem. Decomposition of complex systems is a key part of the formulation performed in the systems design phase. In complex systems, relationships among the elements are key. Connections or relationships define how complex systems work.

A hierarchical system is composed of interrelated sub-systems, each of them being hierarchical in structure. A hierarchy is an arrangement of elements (factors) in which the elements have one higher-level and one lower-level neighbor (except the top and bottom levels). Hierarchical arrangements consist of relationships between elements, and can be represented by directed

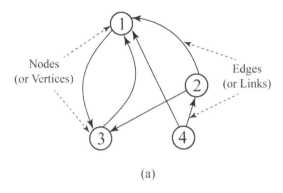
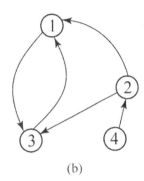

(a)                                              (b)

**Figure 1.9** Directed graph (digraph).

graphs. A directed graph (or digraph), as shown in Figure 1.9, is a set of elements (in this case four elements) that are connected together, where all the links are directed from one node (element) to another. In this figure, the hierarchy shows the four elements by number.

When drawing a digraph, the links are drawn as arrows indicating the direction, as illustrated in Figure 1.9. As shown in the figure, the direction of the linkage arrow is from element 2 to element 1, which indicates that element 2 influences (affects) element 1. As seen in Figure 1.9(a), one may think that a hierarchical representation includes more information than necessary – since element 4 is related to element 2 and element 2 is related to element 1, it follows that element 4 is related to element 1 through the intermediate element 2 (this is called the *transitivity rule*). Therefore, as shown in Figure 1.9(b), we can remove the connection between element 4 and element 1 and still keep all the important relationship conditions of the model.

### 1.8.1 Developing Hierarchy from Matrix

Figure 1.10 shows the relationships among the elements in a system matrix. This square ($n \times n$) matrix has entries in the $i$th row and $j$th column showing the relationship between the elements. If there is no relationship between two elements the entry is 0, whereas if there is a relationship the entry is 1. As shown in Figure 1.10, there is a relationship between elements 2 and 1, but there is no relationship between elements 3 and 2.

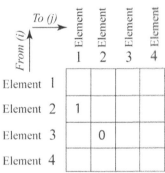

**Figure 1.10** Relationship matrix.

### Example 1.2

Diabetes is a disease in which the person has high blood glucose (blood sugar), either because insulin production is inadequate, or because the body's cells do not respond properly to insulin, or both. Patients with high blood sugar will typically experience polyuria (frequent urination) and will become increasingly thirsty and hungry. The risk related to diabetes can be reduced by (1) controlling the blood sugar level, (2) a diabetic diet, (3) consuming low-carbohydrate food, and (4) exercise.

Using the four elements given above, a relationship matrix and the related digraph can be developed.

**Solution**

As shown in Figure 1.11(a), set all the entries on the matrix diagonal to zero. Next, arbitrarily select four factors (elements) affecting diabetes. Exercise reduces the blood sugar, so there is a relationship between factor 4 and factor 1. Therefore the matrix entry $A_{4,1} = 1$ as shown in Figure 1.11(b). Since exercise does not contribute to a diabetic diet and low-carbohydrate foods, there are no relationships, so we set $A_{4,2} = 0$ and $A_{4,3} = 0$ (see Figure 1.11(b)). The same approach can be used to complete the matrix as shown in Figure 1.11(c).

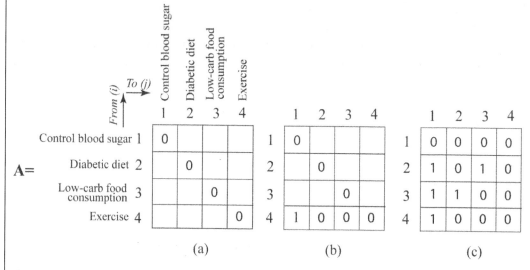

**Figure 1.11** Relationship matrix.

Next, using the relationship matrix, develop a digraph as shown in Figure 1.12. Matrix **A** shown in Figure 1.11 has four factors (elements): hence the hierarchy will show these elements by number. Wherever there is a number 1 in the matrix, there will be a connection (relationship) between the associated factors. For example, since $A_{4,1} = 1$, the direction of the link is from $i = 4$ to $j = 1$. Using same approach, the complete digraph can be developed as shown in Figure 1.12.

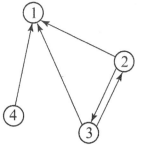

**Figure 1.12** Digraph of associated with matrix **A**.

## 1.8.2 Cycles in Digraph

A directed cycle in a digraph is a string of vertices (elements) starting and ending at the same vertex so that, for each pair of following vertices in the cycle, there is an edge directed from the earlier vertex to the later one – as shown in Figure 1.13, vertices 1 and 4 make a two-element cycle and vertices 1, 2, 3 make a three-element cycle, which means that these two and three elements are coupled, respectively.

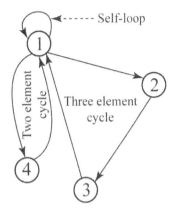

**Figure 1.13** Cycle set.

Figure 1.13 also shows the simplest possible cycle – a cycle called a *self-loop*. In other words, element 1 is related to itself in some way. A digraph that includes no cycles is called a *hierarchical digraph*.

Figure 1.13 shows a cycle set. Any vortex in a cycle set is reachable from any other in the set; hence, all elements of a cycle set are both antecedent to and succedent to any element of the set. It is important to note that as the number of element in a cycle gets larger, analysis become more difficult.

## Example 1.3

a) Does obesity cause obstructive sleep apnea (OSA) or does OSA cause obesity?
b) Explain the chicken and egg cycle.

## Solution

The two-element cycle shown in Figure 1.13 can be explained by Figure 1.14(a). As shown in Figure 1.14(a), possible mechanisms formatting a hurtful cycle where obesity may result in OSA. OSA reduces physical activities, which causes weight gain – OSA and obesity form a cycle. If the question is whether OSA causes obesity or obesity causes OSA, no matter how one thinks, there is no correct answer.

A slightly more complex cycle consisting of three elements is shown in Figure 1.14(b): the chicken and egg cycle. If the chicken came first, then it eventually had to hatch from an egg. If the egg came first, then it had to have a mother to create the egg. If the egg had to have a mother, then the mother should be the chicken – one answer leads to the next, hence making the question impossible to answer. In this case the three elements form a cycle. This situation may create a singularity in design and make the design unstable.

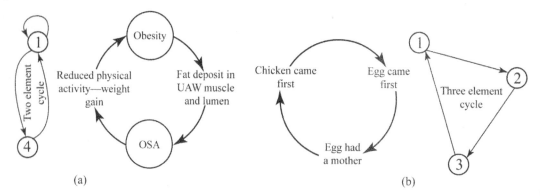

**Figure 1.14** (a) Two-element cycle; (b) three-element cycle.

## 1.9 Axiomatic Design

Axiomatic design (AD) is one example of design theories and tools which can play a major role in transdisciplinary system design. AD provides discipline-independent representations of designs, a general design process, general criteria for effective decision-making, and scalability for complex systems development.[30]

AD was created by Professor Nam Suh of MIT in order to create a science base for design and manufacturing.[31] The AD theory has been further developed by Suh and others.[32,33,34]

AD principles provide a framework that is consistent for all disciplines and at all levels of detail. Designs in AD are modeled using domains and hierarchies. As shown in Figure 1.15, the four domains are called the customer domain, the functional domain, the physical domain, and the process domain. The design elements associated with each domain are:

- customer needs (CNs),
- functional requirements (FRs),
- design parameters (DPs), and
- process variables (PVs).

For example, in the customer domain, suppose that a customer needs to preserve food. There are several ways to achieve this in the functional domain, such as canning, dehydrating, or cooling the food. From these options, the designer selects cooling and then decides on a

30 Tate, D., Ertas, A., Tanik, M., and Maxwell, T.T., *A TD Framework for Engineering Systems Research and Education based on Design and Process*, ATLAS TD Modules, 2006.

31 Suh, N.P., Bell, A.C., and Gossard, D.C., "On an axiomatic approach to manufacturing and manufacturing systems," *Journal of Engineering for Industry*, 100, 127–130, 1978.

32 Nordlund, M., "An information framework for engineering design based on axiomatic design," in Department of Manufacturing Systems. Stockholm, Sweden: The Royal Institute of Technology (KTH), 1996.

33 Suh, N.P., *The Principles of Design*, Oxford University Press, New York, 1990.

34 Suh, N.P., *Axiomatic Design: Advances and Applications*, Oxford University Press, New York, 2001.

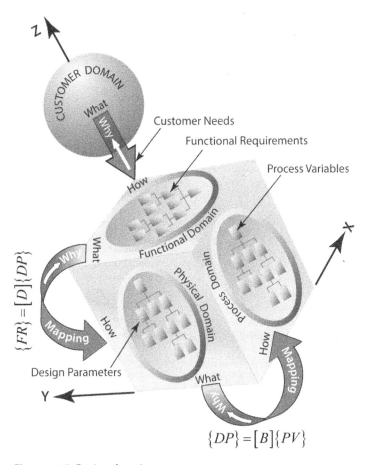

$$\{FR\} = [D]\{DP\}$$

$$\{DP\} = [B]\{PV\}$$

**Figure 1.15** Design domains.

refrigerator in the physical domain. The process domain describes how to manufacture the refrigerator.[35]

Customer needs are a collection of statements extracted in the "voice of the customer" that express customer perceptions of the design task. *Functional requirements* are defined as the minimum set of independent requirements that completely characterize the functional needs of the design solution. These FRs must be specified in a "solution-neutral environment" in terms of the functions to be achieved, not in terms of particular solutions. *Design parameters* are defined as the set of key elements or variables of the design solution that have been chosen to satisfy the FRs. *Process variables* are the key variables that characterize the ways to produce or realize the DPs.

35 Gumus, B., "Axiomatic product development lifecycle," PhD dissertation, Department of Mechanical Engineering, Texas Tech University, Lubbock, TX, 2005.

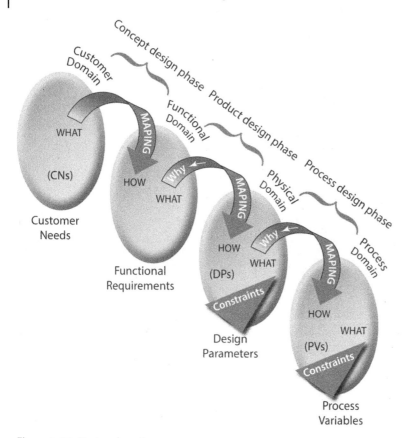

**Figure 1.16** Design domains.

AD provides a model, namely the *design matrix* that shows the engineering links between parts of a design. The design matrix shows the relationships between the FRs and DPs at one level of the design hierarchy.

A simplified model of design domains is given in Figure 1.16. Each domain to the right of an adjacent domain answers the question of *what* the designers want to achieve and *how* they will achieve it. In other words, the customer needs are mapped into functional requirements (concept design), the functional requirements are mapped into design parameters (product design), and the design parameters are mapped into process variables (process design).

Each domain has its own set of elements. The mapping between domains can be described by a set of matrices as:

$$\{CN\} = [\mathbf{R}]\{FR\}, \tag{1.1}$$

$$\{FR\} = [\mathbf{D}]\{DP\}, \tag{1.2}$$

$$\{DP\} = [\mathbf{B}]\{PV\}, \tag{1.3}$$

where $\mathbf{R}$ is the requirement matrix, $\mathbf{D}$ is the design matrix, and $\mathbf{B}$ is the component matrix.

### 1.9.1 Uncoupled, Decoupled, and Coupled Design

Design is concerned with what function a design will achieve (functional requirements) and how it will achieve it (design parameters). Two AD axioms provide a rational basis for evaluation of given solution alternatives. The two axioms can be defined as follows:

*Information content axiom.* Minimize the information content of the design.
*Independence axiom.* Each functional requirement should be satisfied by its corresponding design parameters without affecting the other functional requirements.

As defined in the following equation, the design matrix **D** shows the relationships between functional requirements and design parameters:

$$\{FR\} = [\mathbf{D}]\{DP\}. \tag{1.4}$$

As shown in Figure 1.17(a), each functional requirement of a design is individually satisfied by a design parameter without affecting the other functional requirements. This is called an uncoupled design matrix. In an uncoupled design there is one-to-one relationship between the FRs and DPs. Of course, this is an optimal design solution. However, most design solutions will not be that ideal. When design parameters are constrained by weight, size, cost, etc., they may have secondary effects on the other functional requirements as shown in Figure 1.17(b): DP1 is affecting FR1 and FR2, and DP2 is affecting FR2, FR3 and FR4. This is called decoupled design (triangular matrix). Figure 1.17(c) is a coupled design as it has two cycles shown with dashed lines. In other words, the relationship between the design parameters and their functional requirements is circular – DP1 affects FR1 and FR2, and similarly DP2 affects the same functional requirements (shown in square dashed lines). The other cycle is between elements 2, 3, and 4.

It is important to understand the nature of cycles in societal problems. If a cycle can be replaced by substitute elements then the design map becomes hierarchical, hence the hierarchy of a design

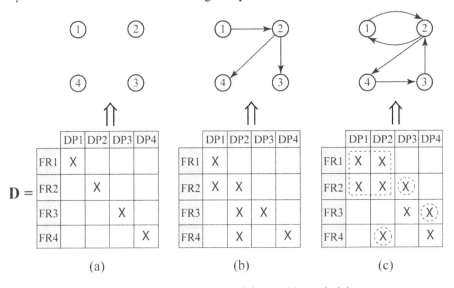

**Figure 1.17** (a) Uncoupled design; (b) decoupled design; (c) coupled design.

map could be better understood. If the cycles cannot be replaced by substitute elements, then each cycle in the design map should be understood individually. Otherwise, the design map becomes complex and the likelihood of understanding it will be very low.

## Example 1.4

Use the independence axiom for a typical water faucet, shown in Figure 1.18.

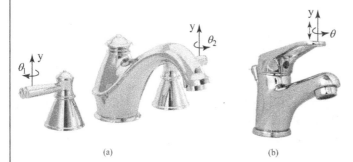

**Figure 1.18** (a) Two-handle water faucet; (b) one-handle water faucet. (Adapted from Frederickson, B., "Holistic systems engineering in product development," *Saab-Scania Griffin*, vol. 1994/95, Saab-Scania AB, Linköping, Sweden, pp. 23–31, 1994.)

(a)                    (b)

**Solution**

The two functional requirements for water faucet shown in Figure 1.18 are:[36]

- FR1: Control flowrate ($Q$) of water.
- FR2: Control temperature ($T$) of water.

As shown in Figure 1.18(a), a two-handle facet has two adjustments – a hot water knob, DP1 ($\theta_1$), and cold water knob, DP2 ($\theta_2$). Both design parameters, DP1 and DP2, affect both functional requirements of flowrate, $Q$, and temperature, $T$. Then, the design matrix will be

$$\begin{Bmatrix} FR1 \\ FR2 \end{Bmatrix} = \begin{bmatrix} X & X \\ X & X \end{bmatrix} \begin{Bmatrix} DP1 \\ DP2 \end{Bmatrix}. \tag{1.5}$$

Substituting FRs and DPs in equation (1.5), we have

$$\begin{Bmatrix} Q \\ T \end{Bmatrix} = \begin{bmatrix} X & X \\ X & X \end{bmatrix} \begin{Bmatrix} \theta_1 \\ \theta_2 \end{Bmatrix}. \tag{1.6}$$

To show the relationship between FRs and DPs, instead of using "1," $X$ is used in equation (1.5). In this design case, the relationship between the design parameters and their functional requirements is circular – DP1 affects FR1 and FR2, and similarly DP2 affects the same functional requirements. This is called a coupled design – the FRs and DPs form a cycle. This is shown in the relationship matrix in Figure 1.19.

|      | DP1 | DP2 |
|------|-----|-----|
| FR1  | X   | X   |
| FR2  | X   | X   |

**Figure 1.19** Coupled design.

36 Frederickson, B., "Holistic systems engineering in product development," *Saab-Scania Griffin*, vol. 1994/95, Saab-Scania AB, Linköping, Sweden, pp. 23–31, 1994.

With a one-handed facet, as shown in Figure 1.18(b), flow rate is adjusted by the vertical motion of the lever and the temperature is adjusted by the angle, $\theta$. The design matrix representing this design is given by

$$\left\{ \begin{array}{c} Q \\ T \end{array} \right\} = \begin{bmatrix} X & 0 \\ 0 & X \end{bmatrix} \left\{ \begin{array}{c} Y \\ \theta \end{array} \right\}. \tag{1.7}$$

As shown in Figure 1.20, the zeros on the off-diagonal of the relationship matrix indicate that each DP affects one FR and not the other. This design is called an uncoupled design, and it satisfies the independence criterion. The design in equation (1.6) is called a coupled design, and does not satisfy the independence criterion.

|     | DP1 | DP2 |
|-----|-----|-----|
| FR1 | X   | 0   |
| FR2 | 0   | X   |

**Figure 1.20** Uncoupled design.

## 1.9.2 Decoupling a Design Matrix

Usually, when a design matrix is initially formed, it happens to be coupled. However, there are simple techniques that can be used to rearrange rows of functional requirements and columns of design parameters for decoupling. As shown in Table 1.4(a), there are two Xs in the upper triangular matrix that need to be moved to the lower triangle to decouple the matrix. The following simple steps are used to decouple matrix (a):

1) Identify the empty row, in this case row 3 (ignore the diagonal terms), swap with the top row (see Table 1.4(b)).
2) Swap the corresponding column with the first column, in this case column 3 (see Table 1.4(c)).
3) Make sure the original matrix relationships will remain the same.

**Table 1.4** Decoupling a design matrix.

(a) Initial matrix

|     | DP1 | DP2 | DP3 |
|-----|-----|-----|-----|
| FR1 | X   | X   | O   |
| FR2 | O   | X   | X   |
| FR3 | O   | O   | X   |

(b) Step 1 (swap FR3 and FR1)

|     | DP1 | DP2 | DP3 |
|-----|-----|-----|-----|
| FR3 | O   | O   | X   |
| FR2 | O   | X   | X   |
| FR1 | X   | X   | O   |

(c) Step 2 (decoupled matrix)

|     | DP3 | DP2 | DP1 |
|-----|-----|-----|-----|
| FR3 | X   | O   | O   |
| FR2 | X   | X   | O   |
| FR1 | O   | X   | X   |

This process of decoupling is not possible in every design. In some cases, part of the design is always coupled. In these cases, a small coupling can be isolated from the rest of the design. In other cases, reevaluate the functional requirements and design parameters to remove the coupling.

## 1.10 Collective Intelligence Management

Collective intelligence management (CIM) was introduced by John Warfield, albeit under the name "interactive management," and it has been used for more than two decades by different organizations involved in a wide variety of problematic situations. CIM is a system of management which applies the science of generic design.[37] The CIM methodology fosters collaboration of group members who share a commitment in solving complex issues within a structure that uses systematic and logical reasoning.

Generic design science incorporates a significant behavioral component and because it is intended to enable complexity to be managed, it falls into a process realization that involves groups and the management of group processes. The following is set of characteristics that distinguish CIM from the traditional design process:[38]

- *Scientifically-based.* It is based on both a philosophy of science and integrated knowledge stemming from behavioral and mathematical sciences.
- *Combining rigidity and flexibility.* It is actually rigid in terms of sub-processes, but it is very flexible in terms of the choice and sequencing to math requirements of the problematic situation.
- *Creates new scientific concepts.* Data taken as a natural part of the constructive effort has been interpreted to produce possible new scientific concepts related to groups – "spreadthink" and "structural thinking".[39]
- *Pattern development.* Team formulation of relational patterns is strongly supported by the method called "interpretive structural modeling" (ISM).

Figure 1.21 shows an example of the management of complexity through CIM. As shown in the figure, CIM involves two closely linked CIM phases – the collective intelligence management workshop (CIMW) and development of collective intelligence (DCI).

---

37 Warfield, J., and Cárdenas, A.R., *A Handbook of Interactive Management*, Iowa State University Press, Ames, 1994.
38 Warfield, J., "A philosophy of design," document prepared for Integrated Design & Process Technology Conference, Austin, TX, 1995, p. 10.
39 Spreadthink is a major difficulty in working together effectively in carrying out the work program of complexity. Structural thinking provides the principal key to working effectively with complexity.

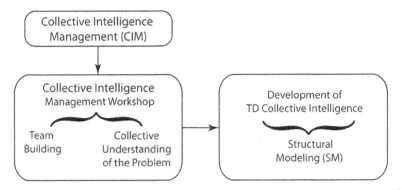

**Figure 1.21** Management of complexity through CIM.

### 1.10.1 Collective Intelligence Management Workshop

Transdisciplinary teams can be developed with distributed leadership – team leadership can change in accordance with the particular expertise required. Most of the time, structuring and understanding a complex problem may become problematic – collaborating team members may not even agree on what the problem is and no solution can be identified that makes everyone happy.[40]

Suppose that there is a complex problem in hand to be structured by engineering under-graduate students. Assume that research sub-project teams of four with specific roles for each sub-project teams are considered. Depending on the assigned task, each sub-project team may have a minimum of three and a maximum of five students. Using any communication platform, a workshop can be organized where sub-project teams will introduce their project proposals (concepts) about the complex problem being explored.

Through dialog, the collective best ideas of sub-project teams will emerge and the incorrect or fuzzy ideas that teams held at the outset will be recognized as incorrect or sharpened to make them useful. Through the CIM process ideas with high interaction will be grouped into clusters.[37] Thus, team members can identify and examine cluster interactions internally. Clusters will be placed in a sequence by using ISM. [41]

Through the CIMW sub-project team members will experience the mutual understanding and learning that are an integral part of the TD research process. Outcomes obtained with CIM include:[37]

- *Learning.* Students engaged in an CIM activity are exposed to a real sharing of ideas and infor-mation, and hence are actively learning about the research project at hand.
- *Commitment.* Research projects are created participatively. Through this kind of approach genuine commitment can be achieved.

---

40 Denning, P.J., "Mastering the mess," *Communications of the ACM*, 50(4), 21–25, 2007.
41 Rittel, H.W.J., and Webber, M.M., "Dilemmas in a general theory of planning" *Policy Sciences*, 4, 155–169, 1973.

- *Documentation*. During the CIM process information generated by team members and decisions taken will be recorded and organized, and will provide the basis for broader diffusion of the outcomes.

### 1.10.2 Structural Modeling

A *structural model* is a diagram which consists of nodes and the links that connect them. In other words, it is a collection of components (elements) showing their relationship. Using structural modeling, one can gain a general understanding of the system as a whole by studying a structural model of the components within the system. Warfield defines structural modeling as:[42]

> Structural modeling is a methodology which employs graphics and words in carefully defined patterns to portray the structure of a complex issue, a system, or a field of study.

Dennis Cearlock defines structural modeling as follows:[43]

> Emphasis is placed on determining if there is a coupling between variables and the relative importance of the coupling, rather than on developing exact mathematical relationships and precisely determining the associated coefficients. As such, pre-programmed, simple functional forms are used for specifying the coupling related between variables composing a system. Thus, structural modeling is concerned with identifying trends and equilibrium states rather than quantitative precision.

The causes of low productivity in complex problems fall within five categories:[6] behavioral, representational, informational, infrastructural, and relational processing. Structural thinking is based on the activity of thinking in regard to relationships and organizing thought in a pattern modeled accordingly. Practitioners of structural thinking should focus on information as a basis for organizing information, explore relationships among system members in detail, specify structural features in generic terms, allow comparisons of relative complexity to be made among relationships, and use computer assistance to develop, organize, and represent relationships.[6]

### 1.10.3 Interpretive Structural Modeling

Interpretive structural modeling will be used for the development of collective intelligence. Figure 1.22 shows a flow chart of the ISM process. A methodology for dealing with complex issues, ISM was proposed by Warfield in 1973. It is a computer-assisted learning process that provides a fundamental understanding of how various parameters (elements, variables, system

---

42 Warfield, J., *Structuring Complex Systems*, Battelle Memorial Institute, Columbus, OH, Battelle Monograph, No. 3, 1974.
43 Cearlock, D.B., "Common properties and limitations of some structural modeling techniques," PhD dissertation, University of Washington, Seattle, 1977.

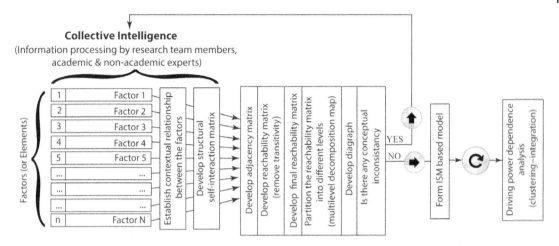

**Figure 1.22** Interpretive structural modeling.

components, etc.) relevant to the problem or issue are interrelated. This helps researchers to structure the parameters in a meaningful way, to develop collective intelligence, and to overcome challenging complex problems.

This interactive learning, information processing, and developing of collective intelligence is especially useful for working in a group to develop a map of the complex relationships between various elements involved in a challenging complex issue. The relationship map includes paths of ideas and threads of thought to help transform unclear and poorly expressed issues into a well-defined and relatively easily solvable model. The fundamental approach of this process is to use academic and non-academic experts with practical experience and knowledge to decompose a complex issue into smaller sub-issues, and build a simpler multilevel structural model. General activities to build an interpretive structural model are shown in Figure 1.22. Note that the sequence of these activities may change from one application to another.

### 1.10.4 Transdisciplinary Collective Intelligence

Transdisciplinary collective intelligence is a new mode of information gathering, knowledge creation, and decision-making that draws on expertise from a wide range of organizations (academic or non-academic), and collaborative partnerships. This technique could add two dimensions to knowledge production – firstly, building collective methods and formulation of practices in addition to knowledge (data, concepts), and secondly, the identification of common problems participatively.

#### 1.10.4.1 Sequence of Activities to Develop an ISM Model

As shown in Figure 1.22, the following steps are used to develop an ISM model:

*Step 1. Team development*

The first step in the ISM process is to organize a group of people with relevant knowledge, skills, and background. This team should consist of academic and non-academic experts from diverse disciplines. A coordinator who will control the ISM process should not only be knowledgeable in the subject matter but also be familiar with the ISM process. At this stage, it is also important clearly identify the problem (issue) which will be studied as explained previously.

*Step 2. Identify factors*

The second step is to identify and define the factors affecting the issue as shown in Figure 1.22. Through brainstorming and consultation with the domain experts, research group members work together to document all the possible factors (elements) whose relationships are to be modeled. Then the most important factors are identified for model development. The nominal group technique (NGT) is an efficient method of generating ideas for defining a set of factors.[44] The five basic steps of the NGT process are as follows:

a) clarification of a trigger question;
b) silent generation of ideas in writing by each group member;
c) round-robin recording of the ideas;
d) ongoing discussion of each idea for clarification and editing;
e) voting to obtain a preliminary ranking of the ideas in terms of significance.

*Step 3. Structural self-interaction matrix*

The next step is to establish the contextual relationship in order to develop a structural self-interaction matrix (SSIM) shown in Figure 1.23. To show the direction of the contextual

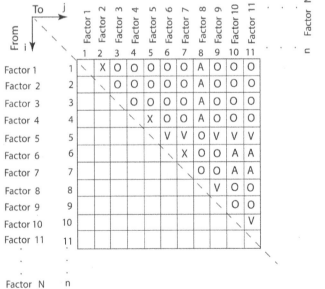

**Figure 1.23** Structural self-interaction matrix.

44 Delbecq, A.L., and VandeVen, A.H., "A group process model for problem identification and program planning," *Journal of Applied Behavioral Science*, 7, 466–491, 1971.

relationship among the factors affecting the issue or problem, the following four symbols are used in the matrix:

- Enter V when the relation is from $i$ to $j$ (in the adjacency matrix the $(i,j)$th entry becomes 1 and the $(j,i)$th entry becomes 0).
- Enter A when the relation is from $j$ to $i$ (in the adjacency matrix the $(i,j)$th entry becomes 0 and the $(j,i)$th entry becomes 1).
- Enter X when the relation is in both directions, from $i$ to $j$ and from $j$ to $i$ (in the adjacency matrix the $(i,j)$th and $(j,i)$th entries become 1).
- Enter O when there is no relationship between the row $(i)$ element and the column $(j)$ element (in the adjacency matrix the $(i,j)$th and $(j,i)$th entries become 0).

*Step 4. Develop adjacency matrix*

Then the adjacency matrix, **A**, shown in Figure 1.24 is developed by transforming the SSIM into a binary matrix by substituting V, A, X, and O by 1 and 0 as per the schema described previously in the matrix that reflects the directed relationships between the elements. This matrix answers the question of "yes" or "no." In other words, starting from factor $i$, can factor $i$ reach factor $j$ to contribute and improve? If so, the $j$th entry becomes 1 and the $i$th entry becomes 0. This means that there is a walk from $i$ to $j$. The main diagonal of adjacency matrix is filled with 1s.

*Step 5. Reachability matrix with transitivity*

The reachability matrix with transitivity, $\mathbf{R_t}$, is shown in Figure 1.25. This matrix is checked for the transitivity rule and will be updated until transitivity is established. The transitivity rule states that, if factor (element) $A$ is related to factor $B$ and if factor $B$ is related to factor $C$, then factor $A$ is related to factor $C$. By following the transitivity rule, a reachability matrix which considers transitivity is developed.

$$
\mathbf{A} =
\begin{array}{r|ccccccccccc}
 & 1 & 2 & 3 & 4 & 5 & 6 & 7 & 8 & 9 & 10 & 11 \;.\;\;.\;\;.\;n \\
\hline
\text{Factor 1}\;\;1 & 1 & 1 & 0 & 0 & 0 & 0 & 0 & 0 & 0 & 0 & 0 \\
\text{Factor 2}\;\;2 & 1 & 1 & 0 & 0 & 0 & 0 & 0 & 0 & 0 & 0 & 0 \\
\text{Factor 3}\;\;3 & 0 & 0 & 1 & 0 & 0 & 0 & 0 & 0 & 0 & 0 & 0 \\
\text{Factor 4}\;\;4 & 0 & 0 & 0 & 1 & 1 & 0 & 0 & 0 & 0 & 0 & 0 \\
\text{Factor 5}\;\;5 & 0 & 0 & 0 & 1 & 1 & 1 & 1 & 0 & 1 & 1 & 1 \\
\text{Factor 6}\;\;6 & 0 & 0 & 0 & 0 & 0 & 1 & 1 & 0 & 0 & 0 & 0 \\
\text{Factor 7}\;\;7 & 0 & 0 & 0 & 0 & 0 & 1 & 1 & 0 & 0 & 0 & 0 \\
\text{Factor 8}\;\;8 & 1 & 1 & 1 & 1 & 0 & 0 & 0 & 1 & 1 & 0 & 0 \\
\text{Factor 9}\;\;9 & 0 & 0 & 0 & 0 & 0 & 0 & 0 & 0 & 1 & 0 & 0 \\
\text{Factor 10}\;\;10 & 0 & 0 & 0 & 0 & 0 & 1 & 1 & 0 & 0 & 1 & 1 \\
\text{Factor 11}\;\;11 & 0 & 0 & 0 & 0 & 0 & 1 & 1 & 0 & 0 & 0 & 1 \\
\end{array}
$$

Factor N    n

**Figure 1.24** Adjacency matrix.

$$R_t =$$

|  |  | 1 | 2 | 3 | 4 | 5 | 6 | 7 | 8 | 9 | 10 | 11 | . | . | . n |
|---|---|---|---|---|---|---|---|---|---|---|---|---|---|---|---|
| Factor 1 | 1 | 1 | 1 | 0 | 0 | 0 | 0 | 0 | 0 | 0 | 0 | 0 |  |  |  |
| Factor 2 | 2 | 1 | 1 | 0 | 0 | 0 | 0 | 0 | 0 | 0 | 0 | 0 |  |  |  |
| Factor 3 | 3 | 0 | 0 | 1 | 0 | 0 | 0 | 0 | 0 | 0 | 0 | 0 |  |  |  |
| Factor 4 | 4 | 0 | 0 | 0 | 1 | 1 | 1 | 1 | 0 | 1 | 1 | 1 |  |  |  |
| Factor 5 | 5 | 0 | 0 | 0 | 1 | 1 | 1 | 1 | 0 | 1 | 1 | 1 |  |  |  |
| Factor 6 | 6 | 0 | 0 | 0 | 0 | 0 | 1 | 1 | 0 | 0 | 0 | 0 |  |  |  |
| Factor 7 | 7 | 0 | 0 | 0 | 0 | 0 | 1 | 1 | 0 | 0 | 0 | 0 |  |  |  |
| Factor 8 | 8 | 1 | 1 | 1 | 1 | 1 | 0 | 0 | 1 | 1 | 0 | 0 |  |  |  |
| Factor 9 | 9 | 0 | 0 | 0 | 0 | 0 | 0 | 0 | 0 | 1 | 0 | 0 |  |  |  |
| Factor 10 | 10 | 0 | 0 | 0 | 0 | 0 | 1 | 1 | 0 | 0 | 1 | 1 |  |  |  |
| Factor 11 | 11 | 0 | 0 | 0 | 0 | 0 | 1 | 1 | 0 | 0 | 0 | 1 |  |  |  |
| Factor N | n |  |  |  |  |  |  |  |  |  |  |  |  |  |  |

**Figure 1.25** Reachability matrix with transitivity.

*Step 6. Final reachability matrix*

The *driving power* and *dependence* of factors are also computed in the *final reachability matrix*, $R_f$, as shown in Figure 1.26. The summation of 1s in the corresponding rows gives the driving power, and the summation of 1s in the corresponding columns gives the dependence. Figure 1.26 is the final form of the relationships of all the factors involved with the complex issue. Assuming that there are 11 factors as shown in Figure 1.26, the total driving power is 38 and the dependence is also 38. The driving power and dependence will be used to classify the factors into four clusters – autonomous, dependent, linkage, and independent.

$$R_f =$$

|  |  | 1 | 2 | 3 | 4 | 5 | 6 | 7 | 8 | 9 | 10 | 11 | Driving power |
|---|---|---|---|---|---|---|---|---|---|---|---|---|---|
| Factor 1 | 1 | 1 | 1 | 0 | 0 | 0 | 0 | 0 | 0 | 0 | 0 | 0 | 2 |
| Factor 2 | 2 | 1 | 1 | 0 | 0 | 0 | 0 | 0 | 0 | 0 | 0 | 0 | 2 |
| Factor 3 | 3 | 0 | 0 | 1 | 0 | 0 | 0 | 0 | 0 | 0 | 0 | 0 | 1 |
| Factor 4 | 4 | 0 | 0 | 0 | 1 | 1 | 1 | 1 | 0 | 1 | 1 | 1 | 7 |
| Factor 5 | 5 | 0 | 0 | 0 | 1 | 1 | 1 | 1 | 0 | 1 | 1 | 1 | 7 |
| Factor 6 | 6 | 0 | 0 | 0 | 0 | 0 | 1 | 1 | 0 | 0 | 0 | 0 | 2 |
| Factor 7 | 7 | 0 | 0 | 0 | 0 | 0 | 1 | 1 | 0 | 0 | 0 | 0 | 2 |
| Factor 8 | 8 | 1 | 1 | 1 | 1 | 1 | 0 | 0 | 1 | 1 | 0 | 0 | 7 |
| Factor 9 | 9 | 0 | 0 | 0 | 0 | 0 | 0 | 0 | 0 | 1 | 0 | 0 | 1 |
| Factor 10 | 10 | 0 | 0 | 0 | 0 | 0 | 1 | 1 | 0 | 0 | 1 | 1 | 4 |
| Factor 11 | 11 | 0 | 0 | 0 | 0 | 0 | 1 | 1 | 0 | 0 | 0 | 1 | 3 |
| Dependence |  | 3 | 3 | 2 | 3 | 3 | 6 | 6 | 1 | 4 | 3 | 4 | $\Sigma = 38$ |

**Figure 1.26** Final reachability matrix.

**Table 1.5** Level I (first iteration).

| Variable | Reachability set | Antecedent set | Intersection set | Level |
|:---:|:---:|:---:|:---:|:---:|
| 1 | 1, 2 | 1, 2, 8 | 1, 2 | I |
| 2 | 1, 2 | 1, 2, 8 | 1, 2 | I |
| 3 | 3 | 3, 8 | 3 | I |
| 4 | 4, 5, 6, 7, 9, 10, 11 | 4, 8 | 4 | |
| 5 | 5, 6, 7, 9, 10, 11 | 4, 5, 8 | 5 | |
| 6 | 6, 7 | 4, 5, 6, 7, 10, 11 | 6, 7 | I |
| 7 | 6, 7 | 4, 5, 6, 7, 10, 11 | 6, 7 | I |
| 8 | 1, 2, 3, 4, 5, 8, 9 | 8 | 8 | |
| 9 | 9 | 4, 5, 8, 9 | 9 | I |
| 10 | 6, 7, 10, 11 | 4, 5, 10 | 10 | |
| 11 | 6, 7, 11 | 4, 5, 10, 11 | 11 | |

*Step 7. Level partition*

The final reachability matrix, $\mathbf{R}_f$, and the antecedents of each element of prospects (see Table 1.5) are used to identify the partition levels.

The previously obtained driving force and dependence help to classify the factors into groups. These groups positions are determined by the separation of the antecedent set and the reachability set. From these two sets an intersection set is determined. The factors common in the reachability set and the antecedent set are included in the intersection set. These three sets help to identify the level of the factors. If all the factors of the intersection and reachability sets of any particular factor are the same, then that factor is identified as the top-level group (level I group) in the ISM hierarchy. Once the top-level factors are identified, they are removed from the set to identify the next level. This iteration process is repeated until all the levels are identified. As shown in Table 1.5, factors 1, 2, 3, 6, 7, and 9 are found to be at level I. Hence, they will be positioned in the top-level group in the ISM hierarchy. This process is repeated till all the levels are determined. The factors along with their reachability set, antecedent set, intersection set and levels, are shown in Tables 1.5–1.9. These levels help to build the digraph and the ISM model.

*Step 8. Formation of digraph and ISM based models*

As shown in Figure 1.27, the association of sets and binary relations through matrices can now be converted into graphical form by using theory of digraphs.[45] If there is a relationship between factors *j* and *i*, the connection between factors will go from *i* to *j*. Finally, the digraph is converted to an ISM based model to give a broad representation of the interrelationship between the factors (see Figure 1.28).

---

45 Harary, F., Norman, R.V., and Cartwright D., *Structural Models: An Introduction to the Theory of Directed Graphs*, Wiley, New York, 1965.

**Table 1.6** Level II (second iteration).

| Variable | Reachability set | Antecedent set | Intersection set | Level |
|---|---|---|---|---|
| 4 | 4, 5, 10, 11 | 4, 8 | 4 | |
| 5 | 5, 10, 11 | 4, 5, 8 | 5 | |
| 8 | 4, 5, 8 | 8 | 8 | |
| 10 | 10, 11 | 4, 5, 10 | 10 | |
| 11 | 11 | 4, 5, 10, 11 | 11 | II |

**Table 1.7** Level III (third iteration).

| Variable | Reachability set | Antecedent set | Intersection set | Level |
|---|---|---|---|---|
| 4 | 4, 5, 10 | 4, 8 | 4 | |
| 5 | 5, 10 | 4, 5, 8 | 5 | |
| 8 | 4, 5, 8 | 8 | 8 | |
| 10 | 10 | 4, 5, 10 | 10 | III |

**Table 1.8** Level IV (fourth iteration).

| Variable | Reachability set | Antecedent set | Intersection set | Level |
|---|---|---|---|---|
| 4 | 4, 5 | 4, 8 | 4 | |
| 5 | 5 | 4, 5, 8 | 5 | IV |
| 8 | 4, 5, 8 | 8 | 8 | |

**Table 1.9** Level V (fifth iteration).

| Variable | Reachability set | Antecedent Set | Intersection set | Level |
|---|---|---|---|---|
| 8 | 8 | 8 | 8 | VI |

Figure 1.27 shows that factors 3 and 9 are isolated since they are not adjacent to any factor – these two factors do not influence system performance. Factors 4, 5, 8, 10, and 11 represent the linear mapping which preserves the acyclicity properties of the system – the system contains no cycle. Factor 8 is the source component since it has only outgoing paths.

*Step 9. MICMAC analysis*

MICMAC (impact matrix cross-reference multiplication applied to a classification) was developed by Duperrin and Godet in 1973 to study the diffusion of impact through reaction paths

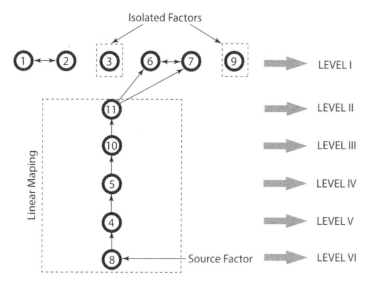

**Figure 1.27** Digraph based on reachability matrix with relation.

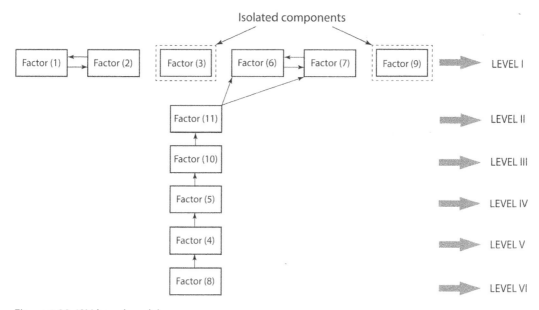

**Figure 1.28** ISM based model.

and loops for developing hierarchies for members of an element set.[46] The purpose of MIC-MAC analysis is to arrange the factors with respect to their driving power and dependence into four clusters:[47]

---

46 Duperrin, J.C., and Godet, M., "Méthode de Hiérarchisation des éléments d'un système," Rapport Economique du CEA, R-45-51, 1973.

47 Mandal, A., and Deshmukh, S.G., "Vendor selection using interpretive structural modelling (ISM)," *International Journal of Operations & Production Management*, 14(6), 52–59, 1994.

1) autonomous,
2) dependent,
3) linkage, and
4) independent factors.

The driving power and dependence of each factor are imported from Figure 1.26. Figure 1.29 shows the driving-power–dependence diagram of factors. As shown in this figure, all factors affecting system performance have been classified into four categories. Cluster I includes six autonomous factors. Many components (1, 2, 3, 9, 10, and 11) have low driving power and low dependence; hence they can be eliminated from the system. This indicates that there are six components disconnected from the system.

Three factors (5, 6, and 7) are placed at the boundaries of the clusters. In such cases, decisions must be made as to which factor should be placed within which cluster. For example, if 5 is an important factor for the system's performance then it should be included within cluster IV. Similarly, cluster II may include two dependent factors (6 and 7) that have low driving power and high dependence. They have a smaller guidance power but it is extremely dependent on the system – These two components may not affect the performance of other components, but they are affected by other components of the system.

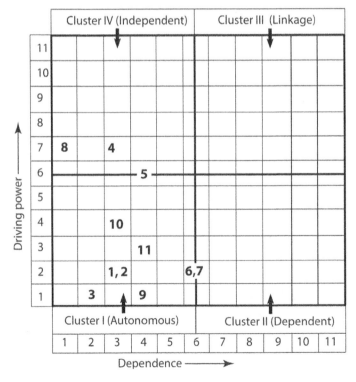

**Figure 1.29** MICMAC analysis.

The factors in the linkage cluster have to be given extreme importance due to their high driving power and high dependence power. It is interesting to note that there are no single components in this cluster; this reduces the complexity of the system performance considerably.

Cluster IV includes independent factors 4, 5, and 8 with a strong driving power but very weak dependence. These components are the key drivers for the system performance. Factors 4, 5, and 8 in this cluster require maximum attention so that the system functionality does not go out of control.

If many factors are affecting an issue, the ISM methodology can get complex. If the number of factors is limited, then the development of an ISM model becomes easier. Structural Equation modeling is also commonly known as the linear structural relationship approach.

### Example 1.5

Figure 1.30 shows the relationship of four components of a system. Develop:

a) the adjacency matrix;
b) the reachability matrix with transitivity;
c) the final reachability matrix;
d) the level partition;
e) the digraph;
f) the MICMAC analysis.

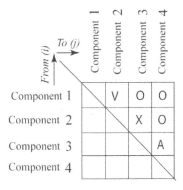

**Figure 1.30** Structural self-interaction matrix.

### Solution

a) The adjacency matrix, **A**, shown in Figure 1.31 is developed by transforming the SSIM into a binary matrix by substituting V, A, X, and O by 1 and 0.
b) The reachability matrix with transitivity, **R$_t$**, shown in Figure 1.32, is checked for the transitivity rule and updated until transitivity is established. In Figure 1.33, component relations imported from adjacency matrix **A** are shown by solid arrows. Dashed arrows show relations through transitivity – if component 1 is related to component 2 and component 2 is related to component 3, then component 1 is related to component 3. The same argument can be made for the relation between component 4 and component 2.
c) The final reachability matrix, **R$_f$**, which includes driving power and dependence, is shown in Figure 1.34. The summation of 1s in the corresponding rows a the driving power of 10, and the summation of 1s in the corresponding

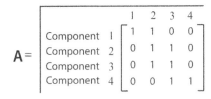

**Figure 1.31** Adjacency matrix.

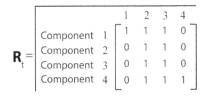

**Figure 1.32** Reachability matrix with transitivity.

columns gives a dependence of 10. Figure 1.34 is the final form of the relationship of all the four components of the system.

d) To identify the level partition, the final reachability matrix, $R_f$, shown in Figure 1.32 is used. For example, consider component 1. For this component, the reachability set is $\{1, 2, 3\}$ (select the 1s from row) and antecedent set is $\{1\}$ (select the 1s from column). The intersection set is the factors common to the reachability set and the antecedent set, $\{1\}$. Since all the factors of the intersection and reachability sets are not same, there is no level for this case. The same procedure will be followed for the other components. As seen from Tables 1.10 and 1.11, this system has two levels.

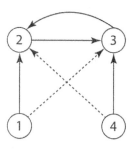

**Figure 1.33** Factor relationships.

$$R_f = \begin{array}{c} \text{Component 1} \\ \text{Component 2} \\ \text{Component 3} \\ \text{Component 4} \end{array} \begin{array}{cccc} 1 & 2 & 3 & 4 \\ \left[ \begin{array}{cccc} 1 & 1 & 1 & 0 \\ 0 & 1 & 1 & 0 \\ 0 & 1 & 1 & 0 \\ 0 & 1 & 1 & 1 \end{array} \right] \end{array} \begin{array}{c} \text{Driving power} \\ 3 \\ 2 \\ 2 \\ 3 \end{array}$$

| | Dependence | 1 | 4 | 4 | 1 | $\Sigma = 10$ |

**Figure 1.34** Final reachability matrix.

e) As shown in Figure 1.35, the digraph can be developed by using the final reachability matrix. Using Figure 1.32, the digraph can be structured as follows:

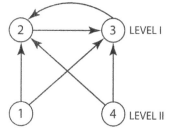

**Figure 1.35** Digraph.

- Component 1 reaches component 2 (arrow should go from 1 to 2) and reaches component 3 (arrow should go from 1 to 3).
- Component 2 reaches component 3 (arrow should go from 2 to 3).
- component 3 reaches component 2 (arrow should go from 3 to 2).
- Component 4 reaches components 2 (arrow should go from 4 to 2) and reaches component 3 (arrow should go from 4 to 3).

f) For the MICMAC analysis, Figure 1.36 can be formed by importing the driving power and dependence of each component from Figure 1.34.

**Table 1.10** Level I (first iteration).

| Component | Reachability set | Antecedent set | Intersection Set | Level |
|:---:|:---:|:---:|:---:|:---:|
| 1 | 1, 2, 3 | 1 | 1 | |
| 2 | 2, 3 | 1, 2, 3, 4 | 2, 3 | I |
| 3 | 2, 3 | 1, 2, 3, 4 | 2, 3 | I |
| 4 | 2, 3, 4 | 4 | 4 | |

**Table 1.11** Level II (second iteration).

| Component | Reachability set | Antecedent set | Intersection set | Level |
|:---:|:---:|:---:|:---:|:---:|
| 1 | 1 | 1 | 1 | II |
| 4 | 4 | 4 | 4 | II |

**Figure 1.36** MICMAC analysis.

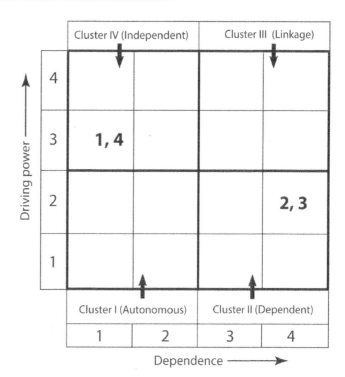

## 1.11 Design Structure Matrix

The design structure matrix (DSM) is the equivalent of the adjacency matrix in graph theory discussed in the previous sections. It is used in systems engineering and project management to model the structure of complex systems or processes. Complexity management using the DSM was originated by Don Steward in 1981. The DSM is an approach for visualizing relations and dependencies within a certain activity.[48] The two main features of DSM are as follows:

- It provides a way to represent a complex system.
- It is flexible for clustering (to facilitate modularity) and sequencing (to minimize cost and schedule risk in processes).

---

48 Danilovic, M., and Börjesson, H., "Managing the multiproject environment," in *The Third International Dependence Structure Matrix (DSM) Workshop, Proceedings.*

A DSM can represent system architecture in terms of the relationships among its essential components. Appropriate decomposition or partitioning is important for managing system complexity. In general, the following three steps are used in system engineering analysis:[49]

1) Decompose the system into elements.
2) Understand and document the interactions between the elements (i.e., their integration).
3) Analyze the potential reintegration of the elements via clustering (integration analysis).[50]

### 1.11.1 Classification of DSMs

As shown in Figure 1.37, four different types of DSM are used for different types of data.[49]

#### 1.11.1.1 Component-based DSM

Component-based DSMs analyze the relations and interactions among components in a complex system architecture. As shown in Table 1.12, interactions between system components can be classified into four types.

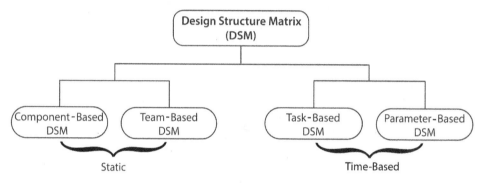

**Figure 1.37** DSM classification.

**Table 1.12** Types of interaction between two components.

| Types | Interaction between two components |
| --- | --- |
| Spatial | Associations of physical space and alignment, needs for adjacency or orientation between two components |
| Energy | Needs for energy transfer/exchange between two components (e.g., power supply) |
| Information | Needs for data or signal exchange between two components |
| Material | Needs for material exchange between two components |

49 Browning, T.R., "Applying the design structure matrix to system decomposition and integration problems: A review and new directions," *IEEE Transactions on Engineering Management*, 48(3), 2001.
50 Pimmler, T.U., and Eppinger, S.D., "Integration analysis of product decompositions," in *Proc. ASME 6th Int. Conf. on Design Theory and Methodology*, Minneapolis, MN, 1994.

#### 1.11.1.2 Team-based DSMs

Team-based DSMs are used for complex organizational analysis to show interactions and information flow among organizational entities, design teams, and team members. Organizational system analysis requires three steps:[48]

1) Break the organization down into elements (e.g., teams) with specific functions, roles, or assignments.
2) Document the interactions between (the integration of) the teams – members of each team are asked to note which other teams their team provides information to and receives information from. Also, the frequency of these interactions can be recorded. This information is used to fill in the rows and columns of the DSM.
3) Analyze the clustering of the teams into "metateams."

#### 1.11.1.3 Task/activity-based DSMs

Task/activity-based DSMs analyze relations among tasks or activities such as project scheduling and process modeling. Process modeling often consumes a lot of effort and takes a long time before results are achieved. Decomposition and integration analysis in process modeling transform the unclear process architecture into a visible and relatively easily solvable model. Process decomposition requires a good understanding of process activities and their interfaces, since the interfaces are what gives a process its added value.[48] The activity-based DSM gives a visual layout for understanding of critical issues related to process modeling. Process modeling requires three steps as follows:

1) Decompose the process into activities.
2) Document the information flow among the activities (their integration).
3) Analyze the sequencing of the activities into a maximum-feed-forward process flow (to be explained in the following sections).

#### 1.11.1.4 Parameter-based DSMs

Parameter-based DSMs analyze relationships among a set of design parameters in a complex system architecture. A parameter-based DSM is built and analyzed similarly to an activity-based DSM.

### 1.11.2 Design Dependencies

A design structure matrix shows design dependencies where the information flow between activities is visualized. Figure 1.38 shows three general dependencies among elements in a system.

In the parallel configuration (Figure 1.38(a), there is no interaction between the system elements – the activity of B is independent of the activity of A, no information exchange occurs between the two activities, and thus they can start anytime without affecting each other. In the sequential configuration (Figure 1.38(b), element A influences the activities of element B – element B of a system is selected based on the system element A. For example, if the activity is a project task, task A must be completed before task B can be started. In the coupled system

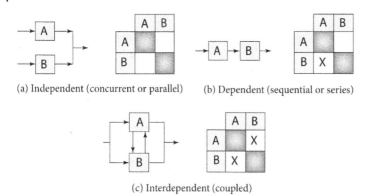

(a) Independent (concurrent or parallel)          (b) Dependent (sequential or series)

(c) Interdependent (coupled)

**Figure 1.38** Dependencies among elements.

shown in Figure 1.38(c), the flow of information is interconnected: element A impacts the activities of element B and element B impacts the activities of element A. This situation can occur if the activity of element A could not be determined without first knowing the activity of element B and activity of element B could not be determined without knowing the activity of element A. Thus, they must be completed together – a chicken and egg situation. This cyclic dependency is referred to as "circuits" or "cycles," as explained in the previous sections.

### 1.11.3   Reading a Design Structure Matrix

A DSM is developed by listing all tasks as rows and columns in an $n \times n$ matrix, where $n$ is the number of tasks. The tasks are listed close to the order in which they should be completed. A DSM can be represented as binary (to show the existence of a relation) or numerically (to represent the strength of a relation). This is also called a weighted DSM. DSMs can also be directional or non-directional. Since diagonal elements have no significance in DSMs, the relation from an element to itself is not considered.

Figure 1.39(b) shows a simple process consisting of four activities that are shown as a flow chart, with a DSM representing that process shown in Figure 1.39(a). The DSM has four activities and the corresponding information exchange patterns – which tasks provide input to others and which tasks are independent and thus can be performed in parallel. For example, in Figure 1.39(a), A provides information to C, and D provides information to A and C, etc. In short, Xs indicate which column tasks provide direct input to corresponding row tasks.

As discussed earlier (see Figure 1.22), one cannot do all data gathering to develop the DSM. Through experts from a wide range of organizations (academic or non-academic), transdisciplinary collective intelligence should be developed for reliable data collection for DSM construction. The quality of the data is the key factor for analysis and interpretation. If possible, a similar process to a CIMW can be used for reliable data collection.

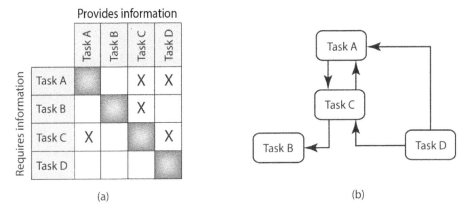

**Figure 1.39** DSM representation of a simple process.

☐ **Example 1.6**

Develop a DSM for the design process shown in Figure 1.40. Discuss the result.

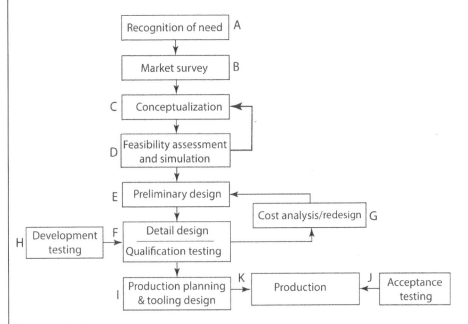

**Figure 1.40** Design process.

**Solution**

The DSM that represents the design process is shown in Figure 1.41. As discussed earlier, Xs indicate column tasks that provide direct input to corresponding row tasks (inputs in row

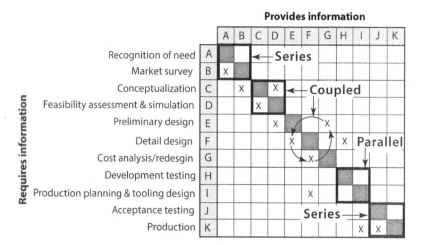

**Figure 1.41** Design structure matrix representing design process.

and outputs in column convention can also be used). Form Figure 1.41, reading across row C (conceptualization), tasks B (market survey) and D (feasibility assessment) provide direct input to task C. Similarly, reading down columns, Xs indicate row tasks that directly receive input from corresponding column task. For example, C and E receive direct input from task D. The flow of information between C and D is interconnected. Tasks C and D require input from and provide input to each other – they are coupled or interdependent. Task C cannot be initiated or completed without input from task D, and task D cannot be determined without the input from task C. The design process shows another interesting loop, E–F–G: F depends on E, E depends on G, and G depends on F. Handling this kind of situation in a DSM will be discussed in Section 1.11.6. As shown in Figure 1.41, tasks in series and parallel are also identified.

### 1.11.4   Domain Mapping Matrix

The DSM illustrated in Figure 1.39(a) can be extended to a domain mapping matrix (DMM) as shown in Figure 1.42(a). DMMs represent relations between elements of different domains. This methodology not only includes one domain at a time but also allows for the mapping between two domains. In this figure, the process consists of four activities and is extended to show who is responsible for which activity. Figure 1.42(b) shows how these persons can be mapped to the activities.

### 1.11.5   Multiple-Domain Matrix

System modeling frameworks using DSMs can be limited in scope for representing the information about entire engineering systems. Often, system complexity arises from the interaction of different domains or disciplines. The modeling of complex systems may require modeling of

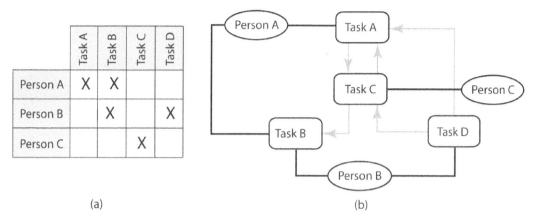

(a)  (b)

**Figure 1.42** Domain mapping matrix.

multiple domains with different types of dependencies, and hence linkages between the domains will be considered in modeling for the entire system stability.

Eppinger and Salminen presented the interactions among different aspects (people, tasks, and components) using additional (non-square) linkage matrices as shown in Figure 1.43.[51] The multiple domain matrix (MDM), which contains DSMs and DMMs, can be used to model

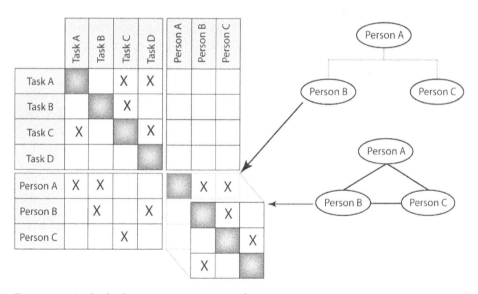

**Figure 1.43** Multiple-domain matrix. (Adapted from Eppinger, S.D., and Salminen, V., "Patterns of product development interactions," International Conference on Engineering Design, ICED '01, Glasgow, August 21–23, 2001.)

---

51 Eppinger, S.D., and Salminen, V., "Patterns of product development interactions," International Conference on Engineering Design, ICED '01, Glasgow, August 21–23, 2001.

entire systems consisting of multiple domains, each having multiple elements, connected by different relationship types.[52]

### 1.11.6 Design Structure Matrix Partitioning

Figure 1.44 shows an example of partitioning for a matrix of 12 tasks.[53] Figure 1.44(a) illustrates the unpartitioned matrix in its original form, whereas Figure 1.44(b) shows the partitioned matrix after all the feedback marks corresponding to required inputs are moved to the lower triangle or as close as possible to the diagonal.

A DSM is partitioned by reordering (i.e., sequencing) DSM rows and columns so that the new DSM becomes a block-diagonal matrix (i.e., lower triangular matrix). However, for complex systems, it is very difficult if not impossible for the simple manipulation of rows and columns to result in a lower triangular matrix.[54] Above the diagonal of a partitioned matrix, X marks denoting feedback or cycles are not desirable – time-consuming iterations are required. As shown in Figure 1.44(b), partitioning generates a decomposition of the process architecture into clusters of independent activities.

After the DSM is partitioned as shown in Figure 1.44(b), tasks in series are identified and executed sequentially, and parallel tasks are found and executed concurrently. Execution of coupled tasks requires iteration by assuming an appropriate initial guess. For example, in Figure 1.44(b), block E–D–H can be executed as follows: task E starts with an initial guess on H's output, E's output is fed to task D, then D's output is fed to task H, and finally H's output is fed to task E. This is the end of the first iteration; therefore, task E compares H's output to the initial guess made. If the error criterion is not satisfied, iteration continues. Otherwise, iteration stops, with the solution converged.[53]

Clusters contain most of the interactions of the DSM marks internally. Clusters may also interact among themselves. For an optimum DSM structure, interactions among the DSM elements (such as team members) within a cluster should be maximized and interactions among separate clusters should be eliminated or minimized.[55] Minimizing feedback loops gets the best results for binary DSMs, but not always for numeric DSMs.

The first step in partitioning is topological sorting (identifying empty rows and columns and reordering them in the DSM) before identifying loops/circuits using the path searching procedure. Information flow is traced either backwards or forwards until a task is encountered twice. All tasks between the first and second occurrences of the task constitute a loop of information flow. The steps for the partitioning procedure are given as follows (see flowchart in Figure 1.45):

---

52 Maurer, M., "Structural awareness in complex product design," Dissertation, Technische Universität, München, 2007.
53 Yassine, A., and Braha, D., "Complex concurrent engineering and the design structure matrix method," *Concurrent Engineering*," 11(3), 165–176, 2003.
54 Moraes, N.A., "Compreensão e visualização de projetos orientados a objetos com matriz de dependâncias," Universidade Federal da Bahia, 2007.
55 Braha, D., "Partitioning tasks to product development teams," in *Proceedings of the ASME 14th International Conference on Design Theory and Methodology*, Montreal, 2002.

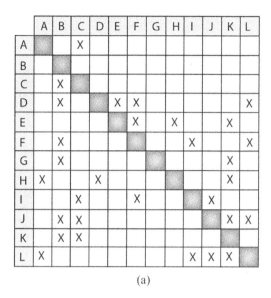

(a)

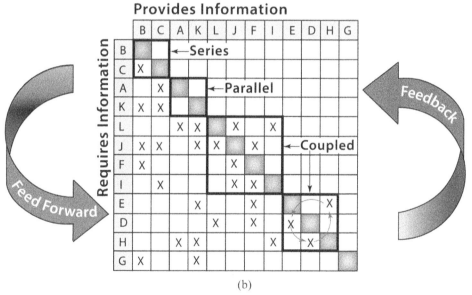

(b)

**Figure 1.44** DSM overview: (a) base DSM; (b) partitioned DSM. (From Yassine, A., and Braha, D., "'Complex concurrent engineering and the design structure matrix method," *Concurrent Engineering*," 11(3), 165–176, 2003. Used with permission.)

1) Check for the existence of any X mark along the upper diagonal of the DSM matrix. If there are no marks along the upper diagonal, then the matrix is partitioned. Hence, stop the procedure.
2) Check for empty rows in the DSM matrix and move all those empty rows to the top of the matrix and the corresponding columns to the left of the matrix and remove them from further consideration.

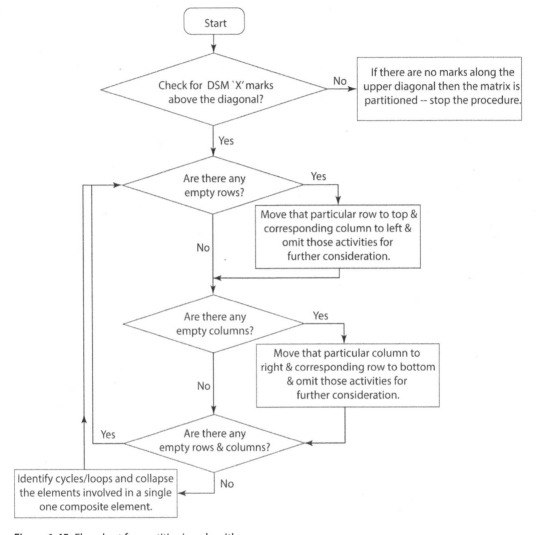

**Figure 1.45** Flowchart for partitioning algorithm.

3) For the remaining matrix, check for any empty columns and move all those empty columns to the right and the corresponding rows to the bottom of the DSM, and remove them from the DSM for further consideration.

4) Repeat steps 2 and 3 until there are no empty rows and columns in the DSM matrix.

5) Using the path searching method, identify cycles/loops, collapse the elements involved into a single composite element, and then go to step 2.

An Excel macro can be used for DSM partitioning and simulation.[56]

---

56  See DSMweb.org.

## Example 1.7

Using topological sorting and the path searching method, partition the DSM matrix shown in Figure 1.46.

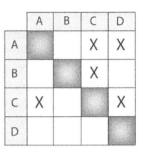

**Figure 1.46** Unpartitioned DSM matrix.

### Solution

As shown in Figure 1.46, there are four tasks in the upper triangular matrix. The following steps are used to partition the DSM matrix (see the flowchart in Figure 1.45):

1) As shown in the unpartitioned DSM matrix shown in Figure 1.47(a), task D in row D is empty – this task does not require any information from any other tasks shown in the DSM matrix. Therefore, move task D to the top of the matrix and the corresponding column task D to the left of the matrix, and removed from further consideration (see Figure 1.47(b)).

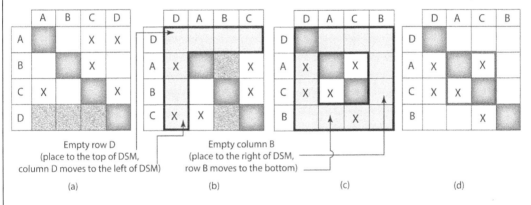

**Figure 1.47** DSM partitioning.

2) Now Figure 1.47(b) shows that task B in column B is empty – this task does not provide information to any other task in the DSM matrix. So move task B to the right and the corresponding row B to the bottom of the DSM matrix and remove them from further consideration (see Figure 1.47(c) – column B and row B are blocked).

3) As seen in Figure 1.47(c), there are no empty rows and empty columns left. There is a loop between task A and task C – task C provides information to task A and task A requires information from task C. Since such a loop is identified and clustered on the diagonal of the DSM, the partitioning process stops. The final partitioned DSM matrix is shown in Figure 1.47(d). When rows and columns are reordered, make sure that all the relationships are kept the same.

## Example 1.8

Using topological sorting and the path searching method, partition the DSM matrix shown in Figure 1.48.

## Solution

As shown in Figure 1.48, there are six tasks in the upper triangular matrix.[57] The following steps are used to partition the DSM matrix (see Figure 1.45):

1) The unpartitioned DSM matrix in Figure 1.49(a) shows that task F in row F is empty. Move task F to the top of the matrix and the corresponding column to the left of the matrix, and removed task F from further consideration (see Figure 1.49(b) – row F and column F are blocked).

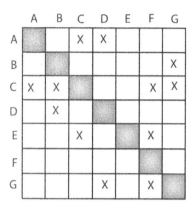

**Figure 1.48** Unpartitioned DSM matrix (from reference [55]).

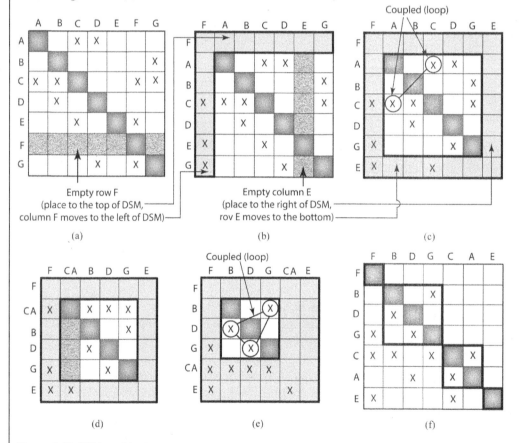

**Figure 1.49** DSM partitioning.

---

57 Adapted from Gebala, D.A., and Eppinger, S.D., "Methods for analyzing design procedures," *ASME Conference on Design Theory and Methodology*," pp. 227–233, 1991.

2) Figure 1.49(b) now shows that task E in column E is empty. Move task E to the right and the corresponding row E to the bottom of the DSM matrix, and remove from further consideration (see Figure 1.49(c) – column E and row E are blocked).

3) As seen in Figure 1.49(c), there are no empty rows and empty columns. However, there is a loop between task A and task C – task A provides information to task C and task C provides information to task A. Since task A and task C are in a loop, collapse both tasks into a single one task as shown in Figure 1.49(d).

4) But now task CA has an empty column suggesting that it is not a part of any other loop in the DSM matrix. Move task CA to the right and the corresponding row CA to the bottom of the DSM matrix, and remove from further consideration (see Figure 1.49(e) – column CA and row CA are blocked).

5) Find dependencies starting with unscheduled tasks – there are three unscheduled tasks: B, D, and G. The dependencies among them are: D depends on B, B depends on G, and G depends on D. Therefore, B–D–G is the final loop which includes all the remaining unscheduled tasks. Since all the loops are identified and clustered as a block on the diagonal of the DSM, the partitioning process stops. The final partitioned DSM matrix, with all the relationships kept the same as in the original matrix (see Figure 1.48), is shown in Figure 1.49(f).

### 1.11.7 Design Structure Matrix Tearing

When the partitioning process is completed and the loops are identified, each loop can be subjected to a further level of analysis, design structure matrix *tearing*. The objective of tearing is to reorder coupled tasks within the groups (blocks) to find an initial ordering to start the iteration process.[58]

There is no universally accepted method in existence for tearing, but when making tearing decisions the following criteria are recommended:[59]

- *Minimal number of tears.* Since tears represent an approximation or an initial guess, it is advisable to reduce the number of guesses used in the tearing process.
- *Limit tears to the smallest blocks along the diagonal.* It is desirable to limit the inner iterations to a small number of tasks.

The concept of tearing is based on guesswork and assumptions. Therefore, tearing should not be done unless absolutely necessary.[60]

---

58 Kron, G., "Diakoptics: The piecewise solution of large scale systems of equations," PhD thesis, University of Texas, Austin, 1963.

59 Yassine, A., "An introduction to modeling and analyzing complex product development processes using the design structure matrix (DSM) method," *Urbana*, 51(9), 1–17, 2004.

60 Bartolomei, J., Cokus, M., Dahlgren, J., de Neufville, R., Maldonado, D., and Wilds, J., "Analysis and Applications of design structure matrix, domain mapping matrix, and engineering system matrix frameworks," Massachusetts Institute of Technology Engineering Systems Division, 2007.

### 1.11.8  DSM Summary

- A DSM can be used by management and engineers for designing, developing, and managing complex systems or processes. It can help them to understand which tasks are constraining and will be performed in series, and which tasks will be performed in parallel. It is equivalent to an adjacency matrix in graph theory.
- The DSM can be reordered (partitioning) to eliminate or reduce X marks from the upper triangle. In short, the goal of DSM partitioning is to reorder the design tasks in order to maximize the availability of information required at each stage of design process.
- The procedures of the partitioning algorithm include scheduling independent design tasks as early as possible. Recall Figure 1.49(a), where row F was empty (meaning an independent design task) and was moved to the top of the DSM to be executed first, as shown in Figure 1.49(b).
- Feedback marks correspond to required information that is not available at the time of executing a task.[59] In another words, not all the information required may be available when it is needed. This implies a circular path of information flow (loop). Partitioning identifies such tasks in a loop and clusters them as a block on the diagonal of the DSM as shown in Figure 1.49(f).[55]
- The objective of partitioning is to reorder the design task to maximize the availability of information required at each stage of the design process and minimize the number of iterative loops within the process – remove the feedback marks from the upper triangle of the matrix to the lower triangle.[52, 55]
- The objective of tearing is break the loops to initialize the iteration.
- DSM banding is an alternative to DSM partitioning to identify the sets of independent (i.e. parallel or concurrent) elements.

## 1.12  Metrics of Complexity

*Computational complexity in design.* One metric used to assess complexity in a problem, product, or process is computational complexity. This is simply the worst-case running time for the best available algorithm to solve a given problem. This metric has its roots in computer science, but given the ability to represent many problems graphically, this method can be applied in physical design approaches. Additional variants for computational solutions of problems include counting logic equations, product terms, and the number of logic functions. These alternatives provide measures for complexity without the cost of iteratively generating computed solutions.

*Information complexity in design.* Another metric for complexity is the information content of a problem. Complexity is reduced by minimizing the number of coupled components (functional requirements and design parameters), and requiring the least information necessary to achieve a desired result from the system examined. In this way, a more complex problem requires more information to be known by the designer in order to produce the desired outcome. This metric is dependent on the knowledge of the designer, and the ability to assess the probability of success, creating an additional component to the measure called imaginary complexity.

*Empirical measures of complexity in design.* Empirically derived measures of complexity exist that are premised on the traditional understanding of complexity: specifically, decomposability. Common complex problems are often characterized by the development of functions of a system incommensurate with the sum of the individual components, or emergence. Typically, these systems consist of multiple levels of combinations, and functional behaviors can arise at any of the levels. An example arises in nature when molecular bonding occurs that differs from the possible atomic bonding.

*Cyclomatic complexity.* Cyclomatic complexity is a software metric used to measure the complexity of a program. This metric was developed by Thomas J. McCabe in 1976 and is based on a control flow representation of the program. Control flow represents a program as a graph which consists of nodes and edges. Similar to digraph as shown in Figure 1.50, nodes represent processing tasks while edges represent control flow between the nodes. Mathematically, the cyclomatic complexity, $M$, is calculated by

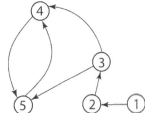

**Figure 1.50** Digraph with five nodes.

$$M = E - N + 2P, \tag{1.8}$$

where $E$ is the number of edges of the graph, $N$ is the number of nodes of the graph, and $P$ is the number of connected components. In the example in Figure 1.50 the number of edges is 6, the number of nodes is 5, and the number of connected components is 1. So the cyclomatic complexity of the digraph is

$$M = 6 - 5 + 2 \times 1 = 3. \tag{1.9}$$

A higher cyclomatic complexity means that the system will be difficult to understand. The question then arises: what is the threshold limit value of cyclomatic complexity? In other words, what is the minimum acceptable cyclomatic complexity number? McCabe stated: "The particular upper bound that has been used for cyclomatic complexity is 10 which seems like a reasonable, but not magical, upper limit."[61] When a digraph has a complexity over 10, we should reconsider the relationships of the factors in the structural self-interaction matrix.

Meaningful measurement of complexity can serve to improve our understanding and ability to work with complex systems. With the help of complexity metrics, it will be possible to track complexity changes over several product generations. We may also be able to benchmark one company's product or process complexity with respect to its competitors. Although complexity metrics are an interesting area for future research, the following factors may be considered for complexity measures:

- the number of decomposed elements (components, tasks, or teams);
- the number of interactions to be managed across the elements;

---

61 McCabe, T.J., "Describing cyclomatic complexity," *IEEE Transactions on Software Engineering*, 2(4), 308, 1976.

- the uncertainty of the elements and their interfaces;
- the patterns of the interactions across the elements (density, scatter, clustering, etc.);
- the alignment of the interaction patterns from one domain to another.

---

**CASE STUDY 1.1  A Transdisciplinary Approach to Unemployment in the United States Petroleum Industry**

Unemployment in the petroleum industry is a complex issue with social implications such as poverty and economic deprivation. With the industry supporting 9 million jobs within the United States alone, it can have far-reaching impacts for the whole economy. Approaching this issue from a transdisciplinary perspective allows for an inherently holistic view of both the problem and solution spaces. In this case study, using a TD approach, the factors involved in unemployment within the US petroleum industry will be evaluated.[a]

**Background**

The petroleum industry is present in all 50 US states and accounts for a large share of the US workforce and gross domestic product (GDP). When considering the petroleum industry, this includes exploration, extraction, refining, transport, and marketing of petroleum products. According the American Petroleum Institute, in 2011, the petroleum industry employed 5.6 percent of the US workforce, and accounted for 8 percent of the national GDP.[62] With such a strong contribution to the US economy, a large change in the employment rate for the petroleum industry can impact the overall US unemployment rate. If the petroleum industry were stable, this might not be of much concern, but history has shown it to be a cyclic industry. Unemployment is a problem that can have far-reaching effects for any society, including poverty and economic deprivation.

**Case Study Approach**

When a transdisciplinary approach is used, the benefits of having both academically collaborative research and industry-specific expertise allow the synthesis of the potential solution set to occur more rapidly, with more robust solutions.[63]

Social and economic concerns mean that it is necessary for policy-makers to define the factors that are affecting the unemployment rate the most. There are a number of factors which can impact unemployment in the petroleum industry. A CIMW was conducted in order to engage in team-building activities and come to a collective understanding of the problem. The CIMW included members of the research team, education professionals, working professionals, and researchers in related fields.

---

62 American Petroleum Institute, "Economic impacts of the oil and natural gas industry on the US economy in 2011," July 2013. Available at http://www.api.org/news-policy-and-issues/american-jobs/economic-impacts-of-oil-and-natural-gas.

63 Ertas, A., "Transdisciplinary collaboration as a vehicle for collective intelligence: A case study of engineering design education," *International Journal of Engineering Education*, 31(6a), 1526–1536, 2015.

A research team conducted brainstorming activities and reached a team consensus in identifying the root causes of and key issues for understanding the problem. The following original list of potential factors affecting unemployment in the petroleum industry was developed:

- Availability of new fields for exploration
- Emergence of new technology (e.g., fracking)
- Issues with lease/land management agreements between companies and land owners
- Market share and competing technologies
- Effect of green energy
- Impact of the coal industry
- Profit margins for oil and gas companies
- Price of oil, international supply/demand of petroleum
- Organization of the Petroleum Exporting Countries (OPEC) output rates
- International demand
- Exploration and production costs
- Automation technology
- Research and development activity levels
- Front-end engineering design activity levels
- Policy and regulations for the petroleum industry
- Export tariffs
- Domestic social issues
- Environmental concerns
- Individual employee adaptability (such as personal willingness to relocate/move family)
- Education levels/skill levels.

Although the initial efforts of the TD research team brought about a good discussion, the team reached out to experts in the petroleum industry who could help guide the TD team in reducing the initial list and preparing for a larger engagement with additional subject-matter experts in the petroleum industry.

As a result of the involvement of a 38-year veteran and former vice president within the petroleum industry, a reduced and more concise list of potential factors was created. The list included cost of labor and benefits (taxes, insurance, pension, etc.), social issues (e.g., climate change activism), individual employees' education/tradecraft skill level, regional job market competition in the energy sector (wind, solar, coal, etc.), US regulations (EPA, OSHA, FERC, etc.), lease operating expenses, price of oil, international tariffs on US oil and gas exports, increase in the use of automation and new technology, OPEC production and output amounts, international political climate and stability, individual employees' adaptability (e.g., willingness to relocate), and US import/export tariffs.

That list of factors became the basis for a formal *Kano survey* (see Chapter 3). This survey was sent to more than 15 subject-matter experts who serve as decision-makers in the petroleum industry to develop a final list of factors potentially linked to unemployment in the petroleum industry. As a

*(Continued)*

---

**CASE STUDY 1.1 (Continued)**

---

result of the inputs received from issue experts and the Kano survey, the list of factors was reduced to seven for further research:

1) Operating expenses
2) Cost of labor and benefits (taxes, insurance, pension, etc.)
3) Automation and new technology
4) Price of oil
5) US regulations (EPA, OSHA, FERC, etc.)
6) Individual employees' education/tradecraft skill level
7) OPEC production and output amounts.

Social issues, although not identified by the experts in the Kano survey as a "must" for inclusion as a part of the research, were maintained as a factor for additional research based on the input from the entire TD team. Furthermore, the OPEC production and output amounts factor was consolidated into the oil price factor based on issue expert suggestions and inputs upon review of the Kano survey results.

**Application of Interpretive Structural Modeling**

Based on the influential factors discussed above, an SSIM was created for this complex problem (see Figure 1.51). The adjacency matrix shown in Figure 1.52 was developed by transforming the SSIM into a binary matrix by substituting 1 and 0 for V, A, X, and O. The final reachability matrix, which includes driving power and dependence, is shown in Figure 1.53. Finally, the digraph is derived and MICMAC analysis performed by using the final reachability matrix (see Figures 1.54 and 1.55).

|  |  | Price of oil | Operating cost (lease) | Operating costs (labor and benefits) | Regulations | Automation & new technology | Social issues | Employee education and trade skills |
|---|---|---|---|---|---|---|---|---|
|  |  | 1 | 2 | 3 | 4 | 5 | 6 | 7 |
| 1 | Price of oil |  | A | A | A | O | O | O |
| 2 | Operating cost (lease) |  |  | O | A | O | O | O |
| 3 | Operating costs (labor and benefits) |  |  |  | A | O | O | O |
| 4 | Regulations |  |  |  |  | A | A | O |
| 5 | Automation & new technology |  |  |  |  |  | V | X |
| 6 | Social issues |  |  |  |  |  |  | O |
| 7 | Employee education and trade skills |  |  |  |  |  |  |  |

**Figure 1.51** Structural self-interaction matrix.

**Figure 1.52** Adjacency matrix.

| | | Price of oil | Operating costs (lease) | Operating costs (labor and benefits) | Regulations | Automation & new technology | Social issues | Employee education and trade skills |
|---|---|---|---|---|---|---|---|---|
| | | 1 | 2 | 3 | 4 | 5 | 6 | 7 |
| 1 | Price of oil | 1 | 0 | 0 | 0 | 0 | 0 | 0 |
| 2 | Operating costs (lease) | 1 | 1 | 0 | 0 | 0 | 0 | 0 |
| 3 | Operating costs (labor and benefits) | 1 | 0 | 1 | 0 | 0 | 0 | 0 |
| 4 | Regulations | 1 | 1 | 1 | 1 | 0 | 0 | 0 |
| 5 | Automation & new technology | 0 | 0 | 0 | 1 | 1 | 1 | 1 |
| 6 | Social issues | 0 | 0 | 0 | 1 | 0 | 1 | 0 |
| 7 | Employee education and trade skills | 0 | 0 | 0 | 0 | 1 | 0 | 1 |

**Figure 1.53** Final reachability matrix.

| | | Price of oil | Operating costs (lease) | Operating costs (labor and benefits) | Regulations | Automation & new technology | Social issues | Employee education and trade skills | Driving power |
|---|---|---|---|---|---|---|---|---|---|
| | | 1 | 2 | 3 | 4 | 5 | 6 | 7 | |
| 1 | Price of oil | 1 | 0 | 0 | 0 | 0 | 0 | 0 | 1 |
| 2 | Operating costs (lease) | 1 | 1 | 0 | 0 | 0 | 0 | 0 | 2 |
| 3 | Operating costs (labor and benefits) | 1 | 0 | 1 | 0 | 0 | 0 | 0 | 2 |
| 4 | Regulations | 1 | 1 | 1 | 1 | 0 | 0 | 0 | 4 |
| 5 | Automation & new technology | 1 | 1 | 1 | 1 | 1 | 1 | 1 | 7 |
| 6 | Social issues | 1 | 1 | 1 | 1 | 0 | 1 | 0 | 5 |
| 7 | Employee education and trade skills | 0 | 0 | 0 | 1 | 1 | 1 | 1 | 4 |
| | Dependence → | 6 | 4 | 4 | 4 | 2 | 3 | 2 | $\sum$ 25 |

(Continued)

**CASE STUDY 1.1 (Continued)**

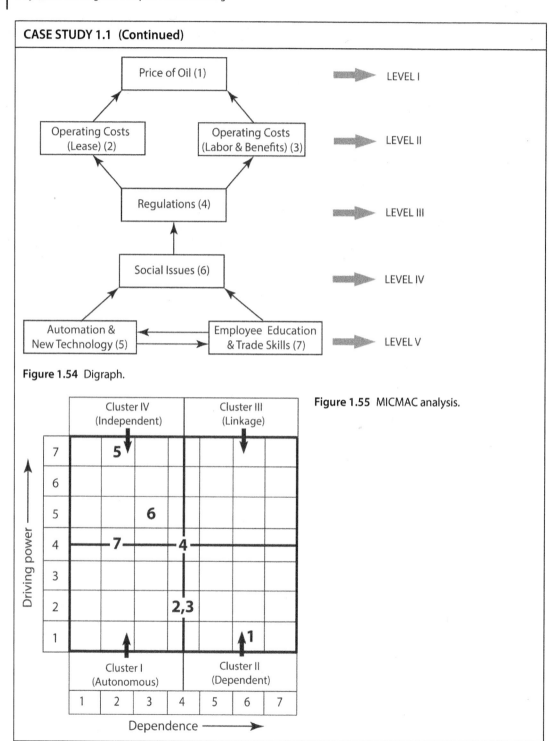

Figure 1.54 Digraph.

Figure 1.55 MICMAC analysis.

The digraph shown in Figure 1.54 reveals the interrelation of the issues impacting unemployment in the petroleum industry. Through the digraph, the cause-and-effect relationships between the factors (root causes and key issues) can be evaluated. At level V, automation (5) and education (7) are the drivers which have high leverage for the issue of unemployment in the petroleum industry. These two factors can lead to a huge increase in productivity, benefiting the whole economy. Both factors (5) and (7) are causes of other factors. Directly or indirectly improving these two factors will impact the other factors. However, it is important to note that jobs are being lost due to automation of tasks that were once performed by manual labor. While many jobs are lost due to this (automation alone is expected to cut nearly 40 percent of oil drilling jobs in Texas),[64] a smaller number of new jobs are being created. These new jobs, which require a different skill set, are focused on the information technology used to monitor and repair the automated technology.

Social issues placed at level IV (human rights, environmental protection, and anti-corruption) can have an impact on the petroleum industry, but mostly through the economic facets. These social issues can lead to laws and regulations (level III), which could then in turn impact operating costs (level II). Hydraulic fracturing (fracking) may be a good example. Fracking is a practice that has been used by the petroleum industry since the 1960s, but it has recently become a focal point of public concern, rising to the status of a social issue as people become worried about environmental contamination and public health and safety concerns. In response to this social issue, some municipalities have passed laws and regulations banning the practice.

As shown in Figure 1.55, factors affecting unemployment in the petroleum industry have been classified into four categories or clusters. It is interesting to note that four factors (2, 3, 4 and 7) are placed at the boundaries of the clusters. In such cases, a decision must be made as to which factor should be placed within which cluster. Automation/new technology is an important factor because it has high leverage for the issue of unemployment. Hence it should be considered in cluster IV. Since factors 2 and 3 are less important, they should be located within cluster I. Cluster I includes autonomous factors. These have low driving power and low dependence, hence they can be eliminated from consideration. Since factor 4 (regulations) can affect and also be affected by other factors, it should be placed in linkage cluster III.

Cluster II includes the oil price factor which has low driving power and high dependence. As seen from Figure 1.55, it has lower driving power but is extremely dependent on the other factors. Being at the top of hierarchy (see Figure 1.54), the price of oil has a direct impact on the ability of oil industry companies to retain skilled labor and therefore it is an important factor for unemployment rates. According to one of the experts who participated in the TD team, when oil prices are low there is an employment ripple effect all the way up and down the support services and product line within the energy sector.

Cluster IV includes three independent factors (5, 6 and 7) with high driving power but very weak dependence. These factors are the key drivers for unemployment in the petroleum industry.

*(Continued)*

---

64  Blum, J., "Fewer jobs in oil patch as automation picks up," *Houston Chronicle*, December 21, 2016.

**CASE STUDY 1.1 (Continued)**

The cyclomatic complexity of the digraph is $M = E - N + 2P = 9 - 7 + 2 = 4$, which means that complexity of this issue is not hard to understand and manage.

### Application of Design Structure Matrix

As shown in Figure 1.56, following the information given in Example 1.8, a design structure matrix was developed to understand and document the interactions between the factors affecting unemployment. The DSM analysis started with the adjacency matrix (see Figure 1.52), which was initially used for evaluation of ISM. As seen from this figure, only one coupling is noted, between 'automation/new technology', and 'employee education and trade skills'. This coupling is also shown in Figure 1.54.

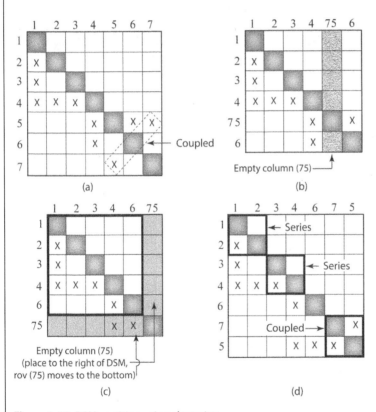

**Figure 1.56** DSM partition using clustering.

The DSM shown in Figure 1.56 represents the process of improving the planning and control of unemployment in the petroleum industry. DSM partitioning generates a decomposition of the process architecture into clusters of independent activities. After the DSM is partitioned as shown in Figure 1.56(d), two tasks in series are identified to be executed sequentially. For example, the oil price (1) must be known before we make a decision on operating costs (2). The execution of coupled tasks (5 and 7) requires iteration by making an appropriate initial guess. While these two factors influence each other, their combined impact on factors like regulations causes a large ripple throughout the rest of the significant factors identified.

In summary, the oil price controls the unemployment rate in the petroleum industry. For example, companies in the energy sector can be expected to respond to low oil prices by eliminating jobs and reducing investment in their future projects.

This case study considered how ISM and DSM can be used in combination to provide different but complementary perspectives on the complex issue of the unemployment problem in the oil industry. The DSM is a useful tool for clustering (integration analysis) and sequencing the factors (tasks), whereas ISM is an effective tool for decomposing a complex issue into smaller sub-issues and provides an understanding of how various factors are relevant to the problem.

[a]This case study is adapted from the TD project assigned to PhD students, Christopher Kreger, Daniel Moran, and Stacie Therson (submitted to Dr. A. Ertas), 2017.

# Bibliography

1 ARORA, S., and BARAK, B., *Computational Complexity: A Modern Approach*, Cambridge University Press, Cambridge, 2009.

2 AXELROD, R., *The Complexity of Cooperation: Agent-Based Models of Competition and Collaboration*, Princeton University Press, Princeton, NJ, 1997.

3 BERGENDORFF, S., *Simple Lives, Cultural Complexity: Rethinking Culture in Terms of Complexity*, Penguin, New York, 2012.

4 CROWELL, D.M., *Complexity Leadership*, F.A. Davis Company, Philadelphia, 2016.

5 GHARAJEDAGHI, J., *Systems Thinking: Managing Chaos and Complexity: A Platform for Designing Business Architecture*, Morgan Kaufmann, Burlington, MA, 2011.

6 JENKINS, D.P., STANTON, N.A., SALMON, P., and WALKER, G.H., *Cognitive Work Analysis: Coping with Complexity*, Ashgate Publishing, Farnham, 2009.

7 KOEN, F., *Innovation, Evolution and Complexity Theory*, Edward Elgar Publishing, Cheltenham, 2005.

8 RZEVSKI, G., and SKOBELEV, P., *Managing Complexity*, WIT Press, Boston, 2014.

9 WOOD, R., *Managing Complexity*, Profile Books, London, 2000.

10 ZELINKA, I., and ADAMATZKY A., *Emergence, Complexity and Computation*, Springer, Berlin, 2012.

## CHAPTER 1 Problems

**1.1**  What do you think "transdiscipline" is? Why?

**1.2**  What do you think "complexity" is? Why?

**1.3**  Do you think complexity is inherently transdisciplinary? Why or why not?

**1.4**  Does reducing social complexity even make sense, especially in an ostensibly engineering context? Why or why not?

**1.5**  As a student, do you think your homework project assignments are becoming more complex? Why or why not? Give evidence to support your conclusion.

**1.6**  Create your own complex system scenario and discuss (see Example 1.1).

**1.7**  Develop a digraph of the matrix shown in Figure P1.7.

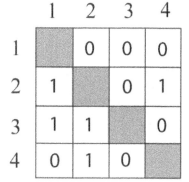

**Figure P1.7** Relationship matrix.

**1.8**  Which elements are coupled in the matrix shown in Figure P1.8?

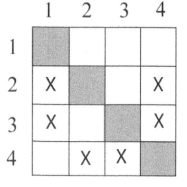

**Figure P1.8** Relationship matrix.

**1.9** What is the DSM representation of the task-based process given in Figure P1.9?

**Figure P1.9** Domain mapping matrix (DMM).

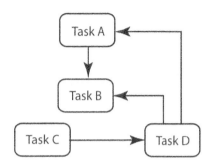

**1.10** For the domain mapping matrix given in Figure P1.10, show how these persons can be mapped to the activities. Use the result of Problem P1.9.

**Figure P1.10** Domain mapping matrix (DMM).

|  | Task A | Task B | Task C | Task D |
|---|---|---|---|---|
| Person A |  | X | X |  |
| Person B | X |  |  | X |
| Person C |  |  |  | X |

**1.11** Develop an adjacency matrix using the SSIM shown in Figure P1.11.

**Figure P1.11** Structural self-interaction matrix.

|  |  | Factor 1 | Factor 2 | Factor 3 | Factor 4 | Factor 5 | Factor 6 | Factor 7 | Factor 8 |
|---|---|---|---|---|---|---|---|---|---|
|  |  | 1 | 2 | 3 | 4 | 5 | 6 | 7 | 8 |
| 1 | Factor 1 |  | A | A | A | O | A | O | O |
| 2 | Factor 2 |  |  | O | A | O | O | O | O |
| 3 | Factor 3 |  |  |  | A | O | O | O | O |
| 4 | Factor 4 |  |  |  |  | A | O | A | O |
| 5 | Factor 5 |  |  |  |  |  | O | V | X |
| 6 | Factor 6 |  |  |  |  |  |  | O | O |
| 7 | Factor 7 |  |  |  |  |  |  |  | O |
| 8 | Factor 8 |  |  |  |  |  |  |  |  |

**1.12** Develop a final reachability matrix using the SSIM shown in Figure P1.11. Show the dependence and driving power of each element.

**1.13** Develop a digraph using the SSIM shown in Figure P1.11. Discuss the relationships of the factors.

**1.14** Perform and discuss the MICMAC analysis of the SSIM shown in Figure P1.11.

**1.15** Develop the design structure matrix of the SSIM shown in Figure P1.11. Identify couplings between the factors if there are any.

**1.16** Patition the design structure matrix (DSM) of Problem P1.15 using clustering. Discuss the result.

**1.17** What is the cyclomatic complexity of the digraph given in Figure P1.17?

**Figure P1.17** Digraph.

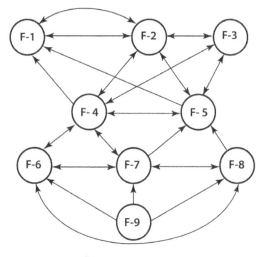

**1.18** Identify the cycles of the DSM given in Figure P1.18.

**Figure P1.18** Design structure matrix.

|   | 1 | 2 | 3 | 4 | 5 | 6 |
|---|---|---|---|---|---|---|
| 1 |   |   |   |   |   |   |
| 2 | X |   |   |   | X |   |
| 3 | X |   |   | X |   |   |
| 4 |   |   |   |   |   | X |
| 5 |   |   | X |   |   |   |
| 6 |   |   |   | X |   |   |

**1.19** Identify the cycles of the DSM given in Figure P1.19.

**Figure P1.19** Design structure matrix.

|   | 1 | 2 | 3 | 4 | 5 | 6 |
|---|---|---|---|---|---|---|
| 1 |   |   | X |   |   |   |
| 2 | X |   | X |   | X |   |
| 3 | X |   |   | X |   |   |
| 4 |   |   |   |   |   |   |
| 5 |   |   |   |   |   | X |
| 6 |   |   |   |   | X |   |

**1.20** Optimize the DSM given in Figure P1.20 by clustering.

**Figure P1.20** Design structure matrix (DSM).

|   | 1 | 2 | 3 | 4 | 5 | 6 | 7 |
|---|---|---|---|---|---|---|---|
| 1 |   |   |   |   |   |   |   |
| 2 | X |   |   |   |   |   |   |
| 3 |   |   |   |   |   |   |   |
| 4 | X |   | X |   | X |   |   |
| 5 |   | X |   |   |   |   | X |
| 6 |   |   |   | X |   |   | X |
| 7 |   |   |   |   | X |   |   |

**1.21** Optimize the DSM given in Figure 1.39 (see Example 1.6) by clustering.

# 2

# Transdisciplinary Design Process

What we need are new choices – new products that balance the needs of the individuals and of society as a whole; new ideas that tackle the global challenges of health, poverty, and education; new strategies that result in differences that matter and a sense of purpose that engages everyone affected by them.

**Tim Brown, 2009**

## 2.1  Introduction

Recent studies suggest engineering education must evolve to teach a more holistic approach to problem solving in order to better prepare students to handle the growing complex problems inherent in today's society.

Definitions of TD research go back to the early 1970s.[1] According to Ertas et al.:

> Transdisciplinarity can be defined as the practice of acquiring new knowledge through education, research, design, and production with a broad emphasis on complex problem solving and the use of knowledge and techniques from multiple scholarly disciplines. TD methods are unique in their ability to bring discipline-specific knowledge together holistically in order to clove complex problems. The goal of transdisciplinary practice is to improve students' understanding of complex issues by extracting the valuable aspects of typical academic disciplines and thereby generating both a more integrative and universal solution to support an issue of importance to society.[2]

---

1  Nicolescu, B., "Methodology of transdisciplinarity – levels of reality, logic of the included middle and complexity," in *ATLAS T3 International Conference Proceedings*, Georgetown, TX, 2010.
2  Ertas, A., Frias, M.K., Tate, D., and Back, M.S., "Shifting engineering education from disciplinary to transdisciplinary practice," *International Journal of Engineering Education*, 31(1-A), 2015, 94–105, 2015.

In another paper, Ertas et al. say:

> TD research and education also focus on leveraging intellectual diversity, collaborative effort, cross-pollinating knowledge, and critical thinking to address real-world problems. The tools of TD – the mindset, the content knowledge, the social-impact-based design process – are all geared toward providing students with the preparation they need to see complexity, and to address those complex problems in innovative ways.[3]

In this chapter, the traditional design methods and design process will be discussed before the transdisciplinary design process is covered. They are both similar inasmuch as both design processes emphasize the creation of good-quality, safe, and more reliable end products or systems. Since both approaches are somewhat different, they can complement each other because they are in many ways alike; they can be combined to create a new design process.

## 2.2 Design

> Design, so construed, is the core of all professional training; it is the principal mark that distinguishes the professions from the sciences. Schools of engineering, as well as schools of architecture, business, education, law and medicine, are all centrally concerned with the process of design.
>
> **Simon, 1996**

What does the term "design" mean? According to one definition, design is a process of logical deduction about a set of requirements and constraints. To say, "I would like to design an artifact" does not provide any understanding of the proposed design. Specific things must be known about the artifact before the design process can be initiated. For example, we might require the artifact to function in a certain way and to a certain efficiency, and to be dependable when operating at specified condition. Several definitions of the concept of design have been put forward by other authors:

> Engineering design is the process of applying various techniques and scientific principles for the purpose of defining a device, a process, or a system in sufficient detail to permit its physical realization.
>
> **Taylor, 1959**

---

3 Ertas, A., Greenhalgh-Spencer, H., Gulbulak, U., Turgut, B.B., and Frias, M.K., "Transdisciplinary collaborative research exploration for undergraduate engineering students," *International Journal of Engineering Education*, 33(4), 1242–1256, 2017.

Engineering design is a purposeful activity directed towards the goal of fulfilling human needs, particularly those which can be met by the technology factors of our culture. And decision making, in the face of uncertainty, with high penalty of error.

**Asimow, 1962**

Finding the right physical components of a physical structure.

**Alexander, 1963**

The performing of a very complicated act of faith.

**Jones, 1966**

Engineering design is the use of scientific principles, technical information and imagination in the definition of a mechanical structure, machine or system to perform pre-specified functions with the maximum economy and efficiency.

**Fielden, 1963**

Design is the process of devising a system, component, or process to meet desired needs.

**Ertas and Jones, 1996**

Design involves a prescription or model, the intention of embodiment as hardware, and the presence of a creative step.

**Archer, 1964**

The creation of a synthesized solution in the form of products, processes or systems that satisfy perceived needs through mapping between the functional requirements (FRs) in the functional domain and the design parameters (DPs) of the physical domain, through proper selection of the DPs that satisfy the FRs.

**Suh, 1989**

Design is concerned with how things ought to be, with devising artifacts to attain goals.

**Simon, 1996**

Different than the above quotes, a new definition of design from the transdisciplinary point of view is as follows: Transdisciplinary design is an activity of collective intelligence bringing imagination into reality to transformationally benefit society. Functional social artifacts, cultural artifacts, and technological artifacts are the end results of TD design activity.

The above definition of transdisciplinary (TD) design indicates that design is more than designing technological (product) artifacts. Factors affecting for TD design are:

- cost;
- environment and sustainability;
- complexity;
- present technology and consciousness;
- political, social, cultural, and psychological issues.

The best engineering solutions are not the best political solutions; conversely, the best political solutions may not be the best engineering solutions. For example, the decision to build a 230-story tower in order to have the tallest building in the United States is definitely not the best engineering approach because high costs and inefficient energy utilization are inevitable. Similarly, when environmental issues are taken into account during a design, the solution may not be the most creative, economically favorable, or aesthetically attractive option. Most advanced technological artifacts may not be the best products if they do not consider the human minds and actions intertwined within consciousness technology.

### 2.2.1 Important Features of Design

The following model highlights important features of the design process:[4,5]

1) Design is an investigative process (research). In general, the initial step of the design process is to communicate and understand the target customer's needs and expectations. In most cases, meeting the needs of customer can be a complex issue. Using design techniques and methods, identifying customer needs and product requirements based on findings from initial user research activities, which include the feedback collected from potential customers and stakeholders for iterative design and testing of a product, past failures and successes should be investigated to seek the solution to the problem at hand.

2) Design is a creative process (art). The question whether design is an art or science has been discussed in many articles published in the past. Most authors recognize the significance of intuition in developing creative solutions to design problems, in particular during the concept formulation stage. The intuition of a well-qualified designer is a trait developed over a period of time during which the designer is *doing design*. These accumulated design experiences are stored in the brain and, in some miraculous way, are accessed during the process of developing solutions to new design problems so that (ideally) the appropriate elements of each experience are synthesized to formulate a concept. This justification makes a convincing argument that design can only be taught by *doing design* and that design is more an art than a science. However, there is another aspect of design that offers itself to the scientific method (analysis). Analysis begins almost immediately after the design concept has been formulated.

---

4 Rzevski, G., On the design of a design methodology, in R. Jacques and J. Powell (eds), *Design Science Method – Proceedings of the 1980 Design Research Society Conference*, Westbury House, Guildford, 1981.
5 Ertas, A., and Jones, J.C., *The Engineering Design Process*, John Wiley & Sons, Inc., New York, 1996.

The analytical method is applied to analyze requirements, establish specifications, evaluate test data, define interfaces, select components, validate performance, etc. The design literature describes many of the structured (or systematic) approaches that can be applied to this analytical effort.

3) Design is a rational process (logic-based). Selection of design attributes is an integral part of the logic process. Logic provides insight into the design process. In general, developing design solutions involves logical reasoning, mathematical analysis, computer simulation, laboratory experimentation, prototype field testing, etc. Making logical deductions about a set of requirements is a rational process which can be performed with the help of well-known scientific methods.

4) Design is a decision-making process (value-based). The design of alternative courses of action is an essential part of the decision-making process. Product design is a knowledge-intensive process and involves large quantities of decisions. Therefore, the selection of design parameters for a product involves considerable uncertainty. Obtaining an optimum solution to a design problem is the key to the design task. By estimating the value which clients are likely to place on different alternatives, one can arrive at an appropriate decision.

Note that not every design problem will demand all four types of knowledge and skills described above.

### 2.2.2   Design Methods

Formal or informal procedures, rules, techniques, and tools for designing are called design methods. Over the last several decades, intensive effort has resulted in a significant increase in scientific design methods; this has led to a large body of research on new conventional and unconventional methods being published in the literature. Tables 2.1 and 2.2 present some of these design methods and their purpose.[6,7]

## 2.3   Design Process Models

Since the 1960s, numerous design process models have been developed by researchers. In general, the design process begins with an identified need and concludes with satisfactory qualification and acceptance testing of the product. Of course, generating a concept early in the design process is important. This initial concept is then subjected to analysis, synthesis, refinement, and finally development.

Design process models can be divided into three categories: those that are based on the essential design activities, those that are based on the phases of evolution of the design object, and those that are based on knowledge.

---

6  Adapted from Cross, N., *Engineering Design Methods*, John Wiley & Sons, Ltd., Chichester, 1994.
7  Adapted from Jones, J.C., *Review of: Design Methods*, John Wiley & Sons, Inc., New York, 1980.

**Table 2.1** Design methods.

| Design methods | Methods | Aim |
|---|---|---|
| *Methods of exploring design situations* | Stating objectives | To identify external conditions with which the design must be compatible |
| | Literature search | To find published information that can favorably influence the designers' output and that can be obtained without unacceptable cost and delay |
| | Searching for visual inconsistencies, interviewing users | To find published information that can favorably influence the designers' output and that can be obtained without unacceptable cost and delay |
| | Questionnaires | To collect usable information from the members of a large population |
| *Methods of searching for ideas* | Brainstorming | To stimulate a group of people to produce many ideas quickly |
| | Synectics | To direct the spontaneous activity of the brain and the nervous system toward the exploration and transformation of design problems |
| | Removing mental blocks | To find new directions of search when the apparent search space has yielded no wholly acceptable solution |
| | Morphological charts | To widen the area of search for the solution to a design problem |

### 2.3.1 Activity-Based Design Process Model

The three-stage model is the simplest example of an activity-based design process, and is based on three essential design activities – analysis, synthesis, and evaluation, as shown in Figure 2.1. The first activity involves understanding the problem and generating an explicit statement of the design process goals. The second activity involves generating possible solutions. During the third activity the validity of possible solutions is tested with respect to the goals (or objectives), requirements, constraints, and finally, the best solution among alternatives is selected. As shown in the Figure 2.1, the problem solution can be refined by reexamining the analysis. In other words, the design process is iterative in the sense that the sequence may be repeated until the problem solution converges to a satisfactory conclusion. Asimow pointed out that each cycle is progressively less general and more detailed than the proceeding one.[8] Although the three-stage design process is the simplest model, there has been widespread agreement on the process shown in Figure 2.1 (Asimow, 1962).

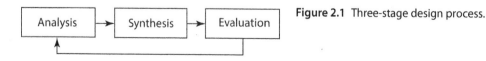

**Figure 2.1** Three-stage design process.

---

8 Asimow, I.M., *Introduction to Design*, Prentice Hall, Englewood Cliffs, NJ, 1962.

**Table 2.2** Design methods.

| Design Methods | Methods | Aim |
|---|---|---|
| *Methods of evaluation* | Checklists | To enable designers to use knowledge of requirements that have been found to be relevant in similar situations |
| | Selecting criteria | To decide how an acceptable design is to be recognized |
| | Ranking and weighting | To compare a set of alternative designs using a common scale of measurement |
| | Specification writing | To describe an acceptable outcome for designing that has yet to be done |
| *Methods of exploring problem structure* | Interaction matrix | To permit a systematic search for connections between elements within a problem |
| | Interaction net | To display the pattern of connections between elements within the design problem |
| | Analysis of interconnected decision areas | To identify and to evaluate all the compatible sets of sub-solutions to a design problem |
| | System transformation | To find ways of transforming an unsatisfactory system so as to remove its inherent faults |
| | Functional innovation | To find a completely new design capable of creating new patterns of behavior and demand |
| | Alexander's method of determining components | To find the right physical components of a physical structure such that each component can be altered independently to suit future changes in the environment |
| | Classification of design information | To decompose design problem into manageable parts |

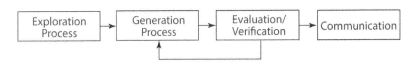

**Figure 2.2** Four-stage design process.

As shown in Figure 2.2, Cross[6] describes design as a four-stage process which is slightly different than the three-stage design process. This simple four-stage model assumes that the final stage of the design process is the communication of the design – that is ready for the production. The design activity prior to the communication stage concerns judging the validity of problem solutions against objectives, requirements, and constraints. The problem solution itself arises from the generation of a concept by the designer, usually after some initial exploration of the ill-defined problem space. In Figure 2.2, an iterative feedback loop for design refinement is shown between the evaluation and generation stages.

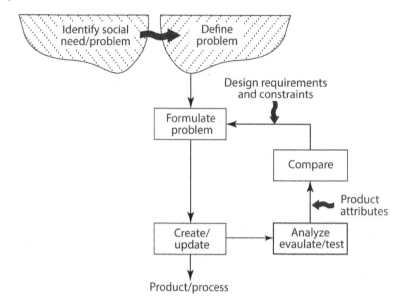

**Figure 2.3** The phase-based design model.

### 2.3.2 Phase-Based Design Process Model

In the phase-based design model, phases can be expanded with more specific activities or steps as in activity-based models. Figure 2.3 presents an example of an idealized phase-based design process model.

## 2.4 Typical Steps in Engineering Design Process

Typical steps in the engineering design process are shown in Figure 2.4.[5] The process shown in the figure is considered to be generally applicable to most design efforts, but the reader should recognize that individual projects will often require variations, including the elimination of some steps. This is especially true for design efforts accomplished within small organizations in which the design process is less formal.

The design process begins with an identified need which can be satisfied by the defined design requirements and constraints. A design process can be initiated based on an idea for a solution to an existing or identified need or from an idea for a product or process for which it is thought a need can be generated. Many toys are products for which the idea precedes the identified need.

### 2.4.1 Recognition of Need

Recognition of customers' needs is the first step in the design process. Good understanding, as well as accurately and completely identifying customers' needs will better enable one to provide

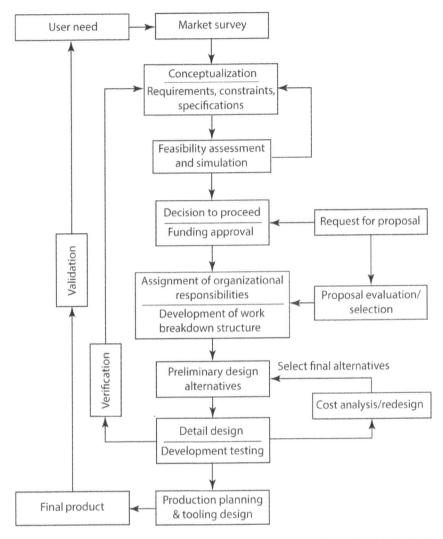

**Figure 2.4** The design process. (From Ertas, A., and Jones, J.C., *The Engineering Design Process*, John Wiley & Sons, Inc., New York, 1996. Reproduced with permission of John Wiley & Sons, Inc.)

them with the products and services that they desire. The process of identifying customer needs and expectations will be discussed in detail in Chapter 3.

### 2.4.2 Process of Concept Development and Creativity

A generic concept development process is shown in Figure 2.5. Concept development is a process of collaborative study and application of innovation driven by a set of customer requirements and target product specifications. The first action in concept development is to bring a group of

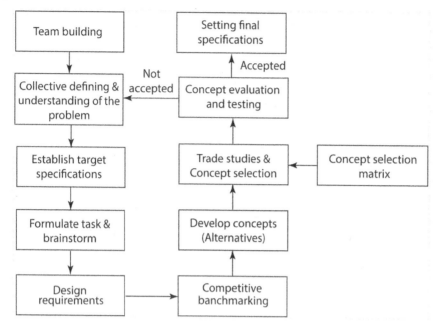

**Figure 2.5** Process of concept development.

individuals together with a diverse range of strengths to create a team. An effective design concept development team should consist of individuals with different and complementary skills, working together towards a shared project goal.

After the problem has been completely understood and defined, the creative process of developing ideas can proceed with ideation, where the design team visualizes concepts and defines key features that contribute to a possible solution to the problem in hand. In general, methods of ideation include: conducting brainstorming sessions, examining existing solutions, creating and using analogies, and collaborative sketching. Through this process, viable solutions need to be identified from which the optimum approach can be selected.

Concept evaluation implies both comparison and decision-making. Concepts should be evaluated against similar current products and against each other to improve the concepts and to determine target specification and market value.

Brainstorming, in which a group of participants try to generate as many possible solutions as they can before choosing the best concept, is one technique that has been used to assist in identifying alternate solutions to problems. The purpose of brainstorming is not to discover the perfect design for the project. Instead, the aim is to create many concepts that approach the project from a wide variety of angles. Since individuals with different and complementary skills are involved in the process, this approach may have some merit since the ideas suggested by one person may trigger ideas for better solutions on the part of another. Note that in the early stages of concept development, group brainstorming should also include customers to document design requirements that drive design.

In a competitive benchmarking study, one can investigate what features and limitations the other existing products have. After reviewing the competitive landscape, it is time to develop several good concepts (alternatives).

After developing several concepts, it is time to select the best option; concept selection matrix is used to evaluate various alternatives. Using measures defined by the criteria, alternatives are compared and the concept to be evaluated is selected. The Pugh matrix can be used to select among the alternatives.

Concept testing is the process of evaluating possible customer responses to a product concept before its introduction into the market. The goal is to prove that the product concept is better than competing solutions – how successful the product will be in the market. As shown in Figure 2.5, if the concept testing fails, the concept development can be refined by reexamining the process until the the final concept selection is made.

### 2.4.2.1 The Pugh Decision Matrix

A decision matrix is a simple tool that can be very helpful in making decisions, especially in cases where there are several alternatives and many criteria of varying importance to be considered. A *Pugh matrix* (PM), also known as a *criteria-based matrix*, is used to compare a number of design concepts in order to identify which best meets a set of criteria. The list of the most important criteria in the decision is developed and then the alternatives are compared according to their positives and negatives against a reference concept called the datum (base concept).

### Example 2.1

When we are traveling, most of us want to get to our destination as quickly as possible and in comfort. Oftentimes, our destination dictates what mode of transportation we use – plane, train, bus, or car. Suppose we are deciding between three alternatives (plane, train, and bus) to travel within the state of Texas. Suppose also that decision will be made against the baseline of using your car to travel.

### Solution

First, decide what are the most important criteria that will be evaluated against alternatives. Suppose that they are: (1) hassle-free journey, (2) comfort, (3) cost, and (4) short travel time. Next, apply a weighting score to the criteria based on their relative importance as shown in Table 2.3 (cost being the most important criterion in this example).

Now score each option in each category based on whether it is better or worse than the baseline: +1 being better than, −1 being worse than, and 0 being equal. For example, consider traveling by plane. In relation to criterion 1, we consider that traveling by plane is worse than traveling by your own car (baseline). Hence we score −1 for this option.

Finally, multiply each score by the weighting factor, then add up the scores for each option. For example, criterion 2 has a weighting of 2. Thus, all the numbers to the right of it are multiplied by 2 (see Table 2.4).

**Table 2.3** Pugh decision matrix.

| | | Baseline: your car | Travel by plane | Travel by train | Travel by bus |
|---|---|---|---|---|---|
| **Criterion** | Weight | | | | |
| Hassle-free | 1 | 0 | 0 | 0 | |
| Comfort | 2 | +1 | −1 | −1 | |
| Cost | 4 | −1 | −1 | +1 | |
| Short time | 3 | +1 | −1 | −1 | |
| | **Totals** | | | | |

**Table 2.4** Pugh decision matrix.

| | | Baseline: your car | Travel by plane | Travel by train | Travel by bus |
|---|---|---|---|---|---|
| **Criterion** | Weight | | | | |
| Hassle-free | 1 | 0 | 0 | 0 | |
| Comfort | 2 | +2 | −2 | −2 | |
| Cost | 4 | −4 | −4 | +4 | |
| Short time | 3 | +3 | −3 | −3 | |
| | **Totals** | +1 | −9 | −1 | |

As shown in Table 2.4, traveling by plane can be selected because it receives the highest overall score. Note that, depending on the number of criteria and the weighting used, we may obtain different results. We can also use different scoring in each category based on whether it is better or worse than the baseline, for example, +2 for much better than, +1 for better than, 0 for equal, −1 for worse than, and −2 for much worse than.

De Bono describes 60 *tools* that can be used to enhance a person's thinking skills. One of the simplest tools presented is named PMI.[9] PMI is an attention-directing tool in which a person's attention is first directed toward the Plus (P) or good points in a suggestion, then toward the Minus (M) or bad points, and finally toward the Interesting (I) points. The benefit of PMI is that it forces the direction of attention onto goals (getting as many P, M or I points as possible) rather than focusing on the single solution that normal prejudices would lead to. The PMI tool provides a good exercise for students in capstone design courses. Students can be divided into groups to do a PMI and the number of plus, minus, and interesting points can be compared, or the entire class can do the exercise together.

---

9 De Bono, E., *De Bono's Thinking Course: Revised Edition*, Facts on File, New York, 1994

#### 2.4.2.2 Creativity

Creativity is a mental characteristic allowing a person to think outside of the box, which results in innovative or different approaches to solving problems. If creativity is a characteristic that is both inherited (intelligence) and capable of being learned (thinking), people with average creative potential can sharpen their abilities with appropriate training. Some people are born with a high potential for creativity, and with training can become exceptional. Several great composers and scientists of the past come to mind. The following list includes several personality characteristics thought to correlate with high creativity:[10]

1) Good guessing
2) Risk taking
3) Challenging authority and procedures
4) Preferring the complex and difficult
5) Being sensitive emotionally, having a sense of beauty
6) Having a vivid imagination
7) Divergent thinking
8) Desiring honesty and frankness
9) Being curious
10) Having high self-esteem.

People who feel that they have very little inherent creative ability need only to remember the comment by Thomas Alva Edison that invention is 99 percent perspiration and 1 percent inspiration. If an individual is committed to achieving a goal, believes that she has it within herself to be successful, and is willing to perspire, she can realize almost any vision. Design concepts are not envisioned by applying mathematical equations and formulas but by drawing on the memories and experiences of that miraculous computer called the mind. The mind not only reviews the contents of our memory to draw useful bits of data together that can be used in formulating conceptual solutions but also uses this data to form new or modified images that can be disassembled, assembled and manipulated to form devices that do not exist.[11]

### 2.4.3 Feasibility Assessment

Assessment of the feasibility of the concept(s) selected is carried out, after which schedules, resource plans, and estimates for the next phase are developed. The feasibility study is an assessment and analysis of the project to support the process of decision-making. The purpose of a feasibility assessment of the concept is to determine whether or not the project can be advanced to the final design phase and production stage with a concept that is achievable technically and within cost constraints. Key issues in the feasibility study are technical, economic, legal, environmental issues, capital and operating costs estimates, and production schedule. It is important to

---

10 Based on communication with colleagues during presentation on "Creativity in Education," Texas Tech University, 1983.
11 Ferguson, E.S., *Engineering and the Mind's Eye*, MIT Press, Cambridge, MA, 1992.

have engineers with experience and good judgment involved in the feasibility assessment phase of the design process.

### 2.4.4 Establishing and Writing Good Design Requirements

Establishing and managing design requirements is one of the most important elements in the design process. This is a task that will be accomplished after the concept has been selected. It is often accomplished during the feasibility study after the design concept has been defined, or it may be accomplished early during the design effort.

Design requirements are specific documented physical and functional needs that a particular design, product or process must be able to perform. They identify capabilities, conditions and attributes. The first step toward developing accurate and complete specifications is to determine correct requirements. This is not an easy task – it is more of an art than a science. Requirements are used throughout the system development life cycle activities – from beginning to completion:

- Requirements capture the essence of a concept.
- Requirements drive a design.
- Requirements provide a means to verify an implementation.

The following steps are used for developing requirements:

1) *Define functions.* Functions describe what the product is going to accomplish.
2) *Establish attributes.* Attributes are characteristics desired by the customer. It is important to note that while two products can have similar functions, they may have completely different attributes.
3) *Determine constraints.* After all the attributes have been clarified and attached to functions, determine the constraints on each of the attributes.
4) *Determine expectations.* Determine what the customer's expectations are and, finally, test on a regular basis to determine if a customer will be satisfied with a product.

The following steps are useful in writing good design requirements:

- Design requirements must be defined as clearly as possible – uncertainty leads to confusion and unhappiness. Although it should be self-evident, it is absolutely essential that all project team participants have a complete and agreed-upon understanding of the requirements. A good example of what can happen as a result of a misunderstanding of the design requirements occurred during a large (and costly) missile program in the late 1950s.[5] In this case, there was confusion between disciplinary organizations as to the meaning of one of the major system design requirements. Unfortunately, this confusion did not become apparent until late in the design process when the overall systems checkout and test procedure was being prepared. The confusion was related to the meaning of the terms *automatic, semi-automatic,* and *manual.* At significant cost, the mechanical design group had provided the capability for all three modes of system operation based on its understanding of these terms, but the control system design group had a different interpretation of these terms and thus did not provide this capability. By the time this misunderstanding was discovered it was impossible to correct the design and one of the major design requirements could not be satisfied.

- Define one requirement at a time – each requirement should be short. To avoid confusion try not to use conjunctions such as *and, or, also, with*. If extensive conjunctions is necessary for clarity, the sentence should be decomposed into a shorter sentences.
- Each requirement should have a complete sentence with no buzzwords or acronyms. Vague and unverifiable terms (e.g., user-friendly, versatile, robust, approximately, minimal impact, easy, sufficient, flexible, adequate, fast, large, small) should not be used. This will cause difficulties in defining the requirement's test cases.
- Each statement that defines a requirement must contain the word "shall." Use positive statements such as "The system shall…," instead of "The system shall not…." "Will" should be used only for statements that provide information and "should" be used to represent a goal to be achieved. The following are some examples using correct terms:
  1) The engine shall operate at a power level of 425 hp.
  2) The tower shall withstand wind loads of 100 mph.
  3) The cruise ship shall travel at a speed of 22 knots.
  Following are some examples using incorrect terms:
  1) The engine shall not operate at less than a power level of 425 hp.
  2) The tower will withstand wind loads of 100 mph.
  3) The cruise ship should travel at a speed of 22 knots.
- Requirements must be realistic and allow acceptable solutions – they should state *what* is needed, not *how* to provide it. The difference between requirements and design is that the requirements represent what the system is supposed to have/do, whereas the design is how the system will accomplish what it is supposed to have/do.
- Requirements must be quantifiable and verifiable.
- Do not use "to be determined" and "to be resolved." Use the current best estimate within brackets in the requirement, and state why the value is still an estimate. State tolerances for qualitative values (e.g., less than, greater than or equal to, plus or minus, 3 sigma).

Establishing the design requirements is one of the most difficult activities in the design process. The requirements are so critical to the final design capability and its cost. Figure 2.6 shows the cost of change with respect to requirement changes, due to being missed or misunderstood, during the product development life cycle (waterfall). As seen from this figure, any design changes near the end of the product development phase will entail a much bigger cost than an earlier design change. For example, if a designer makes a requirement error during the requirement phase, it can be corrected inexpensively, say $1 unit. If this error is found during the analysis and design phase, to make design changes to fix this mistake may cause a cost increase of $5 units. If the designer cannot find the problem until the build stage, design and analysis parts need to be changed. Because of all the wasted development time, the cost increase

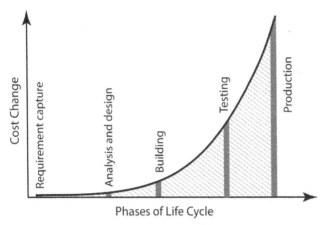

**Figure 2.6** Cost of design requirements.

may go as high as $10 units. Likewise, if the error detection is found during the testing phase the cost increase can be of the order of $70 units. Finally, if the error is captured during the production stage, the cost increase can be drastically high. Factors that lead to this state include the following:

- requirements not reflecting the actual needs of the system stakeholders;
- inconsistent requirements;
- incomplete requirements;
- conflicting requirements;
- misunderstood requirements;
- ambiguous or vague requirements;
- spurious requirements;
- unintended consequences.

To ensure that a common understanding of the design requirements exists, it is essential that adequate coordination be maintained. Some sort of multidisciplinary coordination group needs to be established to discuss the design effort ongoing in each discipline and how the design requirements are being interpreted and applied. One technique that may be beneficial in this regard is to pick a key individual from each appropriate discipline to start preparation of system (or product) checkout and test procedures (or maintenance and use procedures) early in the design process. In this way interdisciplinary coordination is forced to focus on a task that ensures proper sub-system integration and uniform understanding of design requirements. If checkout and test procedure preparation had been initiated early in the design process of the missile program referred to above, the design requirements would have been interpreted uniformly by all participating disciplines and an embarrassing and serious oversight averted. Adequate coordination *within* individual disciplines must be ensured by supervision and management within the discipline. Intradisciplinary coordination is, to a great extent, enhanced by the fact that the personnel involved work together on a day-to-day basis and should normally be aware of each other's actions and plans.

There are three main sources of information concerning the design requirements:

- *Customer.* One of the primary jobs facing the design team is to determine what is to be done for the customer. What does the customer really need? If the design team fully describes the product in terms of requirements and constraints, then they have accomplished one of the most important steps in the design process.
- *Company for which the design is being developed.* The company (customer) may impose restrictions on the design team because of company limitations. For example, a company may not have a component available in the size needed for the proposed design and the design team is expected to use one of the available component sizes, if possible.
- *Organizations having authority to impose restrictions by standards and laws to protect the public from injury or loss of property.* There are numerous organizations, governmental and private, that exist to set standards. These organizations generate and promulgate a large number of standards which affect the design, manufacture, and utilization of essentially all products. In addition, international standards are becoming more important, such as ISO 9000. ISO 9000 is an international standard that documents quality certification and audit requirements. It consists of a series of 20 standards that have been adopted by the European Union as international standards specifying how quality control is to be established and maintained by industry.

It is essential to have well-defined and well controlled requirements, but it is also important to minimize requirement changes as the systems are implemented. Changes to the design requirements during the design process can have a profound and detrimental impact on the final quality of the design. In many high-tech companies, the aforementioned problems may cause long lead times for the design effort. On large-scale complex systems, it is common to experience requirements problems late in the design process.

### 2.4.5 Requirements Decomposition

Not all requirements are equally important, and they can be arranged hierarchically. Higher-level requirements may be vague and should be decomposed into a number of sub-requirements. Decomposition of the requirements should be carried out to a *simple* level so that the requirement is *self-evident* at the lowest level. In other words, requirements have operational status.[12] As a general rule, decomposition should be done from the highest level (most abstract, least concrete) to the lowest level (least abstract, most concrete).

Figure 2.7 shows a requirements decomposition for the transportation of a person. Through brainstorming and discussion, the team can decompose the requirements down to the third level for safety (health and injury), time (layovers, rate of travel), economy (cost and value) and comfort (noise, luxury). Moreover, health can be decomposed down to a fourth level, mental and physical. This exercise illustrates the value of creating a graphical representation of the requirements which helps break down higher-level requirements into multiple related branches.

---

12 Alexander, C., *Notes on the Synthesis of Form*, Harvard University Press, Cambridge, MA, 1964.

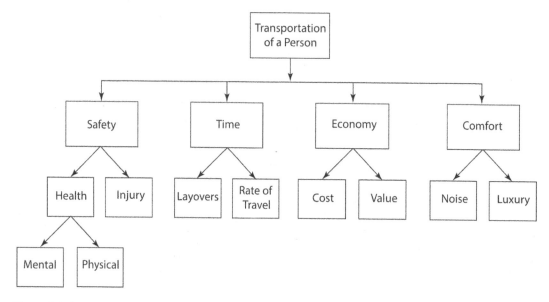

**Figure 2.7** Requirements decomposition.

### 2.4.6 Requirements Traceability[13]

Requirements traceability (RT) is generally practiced in software development life cycles and in the manufacture of high-reliability products and systems such as medical equipment and aerospace components. However, this important practice is not widely known and implemented in other design disciplines. It should be a vital part of any system development life cycle to make sure that customer needs, and in turn the functional requirements and constraints of the design, are considered during the development phases to make the final product/service fully compliant with the original requirements. RT provides stakeholders with the means to show compliance with requirements. Maintaining the system design rationale, establishing change control, and controlling maintenance mechanisms are a few of the many benefits of this approach.[14] In other words, RT is used to ensure continued alignment between stakeholder requirements and various outputs of the system development process.[15] RT is an important part of requirements management and engineering practice. It provides a useful tool to establish product quality with maximum customer satisfaction.

Requirements are features, functions, capabilities, or properties that a system must possess. Requirements state the customer or end-user needs and solution constraints. The traditional way of distinguishing requirements from design is that the requirements represents what the

---

13 Adapted from Gumus, B., Ertas, A., Unuvar, B., and Doganli, M., "Requirements traceability throughout system development lifecycle using axiomatic design approach," IDPT Conference, Vol. 1, CD, 2002.
14 Gotel, O., and Finkelstein, A., "An analysis of the requirements traceability problem," in *Proceedings of the First International Conference on Requirements Engineering*, Colorado Springs, CO, pp. 94–101, 1994.
15 Ramesh, B., Powers, T., Stubbs, C., and Edwards, M., "Implementing requirements traceability: A case study," *Proceedings of the Second IEEE International Symposium on Requirements Engineering*, York, UK, 1995.

system is supposed to have or do (i.e., the whats), whereas the design is how the system will accomplish the whats (i.e., the hows).

The system requirements should be documented in a requirements specification document and the requirements should be communicated and agreed upon by all stakeholders. The design activities and decisions as well as test activities are based on this requirements specification document since it states what the system is supposed to do. Requirements management can be defined as the process of eliciting, documenting, organizing, and tracking changing requirements and communicating this information across the stakeholders.[16]

According to a widely accepted definition, RT is the "ability to follow the life of a requirement, in both forward and backward direction, i.e., from its origins, through its development and specification, to its subsequent deployment and use, and through periods of ongoing refinement and iteration in any of these phases."[17] RT can be divided into two parts:[17]

1) *Pre-requirements traceability* (pre-RT) refers to the ability to describe and follow those aspects of a requirement's life prior to its inclusion in the requirement specification document (system/subsystem specifications, software requirement specifications).
2) *Post-requirements traceability* (post-RT) refers to the ability to describe and follow those aspects of a requirement's life that result from its inclusion in the requirement specification document (requirements deployment and use).

During requirement allocation, all system components (hardware, software, human-ware, manuals, policies, and procedures) created at various stages of the development life cycle are linked to requirements. Therefore, tracing requirements allows developers to easily ascertain the impact of any changes.

There are many different views of traceability depending on the stakeholder's view of the system. To the customer, traceability could mean being able to ascertain that the system requirements are satisfied. The developer's concern with traceability may be how a change in a requirement will affect the system – what modules are directly affected, and what other modules will experience residual effects. To a test engineer, traceability means making sure that each requirement is being tested. Full requirements test coverage is very difficult to achieve without RT.[16]

Many organizations consider RT as a mandate, a contractual requirement to be satisfied. Some organizations, on the other hand, view traceability as an important process in the development of a quality system. Some consider it a must for survival.[14]

Thus, at this point, it is important to mention about the general definition of "system". A system is "any portion of the material universe which we choose to separate in thought from the rest of the universe for the purpose of considering and discussing the various changes which may occur within it under various conditions" (J. Willard Gibbs, 1839–1903). This definition is not restricted to particular types of system – control, computer, mechanical, biological, and so on. This interpretation of system focuses on the important aspect of *change*, which is the motivation for all design activity.

---

16 Suh, P.N., *Axiomatic Design: Advances and Applications*, Oxford University Press, New York, 2001.
17 Davis, A.L., and Leffingwell, D.A., "Making requirements management work for you," *Crosstalk: Journal of Defense Software Engineering*, pp. 10–13, 1999.

### 2.4.6.1 Design Constraints

A design constraint refers to rules or limitations through which design is conceived and created. The constraints are imposed externally by the customer, by industry standards, or by government regulations, and they set limits for acceptable design parameters or for acceptable product performance.

Requirements are the desired functions that the product is expected to provide, whereas the constraints are the restrictions that the product must comply with while providing the desired functions. Table 2.5 shows the different types of constraints that are considered in the design of products, processes, or systems.

Physical constraints define choice of specific design parameters such as acceptable materials, weight and volume limitations, transportation and storage requirements, etc., applicable to the system elements.

Performance/functional constraints determine performance limits of system elements such as valid range of frequency, temperature range, and operation range – what is expected of the product. For example, if the customer needs a "control system operating pressure between 400 psi and 600 psi in a pressure vessel of 10 ft³ volume," then "control system operating pressure between 400 psi and 600 psi" is a functional constraint and "10 ft³ volume" is physical constraint.

As shown in Figure 2.8, it is important to identify the most important limiting factors (i.e., constraints) which prevent achievement of the design goal and then to systematically improve constraints until they are no longer the limiting factors – through an iterative approach leading to better design.

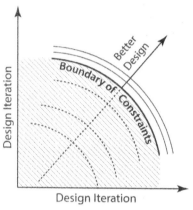

**Figure 2.8** Design improvement.

**Table 2.5** Design constraints.

| Constraints | | | |
| --- | --- | --- | --- |
| **Technical** | **Social** | **Environmental** | **Economical** |
| Physical | Legal/Ethical | Manufacturing | User needs |
| Performance/ | Aesthetic | Ecological | Ergonomic design |
| Functional | Political | Transport | Cybernetic design |
| Timing | Health | Operation | |
| Regulations | Security | Storage | |
| Standards | Safety | Regulations | |
| Codes | | Standards | |
| | | Codes | |
| | | Tests | |

### 2.4.7 Design Specifications

When the design requirements are defined, designers will start writing product design specifications. Note that design requirements define necessary design objectives, whereas specifications define how to meet the design objectives. The product design specification is a very detailed living document which determines the quality of the product. It is a statement of how a design is made, what it is intended to do, and how much it fulfils with the design requirements.

Specific parameters should be quantifiable with target goals (e.g., not "short length" but rather "length shall be less than 20 inches").

### 2.4.8 Preliminary Design[18]

This phase of the design process bridges the gap between the design concept and the detailed design phase of the effort. The design concept is further defined during preliminary design and, if more than one concept is involved, an evaluation leading to selection of the best overall solution must be conducted. As the design concept is refined during the preliminary design phase the overall project cost estimate will become more realistic. The overall system configuration is defined during this phase and a schematic, diagram, layout, definition drawing, or other engineering documentation (depending on the project) should be developed to provide early project configuration control. This documentation will assist in ensuring interdisciplinary and intradisciplinary integration and coordination during the detail design phase. System-level and, to the extent possible, component-level design requirements should be established during this phase of the effort. Establishing requirements is a function involving computation and analysis, literature search, vendor equipment evaluation, previous experience, discussion with experts in the particular field, good judgment, and testing (when absolutely necessary). It is costly to run test programs and thus it is desirable to minimize testing during the preliminary design phase of the project.

The requirements established during the preliminary design phase of the project form the basis for the component specifications developed during the detailed design phase. If the requirements are too stringent, the project cost will escalate and (possibly) no supplier will be found willing to bid on the item in question. If the requirements are too lax, the overall system requirements may not be met, which could lead to dire consequences for the overall project. An additional problem with loose requirements is that they must be tightened up later, sometimes after the initial contract for the component has been consummated, with increased cost. The importance of establishing valid design requirements at the outset is thus apparent. Preparation of system test, checkout and maintenance procedures at an early stage in the design often helps in that regard. The process of thinking these procedures through may help in quantifying the various parameters and thus provide a valid basis for component design.

---

18 Adapted from Ertas, A., and Jones J.C., *The Engineering Design Process*, John Wiley & Sons, Inc., New York, 1996.

### 2.4.9 Detailed Design[19]

The focus of the detailed design phase is to translate the high-level conceptual design into engineering documentation (including schematics, specifications, source code, computer-aided design (CAD) files, assembly drawing, etc.) and, subsequently, test and verify the design for production. At this design phase all the various disciplinary organizations are actively involved in the synthesis/analysis process, resolving the system design concept into its component parts, evaluating components to validate previously established requirements and specifying those design requirements left undefined, and assessing the effect of the component requirements on the overall system requirements. As component requirements are finalized, specifications can be prepared for those components that will be manufactured.

Specifications must include all of the requirements for the component, including the operating parameters, operating and nonoperating environmental stimuli, test requirements, external dimensions and interfaces that must be controlled, maintenance and testability provisions, materials requirements, reliability requirements, external surface treatment (if any), design life, packaging requirements, external marking and any other special requirements such as special lubricants. A good specification will minimize problems of interpretation that could surface later and result in disagreement with the supplier, possibly with negative impact on the entire project.

For components and other system elements that are manufactured internally, detail drawings are prepared that specify the necessary dimensions, the materials of construction, the surface treatments, machining techniques, test requirements, and any other information necessary for fabrication. Specification requirements for these components are often included on the face of the drawing as *notes*, but also may be tabulated on a separate sheet referred to as a specification sheet. The detail drawing must include all the information necessary to produce, test and package the part for shipment (if the part is shipped as a separate unit). The detail drawing must also include adequate orthographic views of the part so that a complete description of its size and shape is furnished to the manufacturing group. Figure 2.9 shows a typical detail drawing.

An assembly drawing is essential for all products that have more than one part. It shows the detail part or component as it is interconnected to other parts and subassemblies. Assembly drawings provide the necessary information for the individual detail parts and components shown on the drawing to be interconnected. Dimensions and other information specified on the drawing are limited to those required to be complied with during assembly. The *bill of materials*, usually located at the lower right corner of the drawing, lists all the elements, components, and other parts that make up the assembly. The name, number required, and part number for each component is given and, if the component is a purchased item, the manufacturer will also be identified. Special assembly instructions and test requirements are also often included in this area of the drawing as well. As with detail drawings and component specifications, subassembly drawings also have their *next assembly* identified in the title block. This procedure (referencing of the *next assembly*) is followed up to the *final* or *top assembly drawing* which depicts the deliverable end item (or product). Assembly drawings do not usually include detail part dimensions,

---

19 Adapted from Ertas, A., and Jones J.C., *The Engineering Design Process*, John Wiley & Sons, Inc., New York, 1996.

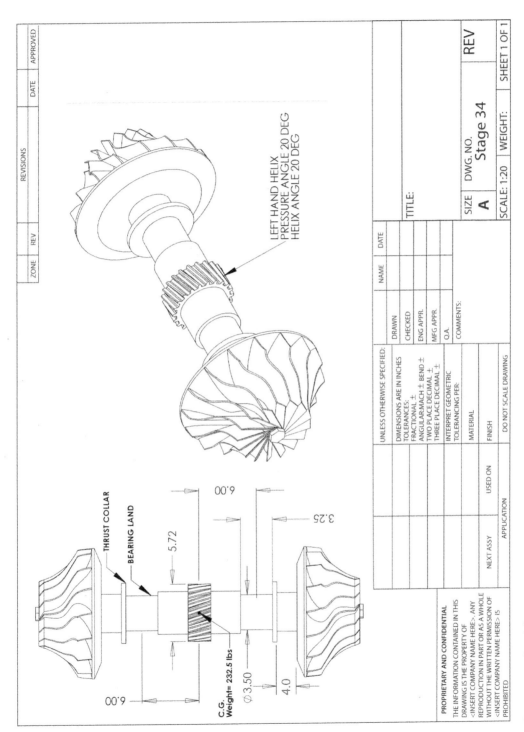

**Figure 2.9** A detail drawing.

LEFT HAND HELIX
PRESSURE ANGLE 20 DEG
HELIX ANGLE 20 DEG

THRUST COLLAR

BEARING LAND

C.G.
Weight= 232.5 lbs

Ø3.50

6.00

5.72

3.25

6.00

4.0

| REVISIONS | | | |
|---|---|---|---|
| ZONE | REV | DESCRIPTION | DATE | APPROVED |

| | | |
|---|---|---|
| UNLESS OTHERWISE SPECIFIED: | | |
| DIMENSIONS ARE IN INCHES | | |
| TOLERANCES: | | |
| FRACTIONAL ± | | |
| ANGULAR: MACH ± BEND ± | | |
| TWO PLACE DECIMAL ± | | |
| THREE PLACE DECIMAL ± | | |
| INTERPRET GEOMETRIC | | |
| TOLERANCING PER: | | |
| MATERIAL | | |
| FINISH | | |
| DO NOT SCALE DRAWING | | |

| | NAME | DATE |
|---|---|---|
| DRAWN | | |
| CHECKED | | |
| ENG APPR. | | |
| MFG APPR. | | |
| Q.A. | | |
| COMMENTS: | | |

TITLE:

| SIZE | DWG. NO. | REV |
|---|---|---|
| A | Stage 34 | |

SCALE: 1:20 | WEIGHT: | SHEET 1 OF 1

NEXT ASSY | USED ON
APPLICATION

but if the part is not complex it may be detailed on the assembly drawing and identified with a dash number. In this case the part number of the detail part will be identical to the assembly drawing number with the dash number added (e.g., 1234-3). A typical assembly drawing, sometimes referred to as a working assembly drawing, is shown in Figure 2.10.

An installation drawing is another type of assembly drawing which is used to specify dimensions that must be maintained during installation of a deliverable end item. In this sense the term *installation* differs from *assembly* in that the end item installation is not accomplished inplant with all the controls, processes and personnel used for an inplant assembly. An installation drawing often depicts the installation of an item in a facility or in an existing system distant from the location where the item was manufactured. It also identifies all the assemblies and parts that make up the installation and often includes the installation procedure, when a fairly simple procedure is involved. If the installation procedure is complicated (lengthy) it will be prepared as a separate document and be referenced on the installation drawing. Installation drawings are often included in parts catalogs and overhaul and repair procedure manuals, as are exploded pictorial assembly drawings.

Development testing is an essential element in most detailed design processes. Development testing is used to determine design requirements, evaluate new concepts, obtain information on the performance of proposed components and subassemblies, validate computational models as well as the design itself, and investigate interdisciplinary integration problems. The value of a well-planned and well-executed development test program can hardly be overstated. It is important to remember that testing is costly and thus needs to be thoroughly planned and diligently controlled. The use of innovative cost control techniques should be considered and applied, where appropriate. As an example of successful and innovative cost control techniques, expenditures during the detail design phase of the Space Shuttle Orbiter were minimized by recycling several major test articles back into the production line after testing was completed. Although this is a somewhat unorthodox approach, it resulted in considerable savings for this project. Care must be exercised when using such techniques to ensure that the probability of success and the potential savings justify the decision to proceed. It should be obvious that someone assigned to the project who has both technical and budgetary responsibility must have the final say as to the level of development testing accomplished. This is normally the project manager or his designee.

## 2.4.10  Design Verification and Validation

Design verification and validation activities are distinctly different and important steps in a design process. Verification and validation activities are performed during and at the end of a product development to continuously verify and validate the product or the product components.

Verification is the process of checking that the product meets the specified requirements – making sure that no requirements are missed in the product design. As shown in Figure 2.4, verification is an iterative process which will continue until all the allocated requirements and constraints are satisfied to eliminate technical and manufacturing risks prior

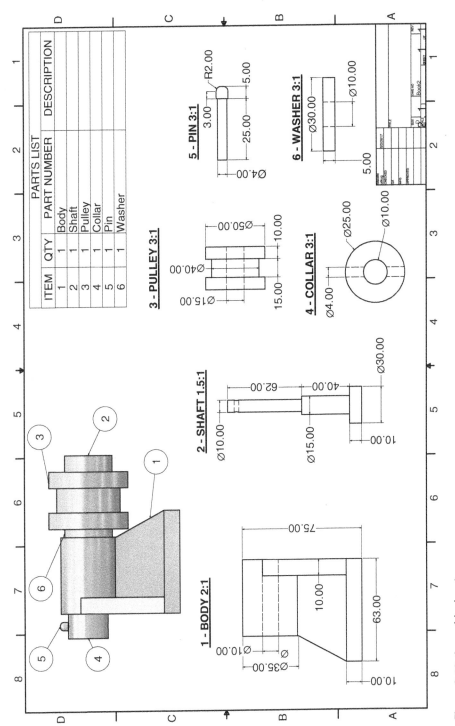

**Figure 2.10** Assembly drawing.

to high-rate production or delivery – are we building the product right? Validation is the process of checking whether the specified requirements capture the customer's needs – are we building the right product?

## 2.5 Design Review

Design reviews are important for product safety management and product risk prevention. They are also a way to ensure that you have designed a safe product that meets user needs and requirements. In other words, a design is evaluated against its requirements during the entire design process activity. It is a continuous process to verify the outcomes of previous design activities and identify issues before committing to the next phase of a design activity. This means that these design reviews are placed at a point during the design process where there has been some design development that should be assessed before the project continues. Product design reviews are essential from an economic perspective. Catching requirement errors during the design phases is orders of magnitude less expensive to fix than catching them during the test or production phases (see Figure 2.6).

Reviews are categorized as formal or informal depending on who participates. An informal review involves those personnel directly involved in the design project. Formal reviews include subject-matter experts who are not directly involved in the product design. Although performing a formal design review looks different for every organization (whether large or small), design reviews are performed with a multidisciplinary design review team from different backgrounds and experienced personnel who are familiar with the technology involved in the product and its related technical risks. If the design is being done for a customer, a representative from the customer should be invited to participate in the design review process. Formal design review team members are selected for their skills and expertise to tie into the requirements of the project to provide an independent assessment.

The following typical information should be provided to the reviewers during the review process:

- current product development specification;
- analysis and test results;
- drawings, schematics, layouts;
- cost and schedule status and projections;
- assessments of alternative designs;
- project risk analysis;
- copies of standards and regulations applicable to the design.

## 2.6 Redesign

Once products have been on the market for some time, they may need to be redesigned because of design defects or new requirements. For example, more and more automobile manufacturers are recalling vehicles for redesign because of safety problems and defects. In 2014, a number of very well-known vehicle manufactures experienced damaging product recalls.

Products may also need to be redesigned to reduce costs, improve quality, or reduce environmental impacts. Therefore, redesign is an important part of the product design and development process. A redesign approach to develop new products has many advantages; usually, it improves the quality and efficiency of the product design process.[20] Redesign approaches are usually more feasible and reliable, since they have already been used and tested successfully.[21] Reusing prior design information also reduces product costs, required design resources, and cycle time.[22]

### Example 2.2

The sock aids shown in Figure 2.11 are devices designed to ease the process of putting on socks or stockings. These devices are helpful for people who struggle to bend at the waist. Although there are a number of designs available, Figure 2.11 shows the more popular styles of sock and stocking aids found in the market. Using the Pugh decision matrix, taking the sock horse device (the easiest and fastest sock aid tool; Figure 2.11(a)) as baseline, decide whether the redesigned device has improved by considering the criteria of cost, manufacturing, and functionality/performace.

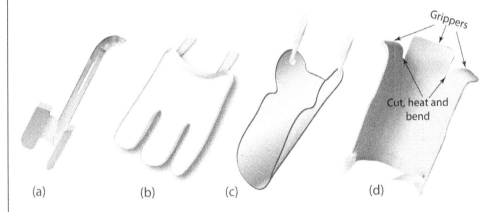

(a)  (b)  (c)  (d)

**Figure 2.11** Sock aid devices.

### Solution

The designs in Figure 2.11(b) and (c) are existing devices in the market. A redesign to existing devices, incorporating improvements, is shown in Figure 2.11(d).

20  Li, Z.S., Kou, F.H., Cheng, X.C., and Wang, T., "Model-based product redesign," *International Journal of Computer Science and Network Security*, 6(1), 99–102, 2006.
21  Han, Y.H., and Lee, K., "A case-based framework for reuse of previous design concepts in conceptual synthesis of mechanisms," *Computers in Industry*, 57(4), 305–318, 2006.
22  Smith, S., Smith, G., and Shen Y.-T., "Redesign for product innovation," *Design Studies*, 33(2), 160–184, 2012.

*Materials.* Design (b) is made with fabric to enable device flexibility. Very high flexibility is a disadvantage when putting on socks: because the device can easily come out of the sock before the process of putting it on ends. Design (c) is made of thermoplastic, which has less flexibility than design (a). Design (d) is made of PVC pipe. This design has no flexibility. No flexibility is not an advantage, but provides 100 percent good performance with no slip from the socks.

*Manufacturing and cost.* When a mold is ready, through injection molding, the manufacturing of and labor time for design (c) are a lot less than for design (b). However, the initial investment in the molding process makes design (c) costly. The redesigned device shown in Figure 2.11(d) has the least manufacturing time. To manufacture this device, buy 5 inches of PVC pipe, and cut two places symmetrically from the center to create grippers by heating (see Figure 2.11(d)). This design also costs less than the two existing other designs. The functionality of redesigned device is greater compared to other two designs because of the catching arms shown in Figure 2.11(d).

First, apply a weighting score to the criteria based on relative importance as shown in Table 2.6 (functionality/performance will be the most important criterion for this example). As shown in Table 2.7, redesign (d) is selected because it receives the highest overall score.

**Table 2.6** Pugh decision matrix.

| | Baseline Sock Horse | Design (b) | Design (c) | Design (d) |
|---|---|---|---|---|
| **Criterion** | Weight | | | |
| Manufacturing | 3 | −1 | 0 | +1 |
| Functionality | 4 | −1 | −1 | +1 |
| Cost | 2 | +1 | 0 | +1 |
| | Totals | | | |

**Table 2.7** Pugh decision matrix.

| | Baseline Sock Horse | Design (b) | Design (c) | Design (d) |
|---|---|---|---|---|
| **Criterion** | Weight | | | |
| Manufacturing | 3 | −3 | 0 | +3 |
| Functionality | 4 | −4 | −4 | +4 |
| Cost | 2 | +4 | 0 | +4 |
| | Totals | −3 | −4 | +11 |

There are several types of redesign activities, varying based on the reason for redesigning products. The following are some examples of product redesign types:[22]

### 2.6.1 Redesign for Conflict Resolution

In general, the main reason for redesigning a product is to correct design defects/errors. Design defects (conflicts) appear when a final product does not meet design requirements or when customers change their existing requirements after a product has been introduced. Conflicts are resolved by changing component attributes, the component itself, or the process of the original design. The focus of this approach is on making modifications to a specific existing product, rather than on developing a new product from prior design information.

### 2.6.2 Redesign for Cost Reduction

The ratio of product quality to product price is an important parameter for customers purchasing a product. If the goal is to reduce the cost of a product, three fundamental changes must take place:[22]

- The product must be redesigned to make it simpler and cheaper.
- The business system must be redesigned to make the product simpler and cheaper to produce and deliver, and to provide protection for the firm against imitators.
- The business must be scaled.

### 2.6.3 Redesign for Product Family Creation

Redesigning products into a product family allows component reuse between products, which in turn reduces product costs, improves product quality, and the product design process overall is less troublesome. The goal of this method is to increase component commonality, reduce design effort, and restore profitability and growth without changing the overall functionality of any of the products.

## 2.7 Other Important Design Considerations[23]

In addition to the design process structure described previously there are several other considerations that are important when initiating a new design project. These are briefly discussed in this section.[5]

### 2.7.1 Design for Regulatory Compliance and Standards

Considering regulatory compliance early in the product design phase and establishing regulatory compliance as a critical requirement will reduce both time and cost prior to product release. Failure to understand and incorrect implementation of regulatory compliance and standards can result in failure of the product release plan and necessitate expensive reworking. The goals

---

23 Adapted from Ertas, A., and Jones J.C., *The Engineering Design Process*, John Wiley & Sons, Inc., New York, 1996.

of the regulatory process is to eliminate surprises late in the product development and ensure that safe and operational products are delivered to the market place on time.

Companies can involve people with extensive regulatory experience in their design teams to ensure that regulatory concerns and requirements are addressed in planning and subsequent design phases. This approach helps the team members to use their experience along with the subject-matter experts to design products and test programs that will allow the creation of regulatory-ready product.

### 2.7.2  Product Distribution and Use

Although the design engineer may not be directly involved in the product distribution and marketing tasks, the considerations and problems that arise during these efforts have an impact on the design of the product. To ensure that the product is delivered to the point of sale and stocked appropriately, packaging must be designed to protect the product from damage during transport and storage. The product designer must specify special shipping and warehousing requirements and often may be involved in the design of shipping containers that must be inexpensive, easy to handle and reuse or recycle, and that effectively protect the product. Conditions such as vibration, temperature and humidity are factors that must often be considered during the shipping container and warehouse systems design and during the product design process as well.

Design engineers are also increasingly involved in product promotional activities. A design engineer can interpret a customer's questions and criticisms relative to the product and can understand the implications they may have for the product design. As the designer becomes more concerned with the customer's use of the product he/she must consider such important product attributes as ease of maintenance, reliability, safety, user friendliness, aesthetics, operational economy, and product life and retirement. Involving the design engineer in the product use phase of the design process is another example of the trend towards further integration of the design process to improve overall product quality, increase customer satisfaction, and shorten the development time.

A phrase often applied to sales activities is that *the customer is always right*. With the current emphasis on profitability, short investment payback periods, shortened product development times, early product obsolescence and intense competition, it often seems that US consumer products are developed under a philosophy that *the customer will buy the cheapest product available*. This is hopefully not true for the more technical engineering design products where the customer is more sophisticated and will normally demand performance, reliability, serviceability, reasonable cost, safe operation and good human engineering (ergonomics). Industries that are renewing their emphasis on quality as a result of competitive pressures (e.g., the US automobile industry) are returning to a *customer is always right* approach and using a disciplined and structured method of determining customer expectations. These customer expectations are compared to the various product design characteristics to identify areas that are not addressed and to develop an optimized product plan incorporating design modifications accommodating the dominant expectations.

Product liability in the USA is a design consideration that is growing in importance. Users of products are increasingly aware of their opportunities to sue manufacturers and court sentiment often leans toward the consumer even when product misuse is involved. This situation has gotten so bad that some products are no longer manufactured in the USA and the availability of some services is limited. Small airplane manufacturing in the USA has all but ceased and women with high-risk pregnancies have a difficult time finding obstetricians willing to perform deliveries. The cost of this litigation in time and money is significant. One expert estimates the overall cost to be $300 billion a year on a national basis.[24] Officials in both large and small businesses spend an exorbitant amount of their time defending their firms against product and design liability claims instead of managing. The cost of this wealth-consuming activity to the USA is enormous in time and money as well as in reduced innovation, reduced economic and career opportunities, and inability to compete on an international basis.

### 2.7.3  Design Life

Many engineering products are designed for a specific installation or assembly, and it is normally assumed that they will remain in this application for their useful life. For these instances the product specification will dictate the design life in terms of cycles or hours of operation, and it is a fairly straightforward task to design and test the product to meet these requirements. An aspect of design that has not received a lot of attention is the problem that arises when the product is removed from its original installation and is put to some *second use*, either before or after it has fulfilled its initial design life. The problem is that since this second use is not included in the specification it is not accounted for in the design, and the result may be failure and personal injury leading to product liability litigation. The fact that the product was used in a way never intended by the original design may not have much influence on the court. This is an extremely difficult problem and there is no real satisfactory solution. The courts seem to focus on whether the failure was *foreseeable* and not on whether there was negligence or ignorance. About the best that the designer can do is to try to foresee both use and misuse and make provision in the design to provide for *credible* failures. Another technique that has seen some limited use is to permanently identify on the product its intended use and limited life, but there is no absolute assurance that the courts will consider this to be adequate warning. When tightly controlled operation of a product by a technically qualified user can be ensured, stringent life requirements can be enforced, for example critical pressure vessels (as in spacecraft use) can be designed for a specified number of pressure cycles and be required to be recertified by retesting before additional use.

Some thought needs to be given to retirement of the product after completion of its design life. If the environment is to be treated as *surroundings* (in a thermodynamic context), the product must be capable of being refurbished and reused or the materials of construction must be recyclable or biodegradable. Some states have imposed a relatively high deposit on soft drink bottles to ensure that they are returned for subsequent reuse, and almost every community has

---

24  Ritter, D., "Litigation pollution,"*Engineering Times*, July 1990.

a program for recycling aluminum soft drink cans, glass, and certain plastic products. Plastic diaper manufacturers have developed a product that is supposedly biodegradable as a result of unfavorable publicity associated with diaper disposal in public landfills. Automobile salvage yards perform a service in recycling automobile parts. After the marketable components are removed from the vehicle chassis the shell is sold for scrap, crushed, and subsequently used in new steel production. The automobile life cycle is a little like the old saying about pigs: "everything gets used but the squeal." The tire casings constitute a problem that has not been satisfactorily solved, however. Some states have implemented tire shredding programs, but this is not a total solution to the problem since the market for shredded tires is not great enough to consume all of the available product. Vehicle and tire manufacturers should be concerned about the retirement of used tires and should develop or encourage second uses or satisfactory recycling processes. Manufacturers of products that do not biodegrade in a reasonable period of time and for which no completely satisfactory recycling program has been developed need to face the problem of product retirement. Some of the considerations that should be factored into the design to account for product retirement are as follows:

1) Design to reduce the rate of obsolescence by anticipating the rapid advancement of technology.
2) Design so that the physical life of the product matches the anticipated service life.
3) Design for more than one level of use so that when the initial service life has been completed the product can be used for a less demanding application.
4) Design the product so that it can be *cannibalized* upon retirement, that is, reusable materials and long-life components can be recovered.
5) Evaluate service-terminated products to obtain useful product design information.

### 2.7.4 Human Factor Considerations

Designers must consider the users of devices, components or systems under development and strive to produce *user-friendly* products. In this sense the term "user" must include the person(s) maintaining and repairing the product as well as the person operating it. A significant amount of anthropomorphic information is available which can be used to determine the size and location of manually actuated devices on the product as well as the optimum location of visual components. Arm and leg actuation force data is available also. Certain general rules pertaining to such things as the meaning of the color of indicator and warning lights, direction of rotation of valves and electrical components, mounting of switches, and the like should also be followed. An individual's capacity for doing work should be considered in the design and, for repetitious tasks, a system layout should be selected that minimizes boredom but does not overload the individual's capacity to comprehend. Considerations pertaining to involuntary reaction as contrasted with intentional action must be accounted for in some designs, as well.

Logic should be applied in all control systems layouts to maximize the efficiency of the operator. Switches, indicators, and the like should be located in a logical sequence simulating the system being controlled. Indicators and devices requiring actuation should be located such that the operator can react appropriately, almost without thinking. A good systems control design philosophy may be to think of the operator as an extension of the system itself and extend the systems design logic to include the operator's functions. The operator must also feel comfortable at his work station and must not be subjected to undue fatigue. High-maintenance items must be accessible (within reason). For example, it should not be necessary to remove half the components under the hood of an automobile to change the spark plugs.

Finally, the product should be aesthetically appealing. Many technical products are purchased primarily because of their customer appeal. The outward appearance of a product may be the feature that initially interests the potential buyer and opens the door to discussion of the technical features of the product.

### 2.7.5  Knowledge-Based Design Process[25]

Knowledge-based design (KBD) is a new product development process evolved from artificial intelligence (AI). KBD allows designers to capture and deploy the knowledge and experiences stored in database with manufacturing best practices, legislation, costing, and other rules. KBD presents a great opportunity to improve the speed and effectiveness of product development process, to reduce the time to create a new design, to refine an existing design, and it provides a significant savings.

> The approaching 21st century is bringing pressure toward more integration, interdisciplinary cooperation, and international collaboration. We have great pressure to convert the integrated-artifact-design process *from a thing of chance and adventure to a regular understood business.* The various entities of our society are redefining themselves to be able to play an effective role in this new order. As engineers, system designers, system integrators, engineering managers, and engineering executives, we are constantly on the outlook for ways and methods of materializing the integration objectives in the hope of delivering a more robust, high-quality line of products.[26]

Although the integrated-artifact-design process using integrated tools has not reached maturity, achievements during the last several decades should not be underestimated. The Boeing 777 was designed entirely digitally, utilizing 3-D solid modeling technology. Boeing invested more than $1 billion in CAD infrastructure for the design of the 777. Such highly sophisticated packages as the CAD/CAM/CAE system, CATIA (Computer-Aided Three-dimensional Interactive Application), EPIC (Electronic Pre-assembly Integration), and ELFINI (Finite Element

---

25  Adapted from: Ertas, A., Gumus, B., and Unuvar, K.B., "Framework of knowledge-based design process: Transdisciplinary approach," in *Proceedings of the Sixth Biennial World Conference on Integrated Design and Process Technology*, SDPS, Pasadena, California, June 23–28, 2002, CD (abstract, p. 22, 2002).
26  Ertas, A., and Tanik, M.M., Note on final program brochure of the Integrated Design & Process Technology Conference, 1995.

Analysis System) were used during the design, modeling, and development of the 777. Through interdisciplinary team management, 238 teams, (3,500 people) were integrated to provide knowledge-sharing concurrently rather than sequentially. All countries and corporations involved were linked by a computer network which consisted of mainframe and work station installations in the Puget Sound area, Japan, Wichita (Kansas), Philadelphia, and other locations. Over 3 million parts were developed and organized in an integrated database that allowed designers to provide a complete 3-D virtual mockup of the aircraft. Comparing the 777 with earlier aircraft (757 and 767) designs, it is clear that Boeing achieved the following improvements:[27]

- Elimination of more than 3,000 assembly interfaces, without any physical prototyping.
- A 90% reduction in engineering change requests (6,000 to 600).
- A 50% reduction in cycle time for engineering change requests.
- A 90% reduction in material rework.
- A 50-fold improvement in fuselage assembly tolerances.
- Reduction of development time from 12 years to 5 years.
- Enhanced reliability and productivity at lower cost.

The Boeing 777 is the first commercial large-scale complex system design using the concept of a knowledge-based engineering platform. Knowledge pertaining to design can be organized in a computer database in terms of design objects. Such information may consist of statements about the physical aspects of the artifacts such as function, data, and control. This information is called *technical knowledge* pertaining to design objects which have been created and produced to accomplish specific goals. Although a vast amount of technical knowledge exists about design objects today, it is not currently in an organized form in a computer database. Obviously, a key part of any organized integration is to agree on the standards for such databases.

In the past, obtaining agreement on standards between corporations was very difficult. But due to the widespread use of computers and cost-cutting pressures from internationalization, the environment is now conducive to more complete standardization. For example, OASIS is a non-profit, international consortium dedicated to accelerating the adoption of product-independent formats based on public standards. OASIS standards include XML, SGML, and HTML, as well as others that are related to structured information processing. Members of OASIS are providers and users of and specialists in the technologies that make these standards work in practice.

XML is one such tool that can be used to improve communication between computer-based systems. The XML language consists of rules to create a markup language that leverages existing infrastructure. The rules ensure that a single compact program, an XML parser, can process all these new languages. Due to its extensible and meta-level properties, XML potentially enables the exchange of information not only between different computer systems, but also across national and cultural boundaries.

---

27 Baba, Y., and Nobeoka, K., "Towards knowledge-based product development: the 3-D CAD model of knowledge creation," *Research Policy*, 26, 643–659, 1998.

### 2.7.6  Design Repositories

Traditionally, design information, schematics, drawings, specifications, requirements, etc., were stored in paper repositories. The use of CAD/CAE processes has moved some of this information storage to electronic format; however, designers and engineers must have access to many types of information other than CAD/CAE files to support the design of complex systems. Engineering knowledge, including previous designs of related components and systems, must be effectively reused in new designs to provide the most cost-effective and best overall solutions. *How* and *why* are as important as *what* to the design team. Thus, electronic repositories in which all types of design-related information and knowledge from across the disciplines can be stored and then rapidly and easily retrieved by designers are necessary. Such design repositories greatly enhance knowledge-based design by facilitating the representation, capture, sharing, and reuse of corporate design knowledge. Companies typically refer to their store of corporate knowledge as design databases; however, these databases are beginning to transition into databases more like the design repositories indicated above. This need for industry to reform its knowledge databases is being addressed by emerging research related to design repositories.

Design databases are static, homogeneous, and data oriented while design repositories must be dynamic, heterogeneous, and knowledge oriented. Databases are used to store limited representations of artifacts (e.g., drawings, CAD models, CAE models, and results). Design repositories provide not only design-specific information such as requirements, specifications, and limitations, but also more general knowledge and data from across disciplines, design rules, simulation models, information related to previously designed components and systems. Not only will images, CAD models, and unstructured text be provided in design repositories, but they will also include mathematical simulation models, structured text, animations, video, and other types of information. Design repositories will also provide such capabilities as a search for information on existing components and systems that satisfy required functions, performance and behavior simulations, automated reasoning and comparison of design options, and more explicit decompositions of physical and functional aspects of a design with cross-mappings. Enough information, knowledge, design and analysis procedures, and active models will be included in design repositories to allow a designer to select components for a sub-system and run simulations to determine performance, or otherwise test the proposed sub-system. Clearly, there will be a need for many design repositories to house all of the information required and there will also be a need for standard formats and protocols to represent the form, function and behavior of components, systems, and other artifacts. Just as clearly, it will take some time for these design repositories to be generated and mature to a point where they can support the design of complex systems.

### 2.7.7  Collaborative Design Framework

Because engineering systems continue to increase in complexity, the membership of design teams must increase in number and in diversity to include not only designers, but also process planners, manufacturers, clients, etc., as well. The ability to collaborate and communicate seamlessly among the members of a design team is particularly critical to the success of complex system design. The environments in which design teams function are sure to become more geographically and functionally distributed while providing the capability to support more direct and more effective interaction and coordination. Numerous knowledge and data repositories, likely distributed and frequently discipline oriented, will provide specifications, design rules, constraints, and rationale to designers. Evolving designs along with design artifacts and rationale will be stored in databases accessible to all members of the team. These databases will provide the capability to generate various snapshots of the evolving design objects at various levels of abstraction. The design environment must also include a work flow management capability to coordinate and manage the overall design process.

The high-speed, distributed computing capabilities required to support such new engineering design processes are being developed and some are already available. Because members of the design team will likely utilize systems, applications, models, data structures, etc., with form and content typical of their specific disciplines, the high-bandwidth network which connects the team members must provide effective exchange protocols to maximize the potential of the collaborative information sharing process. Design applications utilized by the design team must communicate with other applications used by manufacturing, production, process planning, management, etc., through various networks.

### 2.7.8  Rethinking Design Process

The design process is an information processing activity for selection of a means to satisfy identified needs and to solve defined problems. It is a process of gathering information (knowledge) about the problem structure and using this knowledge in discovering potential problem solutions. As shown in Figure 2.12, design knowledge, which defines the information space, is input to the design process. Information used in design may apply to different abstractions of the design process and to different abstractions of design descriptions.[28] For example, knowledge about how components interact in design as well as knowledge for different sub-components

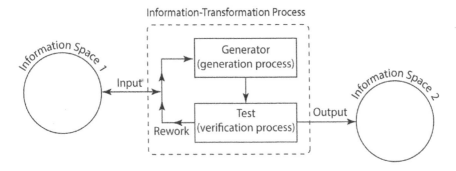

**Figure 2.12** Information- transformation process.

---

28 Coyne, R.D., Rosenman, M.A., Radford, A.D., Balachandra, M., and Gero, J.S., *Knowledge-Based Design Systems*, Addison-Wesley, Reading, MA, 1989.

of complex designs may be included in the information space. Knowledge about political, environmental, social, and physiological issues may also be included to help to make design decisions. Knowledge in axiomatic form, mathematical formulations, rules, and laws may also exist. Information input from the information space is typically manipulated through an information-transformation process to filter solutions that meet the objective of the design.

The basic scheme of design evolution depends on two processes:[29] the generation process and the testing process As shown in Figure 2.12, through the information-transformation process, the generator produces a variety of new forms of knowledge which did not previously exist. The test process extracts generated ideas and knowledge and filters them so that only those that are well fitted to the environment will pass. Objectives of the design can be met, provided that adequate relevant knowledge and enough process direction (defining goals, managing, leading, and controlling towards achieving these goals) are available, and the influences of the environment do not exceed certain constraints and limiting resources.[30]

Consider, a problem structure which may lead to a solution that meets the design objective. The search for problem solutions begins along the possible paths of a tree. Knowledge-transfer processes can be used for each path as processes for seeking a solution for the design problem. In other words, there are several processes (as many as the number of paths and branches) for gathering knowledge about the problem structure that will be valuable in discovering a solution to a design problem. Figure 2.13 shows a typical example. Of course, a partial path solution is not a solution to the design problem and, regardless of the value of the partial path, the total path results in a zero value if it does not lead toward a problem solution. A similar approach can be used for a decomposed complex system.

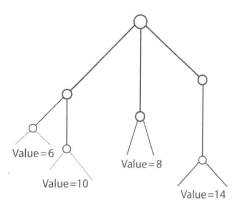

**Figure 2.13** Path tree for design solution.

### 2.7.9 Proposed Knowledge-Based Design Process

As shown in Figure 2.14, proposed knowledge-based design processes can be divided into three domains: the first defines the information space, which includes all the available information necessary to develop the requirements; the second defines the problem space, which stores requirements derived from the information space; and the third defines the solution space, which includes design objects developed through the design object development process. Design objects, needed to satisfy a given set of requirements, can be retrieved from design repositories. It is also important to note that any new or modified design objects developed will

29  Simon, A.H., *The Sciences of the Artificial*, third addition, MIT Press, Cambridge, MA, 1999.
30  Hubka, V., Eder, W.E., Melezinek, A., and Hosnedl, S., *Design Science: Introduction to the Needs, Scope and Organization of Engineering Design Knowledge*, Springer, Berlin, 1996.

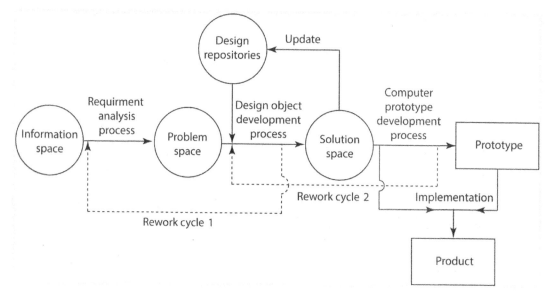

**Figure 2.14** Proposed knowledge-based design life cycle.

be used to update design repositories. As shown in this figure, rework cycle 1 is used to capture additional information to refine the design requirements.

After the solution space is completely developed a computer model is created to test the design before the physical implementation phase starts. During the computer model development process, if there is a need for rework, rework cycle 2 is used. The prototype and solution space are the input to the implementation phase to produce a new product.

## 2.8 Transdiscipline

Convergence: facilitating transdisciplinary integration of life sciences, physical sciences, engineering, and beyond is an approach to problem solving that cuts across disciplinary boundaries. It integrates knowledge, tools, and ways of thinking from life and health sciences, physical, mathematical, and computational sciences, engineering disciplines, and beyond to form a comprehensive synthetic framework for tackling scientific and societal challenges that exist at the interfaces of multiple fields. By merging these diverse areas of expertise in a network of partnerships, convergence stimulates innovation from basic science discovery to translational application. It provides fertile ground for new collaborations that engage stakeholders and partners not only from academia, but also from national laboratories, industry, clinical settings, and funding bodies.

**National Research Council of the National Academies**[31]

---

31 *Convergence: Facilitating Transdisciplinary Integration of Life Sciences, Physical Sciences, Engineering, and Beyond*, National Academies Press, Washington, DC, 2014. Retrieved from: http://www.nap.edu.

In the German-speaking countries the term "transdisciplinarity" is used for integrative forms of research. Transdisciplinary education and research programs take collaboration across discipline boundaries a step further than do multidisciplinary and interdisciplinary programs. The transdisciplinarity concept is a process by which researchers representing diverse disciplines work jointly to develop and use a shared conceptual framework to solve a common problem. "A transdisciplinary approach … seeks not only to go beyond boundaries but also to create a fusion. That is, the walls between disciplines must be removed and a new, unified direction established with a focus on solving problems."[32] The terms multidisciplinary, interdisciplinary, and transdisciplinary are often defined differently among researchers and educators.[33]

Nicolescu stated that transdisciplinarity concerns that which is at once between the disciplines, across the different disciplines, and beyond all disciplines.[34]

Klein defined the terminology of multidisciplinary, interdisciplinary and transdisciplinary approaches as follows:[35]

> Multidisciplinary approaches juxtapose disciplinary/professional perspectives, adding breadth and available knowledge, information, and methods. They speak as separate voices, in encyclopedic alignment…
> Interdisciplinary approaches integrate separate disciplinary data, methods, tools, concepts, and theories in order to create a holistic view or common understanding of complex issues, questions, or problem… Theories of interdisciplinary premised on unity of knowledge differ from a complex, dynamic web or system of relations…
> Transdisciplinary approaches are comprehensive frameworks that transcend the narrow scope of disciplinary world views through an overarching synthesis, such as general systems, policy sciences, feminism, ecology, and sociobiology…
> All three terms evolved from the first OECD international conference on the problems of teaching and research in universities held in France in 1970.

Transdisciplinary research can be defined as exchanging information, altering discipline-specific approaches, using shared methods, tools, resources, and integrating disciplines to achieve a common scientific goal,[36] and represents the highest degree of cross-disciplinary collaboration. "Transdisciplinary research includes cooperation within the scientific community and a debate between research and the society at large. Transdisciplinary research therefore

---

32  Takeuchi, K., "The ideal form of transdisciplinary research as seen from the perspective of sustainability science, considering the future development of IATSS," *IATSS Research*, 38(1), 2–6, 2014.

33  Ertas, A., Greenhalgh-Spencer, H., Gulbulak, U., and Baturalp, T. B. "Transdisciplinary collaborative research exploration for undergraduate engineering students," *International Journal of Engineering Education*, 33(4), 1242–1256, 2017.

34  Nicolescu, B., *Toward Transdisciplinary Education and Learning, Science and Religion: Global Perspectives*, CNRS, University of Paris, 2005.

35  Klein, J.T., "Disciplinary origins and differences," Paper presented at the Fenner Conference on the Environment: Understanding the Population-Environment Debate: Bridging Disciplinary Divides. Canberra, 2004. http://www.science.org.au/events/fenner /klein.htm, accessed Sept. 28, 2010.

36  Rosenfield P.L., "The potential of transdisciplinary research for sustaining and extending linkages between the health and social sciences," *Soc Sci Med.*, 35, pp. 1343–57, 1992.

transgresses boundaries between scientific disciplines and between science and other societal fields and includes deliberation about facts, practices and values."[37]

Cronin writes:[38]

> There is a need for transdisciplinary research (TR) when knowledge about a societally relevant problem field is uncertain, when the concrete nature of problems is disputed, and when there is a great deal at stake for those concerned by problems and involved in dealing with them. TR deals with problem fields in such a say that it can: a) grasp the complexity of problems, b) take into account the diversity of life world and scientific perceptions of problems, c) link abstract and case specific knowledge and d) constitute knowledge and practices that promote what is conceived to be the common good.

In summary, transdisciplinary research includes the key components of interdisciplinarity, along with the incorporation of external non-academic knowledge, applied to solve complex problems.[38]

The transdisciplinary design process is the integrated use of the tools, techniques, and methods from various disciplines.[39] The expected result of TD education is a creative potential for cross-disciplinary collaborative research – ways to solve challenging contemporary social issues. The TD approach teaches us to seek collaboration outside the bounds of our professional experience to make new discoveries, explore different perspectives, express and exchange ideas, and gain new insights.

---

**CASE STUDY 2.1  Multidisciplinary, Interdisciplinary, and Transdisciplinary Research**

Wind power promises a clean and inexpensive source of electricity. It promises to reduce our dependence on imported fossil fuels and the output of greenhouse gases. Many countries are, therefore, promoting the construction of vast wind "farms" and encouraging private companies with generous subsidies.[a]

**Wind Turbine History**

The history of wind power shows a general evolution from the use of simple, light-weight devices to heavy, material-intensive drag devices and finally to the increased use of light-weight, material-efficient aerodynamic lift devices in the modern era. During the winter of 1887–88, Charles F. Brush built the first automatically operating wind turbine for electricity generation. It was the world's largest wind turbine with a rotor diameter of 17m (50 ft) and 144 rotor blades

---

37 Wiesmann, U., Biber-Klemm, S., Grossenbacher-Mansuy, W., Hadorn, G.H., Hoffmann-Riem, H., Joye, D., Pohl, C., and Zemp, E., "Enhancing Transdisciplinary Research: A Synthesis in Fifteen Propositions," in G.H. Hadorn et al., *Handbook of Transdisciplinary Research*, Springer, Dordrecht, 2008.

38 Cronin, K., "Transdisciplinary research (TDR) and sustainability," Environmental Science and Research (ESR) Ltd., 2008.

39 Ertas, A., Maxwell, T., Rainey, V.P., and Tanik, M.M., "Transformation of higher education: The transdisciplinary approach in engineering," *IEEE Transactions on Education*, 46(2), 289–295, 2003.

made of cedar wood. The turbine ran for 20 years and charged the batteries in the cellar of Brush's mansion.[40]

Wind has been an important source of energy in the USA for some time. Over 8 million mechanical windmills have been installed in the USA since the 1860s. It is interesting to note that some of these units have been in operation for more than 100 years.[41]

## Wind Turbine System Design

A wind turbine system design consists of sub-systems to catch the energy of the wind, to point the turbine into the wind, and to convert mechanical rotation into electrical power; there are also systems to start, stop, and control the turbine. To design today's impressive and giant wind turbine structures, many researchers from different disciplines work collaboratively. Mechanical engineers work on gear design, civil engineers work on structure design, material engineers work on the most suitable material selection for the application, electric engineers work on power transmission and control system design, and wind engineers work on rotor blade design, etc. A simple methodology could be to create a collaborative research team to design wind turbines efficiently. Of course, the collaborative effort can be organized in many different ways.

## Research Approaches

The first approach to building a wind turbine that comes to mind could be the *multidisciplinary* research team from diverse disciplines. Multidisciplinary activities involve researchers from various disciplines working essentially independently, each from his/her own discipline specific perspective, to address a common problem (see Figure 2.15). As mentioned earlier, the job responsibilities of the members of a multidisciplinary product development team will vary based on the wind turbine project scope and deliverables. Although multidisciplinary teams do cross discipline boundaries, they remain limited to the framework of disciplinary research (see Figure 2.15). In other words, each team member has specific roles in delivering their sub-product design independently for the final functional system design. Once the wind

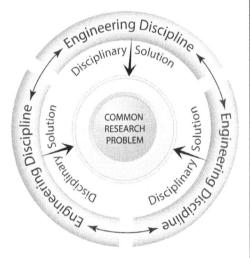

**Figure 2.15** Multidisciplinary research approaches.

*(Continued)*

---

40  Green Energy Ohio, 2009. Danish Wind Industry Association. http://www.greenenergyohio.org/page.cfm?pageId=341, accessed September 7, 2010.
41  Wind Turbine History, Alternative-Energy Resources. 2009. http://www.alternative-energyresources.net/wind-turbine-history.html, accessed August 28, 2010.

**CASE STUDY 2.1  (Continued)**

turbine is functional, the key question we want to answer is whether we have the optimum design. The answer is obviously no! Our wind turbine system design has been carried out independently without having any connections and without any integrating synthesis. Perhaps better collaboration and organization are necessary for this kind of complex system design.

If the research approach is *interdisciplinary*, as shown in Figure 2.16, researchers from different disciplines start communicating, transferring methods and techniques from one discipline to another, and collaborating with each other to optimize their sub-component design, considering the design requirements of the whole system. Once the compatibility and reliability of the sub-components are ensured, then they are delivered for assembly of the system. Although such an integrated system design would provide the optimum solution, the social acceptability of this approach has not been explored. Sustainability is an important issue to consider in design, not only due to economic matters but also due to social issues and environmental concerns.

Although wind power promises a clean and inexpensive source of electricity, it can raise environmental and community concerns:

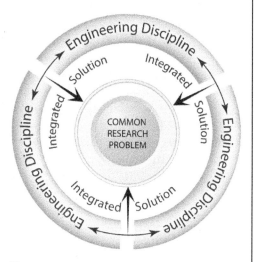

**Figure 2.16** Interdisciplinary research approaches.

- Noise and vibrations caused by wind turbines may cause sleep disruptions and other health problems among people who live nearby.
- They can be visually intrusive for residents living near them.
- They can disturb wildlife habitats and cause injury or death to birds.
- Turbulence from wind farms could adversely affect the growth of crops in the surrounding countryside.
- They may pose significant threats to migrating birds and have impacts on wildlife, habitat, wetlands, dunes, and other sensitive areas such as water resources, soil erosion, and sedimentation.
- Having huge wind turbines, each standing taller than a 60-story building and having blades more than 300 feet long, may disturb the community residents.

In the late 1980s, the California Energy Commission reported that 1,300 birds were killed by wind turbines, including over 100 golden eagles at Altamont Pass, CA. There are many other areas of strong concern: interference with TV reception, microwave reception interference, depreciating property values, increased traffic, road damage, cattle being frightened from rotating shadows cascading from the blades in a setting sun, rotating shadows in nearby homes, concerns about stray voltage, concerns about increased lightning strikes and many others. Currently, all of these issues are being raised in states where wind farms have been introduced.

As shown in Figure 2.17, the TD research process starts by identifying the research problem. This process involves not only crossing disciplinary boundaries (engineering, science, etc.) but also includes guidance of subject-matter experts (academic and non-academic), aspects of practical contexts and values or normative judgments (sustainability, ethical practice), as well as feeding back results into practical actions (societal players).[42] Social sciences and the humanities bring an abundance of knowledge on cultural, economic and social growth and advancement as well as on social systems. Therefore, they provide an important input to decisions being made relative to current problems and challenges. The humanities play an important role in putting to beneficial use new findings in engineering and the natural sciences. For example, natural scientists work together with researchers in the humanities to discover archeological objects and determine their age.

**Figure 2.17** Transdisciplinary research aproaches.

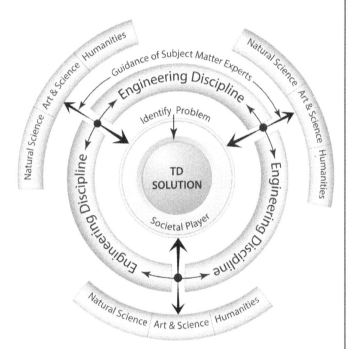

In the case of wind turbine design, researchers from environmental science should undertake an environmental assessment of the site and a comprehensive consultation exercise with local community and environmental bodies in terms of the development of wind turbine farms. Engineers should work with researchers from social science, natural science, and humanities

*(Continued)*

---

42 Baumgärtner, S., Becker, C., Frank, K., Müller, B., and Quaas, M., "Relating the philosophy and practice of ecological economics: The role of concepts, models, and case studies in inter- and transdisciplinary sustainability research," *Ecological Economics*, 67, 384–393, 2008.

---

**CASE STUDY 2.1 (Continued)**

to understand the impact on the environment and nearby communities of people to guide reiteration of their design.

Through the TD research process, researchers can plan early and have frequent consultations with the affected communities. This allows them to identify and address the most serious issues before substantial investments are made. In other words, designers should make reasonable efforts to "design out" or minimize hazards and risks early in the design process.

Further, researchers should collaborate with the required utility agencies, government agencies, environmental organizations, policy-makers, and developers to ensure that such complex problems will be under control.

Continuous education and encouragement are required in order to develop a spirit of collaboration among research team members in order to solve complex problems. Through educational activities that focus on such areas as research team management, problem solving, establishing research goals, optimizing the use of resources, and supporting each other, members of the research team learn to work together more effectively. In other words, team members provide mentoring and support to each other.

[a]From Ertas, A., "Understanding of transdiscipline and transdisciplinary process," *Transdisciplinary Journal of Engineering & Science*, 1, 48–64, 2010 (used with permission).

---

## 2.9 Transdisciplinary Domain

The growing understanding of the importance of technological innovation to economic competitiveness is challenging a new urgency for applied basic engineering research on extremely difficult technical, medical, social, and cultural problems. The lack of a knowledge base to solve these problems is becoming more and more widespread. Engineering education must therefore produce well-trained engineers with the skills required for civic participation, creative thinking, and the ability to interface with other sectors of society to solve such problems. Civic participation and creative thinking around scientific issues can be most richly learned through involvement with the arts and humanities. Transdisciplinary cultures, supporting team-based science and engineering, can be fostered by revising science, technology, engineering, and mathematics (STEM) education and training.

Figure 2.18 describes the attempt to integrate arts and humanities into STEM education. STEAM-H is not just science, technology, engineering, and math education – It is a transdisciplinary applied approach that is linked with real-world problem-based learning. Connections between STEM and sister disciplines such as the arts and humanities occur at the boundary of the transdisciplinary domain – intersection of arts and science, technology, engineering, math and humanities. The six discrete disciplines are no longer separate but "interconnected" through an acronym, STEAM-H. Interconnection goes beyond integration. STEAM-H is a

**Figure 2.18** Transdisciplinary domain.

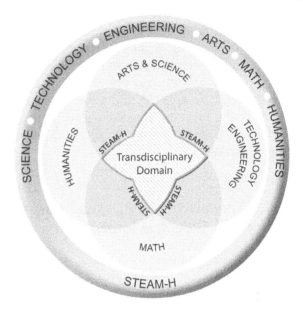

system of interconnected disciplines without boundaries within the TD domain which creates a cohesive teaching and learning paradigm. Transdisciplinary STEAM-H integrates knowledge, tools, methods, and cognitive skills and thinking from widely diverse fields to form an inclusive framework to solve challenging complex contemporary social issues that exist at the interfaces of multiple disciplines.

The unique contributions of the humanities to STEAM-H education may be considered as follows:[43]

- an appeal to an autonomous self with the right and capacity to make independent decisions and interpretations;
- indeterminacy in the subject matter of these decisions and interpretations;
- a focus on meaning, in the context of human responses, actions, and relationships, and particularly on the ethical, aesthetic, and purposive; and
- the possibility of cohesion in standards of decisions and interpretation.

The arts play an important role in science; they also hold the knowledge and skills a person needs to participate actively in civic life. The arts can provide ways for both scientists and engineers to broaden their understanding of concepts from diverse disciplines and generate creative, innovative solutions to unstructured problems. In particular, the arts can help people develop skills such as visual thinking; recognizing and forming patterns; modeling; getting a

---

43 Donnelly, J.F., "Humanizing science education," *Science Education.* 88(5), 762–784.

"feel" for systems; and the manipulative skills learned by using tools, pens, and brushes – thsee are all demonstrably valuable for developing STEM abilities.[44] The arts provide students with problem-solving skills, innovative mindsets, communicative attitudes and motivation. There have been experimental studies which indicate that intense exposure to art develops superior spatial-visual coordination and other basic skills.[45]

Business leaders and economists emphasizing that both the arts and humanities provide the creative and critical thinking skills workers need in a new technology-driven economy that emphasizes multidisciplinary pursuits such as biotechnology, nanotechnology, green energy, clean technologies, and digital media.[44]

## 2.10 Transdisciplinary Design Process: Social Innovation through TD Collective Impact

The transdisciplinary design process teaches students new skills aimed at creativity, innovation, and working across knowledge fields to achieve measurable effects on major social complex issues; it offers an approach that synthesizes methodologies from multiple fields; it teaches the ability to collaborate across multiple spheres of knowledge and practice; it prepares students to design, develop, and deliver systems that qualify them to be workforce ready: and provides a head start on careers in science, engineering, and technology research. TD appropriately emphasizes collaborative, cross-disciplinary, team-based research projects.

Figure 2.19 shows the proposed TD design process model representing social innovation through TD collective impact. "Social innovation is a complex process of introducing new products, processes or programs that profoundly change the basic routines, resource and authority flows, or beliefs of the social system in which the innovation occurs. Such successful social innovations have durability and broad impact."[46]

### 2.10.1 Collective Impact

Collective impact is a new approach to solving complex social problems through cross-sector collaboration. It is an innovative TD approach to tackling unstructured problems by crossing professional and societal domains. We introduce the concept of "collective impact" through a case study.

---

44 Root-Bernstein, R., and Root-Bernstein, M., "Turning STEM into STREAM: Writing as an essential component of science education," 2001. Retrieved from http://www.nwp.org/cs/public/print/resource/3522.

45 Shuster, D.M., "The arts and engineering," http://soliton.ae.gatech.edu/labs/ptsiotra/misc/Arts%20and%20Engineering%20(Shuster).pdf, accessed June 27, 2014.

46 Westley, F., and Antadze, N., "Making a difference: Strategies for scaling social innovation for greater impact," *The Innovation Journal: The Public Sector Innovation Journal*, 15(2), p. 2, 2010.

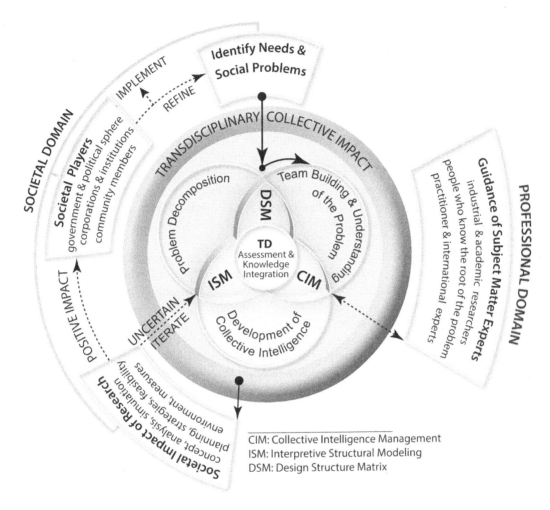

**Figure 2.19** Transdisciplinary design process.

---

**CASE STUDY 2.2**

The environmental cleanup of the Elizabeth River in Virginia was achieved by highly structured collaborative efforts that had a substantial impact on a large-scale social problem.

**Goal**

The aim of the project was to make the Elizabeth River safe for swimming and eating oysters by 2020; this was to be achieved through government, business, and community

*(Continued)*

---

**CASE STUDY 2.2 (Continued)**

---

partnerships. Elizabeth River Project executive director Marjorie Mayfield Jackson helped launch the Elizabeth River Project in 1993 to "restore the river to the highest practical level of environmental quality."

**History**

The Elizabeth River, a 6-mile-long stream of the James River in southeastern Virginia, was named after Princess Elizabeth Stuart. The James River, 23 miles of estuary on the southern end of Chesapeake Bay, has one of the busiest commercial ports in the world. However, industrial pollution contributed to residue contamination and "toxic hotspots" along the river. Twenty-nine species of fish tested positive for cancer as a result of toxins in the Elizabeth. The research study showed that the cancer rate was 38 percent and precancerous lesions 83 percent among the small mummichog fish.

Jackson said the Elizabeth, one of the main sources of the Chesapeake Bay, has a 400-year-old history of pollution and harm. The river, she said, has been integral to industrial and economic growth in Hampton Roads and is home to some of the oldest shipyards in the United States.

As executive director, Marjorie spearheaded the cleanup of Money Point, collaborating with industry, citizens, government projects and the military. Their joint efforts helped to lower cancer rates among the mummichog to less than 7 percent in the initial cleanup area. Stakeholders have worked collaboratively on dozens of other projects, including reducing future pollution and restoring wetlands, setting a goal for the entire Elizabeth River to be both swimmable and fishable by 2020.[47]

---

As shown in Table 2.8, five key elements of collective impact were identified by John Kania and Mark Kramer and published in the *Stanford Social Innovation Review* in 2011.[48] These five main conditions distinguish collective impact from other types of collaboration.

### 2.10.1.1 Shared Measurement

Of these five conditions shown in Table 2.8, one of the most challenging to achieve is shared measurement which identifies how multiple organizations could create uniform ways of measuring their progress toward a common goal.[49]

Shared measurement has been defined as the "use of a common set of measures to monitor performance, track progress towards outcomes and learn what is and is not working in the group's collective approach" (John Kania, FSG). Identifying shared measurement for tracking progress toward a common agenda across organizations allows:[50]

- improved data quality;
- tracking progress toward a shared goal;

---

47  Source: http://intersector.com/case/elizabethriver_virginia/, accessed May 21, 2017.
48  https://ssir.org/articles/entry/collective_impact, accessed, October 2, 2017
49  FSG, http://www.fsg.org/ideas-in-action/collective-impact, accessed May 27, 2017 (used with permission).
50  http://www.collaborationforimpact.com/collective-impact/shared-measurement/

**Table 2.8** Five conditions of collective impact.

| Conditions | Description |
| --- | --- |
| Common agenda | All participants have a shared vision for change, including a common understanding of the problem and a joint approach to solving it through agreed upon actions. |
| Shared measurement | Collecting data and measuring results consistently across all participants ensures efforts remain aligned and participants hold each other accountable. |
| Mutually reinforcing activities | Participant activities must be differentiated while still being coordinated through a mutually reinforcing plan of action. |
| Continuous communication | Consistent and open communication is needed across the many players to build trust, assure mutual objectives, and create common motivation. |
| Backbone support | Creating and managing collective impact requires a separate organization(s) with staff and a specific set of skills to serve as the backbone for the entire initiative and coordinate participating organizations and agencies. |

- enabling coordination and collaboration;
- learning and course correction;
- catalyzing action.

The three phases of the development of a shared measurement system shown in Figure 2.20 can be summarized as follows:[50]

1) Design phase
   - Shared vision for the system and its relation to broader goals, theory of change or roadmap
   - View of current state of knowledge and data
   - Governance and organization for structured participation
   - Identification of metrics, data collection approach, including confidentiality/transparency

**Figure 2.20** Developing shared measurement system.

2) Develop phase
   - Shared vision for the system and its relation to broader goals, theory of change or roadmap
   - View of current state of knowledge and data
   - Governance and organization for structured participation
   - Identification of metrics, data collection approach, including confidentiality/transparency
3) Deploy phase
   - Learning forums and continuous improvement
   - Ongoing infrastructure support
   - Improve system based on a pilot, review, refinement, and ongoing evaluation of usability

Shared measurement systems encourage trans-sectors to align their efforts on shared outcomes, enable them to collectively track and evaluate their collective progressive results and

**Table 2.9** Key success factors in the development of shared measurement systems.[49]

| Key factors | Descriptions |
|---|---|
| Effective relationship with funders | • Strong leadership and substantial funding (multi-year); <br> • Independence from funders in devising indicators, managing system |
| Broad and open engagement | • Broad engagement during design by organizations, with clear expectations about confidentiality/transparency; <br> • Voluntary participation open to all organizations |
| Infrastructure for deployment | • Effective use of web-based technology; <br> • Ongoing staffing for training, facilitation, reviewing data accuracy |
| Pathways for learning and improvement | • Testing and continually improving through feedback; <br> • Facilitated process for participants to share data and results, learn, and better coordinate efforts |

provide them with chances to benchmark their results against others. The key success factors in the development of shared measurement systems are shown in Table 2.9.

### 2.10.2 Four Phases of TD Design Process

The TD design process is conceptualized in four phases (see Figure 2.19):

1) Identifying needs and social problems
2) Team building and collaboratively defining and understanding the research problem
3) Development of collective intelligence to solve complex problems in question
4) Assessment and knowledge integration.

The starting point for the TD design process is identifying pressing social needs and problems. After these have been identified, a cross-team will be developed. To achieve good productive team collaboration the following important steps can be implemented:[51]

- Cross-teams will be developed with distributed leadership (team leadership will change in accordance with the particular expertise required)
- Continual communications system among collaborators to discuss team issues
- Periodic video conferencing among student team members
- Cross-team project progress will be monitored continuously.

Once the cross-team has been developed with a common goal, it meets to collectively define and understand the problem and create a shared vision to solve it. After defining the goals and barriers the research team, with the help of the expert matter guidance team, will investigate the exact nature of the problem and establish shared measurement to track progress that

---

51 Parker, G.M., *Cross-functional Teams*, Jossey-Bass, San Francisco, 2003.

allows for continuous improvement (refinement). Now it is time to organize a collective intelligence management workshop to develop *collective intelligence* for identifying the main *factors* affecting the problem solution. For more information on collective intelligence management, see Section 1.10.

> Given the complexity of the change management needed, businesses will need to realize that collaboration on talent issues, rather than competition, is no longer a nice-to-have but rather a necessary strategy. Multi-sector partnerships and collaboration, when they leverage the expertise of each partner in a complementary manner, are indispensable components of implementing scalable solutions to jobs and skills challenges. There is thus a need for bolder leadership and strategic action within companies and within and across industries, including partnerships with public institutions and the education... These efforts will need to be complemented by policy reform on the part of governments.[52]

Transdisciplinarity provides a good framework and adds to the current approaches for collective intelligence. TD collective intelligence, being a part of collective impact, is a structured approach for knowledge creation and decision-making through organizations from trans-sectors (diverse sectors). By bringing trans-sectors together, common difficult problems will be solved as different participants respond to different aspects of the problem in different ways, but all still with the same common research agenda in mind. This type of collaboration can spawn innovative solutions that could have an impact on the problem solution.

The next step is to implement interpretive structural modeling (ISM), a methodology for dealing with complex issues. Through problem decomposition ISM provides a fundamental understanding of how various elements relevant to the problem interact, helpubf researchers to structure the elements in a meaningful manner to develop collective intelligence. If the ISM results reveal that there is a conceptual inconsistency or that the societal impact of research is uncertain, then the complex problem should be iterated for problem reformulation (see Figure 2.19).

The design structure matrix (DSM) will be applied to "assignment of organizational responsibilities" and "development of work breakdown structure" shown in the traditional design process (see Figure 2.4). DSM analysis will assess the communication needs across team members, especially for large-scale complex problem solutions.

Transdisciplinary knowledge integration includes integrating people (social), artifacts (technical), and knowledge (cognitive) associated with different scientific and non-scientific knowledge domains[53] into an appropriate methodology for research.[54]

---

52 World Economic Forum, "The Future of Jobs Employment, Skills and Workforce Strategy for the Fourth Industrial Revolution," Global Challenge Insight Report, 2015.

53 Becker, E., Jahn, T., Schramm, E., Hummel, D., and Stiess, I., "Social-ecological research. Conceptual framework for a new finding policy" (Synopsis of the Report for the German Federal Ministry of Education and Research). Frankfurt/M. (ISOE), 2000.

54 Hinkel, J., *Transdisciplinary Knowledge Integration: Cases from Integrated Assessment and Vulnerability Assessment*, PhD thesis, Wageningen University, Wageningen, The Netherlands, 2008. ISBN 978-90-8504-825-1.

It is very rare that an idea developed through TD collective impact emerges fully formed. Refinement may continue until the agreed success measures are satisfied. Then the implementation phase can start to realize the product – that is, the traditional design phase begins.

## 2.11 Generic TD Hybrid Design Process

As shown in Figure 2.21, a generic hybrid design process is proposed as a combination of the TD design process and the traditional design process. The TD design phase can be seen as TD activities representing the initial design stage of the traditional design process shown in Figure 2.4. The TD design process will be followed by the traditional design process, starting with embodiment design, where the idea created during the TD design phase can now be firmed up with overall system configurations and working principles – starting to think about more specific engineering considerations, refining and detailing conceptual solution with a significant number of corrective steps: this is the most time-absorbing part of the design process. Then, as discussed in Section 2.4.9, detail design can lead to production.

Both design processes complement each other. Note that TD design methodology can also be implemented, as applied, to some of the other traditional design phases.

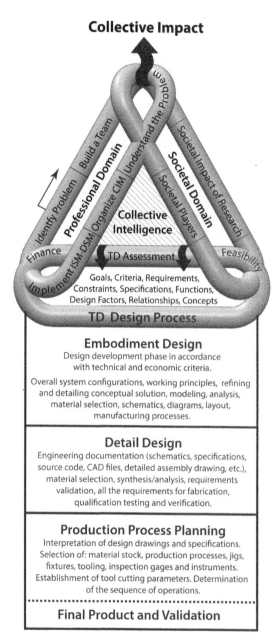

**Figure 2.21** TD hybrid design process.

## 2.12  Transdisciplinary Research Process

The transdisciplinary research (TDR) process can be developed in direct analogy with the system of systems approach:

> System of systems is a collection of task-oriented or dedicated systems that pool their resources and capabilities together to obtain a new, more complex, 'meta-system' which offers more functionality and performance than simply the sum of the constituent systems.[55]

A similar analogy can be used for the TDR process. This process pools resources and capabilities of researchers from diverse disciplines together to obtain a new shared common collection of knowledge to enable the discovery of new approaches that lead to alternative solutions for the growing complexity of today's grand challenges.

The TDR process depends on how researchers from diverse disciplines work together, not how each of the researchers performs when considered separately. Selectively and collectively, organization of the TDR process structure can provide a research capability that is greater than sum of the contributing individual researchers.

### 2.12.1  Transdisciplinary Interoperability

Interoperability refers to the ability of two or more diverse entities (systems, components, organizations, and sectors) to communicate, exchange data, and use the information that has been exchanged. Transdisciplinary interoperability is the ability of TD researchers to communicate with each other and exchange information to enable them to work effectively together (interoperate) to address a common goal by taking into account social, political, and organizational factors that impact the TDR process performance. Four levels of TDR process interoperability execution are given in Figure 2.22.

As shown in Figure 2.22, TDR process interoperability execution is a multiple-step process where the steps are interlinked with each other. Successful TDR process management requires reaching across disciplinary boundaries to establish an end-in-mind set of goals with identified resources. Experienced managers are needed in TDR process activities which requires significant flexibility to resolve the conflict among the researchers from diverse disciplines.

Through connected interoperability, researchers exchange information and share methods and techniques. This TD collaboration lights the fuse of new knowledge generation.

---

55  Broadman, J., DiMario, M., Sauser, B., and Verma, D., "System of Systems Characteristics and Interoperability in Joint Command and Control," 2nd Annual System of Systems Engineering Conference, July 25-26, 2006.

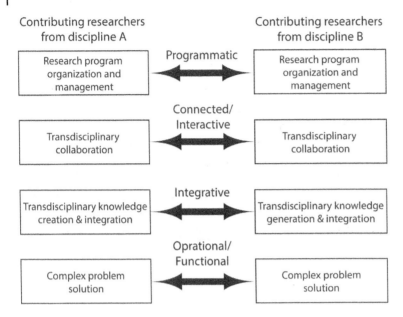

**Figure 2.22** TDR process interoperability execution.

Integration enables TD knowledge creation and ensures synergy of the participating researchers – syntactic (ability to exchange information among the researchers) and semantic (ability to use and understand the information among the researchers) interoperability. Although technical interoperability is essential, it is not sufficient to ensure an effective TDR process. Researchers at all levels of the TDR process must understand each other's capabilities and constraints that unify the participating researchers to achieve desired holistic behavior. A fully integrated TDR process is essential for creation of TD new knowledge which is interlinked with the operational and functional problem solution.

Important characteristics of the TDR process can be summarized as follows:

- Related to real-life complex problems and specific problem solving
- Eliminate disciplinary boundaries for strong collaboration
- Participation of (non-academic) stakeholders
- Acceptance of diverse perspectives, problem framings and interpretations
- Holistic (non-reductionistic) approach
- New knowledge generation for solving specific problems (transformative knowledge).

For a fully integrated successful TDR process, there is a need for:

- Strong leadership and commitment
- Expertise and effective process and control
- A culture to survive and be effective
- Bi-directional collaboration
- Trust and information flow between researchers.

The requirements of the TDR process can be summarized as follows:

- Correct information which leads to a research need
- Information storage and availability
- Information processing and analysis for transdisciplinary knowledge generation (shared concepts, methods and tools required for transdisciplinary research)
- Interconnectivity and communication among the researchers from different disciplines.

In TDR activities, depending on the nature of the research, researchers will interact by exchanging and sharing information, expertise, and resources. Technology like the World Wide Web can significantly enhance this. The TDR process must be carried out rigorously, with issued relating to the research problem evaluated through an iterative process. Research process improvement through such an iterative process allows the researcher to gradually broaden the research scope and in consequence add generality to the research findings.

It is important to note that numerous complexities are involved in this approach. An example is when researchers working together from diverse disciplines are geographically dispersed. How such scientific research teams become organized and managed and address the complexity of collaboration due to the diversity of viewpoints is a great challenge. There are certainly several technical challenges that exist when implementing the TDR process. Some of them are as follows:

- Understanding of research needs
- Establishment of research requirements (dynamically changing requirements increase uncertainty)
- Diverse viewpoints increase complexity of collaboration (interconnectivity is a major issue)
- Achievement of the research goal is put at risk when the various disciplines are not well integrated.

Researchers involved in the TDR process should have the following abilities:

1) The ability to use "outside" knowledge and apply it to engineering problems
2) The ability to create new and more socially attuned engineering questions and solutions
3) The ability to be creative and innovative in engineering design
4) The ability to collaborate across multiple spheres of knowledge and practice (both within and outside of the field of engineering).

Transdisciplinary collaboration is essential in order to solve society's "wicked" problems, for example, those with interconnected influences and requirements (climate change, response to natural disasters; enhanced national security) such that the proposed solution causes other problems. Transdisciplinary collaboration has great potential to speed the rate at which research can contribute to the understanding of the problem, accelerate the pace of new discoveries, and expand human knowledge. Table 2.10 identifies criteria for determining TD skills.

**Table 2.10** Indicators of transdisciplinarity.

| Indicators of transdisciplinarity | Degree of indication |
| --- | --- |
| Deeper understanding of the material | Check: midterm and final exams (individual); modular research projects to what degree and how correctly methods and fundamental concepts are used (individual and team levels) |
| Transdisciplinary skills | Check: problem-solving, conducting research, communication, and self-management skills, self-efficacy, survey of student attitudes and interest |
| Knowledge integration | Check: whether content of research outcomes (final integrated modular research projects) reflect knowledge integration; diversity of knowledge sources; knowledge sharing from different sources; number of the integrative steps set out and how well the steps were carried out |
| Generation of new knowledge that transcends discipline boundaries | Check: content of the research outcome (final integrated modular research projects); what kinds of existing data and information are used to transform them into new knowledge; knowledge assets such as intellectual capital (frequency count) |
| Collaboration and team processes | Check: practice of collaboration of the project teams with different disciplines; transdisciplinary behavioral patterns of project team members; use of external experts |
| Innovation | Check: capture of new physical phenomena; adaptation of existing technologies; use of disruptive technology |
| Creativity | Check: concepts generated (fluency); diversity of concepts (flexibility); originality of concepts (originality); amount of detail (elaboration); technological creativity (invention); economic creativity (entrepreneurship) |
| Management, leadership, and networking | Check: how well the organizational structure fosters communication; networking among group members and teams; joint work activities and shared decision-making; leadership tasks |
| Research and bibliometric indicators | Check: literature search, diversity of the references, content of the research outcome, possible paper publications resulting from project; research benefit to society |

---

**CASE STUDY 2.3**

The objective of this case study is to develop a research proposal to design a high-speed train system line network among Houston, Dallas, and Austin, TX to create a cost-effective, comfortable, safe, and reliable alternative to current transportation systems in order to reduce emissions and traffic loads.[a]

*Research questions*:

1) What are the challenges associated with high-speed train system design?
2) What are the limitations associated with high-speed train system design?
3) What are the federal and state regulations and policies associated with high-speed train system design?

4) How would a high-speed train system impact the society and environment in Texas?

5) What are the economic impacts of having a high-speed train in Texas?

## Introduction

In the fall semester of 2015, the Mechanical Engineering Department at Texas Tech University initiated a two-semester transdisciplinary senior design course sequence emphasizing collaborative, cross-disciplinary team-based research efforts to engage and motivate students through hands-on TD learning experiences.[33]

This case study is a summary of student's first-semester proposal report. Texas is a region with some of the largest cities in the USA. There is a high demand for a fast, efficient, inexpensive way to travel in this region. This is an opportunity to provide a positive impact on society and environment. The Texas Eco Train will not only create countless jobs, but also, by taking cars off the road, reduce fossil fuel emissions and increase the longevity of the Texas highway system.

## Team Building

To effectively solve the complex problem, utilizing collective intelligence management, four student research teams (social, economic, mechanical, and electrical) with relevant knowledge, skills, and background were organized.

## Technology Embedded Learning

Social media and video platforms have connected the TD classroom with others across the world to broaden students' horizons – to explore a new field of knowledge, to become cognizant of possibilities outside of one's respective discipline. In this new TD design course, to provide students with the opportunity to create, collaborate, and communicate, and to enhance learning and knowledge sharing for problem solving, the following educational technologies were used.

- Email
- Video and audio conferences (e.g., Blackboard Collaborate)
- In-class video recordings
- Discussion boards (e.g., Blackboard forums)
- Face-to-face recorded meetings (recorded by students in video or Blackboard format)
- Tablet and laptop
- Smartphone
- File sharing software and document editing (OneDrive, Dropbox).

Technology embedded learning platforms provided students with the ability to:

- interact with "research groups" through interactive discussions;
- interact with the teacher/expert through an integrated video conference along with a shared, interactive whiteboard;

*(Continued)*

**CASE STUDY 2.3 (Continued)**

- form teams with other learners, share files, engage in active discussions, and conduct a group online video conference with a shared, interactive whiteboard;
- form "chat" connections with team members, saving the chat discussions.

**Creativity Tools and Techniques**

In this TD class, students learned skills and techniques needed to be highly creative when they take a job after graduation. Through the use of TD methods, students can learn how to become more creative, discover a range of innovation techniques for producing creative ideas, how to decompose complex problems to understand how various parameters relevant to the problem are interrelated, how to collaborate and share ideas on achieving collective results, how to hold each other accountable for delivery according to their plans, how to openly discuss conflicting ideas, how to embrace critical dialog and debate, and how to trust each other.

Some of the following creativity tools and innovation techniques were used in this TD class: ISM, DSM, axiomatic design, objectives tree method, Kano model analysis, critical to quality characteristics, the KJ method, total quality management, Six Sigma, quality function deployment, house-of-quality, theory of inventive problem solving, robust design, statistical decision-making, Taguchi methods, and design of experiments.

**Implementing TDR Process**

To obtain the desired research outcome for a system design, the proposed TDR process model is used, conceptualized in five steps (see Figure 2.19):

1) Identifying social issues
2) Building a collaborative research team and collective understanding of the problem
3) Developing collective intelligence and producing new transferable STEM knowledge through collaborative research to solve the societal problem in question
4) Problem decomposition
5) Knowledge creation and integration.

The process started with team building and identifying the societal problem as follows. At the beginning of the class, 17 students were assigned to four preliminary sub-project teams to develop their own independent project concept. Teams were randomly selected, so team members often had no previous working experience with others in their team. Using video and audio conferencing, email, forums and chat as communication platforms (Blackboard Collaborate recording), collective intelligence management workshops were organized where sub-project teams introduced their project proposals (concepts) about the project that they would be exploring. Research teams were allowed to generate the following four different project concepts:

1) Texas Eco Railways (high-speed train system design)
2) Tidal power
3) Water crisis
4) Lubbock weather.

Each idea was discussed in detail in class meetings. The advantages and disadvantages of each project concept were identified. Each concept was voted on and ranked. Finally, two closely ranked project concepts were considered as candidates for online voting by the students. The high-speed train system was selected as the final design project concept. This research project was decomposed into four main sub-systems to create expertise groups:

1) Economic modeling
2) Mechanical design
3) Electrical design
4) Social issues.

After the final research project concept was created, depending on the students' interest and area of expertise, sub-project teams were reorganized to develop the final proposal for designing a high-speed train system.

### Development of TD Collective Intelligence

Through brainstorming and consultation with the domain experts, student research group members worked together to document all the possible factors (elements) whose relationships were to be modeled. Then the most important factors were identified for the model development (see Figure 2.23). The next step was to establish contextual relationships to develop the structural self-interaction matrix (SSIM) shown in Figure 2.24. Using expert opinions along with student research team members, an SSIM was developed which shows the direction of contextual relationships among the factors affecting the research problem.

### Complexity Decomposition

The first step in complexity decomposition is to develop an adjacency matrix. The SSIM was transformed into a binary matrix by substituting V, A, X, and O by 1 and 0 in the matrix to reflect the directed relationships between the elements. This matrix was checked for the transitivity rule and updated until transitivity was established. After driving power and dependence of factors were computed, the final reachability matrix, $\mathbf{R}_f$, shown in Figure 2.25 was obtained.

The association of sets and binary relations through matrices can now be converted into graphical form by using theory of digraphs (see Chapter 1). As shown in Figure 2.26, the complex high-speed train system design is decomposed into eight levels, showing how the factors affecting system design are interrelated.

The four most important factors are locomotive design, speed, comfort, and rails (Figure 2.26). These need to be analyzed before other factors as most of the other factors depend on them. For a high-speed train system, the main goal must be to balance consumer travel needs with the project's economic need to be profitable. However, in some areas, laws and regulations can also be major barriers to overcome.

*(Continued)*

**CASE STUDY 2.3 (Continued)**

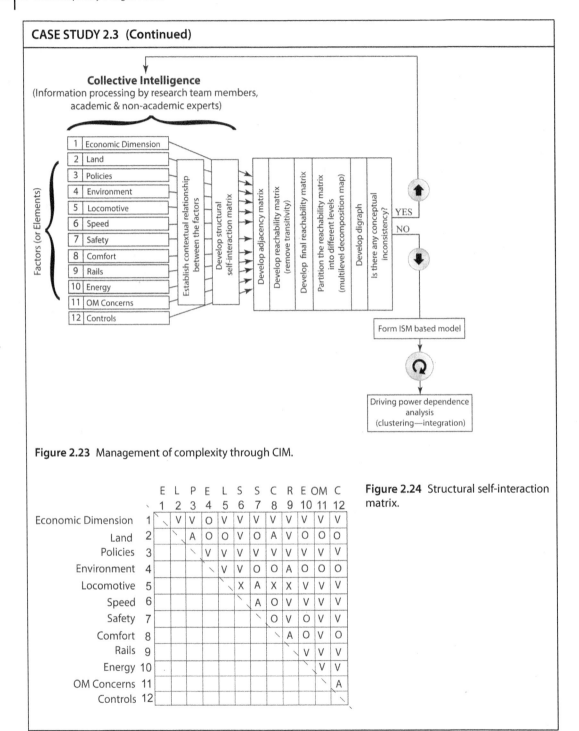

**Figure 2.23** Management of complexity through CIM.

|  | | E | L | P | E | L | S | S | C | R | E | OM | C |
|---|---|---|---|---|---|---|---|---|---|---|---|---|---|
|  | | 1 | 2 | 3 | 4 | 5 | 6 | 7 | 8 | 9 | 10 | 11 | 12 |
| Economic Dimension | 1 | | V | V | O | V | V | V | V | V | V | V | V |
| Land | 2 | | | A | O | O | V | O | A | V | O | O | O |
| Policies | 3 | | | | V | V | V | V | V | V | V | V | V |
| Environment | 4 | | | | | V | V | O | O | A | O | O | O |
| Locomotive | 5 | | | | | | X | A | X | X | V | V | V |
| Speed | 6 | | | | | | | A | O | V | V | V | V |
| Safety | 7 | | | | | | | | O | V | O | V | V |
| Comfort | 8 | | | | | | | | | A | O | V | O |
| Rails | 9 | | | | | | | | | | V | V | V |
| Energy | 10 | | | | | | | | | | | V | V |
| OM Concerns | 11 | | | | | | | | | | | | A |
| Controls | 12 | | | | | | | | | | | | |

**Figure 2.24** Structural self-interaction matrix.

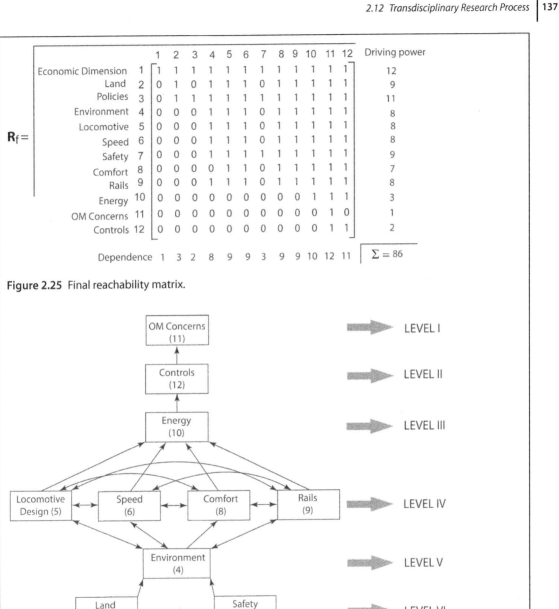

**Figure 2.25** Final reachability matrix.

**Figure 2.26** Directed graph.

(Continued)

---

**CASE STUDY 2.3 (Continued)**

---

As shown in Figure 2.26, factors such as energy, controls, and operation maintenance concerns are positioned at the top of the hierarchy. These are also very significant measures for the development of successful high-speed train systems. Of course, one of the main reasons for creating a high-speed train system is to have a more energy-efficient means of transportation. Controls and operational concerns are two of the most important factors ensuring continued success. These three factors are strongly interrelated. These higher-level factors have greater influence on the high-speed train system. Hence, when designing and developing such a project, these three factors should be evaluated first, and then kept constant as much as possible during the performance improvement process. In other words, we needed to first make sure that this project would be energy efficient, and then that the control systems would allow for a consistent high-speed train network, and then we could work on iterating and evolving the other factors toward success.

**MICMAC Analysis**

The purpose of MICMAC analysis is to arrange the factors with respect to their driving power and dependence in four clusters: (1) autonomous, (2) dependent, (3) linkage, and (4) independent factors (see Figure 2.27). The driving power and dependence of each of the factors were imported from Figure 2.25.

For this particular case, no factors can be identified as autonomous. This indicates that there is no factor disconnected from the system of systems. Dependent factors with low driving power and high dependence are contained in cluster II. As seen in Figure 2.27, energy (factor 10), controls (factor (11) and OM Concerns (factor 12) have a smaller driving power, but they have high dependency. These factors are affected by other factors of the systems, but they may not affect other factors.

Cluster III includes the factors that have high driving power as well as high dependency. These factors include environment issues (4), locomotive design (5), speed (6), comfort (8), and rails (9). Due to the nature of linkage factors, they not only affect but also depend on other factors. This creates a precocious and complex system that must focus on balancing these five factors in order to incorporate them into the system, without putting the whole project in peril.

Finally, the independent factors in cluster IV have very high driving power and minimal dependency. This cluster includes economic dimension (1), land (2), policies (3), and safety (7), and each of these factors has the potential to make or break the project and is influenced largely by external factors. If the project is not economically sound, it will never go beyond the planning stage. Failure to follow policies set by parties in authority can eliminate the project at any phase. Without rights on continuous land, there is no route that a train can follow. Lastly, if the safety of the customer is not guaranteed, legal and financial trouble, as well as a loss of credibility, would cause the project to fail. Even though they are critical to the success of the system, their low dependency means that once the factors have defined criteria they can be ignored as long as the criteria are met.

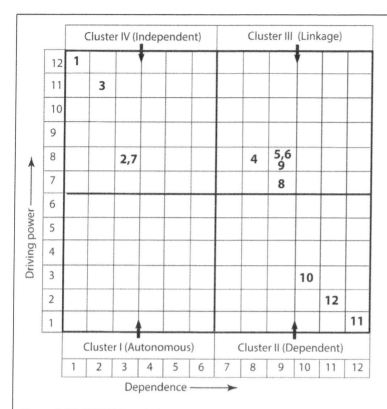

**Figure 2.27** MICMAC analysis.

## System Complexity

As seen from Figure 2.26, while factors 1, 2, 3, 4, 7 and 10, 11, 12 represent the linear mapping which preserves acyclicity properties of the system (system contains no cycle), level IV is the most complex system level. Factor 1 is the source component since it has only outgoing paths.

The number of edges shown in Figure 2.26 is $E = 21$, the number of nodes $N = 11$, and the number of connected components $P = 1$. Therefore, the cyclomatic complexity of the digraph in Figure 2.26 is

$$M = E - N + 2P = 21 - 11 + 2 \times 1 = 12.$$

Since this is higher than 10, the complexity of this system will make it difficult to understand.
Note that MATLAB code for ISM matrix solutions is included in Appendix A.1.

[a]Proposal submitted to Dr. A. Ertas by Mechanical Engineering student for ME 4370 class "Transdisciplinary Pilot Design Course" at Texas Tech University, December 1, 2015.

# Bibliography

1 CLAUSING, D., *Total Quality Development*, ASME Press, New York, 1994.

2 CROSS, N., *Engineering Design Methods*, John Wiley & Sons, Ltd., Chichester, 2008.

3 DIETER, G.E., and SCHMIDT, L.C., *Engineering Design*, McGraw-Hill, New York, 2009.

4 EPPINGER, S.D., and BROWINING, T.R., *Design Structure Matrix Methods and Applications*, MIT Press, Cambridge, MA, 2012.

5 ERTAS, A., and JONES, C. J., *The Engineering Design Process*, John Wiley & Sons, Inc., New York, 1996.

6 LOCHNER, R.H., and MATAR, J.E., *Design for Quality*, ASQC Quality Press, White Plains, NY, 1990.

7 WARFIELD, J.N., and CÁRDENAS, A. R., *A Handbook of Interactive Management*, Iowa State University Press, Ames, 1994.

## CHAPTER 2 Problems

**2.1**  As an exercise demonstrating the difficulty of establishing requirements early during a design project, assign student groups to a review of the curriculum in their engineering discipline to determine the courses that should be included as well as the total number of student credit hours that should be required for graduation. Justification for changes from the current published curriculum should be documented and rationale for the total number of student credit hours for graduation should be defined. Individual groups can then present their recommendations to the class and be prepared to defend their position.

**2.2**  Prepare a specification drawing for one of the devices listed below using an A-size drawing ($8\frac{1}{2} \times 11$ inches). Be sure to include all of the appropriate requirements listed in Paragraph 1.3.8 Detailed Design.
a)  Swiss army knife
b)  Calculator for use by engineering students
c)  Wood bit brace
d)  A simple mechanical press
e)  A car jack
f)  A lug wrench.

**2.3**  Perform a PMI on any of the subjects listed below.
a)  What do you think of curfews as a juvenile crime prevention technique?
b)  Should television sets include a special chip that limits certain objectionable programs?
c)  What do you think of the practice of placing troublesome older people in nursing homes rather than keeping them at home with their family? How about mentally handicapped younger people?
d)  Should university students who are caught cheating on an exam be given a fail grade on the exam? In the course? Expelled from the university?
e)  If you observe another student cheating on an exam should you report it to the teacher? Talk to the student who was cheating about it? Talk to other students about what you should do?

**2.4**  Create your own scenario as in Example 2.1 and make a decision using a Pugh decision matrix.

**2.5**  Although many improvements have been made in the manufacture of automobiles during the past 10 years, the windshield wiper is one component that has not changed dramatically. Windshield wipers are limited in effectiveness during heavy rain, and they pose a problem for further improvement in vehicle aerodynamics. Organize into groups and use brainstorming techniques to develop other solutions for keeping the windshield clear during inclement weather.

**2.6** Analyze the feasibility of using of natural gas, propane, or methanol for fueling yard and garden maintenance equipment. Consider the problems associated with refueling, equipment conversion or purchase, time between fuel tank refuelings, fuel economy, equipment maintenance, and the like. Decide which fuel you think is best for this application and justify your conclusion.

**2.7** Which terms are correct from the following requirement examples?
   a) The cargo ship shall sail at a speed of 15 knots.
   b) The tower will withstand wind loads of 100 mph.
   c) The cruise ship shall not travel at more than a speed of 22 knots.

**2.8** When drivers refuel, gas vapors can escape and contribute to smog and harmful air pollution. As shown in Figure P2.8, to eliminate the release of these vapors, most gas stations have installed special gas pump nozzles that include a rubber boot to block vapors from escaping.

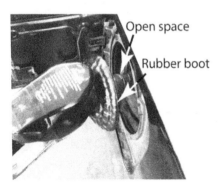

Open space

Rubber boot

**Figure P2.8** Gas pump nozzle with rubber boot.

Motorcyclists and some recreational vehicle drivers have long complained that gas pump nozzles that include a rubber boot are incompatible with their vehicles. Therefore the EPA has removed the rubber boots on gas pump handles. Organize into groups and, considering regulatory compliance, design a new concept and develop new requirements for a pump nozzles that will be compatible with almost every car.

## Team Projects

**2.9** Provide a complexity analysis of the feasibility and challenges associated with the commercial development and deployment of unmanned aerial vehicle drone technology as a method for large-volume product transit and delivery within the urban regions of the United States.

**2.10** Unemployment in the petroleum industry is a complex issue with social implications, such as unemployment, poverty, and economic deprivation. With the industry supporting 9 million jobs within the United States alone, it can have far-reaching impacts on the whole economy. Identify the factors which can impact unemployment in the petroleum industry and provide a complexity analysis and challenges associated with it.

**2.11** History shows that city management is not always aware of or prepared for the unexpected effects of a significant upset or the unintended consequences of their responses to those effects. This is understandable because a city, its people, its essential services, and its environment form a complex evolving system of systems. Although management's goal is to efficiently recover from an upset, at any point in time system interactions may not be known or visible or controllable. There is a need for a management tool or model that captures and reveals system interactions. Identify key vulnerabilities (factors) of the city systems to support development of mitigation strategies to resolve the complexity-induced fragilities. Provide a complexity analysis and challenges associated with it.

**2.12** The Flint water crisis is an ongoing drinking water contamination problem in Flint, Michigan that started in April 2014 when Flint changed its water source from treated Detroit Water and Sewerage Department water to the Flint River. Carry out research on the social impact of the Flint water crisis, and on relevant governmental regulations, and identify the factors that impacted the Flint water crisis. Provide a complexity analysis and challenges associated with it.

# 3

# Project Management and Product Development

A lot of companies have chosen to downsize, and maybe that was the right thing
for them. We chose a different path. Our belief was that if we kept putting great
products in front of customers, they would continue to open their wallets.

**Steve Jobs**

## 3.1 Introduction

Transdisciplinarity is a method of acquiring new knowledge through the rethinking of current
methods and processes, rearranging our fundamental science, and engineering knowledge to
create innovative technological products and services to confront society's complex issues.

For example, steam engines powered all early locomotives (see Figure 3.1), steamboats and
factories, and acted as the foundation of the industrial revolution. During the industrial revolution, steam engines became the main source of power, remaining so until the technological
advances in the design of internal combustion engine. An early reciprocating steam engine is a
heat engine that performs mechanical work using steam as its working fluid. Steam engines are
external combustion engines where the working fluid is separate from the combustion engine.
The ideal thermodynamic cycle used to analyze this process is called the Rankine cycle. In the
cycle, water is heated in a boiler and produces high-pressure steam which is then expanded in
the piston cylinder to create mechanical power. The reduced-pressure steam is then condensed
and pumped back into the boiler.

In contrast to the external combustion steam engine, an internal combustion engine is an
engine where the combustion of fuel occurs in a combustion chamber that is an integral part
of the working fluid. Rearranging our basic science knowledge and rethinking thermodynamic
processes, the "Otto cycle" was developed to build a successful four-stroke internal combustion
engine to replace the steam engine – a profound transformation methodology.

## 3.2 Project Management

A project is not a routine operation but a unique specific set of activities designed to accomplish
a particular goal. Projects have a clear-cut start and finish and a time frame in which the activity
must be completed.

**Figure 3.1** Locomotive powered by steam engines.

Project management is the process of initiating, planning, and coordinating human and material resources throughout the project life cycle to achieve specific goals and meet specific success criteria. The Project Management Institute (PMI) defines project management as "The application of knowledge, skills, tools and techniques to project activities to meet the project requirements."

A project team often includes people from their own organization and sometimes from different organizations and across multiple disciplines. Project resources are well defined and limited, requiring positive and effective management over the various phases and sequences of activities. The project management should bring a unique focus shaped by the goals, making sure that they are specific, well defined, and measurable. A project team must understand and agree with the goals and believe that they are realistic. In achieving the goals, the project management must keep project team members focused on the relation between their objectives and the overall project goals.

As shown in Figure 3.2, the five primary processes involved in project management are as follows:

1) Initiating
2) Planning
3) Executing
4) Controlling
5) Closing

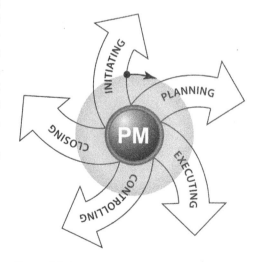

**Figure 3.2** Project management.

Many companies use variations of these project stages and it is common for the stages to be renamed in order to better fit the company functions and products.

### 3.2.1 Initiating

This phase is the first step in the project management life cycle, as it involves starting up a new project. During this phase, project objectives and goals are established and deliverables to be produced are defined. Also, a decision-making team will establish whether the project can be realistically completed.

### 3.2.2 Planning

The project management process begins with planning. Project planning is a continuous task of allocating resources to determine what needs to be done in what period of time and who needs to be involved. Project planning will be executed at the beginning of the project and will continue throughout the project life cycle. When the actual project starts (with the award of a contract), project planning effort takes place parallel to requirement analysis to further plan the activities and resources allocated to the project.

In short, in this phase of the project management, a decision-making team should prioritize the project, calculate a budget and schedule, and determine what resources are needed.

#### 3.2.2.1 Integrated Master Plan

As shown in Figure 3.3, a process driven integrated master plan (IMP) is an event-based contractual document, a relatively top-level plan consisting of a hierarchy of program events. Each event is decomposed into certain accomplishments and each accomplishment is decomposed into specific criteria. The IMP is a critical and important program management tool that provides significant assistance in the planning and scheduling of projects. As seen in Figure 3.3, IMP includes: (1) a list of significant project events; (2) major accomplishments to be attained in order to complete each event; and (3) criteria for each event providing clear evidence that a specific accomplishment has been completed.[1] The five steps in developing the IMP are as follows:[1]

1) Determine the IMP structure and organization.
2) Identify events, accomplishments, and criteria.
3) Prepare the introduction and narrative section.
4) Complete the numbering system.
5) Iterate events, accomplishments, and criteria with the integrated product team during integrated master schedule (IMS) development.

The IMP events do not include calendar dates: each IMP event is completed when its supporting accomplishments are completed and satisfied by supporting criteria. The IMP is usually put on contract and becomes the standard execution plan for the project.

---

1 Adapted from DoD Integrated Master Plan and Integrated Master Schedule Preparation and User Guide, October 21, 2005.

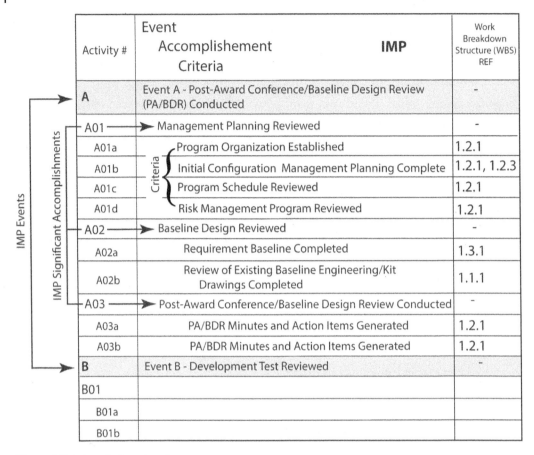

| Activity # | Event Accomplishement Criteria | IMP | Work Breakdown Structure (WBS) REF |
|---|---|---|---|
| A | Event A - Post-Award Conference/Baseline Design Review (PA/BDR) Conducted | | - |
| A01 | Management Planning Reviewed | | - |
| A01a | Program Organization Established | | 1.2.1 |
| A01b | Initial Configuration Management Planning Complete | | 1.2.1, 1.2.3 |
| A01c | Program Schedule Reviewed | | 1.2.1 |
| A01d | Risk Management Program Reviewed | | 1.2.1 |
| A02 | Baseline Design Reviewed | | - |
| A02a | Requirement Baseline Completed | | 1.3.1 |
| A02b | Review of Existing Baseline Engineering/Kit Drawings Completed | | 1.1.1 |
| A03 | Post-Award Conference/Baseline Design Review Conducted | | - |
| A03a | PA/BDR Minutes and Action Items Generated | | 1.2.1 |
| A03b | PA/BDR Minutes and Action Items Generated | | 1.2.1 |
| B | Event B - Development Test Reviewed | | - |
| B01 | | | |
| B01a | | | |
| B01b | | | |

**Figure 3.3** Integrated master plan.

### 3.2.2.2 Integrated Master Schedule

The IMS emerges directly from the IMP and complements it with additional levels of detail. As shown in Figure 3.4, the IMS is a task- and calendar-based day-to-day detailed document which incorporates all of the IMP events, accomplishments, and criteria: it includes a networked schedule containing all the detailed task packages, their duration and timing, key events and milestones, and dependencies among tasks necessary to support the events, accomplishments, and criteria of the IMP.

The IMS is a continuously updated document to reflect the progress of the project. The IMS should:[1]

- maintain consistency with the IMP;
- illustrate the relationships among the events, accomplishments, criteria, and tacks;

| Activity # | Event Accomplishement Criteria                                             **IMP** | WBS            |
|------------|------------------------------------------------------------------------------------|----------------|
| A          | Event A - Post-Award Conference/Baseline Design Review (PA/BDR) Conducted           | -              |
| A01        | Management Planning Reviewed                                                        | -              |
| A01a       | Program Organization Established                                                    | 1.2.1          |
| A01b       | Initial Configuration  Management Planning Completed                                | 1.2.1, 1.2.3   |
| A01c       | Program Schedule Reviewed                                                           | 1.2.1          |
| A01d       | Risk Management Program Reviewed                                                    | 1.2.1          |

| Activity #     | Task Name                     **IMS** | Duration | Start   | Finish  | January, 2016 | |
|----------------|---------------------------------------|----------|---------|---------|---------|---------|
|                |                                       |          |         |         | 1/1-1/7 | 1/8-1/15 |
| A01            | Management Planning Review            | 14 Days  | 1/1/16  | 1/15/16 |         |         |
| A01a           | Establish Program Organization        | 3 Days   | 1/1/16  | 1/4/16  |         |         |
| A01a01-1.2.1   | Identify Integrated Product Team      | 3 Days   | 1/4/16  | 1/7/16  |         |         |
| A01a02-1.2.1   | Customer Survey Complete              | 5 Days   | 1/7/16  | 1/12/16 |         |         |
| A01a03-1.2.1   | Identify customer needs               | 3 Days   | 1/12/16 | 1/15/16 |         |         |

Adapted from Reference 1.

**Figure 3.4** The IMP within an IMS.

- indicate the start and completion dates and duration for all events, accomplishments, criteria, and tacks;
- provide schedule updates on a regular basis that indicate completed actions, schedule slips, and reschedule actions and includes the previous schedule for reference;
- provide the ability to sort schedules in multiple ways (e.g., by event, by integrated product team, by work breakdown structure (WBS), by earned value management (EVM) system, by statement of work, or by contract WBS (CWBS));
- maintain consistency with the work package definitions and EVM systems;
- be traceable between the WBS items supported by each IMS task;
- be vertically and horizontally traceable to the cost and schedule reporting instrument (e.g., contract performance report).

Note that, since IMP activities designate issues associated only with completed effort, to structure the IMP activities past tense verbs are used. On the other hand, since the IMS tasks express what the team needs to do to demonstrate the criterion has been satisfied, present tense verbs are used.

### 3.2.2.3 Work Breakdown Structure[2]

Once the project becomes feasible, a systematic structural breakdown of the effort is required. For most government projects, the preliminary WBS is incorporated into the RFP and is finalized when the contract is consummated. A WBS is a key project deliverable chart in which the critical project tasks are illustrated to describe their relationships to each other and to the project as a whole – it is a project management tool used to break projects down into manageable pieces. The WBS is developed by starting with the end objective and subdividing it into manageable elements in terms of size and complexity, with each successive level reducing the scope, complexity, and dollar value of the WBS elements. As an example, the WBS shown in Figure 3.5 starts with the Space Shuttle program, which is divided into two phases, Shuttle System development and production and Shuttle operations. The next level is the project level which includes such elements as the Orbiter vehicle, the solid rocket motor, external tank, flight operations, etc. Level 4 corresponds to the systems level and includes structures, propulsion, power, avionics, etc. Finally, level 5 corresponds to the subsystem level, which includes a further breakdown of level 4 systems. Using the propulsion system as an example, level 5 divides this system into sub-systems such as the main, reaction control and orbital maneuvering propulsion subsystems. Further breakdown of effort below the sub-system level produces what are often referred to as the component, task, sub-task and work element levels.

The WBS must be compatible with the organizational structure of the work-performing entity to ensure that management accountability is maintained and that diffusion of responsibility across organizational lines is minimized. The WBS must reflect and represent the way in which the work is organized, managed and accounted for so that there is a clear relationship (interface) between each organizational element and the WBS element(s) it is responsible for.[3]

The degree and level to which any design process is sub-divided for work assignment and management responsibility is a function of the complexity and duration of the effort, the overall cost of the project, the organizational structures of the contractor and customer, the number of contracts and their relationships, and the needs of contractor and customer management. Small consumer products and similar design and development efforts obviously do not require the WBS complexity of a project such as the NASA Space Station, for which the preliminary WBS had 80 major systems. In fact, some small projects may not even warrant the preparation of a formal WBS. The point to remember in planning and organizing the work is to sub-divide the project into as many tasks and levels as considered necessary for good management control and organizational element responsibility, taking into account the above considerations.

### 3.2.2.4 Earned Value Management

Earned value management is a systematic project management process used to measure project performance; it helps project managers to assess cost, schedule, and technical progress on programs, and to support proactive decision-making. EVM was developed by the Department of

---

2 Adapted from Ertas, A., and Jones J.C., *The Engineering Design Process*, John Wiley & Sons, Inc., New York, 1996.
3 National Aeronautics and Space Administration, "*Handbook for Preparation of Work Breakdown Structures,*" NHB 5610.1, NASA, Washington, DC, February 1975.

Defense (DoD) to track its programs during the 1960s and is now a mandatory requirement of the US government.

A correct implementation of EVM will provide internal controls and formal program management processes for managing any acquisition within the organization. These controls and processes will ensure both contractor and program managers, as well as other stakeholders, receive contract performance data that:[4]

- relates time-phased budgets to corresponding scope of work;
- objectively measures work progress;
- reflects achievement of program objectives within budget, on schedule, and within technical performance parameters;
- allows for informed decisions and corrective action;
- is timely, accurate, reliable, and auditable;
- allows for estimation of future costs;
- supplies managers at all levels with appropriate program status information;
- is derived from the same EVM the contractor uses to manage the contract.

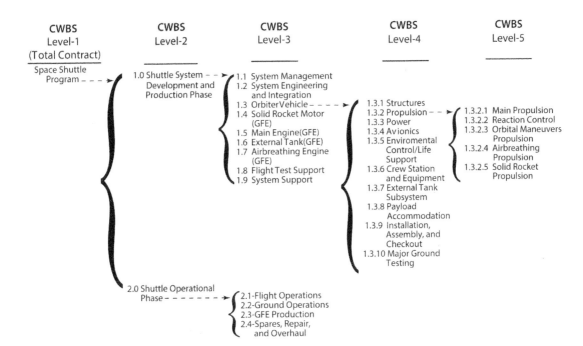

**Figure 3.5** Space Shuttle program work breakdown structure (from *Handbook for Preparation of Work Breakdown Structures*, NHB 5610, NASA, Washington, DC, p. 23, 1975).

---

4 Kranz, M., G. and Bliss, R.G., "Department of Defense Earned Value Management System Interpretation Guide," OUSD AT&L (PARCA), February 18, 2015.

**Table 3.1** EVM Measures.

| Terms | Formula | Description |
|---|---|---|
| **PERFORMANCE PARAMETERS** | | |
| **Budget for completion (BAC)** | | The total overall project budget allocated to the project (projected cost to complete the project) |
| **Planned value (PV)** | | Actual spending at measured points in the project |
| **Earned value (EV)** | $EV = \sum\limits_{begining}^{current} PV$ (completed) | The value of the portion of the work that is completed at a certain point in the project |
| **Actual cost (AC)** | | The actual cost of the work at measured points in the project |
| **PERFORMANCE MEASURES** | | |
| **Schedule variance (SV)** | $SV = EV - PV$; $SV < 0 \Rightarrow$ behind schedule, $SV > 0 \Rightarrow$ ahead of schedule | SV indicated the difference between actual value earned and value that was planned to be earned at a given point in the project. |
| **Schedule performance index (SPI)** | $SPI = EV/PV$; $SPI < 1 \Rightarrow$ behind schedule, $SPI > 1 \Rightarrow$ ahead of schedule | SPI is the amount that the task is ahead or behind schedule, expressed as a percentage of the task. |
| **Cost variance (CV)** | $CV = EV - AC$; $CV < 0 \Rightarrow$ over budget, $CV > 0 \Rightarrow$ under budget | CV indicates the difference between the budgeted cost and the actual cost at a point in the project. |
| **Cost performance index (CPI)** | $CPI = EV/AC$; $CPI < 1 \Rightarrow$ over budget, $CPI > 1 \Rightarrow$ under budget | CPI compares expenditures to actual value at a point in the project. |

Adapted from "Department of Defense Earned Value Management System Interpretation Guide," February 2015, and other sources.

The meaning and calculations of basic EVM terms are given in Table 3.1. The other EVM terms are explained below.

### Estimate to Complete

The estimate to complete (ETC) tells the project manager how much money needed from this point forward till project completion. If the project is expected to continue with the same performance in the future as the past, then the ETC can be calculated as

$$ETC = EAC - AC. \tag{3.1}$$

If the new estimate is required, then $ETC$ = new estimate.

**Estimate at Completion**

The estimate at completion (EAC) reflects, based on today's situation, how much the total project would cost. The following four methods can be used to calculate EAC:

1) If we assume that the project will likely to continue to perform the way it has until now, then

$$EAC = BAC/CPI, \qquad (3.2)$$

where BAC is the total project budget.

2) If the budget estimate is not likely to continue as it did in the past but the remaining project can be completed as planned, then

$$EAC = AC + (BAC - EV). \qquad (3.3)$$

3) If the project is over budget and behind schedule and the customer demands that the project should be completed on schedule, then

$$EAC = AC + (BAC - EV)/(SPI \times CPI). \qquad (3.4)$$

4) If the initial project assumptions were not correct and re-estimating the remaining work is required, then

$$EAC = AC + ETC. \qquad (3.5)$$

**Variance at Completion**

The variance at completion (VAC) tells the project manager, as of now, how much over and under budget we expect the project to finally be compared to the original estimates. The VAC can be calculated as

$$VAC = EAC - BAC. \qquad (3.6)$$

If the VAC is negative, you need that much more money to complete the project. If the VAC is positive, you will finish the project with that much of a surplus.

**To Complete Performance Index**

The to complete performance index (TCPI) is the estimate of future cost performance that is necessary to finish the project within the approved budget. the TCPI can be calculated in the following two ways:

1) If the project is within budget, the remaining funds are $BAC - AC$. Hence,

$$TCPI = (BAC - EV)/(BAC - AC). \qquad (3.7)$$

2) If the project is over budget, remaining funds are $EAC - AC$, since the EAC is the revised end-of-project cost. Hence,

$$TCPI = (BAC - EV)/(EAC - AC). \qquad (3.8)$$

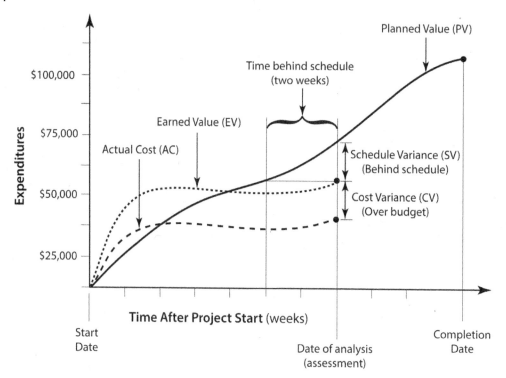

**Figure 3.6** Earned value management.

Figure 3.6 shows a typical project performance monitoring exercise showing whether the project is on, ahead of, or behind schedule and on, under, or over budget. We can estimate the amount of time (on any date of analysis) that we are behind or ahead of the original schedule by drawing a line from the intersection of the EV and date of analysis line parallel to the horizontal axis of the PV curve. As seen in Figure 3.6, the project is about 2 weeks behind schedule. The difference between planned and actual expenditures at the time of analysis is the result of both a schedule delay and being over budget.

### Example 3.1

Assume that a project has a total budget of $8 million and is scheduled for 8 months. It is assumed that the total budget will be spent equally each month until the eighth month is reached. After 2 months, the project manager finds that only 15% of the work is finished and a total of $2.5 million spent. Perform an EVM analysis.

### Solution

Inputs are:

- Planned value (PV) = ($8M/8 months) × 2 = $2 million
- Earned value (EV) = $8M × 0.15 = $1.2 million

- Actual cost (AC) = $2.5 million
- Project total budget (BAC) = $8 million

Outputs are:

- Schedule variance (SV) = EV − PV = $1.2M − $2M = −$0.8 million (since SV is negative, the project is behind schedule).
- Schedule performance index (SPI) = EV/PV = $1.2M/$2M = 0.6 (because SV is negative and SPI < 1, the project is considered behind schedule).
- Cost variance (CV) = EV − AC = $1.2M − $2.5M = −$1.3 million (since CV < 0, the project is over budget).
- Cost performance index (CPI) = EV/AC = $1.2M/$2.5M = $0.48 (since CPI < 1, the project is over budget).
- Estimated at completion (EAC) = BAC/CPI = $8M/0.48 = $16.67 million (note that we assume that the project will likely to continue to perform the same).

*Summary.* Because CV is negative and CPI is less than 1, the project is considered to be over budget. At the time of the analysis (assessment) 15% of the work is finished and a total of $2.5 million spent. This is 31.25% of the total budget (2.5M/8M) but we are 15% of the way through the project. If the project continues at this pace, then the total cost of the project (EAC) will be $16.67 million, as opposed to our original budget of $8 million.

Although EVM serves as a good early-warning system, it represents a single objective data point – it can change quickly and actual costs and project progress seldom occur as budgeted and scheduled. Therefore, assessment of the project through earned value should be performed regularly to identify schedule and budget problems early.

### 3.2.2.5 The Gantt Chart

The Gantt chart or bar chart used today was developed in the early 1900s by Henry Gantt. The simple bar chart shown in Figure 3.7 is probably the most widely used visual presentation of a project scheduling technique. The Gantt chart is a two-dimensional chart. The $x$-axis shows the project timeline. The $y$-axis is a list of specific activities that must be accomplished to complete the project. A similar chart used for displaying events is called a milestone chart (see Figure 3.8). Figures 3.7 and 3.8 actually show the same effort, only the emphasis is different. The content of the bar chart shows the planned start and end times for each task by the length of the bar, whereas the milestone chart normally only identifies the beginning and completion of the task.

In practice, the bar chart and milestone chart are normally combined to show task duration as well as significant milestones.[5] Many available software packages can be used in preparing and updating Gantt, milestone, and other types of charts for presentation purposes.

### 3.2.2.6 PERT Networks[2]

Unfortunately, bar charts have one significant shortcoming: for complex projects with interrelated tasks, bar charts do not reflect the dependence of one task on the completion of another.

---

5 Fleming, Q.W., Bronn, J.W., and Humphreys, G.C., *Project and Production Scheduling*, Probus Publishing, Chicago, 1987.

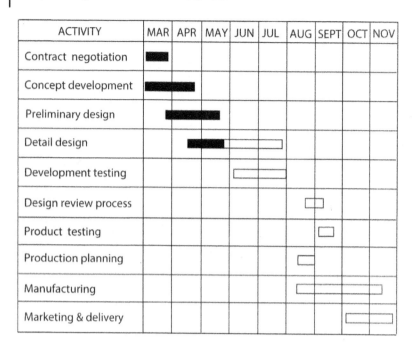

| ACTIVITY | MAR | APR | MAY | JUN | JUL | AUG | SEPT | OCT | NOV |
|---|---|---|---|---|---|---|---|---|---|
| Contract negotiation | ■ | | | | | | | | |
| Concept development | ■■ | | | | | | | | |
| Preliminary design | | ■■ | | | | | | | |
| Detail design | | ■ | ▭ | | | | | | |
| Development testing | | | | ▭ | | | | | |
| Design review process | | | | | | ▭ | | | |
| Product testing | | | | | | ▭ | | | |
| Production planning | | | | | | ▭ | | | |
| Manufacturing | | | | | | ▭▭ | | | |
| Marketing & delivery | | | | | | | | ▭ | |

**Figure 3.7** Typical Gantt chart schedule.

| ACTIVITY | MAR | APR | MAY | JUN | JUL | AUG | SEPT | OCT | NOV |
|---|---|---|---|---|---|---|---|---|---|
| Contract negotiation | ▲ | | | | | | | | |
| Concept development | | ▲ | | | | | | | |
| Preliminary design | | | ▲ | | | | | | |
| Detail design | | ▲ | | | △ | | | | |
| Development testing | | | | △ | △ | | | | |
| Design review process | | | | | | △ | | | |
| Product testing | | | | | | | △ | | |
| Production planning | | | | | | △ | | | |
| Manufacturing | | | | | | △ | | | △ |
| Marketing & delivery | | | | | | | | △ | △ |

**Figure 3.8** Typical milestone schedule.

It was, at least partially, because of this inadequacy that the development of scheduling methods generally known as logic networks came about. One of the early logic network scheduling methods, the program evaluation review technique (PERT), was developed by the US Navy in the 1950s to manage the Polaris submarine missile program. Although PERT is given considerable credit for the success of this program, many of the early efforts to use this technique proved to be cumbersome and difficult. Any scheduling method used in a program should be a tool to assist in getting the job completed. If the effort required to make the tool useful approaches the benefit that the tool provides to the program, then utilization of the tool in that application is questionable. This may have been the problem with many of these early attempts to use PERT. PERT, as originally defined, is no longer used by industry; however, the term PERT has become synonymous with a wide variety of network scheduling techniques. Thus, when a company is said to be using PERT, it often means that they are using network scheduling techniques that employ critical path methodology. Usually one of two acceptable network types is used: the arrow diagram method or the precedence diagram method.[5]

### 3.2.2.7 The Critical Path Method[2]

The critical path method (CPM) is a project management technique created in the 1950s for process planning that identifies critical and non-critical tasks with the goal of eliminating time-frame problems and possible process difficulties. In any given project, if we have many tasks and dependencies, it will be very difficult to identify the most critical tasks. If the most critical tasks are missed, that may immediately impact the success of the whole project.

The main difference between PERT and CPM is the manner in which time estimates are made for activities. CPM relies on a single *most likely* time estimate for each activity, whereas PERT uses statistical uncertainty compensated expected time estimates.

In the late 1950s, CPM began to be referred to as the Arrow diagram method (ADM). Arrow diagrams use an arrow to represent activities in a project and nodes to illustrate when activities precede or follow other activities. Nodes are placed at the start and end of each activity arrow (see Figure 3.9). An activity is a task that occurs over time and consumes resources such as labor, materials, and equipment.

Another variation of the CPM network is the so-called precedence diagram method (PDM). In contrast to the CPM/ADM/later PERT, the PDM identifies the activity in the node, which is typically represented by a rectangular box. Lines tie the nodes together and thus depict the relationship between the activities. Figure 3.10 shows the basic difference between PERT (original), ADM, and PDM networks. An example of the PDM networks method is shown in Figure 3.11.

Once the scheduling method has been selected, several steps must be taken to prepare the project schedule. All of the project activities and events (or a subset thereof) must be identified and their relationships must be determined. Also, during this process, an important question to ask is "how long does each task take before we can finish the project?" Time spans for all the activities must be estimated and the critical path (the longest path in terms of time) must be located. This can best be accomplished by adding all the time durations in all of the possible paths in the network and identifying the one with the longest total time.

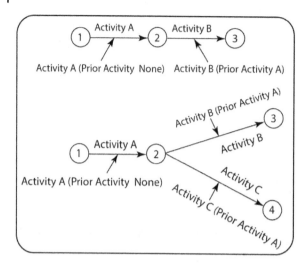

**Figure 3.9** Representation of arrow diagram method.

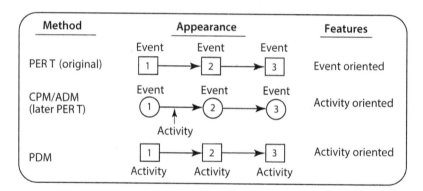

**Figure 3.10** A summary of network types.

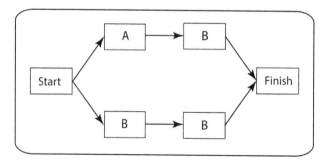

**Figure 3.11** Precedence diagramming method.

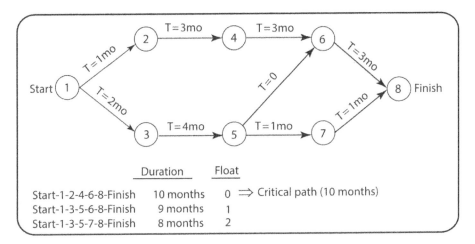

**Figure 3.12** ADM schedule.

Another method that is used to analyze the network for the critical path is to determine the earliest time and latest time that an activity can be accomplished without adversely impacting the total project schedule. Once the critical path has been defined, effort can be concentrated on completing the project in less time. Applying additional resources to critical path activities will allow early project completion, but increased effort on non-critical path tasks will not reduce the overall project schedule time.

Figure 3.12 illustrates a typical ADM schedule with a critical path (the sequence of activities with the longest duration) of 10 months. A delay in any of these activities will result in a delay for the whole project. Float (slack time) for the two other paths is 1 and 2 months. Free float is the time that an activity can be delayed from its earliest start time until it begins to interfere with the earliest start time of the succeeding activity. The duration of each activity is listed in Figure 3.12.

## Example 3.2

Analyze the PERT network schedule shown in Figure 3.13.[a]

**Figure 3.13** ADM schedule.

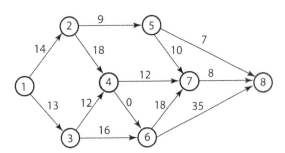

## Solution

Figure 3.13 shows a PERT network schedule. A typical way in which these networks are analyzed is shown in Table 3.2. Starting at the final event, 8, the latest time $T_L$ that event 8 can occur is determined to be 67 time units. The longest path from event 1 to event 8 is noted to be along path 1–2–4–6–8. The earliest time $T_E$ that event 8 can be completed through predecessor event 5 is determined to be 30 time units along path 1–2–5–8. The float or slack time ($T_L - T_E$) along this path is determined to be 37 time units. The time required for activity 5–8, $t_e$, is noted to be 7 time units. For the path including events 7 and 8, $T_E$ is noted to be 58 time units along path 1–2–4–6–7–8. Note that the longest schedule time path to event 7 is used to calculate $T_E$ for the path through activity 7–8. This is because activity 7–8 must wait for the last of the predecessor events to be completed. $T_L$ for successor event 8 remains the same, 67 time units, and the slack time is 9 time units. For successor event 5 the latest time that this event can occur without impacting the project completion date is the $T_L$ for event 8 minus the time for activity 5–8, $67 - 7 = 60$ time units. The longest time to event 5 through predecessor event 2 is 23 time units and the slack for this path is 37 time units. When determining $T_L$ for intermediate events,

**Table 3.2** PERT activity time estimate.

| Activities | | Estimates | | | |
|---|---|---|---|---|---|
| Successor Event | Predecessor Event | $t_e$ | $T_E$ | $T_L$ | Slack $T_L - T_E$ |
| 8 | 5 | 7 | 30 | 67 | 37 |
| 8 | 7 | 8 | 58 | 67 | 9 |
| 8 | 6 | 35 | 67 | 67 | 0 |
| 5 | 2 | 9 | 23 | 60 | 37 |
| 7 | 6 | | | | |
| 7 | 5 | | | | |
| 7 | 4 | | | | |
| 4 | 2 | 18 | 32 | 47 | 15 |
| 6 | 4 | | | | |
| 6 | 3 | | | | |
| 4 | 3 | | | | |
| 4 | 2 | | | | |
| 3 | 1 | | | | |
| 2 | 1 | | | | |

Adapted from *Fundamentals of PERT*, Penton/IPC, Cleveland, OH, 1982.

subtract the time associated with the longest path between the intermediate event and the final event from the $T_L$ for the total network. For example, the $T_L$ for event 4 is determined as follows: $67 - 8 - 12 = 47$. $T_E$ for activity 2–4 is determined by identifying the longest path from the starting event 1 to event 4 through event 2. Thus, $T_E$ for activity 2–4 is $14 + 18 = 32$ time units. Note that there is only one path from the start to event 4 through event 2 in this example.

Complete the information for the remaining paths in Table 3.2 and determine the critical path time. The calculations in this exercise provide an easy way to analyze slack time. The slack time column indicates how important each activity is to timely project completion. If extra resources can be added to activities with zero slack, the project critical path can be shortened. The only activity that cannot be shortened by increasing resources is activity 4–6.

*Summary.* With the graphics capabilities of present-day computers, CPM network planning, scheduling, and analysis are considerably more feasible for projects today than they were when they were first introduced in 1957. With the project management software available today, projects can be planned and tracked, adjustments can be made based on actual performance, and status reports can be generated. CPM networks have proved their worth for the planning and management of one-time-only jobs such as design projects, but they are of limited value for repetitive work such as production scheduling, where variations of the bar chart and other methods are more appropriate. For management presentations, CPM network schedules should be converted to a more understandable form, such as a Gantt chart or milestone display. Most project management software packages include the capability for developing PERT, CPM, and bar chart schedules as well as for displaying data in various graphic formats to track task interdependence and timelines, resources, milestones, and cash flow.[6]

[a]Adapted from *Fundamentals of PERT*, Penton/IPC, Cleveland, OH, 1982.

### 3.2.3  Executing

The project execution phase is the third phase in the project life cycle. The project execution phase is usually the longest phase in the project life cycle and consumes the most resources and time effort. This phase involves achieving objectives by implementing the plans created during the project planning phase. Specifically, this phase includes project management planning and execution, task assignments and execution, team development, resources assignment, updating the project schedule, and modifying project plans as needed.

While the project monitoring and controlling phase has a different set of requirements, the project execution phase goes hand in hand simultaneously with the monitoring and controlling phase.

---

6 K. Landis, "Critical paths," *MacUser*, October 1989.

### 3.2.4 Monitoring and Controlling

The main purpose of the monitor and control phase is to compare and verify project deliverables against the project plan and the requirements. This phase also includes:

- Identify changes
  - Establish the processes to be used to manage and control change
  - Formally record and track all changes requested
  - Estimate the effort, time, cost, and benefits of adopting the changes
  - Define the impact of approving the changes.
- Identify risks
  - Log and track all critical and non-critical risks, including those that have been resolved
  - List preventative actions for reducing risk likelihood
  - Work closely with customers on areas of high uncertainty to minimize risk
  - Evaluate the project in logistics and risk management:
    * Are major risks identified?
    * Is a contingency plan in place to deal with them?
    * Are there areas that can be simplified or automated?
- Perform quality management
  - Make sure that quality reviews are conducted and meet specific parameters as indicated by the project and customer requirements
  - Monitor project deliverables and performance goals to make sure that the requirements are satisfied
  - Monitor regression, functional, and stress testing
  - Monitor implementation strategies and make sure that user acceptance is received.
- Time management
  - Monitor and control all time spent on the project
  - Record the activities performed (the amount of time team members spend working on each activity and monitor the amount of overtime spent on one task)
  - Request approval for time spent on specific tasks
  - Raise and resolve any time management issues related to the project
  - Determine whether the project is currently on track.
- Cost management
  - Identify the expenses required for each task within the project
  - Ensure that expenses are approved before purchasing and kept on an expenses register
  - Keep a central record of all costs incurred
  - Control the overall cost of the project.
- Procurement management
  - Categorize the supplies and services to procure for the project
  - Fill out purchase orders and issue to suppliers
  - Monitor the delivery of supplies and services from providers
  - Review and accept the items obtained from providers
  - Approve provider payments.

### 3.2.5 Closing

This is the last phase in the project life cycle. In this phase, the project is formally closed and its overall level of success is reported to the sponsor. Most of the project management activities in this phase are administrative.

## 3.3 Technical Management

Technical management is the balancing act of managing technical processes and teams along with using technical skills to plan, direct, and control the technical resources of an organization in order to achieve a specific purpose. The management of any effort can be broken up into two main activities: planning and control. The planning phase of management is the most essential element of any activity or project: planning forms the basis and sets the stage for the control phase of operations. Management involves the application and control of resources in order to meet customer expectations. This effort usually involves a difficult balance of time, money, risk, and performance.

Project resources are well defined and limited, requiring positive and effective management over the various phases and sequences of activities. The project manager must align project focus with the corporate vision and get project members to understand, support, and believe that they are realistic. The project manager must build trusting relations, alignment, win–win outcomes, teamwork, and create an environment for happier workers and inspired teamwork. The project manager must consider how to minimize risks (simplify, automate, standardize, rapid prototype) for dependable and sustainable project success. One of the project manager's most important tasks is to keep the project constantly moving toward successful completion and to ensure that project personnel are continuously mindful of the goals. If the project manager's conduct, either technically or ethically, does not engender respect and honor, the goals and objectives he/she has established will not garner the dedication and commitment needed for a successful project.

As depicted in Figure 3.14, technical management is categorized into eight main elements recognized to be highly interrelated, but presented as distinct management components: planning, organization, delegation, decision-making, control, reporting, coordination, and development. These eight elements must be applied consistently with a common purpose. The eight primary elements are first considered to establish a soft baseline for their application within a project as integrated management function. The first three functions can be combined as "planning" and the last five can be combined as "management."[7] Note that there are numerous ways to look at and rearrange the dependencies among the elements shown in Figure 3.14 – this is just a simplified management function interrelationships framework.

---

7 This section is adapted from "Technical Management: Integrated Process," special project in Transdisciplinary Design and Process master's program by K. Himmelreich, submitted to M. Smith and A. Ertas, 2002.

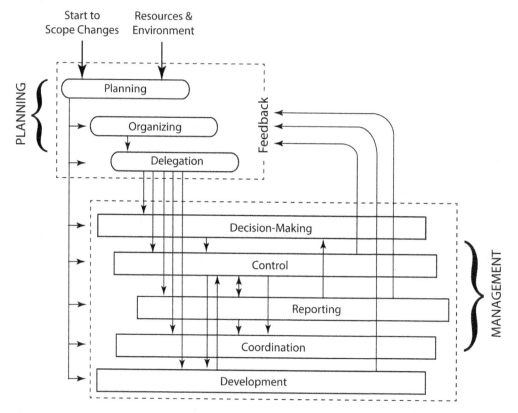

**Figure 3.14** Management function interrelations.

### 3.3.1 Integrated Management System

An integrated management system involves the integration of five management functions: planning, control, reporting, delegation, and development. Planning, control, and reporting are required as the core assignment. Delegation and development will be addressed primarily as they relate to project planning and control. Although some interactions among organizing, decision-making, and coordination exist, we will not include these management functions in the integrated framework. As illustrated in Figure 3.15, planning lays the foundation for all the other management functions. The other end of the spectrum is control, which manages the program. In an oversimplified concept such as this one, the plan feeds all elements, which feed control, which feeds reporting, which then feeds back to the plan. Delegation and development are intermediate control functions, as shown in Figure 3.15.

*Planning.* The project manager is usually the person responsible for laying out the groundwork and getting the project started by following the contract with the customer. Proper planning includes the setting of goals, the approach to be taken, describing tasks at an appropriate level, and the allocation of resources to attain the goal. Program planning is done within internal and

**Figure 3.15** Five management function interrelations.

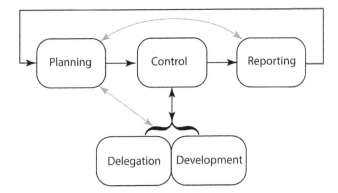

external constraints, and includes consideration of obstacles and potential changes. A careful set of plans serves as an effective program to provide an early look at the likelihood of project success. Each step of the planning is visualized and coordinated with every other step to check cost, schedule, and risk. Trades are made to minimize execution risk while satisfying program objectives (performance, cost, schedule, knowledge, etc.). Milestones are set to provide scheduled points for progress checks during execution.

As shown in Figure 3.16, the process of planning typically include the following outputs:

- objectives;
- organization;
- constraints;
- budget;
- tasks and responsibilities;
- contingencies; and
- schedule

Control standards and reporting mechanisms, shown in Figure 3.16, are also a part of the overall plan. Outputs, as shown in the figure, establish the inputs to other elements that define the

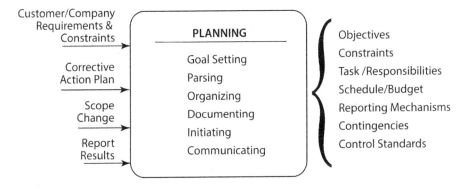

**Figure 3.16** Planning inputs and outputs.

program. Planning has inputs at the start as well as during execution to adapt to circumstances that change during the program. Startup inputs include many of the requirements, constraints, and other programmatics from both the customer side and the internal culture. Execution inputs include scope changes, progress/control reports, and corrective actions.

*Control.* In general, control involves the use of collected data against preset goals to identify problems early enough for improvement. Measurements might include the physical weight of a product, employee satisfaction, and cost. Regardless of the item under analysis, controls must be measurable, predictable, supportive of risk mitigation goals, and within resource constraints. Control is a continuous, iterative process where absolute deviations are identified as well as trends. Variances are evaluated for cause and significance, and appropriate actions are identified and taken. Well-designed controls, which consider the risks that may prevent successful achievement of the goal, may include some degree of pre-planned activity that will occur when a variance from the predicted path takes place.

The process of controlling consists of five distinct but interrelated steps: evaluation, measurement, validation, corrective action, and communication. As shown in Figure 3.17, control of a program involves the following inputs:

- opinion/buy-in of those subject to control;
- control (measurement) standards, including allowances for personal time, etc.;
- actual (measured) progress, including employee development and delegation;
- other planning information.

Outputs of control include control variances, corrective action identification, and reporting of results.

*Reporting.* Performance reports should be published frequently to monitor progress and to maintain buy-in from employees. Negative results must be handled carefully to avoid future problems. The manager must also be subject to an external reporting chain, whether to higher-level managers, stockholders, customers, government agencies (e.g., IRS), or other groups. The external reporting chain is any reporting of progress towards the goal that goes to others

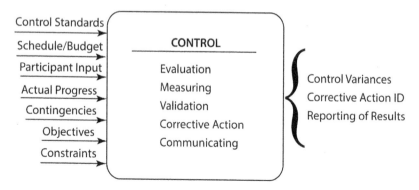

**Figure 3.17** Control inputs and outputs.

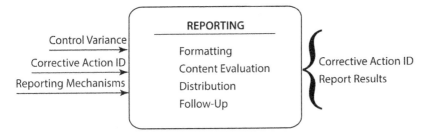

**Figure 3.18** Reporting inputs and outputs.

outside the manager's scope of control. The scope of control, the organizational structure, and the group objectives define the nature and format of external reports. Figure 3.18 shows reporting inputs and outputs. Report outputs include the who, what (status/problems), how, where, when, and why about a particular topic (labeled "Report Results" for simplicity) and specific corrective action plans addressing the problems.

*Delegating.* Managers have a variety of authority and responsibility that includes the actions of their group. In complex organizations, high-level managers delegate authority and responsibility to lower-level managers, each of whom receive resources and tasks to perform and contribute to the overall goal of the program.

Delegation is an essential aspect of the management function of organizing; it organizes concerns and establishes and maintains meaningful relationships among all personnel, regardless of their assigned department or functional field.[8] Delegation allows the authority chain to be broadened as appropriate for the size of the program and existing resources. This is necessary for both project success and employee development. A manager delegates the authority for accomplishing a given task to a subordinate.[9] Without delegation there is no organization.[8] Delegation must be clearly and consistently defined. In delegating, the manager should establish the means of communication of progress, problems, needs, and metrics, and outline the general terms under which the delegation might be revoked. Failures related to delegation can usually be classified as one of the following:

- poor definition of responsibility;
- poor structure of organization;
- poor communications;
- conflicting authority or responsibility.

Figure 3.19 shows inputs and outputs of delegation. Inputs include reporting, control standards, and the assignment of responsibility and authority. Outputs include the goals for employee development, and actual progress for monitoring the authority chain. Rewards may be included

8 Stickney, F.A., and Johnston, W.R., "Delegation and a sharing of authority by the project manager," *Project Management Quarterly*, 14(1), 42–53, 1983.
9 Sisk, H.L., *Management and Organization*, Southwestern Publishing Company, Cincinnati, OH, 1977.

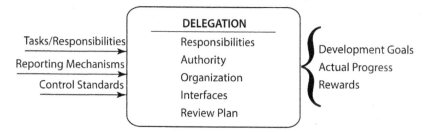

**Figure 3.19** Delegation inputs and outputs.

as outputs of delegation, or they could be part of control. Within the delegation function (as with many others), participant ownership and buy-in are critical to execution.

*Development.* The focus of development for this section is employee development. However, this could be expanded to include broader elements like the refinement of processes or technology for reuse on future programs. Employee development often involves challenging the employee at or beyond their proven abilities and upgrading their technical skills. While there may be formal training involved in development, most workers learn by actually practicing the skills in what is often called on-the-job training. The manager's role in development is to increase the value of his group by making it more productive through increasing its efficiency through individual and group development. Group development comes with practice and is based on standards, policies, personal interaction, and communication. Individual development of skills increases through the ability to take on increased responsibilities, and applies to the manager as well as those within her or his area of control.

Inputs for employee development include goals and specifics from planning and delegation. There is also feedback review of results from control (see Figure 3.20). Actual progress outputs are provided to control and monitor progress, along with more traditional program performance metrics, including rewards.

### 3.3.1.1 Integration Approach

In integrating the management functions, the mission and vision of the company must always be the common thread. Also, the key attributes of program leadership and the company culture

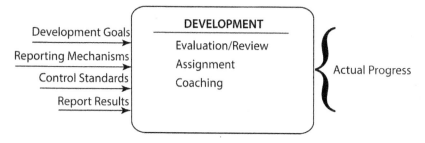

**Figure 3.20** Development inputs and outputs.

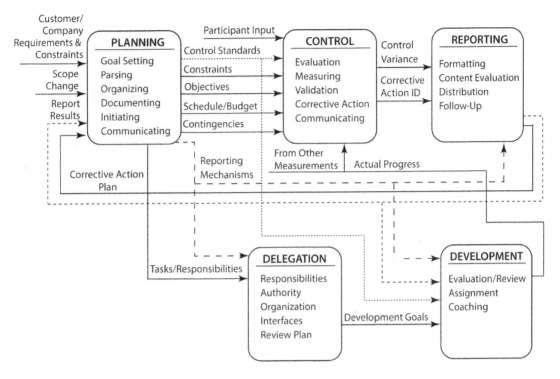

**Figure 3.21** Integrated management functions.

must not be contradicted. Figure 3.21 provides a top-level illustration of the integration of the five management functions described above. This simplified framework is desirable for implementation. The real heart of technical management for a given project lies within the function blocks, and success or failure is usually dependent on the plans.

## 3.4 Clarifying the Project Goals and Objectives

The critical first step in a successful project is to set a clear and compelling goal. The clarification of goals and objectives is vital. Research teams should never assume that project goals and objectives are clear or realistic at the beginning, rather they should analyze and define the goals and objectives so that the project team has a clear idea about what is to be accomplished and an understanding of the subordinate objectives in concrete form for the overall project. In analyzing and defining the project goals and objectives a research team should ask the following questions: What is the anticipated end result of your project? What problem will this project resolve? What need will it meet? How will this project change the way we handle our business? This effort will pay off in several ways:

1) A focus for the project team as the activity level increases and new team members not familiar with the project background are brought into the effort

2) A means to attain a more comprehensive agreement and understanding with the customer and/or upper management about the goals and objectives of the project and the target(s) of the end product
3) A reference point if the project goals and objectives need to be changed over the duration of the project.

### 3.4.1 Objectives Tree Method

The objectives tree method is a useful tool that enables the project team to gain a clear understanding and a more comprehensive agreement of the customer's needs and objectives as well as a baseline from which to proceed. It illustrates in graphic form all the identifiable objectives and their interrelationships organized in a hierarchical pattern of objectives and sub-objectives. The general steps followed in developing objectives trees are as follows:[10]

1) Identify and list the design objectives from the design problem statement, from discussions with the client and/or upper management, and from discussions within the design team.
2) Organize the list into higher- and lower-level objectives and group them according to importance (into hierarchical levels).
3) Draw a diagrammatic tree of the objectives showing the hierarchical relationships and interconnections.

As an example, the design of the Retinal Scanning Identification Device (RSID) objective tree is shown in Figure 3.22. Since the tragic events that occurred on September 11, 2001, airport security has been at the forefront of homeland defense. According to numerous articles and experts, the biggest security risk at airports today is access to airport secure areas such as runways, control towers, gates and terminals. Biometrics is being introduced at airports to mitigate this risk. Biometrics is the science and technology of measuring and statistically analyzing biological data. In information technology, biometrics usually refers to technologies for measuring and analyzing human body characteristics such as fingerprints, eye retinas and irises, voice patterns, facial patterns, and hand measurements, especially for authentication.[11]

To develop the objective tree shown in Figure 3.22, the design team interviewed customers and end users such as airport security and airport personnel, including flight attendants, pilots, and maintenance staff and established a list of customer needs (CNs), grouping the detailed CNs into a more abstract list of CNs. Then, along with the customer, the design team identified the main objective of the RSID, which is to verify the person trying to gain access to a secured area.

Then a list of sub-objectives was compiled and organized into seven categories: timely; secure; accurate; continuous and autonomous; adjustable; failure notification; and provide access or alert security. An objective tree was constructed, reviewed, and modified until

---

10 Cross, N., *Engineering Design Methods*, John Wiley & Sons, Ltd., Chichester, 1989.
11 Adapted from Parker, J. Brown, B., and Moore, H., 2002. "Design of the Retinal Scanning Identification Device," Report, ATLAS Publications, 2002.

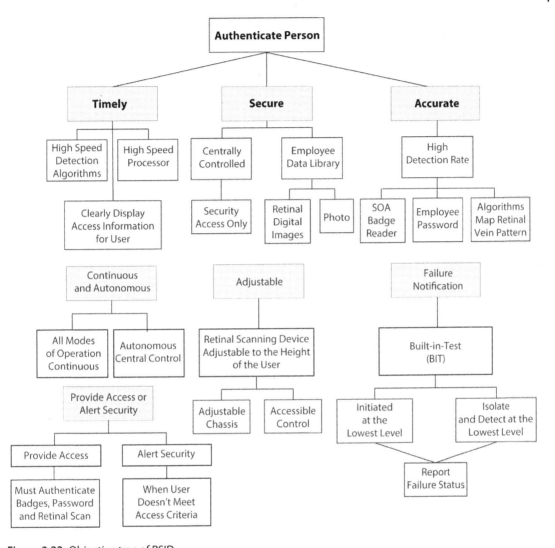

**Figure 3.22** Objective tree of RSID.

the entire project team, along with the customer, understood and agreed on the objectives (see Figure 3.22).

As shown in Figure 3.22, sorting out the objectives and sub-objectives and their relationships early in a design process can be a driving force in clarifying the meaning of the objectives and how they will be accomplished. Perhaps the most important benefit of this effort is the uniform understanding of the objectives accomplished among the various disciplinary groups working on the project. After understanding the customer needs and completing the objectives tree, quality functional deployment (see Section 3.7.2) can be used to provide a structured way of designing quality into the design process while establishing product requirements.

---

**CASE STUDY 3.1  Clarifying Objectives and Planning a Design Project[a]**

**Background**

A state school for physically and mentally handicapped students has a program in which students are trained to perform simple tasks while working on contracts negotiated with local firms. One of these programs involves an automated gravel bagging system used to package 40 lbs of gravel in plastic bags which are then used by construction companies in securing temporary traffic signs and for erosion control. Dump trucks deliver the gravel to a concrete pad located in front of the facility from where it is pushed into a concrete pit by a front-end loader. The gravel is of a small size and is commonly referred to as *chad*. When gravel is being packaged it is fed to a vertical conveyor belt elevator by gravity. The elevator lifts the gravel from the pit to an elevated (10 ft above the floor level) hopper from which it flows into a weighing mechanism. After 40 lbs of gravel has been transferred into the weighing hopper the mechanism dumps the gravel into a plastic bag on a conveyor belt below the weighing hopper. The bag is then sewn shut and transferred to a storage area outside the facility where it is picked up by the customer.

**The Problem**

At times the gravel is delivered in a wet condition. Also, it can pick up moisture while it is located on the concrete pad awaiting movement into the gravel pit. When the gravel is wet it tends to form clumps and will not flow readily from the pit into the vertical elevator. A possible solution suggested by school personnel is to provide a means of drying the gravel on the concrete pad before loading into the pit.

**The Solution**

Develop a design concept that will provide the best solution for this problem. Make sure that you have a thorough understanding of the problem before you start developing the concept. Do not confine your solution to the one proposed by school personnel. Use the PMI method or brainstorming to help in understanding the problem and in widening your perception of proposed solutions. After you have developed a viable design concept, prepare an objective tree to further clarify the objectives and to identify sub-objectives. Finally, identify all of the tasks that must be accomplished to complete the project and schedule them using the Gantt chart and CPM methods (see Example 4.2).

[a] From Ertas, A., and Jones J.C., *The Engineering Design Process*, John Wiley & Sons, Inc., New York, 1996.

### 3.4.2  Coordination with Client and Management[2]

It is no less important that the client and/or upper management agree with the design team's interpretation of the objectives than it is for individual members of the design team to understand and agree on their meaning. Once the objectives have been established, work will begin on the design and resources will begin to be consumed at an accelerated rate. After this point in the design process any effort expended on misunderstood or invalid objectives is wasted. Therefore, it is essential that a mutual and agreed-upon understanding of the objectives exist between the client, upper management, and the design team. If the project is being accomplished for a client, agreement should first be obtained in-house with upper management, followed by coordination with the client. An effective way in which to obtain the concurrence of management, and subsequently the client, is to schedule meetings at which the objectives are identified and described and initial plans are presented as to how these objectives will be accomplished. This presentation should be made by the project manager and key design team members so that management and the client can get to know these key people and develop some confidence in their abilities.

Presentation of the project objectives and initial plans to the customer and/or management is a very significant milestone in the evolution of the project even though it comes at a very early time in the overall effort. It is absolutely essential that the project manager and the team members have a thorough and in-depth understanding of the objectives, and agreement on their meaning, before this presentation is scheduled. From this meeting upper management and the client must develop a feeling of confidence in the knowledge of the team members and in the fact that they can work as a team. Possibly even more important is for the presenters to have the knowledge and capability to effectively counter suggestions that come up during the presentation that are not well thought out, as well as questions that lead to unprofitable discussions. This may be the first time that management and the client have been forced to clarify their ideas about the project. The presenters must thus be able to keep the meeting properly focused so that valid observations can be thoroughly discussed and the others satisfactorily responded to, respectfully, but with minimum discussion.

## 3.5 Decision-Making

Design is a process involving constant decision-making. Decisions are made on organization, objectives, requirements, materials, testing, systems, sub-systems, components, suppliers, personnel assignment, work breakdown, manufacturing, and cost, to name just a few. There are also many ways in which decisions can be made in the design organization. Some of these methods are very disciplined and use well-known and well-defined tools that are available to the decision maker. Decision matrices and decision trees fall into this category. When these methods are used, they are generally applied to the higher-level and more significant project decisions, sometimes as a check on the normal decision-making process. The principal decision-making approach used in design projects works as a function of the organizational structure itself, using formal and informal coordination and approval for decisions. In this type of decision-making an individual is assigned to an area of design responsibility for which he/she has decision authority within guidelines and company policy, subject to supervisory approval. A case study of this type of decision-making follows.

---

**CASE STUDY 3.2  Designing a test enclosure for a missile propellant-loading system.**[a]

### Requirements

A designer on a defense-related project is assigned to design a test enclosure for testing liquid and pneumatic components used on a missile propellant-loading system. The design shall be able to withstand a component failure within the enclosure without injury to the operator. Appropriate pneumatic and liquid connections, component supports, instrumentation, and controls shall be provided for. The enclosure shall include the capability for visual access so that the operator can observe component operation within. The test enclosure shall be portable and capable of operation by one technician. The designer reports to a first-level supervisor and analytical support is available from another group at the same organizational level. A small technical advisory group is also available, made up of experienced designers.

### Solution

In this case study, the designer must first identify all the components in the missile system that will require testing in the enclosure. He must also determine the safety, personnel, and other procedures that apply to the design as well as the design philosophy. To do this, the designer talks to the knowledgeable people within the division in which his branch is located. To assess the safety, quality control, and systems cleanliness requirements the designer must talk to organizations outside the division but within the missile department. To obtain information on the enclosure window the designer decides to talk to the aircraft people who are located in another department of the company. The designer then lays out an initial concept for the enclosure and asks the analytical group to do some calculations supporting the design. Detail design is subsequently initiated and material selection begins. During this period of time the designer meets with his immediate supervisor several times to review the task progress and to ask questions as they arise.

In this assignment the designer is responsible for the following tasks that require decisions:

1) Ensuring that a common understanding of the requirements exists between himself and his supervisor
2) Deciding who needs to be contacted for information
3) Determining that the objectives are defined and well understood
4) Developing a viable design concept
5) Ensuring that detail design requirements are optimally based on objectives
6) Selecting components and detailed parts
7) Selecting the construction materials
8) Deciding whether analytical support results are valid
9) Deciding whether to perform the structural analysis himself
10) Detail design decisions
11) Determining what testing is required to support decision-making
12) Assembly design decisions
13) Procedural design decisions

14) Determining when he needs to talk to his supervisor
15) Determining when the design is completed and ready for review.

In this case study, decision-making is based on experience, judgment, test results, and analysis. If the designer does a good job of coordination and consulting the design will represent the combined experience of the organization. Judgment exercised will be that of the designer and his supervision since they are responsible for making the design decisions. The quality of the analysis is a function of the performance of the analytical group and the judgment of the designer and supervisor.

The importance of being able to work with other experts in accomplishing this task is very apparent. A designer must be able to obtain the enthusiastic support of other employees in providing information and advice. The designer must engender a desire to help on the part of the contacts he makes.

[a]From Ertas, A., and Jones J.C., *The Engineering Design Process*, John Wiley & Sons, Inc., New York, 1996.

## 3.5.1  Decision Matrices[2]

A decision matrix can vary from a simple chart consisting of rows and columns that allows the evaluation of alternatives relative to various design criteria, to a complex array such as the house of quality that allows consideration of the requirements, analysis of the requirements relative to design, manufacturing, and marketing characteristics, evaluation of solutions relative to competitors' as well as target values, technical difficulty associated with the solution, and important control items impacting or resulting from the various solutions. Design criteria do not usually have the same degree of importance; therefore, some sort of weighting scale is normally assigned to account for this variation. The resulting grade for each alternative is, of course, a function of the weighting scheme adopted. For a situation such as that depicted in Table 3.3, in which various materials are evaluated for use in fabricating the test enclosure referred to in the previous case study, the design criteria are given equal weight and a simple utility score, as shown in Table 3.4,

**Table 3.3** Decision matrix for test enclosure material selection (from Ertas, A., and Jones, J.C. *The Engineering Design Process*, John Wiley & Sons, Inc., New York, 1996, used with permission of John Wiley & Sons, Inc.).

| Material | Strength–density ratio | Weldability | Machinability | Corrosion resistance | Availability | Cost | Score |
|---|---|---|---|---|---|---|---|
| Low-alloy steel | 122/3 | 4 | 2.5 | 0 | 4 | 4 | 17.5 |
| Stainless steel | 113/3 | 3 | 2.5 | 4 | 3 | 2.5 | 18.0 |
| Aluminum 2011 | 60/2 | 2 | 3 | 3.5 | 3 | 2 | 15.5 |
| Aluminum 7075 | 194/4 | 2.5 | 3 | 3.5 | 2.5 | 2 | 17.5 |
| Titanium 6A14V | 205/4 | 2 | 1 | 4 | 0 | 0 | 11.0 |

**Table 3.4** Scheme for evaluating design alternatives (from Ertas, A., and Jones, J.C., *The Engineering Design Process*, John Wiley & Sons, Inc., New York, 1996, used with permission of John Wiley & Sons, Inc.).

| Scoring Category | Value |
| --- | --- |
| Far below average | 0 |
| Below average | 1 |
| Average | 2 |
| Above average | 3 |
| Far above average | 4 |

is assigned. Each material is evaluated for each criterion and the score for each material is totaled in the last column. In this case, stainless steel received the highest score and would therefore be a reasonable choice for this application. Aluminum 7075 would also be an acceptable material and, if the strength to density ratio is important enough, should be given serious consideration. Low-alloy steel also received a high grade in this evaluation, but the very poor rating in corrosion resistance would probably eliminate it from consideration. The important point to remember in this type of analysis is to use the evaluation to help make the proper decision. The final decision should not be made solely on the basis of a high decision matrix score but should always reflect the informed judgment of the designer, all factors considered.

The decision matrix can be an important tool in the overall design process, but it should be kept in mind that nothing takes the place of common sense and good judgment. For devices like the decision matrix to be viable, estimates must be made about the relative importance of the different evaluation factors. It is important to realize that these estimates have a certain built-in uncertainty that may result in erroneous conclusions as to the best or most effective solution or alternative. Therefore, it is always prudent to perform the evaluation of alternatives using at least two independent methods. If two separate evaluation methods result in the same conclusion, and if the result seems to be logical based on experience and good judgment, proceeding with that approach is probably warranted. However, a good designer will maintain a questioning attitude, always seeking further confirmation that the decision was correct as the design process evolves.

### 3.5.2 Decision Trees[2]

The decision tree is another method that can be used to evaluate different alternatives. Decision trees are often used in evaluating business investment decisions by providing a structured analysis that takes into account the outcome of possible future decisions including the effect of uncertainty, and allows the benefit of varying levels of present and future profit to be weighed against the concomitant commitments. Decision trees can also be used in the design process to evaluate different design alternatives. After a number of alternative design approaches have been identified, the designer selects the one that appears to offer the optimum solution. This approach is then investigated at a more detailed level, revealing other options that again are evaluated and the best one selected. This process continues to some logical end point that may be a certain critical period of time, (e.g., in the case of financial decisions) or to the lowest level of detail in a design decision. This top-down approach is more commonly used in design, but the process can also be started at the lowest level, building up to the overall concept. Although the decision tree is based on selection of the best option at each branch or level of the tree, this does not ensure that the best possible overall concept will be obtained during the first iteration. Decisions made at any particular level may turn out to be less than optimal in the light of information disclosed at subsequent levels. The impact of later decisions on earlier decisions is important in the effective use of decision trees, and thus considerable iteration is necessary. It is important for the decision tree to be regarded as a dynamic device, one in which new (and better) information is integrated as it becomes available.[10]

---

### CASE STUDY 3.3  Plant Investment Decision using a Decision Tree.[a]

#### Description

A company is trying to decide whether to build a conventional or automated plant to produce a new product with an expected life of 10 years. The decision must be based on the size of the market for the product. Based on projections by the company, demand could be any of the following:

1) High during the first 2 years and, if the product is found unsatisfactory by users, low thereafter
2) High over the 10-year period
3) Low over the 10-year period.

#### Approach

If the company builds an automated plant and significant demand does not develop, it will be saddled with an expensive non-productive facility. If a conventional plant is constructed, management has the option of upgrading the plant by partial automation after two years to accommodate high sustained demand. If the demand is low the company can sustain operations in the conventional plant profitably at the lower volume. Company management is concerned about the possibility that high demand may not develop and leans toward construction of the conventional plant, recognizing that later upgrades for automation to accommodate high demand will be more costly and result in a less efficient operation.

   On the basis of data available, company management believes that the following prognosis is appropriate:

- Demand is expected to develop according to the following probabilities:

   Initially high demand, low long-term demand = 15%
   Initially low demand, high long-term demand = 0%

**CASE STUDY 3.3 (Continued)**

> Initially high demand that is sustained = 60%
>
> Initially low demand that is sustained = 25%.

- The chance that initial demand will be high is 75% (60% + 15%). If initial demand is high, the company estimates that the likelihood that it will continue at a high level is 80% (60% ÷ 75%). A high initial level of sales will affect the probability of high sales in the future. If initial demand is low, the probability is 100% (25% ÷ 25%) that sales will be low in the future. The level of initial sales is thus an accurate indicator of the level of sales in future periods.
- Annual income is projected as follows:
  1) Automated plant, high demand yield = $1.0 million annual cash flow
  2) Automated plant, low demand yield = $100,000 annual cash flow
  3) Conventional plant, low demand yield = $300,000 annual cash flow
  4) Conventional plant, high demand yield = $400,000 annual cash flow
  5) Conventional plant upgrading to meet high demand yield = $600,000 annual cash flow
  6) Conventional plant upgrading without sustained high demand yield = $50,000 annual cash flow.
- Construction estimates for the various options indicate that it will cost $5 million to put the automated plant in operation, $2 million for the conventional plant and $2 million to partially automate the conventional plant.

The decision tree based on this information is shown in Figure 3.23. Note that no information is included that the company's executives did not know before developing the decision tree. However, the value of decision trees in *laying out* information in such a way that systematic analysis is enhanced provides an invaluable tool that improves decision-making.

When management is in the process of deciding whether to build a conventional or automated plant (decision 1), it does not have to concern itself with decision 2. However, it is important for management to be able to place a monetary value on decision 2 in order to compare the gain from taking the "build conventional plant" branch with the "build automated plant" branch at decision 1. The analysis shown in Table 3.5 indicates that the total expected value of the upgrading alternative is negative ($1,920,000 − 3,040,000 = −1,120,000) over the eight-year life remaining after decision 2. Thus, management would not choose to expand the conventional plant if faced with decision 2 (considering financial return only).

Yields for the various decision branches are shown on the right-hand side of Figure 3.23. If these yields are adjusted for their corresponding probabilities less investment cost the following comparison can be made:

$$\text{Build automated plant} = (\$10M \times 0.60) + (\$2.8M \times 0.15) + (\$1M \times 0.25) - \$5M$$
$$= \$1,670,000$$
$$\text{Build conventional plant} = (\$3.84M \times 0.75) + (\$3M \times 0.25) - \$2M$$
$$= \$1,630,000$$

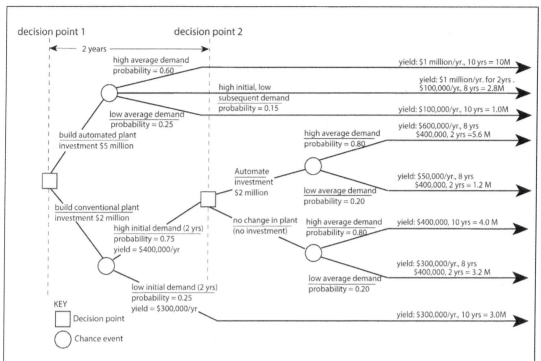

**Figure 3.23** Decision tree with financial data (from Ertas, A., and Jones, J.C., *The Engineering Design Process*, John Wiley & Sons, Inc., New York, 1996, used with permission of John Wiley & Sons, Inc.).

**Table 3.5** Decision analysis using maximum expected total cash flow as a criterion (from Ertas, A., and Jones, J.C., *The Engineering Design Process*, John Wiley & Sons, Inc., New York, 1996, used with permission of John Wiley & Sons, Inc.).

| Choice | Chance event | Probability (1) | Total yield, 8 years (thousands of dollars) (2) | Expected value (thousands of dollars) (1) × (2) |
|---|---|---|---|---|
| Upgrading | High average demand | 0.80 | 4,800 | 3,840 |
| | Low average demand | 0.20 | 400 | 80 |
| | | | Total | 3,920 |
| | | | Less investment | 2,000 |
| | | | Net | 1,920 |
| No upgrading | High average demand | 0.80 | 3,200 | 2,560 |
| | Low average demand | 0.20 | 2,400 | 480 |
| | | | Total | 3,040 |
| | | | Less investment | 0 |
| | | | Net | 3,040 |

*(Continued)*

---

**CASE STUDY 3.3 (Continued)**

---

The choice that maximizes the expected total cash yield at decision 1 is to build the automated plant. Note that the profit margin in favor of building the automated plant is not large and that this analysis does not account for the time value of money. Also, a small error in projecting the probabilities used for this analysis could change the result; therefore, it would be wise to evaluate this investment decision by an alternative means in order to validate the result.

[a]From Ertas, A., and Jones J.C., *The Engineering Design Process*, John Wiley & Sons, Inc., New York, 1996.

---

## 3.6  Process of Defining Customer Needs

The general process of defining customer needs is shown in Figure 3.24. Customer needs or requirements are product characteristics desired or needed by the customer. Just asking what is important to the customer may not give you enough information to make a decision on the engineering specifications (requirements). Customers may provide you with poor and

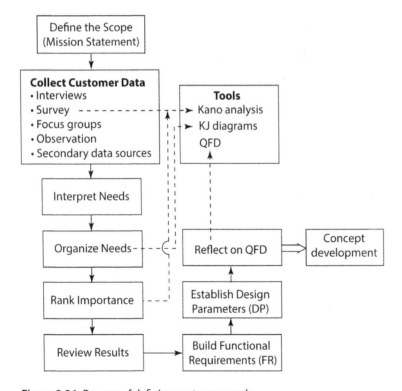

**Figure 3.24** Process of defining customer needs.

misleading or unrealistic information. Understanding customer needs is not an easy task. Five important concerns to define customer requirements are as follows:

1) Customers may have hidden needs that they do not even know they have.
2) Customers may struggle to express their needs.
3) Customers' needs may change quickly over time.
4) Customers may not know what they want until they see it.
5) It is not possible to know all the customers' needs.

### 3.6.1 Customer Data Collection

The following techniques and guidelines can be used to collect data on customer needs.

#### 3.6.1.1 Interviews

*Interviews* are discussions, usually one-on-one between an interviewer and an individual, meant to gather information on a specific set of topics. Interviews can be conducted in person, over the phone, or by submitting a proposal.

#### 3.6.1.2 Observation

*Observation* is data collection in which the researcher does not participate in the communications, for example, observing learning and teaching activities (lectures, seminars, lab classes).

#### 3.6.1.3 Focus Groups

A *focus group* is a dynamic group discussion of approximately six to twelve people who share similar characteristics or common interests. A facilitator leads the group based on a prearranged set of topics. The facilitator creates an environment that help participants to share their points of view. Focus groups are a qualitative data collection method – data is descriptive and cannot be measured numerically. One-on-one interviews may be more efficient than focus groups.

#### 3.6.1.4 Secondary Data Sources

*Secondary data sources* are data sets that are already in existence. Researchers may select parameters to use in their analysis from one secondary data source or may mix data from different sources to create new data sets.

#### 3.6.1.5 Surveys

*Surveys* are fixed sets of questions that can be administered by paper and pencil, as a web form, or by an interviewer following a strict script. How many respondents are really enough to ensure accuracy in your survey's results? Griffin and Hauser conducted a survey interviewing 30 customers to uncover 220 needs.[12] They showed that 15–20 respondents should be enough to get reasonably good understanding of their customers' needs. Of course, depending on the degree of variability, this number can change but at least this is a good rule of thumb. Surveys will be discussed in more detail under the rubric of the Kano model below.

---

12 Griffin, A., and Hauser, J.R., "The Voice of the Customer," *Marketing Science*, 12(1), 1993.

### 3.6.2  Kano Model: Understanding Customer Needs

The Kano model is a method developed in the 1980s by Professor Noriaki Kano. It is used to classify customer needs, define functional requirements, and determine appropriate levels of innovation for product development. The main objective of the Kano method is to help research teams uncover and classify customer needs and attributes into five categories as shown in Table 3.6 and Figure 3.25.

**Table 3.6**  Classification of customer needs and attributes.

| Requirement type | Definition |
|---|---|
| Must be (expected quality) | These requirements are expected. If they are not fulfilled, customers will not be satisfied. |
| One-dimensional (linear) | These requirements result in satisfaction when they are fulfilled and dissatisfaction when not fulfilled. The more of these requirements are fulfilled, the more a customer is satisfied. |
| Excited quality | These qualities are not expected but will delight customers if present. If the requirement is absent, it does not cause dissatisfaction. |
| Indifferent quality | Customers do not care if these requirements are fulfilled or not fulfilled. |
| Reverse quality | These qualities in general cause dissatisfaction. |

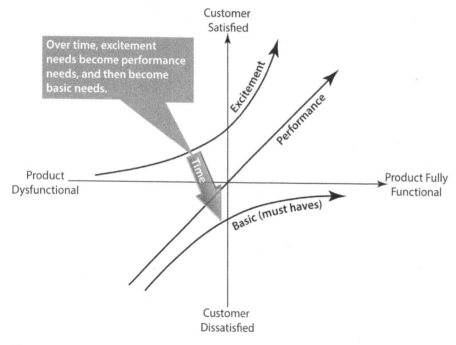

**Figure 3.25**  Kano model.

Basic needs (attributes) are necessary requirements that are expected (obvious, self-evident). A competitive product must meet basic attributes. Increasing the performance of these basic attributes causes diminishing returns in terms of customer satisfaction. However, absence of these attributes cause extreme customer dissatisfaction. Note that basic attributes are either satisfied or not satisfied. An example of a basic attribute would be internet connection in a hotel room. Lack of internet connection makes customers unhappy. If the hotel room has two extra internet connections, it little or no difference to the customer.

Performance attributes are one-dimensional (linear) requirements. Strong performance attributes are usually better and will improve customer satisfaction. On the other hand, an absent or weak performance attribute reduces customer satisfaction. These are measurable technical attributes, and most customer needs will fall into this category. As an example, more gas mileage per gallon leads to high levels of customer satisfaction. As another example, less traffic would result in greater satisfaction.

Excitement attributes (attractive requirements) are not expressed and not expected by customers but can result in high levels of customer satisfaction. However, their absence does not lead to dissatisfaction. For example, internet access when traveling by bus is not expected and the customer will not be dissatisfied if it not provided, but may well be delighted if it is available. Over time, excitement needs become performance needs and then basic needs.

### 3.6.2.1 Kano Model Analysis

A relatively simple way to apply the Kano model analysis is to develop a set of survey questions to send to a group of potential customers. In order to eliminate bias/inconsistency each set will have a positive statement/question and a negative statement/question about each requirement. An example of two simple questions to ask customers on each attribute is as follows:

1) Rate your satisfaction from 1 to 5 (1 being less satisfied and 5 being most satisfied) if the product has this feature (positive statement).
2) Rate your satisfaction from 1 to 5 (1 being less satisfied and 5 being most satisfied) if the product does not have this feature (negative statement).

Here is another example (see Table 3.7):

1) How do you feel if this feature attribute is present? (positive question)
2) How do you feel if this feature attribute is not present? (negative question)

Customers should be requested to answer with one of the following responses:

a) I like it
b) It must be that way
c) I am neutral
d) I can live with it
e) I dislike it

As shown in Table 3.7, each attribute's category can be determined using two paired functional/dysfunctional questions.

**Table 3.7** Example of Kano questionnaire.

| | |
|---|---|
| How do you feel if this feature attribute is present? | (a) I like it |
| Positive question (*functional*) | (b) It must be that way |
| | (c) I am neutral |
| | (d) I can live with it |
| | (e) I dislike it |
| How do you feel if this feature attribute is not present? | (a) I like it |
| Negative question (*dysfunctional*) | (b) It must be that way |
| | (c) I am neutral |
| | (d) I can live with it |
| | (e) I dislike it |

### 3.6.2.2 Evaluating Kano Questionnaire

To decode the survey results, the category of each attribute can be identified through Kano's paired functional/dysfunctional questionnaire (Figure 3.26(a)) and its evaluation matrix (Figure 3.26(b)).

As shown in Figure 3.26, suppose a customer likes warm bread and butter before dinner. He would likely give answer (a) to the positive question and (e) to the negative question. Following the evaluation matrix, this requirement will fall into "linear" requirement L. This means that though the requirement is not a must-have, the more of this kind of requirement the restaurant has, the more satisfied the customer will be. It is clear that the negative question in the Kano questionnaire serves as a consistency check. The combination of customer responses to the two questions will help to determine the type of each requirement.

### 3.6.2.3 Kano Analysis of Results

Let us again consider a restaurant with features evaluated by customers shown in Table 3.8. Since we have to send a questionnaire to many customers, we need to total the results to determine how the majority of customers express their requirements. Suppose we have 50 customers.

After the survey results through the Kano evaluation table shown in Figure 3.26 are completed, the final step is to assign a letter score to every feature as shown in Table 3.8. From the Kano analysis chart one can determine the following:

- As seen from Figure 3.26, although the "must be" *acceptable price range* will satisfy the customers, in general, it is true that price is an important factor in the eyes of customer.
- There are two customer populations. Eighteen customers do not require *choice of quality healthy food*, but would like it. Twenty customers require this feature. Poor-quality food results in customer dissatisfaction and eventually loss of business.
- *Pleasant staff* is a feature required by customers – it is a "must-be" requirement.
- *Pleasant atmosphere* is a feature that customers would not have thought of on their own, but they were excited when they see a clean and pleasant atmosphere; this was not expected.

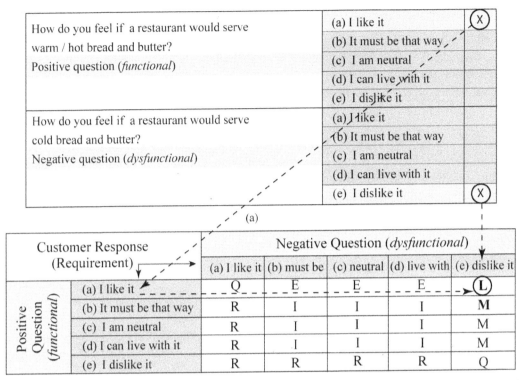

Figure 3.26 Kano model questionnaire evaluation.

Table 3.8 Results of Kano analysis.

| Restaurant features | Exciter | Linear | Must be | Indifferent | Reverse | Questionable |
|---|---|---|---|---|---|---|
| Acceptable price range | 2 | ㉜ | 10 | 3 | 3 | 0 |
| Choice of quality healthy food | 5 | ⑱ | ⑳ | 4 | 2 | 1 |
| Pleasant staff | 4 | 15 | ㉕ | 4 | 1 | 1 |
| Pleasant staff | ㉕ | 8 | 12 | 3 | 1 | 1 |

Note that basic requirements must be implemented. Some or all of the attractive requirements and the performance requirements should be implemented at as high a rating as possible. The indifferent requirements can be safely discarded. The reverse or questionable results would need to be further analyzed in order to properly determine their place. Kano model analysis engenders a better knowledge of the customer and helps one to better address the needs of the customers based on their *critical to quality characteristics*.

### 3.6.3 Critical to Quality Characteristics

A critical to quality (CTQ) characteristic is a measurable characteristic of a product or service that is provided to customers. However, defining quality is not an easy task, and it is easy to overlook factors that are important to customers. This is when CTQ trees are helpful. CTQ helps us to understand what drives quality in the eyes of customers.

CTQ trees help the customer of any company or business to translate the most important needs on products or processes into requirements to ensure their quality. CTQ trees are used to help see the characteristics of a product, or service from a customer point of view.

When developing new products, processes, and services, quality is important. The quality of a product, process, or service is generally defined as the ability to meet the requirements set on them. These requirements are collected or selected from customers and regulations. The CTQ tree is an important tool to translate customer requirements obtained from the voce of the customer into specific measurable activities.

The first step in creating a CTQ tree is to define the needs that are most critical to making the customer happy with the product, process, or service. This first step has many sub-steps, all of which are directly related to what a customer is looking for. Then, for each critical need, define its quality "driver." These are the factors that customers will use to evaluate the quality of the service or product.

The final step is defining measurable performance requirements that each driver must satisfy. Without these requirements, it is difficult, if not impossible, to actually measure the performance and quality of the product. The following example shows how the CTQ tree can be used.

Suppose you are looking for a cozy place to have a dinner with your friends that will satisfy your hunger at an affordable price. The key element is to consider a restaurant. Some of the characteristics (drivers) that can be considered as critical to quality are as follows:

- Service time
- Acceptable price range
- Choice of quality food
- Pleasant server
- Pleasant atmosphere.

Figure 3.27 shows a typical CTQ tree for having a nice dinner (customer need) at a restaurant along with some of the customer voices (at least the major drivers) and requirements. To have a comprehensive list of requirements, the CTQ tree should be completed for each individual critical need that is identified. CTQ trees were originally developed as part of Six Sigma which will be covered in later sections.

### 3.6.4 KJ Method

The KJ method, developed by Jiro Kawakita and commonly known as *affinity diagramming*, is a useful tool used during brainstorming sessions. The concept is a simple but effective method

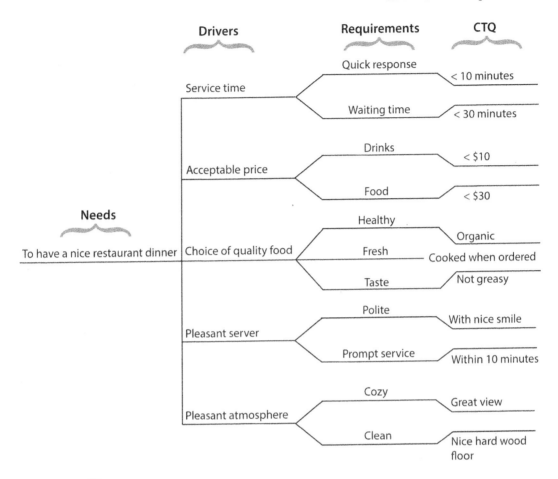

**Figure 3.27** CTQ tree.

for gathering and organizing ideas, opinions, and issues regarding a given specific problem. The tool is especially helpful when working in a team. The goal of the KJ method in concept engineering is to group the qualitative data (in the form of hundreds of voices and images) and extract a useful set of customer requirements. The following simple steps can be used in KJ analysis:

- Collect all ideas without considering their rationality.
- Group the ideas in a manner appropriate to the topic.
- Form hierarchies of groups and sub-groups.
- Evaluate the ideas by ranking and sequencing, and find the best solutions chosen from the groups.

---

### CASE STUDY 3.4  Application of the KJ Method to a Cellular Phone Design

Cellular communications is a rapidly growing and competitive market with great potential for improvements to existing designs. Consider designing a new cellular phone.

#### Background

To design a new cell phone, a survey was sent to a group of customers to define their needs. The survey was sent to approximately 65 people and 21 responses received, a return rate of almost 30%. Raw data from customer responses were examined and customer need statements extracted.[13]

#### Results

As shown in Figure 3.28, a simplified version of the KJ method was used to group customers' need statements. In some cases customers' language were changed to clarify the intent from the context of the overall response. The KJ method was used to compare similar need statements and determine whether they provided unique requirements. For example, several customers wanted the battery to hold a charge longer, so this group of similar need statements was reduced to a single requirement.

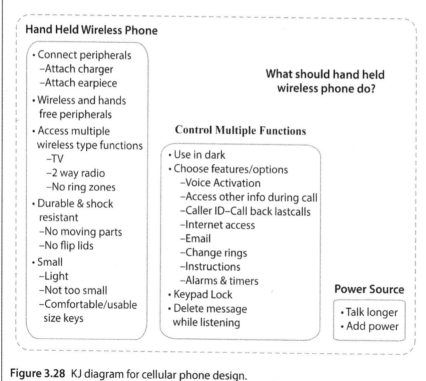

**Figure 3.28** KJ diagram for cellular phone design.

---

13  Survey data were used from Devenport, J., Jernigan, S., Kenny D., and Richardson, W., "Application of axiomatic design project fundamentals of transdisciplinary design and process," Report, ATLAS Publications, 2002.

# 3.7 Techniques and Methods for Product Development and Management

Transdisciplinary integrated product design and development is a complex process that depends on input from many individuals, academic and non-academic experts, and even communities, collaborating to complete the product design and development. Due to the multi-technology and complex nature of modern products, the design process requires intervention or expertise from different disciplines throughout the product's life cycle.[14]

There are a number of techniques and methods currently used in product design and development. In this section, some of these techniques and their relationship each other will be discussed.

## 3.7.1 Total Quality Management

Total quality management (TQM) is a management approach to long-term success through customer satisfaction. It is a management system for a customer-focused organization that involves all employees in continual improvement. It emphasizes that each employee is responsible for the quality of his/her work and for the work of the manufacturing unit to which they are assigned.[2]

Quality is a measure of the extent to which customer needs and expectations are satisfied. It involves developing products or services to meet customer needs and expectations that are economically feasible (*quality of design*) and later manufacturing products or providing services which satisfy previously determined and defined specifications (*quality of conformance*).[15]

*Hoshin planning or management* is a quality tool that is used for organizational improvement developed by the Japanese in the 1960s and 1970s. It is system of planning that is a component of a TQM system and allows an organization to plan and execute strategic organization breakthroughs. Hoshin planning should be used for daily control monitoring of business performance and the existence of cross-functional lines of interaction in an organization. Hoshin planning or management also requires top management leadership to drive the vision throughout the organization since the process begins at policy establishment that can be only done from the top. The Hoshin process is a useful process for managing work towards a key strategic initiative. This approach brings into line all parts of an organization to accomplish an important objective.

## 3.7.2 Quality Function Deployment

Quality function deployment (QFD) was developed in Japan in the early 1970s and has been successfully used in the United States since the 1980s. It is a method for identifying customer

---

14 Ensici, A., and Badke-Schaub, P., "Information behavior in multidisciplinary design teams," in S.J. Culley. B.J. Hicks, T.C. McAloone, T.J. Howard, and P. Badke-Schaub (eds), *The 18th International Conference on Engineering Design: Proceedings*, pp. 414–423, ASM International, Materials Park, OH, 2011.

15 Lochner, R.H., and Matar, J.E., *Design for Quality*, ASQC Quality Press, White Plains, NY, 1990.

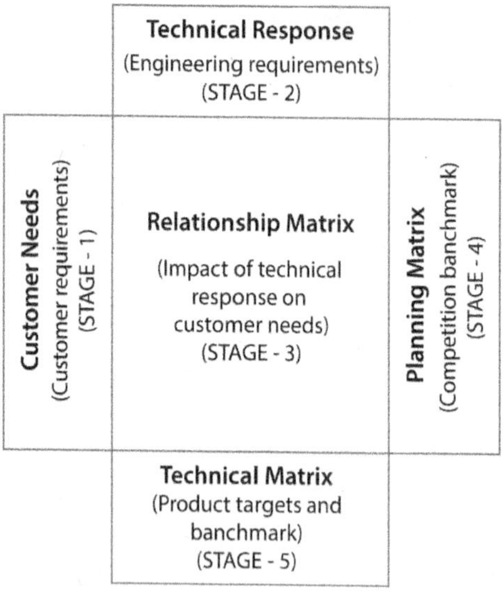

**Figure 3.29** QFD design requirements matrix.

requirements and ensuring that the voice of the customer is integrated into a design process for product development. The five stages of QFD are shown in Figure 3.29. The procedure for documenting information in this chart is as follows:

- *Stage 1* includes customer requirements (needs). Customer needs or requirements are product characteristics desired or needed by the customer. Just asking what is important to the customer may not give you enough information to make a decision on the engineering specifications (requirements). Customers may provide you with poor and misleading or unrealistic information. Understanding customer needs is not an easy task. Five important concerns to define customer requirements:
  1) Customer may have hidden needs that they do not even know they have.
  2) Customers may struggle to express their needs.
  3) Customers' needs may change quickly over time.
  4) Customers may not know what they want until they see it.
  5) It is not possible to know all the customers' needs.
  Establish and practice methods of collecting the voices of customers and converting them into engineering requirements – this is the most important initial part of the project execution.
- *Stage 2* includes the list of engineering requirements. Customers' subjective and non-technical requirements are translated into (corporate language) measurable engineering requirements (or hows). These requirements are developed by design engineers to satisfy the customer requirements. The customer requirements tell us what to do, and the engineering requirements tell us how to do what needs to be done. Both customer requirements and engineering requirements (specifications) are linked to quality.
- *Stage 3* involves developing a matrix of engineering design requirements from customer requirements. This relationship matrix is used to verify and improve the fidelity of the

translation of customer requirements into measurable engineering requirements. By ranking the relationship between two requirements and characterizing the strength of this relationship as strong, medium, or weak, engineering requirements are identified in quantifiable terms. Once all project team participants have an agreed to understanding of stages 1–3, an important initial attempt has been made to improve product planning.

- *Stage 4* is when the proposed design is compared with competitors' products through what is referred to as competition benchmarking. Benchmark design attributes are compared with the list of customer requirements that can be assessed on a five-point scale, with 1 being not satisfied and 5 being completely satisfied. Benchmarking is a remarkable assessment process for improvement to meet or improve on industry best practices. Benchtrending is a continuous potential business practice for predicting and meeting customers' needs, exceeding competitor's performance, and providing industry leadership. Benchmarking is designed to measure processes against others in the industry. It is a strategic-level process to identify and close competitive gaps with best-in-class competitors. Benchtrending focuses on strategic issues and future trends.
- *Stage 5* is when the technical difficulty of each engineering design requirement which needs to be achieved is ranked. Also, target values of each engineering requirement are set. The technical difficulty of achieving each customer requirement listed in stage 1 in terms of the changes defined by the engineering requirement can be assessed on a five-point scale, with 1 being not achieved and 5 being completely achieved.

---

**CASE STUDY 3.5  Quality Function Deployment for a Cellular Phone**

Figure 3.30 shows a simplified QFD model for a cellular phone to provide a structured way of designing quality into the cellular phone while establishing engineering requirements.

**QFD Orthogonal Array**

A QFD orthogonal array is constructed to map the customer's needs to engineering requirements, thus integrating the voice of the customer into the design process (see Figure 3.30). The potential customer's needs (requirements) are listed in rows of the matrix and the engineering requirements are listed in columns of the matrix. At every matrix intersection, the degree of the relationship between engineering requirements and the customer requirements was graded within a three-point range. By ranking the relationships (strong, medium, and weak, receiving a score of 9, 3, and 1, respectively) between the customer's needs and engineering requirements we identified those areas in need of improvement and focus. If the correlation cells are blank, that indicates there is no correlation between two requirements.

As demonstrated in Figure 3.30, hand held (36), internet access/email (30) and inexpensive (60) scored the highest for the customer's needs. Computer/PDA-like features (41), user interface (27), LCD screen (35), short battery charge rate (27), and long battery charge (27) scored the highest for the engineering requirements. Because of the commitment to quality and customer satisfaction,

*(Continued)*

**CASE STUDY 3.5  (Continued)**

Ranking legend:

| Ranking | | |
|---|---|---|
| Strong Relationship | ● | 9 |
| Medium Relationship | ○ | 3 |
| Weak Relationship | ▲ | 1 |

Benchmark ranking legend:

| Ranking | | |
|---|---|---|
| Strong Relationship | ● | 5 |
| Medium Relationship | ○ | 3 |
| Weak Relationship | ◀ | 1 |

QFD matrix — Customer Requirements (rows) vs Engineering Requirements (columns):

| Customer Requirements | Light Weight | Material Strength | Ergonomics | Aesthetic | Computer/PDA-Like Features | User Interface | Multiple RF Functions | LCD Screen | Programmable | Short Battery Charge Rate | Long Battery Life | Low Purchase Price | Bluetooth (or Equivalent) | Score | Competitor 1 | Competitor 2 | Competitor 3 |
|---|---|---|---|---|---|---|---|---|---|---|---|---|---|---|---|---|---|
| Hand held | ● | | ● | ● | | | | ● | | | | | | 36 | 5 | 5 | 5 |
| Hands-free peripherals | | | | ● | | | | | | | | | ● | 18 | 5 | 5 | 5 |
| Crack and shock resistant | | ● | | | | | | | | | | | | 9 | 5 | 5 | 5 |
| Easy to handle | | | ● | | | | | | | | | | | 9 | 3 | 5 | 5 |
| TV | | | | | | ● | ● | ● | | | | | | 27 | 3 | 3 | 1 |
| Voice Activation | | | | | ● | ● | | | | | | | | 18 | 3 | 5 | 1 |
| Access other service during call/ Delete message while listening | | | | | ● | | | | | | | | | 9 | 3 | 5 | 3 |
| Alarms & timers | | | | | ● | | | | | | | | | 9 | 5 | 5 | 5 |
| Internet access/ Email | | | | | ● | ● | | | ○ | ● | | | | 30 | 5 | 5 | 5 |
| Caller ID | | | | | ▲ | | ▲ | | | | | | | 2 | 5 | 5 | 5 |
| Call back last calls | | | | | ▲ | | ▲ | | | | | | | 2 | 5 | 5 | 3 |
| Change rings | | | | | | | ○ | | | | | | | 3 | 5 | 5 | 3 |
| Talk longer | | | | | | | | | | | ● | ○ | | 12 | 3 | 5 | 3 |
| Add power | | | | | | | | | | ● | ● | ○ | | 21 | 3 | 3 | 3 |
| Inexpensive | | | | | ○ | | ● | ● | ○ | ● | ● | ● | ● | 60 | 5 | 3 | 3 |
| Score | 9 | 9 | 18 | 18 | 41 | 27 | 18 | 35 | 3 | 27 | 27 | 12 | 21 | | | | |
| Unit | OZ | PSI | | | | | | Diag.in. | | Hours | Hours | | | | | | |
| Target Value | >10 | 400 | | | | | | 3 | | >2 | <4 | >150 | | | | | |
| Technical Difficulty | 3 | 5 | 3 | 3 | 5 | 5 | 5 | 5 | 5 | 1 | 3 | 5 | 5 | | | | |

Product Targets and Benchmark.

Benchmark ranking legend (bottom):

| Ranking | | |
|---|---|---|
| Strong Relationship | ● | 5 |
| Medium Relationship | ○ | 3 |
| Weak Relationship | ◀ | 1 |

**Figure 3.30** QFD chart for cellular phone.

design effort will focus on improvement in those areas. It is clear that the selling price of the cellular phone is the driving factor for this product's survival in the market.

To understand the market position of the product, benchmarks for each competitor's values are also shown in Figure 3.30. Since the cellular telephone is common, similar products are available on the market. Then the specifications and features for a sample of competitor products via the internet were obtained. Product specifications were matched to the QFD list of customer requirements. Then the degree to which competitors satisfy the requirements were evaluated by rating them on a three-point scale. If no similar product feature is available in the competition, that feature will have a very low benchmark rating.

Finally, the target value of each engineering requirement is defined and target difficulties are shown in Figure 3.30. Each design parameter was evaluated for satisfaction criteria. If the criteria could be measured, a unit of measure was chosen and a target value assigned. For example, the target diagonal size of the LCD Screen was assigned the maximum value of 3 in.

If a design value was not specified in the customer surveys, the design team assigns either a competitive value based on similar products on the market or a reasonable value based on the team's judgment. Finally, each design parameter can be ranked on a three-point scale on its relative difficulty of achievement.

From this QFD analysis, designers can judge the critical areas and potential failure points for the final product design. Critical design points should receive concentrated effort in the design process. Other features can be judged on the degree to which the final product has an advantage or disadvantage relative to the competition. Final design costs, feature choices, and compromises can be based on the QFD.

### 3.7.3 House of Quality

Later, Toyota and its suppliers improved the QFD and introduced the house of quality (HOQ). The HOQ is a transdisciplinary tool or technique which has been used successfully by Japanese manufacturers of consumer electronics, home appliances, clothing, integrated circuits, synthetic rubber, construction equipment, and agricultural engines. Moreover, Japanese designers employed it for different purposes such as swimming schools, retail outlets and planning apartment layouts. The basis of the house of quality is the idea that products should be designed to reflect customers' desires and tastes – thus marketing people, design engineers, and manufacturing staff will work together from the time a product is first envisioned to provide the means for interfunctional planning and communications.

As shown in Figure 3.31, the house of quality matrix, like a house with a *correlation matrix* as its roof, is the most recognized and widely used form of QFD method. It utilizes a planning matrix to relate what the customer wants to how a production company will meet those wants. It translates customer requirements, built on marketing research and benchmarking data, into a proper number of engineering targets to be met by a new product design.

There are several slightly different forms of the HOQ which can be adapted to different kind of problems. There are many advantages of HOQ as indicated below:

- Reduces time involved for planning
- Reduces design changes
- Improves quality
- Reduces design and manufacture costs

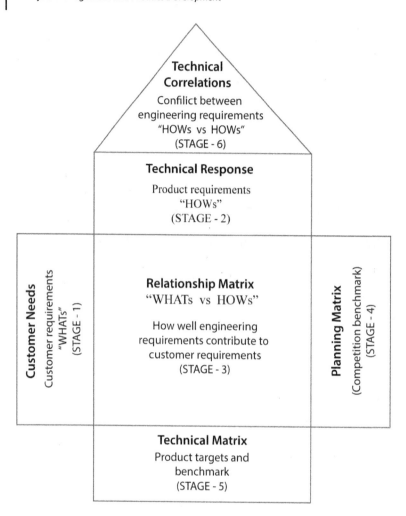

**Figure 3.31** HOQ design matrix.

- Reduces time to market
- Helps in prioritizing different design parameters
- Improves customer satisfaction.

---

**CASE STUDY 3.6  Designing a new SUV car.**

Using team work prepares an orthogonal array (House of Quality) for SUV design.

**House of Quality**

Consider a research team dedicated to design a new SUV car. The research team will prepare an HOQ orthogonal array as shown in Figure 3.32 that lists customers' needs (*WHATs*) for the new SUV

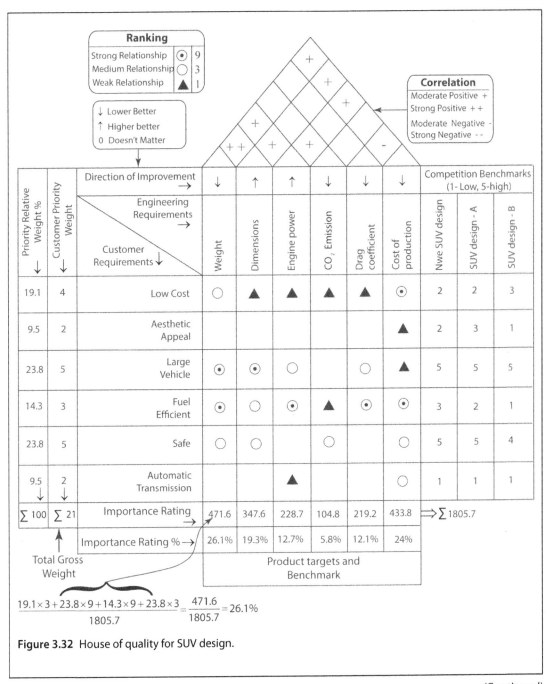

**Figure 3.32** House of quality for SUV design.

(Continued)

---

**CASE STUDY 3.6 (Continued)**

---

design and identifies the corresponding substitute engineering requirements or quality characteristics (*HOWs*). Note that controllable engineering requirements must be measurable.

For the process of identifying customers' needs, the research team should think of as many customer *wants* as possible. Then customers prioritize the importance of each requirement by ranking them on a five-point scale. The *priority relative weight* column is completed by dividing individual gross weights by the total gross weight of 21 as indicated in Figure 3.32. This number will be used later in the relationship matrix.

The next step is the *relationship matrix*. The relationship matrix is where the research team defines the relationship between customer requirements and the company's ability to meet those requirements. As shown in Figure 3.32, the relationship between two requirements is ranked and its strength characterized as strong (9), medium (3), or weak (1). The *importance rating* is calculated by multiplying each relationship rating by the priority relative weight:

$$\text{Importance Rating} = \sum \text{Priority relative weight \% } \times \text{Relationship ranking}$$
$$= 19.1 \times 3 + 23.8 \times 9 + 14.3 \times 3 + 23.8 \times 3$$
$$= 347.6$$

The *Importance rating* % row is completed by dividing the individual importance rating by the total importance rating of 1805.7, giving $347.6/1805.7 = 19.3\%$.

The proposed new SUV design is compared with two competitors' SUVs through what is referred to as competition benchmarking. Benchmark design attributes are compared with the list of customer requirements on five-point scales, with 1 being not satisfied and 5 being completely satisfied.

Engineering (design) requirements in existing designs often conflict with each other. As shown in Figure 3.32, the correlation matrix is used to identify where engineering requirements support or conflict with each other in the new SUV design. For example, Figure 3.32 shows an SUV with higher weight requires higher engine power – these affect each other positively. Similarly, weight and dimensions affect each other strongly positively. The correlation matrix involves a tradeoff between different design parameters – how much one design parameter can be pushed without sacrificing other parameters. If there is a negative or strongly negative impact between engineering requirements, the design must be reiterated or negotiated unless the negative impact can be designed out. Negative impacts can also represent constraints, which may affect the design requirements in two opposite directions. As a result, improving one of the engineering requirements may actually cause a negative impact on the other. One of the main benefits of the HOQ roof is that it identifies these negative relationships so they can be resolved.

The direction of improvement is also shown in Figure 3.32. As the research team defines the engineering requirements, a determination should be made as to the direction of movement for each engineering requirements. For example, if we keep all the other engineering requirements constant, a lower weight is the better for SUV design.

From the HOQ analysis, it is clear that weight, dimensions, and cost of production have the highest importance rating. These critical design points should receive concentrated effort in the design process.

Through the HOQ, one can see where the strongest and weakest, most positive and most negative points lie inside the QFD house. To effectively gain the highest value from the QFD matrix, all key researchers from different collaborating disciplines must communicate as a research team – design engineering, manufacturing, test, customer service, marketing, quality assurance, finance, etc., should focus on meeting customer needs and organizational objectives.

Once the customer's requirements have been identified and prepared in QFD form, we are in a position to develop potential product or process concepts and then select the concept for testing.

If QFD is used for concept development, the weights for the design requirements can be drawn from the QFD analysis. If not, then engineering judgment can be used for the weights. For a simple example, consider the SUV design shown in Figure 3.32. Assume that we have two concepts and one of them will be chosen for further analysis. A selection matrix as shown in Figure 3.33 can be used to quantitatively decide which conceptual design to use for further analysis and eventually as a basis for the final design.

The factors used to measure the effectiveness of each concept were as follows: weight, dimensions, engine power, $CO_2$ emission, drag coefficient, and cost of production. As shown in Figure 3.32, the importance ratings of each of the design (engineering) requirements were determined and used in the concept selection matrix (see Figure 3.33). Each concept is judged according to these criteria. In Figure 3.33, the Greek symbol, $\Pi$ represents multiplication. Three-point ranking (strong, medium, and weak receiving a score of 5, 3, and 1, respectively) is used for the relationships between the design requirements and each concept. Corresponding ranking numbers and importance rating are multiplied together to determine the concept score for each concept. This process is repeated for all six design requirements and for each concept.

As shown in the selection matrix, concept B has the bigger total sum of 412. Hence, concept B will be chosen for further analysis.

| Criteria (Design Requirements) | Importance Rating, % | Concept A | Concept B |
| --- | --- | --- | --- |
| Weight | 26.1 | $\Pi$ ⊙ = 130.5 | $\Pi$ ○ = 78.3 |
| Dimensions | 19.3 | $\Pi$ ⊙ = 96.5 | $\Pi$ ⊙ = 96.5 |
| Engine power | 12.7 | $\Pi$ ○ = 38.1 | $\Pi$ ⊙ = 63.5 |
| $CO_2$ Emission | 5.8 | $\Pi$ ○ = 17.4 | $\Pi$ ○ = 17.4 |
| Drag coefficient | 12.1 | $\Pi$ ▲ = 12.1 | $\Pi$ ○ = 36.3 |
| Cost of production | 24.0 | $\Pi$ ○ = 72.0 | $\Pi$ ⊙ = 120 |
| TOTAL | | 366.6 | 412.0 |

| Ranking | | |
| --- | --- | --- |
| Strong Relationship | ⊙ | 5 |
| Medium Relationship | ○ | 3 |
| Weak Relationship | ▲ | 1 |

**Figure 3.33** Concept selection matrix.

### 3.7.4  Theory of Inventive Problem Solving

"TRIZ" is the Russian acronym for the "theory of inventive problem solving" developed by G.S. Altshuller and his colleagues in Russia between 1946 and 1985. TRIZ is a problem-solving methodology built on the study of the patterns of problems and solutions using logic, data, and research but not on the base of intuitive creativity of individuals or groups.

TRIZ is a powerful empirical method of innovation for creative problem solving which draws on the past knowledge, initiative, and skill of thousands of engineers to accelerate the project team's ability to solve problems creatively in many fields. TRIZ brings reliability, repeatability, and predictability to problem solving with its structured approach. More than 3 million patents have been analyzed to discover the patterns that predict innovative solutions to problems. These problems which have been solved by someone before have been collected and organized within TRIZ.

The range of TRIZ application generally deals not with advancement and development activity (development of technologies, industry, business, art and science, etc.) but with technology for creating something new, inventing methods previously unknown.

In order to create and improve products, services, and systems TRIZ specialists and researchers have produced and apply the following three primary findings:[16]

1) Problems and solutions are repeated across industries and sciences.
2) Patterns of technical evolution are also repeated across industries and sciences.
3) The innovations used scientific effects outside the field in which they were developed.

#### 3.7.4.1  TRIZ Problem-Solving Process

Figure 3.34 describes the TRIZ problem-solving process graphically.

1) The process starts with identifying a specific problem with main functions (e.g., the main functions of a train include transportation, power, speed, weight, safe travel, etc.) that your problem is seeking to achieve.
2) The next step is to reformulate your problem, using main functions to establish contradictions in your specific problem. In other words, deciding what is getting better and what is getting worse (or what is blocking things from improving). For example, engine power will provide a change in train speed, but inertia counteracts this.

   There are two kinds of contradictions: technical (engineering) and physical. Technical contradictions are the classical engineering "tradeoffs." Technical contradictions arise when we want to improve one variable (parameter) in the system, consequently causing another parameter in the system to worsen. For example, consider a cylindrical shaft with a diameter $d$. If $d$ increases (improving parameter) the shaft gets heavier (worsening parameter). If we want to reach to location earlier we go faster (speed is the improving parameter) but increasing speed will cause a decrease in fuel economy (worsening parameter). This contradiction between two different parameters (in this example speed and fuel economy) is usually called a technical or engineering contradiction.

---

16  http://www.triz-journal.com/whatistriz.htm, accessed August 10, 2015.

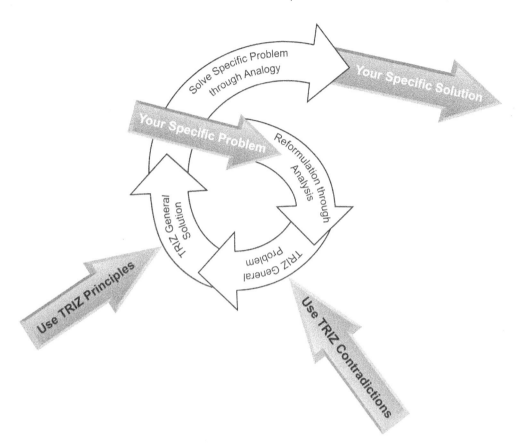

**Figure 3.34** TRIZ problem- solving process.

In contrast to technical contradiction, a conflict within a parameter itself is called a physical contradiction. As an example, if we use a long regular magnetic tool as shown in Figure 3.35(a) to pick up various metal parts, it may not be convenient to carry. If we use a short magnetic tool for the same purpose it is not long enough to reach. In this problem, the length of the tool is a parameter that creates conflict within a parameter (length) itself. The length of the tool should be long enough to reach but it should be short enough to carry – length is the physical contradiction. There

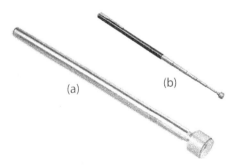

**Figure 3.35** Magnetic pickup tool: (a) regular, (b) telescopic.

are several ways of resolving this contradiction. For example, if we use a telescopic magnetic tool to pick up various metal parts as shown in Figure 3.35(b), it is short enough to carry and if we extend it it will be long enough to reach.

**Table 3.9** Examples of Contradictions.

| Catagory | Good | Bad |
|---|---|---|
| **Technical** | Residential high-rises occupy less precious land area, | but consume more energy. |
| | The material gets stronger, | but the weight increases. |
| | The car with sophisticated shape looks impressive, | but you spend an impressive amount of money to buy it. |
| | A vehicle with higher horsepower | gives you worse gas mileage. |
| | Serving quality food in a timely manner | requires more employees. |
| **Physical** | A system should be *easy* to operate, | but should have many *complex* features. |
| | Friction-reducing shoe soles are *thin*, | but shoe soles should be *thick* for comfort. |
| | A vehicle should have *high* speed to travel in a timely manner, | but it should have *moderate* speed for safety. |

Classical examples are shown in Table 3.9. As can be seen, some parts of the problem solution provide a function which may detract from some other function. For example, the material gets stronger but the weight increases. In this example, "stronger" is a function but that function detracts from the function "weight."[17]

3) Convert your specific problem into a TRIZ general problem – such as contradictions of "power" and "weight." In TRIZ, 39 contradictions have been identified to help us to solve our specific problem. A subset of these contradictions is listed in Tables 3.10 and 3.11.[18]

4) Using TRIZ inventive principles corresponding to TRIZ contradictions, a TRIZ general solution can be obtained. In design, most of the time, there is an underlying physical property which is the root cause of the contradiction between the two different parameters affecting the productivity of the design. For example, consider the contradiction of a large volume of vehicle to improve the carrying capacity of the cargo and the resulting loss of drag coefficient. In this example, we are trying to improve "vehicle volume" (parameter #7) and realize a resulting "loss of energy" (parameter #22) due to the higher drag coefficient. In descending order, possible inventive principles of 7, 15, 13, and 16 have been most common for solving this class of problems, (see Table 3.10).

Now consider another example. If we would like to improve the change of speed (acceleration) of a train with time the intersection of "power" (row) and "weight" (column) as the contradictions will give us the potentially useful TRIZ inventive principles. They are inventive principles 8, 36, 38, and 31 that have been most commonly applied for solving this

---

17 For more examples, see reference http://www.triz-journal.com/contradictions-air-bag-applications/ and http://www.triz-journal.com/innovation-tools-tactics/breakthroughdisruptive-innovation-tools/resolving-contradictions-40-inventive-principles/.
18 http://www.triz-journal.com/39-features-altshullers-contradiction-matrix/

**Table 3.10** TRIZ Contradiction Matrix.

| Improving Feature \ Worsening Feature | Weight of moving object (1) | Weight of stationary object (2) | Length of moving object (3) | Length of stationary object (4) | Area of moving object (5) | Area of stationary object (6) | Volume of moving object (7) | Volume of stationary object (8) | Speed (9) | Force (Intensity) (10) | Loss of energy (22) |
|---|---|---|---|---|---|---|---|---|---|---|---|
| **1** Weight of moving object | + | – | 15, 8 29, 34 | – | 29, 17 38, 34 | – | 29, 2 40, 28 | – | 2, 8 15, 38 | 8, 10 18, 37 | 6, 2 34, 19 |
| **2** Weight of stationary object | – | + | – | 10, 1 29, 35 | – | 35, 30 13, 2 | – | 5, 35 14, 2 | – | 8, 10 19, 35 | 18, 19 28, 15 |
| **3** Length of moving object | 18, 15 29, 34 | – | + | – | 15, 17 4 | – | 7, 17 4, 35 | – | 13, 4 8 | 17, 10 4 | 7, 2 35, 39 |
| **4** Length of stationary object | | 35, 28 40, 29 | – | + | – | 7, 17 10, 40 | – | 35, 8 2, 14 | – | 28, 10 | 6, 28 |
| **5** Area of moving object | 2, 17 29, 4 | – | 14, 15 18, 4 | – | + | – | 7, 14 17, 4 | | 29, 30 4, 34 | 19, 30 35, 2 | 15, 17 30, 26 |
| **6** Area of stationary object | – | 30, 2 14, 18 | – | 26, 7 9, 39 | – | + | – | | – | 1, 18 35, 36 | 17, 7 30 |
| **7** Volume of moving object | 2, 26 29, 40 | – | 1, 7 4, 35 | – | 1, 7 4, 17 | – | + | – | 29, 4 38, 34 | 15, 35 36, 37 | (7, 15 13, 16) |
| **8** Volume of stationary object | – | 35, 10 19, 14 | 19, 14 | 35, 8 2, 14 | – | | – | + | – | 2, 18 37 | – |
| **9** Speed | 2, 28 13, 38 | – | 13, 14 8 | – | 29, 30 34 | – | 7, 29 34 | – | + | 13, 28 15, 19 | 14, 20 19, 35 |
| ... | ... | ... | ... | ... | ... | ... | ... | ... | ... | ... | ... |
| **21** Power | (8, 36 38, 31) | 19, 26 17, 27 | 1, 10 35, 37 | – | 19, 38 | 17, 32 13, 38 | 35, 6 38 | 30, 6 25 | 15, 35 2 | 26, 2 36, 35 | 10, 35 38 |
| ... | ... | ... | ... | ... | ... | ... | ... | ... | ... | ... | ... |
| **39** Productivity | 35, 26 24, 37 | 28, 27 15, 3 | 18, 4 28, 38 | 30, 7 14, 26 | 10, 26 34, 31 | 10, 35 17, 7 | 2, 6 34, 10 | 35, 37 10, 2 | – | 28, 15 10, 36 | 29, 10 23, 35 |

class of problems, in descending order (see Table 3.10). Note that an empty box implies that many of the 40 principles may apply (see Table 3.12) – thus all of them should be considered for the problem in hand.

Suppose there is a need to improve the strength of a light-weight aerospace industry material. Suppose also that the material will be used for an airplane body. Then, for this specific problem, the main functions are strength (improving feature, row 1) and weight (worsening feature, column 14) in Table 3.11. The corresponding cell of the contradiction matrix (see Table 3.11) gives us principles "28, 27, 18, 40." These numbers refer to the TRIZ inventive principles that have the highest probability of resolving our contradiction. A composite material

**Table 3.11** TRIZ Contradiction Matrix.

| | Improving Feature ↓ / Worsening Feature → | Stress of pressure | Shape | Stability of the object composition | Strength | Duration of action of moving object | Duration of action of stationary object | Temperature | Illumination intensity | ⋯ | Productivity |
|---|---|---|---|---|---|---|---|---|---|---|---|
| | | **11** | **12** | **13** | **14** | **15** | **16** | **17** | **18** | **...** | **39** |
| **1** | Weight of moving object | 10, 36 37, 40 | 10, 14 35, 40 | 1, 35 19, 39 | 28, 27 18, 40 | 5, 34 31, 35 | – | 6, 29 4, 38 | 19, 1 32 | ... | 35, 3 24, 37 |
| **2** | Weight of stationary object | 13, 29 10, 18 | 13, 10 29, 14 | 26, 39 1, 40 | 28, 2 10, 27 | – | 2, 27 19, 6 | 28, 19 32, 22 | 19, 32 35 | ... | 1, 28 15, 35 |
| **3** | Length of moving object | 1, 8 35 | 1, 8 10, 29 | 1, 8 15, 34 | 8, 35 29, 34 | 19 | – | 10, 15 19 | 32 | ... | 14, 4 28, 29 |
| **4** | Length of stationary object | 1, 14 35 | 13, 14 15, 7 | 39, 37 35 | 15, 14 28, 26 | – | 1, 10 35 | 3, 35 38, 18 | 3, 25 | | 30, 14 7, 26 |
| **5** | Area of moving object | 10, 15 36, 28 | 5, 34 29, 4 | 11, 2 13, 39 | 3, 15 40, 14 | 6, 3 | – | 2, 15 16 | 15, 32 19, 13 | ... | 10, 26 34, 2 |
| **6** | Area of stationary object | 10, 15 36, 37 | | 2, 38 | 40 | – | 2, 10 19, 30 | 35, 39 38 | | ... | 10, 15 17, 7 |
| **7** | Volume of moving object | 6, 35 36, 37 | 1, 15 29, 4 | 28, 10 1, 39 | 9, 14 15, 7 | 6, 35 4 | – | 34, 39 10, 18 | 2, 13 10 | ... | 10, 6 17, 7 |
| **8** | Volume of stationary object | 24, 35 | 7, 2 35 | 34, 28 35, 40 | 9, 14 17, 15 | – | 35, 34 38 | 35, 6 4 | | ... | 35, 37 10, 2 |
| **9** | Speed | 6, 18 38, 40 | 35, 15 18, 34 | 28, 33 1, 18 | 8, 3 26, 14 | 3, 19 35, 5 | – | 28, 30 36, 2 | 10, 13 19 | ... | – |

(inventive principles of 40 from Table 3.12) would be a solution offered by the matrix to solve this contradiction.

Altshuller's 40 TRIZ *inventive principles* are shown in Table 3.12.[19]

### 3.7.4.2 TRIZ Separation of Principles

Problems of contradictions are resolved through the use of "separation of principles" – one of the most important problem-solving principles and tools within TRIZ. Separation of principles is a TRIZ method to identify where competing parameters or "design contradictions" may not be needed at the *same time*, in the *same space*, on the *same scale*, in the *same direction* or orientation, etc.

The following subset of TRIZ separation of principles will be discussed in this section.

1) Separation of conflicting properties in time: changing a property, response, behavior vs. time. For example, by varying the cargo ship propeller blades to the optimal pitch in time, higher

---

19 For detailed explanation of Table 3.12 see http://www.triz.co.uk/files/U4 8432_40_inventive_principles_with_examples.pdf

**Table 3.12** Altshuller's TRIZ: 40 principles.

| Principles | Principles |
| --- | --- |
| 1. Segmentation | 21. Rushing through |
| 2. Extraction (taking out) | 22. Convert harm into benefit |
| 3. Local quality | 23. Feedback |
| 4. Asymmetry | 24. Mediator (intermediary) |
| 5. Combination (merging) | 25. Self-service |
| 6. Universality | 26. Copying |
| 7. Nesting | 27. Inexpensive short life |
| 8. Counterweight (anti-weight) | 28. Replacement of a mechanical system |
| 9. Prior counteraction | 29. Use pneumatic or hydraulic systems |
| 10. Prior action | 30. Flexible film or thin membranes |
| 11. Cushion in advance | 31. Use of porous materials |
| 12. Equipotentiality | 32. Changing the colour |
| 13. Inversion (the other way round) | 33. Homogeneity |
| 14. Spheroidality (curvature) | 34. Rejecting and regenerating parts |
| 15. Dynamicity | 35. Parameter change |
| 16. Partial, overdone or excessive action | 36. Phase transition |
| 17. Moving to a new dimension | 37. Thermal expansion |
| 18. Mechanical vibration | 38. Use strong oxidisers |
| 19. Periodic action | 39. Inert environment |
| 20. Continuity of useful action | 40. Composite materials |

efficiency can be obtained, thus saving fuel. From Table 3.12 the TRIZ principles most applicable to "separation in time" are 9, 10, 11, 15, 16, 18, 19, 20, 21, 26, 34, and 37.

2) Separation of conflicting properties in space: changing a property, response, behavior vs. special location. From Table 3.12 the TRIZ principles most applicable to "separation in space" are 1, 2, 3, 4, 7, 13, 14, 17, 24, 26 and 30.

3) Separation between parts and the whole: changing a property, response, behavior, etc., to make it different in the sub-system, system, or system of systems. From Table 3.12 the TRIZ principles most applicable to "between parts and the whole" are 1, 5, 6, 8, 12, 13, 22, 23, 25, and 27.

Consider the example of motion-sensitive light switches. They are increasing in use in corporate and public spaces, because they are convenient for employees and shut off after continued inactivity, significantly reducing energy expenditure. The goal is high productivity of the use of lights in such spaces so as not to waste energy. The technical contradiction would be that if lights are always on, the energy use is increased. In other words, two parameters, "lights are always on" and "energy loss," are technical contradictions. Now, consider the physical contradiction – "lights are on" or "not on" (one parameter to consider productivity). When we use separation of principles, the question is: should the 'lights are always on' be separated by time, space or between parts and the whole? The best solution is it should be separated by time – lights should be on when in used and when not in use. Possible applicable principles are 19 (periodic action) and 16 (partial or excessive action); see Table 3.12.

### 3.7.4.3 Ideation TRIZ Methodology

I-TRIZ is a research-based improvement of classical TRIZ science, methodology, tools and applications developed by Ideation International in the USA. The ideation process is a software-based problem-solving process including the ideation TRIZ (I-TRIZ) methodology. The ideation process is designed to help in analyzing a problem situation and forming innovative solution concepts.

I-TRIZ represents an expansion of the TRIZ methodology to *directed evolution* and to non-technical areas (business, management, scientific research, cultural, language, etc.). It provides advanced decision-support knowledge-based integration of classical TRIZ tools and tracks of evolution for higher repeatability, reproducibility and reusability of innovation processes. The Evolution of TRIZ and I-TRIZ is shown in Figure 3.36.[20]

ARIZ (the Algorithm of Inventive Problem Solving) is a part of TRIZ. It is not as widely used as other methods in TRIZ. Anticipatory failure determination (AFD) is an application of I-TRIZ specifically designed for failure analysis and failure prediction. AFD differs from other failure

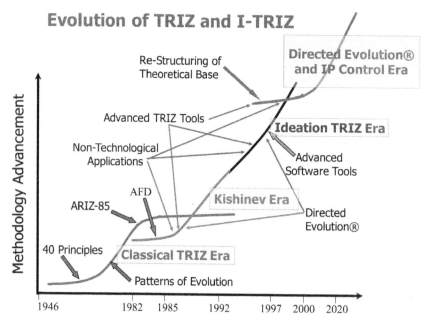

**Figure 3.36** TRIZ evolution (used with permission from Ideation International Inc.)

---

20  http://www.whereinnovationbegins.net

analyses such as failure modes and effects analysis (FMEA) and hazard and operations analysis (HAZOP).

Advanced I-TRIZ methods and tools can be used for improving Six Sigma methodology, especially when Six Sigma methods and tools are inefficient and/or insufficient.

## 3.8 Cascade to Production

Quality can be defined as the characteristic that distinguishes the grade of excellence or superiority of a process, product, or service. Quality is felt after the system is designed and used. Quality is built in by the perception of the designers – designers attempt to satisfy as many stakeholder needs as possible. The aim of satisfying customer needs or requirements puts an emphasis on techniques such as QFD to better understand those needs and produce a product to provide superior value.

In general, one QFD may not be sufficient to explain the voice of the customer (VOC) with regard to actual product design. As seen from Figure 3.37, a four-phase product development process includes four cascaded HOQs where the technical requirements of one house are transformed into the customer requirements of the next. In other words, the first HOQ (product planning) meets the customer requirements as whats against the engineering requirements (the hows) intended to meet the needs. These hows become the whats of the second HOQ (part planning), which charts engineering requirements against hows which are the parts selected to implement them. The parts selected then become the whats of the third HOQ (process planning), plotted against the hows of the process requirements used to create the parts. Finally, the process requirements become the whats of the last HOQ (production planning), where the hows

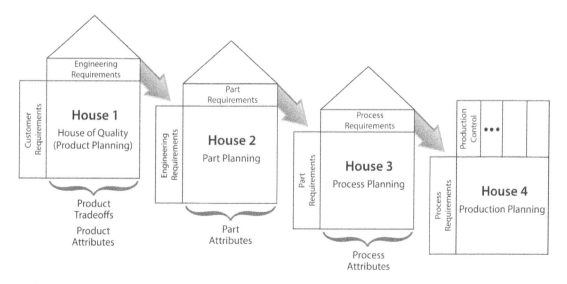

**Figure 3.37** Four-phase QFD approach.

are the process parameters which must be controlled. Thus, the cascaded matrices translate the customer requirements to a set of process parameters to be controlled.

## 3.9  Production Process Planning and Tooling Design[2]

The decline in competitiveness of US manufacturing beginning in the 1960s has been well documented. Various reasons have been suggested for this phenomenon, some of which are not attributable to industry and are beyond its control. One occurrence that is related to this decline is academia's abandonment of courses having to do with manufacturing. With the emphasis on engineering science that began in the 1950s, US universities added courses on the more theoretical subjects and dropped courses having to do with more applied subjects. This resulted in elimination of most of the laboratory courses in machine shop, pattern making, welding and foundry practice that served to introduce the student to manufacturing technology and tended to validate this discipline as an acceptable professional field for graduating engineering students. Lack of emphasis on manufacturing in education undoubtedly contributed to the feeling that many engineering graduates of this era seemed to have careers in this field that were not challenging and held little future. Fortunately, this situation was recognized in the 1980s by many in industry and the federal government. As a result, manufacturing research centers were established in several universities to advance the technology in this important area. These research centers typically involve a cooperative arrangement between a firm associated with manufacturing and a university with the necessary manufacturing facilities and personnel. Startup funding is usually provided by the federal government on the basis of a proposal evaluation. This initiative, though limited, is vitally important for two reasons: first, as indicated, the USA desperately needs an infusion of new ideas and enthusiasm in manufacturing if it is to compete in the world market; and second, 40% of the design engineering workforce is currently involved with production.[21]

Production planning is initiated by review of the design drawings to identify the machines and tooling required and to determine the forming operations to be used. Product design data such as geometrical features, dimensions, tolerances, materials, and surface finishes are evaluated to determine the appropriate sequence of processing operations, given the specific machinery and work stations available on the production line. Detailed production procedures are then prepared based on this evaluation and this documentation is provided to the appropriate work stations as production is initiated. Typical tasks included in production process planning are identified as follows:[22]

1) Interpretation of design drawings and specifications
2) Selection of material stock
3) Selection of production processes

---

21 Kohl, R., "New research shatters stereotypes about engineers," *Machine Design*, February 22, 1990.
22 Zhang, H., and Alting, L., *Computerized Manufacturing Process Planning Systems*, Chapman & Hall, London, 1993.

4) Selection of machines to be used in production
5) Determination of the sequence of operations
6) Selection of jigs, fixtures, tooling, and reference datum
7) Establishment of tool cutting parameters (speed, depth, feed, etc.)
8) Selection of inspection gages and instruments
9) Calculation of processing time
10) Generation of process documentation and numerically controlled (NC) machine data.

Tooling is a term usually taken to mean all of the special equipment required to hold and position the part (jigs), guide the tool in performing the machining operation (fixtures), perform the actual metal removal and forming operations (tools), and check the part for dimensional accuracy (gages). Design engineers, often referred to as tool designers, perform many of these tasks, working with production planning and control and manufacturing personnel. The advantages of concurrent engineering and CAD/CAM, which integrate the detailed design and production planning functions to reduce the overall development time and minimize cost, are apparent. Once the information on the part design is in the computer memory the logical next step is to use this information to program NC machines to accomplish the forming operations. This can be accomplished at work stations which connect to a shop-floor computer which feeds this data to appropriate NC machines, thereby reducing product development time significantly. CAD/CAM also provides a means to store and distribute engineering data for wide use throughout a company or by subcontractors. With adequate database support on materials properties and manufacturing processing information, the part design and/or the manufacturing process can be modified to optimize production. For example, a designer specifies an odd-sized hole in a part based on analysis. Due to the odd size, this hole requires two machining operations, drilling and boring, whereas a slightly smaller or larger nominal sized hole that would satisfy the design requirement requires only one operation, drilling. If this manufacturing process information is available to the designer, the less expensive and time-consuming design approach can be selected.[23] This is known as producibility analysis, and several effective software programs are available to assist in performing this function.

Production planning includes the task of laying out the production line flow (preparing a flowchart). This is a function often performed by industrial engineers who recognize the factors that influence plant layout and can develop a plan that integrates the part manufacturing requirements and ancillary support into a logical sequence with other ongoing production operations. This task was accomplished in the past using templates on a large layout board but can now be performed using CAD, a much more efficient and flexible approach. With the integration of design and production using concurrent design methods the planning and layout of the production line can be initiated early in the design process so that the overall product development time is minimized.

Production control is the function of scheduling the work and providing the materials, supplies, and technical data necessary for the manufacturing operation. It involves the routing of

---

23 Hartnett, J., and Khol, R., "Component design as an engineering discipline," *Machine Design*, November 1989.

parts, components and assemblies; establishing starting and finishing dates for each important component, assembly and end item; and issuing the necessary instructions and other data to ensure a smooth manufacturing operation. Production control is the heart of the production operation, providing all the supplies and parts to the various manufacturing work stations as well as the necessary instructions specifying what to do and how to do it. During the 1980s US industry, under pressure to reduce costs and increase flexibility, began to adopt changes in the production control function. To speed up the preparation and issuance of manufacturing process documentation, some firms turned to a computerized document generation and transmission system. These systems are used to create, maintain, transmit, and store documentation governing essentially all of the various manufacturing operations required in a production line. Computer-integrated manufacturing is a term used to describe systems which integrate the entire manufacturing process. It not only provides for integration of CAD/CAM and all aspects of design engineering/manufacturing but also includes such functions as materials management, cost accounting, shipping and receiving, and personnel training.

An inventory management scheme called just-in-time (JIT) manufacturing has also been adopted by some companies. JIT has been referred to as the twentieth century's most important productivity enhancing management innovation.[24] Using this approach a company attempts to minimize inventory and receive supplies in small lots just as they are needed. With JIT, suppliers must be reliable. A glitch in a supplier's operation can shut down the production line in a short period of time since there is no inventory buffer. With JIT suppliers' operations are, in a sense, an extension of the production line. Suppliers can no longer be dealt with at arm's length as in the past when procurement was based on the lowest bid and satisfactory quality of the supplier's product was ascertained by receiving inspection. Suppliers must be qualified via an in-depth quality review before contract award and some sort of continuous oversight must be maintained to ensure early detection of supply interruptions. The supplier's product quality must be based on a knowledge of its capabilities including facilities, equipment and personnel, and confidence in its operation based on management goals and philosophy. For in-house fabricated parts that are required for subsequent production line assembly, personnel at the working level, and thus in a position recognize potential problems, must feel an identity with, and responsibility for, the effort since every person in the chain is critical in JIT.

Management information is another important aspect of the production control operation. Production control, in conjunction with production supervision and management, develops and maintains the schedule for the work and is thus in a good position to provide the appropriate schedule and other information necessary for management control. If the manufacturing process documentation is prepared using computers with appropriate software, the management information system can use this capability to expedite the availability of information for management. Management can be provided with access to the computerized documentation

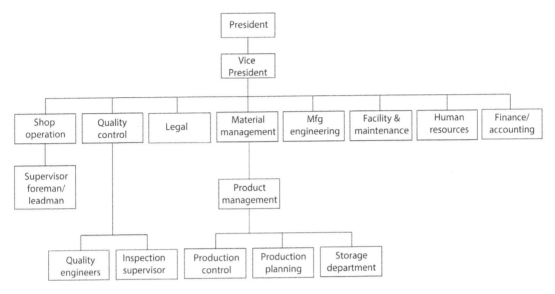

**Figure 3.38** Typical manufacturing organization chart.

system and can select information needed as required. Management often will require special reports depicting specific items of interest. These can usually be prepared relatively easily using the computerized system, but a good rule to follow is to minimize the number of reports generated and to try to use a report format that will satisfy multiple requirements. This is especially true when documentation is prepared without the use of computers.

Figure 3.38 depicts a typical manufacturing corporation organization chart which shows the production control and planning functions and their relative positions in the organization.

### 3.9.1 Production

With the completion of qualification testing and satisfactory acceptance testing of the prototype, the design process is finalized. Support of manufacturing by the design group will still continue, however, since problems usually surface during production that need to be resolved by coordination between design and manufacturing. Product improvement efforts may also be authorized which involve redesign of some element(s) of the product. Acceptance testing should be structured to disclose any discrepancies in the manufacturing process. The basic design has already been proven to be satisfactory by qualification testing; therefore, acceptance testing is used to provide a check on the manufacturing procedures and processes only. Acceptance testing can be performed on every item that comes off the production line or selective testing can be accomplished based on statistical analysis. The approach used will depend on the product and the manufacturing process. For an expensive product such as an automobile some sort of acceptance test will typically be performed on every production article. For a less expensive product that is not as susceptible to manufacturing defects, some sort of statistically based acceptance test program will be instituted.

A quality control inspection plan must also be implemented during the production phase. This is not to be confused with overall product quality assurance which involves other segments of the operation in addition to production. The function involved here is the inspection of manufactured elements during the production process to ensure delivery of a functional product. The inspection may involve the measurement of certain dimensions, the use of a gage to check the dimensional accuracy of some particular feature, the use of instruments to evaluate parameters, or a visual check of the workmanship. Quality control inspectors may be positioned at one location, as in the case of an assembly line or work center, or may be what is referred to as "roaming inspectors," covering larger areas of the operation while watching for unsatisfactory parts as well as poor work practices. Inspection is normally concerned with part characteristics such as dimensions, composition, workmanship, finish, and function. The inspector tries to ensure that the fabricated part conforms to drawings and specifications, evidences quality workmanship, and functions properly. Proper function is normally ascertained via the acceptance test.

### 3.9.2 The Product Realization Process

Many US companies are devoting an increasing amount of effort to the process by which they identify customer needs and product performance; plan the design and manufacturing processes to include the product life cycle, distribution, support, maintenance, recycling and disposal; and plan for product improvement. This process has become known as the *product realization process*. In recognition of the fact that there has been little agreement within industry as to the elements that constitute the product realization process, the National Science Foundation sponsored a study[25] among progressive companies and universities to:

1) define the best practices used by progressive companies in the product realization process;
2) determine the best practices for which knowledge and skills are needed by engineering graduates and experienced engineers;
3) identify model programs for integrating instructional elements for important best practices into mechanical engineering curricula.

Product realization process *best practices* currently used in industry were identified through discussions with individuals from leading companies and through literature search. The 56 different *best practices* which were identified in this study are shown in Table 3.13. Many of these best practices are discussed in this textbook. They are grouped into the following categories:

1) Knowledge of the product realization process
2) Product realization process team skills

---

25 "Integrating the Product Realization Process into the Undergraduate Curriculum," ASME Council on Education, December 1995

**Table 3.13** The consensus set of 56 "best practices" by category (from Ertas & Jones, used with permission of John Wiley & Sons, Inc.).

| Knowledge of PRP | PRP team skills | Design team skills | Analysis & testing skills | Manufacturing testing skills |
|---|---|---|---|---|
| 1. Knowledge of product Realization process | 7. Project Management tools | 18. Competitive analysis | 36. Finite Element analysis | 50. Materials Planning inventory |
| 2. Bench Making | 8. Budgeting | 19. Creative Thinking | 37. Design of experiments | 51. Total quality management |
| 3. Concurrent engineering | 9. Project risk analysis | 20. Tools for "Customer centered" design | 38. Value engineering | 52. Manufacturing processes |
| 4. Corporate Vision and product fit | 10. Design review | 21. Solid modeling/rapid prototyping systems | 39. Mechatronics (mechanisims and control) | 53. Manufacturing floor/ workcell layout |
| 5. Business functions/Mkt'g. Legal, etc. | 11. Information processing | 22. Systems perspective | 40. Process improvement tools | 54. Robotics & automated assembly |
| 6. Industrial Design | 12. Communication | 23. Design for assembly | 41. Statistical process control | 55. Computer-integrated manufacturing |
| | 13. Sketching/drawing | 24. Design for commonality platform | 42. Design standards (e.g., UL., ASME) | 56. Electromechanical packaging |
| | 14. Leadership | 25. Design for cost | 43. Testing standards (e.g., ASTM) | |
| | 15. Conflict management | 26. Design for disassembly | 44. Process standards (e.g., ISO 9000) | |
| | 16. Professional ethics | 27. Design for environment | 45. Product testing | |
| | 17. Team/teamwork | 28. Design for ergonomics (human factors) | 47. Test equipment | |
| | | 29. Design for manufacturing | 48. Application of statistics | |
| | | 30. Design for performance | 49. Reliability | |
| | | 31. Design for reliability | | |
| | | 32. Design for safety | | |
| | | 33. Design for service/repair | | |
| | | 34. CAD systems | | |
| | | 35. Geometric tolerancing | | |

3) Design skills
4) Analysis and testing skills
5) Manufacturing skills.

A glossary of the terms used in this listing is provided in Table C.5, Appendix C.

### 3.9.3 Functional Cost Analysis

Functional cost analysis, or value analysis, is a technique that involves breaking the product down into its component parts and determining the cost-effectiveness of these elements relative to the importance of the functions being provided. A matrix form similar to that shown in Figure 3.39 is used to perform this analysis. Use of this method requires a detailed level of knowledge about the product cost, and thus it is normally applied during the latter stages of the design process after the manufacturing plan has been generated. This technique allows the high-cost areas of the design to be identified so that effort can be concentrated on cost reduction and/or enhanced functionability.[10] Figure 3.39(a) shows the cost analysis of an air valve and highlights the excessive cost of the *valve body* and *connect parts* function as well as the redundancy of some elements. Figure 3.39(b) shows the cost breakdown after redesign. Note the elimination of costs associated with valve elements that were providing no useful function and reduction in the cost of the *connect parts* function.

More recently value analysis has been performed early during the design process at the design concept stage. The returns from applying this analysis technique early in the design process are potentially greater than when applied later since the analysis precedes large expenditures for planning and tooling. Use of this technique early also injects a discipline into the design process that will help in guiding product development. Use of value analysis early in the design process was conceived by a group of General Electric employees in the late 1950s. Only recently have the details of this advanced system, referred to as *value control*, been widely available. A key element in value control is the use of a *market standard*. To establish the market standard it is assumed that the price paid in the marketplace for a class of product reliably measures the worth to the customer of the functions provided by the product. The market standard is then converted into a market standard cost, which is the maximum allowable cost for the product if profit objectives are to be met. A function versus cost diagram is then created to which competitors' products are added. This data, in combination with a thorough search for design alternatives, provides the value standard (lowest cost) for each function. This process has proven to be a powerful tool for generating product value improvements.[26]

---

26 Fowler, T., *Value Analysis in Design*, Van Nostrand Reinhold, New York, 1990.

| Parts | Functions | Stop air | Sense ram air | Sense servo air | Sense cabin air | Connect parts | Provide mounting | Resist corrosion | Provide support | Provide interchangeability | No function | $ total cost | % |
|---|---|---|---|---|---|---|---|---|---|---|---|---|---|
| Banjo assembly | | | 0.2 | | 0.4 | | | | 0.47 | | | 1.07 | 5.5 |
| Valve body | 0.4 | 1.0 | | | 2.82 | 0.8 | 0.2 | 0.8 | | 0.6 | | 6.62 | 34.0 |
| Spring | | | | | | | | | | 0.39 | | 0.39 | 2.0 |
| Diaphragm assembly | 0.6 | 0.1 | 0.1 | 0.1 | 0.94 | | 0.2 | 0.1 | | | | 2.14 | 11.0 |
| Cover | | | 0.4 | | 1.2 | 0.1 | 0.1 | 0.34 | 0.1 | | | 2.24 | 11.5 |
| Lug | | | | | | | | | | 0.1 | | 0.1 | 0.5 |
| Nuts, bolts and washers | | | | | 2.14 | | 0.1 | | 0.1 | | | 2.34 | 12.0 |
| Assembly | | | | | 4.58 | | | | | | | 4.58 | 23.5 |
| Total | 1.0 | 1.1 | 0.7 | 0.1 | 12.08 | 0.9 | 0.6 | 1.24 | 0.67 | 1.09 | | 19.48 | 100.0 |
| % total | 5.1 | 5.7 | 3.4 | 0.5 | 6.2 | 4.6 | 3.1 | 6.4 | 3.4 | 5.6 | | | |
| High or low | | | | | H | | | | | H | | | |

(a)

| Parts | Functions | Stop air | Sense ram air | Sense servo air | Sense cabin air | Connect parts | Provide mounting | Resist corrosion | Provide support | Provide interchangeability | No function | $ total cost | % |
|---|---|---|---|---|---|---|---|---|---|---|---|---|---|
| Cover and connections | 0.15 | 0.25 | 0.50 | 0.10 | 0.25 | 0.30 | | | 0.15 | 0.06 | | 1.76 | 25.5 |
| Body assembly | 0.15 | 0.20 | 0.25 | 0.45 | 0.45 | 0.40 | | | 0.25 | 0.03 | | 2.18 | 31.5 |
| Diaphragm assembly | 0.15 | 0.10 | 0.25 | 0.20 | 0.25 | 0.10 | | | 0.20 | 0.03 | | 1.28 | 18.5 |
| Valve assembly | 0.05 | | | 0.05 | 0.05 | 0.15 | | | 0.31 | 0.05 | | 0.66 | 9.5 |
| Fasteners, nut bolts, etc. | | | | | | 1.04 | | | | | | 1.04 | 15.0 |
| Total | 0.50 | 0.55 | 1.05 | 0.80 | 2.14 | 0.80 | | | 0.91 | 0.17 | | 6.92 | 100.0 |
| % total | 7.2 | 7.9 | 15.1 | 11.6 | 30.9 | 11.6 | | | 13.2 | 2.5 | | | |
| High or low | | | | | H | | | | | | | | |

(b)

**Figure 3.39** Plume of leachate migrating from a sanitary landfill on a sandy aquifer using contours of chloride concentration. (a) Function/cost analysis matrix for an air valve. (b) Function/cost analysis matrix for the redesigned air valve (from Cross, N., *Engineering Design Methods*, John Wiley & Sons, Ltd., Chichester, 1989, p. 132). Reprinted by permission of John Wiley & Sons, Ltd.

## Bibliography

**1** BARCLAY, I., DANN, Z., and HOLROYD, P., *New Product Development*, CRC Press, Boca Raton, FL, 2000.

**2** CREVELING, C.M., SLUTSKY, J., and ANTIS, D., *Design for Six Sigma in Technology and Product Development*, Prentice Hall, Upper Saddle River, NJ, 2003.

**3** HARRISON, F., and LOCK, D., *Advanced Project Management: A Structured Approach*, Gower, Aldershot, 2004.

**4** HORINE, G.M., *Absolute Beginner's Guide to Project Management*, Que, Indianapolis, IN, 2013.

**5** KAYNAK, E., MILLS, N., and BROOKE, M.Z., *New Product Development: Successful Innovation in the Marketplace*, Routledge, New York, 2010.

**6** NAMBISAN, S. (ed.), *Information Technology and Product Development*, Springer, New York, 2010.

# CHAPTER 3 Problems

**3.1** Assume that a year-long project has a total budget of $800,000. After two months, you have completed 15 percent of the project at a total expense of $200,000. Perform an earned value management analysis.

**3.2** A contractor will remove old kitchen counters and install new laminate countertops, then complete the transformation by installing a new sink and faucet for 10 apartment units. Each has a value of $800. The job expected to be completed in 10 days in proportion. By the end of 6 days only 5 units are completed. Perform the earned value management analysis.

**3.3** *Note:* Students will have to be organized into groups to complete this question.

*The situation.* Your organization has just assigned you to a newly formed task team that is to take over a secret project presently being handled by Research and Development. Your team has been assigned responsibility and authority for designing a plan for managing the project and, after top management has reviewed and approved it, for implementing the project. Your team is made up of individuals with experience from a number of divisions because it is thought that a greater range of knowledge and skills is needed to develop the most effective plans. None of the team members have been told anything about the project other than it is expected to grow to sizeable proportions and require additional personnel.

*The task.* Despite the lack of information regarding the project, your team must now design a preliminary plan for managing it. Twenty management activities (A through T) are listed in Table P3.3 in random order. Your task is to rank these activities according to the sequence you think should be followed in managing the project. This ranking will be reviewed by top management before you are given the go-ahead to begin work on the project.

1. Go over the list of activities and, without discussing it with anyone, rank the activities in the sequence you think should be followed in managing the project. Identify the first activity that should be accomplished as "1" and continue through the last activity, "20." Record your ranking under the column heading of "Individual ranking."
2. Now meet with your group and agree on a sequence of activities that should be followed. Record this ranking under the column heading of "Group ranking."
3. "Planning experts ranking" has been recorded already.
4. Enter the difference between columns 1 and 3 in column 4.
5. Enter the difference between columns 2 and 3 in column 5.
6. Total columns 4 and 5 and compare the individual ranking difference with the group ranking difference and discuss. The lower the score the better.

**Table P3.3** Management activities (adapted from Project Management Short Course, TU Electric).

| | Step 1 individual ranking | Step 2 group ranking | Step 3 planning experts ranking | Step 4 difference steps 1 and 3 | Step 5 difference steps 2 and 3 |
|---|---|---|---|---|---|
| A. Find quality people to fill positions. | | | 12 | | |
| B. Measure progress toward and/or deviation from the project's goals. | | | 17 | | |
| C. Identify and analyze the various tasks necessary to implement the project. | | | 8 | | |
| D. Develop strategies (priorities, sequence, timing of major steps). | | | 6 | | |
| E. Develop possible alternative courses of action. | | | 3 | | |
| F. Establish appropriate policies for recognizing individual performance. | | | 20 | | |
| G. Assign responsibility/accountability/ authority. | | | 15 | | |
| H. Establish project objectives. | | | 2 | | |
| I. Train and develop personnel for new responsibilities/authority. | | | 13 | | |
| J. Gather and analyze the facts of the current project situation. | | | 1 | | |
| K. Establish qualifications for positions. | | | 10 | | |
| L. Take corrective action on project (recycle project plans). | | | 19 | | |
| M. Coordinate ongoing activities. | | | 16 | | |
| N. Determine the allocation of resources (budget, facilities, etc.). | | | 11 | | |
| O. Measure individual performance against objectives and standards. | | | 18 | | |
| P. Identify the negative consequences of each course of action. | | | 4 | | |
| Q. Develop individual performance objectives that are mutually agreeable to the individual and supervisor. | | | 14 | | |
| R. Define scope of relationships, responsibilities, and authority of new positions. | | | 5 | | |
| S. Decide on a basic course of action. | | | 5 | | |
| T. Determine measurable checkpoints for the project and variations expected. | | | 7 | | |
| | | Totals (the lower the → score the better) | | | |

**3.4** Develop a Gantt/milestone chart schedule showing the tasks and events identified in Problem 3.3. Use an arbitrary time scale. Note that some of the items identified in Problem 3.3 are neither tasks or events. Suggestion: use Microsoft PowerPoint to prepare your Gantt/milestone chart.

**3.5** Develop an ADM network for the tasks and events identified in Problem 3.3.

**3.6** Develop a Gantt/milestone chart for the design and development of the test enclosure described in Case Study 3.2. Use an arbitrary time schedule.

**3.7** Develop a PDM network for the tasks and events identified in Problem 3.6.

**3.8** Organize students into groups and have each group prepare an orthogonal array as shown in Figure P3.8 that lists customer desires (*What*) for some simple consumer product and identifies the corresponding substitute quality characteristics (*How*). Characterize the relationships as weak, medium, or strong. Think of as many customer *wants* as possible. Groups should be able to identify as many as 12–15. Decide on an appropriate weight for each customer desire and calculate the relationship weighting by multiplying the individual weights by the relationship rating (1, 3, or 9). The total of these scores for each What is then entered in the *gross weight* column. The *% total* column is completed by dividing individual gross weights by the total gross weight. Weights can be established by comparing each desire, in turn, against each of the others, counting the number of times each desire is selected, and multiplying the highest-ranked desire by the total

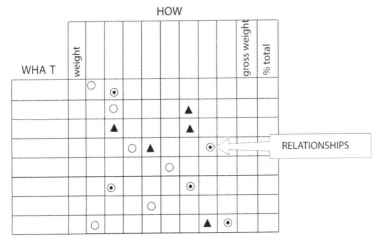

▲  weak relationship    = 1
○  medium relationship = 3
⊙  strong relationship   = 9

**Figure P3.8** "What" versus "How" orthogonal array (from Ford Motor Co. seminar on quality function deployment).

number of desires ($N$), the second-ranked desire by $N - 1$, and so on. Individual group results should be compared and discussed.[1]

**3.9** Complete the table ($T_E$, $T_L$, and Slack) for the network shown in Figure P3.9 and identify the critical path. Because of a critical time constraint, the overall project completion schedule must be reduced to 60 weeks by adding resources. Determine the task to which minimum resources can be added to reduce the overall schedule to 60 weeks and identify the new critical path.

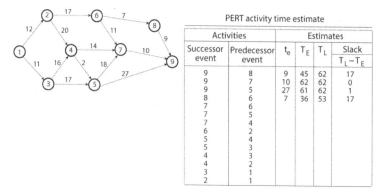

PERT activity time estimate

| Activities | | Estimates | | | |
|---|---|---|---|---|---|
| Successor event | Predecessor event | $t_e$ | $T_E$ | $T_L$ | Slack $T_L - T_E$ |
| 9 | 8 | 9 | 45 | 62 | 17 |
| 9 | 7 | 10 | 62 | 62 | 0 |
| 9 | 5 | 27 | 61 | 62 | 1 |
| 8 | 6 | 7 | 36 | 53 | 17 |
| 7 | 6 | | | | |
| 7 | 5 | | | | |
| 7 | 4 | | | | |
| 6 | 2 | | | | |
| 5 | 4 | | | | |
| 5 | 3 | | | | |
| 4 | 3 | | | | |
| 4 | 2 | | | | |
| 3 | 1 | | | | |
| 2 | 1 | | | | |

**Figure P3.9** PETRI network.

**3.10** Identify the critical paths and slacks shown in Figure P3.10.

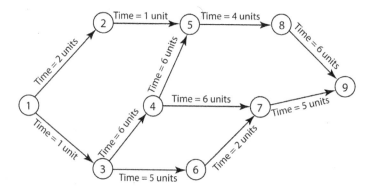

**Figure P3.10** PETRI network.

**3.11** Prepare a decision matrix comparing the use of the following alternative transportation fuels in automobiles and light trucks in the United States. Some of the factors against which the alternatives should be evaluated include conversion cost, infrastructure cost, safety, convenience, environmental effect, effect on vehicle maintenance, US trade balance, and long-term feed stock resource availability.

---

1 Ford Motor Company Seminar on Quality Function Deployment, Detroit, MI, 1988.

1 Compressed natural gas (CNG)

2 Liquid natural gas (LNG)

3 A blend of 85% methanol and 15% hydrocarbon (M85)

4 A blend of 85% ethanol and 15% hydrocarbon (E85)

5 Liquified petroleum gas (LPG)

6 Electricity.

**3.12** Perform a functional cost analysis on the check valve shown in Figure P3.12 using the cost data shown. Use functional cost analysis (described in Section 3.9.3) and Figure 3.39. Identify as many functions for each part as possible.

**Figure P3.12** Check valve assembly.

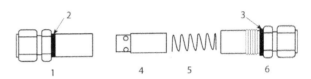

| Parts | Material | Cost $ |
|-------|----------|--------|
| 1. Inlet body | 316 SS | 30 |
| 2. O. ring valve seal | Vito | 2 |
| 3. Body seal gasket | TFE coated 316SS | 5 |
| 4. Poppet | 316SS | 20 |
| 5. Spring | 302 SS | 5 |
| 6. Outlet body | 316SS | 30 |
| | Total cost | $92 |

**3.13** Perform a functional cost analysis on the pressure vessel shown in Figure P3.13 using the cost data shown. Use functional cost analysis (described in Section 3.9.3) and Figure 3.39. Identify as many functions for each part as possible.

**Figure P3.13** Pressure vessel.

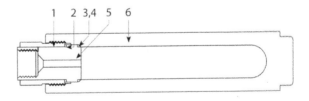

| Parts | Material | Cost $ |
|-------|----------|--------|
| 1. Main nut | 304SS | 15 |
| 2. Thrust ring | 17-4-PH | 2 |
| 3. Gasket | 300SS | 2 |
| 4. Gasket | Copper | 2 |
| 5. Cover | 316SS | 35 |
| 6. Body | 316SS | 60 |
| | Total cost | $116 |

**3.14** Prepare a decision matrix that compares the heating, ventilating and air-conditioning (HVAC) of a house using two alternative approaches: an all electric HVAC system and a semi-passive system using appropriately located double-glazed windows, from walls to store heat, heavily insulated walls and ceiling, collection of roof heat and storage under the slab in a rock bed during winter, and ejection of heat through the roof at night in the summer.

## Team Project

**3.15** Design a toenail clipper for overweight people, seniors, the elderly, and people with back problems or with a limited range of motion. The product must be suitable for both right- and left-hand use.
  a) Perform Kano model analysis using survey results.
  b) Apply the KJ method.
  c) Develop concepts and select one.
  d) Use QFD to develop engineering requirements.
  e) Develop the house of quality.
  f) Use the theory of innovative problem solving (TRIZ) if possible.

# 4

# Transdisciplinary Sustainable Development

Sustainable development is the pathway to the future we want for all. It offers a framework to generate economic growth, achieve social justice, exercise environmental stewardship and strengthen governance.

**Ban Ki-moon**

## 4.1   Introduction

The phrase *sustainable development* gained traction after the 1992 Rio Earth Summit. Sustainable development focuses on the root causes and connections between big issues such as climate change, social inequality, and environmental degradation. Sustainability refers to accepting a duty to seek harmony with other people and with nature; it is not just about the environment. It is about sharing with other people and caring for the Earth.

Although there are many definitions of the term "sustainable development," the most widely accepted one in the literature is from *Our Common Future*,[1] sometimes referred to as the Brundtland definition:[2] "Development which meets the needs of the current generation without compromising the ability of future generations to meet their needs." Nearly two decades before this definition, the United Nations had said that "Unsustainable development leaves behind large areas of poverty, stagnation, marginality, and destruction of the natural environment."

The current economy has increased so much that it already goes beyond/outside the bounds of the ecosystem and social system. Social-ecological systems currently face complex challenges such as the degradation of ecosystems, over-exploitation of natural resources, climate change, wealth inequalities, and human conflicts. These interconnected challenges are threatening the sustainable development of society and cannot be adequately tackled from the content of

---

1 World Commission on Environment and Development, *Our Common Future*, New York: Oxford University Press, New York, 1987.
2 Johnston, A., "Higher education for sustainable development," Final Report of International Action Research Project, March 2007.

specific individual disciplines.[3,4] It is evident that the transdisciplinary research teams must be put together at the beginning of the research project to address such complex issues.

Transdisciplinary research is common in the environmental research area and important in sustainable development research.[5] The transdisciplinary research process takes disciplinary knowledge and implements them in socially relevant issues. It is a process or activity that produces, integrates, and manages knowledge in technical, social and scientific areas.[6,7]

## 4.2 Transdisciplinary Sustainable Development

Transdisciplinary (TD) sustainable development (SD) is a holistic view of sustainable development from transdisciplinary angles of understanding. A clear correlation between SD domains with the concept of transdisciplinarity exists. Domains directly related to SD mainly use a holistic TD approach that integrates economy, ecosphere, and sociosphere.[8] There is value in a holistic idea of sustainability that does not differentiate between social, environmental, and economic sustainability.[9]

As shown in Figure 4.1, the concept of sustainable development has evolved to incorporate three interconnected main domains: economic, social, and environmental. As shown in this figure, sustainability is a multidimensional concept involving environmental equity, economic equity, and social equity. The social attributes of sustainability interrelate with the economic and environmental domains. For example, a community's social sustainability depends on the future of their employment. This includes how a company treats them, how satisfying their job is, or whether they make enough money to support a good quality of life. The environment also has an effect on social sustainability. The environment in which we live in is important for human

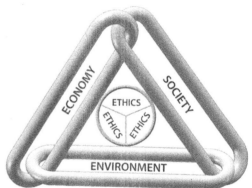

**Figure 4.1** Sustainability and interconnectivity of environment, economy, and society (Ertas, A., *Prevention through Design: Transdisciplinary Process*, ATLAS Publishing, 2010, reproduced with permission).

3 Kates, R.W., and Parris, T.M., "Long-term trends and a sustainability transition," *Proceedings of the National Academy of Sciences of the United States of America*, 100, 8062–8067, 2003.
4 Rockstrom, J., Steffen, W., Noone, K., Persson, A., and Chapin III, F.S., "A safe operating space for humanity," *Nature*, 461(7263), 472–475, 2009.
5 Robinson, J., "Being undisciplined: Transgressions and intersections in academia and beyond," *Futures*, 40(1), 70–86, 2008.
6 Thompson Klein, J., "Prospects for transdisciplinarity," *Futures*, 36(4), 515–526, 2004.
7 Wickson, F., Carew, A., and Russell, A., "Transdisciplinary research: Characteristics, quandaries, and quality," *Futures*, 38, 1046–56, 2006.
8 Cerar, J., "Transdisciplinary sustainable development," master's thesis, Faculty of Economics, University of Ljubljana, 2012.
9 Integrated Network for Social Sustainability, https://clas-pages.uncc.edu/inss/what-is-social-sustainability/, accessed May 31, 2017.

and ecological well-being. As shown in Figure 4.1, ethics is the building block of sustainable development and should be incorporated into design development strategy to ensure long-term sustainability.

### 4.2.1 Economic Sustainability

Economic sustainability involves improving human welfare, primarily through increases in the consumption of goods and services. Traditional development was strongly related to economic growth, which provides economic prosperity for society members. During the early 1960s, the growing numbers of poor in developing countries resulted in considerable attempts to improve income distribution to the poor. As a result, the development paradigm changed towards equitable growth, where social (distributional) objectives, especially poverty alleviation, were accepted to be as important as economic efficiency.

### 4.2.2 Social Sustainability

The social domain focuses mainly on the development of human relationships, success of individual and group goals, and strengthening of values and institutions.[10] Social sustainability seeks to reduce vulnerability and sustain the health of social and cultural systems, and their ability to withstand disasters.[11,12] For example, within a company, social sustainability encompasses such important areas as human rights, fair labor practices, and health, safety, and wellness. It also involves areas such as diversity, equity, work–life balance, and empowerment.

### 4.2.3 Environmental Sustainability

By the early 1980s, there was clear evidence that environmental degradation was a major barrier to development. Hence, protection of the environment became the third major element of sustainable development.[13] In the past several decades, environmentally sustainable development has received more attention than the concept of socially sustainable development. The environmental domain emphasizes the protection of the integrity and resilience of ecological systems.

Considering sustainability, both social and environmental, it is important to note that both require a system of economic activity that is in harmony with the ecological web of life or the social web of life of which we are a part, and upon which we depend for our health, well-being, and quality of life: "Human development and the achievement of human potential require a form of economic activity that is environmentally and socially sustainable in this and future generations."[14]

---

10 Munasinghe, M., "Analysing the nexus of sustainable development and climate change: An overview," Munasinghe Institute for Development (MIND), Sri Lanka, 2002.
11 Chambers, R., "Vulnerability, coping and policy," *IDS Bulletin*, 20(2), 1–7, 1989.
12 Bohle, H.G., Downing, T.E., and Watts, M.J., "Climate change and social vulnerability: Toward a sociology and geography of food insecurity," *Global Environmental Change*, 4(1), 37–48, 1994.
13 Munasinghe, M., "Sustainomics and sustainable development," *The Encyclopedia of Earth*, May 7, 2007.
14 Canadian Public Health Association quoted in Hancock, T., "Social sustainability," http://newcity.ca/Pages/social_sustainability.html, accessed June 1, 2017.

Sustainability seeks justice in the domain of human and nature relationships. In the long-term and essentially uncertain future, there are three crucial relationships:[15,8]

1) Intergenerational justice – justice between humans of different generations
2) Intragenerational justice – justice between different humans of the same generation, in particular the present generation
3) Physiocentric ethics – justice between humans and nature.

Intergenerational and intragenerational justice is expressed as "development that meets the needs of the present generation without compromising the ability of future generations to meet their own needs."[1] The physiocentric ethics aspect is the idea of justice towards nature for its inherent value.[16] This explanation of SD implies nature conservation for its own sake.

Referring to Figure 4.1, a simple example would be that a sustainable building project must not result in harm to the environment during its construction and use. The building must also make economic sense such that, over the long term, the revenues will at least equal the expenses of constructing and operating it. Finally, the building must be socially acceptable such that it will not cause any harm to any person or causes a group of people to experience injustice. What could be more unreasonable than to have workers construct a building that is not as safe to build as it could be? That is to say, is it not fair to design a building to be as safe as possible? A fair construction project is when the designers have made reasonable effort to "design out" or minimize hazards and risks early in the design process. Sustainable construction occurs when design contributes to safety.[17]

The engineering profession is being challenged with a new and forceful set of requirements such as population growth, resource scarcity, and environmental change which appear about to happen. For example, these include apparent changes to the atmosphere, hydrosphere, and biosphere resulting in major shifts from the environmental norms under which the artifacts of our civilization were originally designed. At one time, these aspects of the engineering design could be taken for granted because of the obvious stability of the environment within a narrow, acceptable, and predictable range of change. Including the added interconnectivity and complexity of the environment, shifting requirements from environmental changes will not be easily addressed with methods descended from our industrial age.

Figure 4.2 shows one widely accepted concept of sustainable development: interconnectivity of environment, economy, and society. The environment plays an important role in the well-being of community development. It affects a broad range of social and economic variables which have a vital impact on the quality of community life, human health, and safety.

---

15 Becker, C., "Sustainability ethics and sustainability research" (habilitation thesis), Technical University of Kaiserslautern, 2009.

16 DesJardins, J., *Environmental Ethics* (4th edn), Wadsworth, London, 2005.

17 "Prevention through design, design for construction safety," 2009. http://www.designfor constructionsafety.org/concept.shtml.

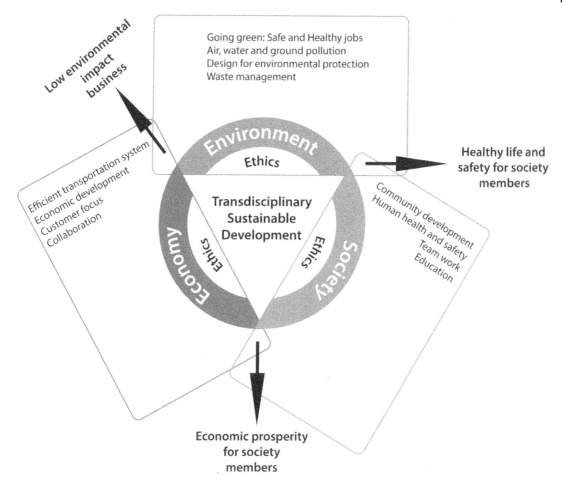

**Figure 4.2** Transdisciplinary sustainable development (Ertas, A., *Prevention through Design: Transdisciplinary Process*, ATLAS Publishing, 2010, reproduced with permission).

A dynamic environment contributes to a healthier society and a stronger economy. Similarly, the environment is affected by economic and social factors.

### 4.2.4 Transdisciplinary Sustainable Development Process

There is no specific process or framework that addresses the problems related to the three inter-acting domains (social, economic, and environmental) of the sustainable development triangle shown in Figure 4.3. This figure shows not only synergies but also a conflict of interest among the interacting domains: the resource conflict between economic development and environmental protection, the property conflict between economic development and equity and social justice,

**Figure 4.3** Sustainable development process (adapted from Campbell, S., "The planner's triangle revisited: Sustainability and the evolution of a planning ideal that can't stand still," *Journal of the American Planning Association*, 82(4), 388–397, 2016).

and the development conflict between social justice and environmental protection. The latter is more abstract, especially if we would like to see these two main concerns go hand-in-hand rather than in opposition. However, in the developing countries where main goal is to industrialize and rise out of poverty often led to rapid resource decrease and environmental disaster.[18]

Tradeoffs exist because they are competing with or exclusive of each other. The three broad social and political institutions (the social welfare state, environmental economics and regulation, and environmental justice) should collaborate and provide feedback to create collective impact for the search for common ground to manage complex conflicting issues. The following three important points should be considered in order to achieve successful sustainable development and mange complexity.

1) Establish equity between three domains.
2) Develop the relationships between three domains.
3) Develop detail level of analysis to make better tradeoff decisions.

---

18 Campbell, S., "The planner's triangle revisited: Sustainability and the evolution of a planning ideal that can't stand still," *Journal of the American Planning Association*, 82(4), 388–397, 2016.

Establishing equity between three domains is a complex issue and this complex issue must be transformed to a map of relationships between interacting domains (social, economic, and environmental) to provide better tradeoff decisions. Because humans and natural systems (social-ecological systems) are strongly coupled, the sustainable development process is highly nonlinear, dynamically changing on going complex process: no single discipline could handle the challenges of complex issues involved with the sustainable development process. To achieve a successful sustainable development, an innovative TD collective impact approach (see Chapter 1 for more information) should be implemented through cross-sector collaboration.

## 4.3 Contaminated Environment[19]

> Over increasingly large areas of the United States, spring now comes unheralded by the return of the birds, and the early mornings are strangely silent where once they were filled with the beauty of bird song.
>
> **Rachel Carson**

Rachel Carson combined her interests in biology and writing as a government scientist with the Fish and Wildlife Service in Washington, DC. Her book, entitled *Silent Spring*, is credited with inspiring much of the late twentieth-century concern with the environment as she documented the effect of pesticides on ecology:

> These sprays, dusts, and aerosols are now applied almost universally to farms, gardens, forests, and homes – non-selective chemicals that have the power to kill every insect, the "good" and the "bad," to still the song of birds and the leaping of fish in the streams, to coat the leaves with a deadly film, and to linger on in soil – all this though the intended target may be only a few weeds or insects. Can anyone believe it is possible to lay down such a barrage of poisons on the surface of the Earth without making it unfit for all life? They should not be called "insecticides," but "biocides."
>
> **Rachel Carson**

The condition of the environment and what can be done to protect it in the future rank high among the concerns of Americans in the twenty-first century. The large amount of environmental degradation that has occurred makes it clear that continued growth in population and economical development makes the correction of past ecological misuse complex and expensive. Hazardous substances, at uncontrolled hazardous waste sites, including chemicals, pesticides, heavy metals, and other toxic substances from industrial processes, refueling facilities and agriculture, have been seeping into the ground and aquifer for many years. Scientists and engineers must begin to recognize the delicate nature of the environment in their endeavors and give it the priority it deserves.

---

19 Global Green New Deal, Policy Brief, 2009. Published by the United Nations Environment Programme as part of its Green Economy Initiative in collaboration with a wide range of international partners and experts.

## 4.4 Groundwater Sustainability[20]

Groundwater sustainability concerns the development and use of groundwater to sustain current and future needs without causing unacceptable consequences. Groundwater is one of the most essential natural resources and degradation of its quality has a major effect on the well-being of people. The quality of groundwater reflects inputs from the atmosphere, from soil and water–rock reactions, as well as from contaminant sources such as mining, land clearance, agriculture, acid precipitation, and industrial waste. The fairly slow movement of water through the ground means that dwelling times in groundwater are generally orders of magnitude longer than in surface water. Groundwater is an important water resource that serves as a source of drinking water for the majority of the people living in the United States. Approximately 75 percent of community water systems and nearly all of rural America uses groundwater-supplied water systems (US EPA, 2000). Contamination from natural and human sources can affect the use of these waters. For example, spilling, leaking, improper disposal, or accidental and intentional application of chemicals on the land surface will result in overspill that contaminates close-by streams and lakes.

Strong competition among users such as agriculture, industry, and domestic sectors is driving the groundwater table lower. The quality of groundwater is severely affected because of the extensive pollution of surface water. The sustainability of groundwater utilization must be assessed from a transdisciplinary perspective, where hydrology, ecology, geomorphology, and climatology play an important role.

Environmental problems are essentially research and development challenges of a different order. These problems can be solved by scientists and engineers working together with political entities that can enact the necessary legislation, obtain the required international cooperation, and provide the necessary funding. The environment can no longer be considered an infinite reservoir in which chemical discharges, toxic material dumping, and harmful stack vapors can be deposited based on the lack of a measurable deleterious effect on the immediate surroundings.

Managing the environment is an international problem that cannot be based on monitoring and control at the local level only. Engineers and scientists must play a key role in providing the essential technology for understanding these global problems and in implementing workable solutions.

For the first time in California's history, the Sustainable Groundwater Management Act allows local agencies to implement groundwater management plans that are tailored to the resources and needs of their communities. Good groundwater management will provide protection against drought and climate change, and contribute to reliable water supplies regardless of weather patterns.[21]

---

20 Adapted from Ertas, A., and Jones, J.C., *The Engineering Design Process*, John Wiley & Sons, Inc., 1996.
21 Sustainable Ground Water Management, http://groundwater.ca.gov/, accessed June 26, 2017.

The primary ways in which groundwater contamination occurs are as follows:[22]

1) *Infiltration.* This is the most common groundwater contamination mechanism. Water that has fallen to earth slowly infiltrates the soil through pore spaces in the soil matrix. As the water moves downward under the influence of gravity, it dissolves materials, including contaminates, with which it comes in contact, forming leachate. The leachate will continue to migrate downward until the saturation zone is reached, after which horizontal and vertical spreading of the contaminates in the leachate will occur primarily in the direction of the groundwater flow, as shown in Figure 4.4.

2) *Direct migration.* Contaminants can migrate directly into groundwater from below-ground sources, such as storage tanks and pipelines that lie within the saturation zone. Significant contamination can occur because of the continually saturated conditions. Storage sites and landfills excavated to near the water table may permit direct contact of contaminants with the groundwater. Vertical leakage through seals around well casings or through improperly abandoned wells may also allow direct entry of contaminants into the groundwater system.

3) *Interaquifer exchange.* Contaminated groundwater can mix with uncontaminated groundwater through a process known as interaquifer exchange, in which one water-bearing unit "communicates" hydraulically with another. This is most common in bedrock aquifers where a well penetrates more than one water-bearing formation. When the well is not being pumped,

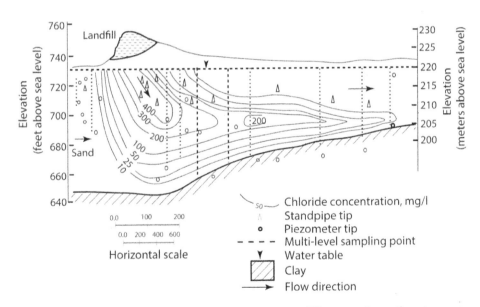

**Figure 4.4** Plume of leachate migrating from a sanitary landfill on a sandy aquifer using contours of chloride concentration (from Freeze, A.R., and Cherry, J.A., *Groundwater*, © 1979, Reprinted by permission of Pearson Education, Inc., New York.

---

22 *Handbook on Ground Water*, Environmental Protection Agency Research Laboratory EPA/625/016, March 1987.

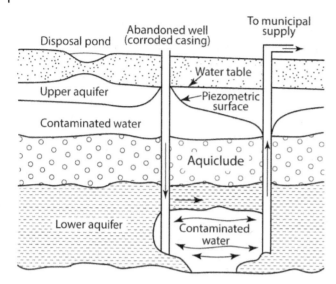

**Figure 4.5** Vertical movement of contaminants along old, abandoned, or improperly constructed well (from Deutsch, M., "Incidents of chromium contamination of ground water in Michigan," US Department of Health, Education and Welfare, Cincinnati, OH, 1961).

water will move from the formation with the greatest potential to formations with less potential. If the formation with the greatest potential contains contaminated water, the quality in another formation can be degraded. Figure 4.5 shows an improperly abandoned well with a corroded casing that was originally used to tap a lower uncontaminated aquifer but now allows migration from an overlying contaminated zone to communicate directly with the lower aquifer. The pumping of a nearby well tapping the lower aquifer creates a downward gradient between the two water-bearing zones allowing contaminated water to migrate through the lower aquifer to the pumping well.

4) *Recharge from surface water.* Normally, groundwater moves toward surface water bodies. Occasionally, however, the hydraulic gradient is such that surface water has a higher potential than the groundwater (such as during flood stages), causing reversal in flow. Contaminants in the surface water can then enter the groundwater system.

Three processes govern the migration of chemical constituents in groundwater:

1) *Advection.* movement caused by the flow of groundwater. The hydraulic conductivity of a geologic formation depends on a variety of physical factors including porosity, particle size and distribution, the shape of the particles, particle arrangement (packing), and secondary features, such as fracturing and dissolution. In general, for unconsolidated porous materials, hydraulic conductivity values vary with particle size. Fine-grained, clayey materials exhibit lower values of hydraulic conductivity, whereas coarse-grained sandy materials normally exhibit higher conductivities.

2) *Dispersion.* movement caused by the irregular mixing of waters during advection. In porous materials, the pores possess different sizes, shapes, and orientations. Similar to stream flow, a velocity distribution exists within the pore spaces such that the rate of movement is greater in the center of the pore than at the edges. Therefore, in saturated flow through these

materials, velocities vary widely across any single pore and between pores. As a result, a miscible fluid will spread gradually to occupy an ever-increasing portion of the flow field when it is introduced into a flow system. This mixing phenomenon is known as dispersion.

3) *Retardation.* chemical and physical mechanisms that occur during advection. Four general mechanisms can retard or slow the movement of chemical constituents in groundwater: dilution, filtration, chemical reaction, and transformation.

Despite increasing awareness that some of the USA's groundwater is contaminated with a variety of toxic metals, synthetic organic chemicals, radionuclides, and pesticides, public policy in this area is still in the formative stages. In spite of increasing efforts by state and federal agencies to protect groundwater, the extent of contamination is likely to appear to increase because more agencies will be searching for contamination, and they will be using increasingly sensitive detection methods.[9] Past excesses and ignorance that (later) result in groundwater contamination will undoubtedly come to light and will add to the staggering projected cost of cleaning up America's groundwater. The present groundwater situation may be satisfactory, but it is probably just a matter of time.

## 4.5   Soil and Groundwater Restoration[23]

A number of techniques are available to contain pollutants and to treat soil and groundwater to clean up contamination. These techniques range from removal of the polluted material and physical, chemical, or biological treatment on the surface, to physical containment and in situ treatment with chemicals or microbes. When selecting soil and groundwater cleanup methods for a particular site, the nature and location of the release, the soil type and geologic conditions, and the required degree of remediation must be evaluated. Liability, cost, period of time necessary to complete the remediation procedure, and any federal or state compliance requirements must also be considered.

The purpose of removing contaminated soil and groundwater associated with a plume of contamination is to treat and/or relocate the wastes to a better engineered and controlled site. The costs of excavation, transportation and new site preparation, soil removal, and reburial are often excessive for a large site, thus this method is used only as a last resort, or in instances of severe pollution where the cost is not significant compared with the importance of the resource being protected. In some cases, removal and reburial in an approved facility is simply transferring a problem from one location to another. Excavation is a technique often used for tank releases when the area of contamination is restricted and well defined. For these cases, excavation costs are usually low compared with other remediation methods, time required for cleanup is short, equipment is generally available, and a wide variety of contaminants can be remediated.

Low-temperature thermal stripping is a process in which soil is excavated and placed in a mobile unit designed to provide heat and drive off contaminants. Heat is usually supplied indirectly as the soil passes through the unit. Contaminants released from the soil are treated in an

---

23 Adapted from Ertas, A., and Jones, J.C., *The Engineering Design Process*, John Wiley & Sons, Inc., 1996.

afterburner, in an activated carbon adsorption system, or condensed for recovery. Soil is treated on site using this method and usually can be accomplished in a relatively short time period. The cost of thermal stripping is higher than landfilling, however, and some contaminants cannot be successfully treated because of the low temperature.[24]

Soil gas extraction, or venting, is a technique frequently used to clean up volatile organic compounds (VOCs) released from storage tanks. This process involves extraction of the contaminate vapor from the soil by using wells and vacuum pumps. Volatile compounds are removed from the area between the soil particles by applying negative pressure to screened wells located in the vadose zone. This process does not require removal of the soil from the site and thus offers certain advantages over extraction. However, cleanup can often take lengthy periods of time, and soils such as heavy clays cannot be readily treated using this method.

---

**CASE STUDY 4.1  Clean up process of contamination from underground gasoline storage tank**

A spill has been detected from an underground gasoline storage tank and abatement procedures have been implemented to prevent any additional leakage. A plan is required that will allow site cleanup in a period of approximately 6 months.[a]

**Solution**

To determine the extent of contamination, organic vapor readings were taken in the basements of all surrounding buildings and utility tunnels in the immediate area. To prevent fire or explosion, venting was initiated in areas where organic vapors were detected.

A hydrogeologic study was then performed to determine the depth to the water table, the local soil strata, and the hydraulic gradient. From these data likely paths of contaminant migration were determined. All water wells in the plume migration path were then sampled for both floating and dissolved hydrocarbons. Surface water near the spill site was also sampled for contamination. Monitoring wells were then drilled, and soil borings were taken at locations radiating out from the spill site. Soil and water samples from these wells were analyzed in a certified laboratory to determine the level of contaminate concentrations. Organic vapor readings were also taken during the drilling process. The resulting data were then assimilated to define the extent of the contamination from the spill. The following determinations were made from these data:

1) No free-floating hydrocarbons were found on the water table.
2) The gasoline contamination was weathered and located vertically in a 2 ft depth region of homogeneous sandy soil.
3) The horizontal migration was contained in a circle of a 36 ft radius with the center located directly below the source of the spill.
4) The total recoverable hydrocarbon concentration was determined to be 1,000 parts per million (ppm).

---

24 *Guidance Manual for Leaking Petroleum Storage Tanks in Texas*, Texas Water Commission, January 1990.

The volume of hydrocarbon contamination within the soil can be approximated as follows. The density $\rho$ of sandy soils can be approximated as 1.5 g/cm$^3$. The volume of contaminated soil is

$$V = hA, \tag{4.1}$$

where $V$ is the volume of soil, $h$ is the zone height, and $A$ is the horizontal area of the zone. Therefore,

$$V = 2 \text{ ft} \times \pi 36^2 = 8{,}143 \text{ ft}^3.$$

The mass $m_s$ of contaminated soil is

$$m_s = \rho V \tag{4.2}$$
$$= (1.5 \text{ g/cm}^3)(8143 \text{ ft}^3)(1728 \text{ in.}^3/\text{ft}^3)(16.387 \text{ cm}^3/\text{in.}^3)$$
$$= 345.88 \times 10^6 \text{ g}.$$

The mass of the contaminant $m_c$ can now be found by using the concentration of the contaminant, which was found to be 1,000 ppm. For every part of soil there is 0.001 parts of contaminant; thus,

$$m_c = 0.001 m_s \tag{4.3}$$
$$= 0.001(345.88 \times 10^6 \text{ g})$$
$$= 345.88 \text{ kg}.$$

The desired cleanup period is determined to be 6 months, which requires an acceptable contaminate removal rate $R_{\text{acc}}$ of

$$R_{\text{acc}} = \frac{345.88 \text{ kg}}{180 \text{ days}}$$
$$= 1.922 \text{ kg/day}.$$

It must now be determined whether this removal rate is attainable. The following equation[25] can be used to determine the estimated contaminant vapor concentration ($C_{\text{est}}$) based on the physical characteristics of the various chemical species:

$$C_{\text{est}} = \sum_i \frac{x_i P_i^v m_{w,i}}{R_u T} \text{ mg/liter}, \tag{4.4}$$

where

$$x_i = \text{mole fraction of constituent},$$
$$P_i^v = \text{pure component vapor pressure, atm},$$
$$m_{w,i} = \text{molecular weight of component, mg/mol},$$

*(Continued)*

25 Johnson, P.C., Stanley, C.C., Kemblowski, M.W., Byers, D.L., and Colthart, J.D., "A practical approach to the design, operation and monitoring of in situ soil-venting systems," *Ground Water Monitoring Review*, 10(2), 159–178, 1990.

**CASE STUDY 4.1 (Continued)**

$R_u$ = universal molar gas constant, L atm/mol K,

$T$ = absolute temperature of soil, K.

Based on the chemical composition of weathered gasoline shown in Table D.1 and the physical properties of these constituents shown in Table D.2 (see Appendix D), $C_{est}$ is determined to be 220 mg/liter. It should be noted that this estimate is valid for vapor concentration at the beginning of venting when the removal rate is greatest. Contaminant concentration declines with time because of changes in composition, lowered contaminant residual levels, and increased resistance to diffusion. This leads to the conclusion that there is a practical limit to the amount of contaminant that can be removed by venting alone. Thus, venting should be considered as only one of the processes used to clean up a contaminated site.

A realistic vapor flowrate must now be determined. Equation (4.5)[25] below can be used to estimate a range of flowrates as a function of soil permeability ($k$). For sandy soils the permeability varies from 1 to 10 darcy (approximately $10^{-8}$ cm$^2$).

$$1 \text{ darcy} = \frac{\dfrac{(1 \text{ centipoise})(1 \text{ cm}^3/\text{s})}{1 \text{ cm}^2}}{1 \text{ atm/cm}} = 0.987(\mu m)^2 = 0.987 \times 10^{-8} \text{ cm}^2,$$

$$\frac{Q}{H} = \frac{\pi \left(\dfrac{k}{\mu}\right)(P_w)\left[1 - \left(\dfrac{P_{atm}}{P_w}\right)^2\right]}{\ln \dfrac{R_w}{R_i}} \tag{4.5}$$

$$= \frac{\pi \left(\dfrac{10^{-8}}{1.8 \times 10^{-4}}\right)(0.9)(1.01 \times 10^6)\left[1 - \left(\dfrac{1}{0.9}\right)^2\right]}{\ln \dfrac{0.0508}{12}}$$

$$= 6.810 \text{ to } 68.10 \text{ cm}^2/\text{s},$$

$$Q = 6.810(60.96) = 415 \text{ to } 4150 \text{ cm}^3/\text{s}$$
$$= 24.9 \text{ to } 249 \text{ liters/min},$$

where

$k$ = soil permeability to air flow, darcy,

$\mu$ = viscosity of air, $1.8 \times 10^{-4}$ g/cm s,

$H$ = height of well screen interval, 60.96 cm (2 ft),

$P_w$ = absolute pressure at extraction well, typically 0.9 to 0.95 atm,

$P_{atm} \approx 1.01 \times 10^6$ g/cm-s$^2$ or 1 atm,

$R_w$ = radius of extraction well, assume 5.08 cm

$R_i$ = radius of influence of vapor extraction well, cm.

$R_i$ can be assumed to be 12 m (40 ft) without significant loss of accuracy.[25] The estimated removal rate $R_{est}$ can now be determined as follows:

$$R_{est} = C_{est}Q \tag{4.6}$$

$$= 220 \text{ mg/liter}(24.9 \text{ liters/min})(1 \text{ kg}/10^6 \text{ mg})(1440 \text{ min/day})$$

$$= 7.9 \text{ to } 79 \text{ kg/day,}$$

where $Q$ is the vapor flowrate, l/min. Since the acceptable removal rate, $R_{acc} = 1.922$ kg/day, the range for the actual removal rate appears to be more than adequate.

States have established target cleanup goals for contaminated groundwater. A maximum allowable residual of 100 ppm total petroleum hydrocarbon has been adopted by many states. For this example, this corresponds to removal of 90 percent of the original 1000 ppm contamination level. From Figure 4.6, which plots the maximum predicted removal rate for weathered gasoline against the volume of air drawn through the contaminated zone per unit mass of contaminant, it can be seen that approximately 100 liters of vapor per gram of contaminant must pass through the soil to remove 90 percent of the contaminant. Recalling that $m_c = 345.88$ kg, the quantity of vapor that must pass through the soil is 345,880 g(100 liters/g) = $345.88 \times 10^5$ liters. Thus the flowrate is

$$Q = \frac{(345.88 \times 10^5 \text{ l})(1 m^3/1000 \text{ l})(35.3 \text{ ft}^3/ m^3)}{180 \text{ days } (1\text{day}/1440 \text{ m})} = 4.71 \text{ cfm.}$$

Because this value is within the level of attainable flowrates, cleanup can be accomplished within the planned six-month period.

**Figure 4.6** Maximum removal rates for weathered gasoline (used by permission from Johnson, P.C., Stanley, C.C., Kemblowski, M.W., Byers, D.L., and Colthart, J.D., "A Practical Approach to the Design, Operation and Monitoring of In Situ Soil-Venting Systems," *Ground Water Monitoring Review*, 10(2), 159–178, 1990).

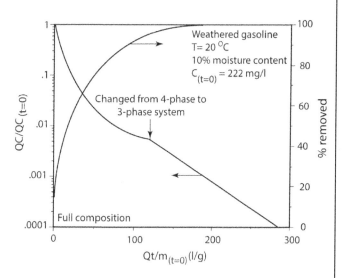

Hydrodynamic controls are often employed to isolate a plume of contamination from the normal groundwater flow regime to prevent the plume from moving into a well field, another aquifer, or surface water. Isolation of the contaminated plume is accomplished when uncontaminated groundwater is circulated around the plume in the opposite direction to the natural groundwater flow. The circulated zone creates a groundwater (hydrodynamic) barrier around the plume. Groundwater upgradient of the plume will flow around the circulated zone while groundwater downgradient will be essentially unaffected.[22]

Withdrawal and treatment of contaminated groundwater is one of the most often used processes for cleaning up aquifers. The type of contamination and the cost associated with treatment determine what specific treatment technology will be used. There are three broad types of treatment possibilities: physical, which includes adsorption, density separation, filtration, reverse osmosis, air and steam stripping, and incineration; chemical, which includes precipitation, oxidation/reduction, ion exchange, neutralization, and wet air oxidation; and biological, which includes activated sludge, aerated surface impoundments, land treatment, anaerobic digestion, trickling filters, and rotating biological discs.[22]

The wide variety of devices that may be used to remove contaminants from water by transfer to air include diffused aeration, coke tray aerators, cross-flow towers, and countercurrent packed towers. In the packed tower, water containing the contaminant flows down through the packing while the air flows upward and is exhausted through the top to the atmosphere or to emission control devices (Figure 4.7). The kinetic theory of gases states that molecules of dissolved gases can readily move between the gas and liquid phases. Consequently, if water contains a volatile contaminant in excess of its equilibrium level, the contaminant will move from the liquid phase (water) to the gas phase (air) until equilibrium is reached. If the air in contact with the water is continuously replenished with fresh, contaminant-free air, eventually all of the contaminant will be removed from the solution. This is the basic operating principal of air-stripping processes. The objective in the design of air-stripping equipment is to maximize the rate of this mass transfer process at minimum cost.

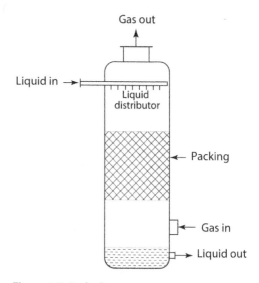

**Figure 4.7** Packed tower.

Another method of cleaning up contamination in groundwater has been proposed by researchers at the Texas Agricultural Experiment Station, College Station, TX. In this approach, aquatic bacteria with a taste for toxic waste are used to break down one of the most common pollutants, TCE. TCE is an industrial solvent used in degreasing operations as well as in dry cleaning procedures, refrigerants, fumigants, and most septic tank cleaning fluids. As indicated previously, TCE is a contaminant in many public drinking water supplies and has been detected in samples from a large number of water wells across the United States. The proposed method,

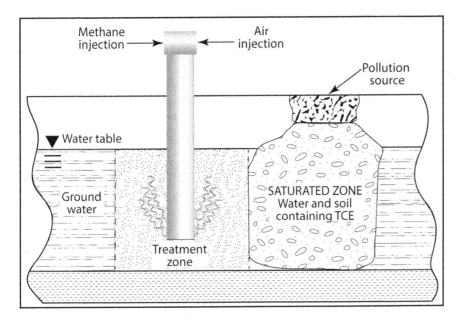

**Figure 4.8** Cleaning groundwater.

which is depicted in Figure 4.8, uses a treatment zone to clean the water like a filter. The filtering bacteria used in the process must be activated by using oxygen and methane as nutrients, which cause the bacteria to grow and to produce the enzyme that breaks down TCE. Bioremediation has some advantages over air stripping and other similar treatments, since it destroys the chemical pollutant. With bioremediation, TCE is converted into microbial cells and carbon dioxide, whereas with air stripping and some other techniques, the pollutant is merely removed from the water and added to the air, which may require subsequent scrubbing processes before release.[26]

### 4.5.1 Design of a Packed Tower[27]

The basic design of packed towers has been well documented in the literature.[28,29] The packed tower offers a cost-effective solution for many scrubbing (change in composition of the air) and stripping (change in composition of the water) operations. The packed tower facilitates high mass transfer rates by incorporating a packing material with a large surface area for the water and air to interact and a configuration that enhances turbulence in the water stream to constantly expose fresh water surfaces to the air. Packings should also have large void areas to minimize pressure drop through the tower. Large-size packing materials are less expensive per unit volume and allow the use of smaller diameter towers for a given water flowrate. Packing materials

---

26 "New method devised to clean up contamination in groundwater," *Lubbock Avalanche Journal*, July 1, 1990.
27 Adapted from Ertas, A., and Jones, J.C., *The Engineering Design Process*, John Wiley & Sons, Inc., 1996.
28 Eckert, J.S., "Design techniques for sizing packed towers," *Chemical Engineering Progress*, 57(9), 54–58, 1961.
29 Kavanaugh, M.C., and Trussel, R.R., "Design of aeration towers to strip volatile contaminant from drinking water," *Journal of American Water Works Association*, 72(12), 684–692, 1980.

of smaller size provide greater mass transfer coefficients and, therefore, allow the use of shorter towers. When the degree of contaminant removal is small, larger packing use is appropriate, whereas for high removal rates smaller size packing may be more economical. As a general rule, the ratio of tower diameter to packing size should be at least 15 : 1 to avoid poor liquid distribution caused by the tendency of the water flowrate to be greater near the walls of the tower.[30] Other desirable features of packing materials include low weight, being chemically inert to fluids being processed, structural strength for easy handling, and low cost.

For towers in which the mass transfer is controlled by the liquid phase resistance, which is the case for most water treatment air stripping operations, the coefficient of interest is $K_L$, the overall liquid phase mass transfer coefficient. After selecting a suitable packing, the product $K_L a$ can be approximated as

$$K_L a = \frac{K_G a H_A}{C_o} \tag{4.7}$$

where $H_A$ is Henry's constant for solute A, and $C_o$ is the molar density of water which is 55.6 kmol/m$^3$. In equation (4.7), $K_G a$ is the product of the overall gas phase mass transfer coefficient and the interfacial surface area per volume of packing over which the mass transfer occurs.

### 4.5.1.1 Sizing the Tower Diameter

The first step in determining the diameter of a packed tower is to assume values of the stripping factor $R$ and to calculate the corresponding gas to liquid flowrate ratio $G/L$ from

$$G/L = \frac{R P_T}{H}, \tag{4.8}$$

where

$G$ = gas flowrate (kmol/m$^2$s),

$L$ = liquid flowrate (kmol/m$^2$s),

$P_T$ = the ambient pressure (atm),

$H$ = Henry's constant for the contaminant (atm),

$R$ = stripping factor (dimensionless).

The flowrate ratio can be converted to SI units (kg/m$^2$s) by multiplying the mol ratio by the gas and liquid molecular weights; thus,

$$\frac{G'}{L'} = \left(\frac{G}{L}\right)\left(\frac{MW_{air}}{MW_{water}}\right). \tag{4.9}$$

After the $G'/L'$ ratio has been determined, Figure 4.9 can be used to establish values for the gas and liquid flowrates. To obtain the value for the abscissa in Figure 4.9, evaluate the expression

$$\frac{L'}{G'}\left(\frac{\rho_G}{\rho_L - \rho_G}\right)^{1/2}, \tag{4.10}$$

---

30 Treybal, R.E., *Mass Transfer Operations*, McGraw-Hill, New York, 1980.

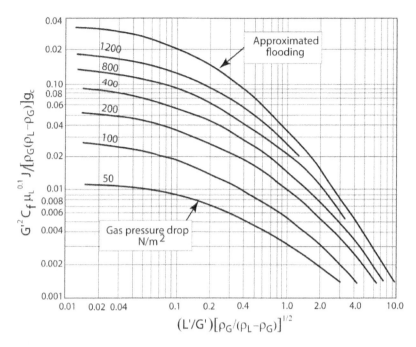

**Figure 4.9** Flooding and pressure drop in random-packed towers (from Treybal, R.E., *Mass Transfer Operations*, McGraw-Hill, New York, 1980, reproduced with permission of McGraw-Hill Education).

where

$\rho_G$ = density of the gas,

$\rho_L$ = density of the liquid.

For a given flowrate the pressure drop of the air rising countercurrent to the water in a packed tower increases approximately in proportion to the square of the gas velocity. At high gas flowrates, entrainment of the liquid may occur, causing a sudden increase in the gas pressure drop. When the liquid flowrate increases at a fixed gas flowrate, the tower may fill with liquid, a condition known as flooding. The tower should be designed to operate at conditions well below flooding. According to Treybal,[30] most stripping towers are designed for gas pressure drops of 200–400 N/m$^2$ per meter of packing depth (0.25–0.50 in. H$_2$O/ft). From the calculated values for the abscissa and a selected pressure drop, values for the ordinate from Figure 4.9 can now be obtained and the following equation can be solved for values of $G'$:

$$G' = \left[ \frac{\rho_G(\rho_L - \rho_G)g_c}{C_f \mu_L^{0.1} J} \times \text{ordinate value} \right]^{1/2}, \tag{4.11}$$

where

$\mu_L$ = liquid viscosity,

$g_c$ = gravity correction factor (for SI units $g_c = 1$, for English units $g_c = 4.18 \times 10^8$),

$C_f$ = packing factor,

$J$ = conversion factor (for SI units $J = 1.0$, for English units $J = 1.502$).

The water flowrate $L'$ can now be determined by using equation (4.10), and the abscissa from Figure 4.9. The tower diameter $\phi$ can then be obtained by solving

$$\phi = \left[\frac{4}{\pi}\frac{Q_L\rho_L}{L'}\right]^{1/2}, \tag{4.12}$$

where $Q_L$ = volumetric liquid flowrate.

### 4.5.1.2 Determining the Height of the Tower

The height of the packing $Z$ required to achieve the desired removal of contaminant is the product of the height of a transfer unit (HTU), and the number of transfer units (NTU):

$$Z = (HTU)(NTU). \tag{4.13}$$

The HTU is inversely proportional to the product of the overall liquid phase mass transfer coefficient and the interfacial area $K_L a$ and is a function of the efficiency of mass transfer from water to air:

$$HTU = \frac{L}{(K_L a)(C_o)}. \tag{4.14}$$

The NTU provides a measure of the difficulty of removing the contaminant from the water and can be calculated as

$$NTU = \frac{R}{R-1}\ln\frac{\frac{X_{in}}{X_{out}}(R-1)+1}{R}, \tag{4.15}$$

where $X_{in}$ and $X_{out}$ are the concentrations of the compound to be removed (percent).

## Example 4.1

Assume, after chlorination, a 50-liter/s surface water supply contains 200 $\mu$g/liter of chloroform. The effluent concentration must be reduced to 40 $\mu$g/liter. Supposing the water and air temperature to be 20°C, determine the dimensions of the packed tower. Assume the following:

- stripping factor, $R = 3$;
- allowable pressure drop in tower is 300 N/m$^2$ per meter.

## Solution

Selection of the packing material: For many years, the most popular packings were Raschig rings and Berl saddles, but these have been almost entirely replaced by Pall rings and Super Intalox made from plastic or ceramics. For this design problem, 2 in. Super Intalox is selected because of its low cost and high mass transfer efficiency. For Super Intalox the packing factor, $C_f = 21$.[31]

---

31 Eckert, J.S., "How tower packings behave," *Chemical Engineering*, 82(70), April 4, 1975.

1) Compute the number of transfer units NTU for $R = 3$.
   The desired removal efficiency $\eta_e$ is

$$\eta_e = 1 - \frac{40}{200} = 80\%.$$

Then

$$\text{NTU} = \frac{R}{R-1} \ln \left[ \frac{\frac{X_{\text{in}}}{X_{\text{out}}}(R-1)+1}{R} \right]$$

$$= \frac{3}{3-1} \ln \left[ \frac{\frac{200}{40}(3-1)+1}{3} \right] = 1.95$$

2) Determine the allowable airflow at 20°C and $P_T = 1$ atm for 300 N/m$^2$ per meter gas pressure drop. Henry's constant for CHCl$_3$ is $H = 170$ atm.[29] Hence,

$$\frac{H}{P_T} = \frac{170}{1} = 170.$$

Knowing

$$\rho_{\text{air}} = 1.205 \text{ kg/m}^3,$$
$$\text{MW}_{\text{air}} = 28.8 \text{ kg/kmol},$$
$$\rho_{\text{water}} = 998 \text{ kg/m}^3,$$
$$\text{MW}_{\text{water}} = 18.02 \text{ kg/kmol},$$

we have

$$R = \frac{H}{P_T}\frac{G}{L} = 170\frac{G}{L}$$

or

$$R = 170\frac{28.8}{18.02}\frac{G'}{L'} = 272\frac{G'}{L'}.$$

Then, for $R = 3$, we find $G'/L' = 0.011$. Figure 4.9 can be used to obtain the value of $G'$ and $L'$. The abscissa of Figure 4.9 is

$$\frac{L'}{G'}\left(\frac{\rho_G}{\rho_L - \rho_G}\right)^{1/2} = 90.56\left(\frac{1.205}{998 - 1.205}\right)^{1/2} = 3.149.$$

For 3.149 and 300 N/m$^2$ per meter pressure drop, from Figure 4.9, the ordinate value

$$\frac{G'^2 C_f \mu_L^{0.1} J}{\rho_G(\rho_L - \rho_G)g_c} = 0.0042$$

is obtained. For SI units $J = 1$, and $g_c = 1$, and $\mu_L = 0.001$ kg/m s. The airflow rate is found by using equation (4.11):

$$G' = \left[\frac{0.0042 \times 1.205 \times (998 - 1.205)}{21 \times (0.001)^{0.1}}\right]^{1/2} = 0.69 \text{ kg/m}^2 \text{ s.}$$

Hence,

$$L' = 90.56G' = 90.56 \times 0.69 = 62.70 \text{ kg/m}^2 \text{ s}$$

3) Determine the tower diameter $\phi$. Using equation (4.12), the tower diameter is

$$\phi = \left[\frac{4}{\pi} \frac{Q_L \rho_L}{L'}\right]^{1/2}$$

$$= \left[\frac{4}{\pi} \frac{0.050 \times 998}{62.70}\right]^{1/2} = 1.013 \text{ m.}$$

4) Compute the height of the packed tower.
   Since the contaminant is being removed from the liquid (water), to determine HTU, the liquid phase mass transfer coefficient $K_L$ controls the process. Estimate $K_L a$ using equation (4.7):

$$K_L a = \frac{K_G a H_A}{C_0} = \frac{0.0094(170)}{55.6} = 0.029 \text{ s}^{-1},$$

$$L = \frac{L'}{MW_{water}} = \frac{62.77}{18.02} = 3.48 \text{ kmol/m}^2 \text{ s.}$$

Note that $K_G a$ is equal to 0.0094 for 2 in. Super Intalox rings.[29] Then

$$\text{HTU} = \frac{L}{K_L a C_o} = \frac{3.48}{0.029 \times 55.6} = 2.16 \text{ m.}$$

5) Compute the height of tower.

$$Z = (\text{HTU})(\text{NTU}) = (2.16)(1.95) = 4.2 \text{ m.}$$

6) Select the appropriate tower dimensions.
   A nominal size tower diameter and height that is readily available should be selected. A reasonable safety factor should also be included. For this application a reasonable choice might be $z = 4.88$ m (16.0 ft) and $\phi = 1.22$ m (4.0 ft). To optimize the tower design, an iterative procedure for various values of stripping factor $R$ is recommended.

### 4.5.2 Importance of Sustainable Air Quality

One of the Earth's most critical natural resources is its atmosphere. Air pollution and concern about air quality are not new. Concern began in the thirteenth century when coal was first used in London. From the middle of the nineteenth century, the atmosphere of the major British cities was constantly polluted by coal smoke in winter.[32]

Breathing is essential for health because it is necessary for life, and so air quality is a fundamental aspect of sustainability. The quality of the layer of air that surrounds the Earth has been degraded to the extent that warnings are issued in many cities when contamination reaches hazardous levels. Joggers are warned about jogging at times of the day when smog levels are elevated, and many metropolitan areas in the world have enacted motor vehicle and other industrial emission controls in an effort to lower air pollution levels. Air pollution in Mexico City is a continuing concern for residents, health experts, and environmentalists. In Mexico City, millions of people live in an atmosphere so foggy that the Sun is obscured, so poisonous that school is sometimes delayed until late morning when the air clears. Air pollution can be prevented by lowering emissions levels from motor vehicles, and changing to more environmentally friendly commercial products. Factories that produce hazardous air pollution should use "scrubbers" or other procedures on their smokestacks to eliminate contaminants before they enter the air outside the plant.

The surrounding air quality, both outdoors and indoors, is crucial to the public's health, productivity, and ability to enjoy life. It is essential to keep the quality of the outdoor air since all life forms depend on it. There is now an increased awareness of the importance of the quality of indoor air because it is dependent on the outdoor air. It is important to note that a recent study reveals that Americans spend 90% of their lives indoors.

### 4.5.2.1 Air Pollution and Toxic Chemical Exposure[33]

Air pollution has been a major concern in the United States since the mid-1950s. This concern was translated into law with the enactment of the 1970 Clean Air Act, and through subsequent amendments to that act in 1977 and 1990. Most states have developed regulatory, enforcement, and administrative programs to reduce air pollution. These include controls on pollution from industries and utilities as well as motor vehicle emission inspection programs. These state programs have been augmented by federal regulations and control measures for reducing emissions from new motor vehicles and certain industrial sources. The US EPA is responsible for coordinating and approving most state air pollution plans and programs. The EPA also provides substantial technical assistance to states which ranges from information on new air pollution control technology to studies of the health effects of air pollution. The overall goal of these programs is to ensure that national air quality standards are achieved and maintained for major air pollutants, including lead, carbon monoxide, ozone, nitrogen oxides, sulfur dioxide, and particulates.

Much public concern has been focused on the harmful effects that acid rain has on fish and other wildlife, lakes, forests, crops, and on human-made objects such as buildings. Certain

---

32 Sustainable Environment, http://www.sustainable-environment.org.uk/Environment/Air_ Pollution.php, accessed May 27, 2017.
33 From Ertas, A., and Jones, J.C., *The Engineering Design Process*, John Wiley & Sons, Inc., 1996.

aspects of the acid rain phenomenon are generally accepted by the scientific community, but many other causes and effects are not well understood. The geographic range of damage from acid rain, the rate at which acidification takes place, and the combination of pollutants that are involved are uncertain. For example, most attention has been focused on the contribution of sulfur emissions to acid deposition, but other pollutants (including oxides of nitrogen) are also contributors.

Another major air quality issue involves the transport of ozone and its precursors. This chiefly involves a process whereby emissions of VOCs are transported over long distances from the source of the release. This phenomenon has complicated the development of control strategies for some areas of the USA and has also compounded the difficulties of achieving air quality standards for ozone. This is a particularly difficult problem for some of the larger urban areas of the country.[34]

Hazardous pollutants also pose serious but primarily localized health problems. This includes hazards to the general public from exposure to refuse and other discarded materials from chemical, metallurgical, agricultural, and other industrial processes as well as exposure in the workplace. The EPA, the Occupational Safety and Health Administration (OSHA), and the Consumer Product Safety Commission share responsibility for issuing and enforcing regulations in this area based on research performed by other government agencies, including the National Institute for Occupational Safety and Health, the National Institute of Environmental Health Sciences, the National Cancer Institute, and the Food and Drug Administration. Although it is probably impossible to determine the magnitude of the toxic chemical risk, it is recognized as a major and growing public health problem. Residents of dwellings in close proximity to hazardous waste landfills are in danger of exposure by direct contact or by inhalation of dusts, fumes, or vapors. Employees working in an environment where chemical and other industrial processes are ongoing can be exposed by accident, through ignorance or carelessness. The difficulty in defining the magnitude of the problem is exacerbated by the long latent period between exposure and the onset of chemically induced disease as well as the relative newness of the science of environmental toxicology.[35]

OSHA has published a document entitled "Air Contaminants – Permissible Exposure Limits" that provides information and recommendations relating to permissible exposure limits (PELs), chemical and physical properties, health hazard information, respiratory protection, and personnel protection and sanitation practices for 600 substances.[36] Substance limits in this OSHA document are provided for direct exposures involving chemical absorption through the skin as well as respiratory exposure whereby chemicals enter the body through the breathing process. Compliance with these limits in the workplace will protect workers against a wide variety of health effects that could cause material impairment of health or functional capacity.

---

34 "Trends in the Quality of the Nation's Air," US Environmental Protection Agency, January 1985.
35 "Health Effects of Toxic Pollution: A Report from the Surgeon General," US Department of Health and Human Services, Serial No. 96-15, August 1980.
36 "Air Contaminants – Permissible Exposure Limits," Title 29 Code of Federal Regulations Part 1910.1000, 1989.

### 4.5.2.2  Particulate Contamination[37]

Particulate contamination contributes to both occupational health and nuisance problems in the workplace. Accordingly, OSHA has promulgated allowable limits that are applicable to various chemical, mineral, and grain dusts for both health concerns and as workplace irritants. These requirements are specified in a US Department of Labor report on air contaminants.[38] Certain dusts also pose explosive hazards and must be controlled for this reason as well. The US Department of Agriculture has published several documents relating to the problems and prevention of grain dust explosions. OSHA PELs for grain dust are less than the minimum explosive concentrations (e.g., the PEL for grain dust is given as 10 mg/m$^3$, whereas the minimum explosive concentration for corn dust is 40 g/m$^3$).

The rate of dust generation during the handling of a material is a function of the material, the material handling rate, and the relative air velocity near the surface of the material. Dust particles less than 125 $\mu$m have surface irregularities that engender Van der Waals surface charges that cause the particles to adhere to each other and to other surfaces.[39] When minimum entrainment velocities are reached, these bonds are broken and the dust is released into the air. If material-handling rates and relative air velocity near the material surface are both held constant, the dust generated by the system will remain essentially constant for each operation.

Dust cloud permanence is related to the settling velocities of the particles, air velocity, turbulence, and flow path. The settling velocity of dust is the terminal velocity that the particle reaches when its drag force is equal to the force of gravity acting on the particle. The terminal settling velocity of a particle as derived by Stokes is

$$V_{ts} = \sqrt{\frac{4\rho_p d_p g}{3 C_D \rho_g}},$$  (4.16)

where

$\rho_p$ = density of the particle,

$\rho_g$ = density of the gas (air),

$C_D$ = drag coefficient for the particle,

$d_p$ = aerodynamic diameter of the particle,

the particle drag coefficient $C_D$ being given by

$$C_D = \frac{4}{3} \frac{\rho_p d_p g}{\rho_g V_{ts}^2}.$$  (4.17)

The solution of $V_{ts}$ from equations (4.16) and (4.17) requires an iterative method. An alternative approach is to introduce the Reynolds number squared:

$$R_e^2 = \left( \frac{\rho_g V_{ts} d_p}{\eta} \right)^2.$$  (4.18)

37  From Ertas, A., and Jones, J.C., *The Engineering Design Process*, John Wiley & Sons, Inc., 1996.
38  "Contaminants – Permissible Exposure Limits," OSHA 3112, 1989.
39  "Review of Literature Related to Engineering Aspects of Grain Dust Explosions, US Department of Agriculture Publication 1375, 1979.

To eliminate the velocity-squared term in equation (4.17), use

$$C_D R_e^2 = \frac{4}{3} \frac{d_p^3 \rho_p \rho_g g}{\eta^2},$$  (4.19)

where $\eta$ is the viscosity of the air.

Now $C_D R_e^2$ can be calculated directly. Equation (4.16) is valid for spherical particles and particles with Reynolds numbers less than 1.0. For small particles (size less than 125 $\mu$m) the Reynolds number is in the range $0.05 < R_e < 4.0$. For this range, an empirical equation was developed by Davies,[40] which permits the direct calculation of $V_{ts}^*$ using $C_D R_e^2$:

$$V_{ts}^* = \frac{\eta}{\rho_g d_p} \left[ \frac{C_D R_e^2}{24} - 2.3363 \times 10^{-4} (C_D R_e^2)^2 \right.$$

$$\left. + 2.0154 \times 10^{-6} (C_d R_e^2)^3 - 6.9105 \times 10^{-9} (C_D R_e^2)^4 \right].$$  (4.20)

The aerodynamic diameter of the particle $d_p$ can be calculated using

$$d_p = d_e \left( \frac{\rho_p}{\rho_0 \chi} \right)^{1/2},$$  (4.21)

where

$$\rho_0 = \text{unit density } (1.0 \text{ g/cm}^3),$$

$$\chi = \text{shape correction factor.}$$

In equation (4.21), $d_e$ is the equivalent diameter of a sphere of the same volume as an irregular particle, and it can be experimentally determined. The shape parameter $\chi$ can be obtained from Table 4.1.[41] The following example demonstrates how the dust concentration in a work area can be determined by applying equation (4.20) to ascertain the quantity of the dust generated that settles out of the air.

Table 4.1 Dynamic shape factors.

| Shape | Shape factor ($\chi$) |
| --- | --- |
| Geometric shapes | |
| *Sphere* | 1.00 |
| *Cube* | 1.08 |
| Sand | 1.57 |

Adapted from Davis, C.N., "Particle fluid interaction," *Journal of Aerosol Science*, 10, 477–513.

40 Davies, C.N., "Particle fluid interaction," *Journal of Aerosol Sciences*, 10, 477–513, 1979.
41 Hinds, W.C., *Aerosol Technology*, John Wiley & Sons, New York, 1982.

## ☐ Example 4.2

Lubbock State School, a school for mentally and physically handicapped students, contacted the Texas Tech University Mechanical Engineering Department regarding a problem it had encountered with the generation of dust during the operation of a gravel bagging system used in vocational training. The gravel bagging system raises gravel from an underground gravel pit through use of a bucket/belt elevator system to a hopper that releases measured amounts of gravel into bags. Large quantities of dust were being produced by the system, raising the dust concentration in the room to a level intolerable to clients working in the area and thought to be in excess of the OSHA limit for nuisance dust in the workplace. The objective of this project was to design and build dust control system.[a]

### Solution

On-site inspection of the bagging operation by a student group interested in the project led to the conclusion that the dust generation was occurring at four different gravel transfer points (Figure 4.10). Standard practices for controlling dust include wetting, ventilation, and the use

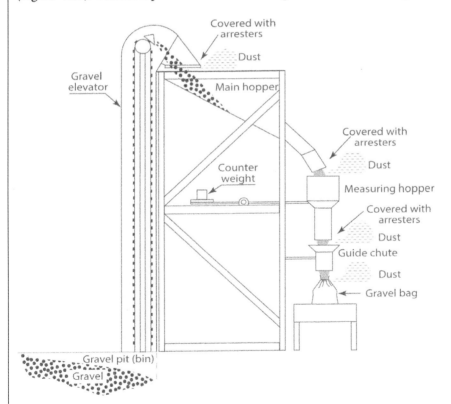

**Figure 4.10** A gravel bagging system (Ertas, A., and Jones, J.C., *The Engineering Design Process*, John Wiley & Sons, Inc., 1996, reproduced with permission of John Wiley & Sons, Inc.).

of fabric dust arresters. An air inlet and exhaust fan had been previously installed in the room to remove the dust but had proved to be ineffective. Based on the on-site inspection the student group decided to undertake the project. The objective established was to develop a dust suppression system to reduce the dust concentration below OSHA nuisance dust requirements.

The concentration of dust in the room is strongly influenced by the terminal settling velocity. To calculate the terminal settling velocity $V_{ts}^*$ for a range of particle sizes, the following parametric values were used:[40]

$$\rho_g = 1.054 \text{ g/cm}^3,$$
$$\rho_p = 2.7 \text{ g/cm}^3,$$
$$\chi = 1.57,$$
$$g = 981 \text{ cm/s}^2,$$
$$\eta = 1.81 \times 10^{-4} \text{ g/cm s}.$$

Assuming an equivalent particle diameter, $d_e = 1$ $\mu$m, the aerodynamic diameter of the particle is

$$d_p = d_e \left( \frac{\rho_p}{\rho_0 \chi} \right)^{1/2}$$

$$= 1 \left( \frac{2.7}{1 \times 1.57} \right)^{1/2} = 1.311 \ \mu\text{m} = 1.311 \times 10^{-4} \text{ cm},$$

and from equation (4.19),

$$C_D R_e^2 = \frac{4}{3} \frac{d_p^3 \rho_p \rho_g g}{\eta^2} = \frac{4}{3} \frac{(1.311 \times 10^{-4})^3 (2.7)(1.054 \times 10^{-3})(981)}{(1.81 \times 10^{-4})^2}$$

$$= 2.560136 \times 10^{-4}.$$

Using $C_D R_e^2 = 2.560136 \times 10^{-4}$ in equation (4.20) gives a settling velocity of 0.014 cm/s. Similar calculations can be performed for other equivalent particle diameter to obtain Table 4.2 for the gravel dust at Lubbock State School.

The dust concentration in the bagging room can be determined by assuming that the dust concentration at any point in time in the room is equal to the dust concentration exhausted out of the room through the ventilation system. This assumption allows the concentration of dust in the room to be calculated knowing the total effective mass of dust suspended in the air and the total effective volume of air. The dust concentration is

$$C = \frac{M_e}{V_e}, \tag{4.22}$$

where

$$C = \text{dust concentration, mg/m}^3,$$
$$M_e = \text{effective mass of dust,}$$
$$V_e = \text{effective volume of air.}$$

**Table 4.2** Settling velocity of gravel dust.

| Equivalent particle diameter $d_e$ ($\mu$m) | Aerodynamic particle diameter $d_p$ ($\mu$m) | Davies settling velocity $V_{ts}^*$ (cm/s) |
| --- | --- | --- |
| 1 | 1.31 | 0.014 |
| 2 | 2.62 | 0.056 |
| 3 | 3.93 | 0.126 |
| 4 | 5.25 | 0.233 |
| 5 | 6.56 | 0.349 |
| 6 | 7.87 | 0.503 |
| 7 | 9.18 | 0.684 |
| 8 | 10.49 | 0.893 |
| 9 | 11.80 | 1.130 |
| 10 | 13.11 | 1.395 |
| 15 | 19.67 | 3.137 |
| 20 | 26.23 | 5.524 |
| 25 | 32.78 | 8.540 |
| 30 | 39.34 | 12.112 |

Source: Ertas, A., and Jones, J.C., *The Engineering Design Process*, John Wiley & Sons, Inc., 1996, reproduced with permission of John Wiley & Sons, Inc.)

The total effective mass of dust $M_e$ in the room is the initial suspended dust mass $M_0$ in the room plus the dust generated during the operation of the bagging machine $M_{in}$ minus the dust that continuously settles out of the air $M_s$ during the time of operation; that is,

$$M_e = M_0 + M_{in} - M_s. \tag{4.23}$$

Because the dust generated in the room during time $t$ includes the dust exiting the room via the ventilation system, it can be formulated as

$$M_{in} = \int_0^t \dot{M}_{in} dt = \dot{M}_{in} t, \tag{4.24}$$

where $\dot{M}_{in}$ is the constant rate of mass of dust generated by the system.

The mass rate of dust generated was estimated by fractionating a sample of the gravel. An American Soil Testing Machine using soil-fractionating sieves was used. The mass distribution for dust particle sizes from 300 to 30 $\mu$m is shown in Figure 4.11. The mass of particles smaller than 30 $\mu$m was determined to be 1 gram for each gravel bag filled. Since the bag processing rate is 1 per minute, this gives a dust generation rate of 1 g/min for particles less than 30 $\mu$m in size. The vertical component of velocity for air moving through the bagging room was measured to be 10 cm/s. Since the settling velocities for particles 30 $\mu$m and larger in size is greater than the vertical air velocity component (see Table 4.2), particles in these size

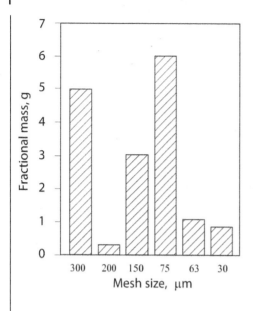

**Figure 4.11** Gravel size distribution (Ertas, A., and Jones, J.C., *The Engineering Design Process*, John Wiley & Sons, Inc., 1996, reproduced with permission of John Wiley & Sons, Inc).

ranges will tend to settle and not contribute to airborne dust contamination within the room. This assumes that personnel working in the bagging room will not be breathing air through which the larger particles are settling (e.g., when the dust generation source is at a level below the worker's head). The mass rate of dust generated, $M_{\text{in}}$, minus the mass rate of the dust that settled, $M_s$, was thus estimated to be equal to 1 g/min.

In equation (4.22), the total effective volume $V_e$ is the volume of the room plus the volume of air exiting the room during time $t$ and can be defined as

$$V_e = V_0 + \int_0^t \dot{Q}dt = V_0 + \dot{Q}t, \tag{4.25}$$

where $V_0$ is the volume of the room and $\dot{Q}$ is the ventilation flowrate. Equation (4.22) can be rewritten in the form

$$C = \frac{M_0 + \dot{M}_{\text{in}}t}{V_0 + \dot{Q}t}. \tag{4.26}$$

The air flowrate of the ventilation system was measured by using a hot-wire anemometer. The air flowrate in the ventilation system was found to be 56.6 m³/min. The volume of the bagging room was measured to be 142.7 m³. Knowing the mass rate of dust generated, $\dot{M}_{\text{in}}$, the ventilation flowrate, $\dot{Q}$, volume of the room, $V_0$, and the initial dust mass, $M_0$, the dust concentration inside the bagging room can now be estimated:

$$C = \frac{M_0 + \dot{M}_{\text{in}}t}{V_0 + \dot{Q}t}$$

$$= \frac{0 + (1)(4 \times 60)}{142.7 + (56.6)(4 \times 60)} = 0.0175 \text{ g/m}^3 = 17.5 \text{ mg/m}^3.$$

The permissible eight-hour exposure limit for inert or nuisance dust is 5 mg/m$^3$.[38] Particles less than 10 $\mu$m are considered to be respirable; thus, particles larger than this size are not hazardous to health.

The bagging machine has four hoppers that generate the dust. It was assumed that each hopper contributed an equal portion of the total dust produced by the gravel machine. With one hopper covered with a fabric dust arrestor, $M_{in}$ = 0.75 g/min, with two hoppers, $M_{in}$ = 0.5 g/min, and so on. Figure 4.12 shows the effect of eliminating individual hopper sources on the dust concentration in the room. It can be seen that reducing dust generation sources to one reduces the dust concentration below the PEL for nuisance dust.

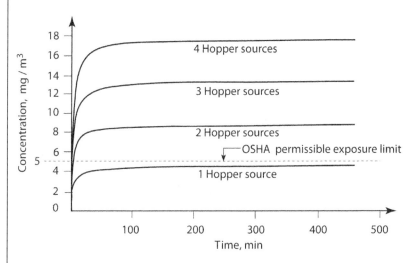

**Figure 4.12** Calculated dust concentration versus time for reduced hopper sources. (Ertas, A., and Jones, J.C., *The Engineering Design Process*, John Wiley & Sons, Inc., reproduced with permission of John Wiley & Sons, Inc.)

### Installation and Testing

The dust suppression method selected was to use fabric dust arrestors at three of the four dust generation points (see Figure 4.10) in the gravel bagging system. The fabric material selected was nylon-reinforced plastic, which is flexible, durable, and impermeable to dust. The main hopper was enclosed by using Velcro around the elevator discharge and a draw rope around the rim of the hopper. This enclosure included a large zipper to allow quick access to the main hopper in case of system clogging. The transfer point between the exit chute of the main hopper and the inlet of the measuring hopper was covered using Velcro hook and loop fasteners at the top and a draw rope at the bottom. The transfer point between the measuring hopper and the guide chute was also covered by using these fasteners at each end.

After installation the air quality in the room was tested using a cascade inertial impactor type sampler which simulates the deposition of particles within the human respiratory system. Air is drawn through the sampler at the rate of 1 cfm by a small vacuum pump. The flowrate can be adjusted by using a bleed-off valve. The flowrate at the beginning and end of each test was measured using a Rockwell gas meter. As air enters the sampler, it is drawn through a series of seven orifices and impaction filters. The incoming particles are deposited at each stage according to size and weight.

The normal bagging rate is one bag per minute, which corresponds to a rate of 600 bags per 10-hour day. To test the system, a sleeve was fitted to the guide chute. The sleeve guided the exiting gravel into a wheelbarrow, allowing multiple bagging operations to be simulated before dumping was necessary. During the simulation of the bagging operation, three air sample tests were conducted for periods of approximately one hour each. The sample time was recorded, and the preweighed collection filters were removed for subsequent gravimetric analysis after each test. Airborne particulate concentrations and size distributions were then determined. The concentration in the room was found to be $1.94$ mg/m$^3$, which is less than the OSHA PEL for nuisance dust ($5$ mg/m$^3$).

## Conclusion

The objectives of this project were satisfied by installation of the fabric dust arrestors. The dust arrestors require no maintenance, are easily removed, and are very durable. The dust concentration level in the bagging room was reduced to less than one half of the maximum level of nuisance dust recommended by OSHA. The operation of the bagging machine was only minimally affected by installation of the arrestors. The customer was pleased with the installation.

[a] Adapted from Ertas, A., and Jones, J.C., *The Engineering Design Process*, John Wiley & Sons, Inc., 1996.

### 4.5.3   Air Contamination from Hazardous Chemical Spills[42]

The sector downwind of the source of a toxic air contaminant release point is defined as the hazard corridor or vulnerability zone. The quantity and rate at which the chemical will actually become airborne depend upon:

1) Total quantity released or spilled
2) Physical state (solid, liquid, gas)
3) Conditions (e.g. temperature and pressure) under which the chemical is stored or handled.

Gases typically become airborne more readily than liquids. Liquids generally become airborne by evaporation. The rate at which liquids become airborne (the rate of vaporization) depends on their vapor pressure, molecular weight, temperature relative to ambient, the surface area of the spill, and the wind speed at the time of the spill. A spilled liquid with a higher vapor pressure will become airborne through evaporation more rapidly than a liquid with a low vapor pressure at the same temperature. Also, a liquid will evaporate faster if the surface area of the spill is increased, if the liquid has a temperature higher than ambient, and if it is exposed to greater wind speeds.

42 From Ertas, A., and Jones, J.C., *The Engineering Design Process*, John Wiley & Sons, Inc., 1996.

Hazardous liquid vapor, released into the atmosphere from near ground level, will expand and mix with the air. As the vapor is carried away by the wind, its density decreases as mixing occurs in a process referred to as turbulent or eddy diffusion. The rate of this diffusion is dependent on complex motions of the air, in both the horizontal and vertical planes. High wind speeds produce a high dilution factor, since the volume of air into which the vapor is introduced is large. A fluctuating wind direction will spread the vapor horizontally about the axis of the mean wind direction. The vapor will also be diffused in the vertical direction, and this is related to the thermal structure of the lowest part of the atmosphere. Terrain, trees, buildings, and diurnal considerations all affect these motions, making accurate predictions difficult. For these reasons, predictions are often made in statistical terms. Figure 4.13 shows the movement of a typical airborne plume following release. Experimental evidence shows that three meteorological parameters need to be considered in determining the rate of atmospheric dispersion of small toxic chemical spills.

1) The mean wind speed $u$ as an indicator of the downwind plume travel from a continuous source
2) The standard deviation $\sigma_y$ in the cross-wind direction of the plume which is a function of wind direction change and is an indicator of the lateral rate of mixing

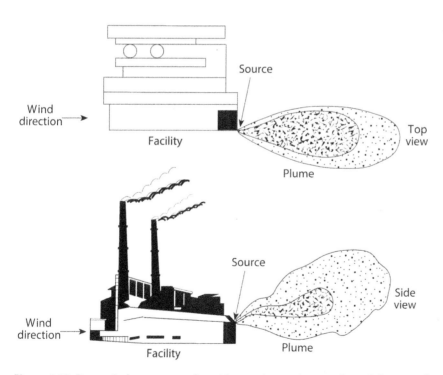

**Figure 4.13** Downwind movement of an airborne chemical vapor plume following release (From "Technical Guidance for Hazards Analysis," EPA, December 1987).

3) The standard deviation $\sigma_z$ of the plume concentration in the vertical direction which is related to the lower atmospheric temperature gradient $\Delta T$, and is an indicator of the vertical rate of mixing.

For decreasing temperatures with increasing altitude, the atmosphere is considered unstable, and for increasing temperatures with increasing altitude, the atmosphere is considered stable. An unstable atmosphere is characterized by significant vertical and horizontal mixing and large contaminant concentration reduction, whereas a stable atmosphere implies little mixing and little reduction in contaminant concentration. To deal with the wide variations in atmospheric conditions meteorologists have introduced stability classes which are used to identify atmospheric conditions. Table 4.3 depicts the equations that have been developed for the above mentioned diffusion deviations based on various atmospheric conditions which are classified from A (very unstable) to F (very stable).[43]

The equations for estimating the extent of the vulnerable zone following the release of a hazardous chemical are given below.[44] The calculations are based on applications of the dispersion model described in Public Health Service Publication No. 999-AP-26, *Workbook of Atmospheric Dispersion Estimates*. Estimates of dispersion distribution parameters are from Department of

**Table 4.3** Formulas recommended by Briggs for $\sigma_y(d)$ and $\sigma_z(d)$ ($10^2 < d < 10^4$ m).

| Pasquill stability type | $\sigma_y$, m | $\sigma_z$, m |
|---|---|---|
| | **Open-country conditions** | |
| A | $0.22d(1 + 0.0001d)^{-1/2}$ | $0.20d$ |
| B | $0.16d(1 + 0.0001d)^{-1/2}$ | $0.12d$ |
| C | $0.11d(1 + 0.0001d)^{-1/2}$ | $0.08d(1 + 0.0002d)^{-1/2}$ |
| D | $0.08d(1 + 0.0001d)^{-1/2}$ | $0.06d(1 + 0.0015d)^{-1/2}$ |
| E | $0.06d(1 + 0.0001d)^{-1/2}$ | $0.03d(1 + 0.0003d)^{-1}$ |
| F | $0.04d(1 + 0.0001d)^{-1/2}$ | $0.016d(1 + 0.0003d)^{-1}$ |
| | **Urban conditions** | |
| A–B | $0.32d(1 + 0.0004d)^{-1/2}$ | $0.24d(1 + 0.001d)^{1/2}$ |
| C | $0.22d(1 + 0.0004d)^{-1/2}$ | $0.20d$ |
| D | $0.16d(1 + 0.0004d)^{-1/2}$ | $0.14d(1 + 0.0003d)^{-1/2}$ |
| E–F | $0.11d(1 + 0.0004d)^{-1/2}$ | $0.08d(1 + 0.00015d)^{-1/2}$ |

43 Briggs, G.A., "Diffusion estimation for small emissions," Atmospheric Turbulence and Diffusion Laboratory, Contribution File No. 79, 1973.
44 "Technical Guidance for Hazards Analysis," EPA, December 1987.

Energy Publication No. DOE/TIC-11223, 1982. The following assumptions apply to calculations made using these equations:

1) Rural flat terrain with no obstacles (hills) that would interfere with the downwind movement of plumes.
2) Ground-level release.
3) Stable atmospheric conditions and low wind speed (3.4 mph).
4) Continuous release with no phase change.
5) The plume is assumed to be at ambient temperature.
6) The substance is released in neutrally buoyant air.
7) A Gaussian distribution of the plume spread is assumed.
8) Gasses are released over a 10 minute period.
9) Liquids are spilled instantaneously from containment onto a flat, level surface forming a 0.033 ft (1 cm) deep pool and are allowed to evaporate at ambient or boiling conditions.

### 4.5.3.1 Estimation of Airborne Quantity Released for Liquids

The rate of release of a chemical is needed for calculation of the radius of the vulnerable zone. It is dependent on the quantity of chemical released, the nature of the release (pool of liquid, release of a pressure relief valve, etc.), and the properties of the chemical released. For spilled pools of chemicals, the rate of release is usually taken to be the evaporation rate (rate of volatization). Using the assumptions presented above, the following equation is used to calculate the rate of release (volatization rate) for liquids to air in lb/min:

$$QR = \frac{0.162 \times MW \times K \times A \times VP}{R(T_1 + 273)},$$

(4.27)

where

$QR$ = rate of release to air (lb/min),

$MW$ = molecular weight (g/g mole),

$K$ = gas phase mass transfer coefficient (cm/s),

$VP$ = vapor pressure of material at temperature ($T_1$),

$T_1$ = temperature at which the chemical is stored (°C).

$K$ can be estimated based on a known value for a reference compound as follows:

$$K = K_{ref} \times (MW_{ref}/MW)^{1/3}.$$

(4.28)

Using water as the reference compound,

$$K_{ref} = K_{water} = 0.25u^{0.78},$$

(4.29)

where $u$ is windspeed (m/s). Combining equations (4.28) and (4.29),

$$K = 0.25u^{0.78} \times (18/MW)^{1/3}.$$

(4.30)

Combining equations (4.27) and (4.30),

$$QR = \frac{0.106 \times (u)^{0.78} \times MW^{2/3} \times A \times VP}{R(T_1 + 273)}.$$
(4.31)

### 4.5.3.2 Calculation of Surface Areas of Spilled Liquids

For diked areas, the surface area is assumed to be the area inside the dike (unless the surface area of the spill is smaller than the diked area). If the area is not diked, in order to calculate the surface area of the spill the density is taken to be 62.4 lb/ft$^3$ (all liquids are assumed to have the same density as water) and the depth of the pool is assumed to be 0.033 ft (1 cm). The surface area of the spilled liquid (ft$^2$) is:

$$A = \frac{QS(\text{lb})}{62.4 \text{ lb/ft}^3 \times 0.033 \text{ ft}} = 0.49 \times QS,$$
(4.32)

where: $QS$ is the quantity spilled (lbs) and $A$ is the surface area (ft$^2$). Substituting for $A$ in equation (4.31), the quantity released to the air per minute ($QR$) can be estimated as

$$QR = \frac{0.633 \times 10^{-3} \times (u)^{0.78} \times MW^{2/3} \times QS \times VP}{(T_1 + 273)}.$$
(4.33)

### 4.5.3.3 Estimation of the Vulnerable Zone

The vulnerable zone radius concentration downwind of a release is given by

$$C = \frac{QR}{\pi \sigma_y \sigma_z u},$$
(4.34)

where

$C$ = airborne concentration, g/m$^3$,

$QR$ = rate of release to air, g/s,

$y$ = horizontal diffusion deviation,

$z$ = vertical diffusion deviation,

$u$ = windspeed, m/s.

Table C.6 (see Appendix C) provides vulnerable zone distances as a function of the *level of concern* and *rate of release*. The level of concern is defined by the EPA as the concentration of an extremely hazardous substance in air above which there may be serious irreversible health effects or death as a result of a single exposure for a relatively short time period. The levels of concern used in Table C.6 have been estimated by using one-tenth of the *immediately dangerous to life and health* levels published by the National Institute for Occupational Safety and Heath. Vulnerable zone data is given for spills in urban areas with atmospheric conditions corresponding to classification F, stable atmosphere with 3.4 mph wind speed.

## Example 4.3

A test facility involved in testing small rocket engines using the propellants nitrogen tetroxide ($N_2O_4$) and unsymmetrical dimethylhydrazine (UDMH) is attempting to establish operational criteria concerning routine and emergency spills. The test facility is to be enclosed by a perimeter fence and, since activities outside of this boundary are not under the control of facility personnel, operating procedures must be conducted in a manner that will ensure that vapor concentration levels at the boundary fence do not exceed allowables. Accordingly, the fence location relative to the facility in which propellant operations are conducted must be such that the level of concern limits established by EPA for these two contaminants are not exceeded.[a]

### Solution

#### Source Strength

The source strength is determined using the following rationale:

1) For large oxidizer spills (emergency spills) the vaporization rate is 30 percent of the total quantity spilled in 10 minutes or 3 percent/min, but not less than 12 lb/min.
2) For oxidizer spills of 1 gallon or less, the total quantity is assumed to evaporate in 1 minute.
3) For bulk fuel spills (emergency) the source strength is 0.25 lb/min ft$^2$ of exposed surface area for UDMH. This source strength is based on a 10-minute spill evaporation time.
4) For fuel spills of 1 gallon or less, the source strength is 0.065 lb/min ft$^2$ of UDMH. One gallon will cover an area of approximately 50 ft$^2$.

#### Operations Involving Nitrogen Tetroxide

For routine (instantaneous) ground releases the determination of vulnerable zone distances for all source strengths of interest can be determined using Table C.6 (EPA 3-13). Based on the size of the spill anticipated (which is a function of the type of operation planned), the quantity released, and the level of concern for the particular propellant, the vulnerable zone distance is determined. For a level of concern of 0.002 g/m$^3$, $QR = 3$ lb/min, and an atmospheric stability classification of F (stable), the vulnerable zone distance is 0.3 miles. For this condition the perimeter fence would have to be located 0.3 miles from the source of the spill to ensure that the level of concern did not exceed the required 0.002 g/m$^3$.

For emergency (continuous) ground releases assuming a level of concern of 0.02 g/m$^3$, $QR = 12$ lb/min, and atmospheric stability classification of F (stable), the vulnerable zone distance is 0.2 miles. Thus, for spills of nitrogen tetroxide the routine ground release governs the location of the perimeter fence at 0.3 miles from the point of the ground release.

*Operations Involving Unsymmetrical Dimethylhydrazine (UDMH)*

For routine (instantaneous) ground releases of UDMH the level of concern is 0.0004 g/m$^3$ and $QR = 3$ lb/min. For a stable atmospheric condition the vulnerable zone distance is 0.8 miles.

For emergency (continuous) ground releases the level of concern is 0.08 g/m$^3$. With an exposed surface area of 100 ft$^2$, and a stable atmospheric condition, the vulnerable zone is 0.2 miles. The UDMH routine spill condition thus governs the distance of the fence from the source of the propellant release for this facility.

---

[a]From Ertas, A., and Jones, J.C., *The Engineering Design Process*, John Wiley & Sons, Inc., 1996.

## 4.6  Occupational Safety and Health[45]

The Occupational Safety and Health Act was signed into law on December 29, 1970. This law is enforced by OSHA, an agency of the Department of Labor. The purpose of the Act is to assure working men and women safe and healthful working conditions and to preserve the USA's human resources. The Act applies to every employer engaged in business affecting commerce. It does not apply to the self-employed or to workers to the extent they are covered by other federal safety and health laws. It is estimated that the Act covers 6 million workplaces employing 75 million workers.[46]

The job safety and health standards included in the Act provide rules for the elimination or avoidance of hazards proven by research and experience to be harmful to personal safety and health. Standards are enforced by OSHA compliance officers who can conduct inspections at any workplace covered by the Act. Employers can require the inspector to provide a warrant. Representatives of employers and employees accompany the inspector during the inspection, which is normally scheduled according to priorities established by OSHA. However, employees can request an inspection if an imminent danger exists that could cause death or serious physical harm. When a violation is found, a written citation is issued requiring abatement within an appropriate time period. Depending on the seriousness of the violation and whether it is willful or repeated, a penalty can be assessed for each violation. Any employer who fails to correct a violation for which a citation has been issued can be penalized for each day the hazard persists beyond the abatement day specified in the citation. Criminal penalties are also included in the Act for making false statements or giving unauthorized advance notice of an OSHA inspection.[47]

The Act encourages states to assume the fullest responsibility for the administration and enforcement of their job safety and health laws through the vehicle of an approved state plan. Up to 50 percent federal funding is provided for these state programs. Once a state receives approval from OSHA to operate its own safety and health plan, OSHA continues to monitor

---

45  From Ertas, A., and Jones, J.C., *The Engineering Design Process*, John Wiley & Sons, Inc., 1996.
46  *OSHA Reference Book*, US Department of Labor, OSHA 3081, 1985.
47  *All About OSHA*, US Department of Labor, OSHA 2056, 1985.

the state's performance and can withdraw approval if the state should fail to continue to meet the criteria established for state plans. The Act also includes provisions for conducting training and education programs designed to assist employers and employees in avoiding unsafe or unhealthy work practices.[48]

The Occupational Safety and Health Act of 1970 established the National Institute for Occupational Safety and Health (NIOSH) as a research agency focused on the study of worker safety and health, and encouraging employers and workers to create safe and healthy workplaces. NIOSH is part of the US Centers for Disease Control and Prevention, in the US Department of Health Services. In 1979, NIOSH sponsored an Engineering Control Technology Workshop which concluded that there was a critical need to include occupational safety and health in the education of engineers. In 1980, the Public Health Service issued a report entitled "Promoting Health/Preventing Disease: Objectives for the Nation." This report included a major objective: "By 1990, at least 70 percent of all graduate engineers should be skilled in the design of plants and processes that incorporate [occupational safety and health] control technologies." Accordingly, NIOSH initiated Project SHAPE to encourage improvements in engineering practice, education, and research. A series of workshops was subsequently conducted to further these goals as well as to establish an engineering school faculty network to ensure that any educational resource material developed by NIOSH was made available to prospective users.

### 4.6.1 Safety and Loss Prevention

Effective management is the key to the control of industrial hazards and thus the primary emphasis of loss prevention is the management system. With the rapid pace of technology and the resulting growth in the number of complex systems and hazardous materials it is no longer possible for business to be conducted in a laissez-faire manner nor to allow hazardous systems operations to evolve by trial and error. The consequences of some potential hazards are so severe that there is zero tolerance for error. The incident in Bhopal, India on December 3, 1984, in which nearly 3,000 people were killed, 10,000 permanently disabled and 100,000 others injured is a good example of a complex system operation with extremely lax safety procedures coupled with gross judgement errors by the plant operators.[49]

Loss prevention is characterized by[50]

1) An emphasis on management
2) A systems rather than a trial-and-error approach
3) A concern to avoid loss of containment resulting in major fire, explosion or toxic release
4) The development of techniques for the quantification of hazards
5) The use of reliability engineering

---

48 *Accident Prevention Manual*, 9th edn, National Safety Council, Chicago, 1986.
49 Martin, M.W., and Schinzinger, R., *Ethics in Engineering*, McGraw-Hill, New York, 1989.
50 Lees, F.P., *Loss Prevention in the Process Industries*, Butterworth, London, 1980.

6) The principle of independence in critical assessments and inspections
7) The planning of emergencies
8) A critique of traditional practices or existing codes, standards, or regulations, where these appear outdated by technological change.

The hazard must be identified if corrective measures and safety procedures are to be developed for mitigation. Several methods have been introduced for identifying hazards, including hazards indices, chemical screening, hazard and operability studies, and plant safety studies. In the early 1970s the author was involved in the aftermath of an accident that occurred as a result of failure to conduct a materials screening test. In this case the operating organization had a screening test requirement in place, but the material under evaluation had been tested several years earlier in a similar environment and it was assumed that no serious hazard existed. The screening test was thus thought to be unnecessary and samples of the material were loaded into a test chamber and exposed to a highly active oxidizer without having been previously screened in a chemistry laboratory test. Unfortunately, a constituent in the material had been changed during the period between the first and second tests and an explosion occurred, seriously injuring the technician assisting in the test activity. Later, it was determined that the material, which was a rubber compound, retained the same identifying number although the resin had been changed to a chemical compound that was not compatible with the oxidizer. The material retained the same identifying number since the specification requirements, which were established by the material blending contractor who was a second-tier supplier, concerned physical characteristics only and did not control the blending constituents used to meet the specification requirements.

After the hazard has been identified it should be categorized by quantifying it in some manner. Quantifying the hazard to provide some measure of its criticality is important in establishing mitigation procedures and in determining the resources that can justifiably be committed to correcting design or other deficiencies related to the hazard. The method used for categorizing hazards is not important; the need for using *some* measure to establish the relative criticality of hazards is. This technique not only forces a certain discipline in the management decision-making process but also provides a record that establishes the rationale for dealing with hazards that may be needed in the future as a result of accidents or litigation.

Another recurring theme in loss prevention is the need to constantly review existing safety rules and procedures and to train employees in following these rules and procedures in the event of an accident. A common safety training procedure in US schools has been the *fire drill*. In a fire drill students are instructed as to what routes to take in exiting the building, what the various alarm signals mean, how to use escape slides and other mechanisms, who is in authority in emergencies, etc. Unfortunately, this level of *drilling* has not been common practice in industry and, as a result, when accidents occur employees often do not know what to do and sometimes panic in trying to evacuate the scene. A little pre-planning, preparation of emergency and evacuation procedures, and training in following such procedures would probably save lives and more than pay for the time lost in training.

### 4.6.2    Making Green Jobs Safe: Integrating Occupational Safety & Health into Green and Sustainability[51]

In 2008 the world experienced the worst financial crisis of our generation, triggering the start of the most difficult recession since the Great Depression. The financial crisis has forced policy-makers to respond powerfully, creatively, and positively to severe financial crises: interest rates have been considerably reduced, a stimulus package for the green economy was signed, hundreds of billions of dollars have been provided to banking systems around the world. A stimulus package was planned to create or save up to 3.6 million jobs and increase consumer spending to stop the recession.

Although many elements of the green economy have value-added benefits for a global economy, we should retrain healthy consciousness of the potential hazards that workers face when performing green jobs.

Schulte and Heidel stated that "There are benefits as well as challenges as we move to a green economy. Defined broadly, green jobs are jobs that help to improve the environment. These jobs also create opportunities to help battle a sagging economy and get people back to work. Yet, with the heightened attention on green jobs and environmental sustainability, it is important to make sure that worker safety and health are not overlooked. NIOSH and its partners are developing a framework to create awareness, provide guidance, and address occupational safety and health issues associated with green jobs and sustainability efforts."[52]

Table 4.4 shows that how our knowledge about old and new hazards intersects with challenges created by new technologies and adaptations of work activities to perform green jobs.[52]

Although many green job programs have the commendable goal of getting young workers into the workforce, it is known that these inexperienced new workers could be the most at risk for job injuries. Moreover, in addition to these green job programs, stimulus package spending on infrastructure projects will also expose thousands of new workers to the myriad hazards encountered in the construction of bridges, highways, and public buildings. Hazards expected to be encountered in green jobs include:[53]

- exposure to lead and asbestos in the course of energy efficiency retrofitting and weatherization in older buildings;
- respiratory hazards from exposure to fiberglass and other materials in re-insulation projects;
- exposure to biological hazards, such as molds, in fixing leaks;
- crystalline silica exposure from fiber-cement materials, which may contain up to 50% silica;
- ergonomic hazards from installation of large insulation panels;
- fall hazards in the installation of heavy energy-efficient windows and solar panels and in the construction and maintenance of windmills (typically 265 feet tall);
- electrical hazards encountered in the course of weatherization projects.

---

51  From Ertas, A., *Prevention through Design: Transdisciplinary Process*, ATLAS Publishing, 2010. Reproduced with permission.
52  Schulte, P., and Heidel, S., "Going green: Safe and healthy jobs," *PtD in Motion*, issue 5, July 2009.
53  Green Jobs – Safe Jobs campaign, National Council for Occupational Safety and Health, 2009. http://coshnetwork.org/node/133

**Table 4.4** Framework for considering green jobs and occupational hazards.

| Job type | Old | New |
|---|---|---|
| *Traditional jobs* | *What we know*<br><br>**Construction.** Creating energy efficient homes could expose workers to hazards such as formaldehyde in insulation and related heatlh effects. | *What we know we don't know*<br><br>**Energy.** Electrical hazards that may arise during installation and maintenance of solar energy panels and generators.<br><br>**Agriculture.** New hazards, equipment, and chemicals linked to green practices. |
| *Green jobs* | *What we don't know we know*<br><br>**Services.** Many of the hazards for recycling workers have been identified in other jobs.<br><br>**Energy and construction.** Hazards and protections for building and maintaining wind energy turbines may be similar to experience in construction trades. | *What we don't know we don't know*<br><br>Unintended consequences of changes in supply chain management practices?<br><br>Unknown health risks with green chemicals and cleaning solvents? |

As another example, solar energy will play an essential role in meeting challenges such as human energy needs, addressing global warming, reducing US dependence on energy imports, creating "green jobs," and helping revitalize the US economy. However, as the solar photovoltaics sector expands, little attention is being paid to the possible environmental and health costs of that fast expansion. The most commonly used solar photovoltaic panels are based on materials and processes from the microelectronics industry and have the capability to create a huge new wave of electronic waste (e-waste) at the end of their useful lives. Recommendations to build a safe and sustainable solar energy industry include:[54]

- reducing and eventually eliminating the use of toxic materials and developing environmentally sustainable practices;
- ensuring that solar photovoltaic panel manufacturers are responsible for the life cycle impacts of their products through extended producer responsibility;
- ensuring proper testing of new and emerging materials and processes based on a precautionary approach;
- expanding recycling technology and designing products for easy recycling;
- promoting high-quality green jobs that protect worker health and safety and provide a living wage throughout the global photovoltaic industry, including supply chains and end-of-life recycling;
- protecting community health and safety throughout the global photovoltaic industry, including supply chains and recycling.

---

54 "Toward a just and sustainable solar energy industry," Silicon Valley Toxics Coalition White Paper, January 14, 2009. http://www.coshnetwork.org/node/266

### 4.6.3   Green During Construction[55]

Green during construction ensures the benefit of the surrounding community, workers, and visitors on the site by reducing emissions, airborne pollution, and toxic gases like CO.

Green building development focuses on energy efficiency and using less toxic products from the perspective of future occupants of a building and also includes air quality issues such as diesel exhaust generated by vehicles (containing nitrogen oxides, sulfur oxides and polycyclic aromatic hydrocarbons) which increases the risk of lung and perhaps bladder cancer; it also includes other health problems such as asthma and cardiovascular diseases. Similar problems can be expected from gasoline-powered vehicles.

Dust is another issue in air quality. Dust consist of small solid particles created by a breakdown of the fracture process, such as grinding, crushing or impact. Particles that are too large to stay airborne settle while others remain in the air indefinitely. General dust levels at considerably elevated concentrations may induce permanent changes to airways and loss of functional lung capacity.

Silica dust is accountable for a major American industrial disaster. Three hundred workers die every year from silicosis, a chronic, disabling lung disease caused by the formation of nodules of scar tissue in the lungs. Hundreds more are disabled and between 3,000 and 7,000 new cases occur each year. Summarizing, high-risk work activities in construction are:[56]

- chipping, drilling, crushing rock;
- abrasive blasting;
- sawing, drilling, grinding concentrate and masonry and products containing silica;
- demonstration of concrete/masonry;
- removing paint and rust with power equipment;
- dry sweeping of air blowing of concrete rock sand dust;
- jackhammering on concrete, masonry and other surfaces.

## 4.7   Prevention through Design: Transdisciplinary Design Process

The NIOSH *Prevention through Design* Initiative is a collaboration with partners in industry, labor, trade associations, professional organizations, and academia. Several national organizations have partnered with NIOSH in promoting this concept of recognizing the hazards of each industry and designing more effective prevention measures. These national organizations are: the American Industrial Hygiene Association, the American Society of Safety Engineers, the Center to Protect Workers' Rights, Kaiser Permanente, Liberty Mutual, the National Safety Council, the Occupational Safety and Health Administration, ORC Worldwide, and the Regenstrief Center for Healthcare Engineering.

---

55  From Ertas, A., *Prevention through Design: Transdisciplinary Process*, ATLAS Publishing, 2010. Reproduced with permission.
56  Barbier, E., "A global green new deal," UNEP-DTIE, 2009.

Work-related injuries are real, devastating, and common. Recent studies indicate that each year in the USA, 294,000 are made sick and 3.8 million are injured, with 55,000 people dying from work-related injuries and diseases. Yearly direct and indirect costs have been estimated to range from $128 billion to $155 billion. Most recent studies in Australia indicate that design is a significant contributor in 37 percent of work-related fatalities; hence, the successful implementation of prevention through design (PtD) concepts can have substantial impacts on worker health and safety.[57]

The PtD concept can be defined as "addressing occupational safety and health needs in the design process to prevent or minimize the work-related hazards and risks associated with the construction, manufacture, use, maintenance, and disposal of facilities, materials, and equipment."[58]

PtD focuses on preventing illness, injury, and fatality by designing out occupational hazards and risks. The PtD process is a collaborative cross-disciplinary activity that relies on the principle that the best way to prevent occupational injuries, illnesses, and fatalities is to anticipate and design out or minimize hazards and risks when new equipment, processes, and business practices are developed.[59]

Integrating PtD into design will require a challenging transformative concept. Transformative changes are broad and can advance to new forms and practices that lead us to safer and more productive environments. PtD, if viewed and practiced with broad vision, should lead to transformative changes that support patient, worker, and environmental safety.[60]

The American workforce is undergoing major changes because of immigration. Immigrants with job opportunities in the USA by and large have lower educational levels, greater poverty, and less income than US-born American citizens. Therefore, the difficulties of developing culturally integrated approaches to workplace safety and health should not to be taken too lightly. As the world becomes increasingly multicultural, PtD process should consider synthesized transcultural theories, models, and research, to facilitate culturally harmonious and capable prevention and control of occupational injuries, illnesses, and fatalities. PtD is a shared concept crossing many diverse disciplines, including agriculture, forestry, and fishing; construction; healthcare and social assistance; manufacturing; mining; services; transportation, warehousing, and utilities; and wholesale and retail trade. A common goal which will be addressed by PtD associated with all the sectors is preventing and controlling occupational injuries, illnesses, and fatalities.

Currently, many business leaders are recognizing PtD as a cost-effective means to enhance occupational safety and health and therefore openly supporting the PtD process to develop management practices to implement them. Besides the USA, many other countries are actively promoting PtD concepts as well. For example, in 1994, the United Kingdom began requiring

---

57 Heidel, D.S., and Schulte, P., "Making the business case for Prevention through Design," NIOSH Science Blog, June 2, 2008.
58 Prevention through Design, National Institute for Occupational Safety and Health, 2009. http://www.cdc.gov/niosh/topics/ptd/
59 *PtD in Motion*, issue 1, February 5, 2008.
60 Fisher, J.M., "Healthcare and social assistance sector," *Journal of Safety Research*, 39, 179–181, 2008.

construction companies, project owners, and architects to address safety and health issues during the design phase of projects, and companies there have responded with positive changes in management practices to comply with the regulations. Also Australia developed the Australian National OHS Strategy 2002–2012, which set "eliminating hazards at the design stage" as one of five national priorities. Consequently, the Australian Safety and Compensation Council developed the Safe Design National Strategy and Action Plans for Australia encompassing a wide range of design areas including buildings and structures, work environments, materials, and plant (machinery and equipment).[58]

Although the Prevention through Design Initiative focuses on design, one should realize the importance of other factors, such as behavior, management, leadership, and personal protective equipment. These factors may interact directly with designs that address occupational safety and health.[61]

### 4.7.1 PtD Design Process

Figure 4.14 shows a general design process for prevention through design: namely, define the work related to product design, then identify and evaluate potential safety hazard and injuries involved with the product, and finally the control hazards that cannot be eliminated. This activity should be implemented throughout the entire design process as shown in Figure 2.4.

PtD must be fully integrated in the early design process of a product design. In other words, during the concept development, a hazard analysis of alternatives to be considered and worker safety and health requirements for the design must be established. The main objective of PtD at the conceptual design phase is to evaluate alternative design concepts, to plan to protect workers safety and health from hazards, and to provide a conservative safety design basis for a chosen concept to carry on into preliminary design. The conceptual design phase offers a key prospect for the safety and health hazard analysis to influence the product design.

PtD efforts during the preliminary design phase are planned to be incremental instead of a complete re-examination of the conceptual design. The hazard analysis will progress from a facility-level analysis to a system-level hazard analysis as design detail becomes available. When the hazard analysis is developed, the selection of controls, safety considerations, and classifications developed during the conceptual design phase must be revisited to make sure they are still appropriate. Decisions made during the preliminary design phase provide the basis for the approach to detailed design and production. During the detailed design phase based on hazards and accident analysis of the final design, a final set of hazard controls will be developed.

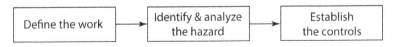

**Figure 4.14** PtD design Process.

---

61 Schulte P.A., Rinehart, R., Okun, A., Geraci, C.L., and Heidel, D.S., "National Prevention through Design (PtD) Initiative," *Journal of Safety Research*, 39, 115–121, 2008.

The National Safety Council has recommended fundamental guidelines for designers to ensure adequate safety and health for products and processes.[62] These guidelines are inclusive, and as many as possible should be considered during product design and use, within the constraints of manpower, cost, and schedule.

1) Eliminate hazards by changing the design, the materials used, or the maintenance procedures.
2) Control hazards by capturing, enclosing, or guarding at the source of the hazard.
3) Train personnel to be cognizant of hazards and to follow safe procedures to avoid them.
4) Provide instructions and warnings in documentation and post them in appropriate locations.
5) Anticipate credible abuse and misuse and take appropriate action to minimize the consequences.
6) Provide appropriate personal protective equipment and establish procedures to ensure that it is used as required.

Engineers must be able to identify hazards associated with their designs and to quantify the relative severity and likelihood of occurrence. Safety hazards normally result in accidents that occur over a relatively brief period of time and for which the acute effects are readily apparent. The effects of health hazards, on the other hand, may not be apparent for some time, often months or years, but the results can be just as devastating.[63] Several techniques have been proposed as aids in the process of recognizing, quantifying, and reducing hazards. Haddon's 10 rules comprise one of the more widely recognized strategies:[64]

1) Prevent the creation of the hazard (e.g., prevent the production of hazardous and non-biodegradable chemicals).
2) Reduce the magnitude of the hazard (e.g., reduce the amount of lead in gasoline).
3) Eliminate hazards that already exist (e.g., ban the use of chlorofluorocarbons).
4) Change the rate of distribution of a hazard (e.g., control the rate of venting a hazardous propellant).
5) Separate the hazard from that which is being protected (e.g., store flammable materials at isolated locations).
6) Separate the hazard from that which is being protected by imposing a barrier (e.g., separate fuel and oxidizer storage areas by using berms or other barriers).
7) Modify basic qualities of the hazard (e.g., use breakaway roadside poles).
8) Make the item to be protected more resistant to damage from the hazard (e.g., use fabric materials in aircraft that do not create toxic fumes when combusted).
9) Counter the damage already done by the hazard (e.g., move people out of a contaminated area).
10) Stabilize, repair, and rehabilitate the object of the damage (e.g., rebuild after a fire).

---

62 Gage, H., "Integrating safety and health into M. E. capstone design courses," *ASEE Southwest Regional Conference*, Texas Tech University, Lubbock, TX, 1989.
63 Ertas, A., and Jones, J.C. *The Engineering Design Process*, John Wiley & Sons, Inc., 1996.
64 Haddon, W., "The basic strategies for reducing damage from hazards of all kinds," *Hazard Prevention*, September/October 1980.

Although the above guidelines and rules do not include every possible safety consideration in a design project, they do provide a checklist against which the design can be evaluated and modified as needed. The designer must develop the habit of continuously evaluating the design for safety, considering not only the product design itself but also the workers involved in fabricating the product, maintaining and repairing the product or system, as well as the end user or purchaser.

## 4.8 Environmental Degradation, Sustainable Development, and Human Well-Being

> There is enough in the world for everyone's need, but not enough for everyone's greed.
>
> **Mahatma Gandhi**

Environmental problems which is the root cause of the environmental degradation are transdisciplinary in nature and the scale of problems varies.[65] Problems like acid rain, climate change, global warming, the greenhouse effect, forest fires, ozone layer depletion, loss of biodiversity, and disappearance of endangered species are global problems.

Environmental degradation is the disintegration of the Earth or deterioration of the environment through consumption of natural resources such as air, water and soil, destruction of ecosystems and the destruction of wildlife. Environmental degradation can lead to a scarcity of resources and is the biggest external threat to our planet.

### 4.8.1 Environment, Climate Change, and Global Warming

#### 4.8.1.1 Environment

The environment is the surroundings of an organism that includes all living (biotic) and non-living (abiotic) things occurring naturally on Earth. The environment consists of the interactions among atmosphere, water, soil, oceans, forests, living species, and non-living things.

All living organisms depend upon environment for their survival. The environment is degrading every day through either natural processes or human activities.

#### 4.8.1.2 Climate Change

Climate change is the increase of average global temperature because of increased greenhouse gas emissions in the Earth's atmosphere (mainly $CO_2$), trapping heat in the atmosphere.

Climate change is one of the most crucial public health threats facing the planet. Children, the elderly, and communities living in poverty are among the most vulnerable to climate change – around the world natural disasters have cost lives and devastated families. Species

---

65 Sankar, U., "The state of environment and environmental policy in India," Dr. S. Ambirajan Eighth Memorial Lecture, March 18, 2009.

are under threat, farmers are hurting, and our resources are being stretched. As our climate changes, the chance of injuries, illness, and death due to heatwaves, wildfires, intense storms, and floods increases.

Tangier Island in Chesapeake Bay has lost two-thirds of its landmass since 1850. The 1.2 square mile island is suffering from flood and erosion and slowly disappearing. The flood was caused by a combination of seasonal high tides and it is believed that the sea level rise is due to climate change. In the Mississippi Delta, trees are withering away because of rising saltwater, creating ghost forests. A gigantic cloud of dust known as a "haboob" comes over Sudan's capital, Khartoum, moving like a thick wall, damaging homes, while increasing evaporation in a region that is struggling to preserve water supplies. According to experts, without immediate intervention, part of Sudan, one of the most vulnerable countries in the world, could become uninhabitable as a result of climate change.

In July 2017, a massive crack in an Antarctic ice shelf grew at a rate of five football fields per day. An accelerating crack in the ice shelf known as Larsen C is now about 120 miles long, and only 8 miles remain until the crack cuts all the way across. A warming climate can contribute to developing a crack: if the ice shelf is exposed to warmer air above and/or warmer water below, then this will cause through-depth (top to bottom) cracks to form in the shelf. Such a crack will cause pieces of ice to break off the front of the shelf. It is important to note that this will be an iceberg about the size of the state of Delaware. The massive ice shelf in Antarctica could break up without warning, say scientists – and it would be a disaster to our coastlines, possibly causing storm surges, groundwater contamination, and farmland loss.

### 4.8.1.3 Global Warming and Greenhouse Effect

When burned, carbon-based substances such as oil, natural gas, and coal produce water and $CO_2$ which enters the atmosphere and creates a sort of thickening blanket, trapping the Sun's heat and causing the planet to warm up. The burning of fossil fuel increases the blanket effect.

To understand global warming, we should first become familiar with the *greenhouse effect* (see Figure 4.15). The greenhouse effect is a natural process by which some of the heat through solar radiation is captured in the atmosphere of the Earth, thus keeping the Earth warm enough to sustain life. The gases that help capture the solar heat, called "greenhouse gases," are water vapor, carbon dioxide, methane ($CH_4$), nitrous oxide ($N_2O$), and other manufactured chemicals resulting from human activities.

As shown in Figure 4.15, in the natural greenhouse effect, solar radiation reaches the Earth's atmosphere while some of the heat escapes back into space through reflection. Then the remaining part of the heat is absorbed by the land, lakes, and oceans, warming the Earth. Next, heat radiates from the Earth to escape into space while some of the heat is captured by greenhouse gases.

The human-enhanced greenhouse effect results in *global warming*. Before the industrial revolution, the greenhouse effect was a natural process. With the increase in industrial pollution, global warming has been increasing at a steady rate. As shown in Figure 4.15, increasing use of vehicles, and the smoke and greenhouse gases released by factories into the air increase the

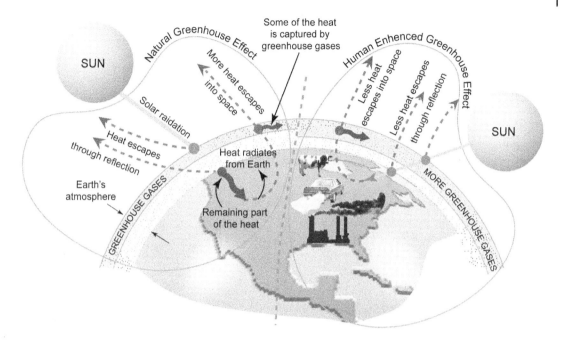

**Figure 4.15** Greehouse effect.

greenhouse effect – humans have changed Earth's atmosphere in dramatic ways over the years, resulting in global warming. Although there are natural causes that contribute to climate fluctuation, industrial practices are behind the recent rapid acceleration in global warming.

A growing world population has led to deforestation, burning fossil fuels, and intensive farming. These activities collectively and selectively produce greenhouses gases in our atmosphere.

### 4.8.1.4  Human Activities Causing Global Warming

1) *Carbon dioxide emissions from fossil fuel burning power plants.* To produce electricity from coal-burning power plants releases enormous amounts of carbon dioxide and other heat-trapping greenhouse gases into the atmosphere.
2) *Carbon dioxide emissions from burning gasoline for transportation.* Together, cars and trucks are responsible for nearly one-fifth of all US emissions, emitting approximately 20 pounds of carbon dioxide and other global-warming gases for every gallon of gas.
3) *Methane emissions from animals and agriculture.* Methane is considered a greenhouse gas, ranking behind carbon dioxide. Both carbon dioxide and methane emissions must be addressed if we want to effectively reduce the impact of climate change.
4) *Deforestation.* Forests remove and store carbon dioxide from the atmosphere. They use carbon dioxide to grow. Deforestation releases large amounts of carbon, as well as reducing the amount of carbon capture on the planet – we do not have as many trees to absorb the extra carbon dioxide.

5) *Usage of chemical fertilizers on croplands.* The increased use of commercial fertilizers has been identified as a significant emitter of greenhouse gases. The high rate use of nitrogen-rich fertilizers has effects on the heat storage of cropland more than carbon dioxide.

#### 4.8.1.5   Consequences of Climate Change and Global Warming

Climate change is a complex issue, and its potential impacts are hard to predict far in advance. It is obvious that certain consequences are likely to occur if current trends continue. Climate change could affect human health, infrastructure, transportation systems, energy, food and water supplies, as well as ecosystems. Some of the most important consequences of climate change are as follows:

- *Higher temperatures.* Heat-trapping greenhouse gases emitted by power plants, automobiles, deforestation, and other sources are warming up the planet. High temperatures are causing an increase in heat-related deaths and illnesses.
- *Changing landscapes.* Rising temperatures and changing precipitation levels (rain and snow) are forcing trees and plants to move toward polar regions and up mountain slopes.
- *Sea level rise.* Sea levels rise due to rising temperatures causing thermal expansion and the Arctic ice to melt, which could lead to the relocation of tens of millions of people.
- *Increased risk of drought, fire, and floods.* As temperatures rise globally, droughts will become more frequent and more severe, with potentially damaging consequences for agriculture, water supply, and human health. These hot, dry conditions will increase the likelihood of wildfires. The costs of wildfires, in terms of risks to human life and health, property damage, and state and federal dollars, could be devastating.
- *Stronger storms and hurricanes.* Climate change will cause hurricanes and tropical storms to become more severe – lasting longer, causing more damage to coastal communities and ecosystems.
- *Ecosystems will change.* Climate has an important environmental effect on ecosystems. Increasing global temperature means that ecosystems will change – some species will move out of their habitats because of changing conditions. For example, they migrate to higher latitudes or higher elevations where temperatures are more convenient for their survival.

Presently, 10 percent of the land area on Earth is covered with glacial ice, including glaciers, polar ice caps, and the ice sheets of Greenland and Antarctica (main ice-covered landmass). Glaciers store about 75 percent of the world's freshwater. If all land ice melted directly into the ocean, sea level would rise approximately 70 meters (230 feet) worldwide.[66] The polar regions are critical drivers of the world's climate. The melting of the polar ice caps is caused by the overall increase in global temperature, and this melting can have serious effects for all organisms on Earth – global sea levels are rising, fresh water flows into the sea, changing ocean currents and the living conditions for marine organisms. A warmer Arctic will also affect weather patterns and consequently food production around the world.

---

66 https://nsidc.org/cryosphere/glaciers/quickfacts.html, accessed July 30, 2015.

Ice at the North Pole is not as thick as at the South Pole. The ice floats on the Arctic Ocean. If it melted, sea levels would not be affected. There is a considerable amount of ice covering Greenland, which would add another 7 meters to the oceans if it melted. It is important to note that the polar ice caps are frozen because the snow they are covered with reflects a large amount of light back into space, thus maintaining and controlling the Earth's temperature. Otherwise the melting of polar ice caps increases the rate of global warming.

Since polar ice caps are made of fresh water, melting of the ice caps will add fresh water into the salty ocean waters, thus making the ocean water less saline. This will cause problems for organisms that are well adapted to ocean waters.

## 4.9 Ecosystems

Figure 4.16 illustrates the levels of organization in an ecosystem – (a) individual, (b) population, and (c) community. A community (see Figure 4.16(c)) of both living or biotic organisms (plants, animals, microorganisms) and non-living or abiotic organisms (soil, air, water, forces (gravity and wind) and conditions (temperature, light intensity, humidity, or salinity)) interacting as a system is called an ecosystem. All the organisms (elements) in an ecosystem are dependent on each other to survive – these elements help to sustain one another in regular patterns. If any part of the ecosystem is damaged it has an impact on all the other elements of the ecosystem – it can be disastrous to all living and non-living organisms within the ecosystem. A sustainable ecosystem is one where all the elements live in balance and are capable of reproducing themselves.

Elements of most ecosystems include water, air, sunlight, soil, plants, microorganisms, insects and animals. Ecosystems can be on land or water (sea, river, or lake). Ecosystems vary in size

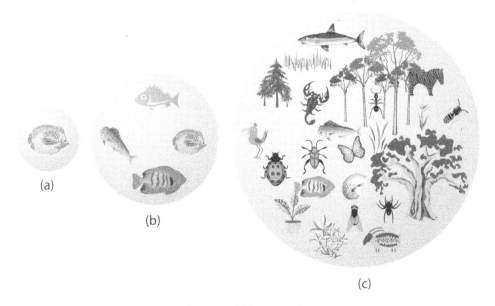

(a)

(b)

(c)

**Figure 4.16** (a) Individual, (b) population, and (c) community ecosystem.

from a small puddle to a huge desert. An ecosystem exists within a larger area called a *biome*. For example, the Sonoran Desert in Arizona is proclaimed to be the most species-rich and biologically diverse dry land on Earth. Some of the desert has stream and creeks with a variety of well-adapted desert species of plants, animals, and birds. This is one type of ecosystem. In contrast, the other part of the desert has little water – only few plants such as cacti, snakes, ground rats, and scorpions can survive. This is another kind of ecosystem within the same biome.

There are three main types of ecosystems on Earth: freshwater, ocean, and terrestrial. Freshwater ecosystems are relatively small and support many species of life, including fish, amphibians, insects, and plants. There are three main types of ocean ecosystems – shallow ocean waters, deep ocean water, and deep ocean surface ecosystems. Terrestrial ecosystems are found only on land and their location is usually dependent on the latitude of the area. The main types of terrestrial ecosystems are forest, desert, grassland, and mountain ecosystems.

### 4.9.1   System of Ecosystems

A *system of ecosystems* is a complex set of interacting ecosystems. If any part of any ecosystem is damaged it may have an impact on the other ecosystems. Unsustainable changes within an ecosystem are likely to lead to other effects on the other ecosystems. The loss of biodiversity can create an imbalance in ecosystems.

### 4.9.2   Biodiversity

Biodiversity or "biological diversity" is the variety of different types of complex webs of life found on land and water – it includes all species (microorganisms, plants, and animals), the genetic information they contain and the ecosystem they form. All living species are interconnected each other. Biodiversity is an essential part of Earth's life-support system – it is a global resource.

There are three levels of biodiversity as shown in Figure 4.17: genetic, species, and ecosystem diversity.

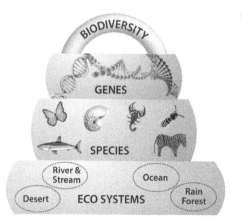

**Figure 4.17** Levels of biodiversity.

### 4.9.2.1 Genetic Diversity

Genetic diversity is the variation of genes within a species. Each species has its own individual genetic composition; species may have different populations, each having different genetic compositions. To preserve genetic diversity, different populations of a species must be preserved. If there are changes in an environment, genetic diversity plays an important role in adaptation to that environment. More genetic diversity in a population means a greater ability for some of the individuals in it to adapt to change in the environment. Genetic diversity can be measured at many different genetic levels, including population, species, community, and biome.[67]

### 4.9.2.2 Species Diversity

Species are a distinct unit of diversity within a habitat or a region. Some habitats, such as rainforests, have many species. Each species can be considered to have a specific role in an ecosystem. Therefore the addition or loss of single species may have effects on the whole ecosystem.

### 4.9.2.3 Ecosystem Diversity

Ecosystem diversity involves the variety of species' functions and their interactions within a region. As already said, an ecosystem can be a large area, such as an entire forest, or a small area, such as a pond. As shown in Figure 4.17, an ecosystem is a community of organisms and their physical environment interacting together.

Biodiversity improves ecosystem productivity where each species, no matter how small, has an important role to play within the ecosystem. For example, the existence of a large number of plant species indicates a large variety of crops and food. They provide shelter – forest products and fibers such as wool and cotton. They are a source of medicine – both traditional medicines and those synthesized from biological resources and processes.

---

**CASE STUDY 4.2 Eco-village system design**

Design an ideal self-sustaining rural eco-village system of systems. Assume that approximately 100 people will live in independent 25 homes.[a]

**Background**

The hypothetical example of an ideal self-sustaining rural eco-village systems of system (SoS) in a multi-project environment given in this case study can be created by optimizing natural resources – using renewable energy, growing as much food as possible through intensive agriculture, community building, sharing common resources, and finally "all for one and one for all." Since a detailed solution will not be provided, refer to Chapter 1 and 2 for the solution process and steps.

---

*(Continued)*

---

67 http://canadianbiodiversity.mcgill.ca/english/theory/threelevels.htm, accessed July 15, 2015.

---

**CASE STUDY 4.2 (Continued)**

### Collective Intelligence Management Workshop

The first step in an interpretive structural modeling (ISM) process is to organize a group of researchers, in this case a student research team, with relevant knowledge, skills, and background. The team included non-academic experts from different disciplines. The student team clearly identified the problem to be studied. Through brainstorming and consultation with the domain experts, the team identified the factors in designing an ideal self-sustaining rural eco-village (see Figure 4.18).

The factors shown in Figure 4.18 have been simplified by lumping together similar factors serving the same purposes. For example, renewable energy may include solar, wind, and hydro energies. Environmental issues may include recycling, air, water, and soil protection. Self-sufficiency may include recycling, and sharing common resources (food production, etc.). Diversity may include age, income levels, culture, color, being from different countries. Social issues may include encouraging unity through diversity, fostering cultural expression, emphasizing holistic health practices, and being committed to living in a community. Designing and developing an ideal SoS is not an easy task. Many subsystems (e.g., energy, green building, transportation, recycling, etc.) must be integrated to achieve an overall optimum SoS solution.

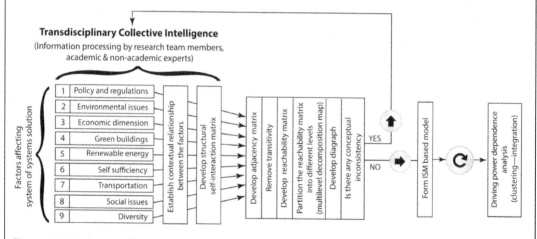

**Figure 4.18** Sequence of activities to develop an ISM model.

### Structural Self-Interaction Matrix

With the help of expert opinion, the research team developed a structural self-interaction matrix (SSIM) which shows the direction of contextual relationships among the factors affecting the issue or problem (Figure 4.19). Note that, in developing the SSIM, if the relationship between factors is weak, it is assumed that there is no relationship.

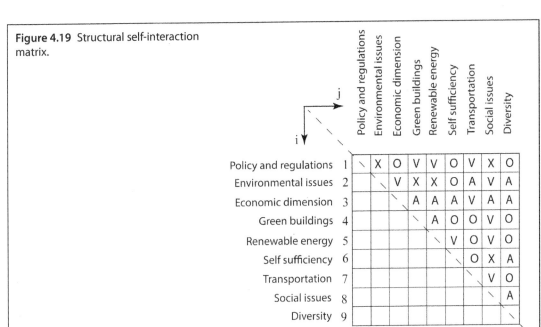

**Figure 4.19** Structural self-interaction matrix.

Then an adjacency matrix, $\mathbf{R}_a$, is developed by transforming the SSIM into a binary matrix by substituting V, A, X, and O by 1 and 0 to reflect the directed relationships between the elements (Figure 4.20). Following the transitivity rule, a reachability matrix which considers transitivity is developed (see Figure 4.21). Then the driving power and dependence of factors are also computed in the final reachability matrix (Figure 4.22). The summation of 1s in the corresponding rows gives the driving power and the summation of 1s in the corresponding columns gives the dependence.

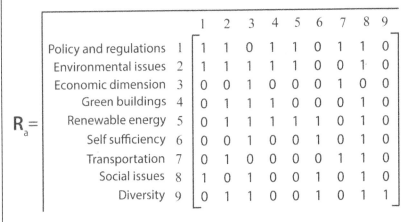

**Figure 4.20** Adjacency matrix.

*(Continued)*

**CASE STUDY 4.2 (Continued)**

|  |  | 1 | 2 | 3 | 4 | 5 | 6 | 7 | 8 | 9 |
|---|---|---|---|---|---|---|---|---|---|---|
| Policy and regulations | 1 | 1 | 1 | 1 | 1 | 1 | 1 | 1 | 1 | 0 |
| Environmental issues | 2 | 1 | 1 | 1 | 1 | 1 | 1 | 1 | 1 | 0 |
| Economic dimension | 3 | 0 | 1 | 1 | 0 | 0 | 0 | 1 | 1 | 0 |
| Green buildings | 4 | 1 | 1 | 1 | 1 | 1 | 1 | 1 | 1 | 0 |
| Renewable energy | 5 | 1 | 1 | 1 | 1 | 1 | 1 | 1 | 1 | 0 |
| Self sufficiency | 6 | 1 | 0 | 1 | 0 | 0 | 1 | 1 | 1 | 0 |
| Transportation | 7 | 1 | 1 | 1 | 1 | 1 | 1 | 1 | 1 | 0 |
| Social issues | 8 | 1 | 1 | 1 | 1 | 1 | 1 | 1 | 1 | 0 |
| Diversity | 9 | 1 | 1 | 1 | 1 | 1 | 1 | 1 | 1 | 1 |

$R_t =$

**Figure 4.21** Reachability matrix with transitivity.

|  |  | 1 | 2 | 3 | 4 | 5 | 6 | 7 | 8 | 9 | Driving power |
|---|---|---|---|---|---|---|---|---|---|---|---|
| Policy and regulations | 1 | 1 | 1 | 1 | 1 | 1 | 1 | 1 | 1 | 0 | 8 |
| Environmental issues | 2 | 1 | 1 | 1 | 1 | 1 | 1 | 1 | 1 | 0 | 8 |
| Economic dimension | 3 | 0 | 1 | 1 | 0 | 0 | 0 | 1 | 1 | 0 | 4 |
| Green buildings | 4 | 1 | 1 | 1 | 1 | 1 | 1 | 1 | 1 | 0 | 8 |
| Renewable energy | 5 | 1 | 1 | 1 | 1 | 1 | 1 | 1 | 1 | 0 | 8 |
| Self sufficiency | 6 | 1 | 0 | 1 | 0 | 0 | 1 | 1 | 1 | 0 | 5 |
| Transportation | 7 | 1 | 1 | 1 | 1 | 1 | 1 | 1 | 1 | 0 | 8 |
| Social issues | 8 | 1 | 1 | 1 | 1 | 1 | 1 | 1 | 1 | 0 | 8 |
| Diversity | 9 | 1 | 1 | 1 | 1 | 1 | 1 | 1 | 1 | 1 | 9 |
| Dependence |  | 8 | 8 | 9 | 7 | 7 | 8 | 9 | 9 | 1 | $\Sigma = 66$ |

$R_f =$

**Figure 4.22** Final reachability matrix.

After removing the transitivities based on the reachability matrix as described in the ISM approach, the structural hierarchy of the SoS factors is developed as shown in Figure 4.23. This figure illustrates visually the direct and indirect relationships between the factors affecting the performance of an ideal self-sustaining rural eco-village. Two main important factors, self-sufficiency (measure 6) and policy and regulations (measure 1) shown in Figure 4.23 are the preliminary factors which need to be analyzed before other factors, as most of the other factors depend on them. These two factors not only affect but also they are affected by the other factors.

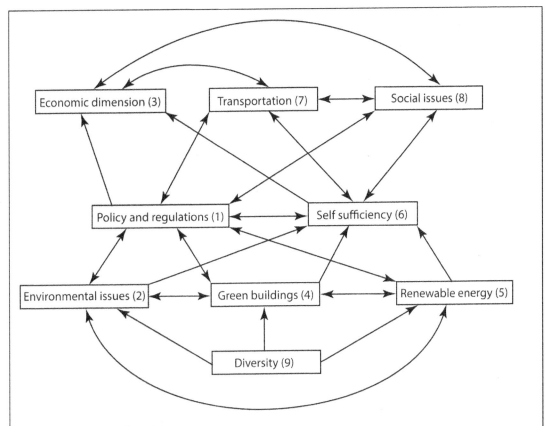

**Figure 4.23** ISM based model.

For a self-sufficient eco-village, the main goal must be a very high level of self-sufficiency and a very small ecological footprint. However, in some areas, laws and regulations can be major barriers for rural eco-village development.

As shown in Figure 4.23, the economic dimension (measure 3), transportation (7), and social issues (8) factors are positioned at the top of the hierarchy. They are also very significant measures for the development of a rural eco-village. Of course, one of the main reasons for creating eco-villages is families (social); also, a rural eco-village must have connections to the outside world (transportation) and a true economy that results from the value of people and benefits from balanced use of land (economic dimension). These three factors are strongly interrelated – affecting and affected by each other. These higher-level factors have greater influence on the success of the eco-village. Hence, when designing and developing such an SoS, these three factors can be first evaluated and kept constant as much as possible during the performance improvement. In other words, first make sure that transportation issues are solved, within certain limitations economic modeling is developed, and social issues are resolved, and

(Continued)

**CASE STUDY 4.2 (Continued)**

then start iteration by changing the other remaining factors to find the best possible solution for the ideal self-sufficient eco-village SoS.

As illustrated in Figure 4.24, all performance measures of factors affecting the SoS have been classified into four categories. In this case, no factor has been identified as autonomous. This indicates that there is no factor disconnected from the SoS. Cluster II includes only the economic dimension (measure 3) which has a smaller guidance power but is extremely dependent on the other factors.

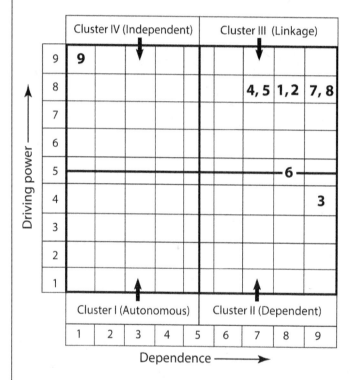

**Figure 4.24** MICMAC analysis.

Cluster III includes linkage factors of policy and regulations (measure 1), environmental issues (2), green buildings (4), renewable energy (5), self-sufficiency (6), transportation (7), and social issues (8) which need to be given extreme importance due to their high driving power and high dependence power.

Any action on any of the factors in cluster III will affect the entire SoS. Therefore, as mentioned earlier, designing such an SoS is a challenging task and seven major factors within this cluster which influence SoS performance must be carefully analyzed and integrated with the whole projected system. Note that self-sufficiency is on the border of the dependent and linkage cluster. This factor not only can be affected by other factors of SoS but also can affect the other factors.

Cluster IV includes the independent factor of diversity (measure 9) with strong driving power but very weak dependence. This factor is the key driver for the eco-village performance. Diversity is the strength and an important element. As diversity has an impact on many factors, eco-village participants have to pay maximum attention to establish a diverse but integrated community for peaceful, healthy, and sustainable life so that issues will not be out of control.

In conclusion, the eco-community determines the course of human interaction, the prospects for the impact on the natural environment, social issues, and sustainable, peaceful, healthy life. Diverse participants in an eco-village may have their personal and collective choices and may have an enormous impact on their interaction in cooperative and competitive modes. See Appendix A.1 for Matlab code.

[a]Ertas, A., Rohman, J., Chillakanti, P., and Baturalp, T. B., "Transdisciplinary collaboration as a vehicle for collective intelligence: A case study of engineering design education," *International Journal of Engineering Education*, 31(6A), 1526–1536, 2015. Used with permission.

## 4.10   Conclusion

It should be obvious that the material presented in this chapter constitutes only cursory treatment of the very broad and important subject of transdisciplinary sustainable development. Some understanding of the relative roles of the EPA and OSHA is important; thus, a brief description of the responsibilities and authority of these organizations has been included. Since the pollution problem involves air, land, and water, some limited discussion of the main considerations affecting these has also been included. A brief concept description of the transdisciplinary PtD process has been given. By means of examples, it has been shown that prevention through design is a transdisciplinary process that involves many transnational and transcultural issues. The examples and design problems also represent only a fraction of the type of pollution abatement techniques in use or under development, but it is hoped that they will serve to introduce the student to this technology and pique his or her interest to delve into this subject in greater depth. There is a vast amount of material available on the subjects mentioned in this chapter, and there are certainly enough challenges to provide a very fertile field for future engineers.

## Bibliography

1 BALL, W.P., JONES, M.D., and KAVANAUGH, M.C., "Mass transfer of volatile organic compounds in packed tower aeration," *Journal (Water Pollution Control Federation)*, 56(2), 127–136, 1984.

2 BOROMISA, A.M., *Green Jobs for Sustainable Development*, Taylor & Francis, New York, 2015.

**3** BOROWY, I., *Defining Sustainable Development for Our Common Future: A History of the World Commission on Environment and Development*, Routledge, Abingdon, 2013.

**4** CARADONNA, J.V., *Sustainability: A History*, Oxford University Press, New York, 2014.

**5** CROWL, D.A., and LOUVAR, J.F. *Chemical Process Safety: Fundamentals with Applications*, Prentice Hall, Englewood Cliffs, NJ, 1990.

**6** ENDERS, J.C., *Theories of Sustainable Development*, Routledge, London, 2016.

**7** EPA, *Handbook on Ground Water*, Environmental Protection Agency Research Laboratory EPA/625/016, March 1987.

**8** EPA, *National Air Quality and Emissions Trends Report*, April 1991.

**9** KAVANAUGH, M.C. and TRUSSEL, R.C., "Design of aeration towers to strip volatile contaminants from drinking water," *Journal of the American Water Works Association*, 72(12), 684–692, Dec. 1980.

**10** LEES, F.P., *Loss Prevention in the Process Industry*, Butterworths, London, 1989.

**11** MOON, B., *The Age of Sustainable Development*, Columbia University Press, New York, 2015.

**12** NATIONAL SAFETY COUNCIL, *Accident Prevention Manual*, 9th edn, Chicago, 1986.

**13** PERRY, H.R. & CHILTON, H.C., *Chemical Engineers' Handbook*, 4th edn, McGraw-Hill, New York.

**14** PORTNEY, K.E., *Taking Sustainable Cities Seriously*, MIT Press, Cambridge, MA, 2013.

**15** TREYBAL, R.E., *Mass Transfer Operations*, McGraw-Hill, New York, 1980.

**16** US GEOLOGICAL SURVEY, "National Water Summary 1986," US Geological Survey Water Supply Paper 2325, 1988.

**17** WILKE, C.R., and CHANG, P., "Correlation of diffusion coefficients in dilute solutions," *AICHE Journal*, 1(2), 264–270, 1955.

**18** Air Contaminants – Permissible Exposure Limits, Title 29 Code of Federal Regulations, Part 1910-1000, 1989.

# CHAPTER 4 Projects

**4.1** As an individual, what can you do to incorporate green living practices into your everyday practices?

**4.2** Write a research paper on how green energy impacts environmental, economic, and social aspects of sustainability.

**4.3** Write a research paper on environmental impacts of renewable energy technologies.

**4.4** Write a research paper on carbon-positive and carbon-negative effects on the environment. Explain how carbon can be recycled for healthy soil and clean water.

**4.5** Write a research paper on how green policies can reduce carbon dioxide emissions and stop high levels of air pollution.

**4.6** *Correction of groundwater contamination.* An aerospace engineering facility is located approximately 20 miles from a city of approximately 60,000 population. A number of organic solvents are routinely used in various design and testing activities. Waste solvents have been discharged in the past into unlined, earthen evaporation ponds for disposal. A review of the test facility's past and current waste management operations by the state environmental agency indicates a high potential for threats to human health and the environment. This has prompted regulatory action and forced the facility to determine if releases of hazardous wastes to the environment have occurred.

Groundwater monitoring wells were subsequently installed downgradient from each evaporation pond. Groundwater samples were collected from each well and analyzed for volatile organic constituents. Results indicated that releases have occurred to groundwater beneath the facility. The following compounds were detected: freon-11 (MF), trichlorethylene (TCE), and tetrachloroethylene (PCE).

The facility then conducted a contamination assessment for the site to determine the lateral and vertical extent of groundwater contamination resulting from past activities. Two hundred monitoring wells were drilled over a three-year period in an effort to define the extent of groundwater contamination. The results indicated that groundwater contaminants would impact public drinking water supplies (located approximately 4 miles downgradient) within one year unless a remediation program was initiated.

The facility plans to drill a series of interception wells to prevent potential contamination of the public water supply. Pumping of these wells will prevent continued migration of contaminants and recover contaminated groundwater for subsequent treatment and restoration.

A corrective measures study was undertaken to determine the most efficient and cost-effective method for treating groundwater containing volatile organic compounds. The results of the study indicated that air-stripping of the volatile organic constituents was the most appropriate method for this site. Accordingly, project hydrogeologists have installed the interception well network designed to capture the contaminant plume. An air-stripper design has not been completed to date.

Given the site-specific information below, design an air-stripper system to treat contaminated groundwater to acceptable standards. The design should include the diameter and height of individual stripper tower(s), the number of towers required, and the size of pipe and air-blower requirements. The type of packing material should also be included in the design. Design and size the system based on the most difficult compound to strip. Assume that no additional pumps are required to transfer the water from the well pump to the air stripping tower. Design the air-stripping system to provide the most efficient and cost-effective approach.

The following information has been provided:

a) Total flow rate from production wells = 150 gpm.
b) Atmospheric pressure = 0.85 atm.
c) Water temperature = 23°C
d) Influent concentration of TCE = 220 $\mu$g/liter
e) Influent concentration of PCE = 20 $\mu$g/liter
f) Influent concentration of MF = 160 $\mu$g/liter
g) Maximum allowable effluent concentration of each contaminant = 0.1 $\mu$g/liter
h) Henry's constant for MF = 3239 atm

The design package should include a schematic of the tower showing significant components and other elements, analysis justifying the tower design and components selected, theory involved in air-stripping of VOCs, and an economic analysis supporting the cost-effectiveness of the design. Since this is an iterative design process, computer printouts or other data should be included to show the number of tower design options considered and to validate the option selected.

**4.7** *Environmental design concept for disposal and release of propellants.* A design concept that will provide for the safe disposal and release of earth-storable hypergolic propellants at an aerospace test complex located within a large research and development facility is required.

*Background.* The test complex consists of five individual test facilities located within a fenced and access-controlled area of approximately 100 acres. This test complex was established to test low-thrust rocket engines, fluid components, fuel cells and batteries, and to provide the capability for hard vacuum testing of propellant system components and subsystems. The complex includes a central office building and supporting laboratories, a warehouse, and the five test facilities listed below:

a) Thermal vacuum test facility
b) Pyrotechnics test facility
c) Propulsion test facility
d) Power systems test facility
e) Fluid systems test facility.

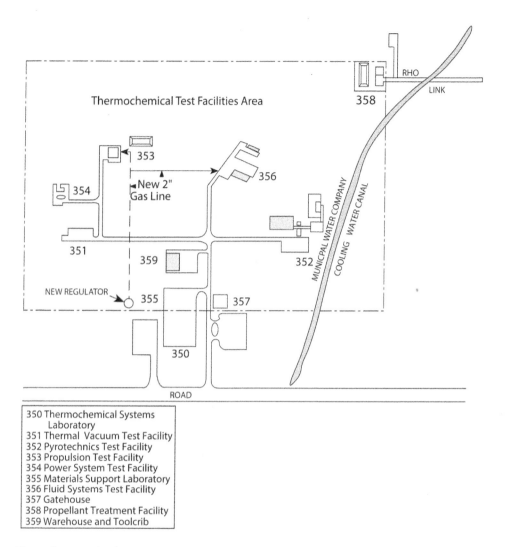

**Figure P4.7** Facility layout.

Figure P4.7 depicts the layout of the facilities and their general relationship. The problems in the handling of propellants in these facilities are the potential exposure of off-site personnel to toxic vapor from propellant releases, both intentional and unintentional, and the potential contamination of ground and surface water from oxidizer, fuel, or other fluid seepage or flow. Personnel assigned to work in the test area are knowledgeable as to the hazards of propellants and the safety procedures necessary to ensure that area personnel are not exposed to toxic vapors in excess of what is allowed. The primary concern is for off-site personnel exposure to air-transported vapor and to contaminated water as a result of propellant seeping into the ground and water table or surface flow into a municipal power company cooling water canal that flows through the area (see Figure P4.7).

Only two of the facilities, the Fluid Systems and Propulsion Test Facilities, require the use of propellants that are of concern (nitrogen tetroxide, $N_2O_4$, and the family of hydrazines including monomethylhydrazine (MMH), unsymmetrical dimethylhydrazine (UDMH), Aerozine-50 (A-50), and neat hydrazine, $N_2H_4$). These facilities include several test cells in which testing is accomplished and in which small amounts of propellant (less than one gallon) must necessarily be disposed of routinely. These cells are constructed of reinforced concrete and are open at one end. Cell floors slope toward this open end so that spills can be flushed down. Testing can be accomplished remotely from the control room, but test cell preparations and pretest operations require that personnel work in the cell when the systems are charged with propellant, but not under pressure. These personnel are well trained and equipped with personal protective equipment. They are also under the control of the test conductor or test engineer when the cell systems are charged. Personnel are also required to work around charged systems when filling run tanks, performing system maintenance or modification, or transferring propellant from the receiver tank.

Oxidizer and fuel storage equipment at these two facilities are separated by location on opposite sides of the facility. Storage tanks are approximately 400 gallons in size and propellant transfer is accomplished by pressurization of the run (storage) tank with gaseous nitrogen. Propellant flows from the run tank to the test cell through stainless steel lines. Normally, the propellant lines are filled during pretest operations up to a valve located in the test cell. This valve, which could be the rocket engine propellant valve, depending on the particular component under test, is controlled during test from the control room.

Following testing, propellant is removed from the system by flowing into a receiver tank and the system is cleared of propellant by blowdown with hot, dry nitrogen. The fuel system often requires flushing with deionized water prior to drying. Since receiver tank propellant can normally be reused, this contaminated water is not transferred to the receiver tank but must be disposed of. Small leaks and spills within the test cell can be decontaminated by flushing with water. After drying, the system is normally maintained under a "blanket" pressure to ensure that leakage flows from the system to the environment.

*Toxic vapor exposure limits.* The limiting oxygen concentrations (LOCs) for the propellants used in this facility are shown in Table P4.7 and are based on the following assumptions:

a) Exposures at these levels will be accidental and not the result of engineering controls designed to yield exposures at these levels.

b) These accidental exposures will be single events, that is, if a man is exposed at these levels further exposure will be prevented until he regains his normal resistance.

c) Personnel who could be exposed under these conditions are not idiosyncratic, hypertensive, or otherwise predisposed to disease from the specific contaminate.

d) The probable severity of injury due to secondary accidents, including those resulting from impairment of vision, judgment, and coordination, must be considered in applying these values.

**Table P4.7** Propellant spill criteria.

| Propellant | LOC mg per m³ | Emergency strength lb per min ft² | Routine source strength lb per min ft² | Wind speed mph |
|---|---|---|---|---|
| Oxidizer ($N_2O_4$) | 0.002* | 30 | 30 | 3.4 |
| MMH | 0.26 | 0.089 | 0.021 | 3.4 |
| Hydrazine ($N_2H_4$) | 1.3 | 0.025 | 0.005 | 3.4 |
| UDMH | 0.65 | 0.25 | 0.065 | 3.4 |

*$N_2O_4$ emergency source strength is 30% of the total quantity spilled in 10 minutes but not less than 12 lb/min.

In addition to the LOCs for these test facilities, Table P4.7 gives the source strengths based on operational procedures. The design concept must be based on allowable contaminant level values for both air and water. The research and development facility is located adjacent to a recreational lake and effluent from these test facilities could well end up in this lake. The water quality criteria for recreational lakes in this locale states that sewage effluents shall not exhibit either acute or chronic toxicity to human, animal or aquatic life to such an extent that lake use is interfered with.

*Required product.* Provide the information necessary to convey an adequate understanding of the proposed design concept, including the necessary discussion, calculations, and sketches, organized in a neat and orderly fashion. The two specific considerations that must be accounted for in the design concept are (1) how to ensure that personnel outside the perimeter fence are not affected in any way by activities (spills or tank venting) within the test complex, and (2) how to ensure that routine and emergency propellant spills are disposed of in a manner that will not result in ground or surface water contamination and will not be a prohibitive operational burden on facility personnel.

**4.8** *Leaking petroleum tank site remediation.* A leaking underground storage tank was discovered in November 1989, while a university was in the process of removing four underground storage tanks at the physical plant. The tanks that were removed consisted of a 2,000 gallon regular gasoline steel tank, a 1,000 gallon regular gasoline steel tank, a 500 gallon regular gasoline steel tank, and a 500 gallon diesel steel tank. A 2,000 gallon unleaded gasoline fiberglass tank remains at the site. As the tanks were removed it was discovered that the 500 gallon regular gasoline steel tank had been leaking an undetermined length of time. Upon inspection of the tank, several small holes, the size of buckshot, were found. The small holes were in an area of approximately 1.5 square feet. The holes were located approximately at or below the centerline of the tank. The cause of the holes is not known.

The four tanks that were removed were installed in 1960, and used to service university vehicles. The remaining 2,000 gallon fiberglass tank, which was installed in 1976, is currently still in use and shows no evidence of leaking. Emergency abatement measures were subsequently taken and an engineering firm was retained to assess the extent of the contamination. Some soil tests were performed and a monitoring well installed before the consulting firm became involved in the project. Soil and water samples were taken during the drilling of the monitoring well and were analyzed by a certified laboratory. In addition, organic vapor meter readings were taken at various depths during the drilling of the monitoring well.

After studying the results of the laboratory work performed on the soil and water samples, the consulting engineer recommended additional monitoring wells and one soil boring to better define the extent of the contamination. Soil samples were taken from these wells and were analyzed. Results of this analysis are shown in Table P4.8.

**Table P4.8** Soil and water analysis.

| Well ID | Sample type | Sample depth (feet) | Sample date | TRPHC (ppb) | MTBE (ppb) | Benzene (ppb) | Toluene (ppb) | Ethylbenzene (ppb) | m, p, & o Xylene (ppb) |
|---|---|---|---|---|---|---|---|---|---|
| 1 | Soil | 30–32 | 1/24/90 | 5,364.0 | 45.3 | 6.5 | 14.0 | <0.2 | 20.2 |
|   | Water |   | 3/06/90 |   | 70.3 | 46.8 | 82.6 | <0.2 | 163.2 |
|   |   |   | 4/14/90 | 519.0 | <0.2 | 238.4 | 62.6 | <0.2 | 144.7 |
| 2 | Soil | 20 | 4/07/90 | <500 | <0.2 | <0.2 | <0.2 | <0.2 | <0.2 |
|   |   | 25 | 4/07/90 | 5,692.0 | 23.4 | 1.7 | 4.4 | 0.2 | 1.4 |
|   |   | 30 | 4/07/90 | 33,843.0 | 65.7 | 7.3 | 108.6 | 11.9 | 710.9 |
|   |   | 33 | 4/07/90 | 295,100.0 | 41.4 | 8.6 | 52.9 | 34.9 | 183.2 |
|   | Water |   | 4/14/90 | 767.0 | 54.5 | 12.2 | 8.7 | <0.2 | 13 |
| 3 | Soil | 20 | 4/07/90 | 10,733.0 | <0.2 | <0.2 | <0.2 | <0.2 | <0.2 |
|   |   | 25 | 4/07/90 | 6,849.20 | 26.2 | 23.7 | 32.6 | 6.4 | 49.7 |
|   |   | 30 | 4/07/90 | 3,682.0 | 6.5 | 22.3 | 21.1 | 15.9 | 56.8 |
|   |   | 33 | 4/07/90 | 262,859.0 | 245.0 | 196.2 | 197.7 | 196.5 | 603.6 |
| 3 | Water |   | 4/14/90 | <200 | <0.2 | 0.2 | 0.9 | <0.2 | <0.2 |
| B1 | Soil | 20 | 4/07/90 | 15,139.0 | <0.2 | <0.2 | <0.2 | <0.2 | <0.2 |
|   |   | 25 | 4/07/90 | 4,500.0 | <0.2 | <0.2 | <0.2 | <0.2 | <0.2 |
|   |   | 30 | 4/07/90 | 7,189.0 | 13.5 | 3.3 | 7.4 | 0.5 | 4.3 |
|   |   | 33 | 4/07/90 | 225,344.0 | 41.8 | 43.8 | 688.4 | 698.6 | 2,729.0 |

From the data provided by the soil and water analysis, the contaminant was determined to have migrated vertically to a depth of 27–32 feet, and horizontally to points just outside of the monitoring wells. Migration was facilitated by the sandy soil comprising much of the lower unsaturated strata. A cross-section of the soil structure is shown in Figure P4.8(a).

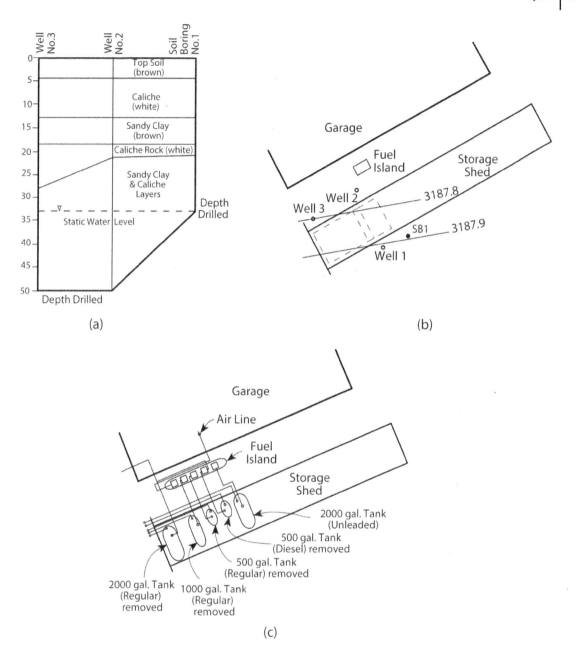

(a)

(b)

(c)

**Figure P4.8** Tank locations.

Figure P4.8(b),(c) depicts the site layout and location of storage tanks and wells. A site remediation plan contemplating the use of soil venting is required. The plan must provide answers to the following questions:

a) What contaminant vapor concentrations are likely to be obtained?
b) Under ideal vapor flow conditions (100–1000 scfm vapor flowrates) is this concentration great enough to yield acceptable removal rates?
c) What range of vapor flowrates can be achieved?
d) Will the contaminant concentrations and realistic vapor flowrates produce acceptable removal rates?
e) What residual, if any, will be left in the soil? What vapor composition and concentration changes will occur with time? How do these values relate to regulatory requirements?
f) Are there likely to be any negative effects of soil venting?

**4.9** Organize students into a research team to create their own research project (use Case Study 4.2 as an example) according to he following requirements:
- Relate the modular research project to: (a) energy, recycling, human health, security, disasters, economics, management; (b) sustainability (addressing environmental, economic, social aspects); (c) transportation and societal issues.
- Integrate planning and technical aspects (design, process, and systems) as well as economic, ecological and social goals.
- Consider functional requirements, ethical, safety, and contemporary issues.

# 5

# Design for Manufacture

Perfection is not attainable, but if we chase perfection we can catch excellence.

**Vince Lombardi**

## 5.1   Introduction

Design for manufacture (DFM) and design for assembly (DFA) are the concepts that have grown out of the need for the United States industry to compete on an international scale, and to minimize the time-to-market for developing new products and introducing them in the market-place. Design for manufacture was developed by Motorola as a means of reducing product cost and development time. It may require additional effort during early product development since approximately 70–80 percent of the manufacturing cost of a product is determined by design decisions, while production decisions account for only 20 percent. Design for manufacture is obviously worthy of serious consideration.

This method emphasizes the proactive design of products to optimize manufacturing so that fabrication time is minimized, and assembly of the product is simplified, thus ensuring the highest quality and reliability, with minimum cost. This can be accomplished by:

- selecting manufacturing processes and materials adaptable to production methods and supply chain variability as early as in the concept phase;
- avoiding unnecessary design features that require extra processing effort and complex tooling;
- organizing transdisciplinary teams that design the product and develop the manufacturing process concurrently – a process referred to as concurrent or simultaneous engineering.

## 5.2   Why Design for Manufacture?

Integrated product design and development that ensures six sigma defect levels in manufactured components are met.

The goal of DFM is to develop integrated product designs that allow manufacturing to minimize waste and deliver high-quality products that meet all specification requirements. The DFM approach will enable the design team to provide:

- integrated product design and development that ensures $\mp 6$ sigma defect levels in manufactured components are met;
- design that ensures a waste-minimizing manufacturing product;
- product design that emphasizes simplicity in the manufacturing process;
- design that ensures low costs with minimum time-to-market.

Companies can lose as much as 20 percent in profits each year due to non-manufacturable product and process designs.

## 5.3   The Six Steps in Motorola's DFM Method

*Step 1.* Identify the physical and functional requirements of the product which are necessary to satisfy the design requirements and customer needs.

*Step 2.* Identify the critical characteristics of the product necessary to meet the requirements. The key critical features essential to quality should be listed so that special care will be taken during the design and manufacturing processes. For example, key critical characteristics of LED bulbs are: energy efficiency (low power consumption), a lifetime in excess of 20 years, giving good light, ease of installation, extreme durability, reducing carbon emissions, etc.

*Step 3.* For each key characteristic, determine whether it is controlled by each product component, the assembly process, or a combination of both.

*Step 4.* For each key characteristic, determine the target value which minimizes the effects of variation on successful product development, and establish the maximum allowable range which can be tolerated by the product design.

*Step 5.* Determine a nominal value and maximum allowable tolerance for each product component and process step.

*Step 6.* Determine the capability of the product components and the process elements that control the required performance.

## 5.4   Lean and Agile Manufacturing

Lean manufacturing is a systematic manufacturing approach for the elimination of waste and minimizing costs – the minimum amount of money is invested in raw materials and inventory at all times. Agile manufacturing responds to customer needs quickly while producing faster high-quality product delivery but also controlling costs. Both lean and agile manufacturing are designed to keep companies competitive in the marketplace. Applying either or both must be decided on early in the manufacturing planning process, as they affect all aspects of the product design life cycle.

Agile manufacturing gives companies' a competitive advantage for design projects where the final goal is not clearly defined – as the project progresses, the design approach is continuously changing to satisfy customers, hence, innovating based on the customer demand. Agile manufacturing emphasizes the need for strong team interaction and collaboration to respond rapidly to emerging crises. Agile manufacturing may not be affective when the product design is simple and has no urgency.

Figure 5.1 shows an exponentially increasing cost of change if design changes are implemented during the product development life cycle time. Whereas when an agile process is used, at first the cost of change reduces because of the fast feedback (customer feedback, unit testing, etc.), but increases again over time if short feedback mechanisms are not ready to provide information about the project.

### 5.4.1   Just-In-Time Supply Chain in Lean Manufacturing

The objective of just-in-time (JIT) is to produce and deliver finished products "just in time" to be sold. JIT is essentially an inventory strategy in supply chain management. Using the JIT philosophy, companies keep increase efficiency and decrease waste by receiving goods only as they are needed in the production process. To reduce inventory costs with this method producers must forecast demand accurately; sophisticated planning and considerable experience are required in this field, as companies will face challenging issues and complexities when implementing the JIT concept in the supply chain.

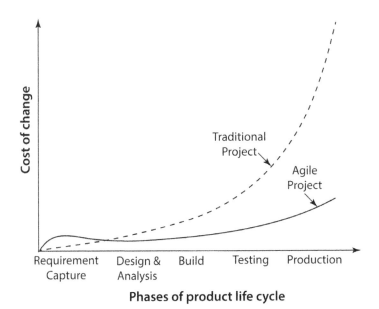

**Figure 5.1** Cost of change over time.

## 5.5 Design for Manufacture and Assembly Guidelines

A common approach to design for manufacturing and assembly is to apply a set of design guidelines. Design for manufacture and assembly guidelines that have been developed over years of design and manufacturing experience are outlined below:[1,2]

1) *Eliminate operations that require skill.* Identify manufacturing tasks that require skill and try to eliminate them. Ask whether the assembly task could be accomplished by a blind person or a person with only one hand.

2) *Minimize part counts.* Reduce the number of parts, thereby reducing the cost of manufacturing – this normally contributes to reduced weight, reduced material requirements, and reduced complexity as well. Design drives the cost of the product and the number of parts drives the design; therefore, challenge the need for every part. Ask questions such as: Does the part move relative to the mating part? Does the part have to be of a different material? Does the part have to be removed for disassembly? Figure 5.2(a) shows an example of reduced production cost by reducing the number of parts.

3) *Use a modular design.* Products made-up of 4–8 modules with 4–12 parts per module can usually be automated most effectively. Maintain a generic product configuration through the assembly process as far as possible and install the specialized modules during the latter stages. Elimination of the need for a separate housing or enclosure is also a good technique (Figure 5.2(b)).

4) *Minimize part variations.* Standardized parts should be used whenever possible. This is especially true in regard to piece-parts such as nuts, bolts, screws, washers, fittings, gaskets, resistors, condensers, etc. (Figure 5.2(c)). Use of standardized parts eliminates the design, fabrication, and inspection of the specialized part as well as the need to carry an additional part number in the inventory system. Standard stock items are widely available and the reliability and quality are usually well established.

5) *Use a multifunctional design.* Design components to perform more than one function. For example, an electrical chassis can be designed to support circuit boards, act as an electrical ground, and function as a heat sink for heat dissipation. An engine cylinder head can be designed to perform as a pressure vessel closure, to allow instrumentation and component mounting, to act as a heat exchanger, and to perform as a structural member. Features can be incorporated into a design to facilitate assembly (e.g., self-alignment guides, detents, or non-symmetric bolt and fastener patterns).

6) *Design parts for multiple uses.* Many parts that perform very basic functions lend themselves to more than one use. A standardized parts program can be established and new part designs that lend themselves to standardization can be included. The standard parts data base can then be consulted as new products are designed.

---

1 Pearse, J. and Stoll, H.W., *Design for Manufacturing*, Tool and Manufacturing Engineers Handbook, SME, Dearborn, MI, 1989.
2 Hock, G.W., Presentation at the IEEE Videoconference Seminars on Design for Manufacturability, May 16, 1985.

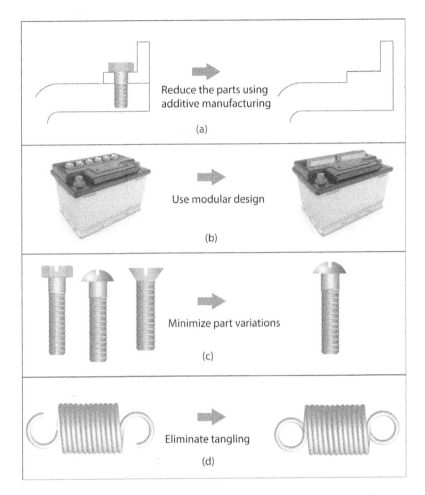

Reduce the parts using
additive manufacturing

(a)

Use modular design

(b)

Minimize part variations

(c)

Eliminate tangling

(d)

**Figure 5.2** Examples of product design improvements (from Hock, G.W., Presentation at the IEEE Videoconference Seminars on Design for Manufacturability, May 16, 1985).

7) *Design to simplify fabrication.* The design of parts should make use of the lowest-cost material that will satisfy the requirements and minimize waste and production processing time. Fabrication processes that avoid surface treatments (painting, plating, buffing, etc.) and secondary processes (grinding, reaming, polishing, etc.) should be used where possible. The design should be based on the use of simple fabrication processes. Design parts to be self-securing so that they do not require complicated and expensive joining operations using threads, welding, brazing, or soldering. Design parts so that they are not too small, too large, too thin, or too delicate.

8) *Use of fasteners.* Fasteners add significant cost to the manufacturing process. The cost of installing screws and other fasteners significantly exceeds the cost of the fastener. If fasteners

must be used, the number and variation in size and type should be minimized. Fasteners and piece-parts that are small and hard to handle during fabrication should not be used. The use of captured washers and self-tapping screws will help in this regard.

9) *Minimize assembly directions.* To minimize assembly time, parts should be assembled from one direction. The best solution in this regard is to assemble parts in a top-down fashion along the z-axis like making a sandwich.

10) *Maximize compliance.* To overcome problems associated with tolerance buildup, alignment during mating of parts, and part insertion during assembly, the design should include the use of generous tapers and chamfers, location points for manufacturing fixtures, guiding features, and generous radii.

11) *Minimize handling.* Positioning of parts during manufacture is costly. Thus, the part should be designed so that proper positioning is easy to attain. This can be achieved by using symmetrical designs or identifying the proper position by marking the part. The number of positions in which the part must placed during fabrication should also be minimized. Deliver parts to the production line in proper orientation when possible and design the part so that the orientation can be determined from geometric features. Design parts so that they will not tangle or nest (Figure 5.2(d)).

12) *Eliminate or simplify adjustments.* Mechanical adjustments add to the cost of fabrication and cause assembly, test, and reliability problems. The need for these adjustments can often be negated by using stopping points, detents, notches, and spring mounted components. If the designer understands why the adjustment has been recommended he can often find a way of eliminating or reducing the need.

13) *Avoid flexible components.* Wiring and other flexible components are difficult to handle during assembly. The use of rigid or process-applied gaskets, plugs or connectors to replace lead wires, and circuit boards instead of electric wiring, will help to minimize this problem.

14) *Minimize testing.* Use design techniques that provide for automatic testing. Internal test points should have parallel external test points and should be accessible with standard probes. Test points should help in guiding and aligning test probes.

## 5.6 Six Sigma

Six Sigma was introduced by Bill Smith as a management strategy in 1986 while working as an engineer at Motorola. In 1995, Jack Welch made it central to his business strategy at General Electric. Now it is used in many industrial sectors. The purpose of Six Sigma is shown in Figure 5.3.

Six Sigma is a measure of quality that tries to develop and deliver near-perfect products and services. It is a methodology for eliminating *defects* in any process – from manufacturing to healthcare, and in products and services. There are two types of product characteristics – a variable characteristic that is measured in physical units and an attribute characteristic that can be counted. A defect is any variation in a required characteristic (variable or attribute) of a product which is far enough removed from its nominal value to prevent the product from satisfying the

**Figure 5.3** Pupose of Six Sigma.

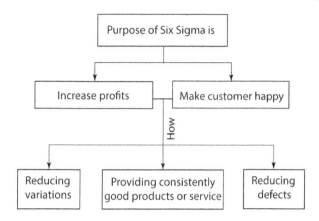

physical and functional requirements of the customer. To achieve long-term six sigma quality, a process must not produce more than 3.4 defects per million opportunities.

Six Sigma is a performance target that applies to a single critical to quality (CTQ) characteristic, not to the total product. For example, when a work station provided by a computer company is described as "Six Sigma," this does not mean that only 3.4 work stations out of a million will be defective. In this case, it means that within a single work station, the average opportunity for a defect of a CTQ characteristic is only 3.4 defects per million opportunities.

It is known that Six Sigma can be costly to implement, and it can take quite a few years before a company starts to see the end results. Six Sigma users claim that its benefits include:

- up to 50 percent process cost reduction;
- cycle-time improvement;
- less waste of materials;
- a better understanding of customer requirements;
- increased customer satisfaction;
- more reliable products and services.

Six Sigma can be discussed from three different viewpoints:

1) *Six Sigma is a philosophy.* (a) Anything less than perfect is an opportunity for improvement. (b) Defects cost money and time. (c) Understanding processes and improving them is the most effective way to achieve lifelong results.
2) *Six Sigma is statistics.* Six Sigma processes will produce no more than 3.4 defects per million opportunities.
3) *Six Sigma is a process.* The define–measure–analyze–improve–control (DMIAC) approach is used to achieve a higher level of performance.

There are two improvement processes in Six Sigma – Six Sigma DMAIC and Six Sigma DMADV.

### 5.6.1 Six Sigma DMAIC Process

Six Sigma DMAIC is a process that defines, measures, analyzes, improves, and controls existing processes that fall below the Six Sigma specification (see Figure 5.4).

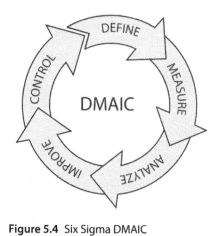

1) *Define phase.* The main goal of this phase is to outline the process requirements of the project – process improvement goals are consistent and aligned with customer demand and project goals.
2) *Measure phase.* This phase includes data collection and review of the types of measurement methods and their key features.
   - Companies collect reliable baseline data to compare against future results.
   - Companies must think about where errors in measurements can occur and impact data collection on a project success.
   - Companies must study the frequency occurrence of defects.

**Figure 5.4** Six Sigma DMAIC methodology.

3) *Analyze phase.* This phase is often linked with the measure phase to understand the nature of the data. In this phase, both the data and the process are analyzed to narrow down and verify the root causes of waste and defects. The project team investigates areas for improvement.
4) *Improve phase.* After the data analysis, this phase investigates the key parameters that cause problems and how to fix them. The improve phase includes the process known as Design for Six Sigma (DFSS) which leads to quality products and services. The project team investigates improvement ideas throughout the project. Project teams use techniques such as design of experiments to eliminate defects.
5) *Control phase.* Finally, in the control phase, a project team develops metrics that help the team members monitor and document continued success. The project team has to ensure that any deviations from target are corrected before they result in defects so that future process performance is controlled.

### 5.6.2 Six Sigma DMADV Methodology

Six Sigma DMADV is a process that defines, measures, analyzes, designs, and verifies new processes or products that are trying to achieve Six Sigma quality (see Figure 5.5).

1) *Define phase.* The main goal of this phase is to outline the design requirements of the project – design requirements are consistent and aligned with customer demand and project goals.
2) *Measure phase.* This phase includes data collection and review of the types of measurement methods and their key features. CTQ characteristics, product capabilities, and risks are identified.
3) *Analyze phase.* In this phase, both the data and the process are analyzed to create a high-level design. Design capabilities are evaluated to select the best design from the design alternatives.

4) *Design phase.* Design tests results are compared with customer requirements and needs. Product design verification encompasses functional performance, operational and usage requirements, reliability, and safety requirements. Manufacturing process verification includes process capability and production costs. Any additional design changes needed are made and a plan is made for design verification.

5) *Verification phase.* The prototype is tested and the design verified. Performance, failure modes, reliability, and risks are assessed. The product or service is released to the customer for review. The last phase in the methodology could be continuous – the process may require adjustment.

Figure 5.6 shows a comparison of the design and improvement processes of DMAIC and DMADV.

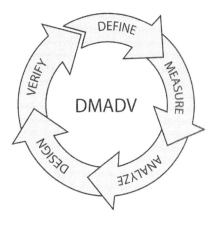

**Figure 5.5** Six Sigma DMADV methodology.

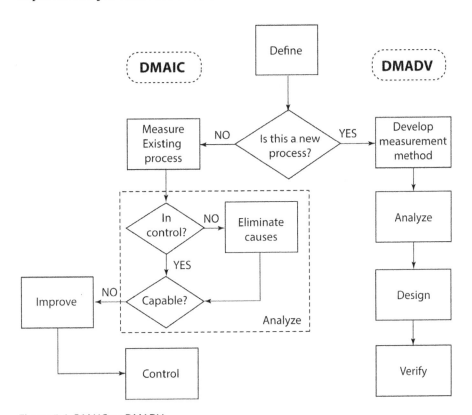

**Figure 5.6** DMAIC vs. DMADV.

### 5.6.3 Statistical Six Sigma Definition

The term "Six Sigma" is derived from the normal distribution used in statistics. As shown in Figure 5.7, in a normal distribution, the interval created by the mean plus or minus 1 standard deviation contains 68.26% of the data points; in the Six Sigma context this is referred to as the *yield*. The number of *defects per million opportunities* (DPMO), the number of data points per million outside of the area created by the mean plus or minus 1 standard deviation, is calculated as

$$DPMO = (1.00 - 0.6826) \times 1{,}000{,}000 = 317{,}400.$$

The *defect rate* (percentage defective) can be calculated as

$$Defect\ rate = \frac{Total\ number\ of\ defective\ units}{Total\ number\ of\ opportunities} \times 100$$

$$= \frac{317{,}400}{1{,}000{,}000} \times 100 = 31.74\%.$$

If the process is a service, the *error per million opportunities* (EPMO) is calculated as

$$EPMO = \frac{Total\ number\ of\ errors}{Total\ number\ of\ processes} \times 1{,}000{,}000.$$

The quantity *parts per million defective* (PPM) is the number of defective units in 1 million units. PPM is usually used if the number of defective products is small – hence, a more

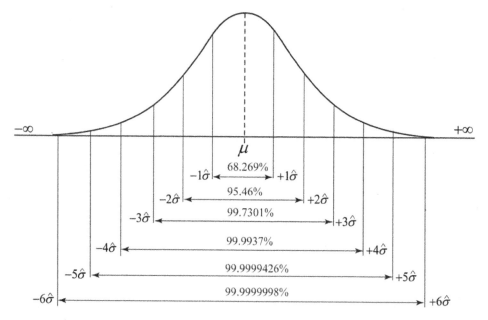

**Figure 5.7** Normal distribution percentages lying within $1\hat{\sigma}, \ldots, 6\hat{\sigma}$ of the mean value of $\mu$.

accurate measure of the defective rate can be obtained than with the percent defective. PPM is calculated as

$$\text{PPM} = \frac{\text{Total number of defective units found in a sample}}{\text{Sample size}} \times 1{,}000{,}000.$$

For example, in a sample of 60 calculators we find that 4 of them are defective. The PPM defective is:

$$\text{PPM} = \frac{4}{60} \times 1{,}000{,}000 = 66{,}667.$$

In a normal distribution, the interval created by the mean plus or minus 6 standard deviations contains 99.9999998% of the data points, which means that 0.002 data points per million are outside the interval $((1 - 0.999999998 = 2 \times 10^{-9}) \times 1{,}000{,}000 = 0.002$ per million opportunities). In other words, statistically a Six Sigma process means 2 defects per billion opportunities. Defect rates for various short-term sigma levels are given in Table 5.1. Note that yield and defect rate sum to 100 percent.

However reliably short-term processes may perform, processes usually do not perform as reliably in the long term as in the short term, and we consider this briefly in the next subsection.

### 5.6.4 The 1.5 Sigma Shift

Motorola has found out, through years of process and data collection, that processes vary and drift over time – what they call *long-term dynamic mean variation*. This variation usually falls between 1.4 and 1.6 sigma. In the short term, only normal process variation needs to be controlled. However, in the long term, special process variations occur due to wear in measurement tools or fatigue in important process elements, leading to shifts in the upper and lower specification limits by $1.5\hat{\sigma}$. Offsetting the normal distribution by 1.5 standard deviations on either side will take into account the special process variations. This results in the process performing at

Table 5.1 Defect rates for short-term sigma levels.

| Sigma ($\hat{\sigma}$) level | DPMO* | Yield | Defect rate |
| --- | --- | --- | --- |
| 1 | 317,400 | 68.2690% | 31.74% |
| 2.0 | 45,400 | 95.4600% | 4.54% |
| 3.0 | 2,699 | 99.7301% | 0.2699% |
| 4.0 | 63 | 99.9937% | 0.0063% |
| 5 | 0.574 | 99.9999426% | 0.0000574% |
| 6 | 0.002 | 99.9999998% | 0.0000002% |

*Defects per million opportunities

**Table 5.2** Defect rates for long-term sigma levels.

| Sigma ($\hat{\sigma}$) level | DPMO* | Yield | Defect rate |
|---|---|---|---|
| 1 | 697,612 | 30.2388% | 69.7612% |
| 2 | 308,770 | 69.1230% | 30.8770% |
| 3 | 66,810 | 93.3190% | 6.6810% |
| 4 | 6,209 | 99.3791% | 0.6209% |
| 5 | 232 | 99.9768% | 0.0232% |
| 6 | 3.4 | 99.99966% | $3.4 \times 10^{-4} \approx 0.0\%$ |

*Defects per million opportunities

6 sigma levels in the short run but at 4.5 sigma levels in the long run. Defect rates for long-term sigma levels are given in Table 5.2.

### Example 5.1

If a company discovered 5 errors while processing 21,000 payment to its employees during the last year, calculate:

a) EPMO.
b) At what level does the company perform?

**Solution**

a)

$$EPMO = \frac{\text{Total number of error}}{\text{Total number of process}} \times 1,000,000$$

$$= \frac{5}{21,000} \times 1,000,000 = 238$$

b) Table 5.3 (with 1.5 sigma shift) shows that the company performs almost at 5 sigma level. The defect rate is approximately 0.023%.

### 5.6.5 Process Capability

Process capability is a method for designing components and products that are robust to process variation. The capability indices are used to measure the quality of the manufactured parts or processes whose characteristic variations must be within lower and upper specification limits and close to a specified target value. There are two kinds of capability index for Six Sigma – short-term and long-term capability indices.

**Table 5.3** Defect rates for sigma levels.

| Sigma level | Without 1.5 sigma shift | | | With 1.5 sigma shift | | |
|:---:|:---:|:---:|:---:|:---:|:---:|:---:|
| | DPMO* | Yield % | Defect rate% | DPMO* | Yield % | Defect rate% |
| 1.0 | 317,400 | 68.2690 | 31.7400 | 697,612 | 30.2388 | 69.7612 |
| 1.1 | 271,332 | 72.8668 | 27.1332 | 660,082 | 33.9918 | 66.0082 |
| 1.2 | 230,139 | 76.9861 | 23.0139 | 621,378 | 37.8622 | 62.1378 |
| 1.3 | 193,601 | 80.6399 | 19.3601 | 581,814 | 41.8186 | 58.1814 |
| 1.4 | 161,513 | 83.8487 | 16.1513 | 541,693 | 45.8307 | 54.1693 |
| 1.5 | 133,614 | 86.6386 | 13.3614 | 501,249 | 49.8651 | 50.1349 |
| 1.6 | 109,598 | 89.0402 | 10.9598 | 461,139 | 53.8861 | 46.1139 |
| 1.7 | 89,130 | 91.0870 | 8.9130 | 421,427 | 57.8573 | 42.1427 |
| 1.8 | 71,860 | 92.8140 | 7.1860 | 382,572 | 61.7428 | 38.2572 |
| 1.9 | 57,432 | 94.2568 | 5.7432 | 344,915 | 65.5085 | 34.4915 |
| 2.0 | 45,400 | 95.4600 | 4.5400 | 308,770 | 69.1230 | 30.8770 |
| 2.1 | 35,728 | 96.4272 | 3.5728 | 274,412 | 72.5588 | 27.4412 |
| 2.2 | 27,806 | 97.2194 | 2.7806 | 242,071 | 75.7929 | 24.2071 |
| 2.3 | 21,448 | 97.8552 | 2.1448 | 211927 | 78.8073 | 21.1970 |
| 2.4 | 16,395 | 98.3605 | 1.6395 | 184,108 | 81.5892 | 18.4108 |
| 2.5 | 12,419 | 98.7581 | 1.2419 | 158,686 | 84.1314 | 15.8686 |
| 2.6 | 9,322 | 99.0678 | 0.9322 | 135,686 | 86.4314 | 13.5686 |
| 2.7 | 6,934 | 99.3066 | 0.6934 | 115,083 | 88.4917 | 11.5083 |
| 2.8 | 5,110 | 99.4890 | 0.5110 | 96,809 | 90.3191 | 9.6809 |
| 2.9 | 3,731 | 99.6269 | 0.3731 | 80,762 | 91.9238 | 8.0762 |
| 3.0 | 2,699 | 99.7301 | 0.2699 | 66,810 | 93.3190 | 6.6810 |
| 3.1 | 1,935 | 99.8065 | 0.1935 | 54,801 | 94.5199 | 5.4801 |
| 3.2 | 1,374 | 99.8626 | 0.1374 | 44,566 | 955434 | 4.4566 |
| 3.3 | 966 | 99.9034 | 0.09660 | 35,931 | 96.4069 | 3.5931 |
| 3.4 | 673 | 99.9327 | 0.0673 | 28,716 | 97.1284 | 2.8716 |
| 3.5 | 465 | 99.9535 | 0.0465 | 22,750 | 97.7250 | 2.2750 |

**Table 5.3** (Continued)

| Sigma level | Without 1.5 sigma shift | | | With 1.5 sigma shift | | |
|---|---|---|---|---|---|---|
| | DPMO* | Yield % | Defect rate% | DPMO* | Yield % | Defect rate% |
| 3.6 | 318 | 99.9682 | 0.0318 | 17,864 | 98.2136 | 1.7864 |
| 3.7 | 215 | 99.9785 | 0.0215 | 13,903 | 98.6097 | 1.3903 |
| 3.8 | 144 | 99.9856 | 0.0144 | 10,724 | 98.9276 | 1.0724 |
| 3.9 | 96 | 99.9904 | 0.0096 | 8197 | 98.1803 | 0.8197 |
| 4.0 | 63 | 99.9937 | 0.0063 | 6209 | 98.3791 | 0.6209 |
| 4.1 | 41 | 99.9959 | 0.0041 | 4661 | 99.5339 | 0.4661 |
| 4.2 | 26 | 99.9974 | 0.0026 | 3467 | 99.6533 | 0.3467 |
| 4.3 | 17 | 99.9983 | 0.0017 | 2555 | 99.7445 | 0.2555 |
| 4.4 | 10 | 99.9990 | 0.0010 | 1865 | 99.8135 | 0.1865 |
| 4.5 | 6 | 99.9994 | 0.0006 | 1349 | 99.8651 | 0.1349 |
| 4.6 | 4 | 99.9996 | 0.0004 | 967 | 99.9033 | 0.0967 |
| 4.7 | 2 | 99.9998 | 0.0002 | 687 | 99.9313 | 0.0687 |
| 4.8 | 1 | 99.9999 | 0.0001 | 483 | 99.9517 | 0.0483 |
| 4.9 | 0.96 | 99.999904 | 0.000096 | 336 | 99.9964 | 0.0336 |
| 5.0 | 0.574 | 99.9999426 | 0.0000574 | 232 | 99.97680 | 0.02320 |
| 5.1 | 0.34 | 99.9999660 | 0.0000340 | 159 | 99.98410 | 0.01590 |
| 5.2 | 0.2 | 99.9999800 | 0.0000200 | 107 | 99.98930 | 0.01070 |
| 5.3 | 0.116 | 99.9999884 | 0.0000116 | 72 | 99.99280 | 0.00720 |
| 5.4 | 0.067 | 99.9999933 | 0.0000067 | 48 | 99.99520 | 0.00480 |
| 5.5 | 0.038 | 99.9999962 | 0.0000038 | 31 | 99.99690 | 0.00310 |
| 5.6 | 0.021 | 99.9999979 | 0.0000021 | 20 | 99.99800 | 0.00200 |
| 5.7 | 0.012 | 99.9999988 | 0.0000012 | 13.35 | 99.99867 | 0.00134 |
| 5.8 | 0.007 | 99.9999993 | 0.0000007 | 8.55 | 99.99915 | 0.00086 |
| 5.9 | 0.004 | 99.9999996 | 0.0000004 | 5.42 | 99.99946 | 0.00054 |
| 6.0 | 0.002 | 99.9999998 | 0.0000002 | 3.4 | 99.99966 | 0.00034 |

*Defects per million opportunities

### 5.6.5.1 Short Term Capability Index

A simple and straightforward indicator of process capability is the short-term capability index. This index measures the capability of a process when the process is centered or not centered.

If the process is *centered*, in other words the process mean $\mu$ is equal to the target or nominal value of the specification as shown in Figure 5.8, the process capability index $C_p$ can be calculated as

$$C_p = \frac{USL - LSL}{6\hat{\sigma}}. \qquad (5.1)$$

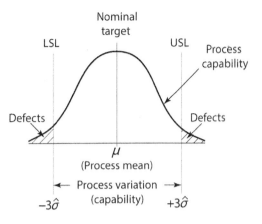

**Figure 5.8** Process centered.

The variation of process capability should be within the tolerance limits of the characteristics. Since the process is targeted at the nominal value, the process mean should be equal to the nominal target value. Unreasonably tight design tolerances can cause manufacturing defects.

In equation (5.1), $\hat{\sigma}$ is the short-term standard deviation and *LSL* and *USL* are the lower and upper specification limits, respectively. Note that, as a measure of process capability, it is customary to take a $6\hat{\sigma}$ spread in the product quality characteristic distribution. This would include about 99.73% of the distribution of output if the process output is approximately normally distributed. As seen from Figure 5.8, the comparison is made by establishing the ratio of the spread between the process specifications (*USL* − *LSL*) to the spread of the process values ($6\hat{\sigma}$ − process width). In equation (5.1), *USL* − *LSL* represents the voice of the customer's requirements and $6\sigma$ represents the characteristic voice of the process.

In equation (5.1), a value of 1 for $C_p$ indicates that 99.73% of the process output meets the specification and 0.27% of the process output will fall outside the tolerance limits. In other words, for a normal distribution approximately 2,700 out of every 1,000,000 production units would be non-conforming. The process is not capable if $C_p < 1$. Note that $C_p$ is a measure of short-term process capability. This value is calculated by using the short-term standard deviation.

### Example 5.2

Assume that a shaft target diameter is set at 20 mm. If the process output has a short-term standard deviation of $\hat{\sigma} = 0.02$ mm, what would be the specification limits to ensure that the shafts manufactured are within the design specification limits when $6\hat{\sigma}$ process spread is used. Assume that the process output is approximately normally distributed.

## Solution

Since $\hat{\sigma} = 0.02$ mm, the process is not capable of making each pin exactly the same diameter, 20 mm. However, if we assume that the process spread is $6\hat{\sigma}$ as shown in Figure 5.9, the process could be capable of manufacturing each shaft within the design specification limits. Assume that the mean is equal to 20 mm. Then the design specification limits are:

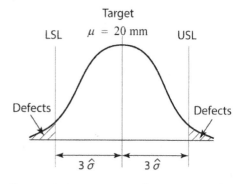

Figure 5.9 Process centered.

$$USL = 20 + 3\hat{\sigma} = 20 + 3(0.02) = 20.06 \text{ mm},$$

$$LSL = 20 - 3\hat{\sigma} = 20 - 3(0.02) = 19.94 \text{ mm}.$$

The process capability index, $C_p$, is

$$C_p = \frac{USL - LSL}{6\hat{\sigma}} = \frac{20.06 - 19.94}{6 \times 0.02} = 1.$$

$C_p = 1$ means that the process is just meeting specifications. This process will produce approximately 2,700 non-conforming production units per million.

If the process is *not centered* in the specification requirements (i.e., the process mean is not at the midpoint of the specification or target value) an amended version of process capability is used to predict the process capability. This is called the *adjusted short-term capability index*, $C_{pk}$. The smaller of the values obtained from equations (5.2) and (5.3) is used for the measure of process capability – how many standard deviations can be fit between the process mean and the closest specification limit. As shown in the equations, since the specification has been split into two pieces, the process spread is also split into two pieces, namely, $6\hat{\sigma}/2 = 3\hat{\sigma}$. To summarize, given

$$C_{pu} = \frac{USL - \mu}{3\hat{\sigma}}, \tag{5.2}$$

$$C_{pl} = \frac{\mu - LSL}{3\hat{\sigma}}, \tag{5.3}$$

we obtain

$$C_{pk} = \min(C_{pu}, C_{pl}). \tag{5.4}$$

Note that, for a sample distribution, $\mu$ and $\hat{\sigma}$ in equations (5.2) and (5.3) should be replaced by $\bar{x}$ and $S$, respectively. The value of $C_{pk}$ should be at least 1.33 for $4\sigma$ to satisfy most customers. As shown in Table 5.4, for the $6\sigma$ level the value of $C_{pk}$ should be at least 2.

**Table 5.4** Defect rates for short term sigma levels.

| Sigma ($\hat{\sigma}$) Levels | DPMO* | Yield % | $C_{pk}$ |
|---|---|---|---|
| 1.0 | 317400 | 68.2690 | 0.33 |
| 1.5 | 109598 | 89.0402 | 0.50 |
| 2.0 | 45400 | 95.4600 | 0.67 |
| 2.5 | 12419 | 98.7581 | 0.83 |
| 3.0 | 2699 | 99.7301 | 1.00 |
| 3.5 | 465 | 99.9535 | 1.17 |
| 4.0 | 63 | 99.9937 | 1.33 |
| 4.5 | 6 | 99.9994 | 1.50 |
| 5.0 | 0.574 | 99.9999426 | 1.67 |
| 5.5 | 0.038 | 99.9999962 | 1.83 |
| 6.0 | 0.002 | 99.9999998 | 2.00 |

*Defects per million opportunities.

## Example 5.3

In Example 5.2, assume that the process has the upper specification limit 20.04 mm, and lower specification limit 19.96 mm. If the observed process mean is 20.005 mm and the short-term standard deviation is 0.012 mm, investigate whether the process is capable of manufacturing shafts within the defined specification limits. Also calculate the total percentage of expected shafts that would be out of specification.

### Solution

Since the process mean, $\mu = 20.005$ mm is not at the midpoint of the specification values $((20.04 + 19.96)/2 = 20.00$ mm$)$, the process is not centered (see Figure 5.10). Hence, equations (5.2) and (5.3) are used to measure the process capability:

$$C_{pu} = \frac{USL - \mu}{3\hat{\sigma}} = \frac{20.04 - 20.005}{3 \times 0.012} = 0.972,$$

$$C_{pl} = \frac{\mu - LSL}{3\hat{\sigma}} = \frac{20.005 - 19.96}{3 \times 0.012} = 1.25.$$

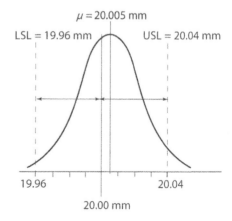

**Figure 5.10** Process is not centered.

Thus, the smallest capability indiex is $C_{pu} = 0.972$. Since this value is less than 1, the process is not capable of manufacturing parts within the defined specification limits when a $6\hat{\sigma}$ (or $3\hat{\sigma}$ for one side) process spread is used.

### 5.6.5.2 Long-Term Capability Index

The equations used to calculate short-term capability indices can also be used to calculate long-term capability indices. These long-term capability indices are designated by $P_p$ and $P_{pk}$. ($P$ stands for "performance.") The only difference in their calculation is that we use the long-term standard deviation. Long-term capability indices are crucial because every process extends out over time to create long-term performance. The equations used to calculate the *process performance* are as follows:

When the process is centered, the process performance, $P_p$, is

$$P_p = \frac{USL - LSL}{6S}. \tag{5.5}$$

When the process is not centered, the process performance, $P_{pk}$ is

$$P_{pk} = \frac{\min{(USL - \bar{x}, \bar{x} - LSL)}}{3S}. \tag{5.6}$$

Here $S$ is the sample standard deviation for all the data (long-term) and $\bar{x}$ is the process sample mean.

### 5.6.5.3 Short-Term and Long-Term Standard Deviations

The *short-term standard deviation* helps us determine how many repetitive measurements should be made at any one time to achieve a required level of precision. The calculated long-term standard deviations are usually smaller than or equal to the calculated short-term standard deviations – this is because the measurement system is more precise over short intervals than over long intervals of time. The short-term standard deviation can be calculated as

$$\hat{\sigma} = \frac{\bar{R}}{d_n}, \tag{5.7}$$

where $\bar{R}$ is the range mean and $d_n$ is Hartley's constant, given in Table 5.5.

The long-term standard deviation, $S$ (also called the *sample standard deviation*), can be calculated in part by summing the differences between the individual data points and the data set's overall average, $\bar{x}$. As shown in Figure 5.11, the long-term data distribution includes all the short-term data distributions.

The long-term standard deviation is computed as

$$S = \sqrt{\frac{1}{n-1} \sum_{i=1}^{n} (x_i - \bar{x})^2}, \tag{5.8}$$

where $n$ is the number of observations. The long-term standard deviation is a measure of reproducibility of the measurement process.

**Table 5.5** $\bar{X}$ and $R$ chart variables.

| $n$ | $D_2$ | $D_4$ | $d_n = d_2$ |
|---|---|---|---|
| 2 | 0.000 | 3.267 | 1.128 |
| 3 | 0.000 | 2.574 | 1.693 |
| 4 | 0.000 | 2.282 | 2.059 |
| 5 | 0.000 | 2.115 | 2.326 |

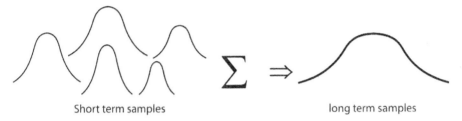

Short term samples          $\Sigma \Rightarrow$          long term samples

**Figure 5.11** Long-term distribution combines all the short-term distributions.

#### 5.6.5.4 Product Design and Process Design

The product designer's objective is to maximize the allowable tolerance while still allowing successful functioning of the product, whereas the process designer's goal is to minimize the variability of the process which produces the characteristic required for functionality of the product and to center the process on the target value of the characteristic. To optimize the capability, $C_p$ requires joint effort from product and process designers.

A high $C_p$ index implies that the process can reproduce the characteristics. A high $C_{pk}$ implies that the process is reproducing the characteristic within the desired design limits.

#### 5.6.5.5 Calculating Sigma Capability Using the Z Score

Instead of the $C_p$ and $P_p$ capability indices, $Z$ score values can be used to define the sigma capability of a process. $Z$ score statistics are standard values that help us to compare process capability. The $Z$ score is the number of standard deviations that can fit between the mean and the specification limits (see Figure 5.12). It is the point on the standard normal distribution such that the area to the right of the point is equal to the average $P$ (the proportion of defective units in a process). The higher the process $Z$ score, the better the process performance. Upper and Lower $z$ scores, $Z_U$ and $Z_L$ respectively, can be calculated from

$$Z_U = \frac{USL - \mu}{S},$$

(5.9)

**Figure 5.12** Normal distribution.

$$Z_L = \frac{\mu - LSL}{S}. \tag{5.10}$$

The short-term $Z_{ST}$ score is calculated using the short-term standard deviation of the process, and the long-term $Z_{LT}$ score is calculated using the overall standard deviation of the process. The total sigma score, called $Z_{\text{bench}}$, is found from the *total yield*, $Y_T$, of the process:

$$Y_T = 1 - \text{total defects}. \tag{5.11}$$

### Example 5.4

Using the $Z$ score, calculate the long-term sigma level of the process if $USL = 4.1$, $LSL = 3.5$, $\mu = 3.839$ and the long-term standard deviation $S = 0.17588$.

**Solution**

The long-term $Z_{\text{bench}}$ is calculated using the overall standard deviation of the process as follows. First,

$$Z_U = \frac{USL - \mu}{S} = \frac{4.1 - 3.839}{0.17588} = 1.48.$$

The $Z$ score shown in Figure 5.13 can be converted into a probability using the normal distribution values shown in Table C.1 (see Appendix C). For a $Z$ score of 1.48, the yield is 0.9306. Thus,

$$\text{Defects} = 1 - \text{Yield} = 1 - 0.9306 = 0.0694,$$

or 6.94%.

Next,

$$Z_L = \frac{\mu - LSL}{S} = \frac{3.839 - 3.5}{0.17588} = 1.93.$$

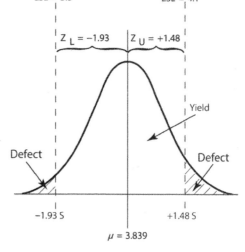

**Figure 5.13** Process with $S = 0.17588$.

For a $Z$ score of 1.93, the yield is 0.9732. Thus,

$$\text{Defects} = 1 - \text{Yield} = 1 - 0.9732 = 0.0268, \quad \text{or} \quad 2.68\%$$

Now the total yield, $Y_T$, is

$$Y_T = 1 - \text{Total defects}$$

$$= 1 - (0.0694 + 0.0268) = 0.9038.$$

From normal distribution table, we have approximately 1.3 sigma level ($Z_{\text{bench}}$) for this process. Note that, since we used the long-term standard deviation in our calculations, this is a long-term sigma level. To determine the short-term sigma level we add 1.5 sigma – the short-term sigma level is $= 1.3 + 15 = 2.8$.

### 5.6.6 Manufacturing Metrics

*Yield* is a key performance indicator in manufacturing. A highly manufacturable product exhibits very high yield, whereas a poorly manufacturable product exhibits very poor yield. The term yield has several meanings:

*Process yield* (PY) is used in manufacturing to measure and control the performance of the process. It is defined as the percentage of products that pass through the conformity check – product key parameters fall within certain range of manufacturing tolerance. That is:

$$PY = \frac{\text{Number of units confirmed}}{\text{Number of units started}}. \tag{5.12}$$

However, process yield only considers the final yield and does not take into rework or retesting. Moreover it does not tell us what processes involve defects.

*First pass yield* (FPY) is an important manufacturing metric for measuring quality and production performance and is often much different than traditional process yield. It is the percentage of products that are manufactured correctly within the specifications first time, without rerun or rework:

$$FPY = \frac{\text{Number of units confirmed to specification}}{\text{Total units entering the process}}. \tag{5.13}$$

*Rolled throughput yield* is the probability that a unit will pass through a series of process steps free of defects. RTY takes rework into account. It uses the number of defects found at each process step. If the number of defective units after each process step is known, then the RTY can be calculated as

$$RTY = \prod_{k=1}^{n} FPY_k \tag{5.14}$$
$$= FPY_1 \times FPY_2 \times \ldots \times FPY_n,$$

where $n$ is the number of process steps.

### Example 5.5

Figure 5.14 shows an automated overhead conveyor wet spray finishing line used in coating metal parts. Several different kind of processes are used to coat metal parts – electro-coating, powder coating, wet spray painting, etc. Regardless of the process used, paint is applied to a part at a certain film thickness. If the film thickness is too small the coating will be light. If film thickness is too great, excess paint will run off. In either case the part will not pass inspection. The wet spray finishing process shown in Figure 5.14 can be divided into six distinct processes:

- Pre-washing process to wash out the dirt, oil, grease, and metal particles.
- Pretreatment process for corrosion resistance and paint adhesion.
- Final washing process to remove the residual emulsion from the pretreatment step. The metal surface must be chemically clean.

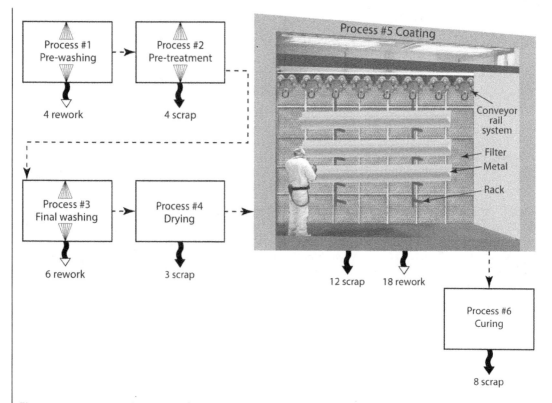

**Figure 5.14** Automated wet spray finishing line process flow.

- Drying process (drying oven) – the metal surface must be dry and free of water particles.
- Coating process – wet spray painting operations are used in spray booths with automatic spray guns or robotics for precision spray painting and even manual spray systems.
- Curing process (bake oven) – the bake oven receives the metal parts after they exit the coating process. The bake oven cures the paint film to ensure maximum performance properties.

Using the data shown in Figure 5.14, calculate:

1) Process yield (assume the system is only one single process)
2) First pass yield
3) Rolled throughput yield.

**Solution**

For this hypothetical problem, in two hours' work 500 aluminum metal frames were processed through the paint line. During the pre-washing process, due to the high-pressure water spray, 4 metal frames fell off the paint racks. Since they were not damaged, they could be reloaded and painted later (reworked). During the pretreatment process, for the same reason, four metal frames fell off. Since the paint line is continuously running, pulling out the metal frames is not practical. In two hours, these metal frames changed color to green due to lying in a chemical solution, meaning they could not be painted (so were scrapped). Six metal frames lost during the final washing process would be painted later (reworked). In the drying oven, air circulation through a fan can also cause metal loss: usually they will be damaged. During the drying process 3 metal frames are scrapped. Because of the excess paint 12 metal frames were scrapped and 18 light paintings were separated for rework. In the curling process, eight metal frames fell of and the paint on the metals were burned (scrapped). During the painting process, 28 reworks and 27 scraps out of 500 metal frames input into the paint line conveyor system were discovered. Two hours of the complete painting process resulted in 28 reworks and 27 scraps out of 500 metal frames.

1) The process yield is

$$PY = \frac{\text{Number of units confirmed}}{\text{Number of units started}} = \frac{500 - (28 + 27)}{500} = 89\%.$$

2) The first pass yield is

$$FPY = \frac{\text{Number of units confirmed to specification}}{\text{Total units entering the process}}$$
$$= \frac{445 - 28}{500} = 83.4\%.$$

3) The rolled throughput yield is

$$RTY = \prod_{k=1}^{n} FPY_k$$

$$= FPY_1 \times FPY_2 \times ...FPY_n$$

Process #1. 500 units and 4 reworks. The yield becomes $(500 - 4)/500 = 99.2\%$.

Process #2. After 4 reworks, the remaining good product is 500 units and 4 scrap. The yield becomes $(500 - 4)/500 = 99.2\%$.

Process #3. After 4 scraps, the remaining good product is 496 units and 6 reworks. The yield becomes $(496 - 6)/496 = 98.8\%$.

Process #4. After 6 reworks, the remaining good product is 496 units and 3 scrap. The result of yield becomes $(496 - 3)/496 = 99.4\%$.

Process #5. After 3 scraps, the remaining good product is 493 units and 12 scraps, 18 reworks. The result of yield becomes $(493 - (12 + 18))/493) = 94\%$.

Process #6. After 12 scraps and 18 reworks, the remaining product is 481 units and 8 scraps. The result of yield becomes $(481 - 12)/481 = 469/481 = 98.3\%$.

Thus

$$RTY = 99.2 \times 99.2 \times 98.8 \times 99.4 \times 94.0 \times 98.3 = 87.9\%.$$

If the number of defects after each step is known, then the first pass yield can also be estimated using the Poisson approximation:

$$FPY = e^{-DPU},$$ (5.15)

where *DPU* is the average *defect per unit* for each process step. This is simply a ratio of the number of defects to the total number of units processed:

$$DPU = \frac{\text{Number of defects}}{\text{Total number of units processed}}.$$ (5.16)

## Example 5.6

Referring to Example 5.5, calculate the defects per units and, using calculated values, determine the FPY for each process.

### Solution

For process #1, we have:

$$DPU = \frac{\text{Number of defects}}{\text{Total number of units processed}}$$

$$= \frac{4}{500} = 0.008$$

and

$$FPY = e^{-DPU} = e^{-0.008} = 99.2\%$$

Using similar calculations, the results of the six processes of DPU and FPY are given in Table 5.6. Note that the FPY results are identical to those found in Example 5.5.

**Table 5.6** Calculation results of DPU and FPY.

| Process # | Scrap | Rework | # Defects | # Units | DPU | FPY |
|:---:|:---:|:---:|:---:|:---:|:---:|:---:|
| 1 | 0 | 4 | 4 | 500 | 0.008 | 99.2% |
| 2 | 4 | 0 | 4 | 500 | 0.008 | 99.2% |
| 3 | 0 | 6 | 6 | 496 | 0.024 | 98.8% |
| 4 | 3 | 0 | 3 | 493 | 0.0061 | 99.4% |
| 5 | 12 | 18 | 30 | 481 | 0.062 | 94.0% |
| 6 | 8 | 0 | 8 | 469 | 0.0171 | 98.3% |

### 5.6.7   Inspection and Testing in Quality Control and Process Planning

Inspection and testing are a vital part of ensuring that quality product will be produced and that defective parts will be removed efficiently. While the need for better quality control is generally expected, in the manufacturing process it is generally accepted that inspections are money wasted and a time-consuming operation that reduces cost-effectiveness. But in software development this is not the case – the value and payoff can be realized by introducing software inspections.

*Defect removal efficiency* (DRE) was identified as the most important measure of inspection and testing quality.[3] Defect removal efficiency can be expressed as a percentage defined by:

$$\text{DRE} = \frac{\text{Total defects found during the process}}{\text{Total defects introduced by the process}} \times 100. \tag{5.17}$$

A typical manufacturing process which includes inspection and testing is shown in Figure 5.15. Here

$D_0$ = defects submitted from previous process,

$D_1$ = defects added in process 1,

$D_n$ = defects added in process $n$

$D_{sn}$ = defects submitted from the $n$th process,

$D_{si}$ = defects submitted from inspection,

$D_i$ = defects detected by inspection and corrected,

$D_t$ = defects detected by testing and corrected,

$D_e$ = defects escaping to next process.

Note that an escape is a defect that was not detected by test teams; instead, the defect was found by customers. The process planning flow given in Figure 5.15 should be revised at any point

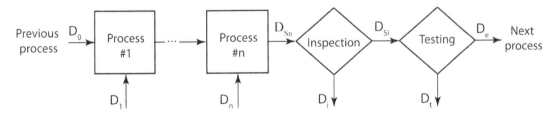

**Figure 5.15** Typical manufacturing process.

---

3 O'Brien, J., "Software testing metrics – defect removal efficiency (DRE)," https://www.equinox.co.nz/blog/software-testing-metrics-defect-removal-efficiency, accessed December 10, 2016.

of process to ensure that submitted defect per unit amounts do not go above the maximum acceptable level.

If the observed defect level and the defect removal efficiency are known, submitted and escaping defects can be estimated. For example, as shown in Figure 5.16, if the testing DRE is 0.9 and the observed defects are $D_t = 0.30$ DPU, then the submitted defects from the previous process are equal to $D_{sn} = 0.30/0.9 = 0.33$ DPU. Escaping defects will be $D_e = 0.33 - 0.30 = 0.03$.

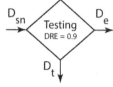

**Figure 5.16** Defects calculations.

### Example 5.7

Insulated double-pane windows (Figure 5.17) are soundproof windows with insulated dual glass and coated aluminum frames. One of the manufacturing defects is a weak seal causing failure of insulation. When it fails, moisture will migrate into the middle of the insulated glass unit and cloud up the window; the solution is to replace the glass or the window. Besides the assembly problems, another manufacturing issue is not having correct tolerances during the aluminum frame cuts.

The defect from the previous process of not having correct tolerances is 0.04 DPU (see Figure 5.18). Assume that the DRE for inspection is $PD_E = 0.90$, $SD_E = 0.80$, and $PA_E = 0.90$, for part defects, seal defects and part assembly errors, respectively. Also, the DRE for testing is $PD_E = 0.0$, $SD_E = 0.80$ and $PA_E = 0.90$.

Also, assume that:

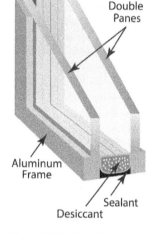

**Figure 5.17** Double-pane window.

- defects created in process #1 through part defects are 0.02 DPU, seal defects 0.06 DPU and part assembly errors is 0.04 DPU;
- defects created in process #2 through part defects are 0.02 DPU, seal defects 0.0 DPU and part assembly errors is 0.08 DPU.

Calculate:

1) The total detected defects
2) The first pass yield for inspection and testing
3) The rolled throughput yield
4) The escapes not detected by inspection and test teams.

### Solution

- $D_1$, total defects created in process #1, is $0.02 + 0.06 + 0.04 = 0.12$ DPU.
- $D_{S1}$, defects submitted in process #2, is $0.12 + 0.04 = 0.16$ DPU.
- $D_2$, total defects created in process #2, is $0.02 + 0.0 + 0.08 = 0.10$ DPU.

- $D_{Si}$, total defects submitted to inspection through PD, SD, and PA, is $(0.06 + 0.02) + (0.06 + 0.0) + (0.04 + 0.08) = 0.08 + 0.06 + 0.12 = 0.26$ DPU.
- $D_i$, total defects detected through inspection, is $(0.08 \times 0.90) + (0.06 \times 0.80) + (0.12 \times 0.90) = 0.072 + 0.048 + 0.108 = 0.228$ DPU.
- $D_{St}$, total defects submitted in testing, is $(0.08 - 0.072) + (0.06 - 0.048) + (0.12 - 0.108) = 0.008 + 0.012 + 0.012 = 0.032$ DPU.
- $D_t$, total defects detected through testing, is $(0.008 \times 0.0) + (0.012 \times 0.80) + (0.012 \times 0.90) = 0.0 + 0.0096 + 0.0108 = 0.0204$ DPU.
- $D_e$, total escaping defects, is $(0.008 - 0.0) + (0.012 - 0.0096) + (0.012 - 0.0108) = 0.008 + 0.0024 + 0.0012 = 0.00116$ DPU.

1) The total detected defects are $0.228 + 0.0204 = 0.2484$ DPU.
2) The first pass yield for inspection is $FPY = e^{-DPU} = e^{-0.228} = 79.6\%$, and for testing is $FPY = e^{-DPU} = e^{-0.0204} = 98\%$.
3) The rolled throughput yield is $RTY = 0.796 \times 0.98 = 78\%$.
4) The total escape not detected by inspection and test teams is $0.008 + 0.0024 + 0.0012 = 0.0116$ DPU.

Results are shown in Table 5.7 and Figure 5.18.

**Table 5.7** Results of process planning of double-pane windows.

| Catagory | Process # 1 | | Process #2 | | Inspection | | Testing | | |
| | Submitted | Created | Submitted | Created | Submitted | Detected | Submitted | Detected | Escaping |
|---|---|---|---|---|---|---|---|---|---|
| PD | 0.04 | 0.02 | 0.06 | 0.02 | 0.08 | 0.072 | 0.008 | 0.0 | 0.008 |
| SD | 0.0 | 0.06 | 0.06 | 0.0 | 0.06 | 0.048 | 0.012 | 0.0096 | 0.0024 |
| PA | 0.0 | 0.04 | 0.04 | 0.08 | 0.12 | 0.108 | 0.012 | 0.0108 | 0.0012 |
| TDPU | 0.04 | 0.12 | 0.16 | 0.10 | 0.26 | 0.228 | 0.032 | 0.0204 | 0.0116 |
| FPY | | | | | | 79.6% | | 98% | |

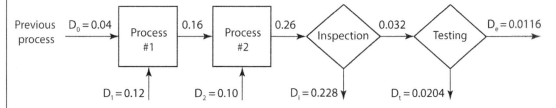

**Figure 5.18** Process planning: effectiveness of inspection and testing.

### 5.6.8 Statistical Process Control

Statistical process control (SPC) is a tool utilized by the Six Sigma process. It is a statistically based method for monitoring process behavior to determine whether the process is performing consistently up to its capability. The SPC process collects data and provides results as the process is occurring, so that when the process is out of control immediate action can be taken. This should help a process, and its quality measures, prevent drifting beyond specification limits, and avoid the production of bad parts. SPC can essentially not only eliminate the production of defective parts but also create visibility of the cause of the problem while a failure is occurring. Process improvement to achieve an acceptable level of product variation is accomplished later, after the process has been brought under control.[4] With real-time SPC we can:

- significantly reduce inconsistency and scrap, thus reducing costs;
- scientifically improve productivity;
- uncover hidden process characteristics;
- immediate react to process changes;
- obtain productivity and quality information about a production process in real time so as to make real-time decisions on the shop floor.

One of the main tools used in SPC analysis is the quality control flowchart originally developed and used by Shewhart. The quality control flowchart shows the trend, with respect to time, of actual values of selected process variables. To understand when to determine whether or not changes are occurring, Shewhart proposed that the data be plotted versus time on a chart showing the upper control limit (UCL) and the lower control limit (LCL) to determine whether the process was or was not *in control*. If the control limits are set to plus or minus three standard deviations about the target value, then the likelihood that the variables would exceed the limits is very small as long as the process is *in control*. A typical Shewhart control chart is shown in Figure 5.19. As seen from this figure, sample 5 is out of control.

The Shewhart control chart is known as an $\overline{X}$ chart and it is often combined with an R chart (see Figure 5.19). The $\overline{X}$ chart has a reference line which shows the sub-group average, $(\overline{\overline{X}})$. The individual samples are also compared to determine their range, R. The control limits for the $\overline{X}$ are calculated by

$$UCL_{\overline{X}} = \overline{\overline{X}} + 3\hat{\sigma}_{\overline{X}}, \tag{5.18}$$

$$LCL_{\overline{X}} = \overline{\overline{X}} - 3\hat{\sigma}_{\overline{X}}, \tag{5.19}$$

---

4 Beauregard, M.R., Mikulak, R.J., and Olson, B.A., *A Practical Guide to Statistical Quality Improvement*, Van Nostrand Reinhold, New York, 1992.

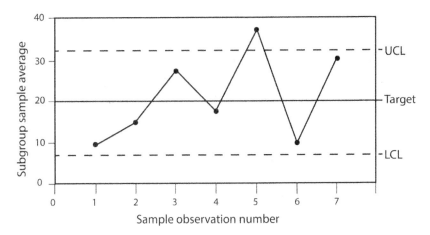

**Figure 5.19** Process planning: Shewhart control chart.

where $\hat{\sigma}_{\overline{X}}$ is the standard error of the mean given by

$$\hat{\sigma}_{\overline{X}} = \frac{\hat{\sigma}}{\sqrt{n}}. \tag{5.20}$$

The control limits for the $R$ chart are calculated by

$$UCL_R = D_4\overline{R}, \tag{5.21}$$

$$LCL_R = D_2\overline{R}, \tag{5.22}$$

where $D_4$ and $D_2$ are constants depending on the sub-group size, $n$ (see Table 5.5).

## Example 5.8

The Food and Drug Administration (FDA) quality system regulations include requirements related to the methods used in, and the facilities and controls used for, designing, manufacturing, packaging, labeling, storing, installing, and servicing of medical devices for human use. This action provides preproduction design controls to achieve consistency with quality system requirements.

The biopsy instrument shown in Figure 5.20 is used to harvest diagnostic quality specimens. Jaw alignment and its precise dimension are crucial to ensure full consistent tissue size extraction and minimize patient discomfort. Cost control is important, as this is a disposable device.

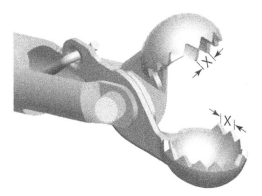

**Figure 5.20** Biopsy instrument.

During the manufacture of this device, a critical measurement, the distance $X$ between two jaw teeth, was made on five devices each day for 16 days. The first device was measured at 8:00 am, the next four hours later, etc. As shown in Table 5.8, the data were kept in time order to make $\overline{X}$ and R charts with the usual $3\hat{\sigma}$ limits, having 16 sub-groups (days) of size five (times of the day). Construct $\overline{X}$ and R charts to establish the statistical control of the manufacturing process.

**Table 5.8** Measured data for jaw tooth distance, $X$.

| Observation Number | Distance, $X$ mm | | | | | Sample Mean $\overline{X}$, mm | Sample Range $R$, mm |
|---|---|---|---|---|---|---|---|
| 1 | 3.9 | 3.8 | 3.9 | 4.0 | 3.9 | 3.900 | 0.2 |
| 2 | 3.6 | 3.8 | 3.7 | 3.9 | 3.7 | 3.740 | 0.3 |
| 3 | 3.6 | 4.1 | 3.9 | 4.0 | 3.9 | 3.900 | 0.5 |
| 4 | 4.1 | 3.7 | 4.0 | 3.8 | 3.9 | 3.900 | 0.4 |
| 5 | 3.8 | 4.1 | 3.9 | 4.1 | 4.0 | 4.000 | 0.3 |
| 6 | 3.6 | 4.0 | 4.0 | 4.1 | 3.9 | 3.920 | 0.5 |
| 7 | 3.6 | 3.9 | 3.6 | 3.7 | 3.6 | 3.680 | 0.3 |
| 8 | 3.6 | 4.0 | 4.0 | 3.9 | 3.6 | 3.820 | 0.4 |
| 9 | 4.1 | 3.6 | 3.8 | 3.6 | 4.0 | 3.820 | 0.5 |
| 10 | 4.0 | 3.6 | 3.6 | 3.6 | 3.7 | 3.700 | 0.4 |
| 11 | 3.6 | 4.1 | 3.7 | 3.6 | 3.7 | 3.740 | 0.5 |
| 12 | 4.0 | 3.7 | 3.6 | 3.6 | 3.9 | 3.780 | 0.4 |
| 13 | 3.6 | 4.0 | 4.0 | 4.1 | 3.8 | 3.900 | 0.5 |
| 14 | 3.8 | 3.6 | 3.8 | 4.0 | 3.8 | 3.800 | 0.4 |
| 15 | 3.9 | 4.1 | 4.1 | 3.9 | 4.0 | 4.000 | 0.2 |
| 16 | 3.6 | 3.8 | 3.9 | 4.0 | 3.8 | 3.820 | 0.4 |
| | | | | | | $\overline{\overline{X}} = 3.839$ | $\overline{R} = 0.3875$ |

## Solution

For observation 1, the sample mean, $\overline{X}$, and range, $R$, can be calculated as

$$\overline{X} = \frac{3.9 + 3.8 + 3.9 + 4.0 + 3.9}{5} = 3.90 \text{ mm},$$

$$R = 4.0 - 3.8 = 0.2 \text{ mm}.$$

Table 5.8 shows the mean and the range for the other observations calculated as indicated above. The sub-group average and the range mean are

$$\overline{\overline{X}} = \frac{3.900 + 3.740 + \cdots + 3.820}{16} = 3.839 \text{ mm,}$$

$$\overline{R} = \frac{0.2 + 0.3 + \cdots + 0.4}{16} = 0.3875 \text{ mm.}$$

The estimated standard deviation of the process can be calculated as

$$\hat{\sigma} = \frac{\overline{R}}{d_2} = \frac{0.3875}{2.326} = 0.1666.$$

For $n = 5$, Hartley's constant $d_2 = 2.326$ is obtained from Table 5.5. The estimated standard error of the mean is

$$\hat{\sigma}_{\overline{X}} = \frac{\hat{\sigma}}{\sqrt{n}} = \frac{0.1666}{\sqrt{5}} = 0.0745.$$

The control limits for the $\overline{X}$ chart are

$$UCL_{\overline{X}} = \overline{\overline{X}} + 3\hat{\sigma}_{\overline{X}}$$

$$= 3.839 + 3 \times 0.0745 = 4.063,$$

$$LCL_{\overline{X}} = \overline{\overline{X}} - 3\hat{\sigma}_{\overline{X}}$$

$$= 3.839 - 3 \times 0.0745 = 3.616.$$

The control limits for the R chart are

$$UCL_R = D_4 \overline{R}$$

$$= 2.115 \times 0.3875 = 0.82,$$

$$LCL_R = D_2 \overline{R}$$

$$= 0.0 \times 0.245 = 0.0.$$

Constants $D_2$ and $D_4$ can be obtained from Table 5.5. The result of the analysis is shown graphically in Figure 5.21. Although it is seen from the figure that the $\overline{X}$ for sub-groups 5 and 15 are barely inside the UCL, both charts exhibit statistical control.

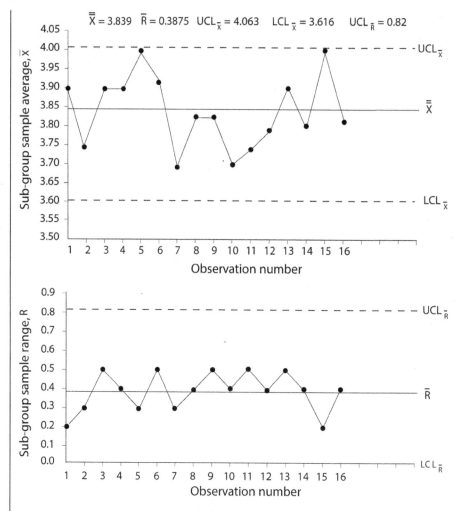

**Figure 5.21** Control chart for jaw tooth distance, $X$

## 5.7 Tolerancing in Design

Tolerancing is a crucial issue in the context of innovative design. Design of tolerances should be within manufacturing capabilities – design should avoid unnecessarily tight tolerances that are beyond the capability of the manufacturing processes. Tight tolerances will require more care to be taken during machining, longer processing time, and higher defects rates, resulting in increased costs of product development.

## 5.8 Geometric Dimensioning and Tolerancing

Tolerances and process capabilities to meet the dimensional limits should be reviewed by the manufacturing team. Where specific fits and functions are required, industry standard dimensioning and tolerancing standard should be utilized. Geometric dimensioning and tolerancing (GD&T) is an international language for defining and communicating engineering drawings. It consists of a well-defined set of symbols, rules, definitions, and conventions that can be used to describe the size, form, orientation, and location tolerances of component (part) features. This section will provide a general overview of GD&T meanings and applications.

### 5.8.1  Definitions of the terms used in GD&T

*Nominal size* is an approximation of the actual size but not necessarily exactly the same. For example a 1/2 inch hole (nominal size) could be 0.495 in. (actual size) on the engineering drawing.

*Basic size* is the nominal size of the hole and shaft from which the tolerances and allowances are established (see Figure 5.22). This is basically same for both hole and shaft.

*Allowance* is the planned deviation between the maximum material limits of mating components. It is the minimum clearance (positive allowance), or maximum interference (negative allowance) between mating components.

*Tolerance* is the acceptable amount of dimensional variation that will still allow the part to function properly. It is the total variation between the upper and lower limits.

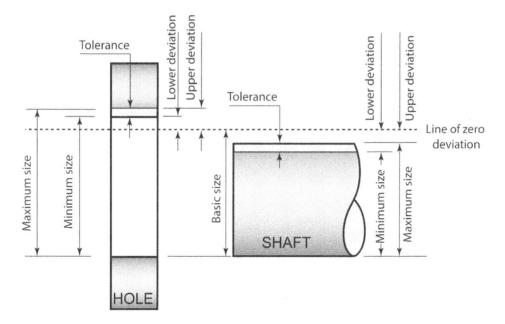

**Figure 5.22**  Hole and shaft definations.

*Limit* is the extreme permissible dimension of the part: the dimensions show the largest and smallest values allowed. Any size between largest and smallest values is acceptable.

*Feature* is a physical portion of a part like a hole, flat surface, point, edge centerline, etc. Also it can be a size, as in the width of a slot or diameter of a cylinder.

The *maximum material condition* (MMC) occurs when a size feature contains the maximum amount of material within the stated limits of size, for example the maximum shaft diameter or the minimum hole diameter.

The *least material condition* (LMC) occurs when a size feature contains the least amount of material within the stated limits of size, for example the minimum shaft diameter or the maximum hole diameter.

*Regardless of feature size (RFS)* is a term that specifies that the feature can be any size within the stated limits.

*Bilateral tolerance* refers to a tolerance method using an equal plus and minus deviation from the specified basic dimension (see Figure 5.23).

*Unilateral tolerance* refers to a tolerance method using a deviation in only one direction, either plus or minus, from the specified basic dimension (see Figure 5.23).

*Fits between mating parts* refers to measures of tightness and looseness between two mating parts. Three three main types of fits between parts are discussed below.

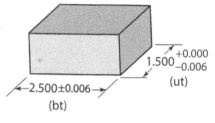

**Figure 5.23** Bilateral and unilateral tolerances.

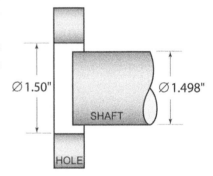

**Figure 5.24** Clearance fit.

### 5.8.2 Fits between Parts

#### 5.8.2.1 Clearance Fit

In clearance fit, mating parts will always leave a space or clearance when assembled. In this case, the maximum shaft diameter is always less than the minimum hole diameter. In the example shown in Figure 5.24 there are at least 0.002 inches of clearance between the parts.

#### 5.8.2.2 Interference Fit

For interference fits, mating parts will always interfere when assembled. In such cases, the diameter of the shaft is always larger than the hole diameter. In the example in Figure 5.25 the smallest shaft is 1.509 in. and the largest hole is 1.504 in., which creates at least 0.005 inches of interference.

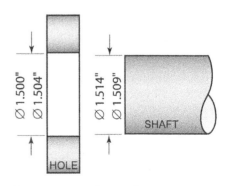

**Figure 5.25** Interference fit.

### 5.8.2.3 Transition Fit

In transition fit, mating parts will sometimes be an inter-ference fit and sometimes be a clearance fit when assem-bled. In the example in Figure 5.26, the smallest shaft diameter (1.502 in.) will fit in the largest hole (1.504 in.) with 0.002 in. clearance fit, and the largest shaft diame-ter (1.510 in.) will have to be forced into the smallest hole (1.500 in.) with 0.01 in. interference fit.

Types of fits and their applications are given in Table 5.9. As an example, the ISO symbol for the hole for a normal running fit with basic size of a 30 mm is 30H7; the 7 designates the International Tolerance (IT) grade 7.

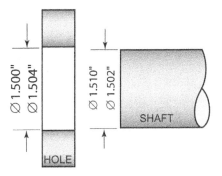

**Figure 5.26** Transition fit.

### 5.8.3 International Tolerance Grade

International tolerance grade (IT) is the classification system representing groups of tolerances which vary depending upon the basic size; these are designated by the symbols IT0, IT1, and IT01 to IT16 – a total of 18 IT grades. The relation of machining process to tolerance grades and practical usage of international tolerance grades are summarized in Table 5.10 and Table 5.11, respectively.

### 5.8.4 Tolerance Symbols

Tolerance symbols are used to specify the tolerance and fits for mating components.

*Hole basis* is the system of fits where the minimum hole size is the basic size. As shown in Figure 5.27(a), the fundamental deviation for a hole basis system is denoted by the upper-case letter "H." The 30 indicates the diameter of the hole. The numbers following the letter H indicate the IT grade.

**Figure 5.27** Hole–shaft system fit.

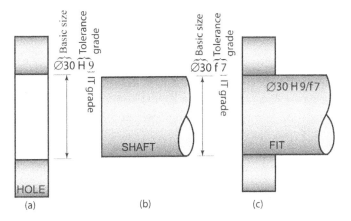

**Table 5.9** Types of fits and their applications (from different sources).

**INTERFERENCE FIT**

| Type of fit | Symbol | Description |
| --- | --- | --- |
| Shrink fit | H8/u8 | Wheel steel tyres, bronze crowns on worm wheel hubs, couplings, etc. |
| Medium drive fit | H7/s6 | For ordinary steel parts or shrink fits on light sections, the tightest fit can be used with cast iron |
| Press fit | H7/r6 | Coupling of shaft ends, bearing bushing in hubs, valve seats, gear wheels |
| Locational interference fit | H7/p6 | For parts requiring rigidity and alignment with good accuracy of location but not special bore pressure requirements |

**TRANSITION FIT**

| Type of fit | Symbol | Examples of application |
| --- | --- | --- |
| Locational clearance fit | H7/h6 | Allows snug fit for locating stationary parts: can be freely assembled and disassembled |
| Locational transition fit | H7/n6 | For more accurate location where greater interference is allowed, such as gears and bearing bushes, shaft and wheel assembly fixed by feather key |
| Force fit | H7/m6 | Used for parts of machine tools that must be dismantled without damage, e.g. gears belt pulleys, couplings, fit bolts, inner ring of ball bearings |
| Push fit | H7/k6 | For accurate location, a compromise between clearance and interference such as belt pulleys, brake pulleys, gears and couplings |
| Easy push fit | H7/j6 | Used for parts which are frequently dismantled, but are secured by keys, e.g. pulleys, hand wheels, bushes, bearing shells, piston on piston rods, change gear trains |

**CLEARANCE FIT**

| Type of Fit | Symbol | Examples of application |
| --- | --- | --- |
| Precision sliding fit | H7/h6 | Used for sealing rings, bearing covers, milling cutters on milling mandrels |
| Close running fit | H7/g6 | This sliding fit is intended to move and turn freely and locate accurately |
| | | Used for sleeve shafts, clutches, movable gears in change gear trains |
| Normal running fit | H7/f7 | Sleeve bearings with high revolution, bearings on machine tool spindles |
| Easy running fit | H8/e8 | Sleeve bearings with medium revolution, grease lubricated bearings of wheel boxes, gear sliding on shafts and sliding blocks |
| Loose running fit | H11/c11 | For wide commercial tolerances on external parts such as sleeve bearings with low revolution |

**Table 5.10** Relation of machining process to international tolerance grades.

| Machining Operation ↓ | IT grades | | | | | | | | | |
|---|---|---|---|---|---|---|---|---|---|---|
| | 4 | 5 | 6 | 7 | 8 | 9 | 10 | 11 | 12 | 13 |
| Lapping & Honing | ■ | ■ | | | | | | | | |
| Cylindrical grinding | | ■ | ■ | ■ | | | | | | |
| Surface granding | | ■ | ■ | ■ | ■ | | | | | |
| Diamond turning | | ■ | ■ | ■ | | | | | | |
| Diamond boring | | ■ | ■ | ■ | | | | | | |
| Broaching | | ■ | ■ | ■ | ■ | | | | | |
| Reaming | | ■ | ■ | ■ | ■ | ■ | ■ | | | |
| Turning | | | | ■ | ■ | ■ | ■ | ■ | ■ | ■ |
| Boring | | | | | ■ | ■ | ■ | ■ | ■ | ■ |
| Milling | | | | | | | ■ | ■ | ■ | ■ |
| Planing & shaping | | | | | | | ■ | ■ | ■ | ■ |
| Drilling | | | | | | | ■ | ■ | ■ | ■ |

**Table 5.11** Practical usage of international tolerance grades.

| Machining Operation ↓ | IT grades | | | | | | | | | | | | | | | | |
|---|---|---|---|---|---|---|---|---|---|---|---|---|---|---|---|---|---|
| | 01 | 0 | 1 | 2 | 3 | 4 | 5 | 6 | 7 | 8 | 9 | 10 | 11 | 12 | 13 | 14 | 15 | 16 |
| Measuring tools | ■ | ■ | ■ | ■ | ■ | ■ | ■ | ■ | ■ | | | | | | | | | |
| Fits | | | | | | | ■ | ■ | ■ | ■ | ■ | ■ | ■ | | | | | |
| Material | | | | | | | | | | ■ | ■ | ■ | ■ | ■ | ■ | ■ | | |
| Large manufacturing Tolerances | | | | | | | | | | | | | | ■ | ■ | ■ | ■ | ■ |

*Shaft basis* is the system of fits where the maximum shaft size is the basic size. The fundamental deviation for a shaft basis system is denoted by the lowercase letter "f" (see Figure 5.27(b)).

## 5.8.5 Geometric Characteristic Symbols

To raise the level of precision, a standardized system of geometric tolerance control symbols is used. Fourteen characteristic symbols used in GD&T are shown in Table 5.12. They are used to describe size, location, orientation, and form. These symbols have uniform meaning, readily adaptable to computer applications, and they are accepted internationally.

**Table 5.12** Characteristic Symbols Used in GD&T.

| Type of tolerance | Geometric characteristics | Symbol | Type of tolerance | Geometric characteristics | Symbol |
|---|---|---|---|---|---|
| FORM | Straightness | — | ORIENTATION | Angularity | ∠ |
| | Flatness | ▱ | | Perpendicularity | ⊥ |
| | Circularity | ◯ | | Parallelism | // |
| | Cylindricity | ⌭ | PROFILE | Profile of a line | ⌒ |
| LOCATION | Position | ⊕ | | Profile of a surface | ⌓ |
| | Concentricity | ◎ | RUNOUT | Circular runout | ↗ |
| | Symmetry | ⩸ | | Total runout | ↗↗ |

GD&T is applied on a drawing through feature control frames. The feature control frame is a rectangle divided into compartments shown in Figure 5.28. The first compartment contains a geometric tolerance symbol that shows the geometric characteristic to which the tolerance is applied, such as location, orientation, or form. In the example in Figure 5.28, the geometric tolerance symbol shows that the specified feature must lie perpendicular within a tolerance zone. The second compartment contains the tolerance value of 0.04 diameter and followed by a material condition symbol. The material condition symbol can then be followed by primary (in this case letter A), secondary, and tertiary datum reference letters, along with the material conditions of each datum.

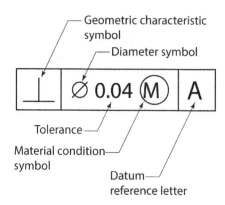

**Figure 5.28** Geometric tolerance control symbols.

Datum reference letters are usually used as reference tolerances to one of up to three perpendicular planes from which a measurement is made. However, datum reference letters can also show an exact point or axis.

### 5.8.6 Methods for Tolerance Stack-up Analysis in Assemblies

Assembly tolerance stack-up analysis predicts the tolerance value of the whole assembly or a specific gap of the assembly from the tolerance values of individual components. For example, if two mating parts have a tolerance of 0.003 in., then the total tolerance would stack up to 0.006 in. in the assembly.

Tolerance stack-up analysis can be performed in many ways such as worst-case analysis, statistical analysis, and Monte Carlo simulation. In this section, for predicting assembly measurement, two types of stack-up analysis will be discussed – worst-case analysis and statistical analysis.

#### 5.8.6.1 Worst-Case Analysis

This method predicts the maximum expected variation of the measurement by summing absolute values of the tolerances of the assembly components. Since the worst case requires tightest component tolerances, the manufacturing process is costly. This method requires 100 percent of the components to assemble and function properly. The simple formula to calculate the worst-case tolerance stacking for an assembly, $WCT_{assem}$, is given by

$$WCT_{assem} = \sum T_i,$$ (5.23)

where $T_i$ is the $i$th component tolerance.

## Example 5.9

Figure 5.29 shows the tolerances of three disks. Calculate the assembly dimension $H$ and its tolerance value by using worst-case analysis. Dimensions are in mm.

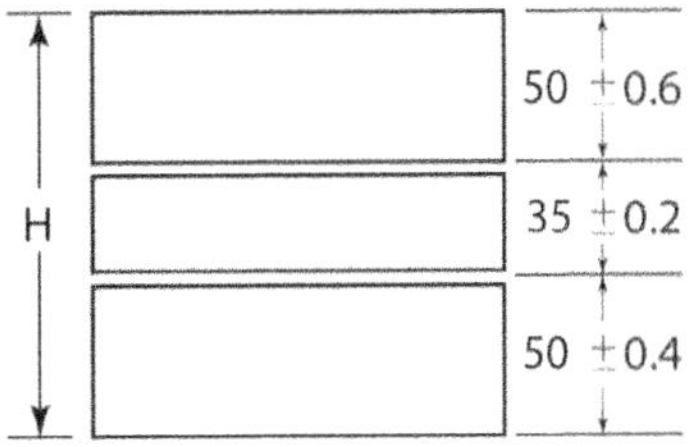

## Solution

Worst-case tolerance for the whole assembly is

$$WCT_{assem} = \sum T_i = 0.4 + 0.2 + 0.6 = 1.2.$$

Nominal hight, $H$, is

$$H = 50 + 35 + 50 = 135.$$

Then the overall dimension of the assembly is

$$H = 135 \mp 1.2.$$

The upper tolerance limit (UTL) is

$$UTL = 135 + 1.2 = 136.2.$$

The lower tolerance limit (LTL) is

$$UTL = 135 - 1.2 = 133.8.$$

**Figure 5.29** Disk assembly.

(The figure shows three stacked disks labeled: 50 ±0.6, 35 ±0.2, and 50 ±0.4, with overall height $H$.)

### 5.8.6.2 Statistical Tolerance Analysis (Root Sum Square)

The worst-case tolerance analysis is simple and parts will always fit. However, if we are producing thousands of parts, making sure that each and every component works properly is costly. The root sum square method works on a statistical approach and assumes that most likely the components fall in the middle of the tolerance zone rather than at the ends.

Assuming that the tolerances have a normal distribution and the half tolerance band is $3\hat{\sigma}$, the total assembly standard deviation, $\hat{\sigma}_{total}$, is given by

$$\hat{\sigma}_{total} = \sqrt{\hat{\sigma}_1^2 + \hat{\sigma}_2^2 + \cdots + \hat{\sigma}_n^2},$$ (5.24)

where $\hat{\sigma}$ is the standard deviation of the individual component. From equation (5.2), tolerance capability is

$$C_p = \frac{UTL - \mu}{3\hat{\sigma}}.$$ (5.25)

Setting $T = UTL - \mu$,

$$C_p = \frac{T}{3\hat{\sigma}} \tag{5.26}$$

or

$$\hat{\sigma} = \frac{T}{3C_p}. \tag{5.27}$$

Substituting equation (5.27) into equation (5.24), we have

$$\frac{T_{\text{total}}}{3C_{\text{assem}}} = \sqrt{\left(\frac{T_1}{3C_{p1}}\right)^2 + \left(\frac{T_2}{3C_{p2}}\right)^2 + \cdots + \left(\frac{T_n}{3C_{pn}}\right)^2}. \tag{5.28}$$

If we assume that the component tolerances in the assembly have an equal quality (the same $C_p$), then equation (5.28) simplifies to

$$T_{\text{total}} = \sqrt{T_1^2 + T_2^2 + \cdots + T_n^2}, \tag{5.29}$$

where

$$T_{\text{total}} = \text{total assembly tolerance,}$$
$$T_1, T_2, \ldots, T_n = \text{individual component tolerances.}$$

### Example 5.10

Referring to Example 5.9, calculate the assembly dimension $H$ and its tolerance value by using statistical analysis.

**Solution**

Using the statistical tolerance analysis approach, the total assembly tolerance from equation (5.29) is

$$T_{\text{total}} = \sqrt{T_1^2 + T_2^2 + \cdots + T_n^2}$$
$$= \sqrt{0.6^2 + 0.2^2 + 0.4^2} = 0.75.$$

Then the overall dimension of the assembly becomes

$$H = 135 \mp 0.75.$$

The upper tolerance limit is

$$UTL = 135 + 0.75 = 135.75.$$

The lower tolerance limit is

$$UTL = 135 - 0.75 = 134.25.$$

### 5.8.6.3  Comparison of Worst-Case and Statistical Approaches

Considering Examples 5.9 and 5.10, the assembly dimension tolerance obtained from statistical analysis is 37.5% smaller than the worst-case tolerance. This is a very significant decrease obtained from the 1.2 mm worst-case analysis. The statistical tolerance analysis approach is used for a given assembly tolerance to increase the component tolerances, hence lowering manufacturing costs. If we increase the tolerance of each component by 50%, statistical analysis will provide approximately 1.12 mm assembly tolerance, which is still below the 1.2 mm worst-case assembly tolerance. We observe a significant difference between two methods. That is why statistical tolerance analysis is used to increase component tolerances which reduces the manufacturing cost.[5]

## 5.9  Future of Manufacturing: Additive Manufacturing

From the 1700s manufacturing was primarily limited to conventional machining methods such as turning and milling. These machining approaches have dominated manufacturing industry for the last several centuries and have shaped the designer's way of thinking about form and function. Even though these conventional machining techniques have greatly improved in precision, cost, and productivity they possess inherent limitations on fabrication of complex parts and shapes. With the advent of additive manufacturing (AM), complexity in design comes "free" as product components are built by melting thin layers of material powder in successive overlapping 2-D shapes creating a 3-D part. Material is added without the need for traditional tooling (see Figure 5.30) and can take very complex shapes and forms. This new method is complementary to traditional metal-removing methods and can be integrated into the existing production workspace.

**Figure 5.30** Additive manufacturing.

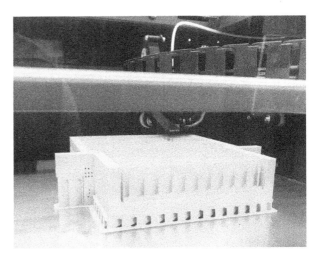

---

5  For more information about this subject, see "Product design and analysis," http://www.pddnet.com/blogs/2013/06/statistical-tolerance-analysis, accessed June 3, 2013.

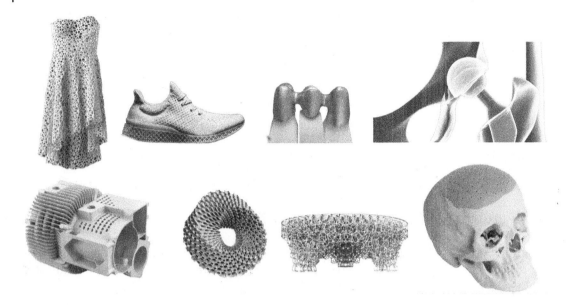

**Figure 5.31** Applications of additive manufacturing (reproduced from different internet sources).

As shown in Figure 5.31, applications of additive manufacturing include not only complex shape component manufacturing but also wide-reaching solutions for industry. AM has quickly taken a foothold in mainstream manufacturing processes, penetrating the aerospace, defense, art and design industries.

Surgeons are now creating hi-tech virtual models of their patients before operating on knee and hip joints, which are then used to create custom precision parts for joints. This could both drastically improve the effectiveness of the surgery and reduce recovery times. In addition to superior design performance, the AM approach has the potential to drastically reduce cost. For example, in 2015, if the orthopedic industry used 1,500 to 2,000 tons of titanium, AM would require less than 50 tons, a simple cost reduction due to minimizing material waste. Cost reduction using AM can also be achieved by combining several components into a single part. This advantage reduces part count, tolerance stack-ups, and assembly time operating costs per unit sold.[6]

Additive manufacturing is also successfully used for implementation of dental parts: patient specific bridges, crowns, and partial dentures are printed using biocompatible cobalt chrome powder. In the automotive industry, advances in AM technologies have opened doors for newer designs; cleaner, lighter, and safer products; shorter lead times; and lower costs.[7]

---

6  Madani, A., "Additive manufacturing status in the orthopedics industry," http://www.odtmag.com/contents/view_online-exclusives/2016-05-04/additive-manufacturing-status-in-the-orthopedics-industry, accessed, January 5, 2017.
7  Cotteleer, M., Holdowsky, J., and Mahto, M., *The 3D Opportunity Primer: The Basics of Additive manufacturing*, Deloitte University Press, 2014.

So far, AM has mainly been used for rapid prototyping within different industries, but in recent years this technology has been used in the fashion industry, for example, in magnificent-looking complex running shoes and dress designs.

It has taken researchers 20 years to improve the AM capability to the point where it is now being used in mainstream processes. Now researchers are investigating how 3-D printing will be able to use cotton, other natural fibers, and biological material. In fact, new research, published in the journal *Science Advances*, demonstrates that it is possible to replicate the heart through 3-D printing.[8]

While design for manufacturing is required to keep the product development process efficient and cost-effective, this approach is no longer necessary when it comes to AM. However, with this new technology in product design, there is a need for design for additive manufacturing (DFAM). Building a product no longer requires multiple parts to be manufactured separately and assembled together, but the product can be manufactured as single system components.

The unique capabilities of AM technologies bring new opportunities for customization, improvements in product performance, multifunctionality, and lower overall manufacturing costs. The following unique capabilities of this new technology enable design freedom and push complexity to new limits:[9]

- **Shape complexity**. Through shape optimization, it is possible to build virtually any complex shape.
- **Material complexity**. Material can be processed one point, or one layer, at a time, enabling product components with complex material compositions and designed property gradients.
- **Hierarchical complexity**. Multiscale structures can be designed and fabricated from the microstructure through geometric mesostructure (> 5 mm) to the part-scale macrostructure.
- **Functional complexity**. Complete functional assemblies and mechanisms can be manufactured at any time.

### 5.9.1 Generic AM Process Flow

The Generic AM process can be summarized in eight steps as shown in Figure 5.32[10].

*Step 1* is the development of CAD solid modeling that describes the external geometry of the product.

*Step 2* is the exporting of the CAD output as an STL file. STL is the standard file type used by most additive manufacturing systems. It is a triangulated representation of a 3-D CAD model.

*Step 3* is transferring the STL file to the AM machine. This step may need some changes in the programming format to make sure that required design parameters (correct size, positions, etc.) are correctly defined.

8 Hinton, T.J., et al., "Three-dimensional printing of complex biological structures by freeform reversible embedding of suspended hydrogels," *Science Advances*, 1(9), e1500758, 2015.
9 David, W.R., "Design for additive manufacturing: A method to explore unexplored regions of the design space," https://sffsymposium.engr.utexas.edu/Manuscripts/2007/2007-34-Rosen.pdf, accessed January 2, 2017.
10 Gibson, I., Rosen, D., and Stucker, B., *Additive Manufacturing Technologies*, Springer, New York, 2015.

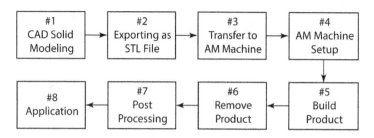

**Figure 5.32** Generic AM process.

*Step 4* is the AM machine setup related to the product building parameters – material constraints, layer thickness, timing, etc.

*Step 5* is automatically building the product by the AM machine. The AM technician must make sure that machine has enough material to finish the job of building.

*Step 6* is removing the finish product from the AM machine. Again, the AM technician must make sure that the operating temperature is sufficiently low and the machine is completely stopped.

*Step 7* is the post-processing step – separation of supports from the product, cleaning, powder removal, etc.

*Step 8* is the application step – any other required steps such as acceptable surface finish, painting, etc. before the product is used.

During the US government sponsored study of European researcher groups, many researchers mentioned that they can see the lack of capable CAD tools as a serious barrier for their research and for the utilization of AM technologies for product manufacturing.[11] But, if suitable CAD and DFM methods and tools can be developed, designers can design components with notably improved manufacturing processes. With the right shape, material, and hierarchical complexity capabilities, DFM can move from an emphasis on cost minimization to a focus on achieving incredible capabilities.[9]

---

**CASE STUDY 5.1**

Just in time (JIT) inventory management is always a challenge for all organizations due to the heavy cost associated with inventory holding and also due to organizations' production processes. Suppose the most important issues of the JIT supply chain we identified as follows: (1) high variance in demand and supply; (2) high machine setup time; (3) lack of flexibility of manufacturing system; (4) supplier capacity constraints; (5) line stoppage; (6) poor product

---

11 Rosen, D.W., Atwood, C., Beaman, J., Bourell, D., Bergman, T., and Hollister, S., "Results of WTEC Additive/Subtractive Manufacturing Study of European Research," *Proceedings of SME Rapid Prototyping & Manufacturing Conference*, paper TP04PUB211, Dearborn, MI, May 10–13, 2004. See also http://wtec.org/additive/welcome.htm for the complete report.

quality; (7) optimization of product variety; (8) high production cost. Assume that using domain expert opinions, contextual relationships of identified challenging issues are given in the structural self-interaction matrix shown in Figure 5.33.[a] Investigate the complexity of the JIT inventory management process.

## Solution

Interpretive structural modeling (ISM) will be used for the complexity analysis of the supply chain in question. ISM helps us to identify the complex relations between the main factors given in Figure 5.33. Assume that, through domain experts and group discussion, a decision is made as to whether and how the factors given in Figure 5.33 are related.

| | | High variance in demand and supply | High machine setup time | Lack of flexibility of manufacturing system | Supplier capacity constraints | Line stoppage | Poor product quality | Optimization of product variety | High production cost |
|---|---|---|---|---|---|---|---|---|---|
| | | 1 | 2 | 3 | 4 | 5 | 6 | 7 | 8 |
| 1 | High variance in demand and supply | | A | A | A | V | O | V | V |
| 2 | High machine setup time | | | V | O | V | O | V | V |
| 3 | Lack of flexibility of manufacturing system | | | | A | V | O | V | V |
| 4 | Supplier capacity constraints | | | | | V | O | V | O |
| 5 | Line stoppage | | | | | | O | O | V |
| 6 | Poor product quality | | | | | | | O | V |
| 7 | Optimization of product variety | | | | | | | | V |
| 8 | High production cost | | | | | | | | |

**Figure 5.33** Structural self-interaction matrix.

*(Continued)*

**CASE STUDY 5.1 (Continued)**

| | | High variance in demand and supply | High machine setup time | Lack of flexibility of manufacturing system | Supplier capacity constraints | Line stoppage | Poor product quality | Optimization of product variety | High production cost | Driving power |
|---|---|---|---|---|---|---|---|---|---|---|
| | | 1 | 2 | 3 | 4 | 5 | 6 | 7 | 8 | |
| 1 | High variance in demand and supply | 1 | 0 | 0 | 0 | 1 | 0 | 1 | 1 | 4 |
| 2 | High machine setup time | 1 | 1 | 1 | 0 | 1 | 0 | 1 | 1 | 6 |
| 3 | Lack of flexibility of manufacturing system | 1 | 0 | 1 | 0 | 1 | 0 | 1 | 1 | 5 |
| 4 | Supplier capacity constraints | 1 | 0 | 1 | 1 | 1 | 0 | 1 | 1 | 6 |
| 5 | Line stoppage | 0 | 0 | 0 | 0 | 1 | 0 | 0 | 1 | 2 |
| 6 | Poor product quality | 0 | 0 | 0 | 0 | 0 | 1 | 0 | 1 | 2 |
| 7 | Optimization of product variety | 0 | 0 | 0 | 0 | 0 | 0 | 1 | 1 | 2 |
| 8 | High production cost | 0 | 0 | 0 | 0 | 0 | 0 | 0 | 1 | 1 |
| | Dependence | 4 | 1 | 3 | 1 | 5 | 1 | 5 | 8 | $\sum 28$ |

**Figure 5.34** Final reachability matrix, $R_f$.

Considering transitivity, from SSIM the final reachability matrix is developed (see Figure 5.34). After level partitioning of the reachability matrix, a digraph is developed with transitivity links removed (see Figure 5.35). This figure shows that there is no coupling between the supply chain factors.

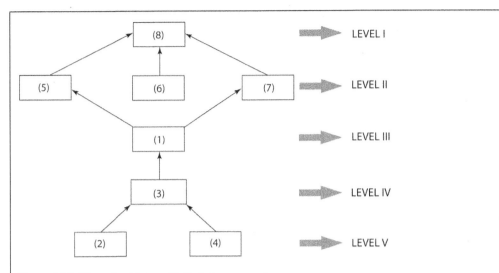

**Figure 5.35** Digraph with transitivity links removed.

**Figure 5.36** MICMAC analysis.

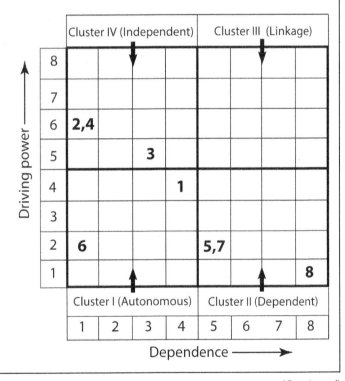

---

**CASE STUDY 5.1 (Continued)**

---

As shown in Figure 5.35, at level V, high machine setup time (2) and supplier capacity constraints (4) are the drivers which have high leverage for the JIT supply chain. As seen in Figure 5.36, factors affecting the supply chain have been classified into four categories. This figure also shows that the highest driving power of the supply chain lies with high machine setup time (2) and supplier capacity constraints (4), being independent factors in cluster IV. In cluster IV, lack of flexibility of the manufacturing system (3) is also the key driving factor for JIT supply chain. Because supply chain efficiency significantly depends on these factors, management should understand the importance of these factors when considering planning and implementation of JIT.

High production cost (8) is classified as a risk highly dependent on supply chain efficiency, followed by line stoppage (5) and optimization of product variety (7). They have minor driving power and are at the top of the ISM hierarchy.

In cluster I, high variance in demand and supply (1) and especially poor product quality (6) have very weak risk supply chain efficiency – they do not have much influence on the system efficiency. It is important to note that there are no factors affecting the efficiency of the supply chain in linkage cluster III.

[a]Adapted from Mahajan, V.B., Jadhav, J.R., Kalamkar, V.R., and Narkhede, B.E., "Interpretive structural modelling for challenging issues in JIT supply chain: Product variety perspective," *International Journal of Supply Chain Management*, 2(4), 50–63, 2013.

---

# Bibliography

1 CLAUSING, D., *Total Quality Development*, ASME, New York, 1994.

2 HARRY, M., and SCHROEDER, R., *Six Sigma*, Doubleday, New York, 2000.

3 KALPAKJIAN, S., and SCHMID, S., *Manufacturing Engineering & Technology*, Pearson, Upper Saddle River, NJ, 2013.

4 LOCHNER, R.H., and MATAR, J.E., *Design for Quality*, ASQC Quality Press, White Plains, NY, 1990.

5 WILSON, L., *How To Implement Lean Manufacturing*, McGraw-Hill Education, New York, 2009.

6 WOMACK, J.P., and JONES, D.T., *Lean Thinking*, Free Press, New York, 2003.

## CHAPTER 5 Problems

**5.1** For the manufactured parts shown in Figure P5.1 suggest changes in the design that will (a) improve the joining process for the spectrophotometer knob, and simplify the handling and orientation procedure for (b)the motor bracket and (c) the link.

**Figure P5.1** Parts for improvement

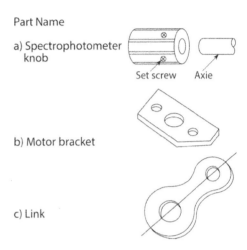

Part Name

a) Spectrophotometer knob

Set screw    Axie

b) Motor bracket

c) Link

**5.2** A company discovered 8 errors while processing 24,000 payments to its employees during the last year. (a) Calculate the EPMO. (b) At what level does the manufacturing company perform?

**5.3** Assume that a pin target diameter is set at 8 mm. If the process output has a short-term standard deviation, $\hat{\sigma} = 0.04$ mm, what would be the specification limits to ensure that the pin manufactured will be within the design specification limits when $6\hat{\sigma}$ process spread is used. Assume that the process output is approximately normally distributed.

**5.4** In Problem 5.3, assume that the process has upper specification limit 8.08 mm, and lower specification limit 7.90 mm. If the observed process mean is 8.04 mm and the short-term standard deviation is 0.03 mm, investigate whether the process is capable of manufacturing pins within the defined specification limits. Also calculate the total percentage of expected shafts that would be out of specification.

**5.5** Using the $Z$ score, calculate the long-term sigma level of the process if $USL = 4, LSL = 3.2$, $\mu = 3.78$, and the long-term standard deviation $S = 0.16$.

**5.6** If the testing defects removal efficiency $DRE = 0.92$ and the observed defects are $D_t = 0.28$, calculate the submitted and escaping defects to customer.

**5.7** A manufacturing company detected 9 errors in processing 30,000 manufacturing data. (a) Calculate the EPMO. (b) At what level does the manufacturing company perform?

**5.8** Consecutive diameter measurements taken every hour from manufacturing precision-made pins are shown in Table P5.8. Construct $\overline{X}$ and R charts to establish the statistical control of the manufacturing process. Discuss the result.

**Table P5.8** Pin diameter measurement data.

| Observation number | Diameter mm | | | |
|:---:|:---:|:---:|:---:|:---:|
| 1 | 30.0 | 29.8 | 30.0 | 30.2 |
| 2 | 30.0 | 29.8 | 29.6 | 29.8 |
| 3 | 30.0 | 29.9 | 29.7 | 30.2 |
| 4 | 29.8 | 29.5 | 29.7 | 29.0 |
| 5 | 30.0 | 30.4 | 30.0 | 30.6 |
| 6 | 30.1 | 30.4 | 30.3 | 30.6 |
| 7 | 29.7 | 29.8 | 29.4 | 29.7 |
| 8 | 30.0 | 29.8 | 29.9 | 30.1 |
| 9 | 30.0 | 30.4 | 30.0 | 29.8 |
| 10 | 29.2 | 29.8 | 29.8 | 30.0 |
| 11 | 29.6 | 30.2 | 29.8 | 29.8 |
| 12 | 29.4 | 29.8 | 29.7 | 30.1 |
| 13 | 30.0 | 30.2 | 30.4 | 30.0 |
| 14 | 29.8 | 29.6 | 30.0 | 30.2 |
| 15 | 30.1 | 30.3 | 30.2 | 30.0 |
| 16 | 30.1 | 29.7 | 30.1 | 30.3 |
| 17 | 30.1 | 30.1 | 29.9 | 29.5 |
| 18 | 30.0 | 30.2 | 30.4 | 30.2 |
| 19 | 30.2 | 30.1 | 30.3 | 30.2 |
| 20 | 30.0 | 29.8 | 29.6 | 29.4 |

**5.9** If the specification limits on the pins given in Problem 5.8 are
a) $29.9525 \mp 0.8000$ mm,
b) $29.9525 \mp 0.75036$ mm,
c) $29.9525 \mp 0.5000$ mm,
discuss the relationship between process variability and specification limits.

**5.10**  Figure P5.10 shows the tolerances of three disks. Calculate the assembly dimension $H$ and its tolerance value by using worst-case analysis. Dimensions are in mm.

**Figure P5.10**  Disk assembly.

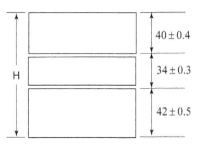

**5.11**  Referring to Problem 5.10, calculate the assembly dimension $H$ and its tolerance value by using statistical analysis.

**5.12**  Referring to Case Study 5.1, develop the adjacency matrix.

**5.13**  Referring to Case Study 5.1, develop the level partitioning tables of the reachability matrix.

**5.14**  Assume that four lean practice main factors and their relationships were identified through literature review and experts opinion as shown in Figure P5.14. Develop the adjacency matrix, final reachability matrix, level partitioning tables, digraph, and MICMAC analysis, and discuss the lean practice with the factors in Figure P5.14.[1]

**Figure P5.14**  Structural self-interaction matrix.

|   |                     | Waste elimination | Delivery reliability | Volume flexibility | Low cost |
|---|---------------------|:---:|:---:|:---:|:---:|
|   |                     | 1 | 2 | 3 | 4 |
| 1 | Waste elimination   |   | V | V | V |
| 2 | Delivery reliability |   |   | A | V |
| 3 | Volume flexibility  |   |   |   | V |
| 4 | Low cost            |   |   |   |   |

---

1  Adapted from Jadhav J.R., Mantha S.S., and Rane, S.B., "Development of framework for sustainable Lean implementation: An ISM approach," *Journal of Industrial Engineering International*, September 2014, 10:72.

# 6

# Design Analyses for Material Selection

Engineering has never been easy. The speed of introduction of new materials, tools, and techniques is increasing. We are approaching a human processing bottleneck for effective use of these inventions in better engineering of cost-effective, timely, useful and reliable artifacts.

## 6.1   Introduction

Materials engineering is a field of engineering that includes a range of kinds of material and how to use them in manufacturing. To convert the basic materials into an engineering product, materials with specific properties should be selected. Material selection is a process that is performed to select the best materials for the specific application and product development. If an appropriate material selection is not performed, the product life tends to be highly unpredictable.

The competitive market mechanism and increase in product consumption makes product designers think more about materials than before. Among the many factors to consider in product design, selecting proper materials and embracing the challenges to explore and develop a variety of manufacturing processes associated with design are two of the most important responsibilities of a design engineer. A primary design requirement in the selection of the proper materials for a specific application is that the material be capable of meeting the design service life requirement at the least cost. For a defined service life, first, the designer assess the suitability of a range of candidate materials. Then, the designer narrows the available choices to a few candidate materials that best meet the mechanical, thermal, electrical, and economic constraints of the product design. The final selection of a particular material can be made based on past experience, accessed through a questionnaire-based selection engine. Selecting materials based on past experience is still popular because the designer feels confident in using a tried and proven material. However, rapidly changing advances in technology are demanding better performance from engineering materials. Hence, selecting the proper engineering material for a given application requires a broad knowledge of the state of the art in materials development.

## 6.2 General Steps in Materials Selection

- Consider functional requirements for an intended use of the product.
- Analyze the material performance requirements.
- Develop alternative solutions for the material selection process.
- Identify candidate materials which satisfy the requirements.
- Make a decision on the appropriate material for the product considering load, shape, size, manufacturability, availability, cost, and environmental issues.

For materials selection, especially during new product development, designers usually consider ways to reduce costs, improve product performance, and improve manufacturing or assembly yields. In other words, select the lowest-cost material that enables the product's expected performance. Think about the large number of different kinds of material that exist from which to select and the vast amount of information available on which processes could be used for product manufacturing. Asking following questions would be appropriate:

- What do we need to know for material selection?
- How do we begin material selection?
- How do we decide which process would be the best to satisfy the functional requirements?

In the following sections, we will take a systematic approach to the material selection process.

### 6.2.1 Interactions between Factors Affecting Material Selection

Materials selection may involve a complex set of relationships among product function, materials, manufacturing process, shape, load, and size. Materials selection for a proposed product may not be an easy task; it involves a map of the complex relationships between various factors affecting the materials selection process. This unclear complex relationship map should be transformed into a visible, well-defined, and relatively easily solvable model – decompose a complex issue into smaller sub-issues and build a simpler multilevel structural model.

Interpretive structural modeling (ISM) methodology can be used to deal with this complex issue; it provides a fundamental understanding of the various elements relevant to the material selection process (for more information on ISM, refer to Chapter 1). As previously mentioned, the six most important factors affecting material selection are identified and a contextual relationship is establish to develop structural self-interaction matrix shown in Figure 6.1. In this matrix, "X" represents a two-way interaction whereas "A" and "V" represent a one-way interaction.

In general, product function dictates the choice of both material and shape. Process is affected by the material selected. Process also interacts with shape as the shape of

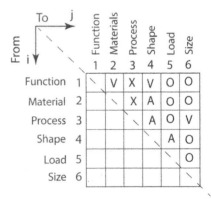

**Figure 6.1** Structural self-interaction matrix for material selection.

the product affects the manufacturing process that will be used. The shape requirement dictates the choice of material and process. However, the requirements of the manufacturing process affect the materials that will be used and the shape of the products they can take. Load has a strong relationship with the shape of the product.[1]

Figure 6.1 reflects all these interactions. Note that in developing structural self-interaction matrix, if the relationship between factors is weak, it is assumed that there is no relationship between factors.

After removing the transitivities based on the reachability matrix as described in the ISM approach in Chapter 1, the structural hierarchy of the performance factors in material selection is developed as shown in Figure 6.2. As shown in Figure 6.2, the unclear complex relationship map among the six factors affecting the material selection process is transformed into a visible, well-defined, and relatively easily solvable model. This figure depicts visually the direct and indirect relationships between the factors affecting the material selection process. As shown in Figure 6.3, a MICMAC analysis to arrange the factors with respect to their driving power and dependence into four clusters is also developed.

There is no factor that has been identified as an autonomous factor in cluster I (see Figure 6.3). This indicates that there is no disconnected factor in the analysis of material selection – they are all important.

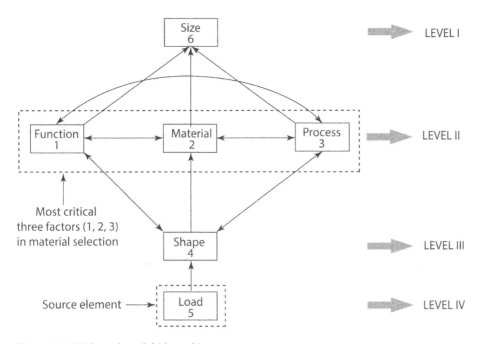

**Figure 6.2** ISM based model (digraph).

1 Ashby, M.F., *Materials Selection in Mechanical Design*, 3rd edn, Butterworth-Heinemann, Oxford, 2005.

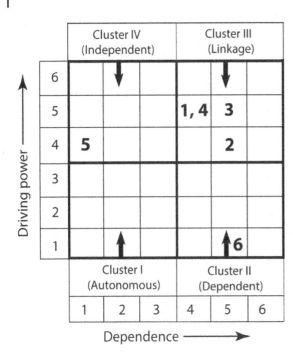

**Figure 6.3** MICMAC analysis.

As shown in Figure 6.3, cluster II includes the dependent factor of size (factor 6) with low driving power and high dependence; size may not affect other factors but it is affected by other factors of material selection process.

The factors in the linkage cluster III (functions, material, manufacturing process, and shape) have to be given extreme importance due to their high driving power and high dependence. Being the most important factors, they affect and depend on other factors in material selection. Therefore, as mentioned earlier, optimum material selection is a challenging task in design and these four major factors within this cluster which influence the material selection performance must be carefully analyzed and integrated with the whole product design.

Factor load (5) in cluster IV is the key driver for the material selection. As shown in Figure 6.3, load is an independent factor with strong driving power but very weak dependence. As load directly and indirectly has impact on other factors, designers have to pay maximum attention to the nature of the load applied on a machine component. As seen from Figure 6.2, load affects the shape of a product to be designed. For example, if the load is bending, an I-section of the beam is desirable. If the load is torsion, a circular tube is better than either an I-section or solid section. However, when the load is axial, the shape of the cross-section is not important – the area of the cross-section is important.[2]

---

2 For more information refer to Ashby, M.F., "Materials and shapes," *Acta Metallurgica et Materialia*, 39(6), 1025–1039, 1991.

Element 6 (load factor) is the source element since it has only an outgoing path: material selection analysis should start from this element. The next step is to design and optimize the shape of the components which will carry the applied mode of loading. Since a large number of lines enter and leave the elements at level III (function, material, and the manufacturing process) as shown in Figure 6.2, they are the most critical factors which need to be analyzed with great care.

## 6.2.2 Interactions between Material Selection Factors

Figure 6.2 provides two types of relationships:

1) interactions between functional, materials, process, and shape (see Figure 6.4);
2) interactions between function, materials, and process (see Figure 6.5).

The interaction between function, material, process, and shape is shown in Figure 6.4. Ashby terms these interactions the central problem of material selection. As seen from the ISM model, there is a walk from shape to material. Since material has less of an effect (weak) on shape, the walk is considered one-way. Note that interactions between other elements are two-way. The manufacturing process dictates the shape and size. To give a certain shape to material, an appropriate manufacturing process is used. Figure 6.4 can be reduced to three critical elements as shown in Figure 6.5. A detailed flow chart for the material selection process representing Figure 6.5 is giving in Figure 6.6.

As depicted in Figure 6.6, once the design requirements and constraints are defined, the designer must consider the product configuration and material requirements. Materials must satisfy the requirements of operating in harsh environments and still maintain required performance, for example, turbine blades in jet engines operate in extreme temperature environments; structural materials subject to radiation and corrosive environments such as nuclear reactors; structural materials subject to extreme loading conditions in corrosive environments such as offshore structures.

The next step in the process is the analysis of the critical performance domain. The critical performance domain has the greatest influence on the material selection process to select the most suitable material for a product. Within this domain, product functionality dictates the choice of material.

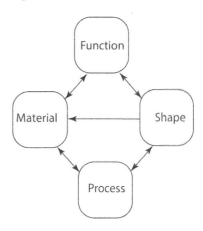

**Figure 6.4** Interractions between function, material, process, and shape.

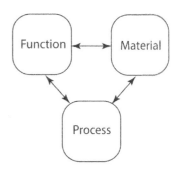

**Figure 6.5** Interactions between function, material, and process.

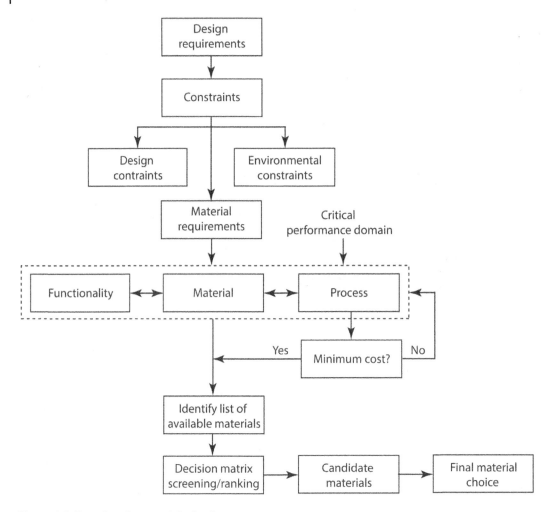

**Figure 6.6** Flow chart for material selection.

For example, functional requirements may require the material to have certain performance characteristics such as a certain stiffness and strength. In many cases, more than one material can satisfy the design performance requirements. Therefore, the various manufacturing processes must be considered which affect the final cost of the product. Cost is always an important consideration in product design and development. The designer should make sure that a selected material will not only satisfy the desired final product performance but also be acceptable to the producer in terms of lower material cost and manufacturing cost. Tradeoffs should be exposed and effort will be made to make an educated decision – for example, in the material selection process, one material could be very costly, but provide a low-maintenance product; and at the same time, if materials can be easily recycled, disposal costs are reduced.

After identifying an available and suitable range of candidate materials, a decision matrix is developed for screening. Through ranking, the final material choice will be decided to achieve the acceptable product performance.

The decision matrix can be an important tool in the overall design process, but it should be kept in mind that nothing takes the place of common sense and good judgment. For methods like the decision matrix to be viable, estimates must be made about the relative importance of the different evaluation factors. Note that these estimates have a certain built-in uncertainty that may result in wrong conclusions as to the best or most effective solution or alternative. Therefore, it is important to perform the evaluation of alternatives using at least two independent methods.

### 6.2.3 Other Factors in Material Selection

The most important factors that should be considered during the selection process include:

1) Availability
2) Manufacturability
3) Repairability
4) Reliability
5) Service environment
6) Compatibility
7) Cost.

Factors to be considered in determining the availability of materials are as follows:

1) Thicknesses
2) Widths
3) Minimum quantities
4) Number of sources.

Failure to factor these considerations into the overall design process will result in serious scheduling difficulties.

The designer should not overlook manufacturability; The designer is responsible for making certain that the design is producible with the selected material. As shown in Figure 6.7,[3] material selection may significantly affect the manufacturing cost. Because high-performance alloys have a low machinability rating, the relative cost of machining alloy steel is high in comparison with that of carbon steels. This is due to alloy steel's high strength, high resistance to shear load, and rapid work hardening. Careful attention is therefore necessary when making

3 Trucks, H.E., *Design for Economical Production*, Society of Manufacturing Engineers Publications Development Department, Dearborn, MI, 1987, p. 2.

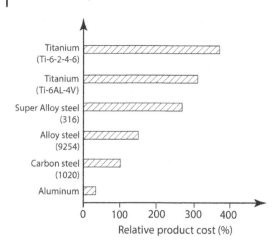

**Figure 6.7** Relative production cost for various types of materials.

the selection of materials in the early stages of design. Once the product is manufactured and is in service, the material of construction must be easily and economically repaired. Part repair must often be done within a short period of time, and procedures must be as simple as possible. Selected material candidates should require a minimum of special processes, tools, and skills for repair. Therefore, repair techniques must be considered during the material selection process.

Other factors that should be considered during the selection process include past experience, service history, mechanical and physical material properties as affected by mill and manufacturing processes, quality control requirements, and correlation of laboratory test data, such as fracture toughness, impact, and stress corrosion, to the service requirements. Prior service history is a very important criterion for material selection. Although some new materials offer significant improvements in physical and chemical properties, designers must be cautious in using them until their performance is proven. Designers should recognize the risk of failure when a new unproven material is selected for an application. Selected materials must perform under the specified environmental conditions; thus, the designer should review the material properties and the predicted performance in the environment in which the product will function. The compatibility of two or more materials used in an application is another important factor in selection. For example, two materials performing in a high-temperature environment may fail if their coefficients of thermal expansion are significantly different.

Perhaps the most important consideration of all in the selection process is the cost of the material for the proposed application. Cost is the most important selection factor in reducing the number of possible material candidates to a manageable level. To finalize the choice, the total cost of a material used for a specific application must be compared with the total cost of alternate materials. Cost and weight can be incorporated into property parameters (strength or stiffness) to facilitate comparisons.

**Figure 6.8** Material selection at design stages (adapted from Ashby[1]).

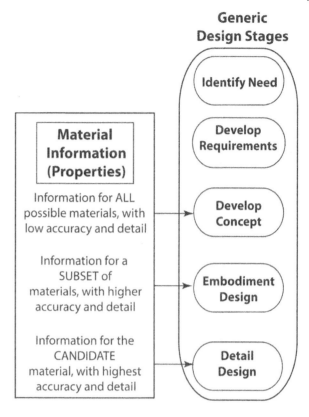

**Generic Design Stages**

Identify Need

Develop Requirements

**Material Information (Properties)**

Information for ALL possible materials, with low accuracy and detail → Develop Concept

Information for a SUBSET of materials, with higher accuracy and detail → Embodiment Design

Information for the CANDIDATE material, with highest accuracy and detail → Detail Design

### 6.2.4 Material Selection at Design Stages

Figure 6.8 illustrates material selection at each design stage.[1] Decisions concerning the choice of materials and processes can occur at varying times during the design process. Although materials are selected during detail design, a variety of candidate materials may be identified during the concept and embodiment design stages of the effort. Material changes may sometimes be required during the early production phases as a result of unanticipated machining and assembly problems. However, the need for material changes this late in the process is decreasing as a result of the integration of design and production with concurrent engineering and other similar concepts. Infrequently, material changes are required after the product has been placed in operation such as recall. For example, an auto recall occurs when a manufacturer determines that a car model has a safety-related defect due to the material used in production.

During the selection of the optimum solution to a design problem, an attempt should be made to determine if the concept is unduly limited by the materials selected. Based on the design requirements, certain classes of materials may be eliminated from consideration, and other classes may be selected as suitable candidates for the final choice because of their characteristics and properties. Conversely, in some instances, it may be advisable to modify the design concept if it is limited by the available materials.

At the detail design stage, candidate materials identified earlier in the design process are subjected to an in-depth analysis. At this stage, analysis and determination of the material requirements should be accomplished, selection and evaluation of candidate materials should be completed, and the materials selection decision should be finalized.

See http://www.matweb.com/search/CompositionSearch.aspx for data and background information on materials' properties of materials.

### 6.2.5 Ashby Charts and Material Selection

Initial selection of candidate materials for a product design can be performed by using Ashby charts (see, for example, Figures 6.9, 6.10, and 6.11). These charts provide physical insight

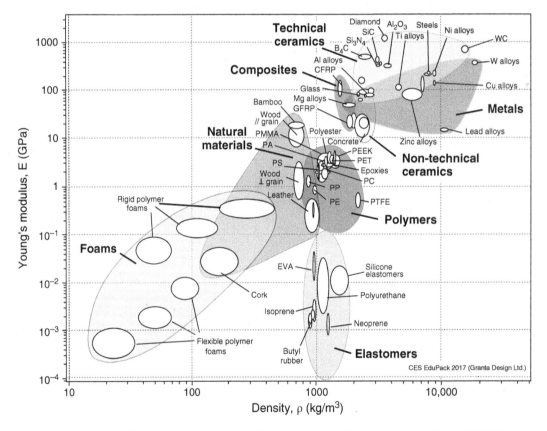

**Figure 6.9** Young's modulus versus density chart (from Ashby, M.F., "Material property charts," CES EduPack, Granta Design, Cambridge, 2017 (used by permission of Granta Design Ltd.).

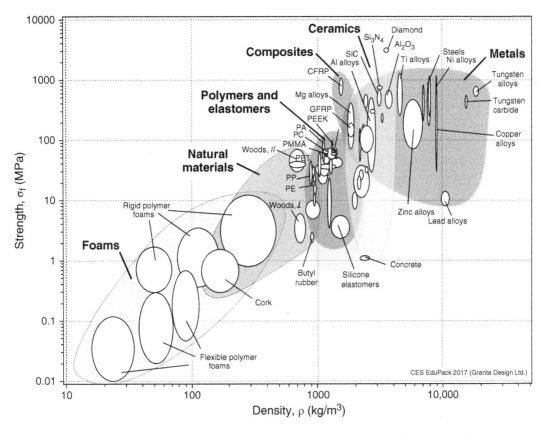

**Figure 6.10** Strength versus density chart (From Ashby, M.F., "Material property charts," CES EduPack 2017 (used by permission of Granta Design Ltd.).

into tradeoffs (usually between product performance and cost) by pairing properties which normally should both be considered at once. This process will help to prevent the need to work with many material property tables. Ashby materials selection charts are a unique graphical way of presenting material property data. They provide materials property data as *balloons* in an easy way to compare and show the relative position for all of the materials being considered for a specific design application. This will help to focus on possible candidates, and also helps to develop physical insight into the relative performance of materials. When data for a given material class such as metals, ceramics, composites, and polymers are plotted, as shown in Figure 6.9, they are grouped in an enclosed field ("balloon"). These distinct balloons which define the ranges of their properties may overlap and may have little balloons within their enclosed region.

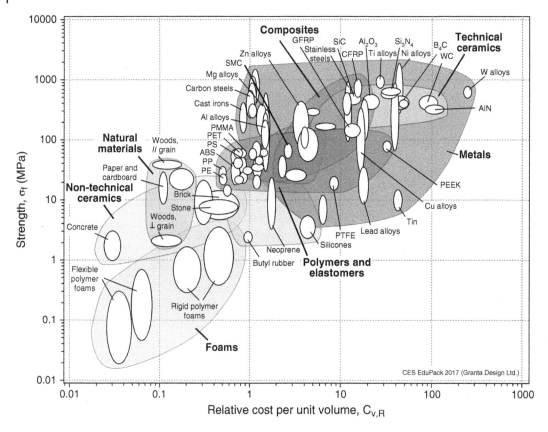

**Figure 6.11** Strength versus relative cost per unit volume chart (From Ashby, M.F., "Material property charts," CES EduPack 2017 (used by permission of Granta Design Ltd.).

Ashby material bar charts, shown in Figures 6.12 and 6.13, can also be used for initial material selections. Figures 6.12 shows the cost per unit mass of engineering materials as a bar chart. Each bar represents a single material. The length of the bar shows the range of cost displayed by that material in its various forms. Note that although material properties such as strength and stiffness do not change with time, the cost of the materials can change depending on the market values at the time of need.

As shown in Figure 6.13, each bar describes one material; its length shows the range of modulus exhibited by that material. As in the case of Figures 6.12, the materials are separated by class. Each class shows a characteristic range: Metals and ceramics have high moduli; polymers have low; hybrids show relatively moderate modulus of elasticity.

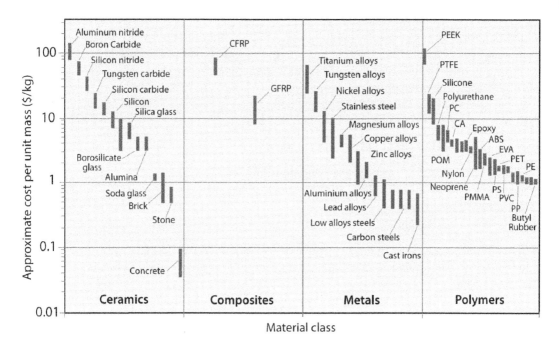

**Figure 6.12** Approximate material costs chart. (Reproduced from Ashby, M.F., "Materials and process selection charts," Granta Design, p. 19, 2010.)

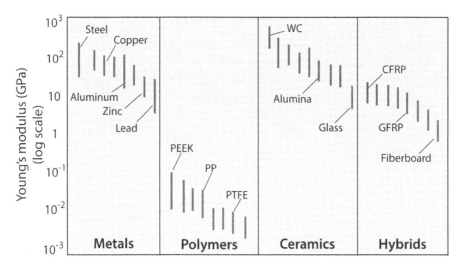

**Figure 6.13** Young's modulus chart. (Reproduced from Ashby, M.F. and Cebon, D., *New Approaches to Materials Education*, Granta Design, Cambridge, 2007.)

## Example 6.1

Let us think of a material selection problem where the specification is for a component that is light, strong, has good vibration damping, and is cost-effective (e.g. the inner core of a snowboard). We wish to narrow the initial material choice.

**Solution**

The problem can be structured as follows:

- Function: ability to glide on snow without vibration
- Objective: minimize cost
- Constraints: light and strong
- Free variable: material choice.

Many materials can be selected for the inner core of a snowboard. The strength–density chart (Figure 6.10) and cost–material class bar chart (Figure 6.12) can help us to narrow down the initial material choice.

Because a snowboard must be light, the density of the material should be less than 3 Mg/m$^3$. We might also select a relatively strong material which has a strength value of 25 MPa. These constraint values provide us a window in Figure 6.10. Within this window, some metals, ceramics, woods, and composites could be considered as the material candidates. We have many wood candidates but the strength of the wood is the lowest compared to other materials within this window. Ceramic is very brittle and not good as a vibration absorber.

The strength of the polymers are low, so most polymers are unlikely to be useful for this application. Steel is very heavy and is not within the selected window. Mg alloys could be a choice but it is relatively heavy, expensive and cannot absorb vibrations well.

The only material left within the region is composite – it is light and strong. Figure 6.12 reveals that cost of the composite is little high but reasonable – composites appear best of all.

## 6.3 Classification of Materials

There are numerous kinds of materials used in the field of engineering. As shown in Table 6.1, materials can be classified into six groups: metals, ceramics, polymers, composites, semiconductors, and new materials. These classes can be further organized into various sub-groups based on their mechanical, physical, and chemical properties.

### 6.3.1 New Materials

New materials are necessary for innovative new design development. Big spikes in design fields are often related with the innovation of new materials that impact not only the feasible solutions but also the very nature of design problems. The development and application of new materials may provide safer, more economical and stringent design.

**Table 6.1** Classification of Materials.

| Material type | Sub-groups |
| --- | --- |
| Metal | **Ferrous metals and alloys** (irons, carbon steels, alloy steels, stainless steels, tool and die steels) |
| | **Non-ferrous metals and alloys** (aluminum, copper, magnesium, nickel, titanium, precious metals, refractory metals, superalloys) |
| Ceramic | Glasses |
| | Glass ceramics |
| | Graphite |
| | Diamond |
| Polymers | Thermoplastics |
| | Thermoset |
| | Elastomers |
| Composite | Polymer-matrix composites |
| | Metal-matrix composites |
| | Ceramic-matrix composites |
| Semiconductors | Silicon |
| New materials | Smart materials |
| | Nanomaterials |
| | Biomaterials |

Material atoms can be combined in new ways to produce new materials with smart properties. For example, a window glass changes color to control the room temperature. New materials can be developed from existing materials by applying high pressure, temperature, electric or magnetic fields. Some examples of new materials are as follows.

### 6.3.1.1 Smart Materials

Smart materials can considerably change their mechanical or other properties in a controllable manner. Based on input and output, the smart materials are classified as follows:

- Shape memory alloys. The field of smart materials is growing rapidly, with one of the most fascinating areas being that of shape memory alloys. A shape memory alloy can undergo significant plastic deformation, and then returns to its original shape by the application of heat. Nickel-titanium-copper, Gold-cadmium, and Nickel-titanium are good examples of shape memory alloys. Examples of applications of shape memory alloys are: micro-actuators, thermostats, electrical circuit breakers, fire dampers, and many medical applications.
- Magnetostrictive materials. Magnetostrictive materials exhibit change in shape when exposed to a magnetic field. Some of the applications of magnetostrictive materials are: actuators,

transducers, magnetostrictive film applications, sensors and many other applications. Magnetostrictive materials include nickel and alloys such as Fe-Ni and Co-Ni.

- Piezoelectric materials. Piezoelectric materials produce an electric current if deformed by mechanical stress. The piezoelectric process is also reversible, creating deformation due to the application of an electrical field. Some of the applications of piezoelectric materials are: mechanical sensor to pick up a mechanical deformation, used as an actuator, used for imaging, mostly in medicine, used as gas lighters. The most commonly known piezoelectric material is quartz ($SiO_2$); other are berlinite ($AlPO_4$), barium titanate ($BaTiO_3$), zinc oxide ($ZnO$), and aluminum nitride ($AlN$).
- Electro-rheological fluids. Electro-rheological fluids change their physical properties when exposed to an electric field. They can be transformed from the liquid state into the solid state like gel in milliseconds by applying an electric or a magnetic field. The change is reversible once the electrical field is removed. Research in this area has produced significant impacts on the automobile industry, bridge and building construction, the aerospace industry, the defense industry, the design of fast actuated hydraulic devices, energy production, and energy conservation.

### 6.3.1.2 Nanomaterials

Nanomaterials can be metals, ceramics, polymeric materials, or composite materials. Their defining characteristic is that they contain nanoparticles, smaller than 100 nanometers in at least one dimension. Nanomaterial properties are different from those of the same materials with micron- or millimeter-scale dimensions and are intentionally produced and designed with very specific properties related to shape, size, surface properties and chemistry. Their relative-surface area is one of the main factors that increase their reactivity, strength and electrical properties. Applications of nanomaterials include, but are not limited to, better insulation and absorption materials, low-cost flat-panel displays, superior cutting tools, high-energy-density batteries, high-sensitivity sensors, automobiles with better fuel efficiency, fuel cells, reliable satellite design, coatings, nano-ball-bearings, nanoscale magnetic materials in data storage devices, and nanostructured membranes for water purification.

Although many potentially beneficial applications exist, many scientific institutions and government organizations across the world have underscored the need to assess their possible health and environmental risks. Among these are worker protection during manufacture, safe disposal of engineered nanomaterials, disposal of contaminated equipment, and contamination of soil, air, and water.

### 6.3.1.3 Biomaterials

Biomaterials are any natural or human-made materials that interact with biological systems. They are used to replace natural functions. Biomaterials are used to fill bony defects, as implants for young children that resorb after the body's bones have healed, used to provide faster healing and reducing complications, used in dental applications, surgery, and drug delivery. Some examples of biomaterials are knee braces, contact lenses, breast implants, heart transplants.

## 6.4 Material Properties

When selecting materials for a design, it is important to understand their properties. Although there are four main material properties (mechanical, physical, chemical, and environmental), discussions in this section will be on some of the mechanical properties of materials.

### 6.4.1 Mechanical Properties

One of the simplest tests for determining some of the main mechanical properties of a material is the tensile test. In this test, a load $P$ is applied along the longitudinal axis of a circular test specimen, as shown in Figure 6.14(a). Two gages are marked at a distance $L_0$ apart. The distance $L_0$ is called the original gage length of the specimen. The original cross-sectional area of the central part of the specimen is denoted by $A_0$.

During an experiment the applied load and the change in gage length are measured and converted to stresses and strains. The resulting stress–strain curve in Figure 6.14(b) (not scaled) gives a direct indication of the material properties. If $L$ is the observed length corresponding to an applied load, $P$, the gage elongation $\Delta L = L - L_0$. The elongation per unit of the initial gage length is called the strain, $\varepsilon$, and given by $\varepsilon = \Delta L / L_0$.

As shown in Figure 6.14(b), as the strain increases, there are four well-defined regions with distinct types of behavior. In the interval OA, the plot is a straight line and materials show linear

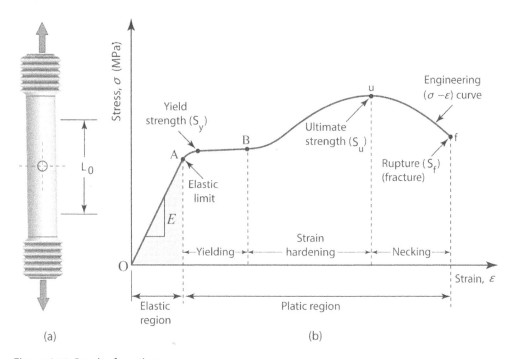

**Figure 6.14** Result of tensile test.

elastic behavior. In this region, deformations are fully recoverable when the applied load is removed. Beyond point A, for ductile materials, deformations are plastic. A plastically deformed material will not return to its original size and shape when the load is removed. This region of the stress–strain curve is assumed to be horizontal. Within the interval Bu, stress increases with increasing strain although at a much lower rate than *E*. This is the strain hardening region. And finally, as the cross-sectional area of the specimen decreases due to plastic flow, at point u necking starts. As shown from the figure, from point *u* stress starts decreasing until rupture occurs.

### 6.4.1.1   Elastic Modules
For elastic materials, the strain is proportional to the applied load and the slope of the line OA in the elastic region is designated by a constant, *E*, and called elastic modulus or Young's modulus (see Figure 6.14(b)).

### 6.4.1.2   Elastic Limit
The elastic limit is the maximum stress within a solid material before the onset of permanent deformation. When stresses are removed, the material returns to its original size and shape. (see Figure 6.14(b)).

### 6.4.1.3   Yield Strength
The yield strength, $S_y$, of a material is defined as the stress at which a material begins to deform plastically. Prior to the yield point the material will deform elastically (leaving 0.2 per cent permanent deformation) and will return to its original shape when the applied stress is removed.

For some materials, such as metals and plastics, the change from the linear elastic region cannot be easily seen. In such cases, as shown in Figure 6.15, to determine the yield strength of the material an *offset method* test is used (see ASTM E8 for metals and D638 for plastics). As shown in Figure 6.15, an off-set, OB, is expressed as a percentage strain. The usual value of the offset is taken to be 0.2 percent for metals and plastics.

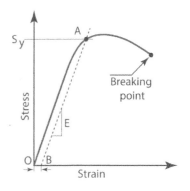

**Figure 6.15** Offset method.

The yield stress $S_y$ is determined by drawing a line from point B parallel to the straight-line portion of the diagram. The intersection point A becomes the yield strength by the offset method.

### 6.4.1.4   Ultimate Tensile Strength
Ultimate tensile strength is the highest stress developed in material before rupture. Usually, changes in area due to changing load and *necking* are ignored in determining ultimate tensile strength.

### 6.4.1.5   Rupture Strength
Rupture strength is the stress developed in a material at the breaking point. It is not necessarily equal to ultimate strength. Since necking is not taken into account in determining rupture strength, it rarely indicates true stress at rupture.

**Figure 6.16** Necking leading to fructure.

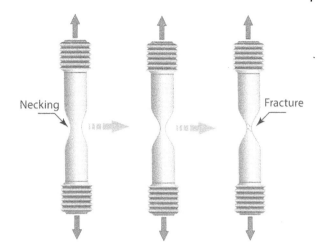

### 6.4.1.6 Necking

Necking is a reduction in diameter in a small region of the material during the tensile deformation (see Figure 6.16). A significant amount of shearing occurs in the necking area. Basically, it is a region of local instability in the material.

### 6.4.1.7 Fracture Toughness

*Fracture toughness*, $K_{IC}$, is a very important material property that defines its ability to resist stress at the tip of a crack. This subject will be discussed in detail in this chapter.

### 6.4.1.8 Brittle and Ductile Behavior

Materials can be classified into two behavior categories: brittle and ductile. In general, steel, aluminum, gold, silver, and copper are typical examples of ductile materials. Glass, concrete, and cast iron are considered as brittle materials. Brittle and ductile materials can be distinguished by comparing the stress–strain curves as shown in Figure 6.17.

As shown in Figure 6.17, ductile materials show larger strains before failure. In general, failure of ductile materials is limited by their shear strength. Ductile materials often have relatively small Young's modulus and ultimate stresses. Brittle materials fail at much lower strains, and material failure is limited by tensile strength. Since ductile materials do not fail without warning as brittle materials do, they are preferable to brittle materials for building structural members.

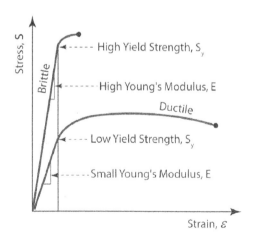

**Figure 6.17** Brittle and ductile material behavior.

### 6.4.1.9  Hardness

Hardness is a property of metals and defined as the resistance of a material to a permanent indentation of particular geometry over a specified length of time. The greater the hardness of the metal, the greater the resistance it has to permanent deformation. Many methods have been developed for hardness testing. Those most often used by industry are Brinell, Rockwell, and Vickers hardnesses. Hardness measures are based on indentation hardness testing. A metal's hardness is measured by the resistance to the penetration of a non-deformable ball or cone. The tests determine the depth of a ball or cone which will sink into the metal, under a specified load, within a certain period of time.

### 6.4.1.10  Impact Resistance

Impact resistance is the ability of a material to withstand a high force or shock applied to it over a short period of time without fracturing. Impact tests are used to investigate the *toughness* of materials. Toughness is the capacity of a material to absorb energy and plastically deform without fracturing.

### 6.4.1.11  Creep Strength

Creep strength is the ability of a material to withstand a load at elevated temperatures over a sufficiently long time.

### 6.4.1.12  Wear

Wear is the material transfer from one surface to another between two sliding surfaces in metal-to-metal contact. Wear can also be the result of impact, erosion, abrasion, oxidation, and corrosion, or a combination of these.

### 6.4.1.13  Fatigue Strength

Fatigue strength is the ability of a material to withstand a time-dependent fluctuating load.

### 6.4.1.14  Damping Capacity

Damping capacity is the ability of a material to absorb vibrational energy. High damping is a desirable property since it minimizes the resonant amplitude, and hence minimizes the operating stresses to which the part is exposed.

Table 6.2 gives the mechanical properties of selected materials at room temperature.

## 6.4.2  Machinability of Metals

Machinability of a material can be defined as the ease or difficulty with which it can be machined. Machinability changes the physical properties and cutting conditions of the material. Materials with good machinability require minimum power to machine, can be cut relatively fast, easily obtain a good finish, and do not wear the tooling as much.

There is no quantitative measurement of the machinability of a material. In general, machinability is expressed as a percentage or a normalized value. The American Iron and Steel Institute

**Table 6.2** Mechanical properties of selected materials at room temperature.*

| Material | Modulus of Elasticity, $E$ | | Modulus of Rigidity, $G$ | | Possion's Ratio, $v$ | Weight Density, $\gamma$ | Mass Density, $\rho$ | Thermal Conductivity $k$ | |
|---|---|---|---|---|---|---|---|---|---|
| | Mpsi | GPa | Mpsi | GPa | — | (lb/in³) | (Mg/m³) | $\dfrac{BTU}{h\,ft°F}$ | $\dfrac{W}{m°C}$ |
| Aluminum alloys | 10.4 | 71.7 | 3.9 | 27 | 0.32 | 0.10 | 2.8 | 100 | 173 |
| Beryllium copper | 18.5 | 127.6 | 7.2 | 49.4 | 0.29 | 0.30 | 8.3 | 85 | 147 |
| Brass, bronze | 16.0 | 110.3 | 6.0 | 41.5 | 0.33 | 0.31 | 8.6 | 45 | 78 |
| Copper | 17.5 | 121.0 | 6.6 | 46.0 | 0.33 | 0.32 | 8.9 | 220 | 381 |
| Iron, gray cats | 15.0 | 103.4 | 5.9 | 40.4 | 0.28 | 0.26 | 7.2 | 29 | 50 |
| Magnesium alloys | 6.50 | 45.0 | 2.4 | 16.8 | 0.33 | 0.07 | 1.8 | 55 | 95 |
| Nickel alloys | 30.0 | 206.8 | 11.5 | 79.6 | 0.30 | 0.30 | 8.3 | 12 | 21 |
| Steel, carbon | 30.0 | 206.8 | 11.7 | 80.8 | 0.28 | 0.28 | 7.8 | 27 | 47 |
| Steel, alloy | 30.0 | 206.8 | 11.7 | 80.8 | 0.28 | 0.28 | 7.8 | 22 | 38 |
| Steel, stainless | 27.5 | 189.6 | 10.7 | 74.1 | 0.28 | 0.28 | 7.8 | 12 | 21 |
| Titanium alloy | 16.5 | 114.0 | 6.2 | 43.0 | 0.33 | 0.16 | 4.4 | 7 | 12 |
| Zinc alloy | 12.0 | 82.7 | 4.5 | 31.1 | 0.33 | 0.24 | 6.6 | 64 | 111 |

*From different sources.

(AISI) has determined that AISI No. 1112 carbon steel has a machinability rating of 100 percent. Machinability of different materials based on 100 percent machinability for AISI 1212 steel are given in Table 6.3.

## 6.5 Analysis of Material Requirements

During the design analysis of material, the maximum and minimum requirements necessary for proper performance of the material must be fully described and understood. It is also important that selection criteria be established for material selection process to provide qualified material advice. Note that the material selection can only be a recommendation, which will require verification by testing.

In some cases, materials selection for a product design can be a challenging task, especially when multiple objectives and constraints such as minimum weight, flexibility, strength, minimum cost, environmental impact, and recyclability are considered.

### 6.5.1 Material Performance Requirements

Material properties and characteristics cover a wide range of parameters and play an important role in meeting the design requirements. Material selection properties and characteristics

**Table 6.3** Machinability of metals.*

| Carbon steel | % | | % | | % |
|---|---|---|---|---|---|
| 1015 | 72 | 1055 | 55 | 1144 | 76 |
| 1016 | 70 | 1060 | 60 | 1144 stressproof | 83 |
| 1020 | 72 | 1117 | 91 | **1212** | **100** |
| 1035 | 65 | 1137 | 72 | 1213 | 136 |
| 1042 | 64 | 1141 | 70 | 12L14 | 170 |
| 1045 | 57 | 1141 annealed | 81 | 1215 | 136 |

| Alloy steels | % | | % | | % |
|---|---|---|---|---|---|
| 2355 annealed | 70 | 4150 annealed | 60 | 6150 | 55 |
| 4130 annealed | 72 | 4340 annealed | 57 | 8620 | 65 |
| 4140 annealed | 65 | 4615 | 65 | 9254 | 45 |
| 4142 annealed | 66 | 4820 annealed | 50 | 9260 | 40 |
| 41L42 annealed | 77 | 5046 | 60 | 9310 | 50 |

| Stainless steels & super alloys | % | | % | | % |
|---|---|---|---|---|---|
| 301 | 45 | 410 | 54 | 15-5PH condition A | 48 |
| 303 | 85 | **416** | **100** | 17-4PH condition A | 48 |
| 304 | 40 | 420 | 45 | A286 aged | 33 |
| 316 | 36 | 430 | 60 | Hastelloy X | 19 |
| 321 | 36 | 431 | 48 | Type 2205 | 45 |
| 347 | 36 | 440C | 40 | Type 2304 | 45 |

| Tool steels | % | | % | | % |
|---|---|---|---|---|---|
| A-2 | 42 | D-3 | 27 | O-2 | 42 |
| A-6 | 33 | M-2 | 39 | – | – |
| D-2 | 27 | O-1 | 42 | – | – |

| Gray cast iron | % | | % | | % |
|---|---|---|---|---|---|
| ASTM class 20 annealed | 73 | ASTM class 35 | 48 | ASTM class 50 | 36 |
| ASTM class 25 | 55 | ASTM class 40 | 48 | – | – |
| ASTM class 30 | 48 | ASTM class 45 | 36 | – | – |

| Aluminum & magnesium alloys | % | | % | | % |
|---|---|---|---|---|---|
| aluminum, cold drawn | 360 | aluminum, wrought | 480 | magnesium, cast | 480 |
| aluminum, cast | 450 | magnesium, cold drawn | 480 | magnesium, wrought | 480 |

*From different sources

**Table 6.4** Material selection properties and characteristics.

| | |
|---|---|
| **1) Static characteristics**<br>  a) Strength<br>    Ultimate strength<br>    Yield strength<br>    Shear strength<br>  b) Ductility<br>  c) Young's modulus<br>  d) Poisson's ratio<br>  e) Hardness<br>**2) Fatigue characteristics**<br>  a) Corrosion fatigue<br>  b) High load<br>  c) Low load/extended life<br>  d) Constant amplitude load<br>  e) Spectrum load<br>  f) Fatigue strength<br>**3) Fracture characteristics**<br>  a) Fracture toughness<br>  b) Flaw growth<br>  c) Crack instability<br>**4) Thermal properties**<br>  a) Coefficient of linear thermal expansion<br>  b) Melting and boiling points<br>  c) Heat transfer coefficient<br>  d) Specific heat<br>  e) Thermal conductivity<br>  f) Thermal shock resistance<br>  g) Heat of sublimation | **5) Manufacturing**<br>  a) Producibility<br>  b) Availability<br>  c) Processing characteristics<br>    Machinability<br>    Weldability<br>    Moldability<br>    Heat treatability<br>    Formability and forgeability<br>    Hardenability<br>  d) Minimum handling thickness<br>  e) Joining techniques<br>  f) Quality assurance<br>**6) Hostile environments**<br>  a) Moisture<br>  b) Temperature<br>  c) Acidity/alkalinity<br>  d) Salt solution<br>  e) Ammonia<br>  f) Hydrogen attack<br>  g) Nuclear hardness<br>**7) Damping characteristics**<br>**8) Anisotropy**<br>**9) Others**<br>  a) Electrical properties<br>  b) Magnetic properties<br>  c) Chemical properties<br>  d) Corrosion properties |

normally considered in design are summarized in Table 6.4 and are discussed (in part) in this chapter.

In general, material strength has three distinct elements:[4]

1) *Static strength.* The ability to withstand a constant load at ambient temperature.
2) *Fatigue strength.* The ability to withstand a time-dependent fluctuating load.
3) *Creep strength.* The ability to withstand a load at elevated temperatures over a sufficiently long time.

Although the strength of the material is important for performance requirements, for minimum-weight design, material density is also important. That is, product performance is not limited by only one property and one should compare materials based on several properties at once. High strength is desired for many design applications; however, it should be remembered that machining and processing of high-strength material are costly. To increase the strength of the product by changing its shape (geometry) may also affect the processing of the material and consequently the cost. Functionality of the final product depends on the choice of the material

---

4 Ertas, A., and Jones, J.C., *The Engineering Design Process*, 2nd edn, John Wiley & Sons, Inc., New York, 1996.

and the shape. All the parameters mentioned in this section for material performance are interrelated. One concludes that materials selection should be based on the following criteria:

- Functionality of the product
- Shape of the product (geometry)
- Properties of the material used (material strength)
- Process used in manufacturing
- Cost
- Reliability and ability to meet design service life requirement.

### 6.5.2   Material Performance Indices

Usually, performance depends on a combination of properties, and then the best material is selected by optimizing one or more material performance indices. A *material performance index* is a group of material properties which governs some aspect of the performance of a product.[1,2]

There are two kinds of effects inherently associated with any product or system:

1) Undesirable effects, such as high cost, excessive weight, large deflections, and vibrations
2) Desirable effects, such as light weight, long useful life, efficient energy output, good power transmission capability, and high cooling capacity.

*Optimum design* can be defined as the best possible design from the standpoint of the most significant effects, that is, minimizing the most significant undesirable effects and/or maximizing the most significant desirable effects.

Application of optimization to a design problem requires the functionality of the product to be designed, the formulation of an *objective function* such as weight, cost, or shape, the expression of design *constraints* as equalities or inequalities, and *design variables* (desirable and free design variables) from which an optimum solution is sought.

Material indices are derived from the objective function of the optimization, as illustrated by the following example.

### Example 6.2

Define the material stiffness performance index for a light, stiff solid cylinder under bending, as shown in Figure 6.18.

### Solution

In this problem, the desirable function is to support the bending load, $P$, the objective is to minimize the mass, $m$, and constraints are length of the solid cylinder and the the stiffness, $SF_b$. The free design variables

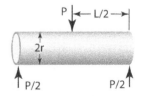

**Figure 6.18** Solid cylinder under bending.

are the cross-sectional area and the material to be used. The objective function to be minimized is

$$m = V \times \rho = L(\pi r^2)\rho, \tag{6.1}$$

where $r$, $\rho$, and $V$ are the radius, density, and volume of the cylinder, respectively. The design constraint stiffness in bending, $SF_b$, is given by

$$SF_b = \frac{P}{\delta} = \frac{48EI}{L^3}, \tag{6.2}$$

where $E$ is the modulus of elasticity, $I$ is the second moment of area, and $\delta$ is the central deflection. The second moment of area for a cylindrical cross-sectional area is given by

$$I = \frac{\pi r^4}{4}. \tag{6.3}$$

The area $A$ and consequently $r$ are the free design variables to be considered to minimize the mass while satisfying the constraints. From above three equations, we obtain

$$m = \frac{\pi}{2(3\pi)^{1/2}} L^{5/2} [S_b]^{1/2} \frac{\rho}{E^{1/2}}. \tag{6.4}$$

For a given bending stiffness $S_b = P/\delta$, the weight of the solid beam is minimized (performance is maximized) when the material performance index, $M_{pi} = E^{1/2}/\rho$, has the largest value. Table 6.5 shows the material indices when stiffness is prescribed.

As shown in Example 6.1, If the desire is to design a light but stiff structure, then we want to choose a material that has a high stiffness to density ratio. In another words, $(E^{1/2}/\rho)$ must be maximized. Let the material performance index, $MP$, be a constant quantity

$$MP = \frac{E^{1/2}}{\rho} = C. \tag{6.5}$$

Rewriting equation (6.5) by taking the base-10 logarithm of both sides gives

$$\frac{1}{2} \log E = \log C + \log \rho \tag{6.6}$$

or

$$\log E = 2 \log C + 2 \log \rho. \tag{6.7}$$

Equation (6.7) represents a straight-line equation

$$Y = a + bx, \tag{6.8}$$

where $b = 2$ is the slope of the straight line. Equation (6.7) provides a family of parallel straight-line plots of $\log E$ versus $\log \rho$ all having a slope of 2. Each line, called a *design guideline*, in the family corresponds to a different material performance index. Formulas for minimum stiffness–weight ratio are shown in Tables 6.5.

**Table 6.5** Examples of material indices for stiffness prescribed.

| Function | Objective | Loading | Maximize | Minimize |
|---|---|---|---|---|
| | Minimum weight | Tension: tie | $\dfrac{E}{\rho}$ | $\dfrac{\rho}{E}$ |
| | Minimum weight | Column: buckling | $\dfrac{E^{1/2}}{\rho}$ | $\dfrac{\rho}{E^{1/2}}$ |
| | Minimum weight | Solid beam: bending | $\dfrac{E^{1/2}}{\rho}$ | $\dfrac{\rho}{E^{1/2}}$ |
| | Minimum weight | Solid beam: torsion | $\dfrac{G^{2/3}}{\rho}$ | $\dfrac{\rho}{G^{2/3}}$ |
| | Minimum weight | Shaft: torsion | $\dfrac{G}{\rho}$ | $\dfrac{\rho}{G}$ |
| | Minimum weight | Solid rectangular beam: bending | $\dfrac{E^{1/3}}{\rho}$ | $\dfrac{\rho}{E^{1/3}}$ |
| | Minimum weight | Pressure vessel: internal pressure | $\dfrac{E}{\rho}$ | $\dfrac{\rho}{E}$ |

☐ **Example 6.3**

Define the material strength performance index for a light, strong solid cylinder under bending, as shown in Example 6.2.

**Solution**

In this problem, the desirable function is to support the bending load, $P$. The objective is to minimize the mass, $m$, and the constraints are the length of the solid cylinder and the bending strength, $\sigma_b$. The free design variables are the cross-sectional area and the material to be used. The objective function to be minimized is

$$m = V \times \rho = L(\pi r^2)\rho, \tag{6.9}$$

where $r$, $\rho$, and $V$ are the radius, density, and volume of the solid cylinder, respectively. The design constraint bending strength in bending, $\sigma_b$, is given by

$$\sigma_b = \frac{Mc}{I} \tag{6.10}$$

where, for the maximum fiber stress, $c = r$, $I = \pi r^4/4$, and maximum bending moment, $M = PL/2$. Then

$$\sigma_b = \frac{2PL}{\pi r^3}. \tag{6.11}$$

The area, $A$, and consequently radius, $r$, are the free design variables to be considered to minimize the mass while satisfying the constraints. Solving $r$ from the bending stress equation and substituting into the mass equation, we have

$$m = (4\pi)^{1/3}(L^5 P^2)^{1/3}\left(\frac{\rho}{\sigma_b^{2/3}}\right). \tag{6.12}$$

For a given $L$ and $P$, the above result shows that the weight of the solid cylinder is minimized when the material performance index, $MP = \sigma_b^{2/3}/\rho$, has the largest value.

As shown in Example 6.2, If the desire is to design a light but strong structure, then we want to choose a material that has a high strength to density ratio. In another words, $\sigma^{2/3}/\rho$ must be maximized. Let the material performance index, $MP$, be a constant quantity

$$MP = \frac{\sigma^{2/3}}{\rho} = C. \tag{6.13}$$

Rewriting equation (6.13) by taking the base-10 logarithm of both sides gives

$$\frac{2}{3}\log \sigma = \log C + \log \rho \tag{6.14}$$

or

$$\log \sigma = \frac{3}{2}\log C + \frac{3}{2}\log \rho. \tag{6.15}$$

Equation (6.15) represents a straight-line equation

$$Y = a + bx, \tag{6.16}$$

where $b = 3/2$ is the slope of the straight line. Equation (6.15) provides a family of parallel straight-line plots of $\log \sigma$ versus $\log \rho$, all having a slope of $3/2$. Formulas for minimum strength–weight ratio are shown in Tables 6.6.

**Table 6.6** Examples of material indices for strength prescribed.

| Function | Objective | Loading | Maximize | Minimize |
|---|---|---|---|---|
| | Minimum weight | Tension: tie | $\dfrac{\sigma_y}{\rho}$ | $\dfrac{\rho}{\sigma_y}$ |
| | Minimum weight | Column: buckling | $\dfrac{\sigma_y}{\rho}$ | $\dfrac{\rho}{\sigma_y}$ |
| | Minimum weight | Solid beam: bending | $\dfrac{\sigma_y^{2/3}}{\rho}$ | $\dfrac{\rho}{\sigma_y^{2/3}}$ |
| | Minimum weight | Solid beam: torsion | $\dfrac{\tau^{2/3}}{\rho}$ | $\dfrac{\rho}{\tau^{2/3}}$ |
| | Minimum weight | Shaft: torsion | $\dfrac{\tau}{\rho}$ | $\dfrac{\rho}{\tau}$ |
| | Minimum weight | Solid rectangular beam: bending | $\dfrac{\sigma_y^{1/2}}{\rho}$ | $\dfrac{\rho}{\sigma_y^{1/2}}$ |
| | Minimum weight | Pressure vessel: internal pressure | $\dfrac{\sigma_y}{\rho}$ | $\dfrac{\rho}{\sigma_y}$ |

For illustration of a cost-performance index, consider again the case of a solid cylinder under bending. The stress due to maximum bending is

$$\sigma = \frac{16PL}{\pi D^3}. \tag{6.17}$$

To facilitate comparison, consider two solid cylinders A and B with different material properties and cross-sectional areas. For the same load, $P$, and length, $L$, the load-carrying ability of two solid cylinders can be written as

$$\frac{\pi D_A^3 \sigma_A}{16L} = \frac{\pi D_B^3 \sigma_B}{16L} \tag{6.18}$$

or

$$\frac{D_B}{D_A} = \left(\frac{\sigma_A}{\sigma_B}\right)^{1/3}. \tag{6.19}$$

The weight, $W$, of the solid cylinder is

$$W = V\rho = L\frac{\pi D^2}{4}\rho. \tag{6.20}$$

Using equation (6.20) yields

$$\frac{W_B}{W_A} = \frac{D_B^2 \rho_B}{D_A^2 \rho_A} \tag{6.21}$$

or

$$\frac{W_B}{W_A} = \left(\frac{\sigma_A}{\sigma_B}\right)^{2/3}\frac{\rho_B}{\rho_A}. \tag{6.22}$$

Suppose that the unit costs of materials A and B are $C_A$ $/lb and $C_B$ $/lb, respectively. The total cost $C_T$ of the solid cylinders to withstand the same load $P$ is given by

$$C_{TA} = W_A \times C_A \quad \text{and} \quad C_{TB} = W_B \times C_B. \tag{6.23}$$

From equations (6.22) and (6.23),

$$\frac{C_{TB}}{C_{TA}} = \left(\frac{\sigma_A}{\sigma_B}\right)^{2/3}\frac{\rho_B}{\rho_A}\frac{C_B}{C_A}. \tag{6.24}$$

From equations (6.12) and (6.24), we conclude that $C(\rho/\sigma^{2/3})$ is the cost per unit of strength of a solid cylinder loaded under bending. This is often referred to as a *minimum-cost criterion*. From equation (6.4), it can be seen that the minimum-cost criterion based on stiffness is $C(\rho/E^{1/2})$. Formulas for *minimum-cost criterion* for different loading conditions are given in Table 6.7.

**Table 6.7** Examples of material indices for minimum cost ($C_m$).

| Function | Objective | Loading | Material index (miximize) | |
|---|---|---|---|---|
| | | | **Strength based** | **Stiffness based** |
| | Minimum cost | Tension: tie | $\dfrac{\sigma_y}{C_m \rho}$ | $\dfrac{E}{C_m \rho}$ |
| | Minimum cost | Column: buckling | $\dfrac{\sigma_y}{C_m \rho}$ | $\dfrac{E^{1/2}}{C_m \rho}$ |
| | Minimum cost | Solid beam: bending | $\dfrac{\sigma^{2/3}}{C_m \rho}$ | $\dfrac{E^{1/2}}{C_m \rho}$ |
| | Minimum cost | Solid beam: torsion | $\dfrac{\tau^{2/3}}{C_m \rho}$ | $\dfrac{G^{2/3}}{C_m \rho}$ |
| | Minimum cost | Shaft: torsion | $\dfrac{\tau}{C_m \rho}$ | $\dfrac{G}{C_m \rho}$ |
| | Minimum cost | Solid rectangular beam: bending | $\dfrac{\sigma^{1/2}}{C_m \rho}$ | $\dfrac{E^{1/3}}{C_m \rho}$ |
| | Minimum cost | Pressure vessel: internal pressure | $\dfrac{\sigma}{C_m \rho}$ | $\dfrac{E}{C_m \rho}$ |

## Example 6.4

Identify material candidates for a light but strong rotating shaft under bending loads for a double over-hung integrally geared centrifugal compression stage, as shown in Figure 6.19. Consider following design criteria:

1) Use design guideline at $MP_1 = 1$ MPa.
2) The strength of the rotating solid bar under bending should not be less than 250 MPa.
3) Consider material cost.

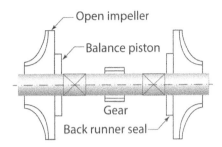

**Figure 6.19** Rotating machinery.

## Solution

The first constraint, material performance index, $MP_1 = 1$ MPa provides a design guide line having a slope of 3/2 (see equation (6.15)) The performance index $MP_1$ for $\rho = 0.1$ and strength 1 MPa is

$$MP_1 = \frac{\sigma^{2/3}}{\rho} = \frac{(1.0 \times 10^6)^{2/3}}{0.1} = 10.0 \times 10^4.$$

Note that, because design lines are for comparison, units are ignored. Strength > 250 MPa and $MP_1 = 10.0 \times 10^4$ define the shaded search region as shown in Figure 6.20. Within this region possible material candidates are:

- carbon fiber-reinforced composite (CFRP);
- aluminum alloy;
- titanium alloy;
- high-strength steel;
- ceramics.

Although composite materials have high strength to weight ratio, compared to metals, for this application composites may not be a practical solution due to the cost consideration. Ceramic materials are more brittle than their metallic counterparts, and carry a greater risk of catastrophic failure when fatigue loading under high speed exists. Therefore, ceramic materials will be eliminated from consideration. As shown in Table 6.8, three remaining materials – aluminum alloy, titanium alloy, and high-strength steel will be evaluated and compared. In this table, strength is denoted by $S$ instead of $\sigma$.

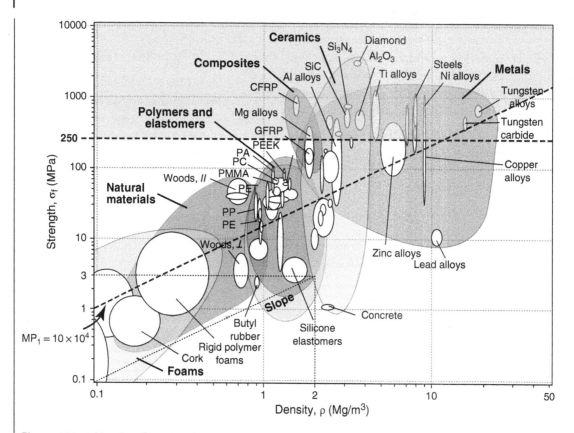

**Figure 6.20** Ashby chart for strength versus density.

**Table 6.8** Mechanical properties of material candidates.

| Material | $\rho$ $Mg/m^3$ | $S$ $MPa$ | $S/\rho$ $MPa/(Mg/m^3)$ | $MP\left(\dfrac{S^{2/3}}{\rho}\right)$ $(MPa)^{2/3}/(Mg/m^3)$ |
| --- | --- | --- | --- | --- |
| High strength steel | 7.7 | 1,700 | 221 | 18.4 |
| Aluminum alloy | 2.7 | 350 | 130 | 18.3 |
| Titanium alloy | 4.5 | 1,100 | 244 | 23.6 |

The results in Table 6.8 substantiate that aluminum alloy can clearly be eliminated from the selection process because of its low strength–weight $(S/\rho)$ ratio. The question is which material to select from the remaining two. Titanium has a high strength–weigh ratio and better material performance index. However, because of the higher cost (twice as much) of titanium compared to steel, the selected material candidate would be steel.

### 6.5.3 Selection and Evaluation of Candidate Material

After the material analysis step is completed, the designer then narrows the candidate list to a few materials. The elimination process is then initiated for the final material selection. The rating system shown in Table 6.9 can be used for this purpose. Two design requirements and three selection factors are adopted. Since not all the design requirements and selection factors are of equal importance, weighting factors $a, b, \ldots$ are used to find the overall rating.[5] The overall rating, $G_i$, can be calculated as

$$G_i = \frac{aR_1 + bR_2 + cR_3 + \cdots}{a + b + c + \cdots} \tag{6.25}$$

A high value of $G_i$ identifies the best material. However, a low design requirement or selection factor value can be an indication of the most suitable material to meet a specific requirement such as cost. In this instance, assuming that $R_5$ is the relative value of cost, then equation (6.25) is modified to

$$G_i = \frac{aR_1 + bR_2 + cR_3 + dR_4 + e(1 - R_5)}{a + b + c + d + e}. \tag{6.26}$$

### 6.5.4 Decision

The previously described screening processes will point to the best material. The decision will be based on selecting the one material that meets the requirements with the best balance of properties.

Table 6.9 Final evaluation of the candidate materials.

| Material Candidates | Design Requirements | | | | Material Selection Factors | | | | | | Overall Rating $\dfrac{aR_1 + bR_2 + \cdots}{a + b + \cdots}$ |
|---|---|---|---|---|---|---|---|---|---|---|---|
| | **1** | | **2** | | **3** | | **4** | | **5** | | |
| | $A_1$ | $R_1$ | $A_2$ | $R_2$ | $A_3$ | $R_3$ | $A_4$ | $R_4$ | $A_5$ | $R_5$ | |
| Material$_1$ | — | — | — | — | — | — | — | — | — | — | $G_1$ |
| Material$_2$ | — | — | — | — | — | — | — | — | — | — | $G_2$ |
| Material$_3$ | — | — | — | — | — | — | — | — | — | — | $G_3$ |
| Material$_4$ | — | — | — | — | — | — | — | — | — | — | $G_4$ |

A: Absolute value.
G: Overall rating value with respect to the largest value.
R: Relative value respect to largest value.
Adapted from Crane, F.A., and Charles, J.A., *Selection and Use of Engineering Materials*, Butterworth & Co., Boston, 1984.

5 Crane, F.A., and Charles, J.A., *Selection and Use of Engineering Materials*, Butterworth & Co., Boston, 1984, p. 213.

## Example 6.5

Table 6.10 shows the materials that can be used for manufacturing the landing gear cylinder shown in Figure 6.21. Considering strength, rigidity, and cost, determine the appropriate material choice from the given alternates.[a]

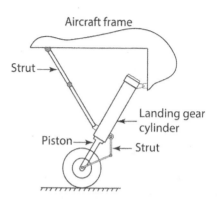

**Solution**

To determine the overall rating, for simplicity, assume that the weighting factors for the design requirements and the selection factor are unity. Also, assume that the landing gear cylinder is a thin-walled pressure vessel under internal pressure. Consequently, Table 6.11 can be constructed by the use of the information provided in Tables 6.6, 6.7, and 6.10. For simplicity, $a$, $b$, and $c$ are assumed to be equal to 1.

**Figure 6.21** Aircraft landing gear mechanism.

**Table 6.10** Mechanical properties of material candidates.

| Material | Yield strength $S_y$ (MPa) | Density $\rho$ (tons/m³) | Elastic modulus $E$ (GPa) | Cost $C$ ($/ton) |
|---|---|---|---|---|
| High-strength steel | 1,700 | 7.7 | 200 | 650 |
| Aluminum alloy | 400 | 2.7 | 71 | 700 |
| Titanium | 1,100 | 4.5 | 120 | 5,000 |

**Table 6.11** Design requirements and selection factors for overall rating.

| Material | $S_y/\rho$ | | $E/\rho$ | | $C \times \frac{\rho}{S_y}$ | | Overall rating ($G$) $\dfrac{aR_1 + bR_2 + c(1-R_3)}{a+b+c}$ |
|---|---|---|---|---|---|---|---|
| | $A_1$ | $R_1$ | $A_2$ | $R_2$ | $A_3$ | $R_3$ | |
| High-strength steel | 221 | 0.91 | 26 | 0.96 | 2.9 | 0.14 | 0.91 |
| Aluminum alloy | 148 | 0.61 | 26 | 0.96 | 4.7 | 0.23 | 0.78 |
| Titanium | 244 | 1 | 27 | 1 | 20.5 | 1 | 0.67 |

The results in Table 6.11 substantiate that the stiffness–weight ratio will not affect the material selection decision, since the values for all choices are close (within 4 percent). Clearly, aluminum alloy can be eliminated from the selection process because of its low strength–weight ratio. The question is which material to select from the remaining two. Titanium has a high strength–weight ratio and low weight compared with high-strength steel. However, because of the cost, the overall rating of steel is considerably higher than of titanium. Although the cost–weight ratio of titanium is relatively high, the desire to reduce the weight of the aircraft may make it the favorable choice over high-strength steel.

Now consider the following typical cost analysis of a landing gear cylinder:[6]

- Net weight

| | |
|---|---|
| Steel | 259 lb |
| Titanium | 188 lb |
| Saving in weight | 71 lb |

- Relative cost (material plus manufacturing)

| | |
|---|---|
| Steel | $4,710 |
| Titanium | $9,900 |
| Total increase in cost | $5,190 |

Hence, the increase in cost per pound of weight saved is

$$\text{Unit increase in cost} = \frac{5190}{71} = \$73.$$

From the above results the choice would be titanium (if the increase in cost can be justified) because of its high strength and light weight. Cost may be a less important factor for a military aircraft than for a commercial aircraft; hence, the choice of titanium might be appropriate for one but not the other. As discussed previously, though, reliability, availability, manufacturability, and maintainability must be studied in detail before a final choice is made.

[a]Ertas, A., and Jones, J.C., *The Engineering Design Process*, 2nd edn, John Wiley & Sons, Inc., New York, 1996.

## 6.6 Design Analysis for Fatigue Resistance

Structures are often found to have failed under the action of cyclic loading even when the applied stresses are below the yield strength of the materials. This type of failure is known as fatigue. A good illustration of fatigue failure is breaking a thin wire with your hands after bending it back and forth several times in the same location. The majority of machine component failures are caused by fatigue. Therefore, engineering requirements during the design phase for avoiding fatigue failure must be considered.

---

6 Verink, E.D., Jr., *Methods of Material Selection,* Gordon and Breach, New York, 1968.

The *S–N* curve method of fatigue life calculation is stress based, and is only fully applicable to cyclic stresses in the elastic range. However, some structural components in some situations can serve under cyclic stresses that exceed the yield stress of the material. High-cycle and low-cycles fatigue are often dealt with as separate regimes. In the low-cycle regime fatigue life models are generally developed for $10^3$ cycles and below. Low-cycle fatigue can be a crucial consideration in the design of products. It is important for situations in which mechanical components go through either mechanically or thermally induced cyclic plastic strains that cause failure within reasonably few cycles. Knowledge gained from low-cycle fatigue testing can be an important information in the establishment of design criteria to protect against mechanical component failure by fatigue.

## 6.6.1 Low-Cycle Fatigue

The low-cycle fatigue region is associated with high loads and short service life. Considerable plastic strain occurs during each cycle. In general, this model is considered for $10^3$ cycles and below. Manson proposed a simplified formula known as the method of universal slopes:[7]

$$\Delta \epsilon_t = \Delta \epsilon_e + \Delta \epsilon_p = 3.5 \frac{S_{ut}}{E} N^{-0.12} + \epsilon_f^{0.6} N^{-0.6}, \tag{6.27}$$

where $\Delta \epsilon_t$ is the total strain range which has two components (elastic strain range, $\Delta \epsilon_e$, and plastic strain range, $\Delta \epsilon_p$), $\epsilon_f$ is the true strain corresponding to fracture in one reversal (*fatigue ductility coefficient*), $E$ is Young's modulus, and $N$ is the number of strain cycles to failure.

A monotonic test and an incremental step test can be used to obtain all the constraints in equation (6.27) for predicting the low-cycle fatigue of a variety of steels. Using equation (6.27), fatigue life, $N$, can be calculated if the total strain amplitude, $\Delta \epsilon_t$, is known. However, calculation of the total strain would be troublesome if discontinuities exist.

## 6.6.2 High-Cycle Fatigue

In high-cycle fatigue situations, the fatigue strength of machine components is analyzed using an *S–N* diagram, obtained in constant amplitude fatigue test by using R.R. Moore's rotating-beam fatigue testing machine, shown in Figure 6.22. Fatigue testing machines apply repeated cyclic loads to test specimens.

As shown in Figure 6.23(b), the rotating-beam testing machine applies cyclic load through the use of dead weights as the specimen rotates. Standard test specimens shown in Figure 6.23(a) are tested at different loads to obtain data points for plotting an *S–N* curve. In general, the fatigue strength of materials is documented by the *S–N* curve obtained from constant stress amplitude, $\sigma_a$, imposed by a pure bending stress test. In other words, the mean stress, $\sigma_m = 0$. A typical *S–N* curve for a steel material constructed using experimental data is shown in Figure 6.24. As seen from the figure, the experiment is repeated at each stress level to verify the results are consistent

---

7 Manson, S.S., "Fatigue: A complex subject – some simple approximations," *Experimental Mechanics*, 5(7), 193, 1965.

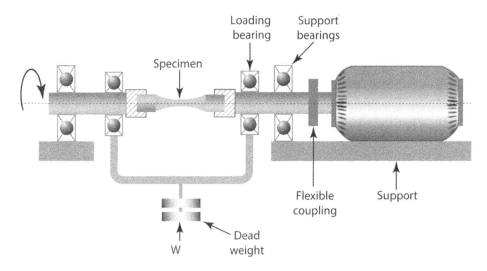

**Figure 6.22** R.R.'s Moore rotating beam fatigue testing machine.

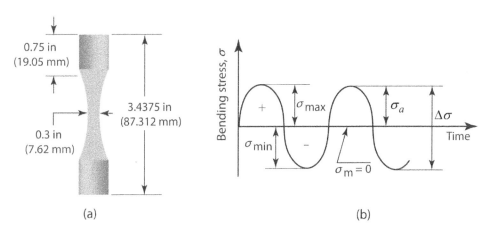

**Figure 6.23** Cyclic loading.

and not just an accident. In this figure, $S_f'$ and $S_e'$ are called uncorrected fracture strength and endurance strength of a material, respectively.

As the specimen is under cyclic stresses, the load acting on the specimen induces fluctuating bending stress. The load on the specimen will be reduced gradually to collect the failure data at each stress level until the failure will not occur. At this point, the corresponding bending stress amplitude will be the endurance strength of the materials. The endurance strength obtained in laboratory by means of R.R. Moore's machine will be unrealistic for practical engineering use since the specimen used for testing is highly polished and homogeneous, with no surface defect. Therefore, the endurance strength is not a material property such as yield strength and ultimate strength. It depends on other parameters which should be incorporated to obtain realistic

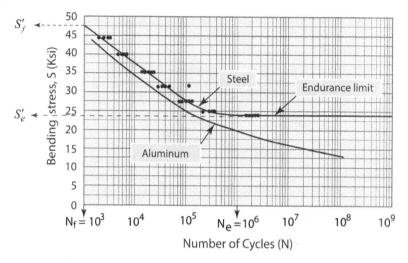

**Figure 6.24** Typical S–N curve.

endurance strength. These parameters are represented as fatigue strength correction factors with values less than unity.

The endurance limit, $S'_e$, shown in Figure 6.24 is a safe range of fluctuating stress value, and below this value it is assumed that failure will not occur. The endurance limit implies that structural members stressed under this limit will have infinite life; $N = 10^6$ cycles. As shown in Figure 6.24, steel and titanium alloys (ferrous alloys such as low-strength carbon and alloy steel; some stainless steels, iron, and titanium alloys; and also some polymers) have an endurance limit, the stress amplitude below which there appears to be no number of cycles that will cause failure. On the other hand, structural metals such as aluminum, copper, magnesium, and nickel do not show an endurance limit and will eventually fail even from small stress amplitudes.

### 6.6.3 Fatigue Strength Calculation

Normally, the fatigue strength of materials is doc-umented by S–N diagrams, obtained from fully reversed constant-amplitude fatigue load tests. As seen from Figure 6.25, the fracture strength, $S'_f$, and endurance strength, $S'_e$, are known if the S–N curve of the material is known. If the S–N curve of the material is not available, then simple equations approximately representing the S–N curve can be used to calculated the fatigue life.

Now consider the simplified S–N curve for a given material shown in Figure 6.25. The approximate

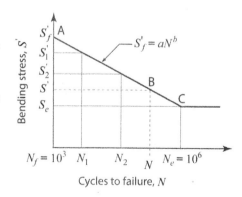

**Figure 6.25** Simplified S–N curve.

equation for the *S–N* curve shown in Figure 6.25 can be written as

$$S' = aN^b,$$ (6.28)

where *b* is the slope of the *S–N* curve. To determine *b*, take logarithms of both sides of equation (6.28):

$$\log S' = \log a + b \log N.$$ (6.29)

The coordinates of points A and C in Figure 6.25 must satisfy equation (6.29). Using point A, we have

$$\log S'_f = \log a + b \log N_f.$$ (6.30)

Similarly, using the coordinates of point C yields

$$\log S'_e = \log a + b \log N_e.$$ (6.31)

From equations (6.30) and (6.31), we obtain

$$b = \frac{\log(S'_f/S'_e)}{\log(N_f/N_e)}.$$ (6.32)

To determine the constant *a*, substitute equation (6.32) into equation (6.31):

$$\log S'_e = \log a + \frac{\log(S'_f/S'_e)}{\log(N_f/N_e)}(\log N_e).$$ (6.33)

Substituting $N_f = 10^3$ and $N_e = 10^6$ into equation (6.33) gives

$$a = \frac{(S'_f)^2}{S'_e}.$$ (6.34)

Similarly, for various values of $N_f$ and $N_e$, constant *a* in equation (6.28) is found to be

$$a = \frac{(S'_f)^2}{S'_e} \quad \text{for } N_f = 10^3 \text{ and } N_e = 10^6,$$ (6.35)

$$a = S'_e\left(\frac{S'_f}{S'_e}\right)^{7/4} \quad \text{for } N_f = 10^3 \text{ and } N_e = 10^7,$$ (6.36)

$$a = S'_e\left(\frac{S'_f}{S'_e}\right)^{8/5} \quad \text{for } N_f = 10^3 \text{ and } N_e = 10^8.$$ (6.37)

The equation to calculate the fatigue life, *N*, for a given stress level, *S*, can be derived from equation (6.28):

$$N = N_f\left(\frac{S'}{S'_f}\right)^{1/b}.$$ (6.38)

**Table 6.12** Values for S–N curve end coordinates.

| Material | $S'_f$ | $N_f$ | $S'_e$ | $N_e$ | Endurance limit in shear, $S'_{S_e}$ Maximum shear | Von Mises |
|---|---|---|---|---|---|---|
| $S_{ut} \le 200\text{kpsi}$ $S_{ut} \le 1380\text{MPa}$ | $0.9S_{ut}$ | $10^3$ | $0.5S_{ut}$ | $10^6$ | $0.5S'_e$ | $0.577S'_e$ |
| $S_{ut} \ge 200\text{kpsi}$ $S_{ut} \ge 1380\text{MPa}$ | $0.9S_{ut}$ | $10^3$ | 100 kpsi 700 MPa | $10^6$ | $0.5S'_e$ | $0.577S'_e$ |
| Cupper alloys | $0.9S_{ut}$ | $10^3$ | $(0.25 - 0.5)S_{ut}$ | $10^6$ | $0.5S'_e$ | $0.577S'_e$ |
| Nickel alloys | $0.9S_{ut}$ | $10^3$ | $(0.35 - 0.5)S_{ut}$ | $10^6$ | $0.5S'_e$ | $0.577S'_e$ |
| Aluminum alloys | $0.9S_{ut}$ | $10^3$ | $0.35S_{ut}$ | $10^6$ | $0.5S'_e$ | $0.577S'_e$ |
| Magnesium alloys | $0.9S_{ut}$ | $10^3$ | $0.35S_{ut}$ | $10^6$ | $0.5S'_e$ | $0.577S'_e$ |
| Titanium | $0.9S_{ut}$ | $10^3$ | $(0.45 - 0.65)S_{ut}$ | $10^6$ | $0.5S'_e$ | $0.577S'_e$ |

The S–N curve for a specific material can be constructed by using the parameters given in Table 6.12.

### Example 6.6

The cantilever beam shown in Figure 6.26 has an ultimate strength of 720 MPa.

a) Estimate the endurance limit, $S'_e$, of the beam material for an infinite life.
b) Estimate the fracture strength, $S'_f$, of the beam material at $10^3$ cycles.
c) Determine the fatigue strength, $S'$, corresponding to a life of $90 \times 10^3$ cycles.

**Figure 6.26** Cantilever beam.

### Solution

(a) From Table 6.12,

$$S'_e = 0.5S_{ut} = 0.5 \times 720 = 360\text{MPa}.$$

(b) From Table 6.12,

$$S'_f = 0.9S_{ut} = 0.9 \times 720 = 648\text{MPa}.$$

(c) From equation (6.28),

$$S' = aN^b,$$

where

$$a = \frac{(S'_f)^2}{S'_e}$$

From Table 6.12, $N_f = 10^3$ and $N_e = 10^6$. Substituting these values into the above equation yields

$$a = \frac{648^2}{360} = 1{,}166.4 \text{MPa}$$

and

$$b = \frac{\log(S'_f/S'_e)}{\log(N_f/N_e)} = \frac{\log(648/360)}{\log(10^3/10^6)} = -0.085.$$

Then the fatigue strength, $S'$, corresponding to a life of $90 \times 10^3$ cycles is

$$S' = (1{,}166.4)(90 \times 10^3)^{-0.085} = 442 \text{ MPa}.$$

For linear systems, the fatigue life, $N$, is directly related to the time, $T$, and the stress, $S$, is directly related to the acceleration, $G$, and to the displacement, $Z$. Hence, equation (6.38) can be rewritten as[8]

$$N = N_f \left( \frac{G}{G_f} \right)^{1/b}, \tag{6.39}$$

$$N = N_f \left( \frac{Z}{Z_f} \right)^{1/b}, \tag{6.40}$$

$$T = T_f \left( \frac{G}{G_f} \right)^{1/b}. \tag{6.41}$$

Equations 6.39–6.41 can be used for accelerated testing to obtain the long-term performance of a product in a shortened test time. Some products will exhibit the same performance in a short time at high stress that they will exhibit in a longer time at lower stress. This method is useful when reliability information must be obtained in a short time period.

### 6.6.4 Fatigue Analysis for Fluctuating Stress

Thus far, fatigue analysis has been explained based on a fully reversed load as shown in Figure 6.27(a). In some cases, stress applied on structural members fluctuates without passing through zero. For example, two unique cases are shown in Figure 6.27(b, c). In these cases (where the applied force does not cause complete reversal), the combination of stress amplitude and mean stress must be considered in the fatigue analysis. Hence, the equivalent bending

---

8 Steinberg, D.S., *Vibration Analysis for Electronic Equipment*, John Wiley & Sons, New York, 1988.

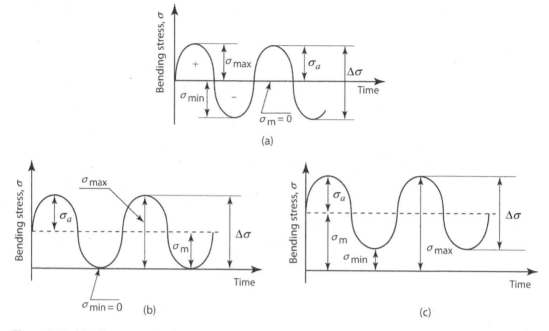

**Figure 6.27** (a) Fully-reversed cyclic load. (b) Repeated cyclic load. (c) Fluctuating cyclic load.

stress, which takes into account the combination of stress amplitude, $\sigma_a$, and mean stress, $\sigma_m$, can be calculated as follows:

- for ductile materials (Soderberg equation),

$$\sigma_{eq} = \sigma_a \left( \frac{S_y}{S_y - \sigma_m} \right) ; \tag{6.42}$$

- for brittle materials (modified Goodman equation):

$$\sigma_{eq} = \sigma_a \left( \frac{S_{ut}}{S_{ut} - \sigma_m} \right) . \tag{6.43}$$

Equation (6.43) can only be used when the mean stress, $\sigma_m$, due to tensile loading is considered. If $\sigma_{min}$ and $\sigma_{max}$ are known, the stress amplitude, $\sigma_a$, mean stress, $\sigma_m$, and stress range, $\sigma_r = \Delta\sigma$ can be determined by

$$\sigma_a = \frac{\sigma_{max} - \sigma_{min}}{2},$$
$$\sigma_m = \frac{\sigma_{max} + \sigma_{min}}{2}, \tag{6.44}$$
$$\sigma_r = 2\sigma_a.$$

Experimental data shows that compressive mean stress has almost no effect on the endurance limit. Then $\sigma_m$ is assumed to be equal to zero and equation (6.43) reduces to $\sigma_{eq} = \sigma_a$. Hence, in the absence of the mean stress, the stress amplitude is used for the fatigue life calculation.

## 6.7 Miner's Rule: Cumulative Fatigue Damage

Miner's rule first proposed by Palmgren in 1924 and was further developed by Miner in 1945. It is the simplest and most widely used cumulative damage model for fatigue life estimation under variable amplitude loading. Miner's rule states that if there are $k$ different stress level of amplitudes, $\sigma_a$, then the damage fraction, $D$, is

$$D = \sum_{i=1}^{k} \frac{n_i}{N_{fi}} = \frac{n_1}{N_{f1}} + \frac{n_2}{N_{f2}} + \cdots + \frac{n_k}{N_{fk}}, \tag{6.45}$$

where $n$ is the number of cycles at stress amplitude, $\sigma_a$, and $N_f$ is the total failure cycle corresponding to that $\sigma_a$. The damage fraction is experimentally found to be between 0.7 and 2.2. For design purposes, we generally assume that when the damage fraction reaches 1, failure occurs.

Miner's (linear damage) rule is used in estimating the cumulative damage of machine components at stress levels exceeding the endurance limit. According to Miner's rule, if a machine component is subjected to a constant amplitude cyclic load and fails after $N_f$ cycles, each cycle expended $1/N_f$ fraction of the component's total life. This can be thought of as determining what proportion of total life is spent by stress reversal at each magnitude, then developing a linear combination of their damage values. For example, if a machine component is subjected to $n_1$ reversed stress cycle at a stress level of $\sigma_1$, $n_2$ reversed stress cycle at a stress level of $\sigma_2$, etc., what is the damage done to the machine component?

### Example 6.7

Assume that the cantilever beam shown in Figure 6.28 has material properties $S_{ut} = 80$ kpsi and $S_y = 66$ kpsi. Suppose the loads $P_1$ and $P_2$ (see Figure 6.28) on the beam produce fluctuating stress of $\sigma_1$ for a total of $n_1 = 6{,}000$ cycles and $\sigma_2$ for a total of $n_2 = 50{,}000$ cycles. Determine the damage and the remaining life of the beam.

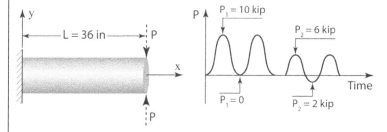

**Figure 6.28** Cantilever beam.

## Solution

The bending stress due to the bending moment, $M = PL$, is

$$\sigma = \frac{M}{I/c}, \quad \text{where } \frac{I}{c} = \frac{\pi D^3}{32} = \frac{\pi (4)^3}{32} = 6.28 \text{in}^3.$$

The maximum and minimum stresses due to $P_1$ are

$$\sigma_{1max} = \frac{10 \times 36}{6.28} = 57.32 \text{ kpsi}, \quad \sigma_{1min} = 0.0 \text{ kpsi}.$$

Then the stress amplitude, $\sigma_a$, and the mean stress, $\sigma_m$, are

$$\sigma_{1a} = \frac{\sigma_{1max} - \sigma_{1min}}{2} = \frac{57.32 - (-0.0)}{2} = 28.66 \text{ kpsi},$$

$$\sigma_{1m} = \frac{\sigma_{1max} + \sigma_{1min}}{2} = \frac{57.32 + (0.0)}{2} = 28.66 \text{ kpsi}.$$

To calculate the equivalent stress, $\sigma_{1eq}$, we choose Soderberg line relationships (ductile material):

$$\sigma_{1eq} = \sigma_{1a} \frac{S_y}{(S_y - \sigma_{1m})} = 28.66 \left[ \frac{66}{(66 - 28.66)} \right] = 50.66 \text{ kpsi}.$$

The equivalent stress, $\sigma_{1eq} = 50.66$ kpsi, includes the effect of the mean stress, $\sigma_{1m}$, for the damage calculation. Using similar calculations, the following results are obtained for the fluctuating load, $P_2$:

$$\sigma_{2max} = \frac{6 \times 36}{6.28} = 34.39 \text{ kpsi}, \quad \sigma_{2min} = \frac{-2 \times 36}{6.28} = -11.46 \text{ kpsi},$$

and

$$\sigma_{2a} = \frac{\sigma_{2max} - \sigma_{2min}}{2} = \frac{34.39 - (-11.46)}{2} = 22.93 \text{ kpsi},$$

$$\sigma_{2m} = \frac{\sigma_{2max} + \sigma_{2min}}{2} = \frac{34.39 + (-11.46)}{2} = 11.47 \text{ kpsi}.$$

The equivalent stress is

$$\sigma_{2eq} = \sigma_{2a} \frac{S_y}{(S_y - \sigma_{2m})} = 22.93 \left[ \frac{66}{(66 - 11.47)} \right] = 27.75 \text{ kpsi}.$$

To estimate the total life corresponding to the stress levels, we develop an S–N curve for the beam material using Table 6.12. The S–N curve can be developed by having a straight line through the points A ($N_f = 10^3, 0.9S_{ut}$) and B($N_e = 10^6, 0.5S_{ut}$), plot an approximate S–N curve as shown in Figure 6.29. From Figure 6.29, the total life, $N_1$, corresponding to a stress level of

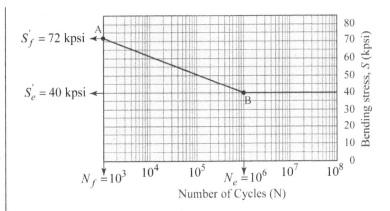

**Figure 6.29** Aproximate *S–N* curve.

$\sigma_{1eq} = 50.66$ kpsi is 100,000 cycles. Similarly, the total life for a stress level of $\sigma_{2eq} = 27.75$ kpsi is infinity. Substituting into the damage equation, we have

$$D = \left( \frac{n_1}{N_1} + \frac{n_2}{N_2} \right) 100 = \left( \frac{6,000}{100,000} + \frac{50,000}{\infty} \right) 100 = 6\%.$$

The remaining life, *RL*, is

$$RL = 100 - D = 100 - 6 = 94\%.$$

## 6.7.1 Offshore Pipeline Fatigue Application

As shown in Figure 6.30, a riser is a long tensioned cylindrical hollow steel tube that connects an offshore floating production structure or a drilling vessel to a sub-sea system for drilling or production purposes. The riser is considered to be the most critical component in an offshore pipeline development, taking into account the dynamic wave and current loads and hostile service conditions it has to withstand. Riser ball joints are used in the upper and lower portion of the riser to provide flexibility. When a riser is used in water depths greater than about 20 meters, it has to be tensioned to maintain stability. The riser not only protects the drill pipe from wave and current forces but also provides a path for the return of the drilling fluid to the drilling vessel. It is evident that the drill pipe undergoes a relatively sharp bend due to the deflection of the ball joints caused by wave and current loads on the riser string. The bend at the ball joints can cause fluctuating stresses and result in fatigue damage to the drill pipe that is inside the riser. In general, to be safe, in drilling from drilling vessels, the allowable ball joint angle should not be more than approximately 5 degrees.

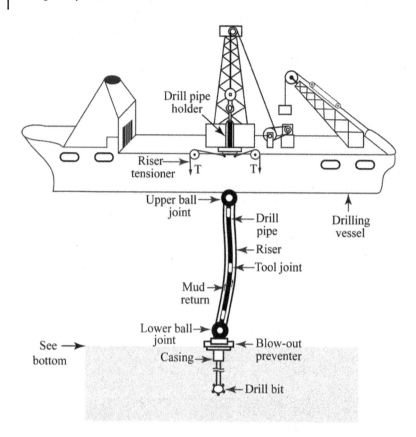

**Figure 6.30** Marine riser system.

### Example 6.8

Using the *S–N* curve approach (see Figure 6.31), find the cumulative fatigue damage inflicted on the drill pipe during one pass through the riser ball joint. One pass is defined as a vertical movement of the drill pipe through 144 inches while being rotated. Assume that the drill pipe is subjected to a constant axial tension load of 100,000 lbf to prevent failure by buckling and is of material with elongation less than 5%.[a]

### Solution

During the drilling operation, the drill pipe is subjected to alternate bending stresses (compression and tension) due to rotation. This bending stress is a maximum at the ball joint and falls off exponentially with distance from the ball joint, both below and above. As a result, when the drill pipe makes one pass through, as each joint of the drill pipe passes through the ball joint,

**Figure 6.31** *S–N* curve for grade E drill pipe.

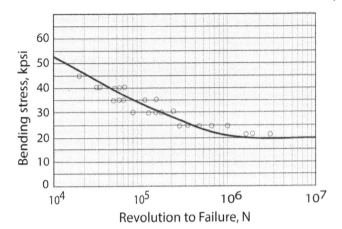

the upper ball joint experiences cumulative fatigue damage from stress cycles of widely varying amplitudes.

An element of the deformed drill pipe is acted on by the forces shown in Figure 6.32, for which the bending moment equation is

$$M(x) = \frac{T\theta}{k}(\cosh\ kx - \sinh\ kx), \tag{6.46}$$

where $\theta$ is the upper ball-joint angle and

$$k = \sqrt{\frac{T}{EI}}. \tag{6.47}$$

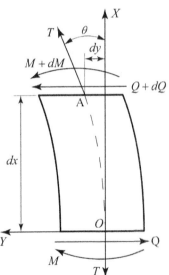

**Figure 6.32** Free-body diagram of a drill pipe.

*Given Design Data*

Modulus of elasticity of drill-pipe material, $E = 30 \times 10^6$ psi
Yield strength of drill-pipe material, $\sigma_{ys}$ = 90 kpsi
Penetration speed of drill pipe, $V$ = 120 in./h
Outside diameter of drill pipe, $D_o$ = 5.00 in.
Inside diameter of drill pipe, $D_i$ = 4.276 in.
Rotational speed of drill pipe, $n$ = 100 rpm
Tension load, $T$ = 100 kips.

The moment of inertia of the drill pipe is

$$I = \frac{\pi(D_o^4 - D_i^4)}{64} = \frac{\pi(5^4 - 4.276^4)}{64} = 14.27 \text{ in.}^4.$$

The value of the constant $k$ for $T = 100,000$ lb and $E = 30 \times 10^6$ psi is

$$k = \sqrt{\frac{T}{E \times I}} = \sqrt{\frac{100,000}{(30 \times 10^6)(14.27)}} = 0.01528 \text{ in.}^{-1}.$$

The cross-sectional area of the drill pipe is

$$A = \frac{\pi(D_o^2 - D_i^2)}{4} = \frac{\pi(5^2 - 4.276^2)}{4} = 5.275 \text{ in.}^2.$$

The mean stress. $\sigma_m$, due to the tension load of 100 kips, is

$$\sigma_m = \frac{T}{A} = \frac{100}{5.275} = 18.96 \text{ kpsi.}$$

Since the drill pipe material is brittle (material elongation less than 5%), we use the modified Goodman equation

$$\sigma_{eq} = \sigma_a \underbrace{\left( \frac{S_{ut}}{S_{ut} - \sigma_m} \right)}_{F_{cr}}$$

The correction factor, $F_{cr}$, which takes into account the mean stress, is

$$F_{cr} = \frac{S_{ut}}{S_{ut} - \sigma_m} = \frac{100}{100 - 18.96} = 1.234.$$

The following steps are performed to determine the cumulative fatigue damage inflicted on the drill pipe during one pass through the upper ball joint:

*Step 1.* Divide a joint of the drill pipe into a number of finite elements. As shown in Figure 6.33, smaller elements are taken close to the higher stress region. As we will learn later, the fatigue damage calculation shows that the bending moment beyond a distance of 72 in. of the ball joint has no significant contribution to the life of a drill pipe.

*Step 2.* Calculate the bending moment using equation (6.46). For $\theta = 1°$ (0.01744 rad), the bending moment, $M$, at $x = 0$ is

$$M(x = 0) = (100,000) \left( 1° \times \frac{\pi}{180} \right) \left( \frac{1}{0.01528} \right)$$
$$\times [\cosh(0.01528)(0) - \sinh(0.01528)(0)]$$
$$= 114.20 \text{ kipsin.}$$

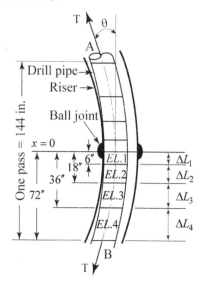

**Figure 6.33** Schematic of a drill pipe in a ball joint.

Similarly, the calculated bending moment at $x = 6$ in. is $M(x = 6) = 104.22$ kipsin. The average bending moment at the first element, $M_{ave}$ (EL.1), can now be calculated as

$$M_{ave}(EL.1) = \frac{114.20 + 104.22}{2} = 109.21 \text{ kipsin.}$$

*Step 3.* Determine the average alternating (reversed) bending stress $\sigma_{ave}(x)$, from the following relation:

$$\sigma_{ave}(EL.1) = \frac{M_{ave}(EL.1) \times D_o}{2 \times I} = \frac{109.21 \times 5}{2 \times 14.27} = 19.13 \text{ kpsi.}$$

*Step 4.* Determine the equivalent stress, $\sigma_{eq}$, by the use of the equation representing the modified Goodman line to account for the mean stress due to nominal tension:

$$\sigma_{eq}(EL.1) = F_{cr}\sigma_{ave}(EL.1) = 1.234 \times 19.13 = 23.61 \text{ kpsi.}$$

This step is performed because a tensile load is superimposed on a drill pipe experiencing alternating being stress, which reduces its endurance limit. Hence, the fatigue effect of bending becomes more severe because of this tensile load.

*Step 5.* Evaluate the number of stress reversals $n_1$ for the first element from

$$n_1 = \frac{\text{rpm} \times 60 \times \Delta L_1}{V} = \frac{100 \times 60 \times (6 \text{ in.} - 0 \text{ in.})}{120} = 300 \text{ cycles.}$$

*Step 6.* Find the number of cycles, $N_1$, for failure at the stress level of 23.61 kpsi for the first element from the *S–N* curve shown in Figure 6.31:

$$N_1 = 500,000 \text{ cycles.}$$

Note that if the stress level is higher than 53 kpsi (again see Figure 6.31), the total life, $N$, should be assumed to be $10^4$ cycles.

*Step 7.* Apply Miner's rule to evaluate the cumulative fatigue damage for the first element:

$$\frac{n_1}{N_1} \times 100 = \frac{300}{500,000} \times 100 = 0.06\%.$$

This procedure is repeated for the remaining elements, and the results are summarized in Table 6.13. Note that Figure 6.31 shows that the life is infinite at stresses below 20 kpsi. Therefore, at stresses below 20 kpsi, the damage is assumed to be zero. The total damage to the drill pipe during its passage through the location beginning 72 in. above and ending 72 in. below (one pass) the upper joint can be calculated as follows:

$$\text{Total damage} = 2 \times \left( \frac{n_1}{N_1} + \frac{n_2}{N_2} + \frac{n_3}{N_3} + \frac{n_4}{N_4} \right) \times 100$$
$$= 2 \times (0.06 + 0 + 0 + 0) = 0.12\%.$$

**Table 6.13** Fatigue damage calculation.

| EL. No. | x (in.) | M(x) (kips-in.) | $M_{ave}$ (kips-in.) | $\sigma_{ave}$ (kpsi) | $\sigma_{eq}$ (kpsi) | n (cycle) | N (cycle) | (n/N) × 100 % |
|---------|---------|-----------------|----------------------|-----------------------|----------------------|-----------|-----------|----------------|
| 1 | x = 0 | 114.20 | 109.22 | 19.13 | 23.61 | 300 | $5 \times 10^5$ | 0.06 |
|   | x = 6 | 104.22 | | | | | | |
| 2 | x = 6 | 104.22 | 95.49 | 16.73 | 20.64 | 600 | inf. | 0.0 |
|   | x = 18 | 86.76 | | | | | | |
| 3 | x = 18 | 86.76 | 76.33 | 13.37 | 16.50 | 900 | inf. | 0.0 |
|   | x = 36 | 65.90 | | | | | | |
| 4 | x = 36 | 65.90 | 51.95 | 9.10 | 11.23 | 1800 | inf. | 0.0 |
|   | x = 72 | 38.00 | | | | | | |

[a] Adapted from Ertas, A., et al., "The effect of tool joint stiffness on drill pipe fatigue in riser ball joints," *Transactions, ASME, Journal of Engineering for Industry*, 111(4), 369–374, 1989.

## 6.8 Fracture Mechanics Based Fatigue Analysis

Fracture mechanics is an alternative approach to fatigue assessment that has evolved over the last several decades. Estimation of fatigue lives of mechanical components is an essential part of engineering design. The fracture mechanics approach can be divided into linear elastic fracture mechanics (LEFM) and elasto-plastic fracture mechanics (EPFM). LEFM gives good results for brittle elastic materials such as high-strength steel, glass, ice, concrete, etc. When the applied load is low enough, LEFM provides good results for ductile materials such as low carbon steel, stainless steel, and some aluminium alloys.

Fracture failure can also be classified on the basis of the ability of a material to experience plastic deformation:

a) Ductile fracture – associated with significant plastic deformation.
b) Brittle fracture – associated with little or no plastic deformation and sudden catastrophic failure.

Brittle fracture, due to the loss of plasticity, was first studied as early as the 1950s. With the development of welding and its application to the construction of large-scale structures, a certain number of catastrophic failures have occurred. The T2 oil tankers, built in large quantities in the United States during World War II, split in two in cold weather. Investigations showed that the tendency of the tankers to split in two was due to poor welding techniques. But later, it was concluded that the steel used during wartime construction had too high a sulfur content that turned the steel brittle at lower temperatures.[9] Again, during World War II, the USA started to build all-welded cargo vessels (Liberty ships). In total, 2,708 Liberty ships were constructed from 1939 to 1945; 1,031 accidents due to brittle fracture were reported by April 1946. More than 200 Liberty ships were sunk or damaged beyond repair. The investigation showed that the failures were caused by the development of brittle crack because of the poor fracture toughness of welded joints. These accidents showed the importance of fracture toughness and eventually would lead to the birth of the fracture mechanics.

Under certain conditions a ductile metal may behave as if it were brittle and a brittle metal may behave as if it were ductile. Ductile materials frequently undergo fracture; however, brittle materials rarely crack in a ductile mode. There are a number of affecting factors that make steel behave in a brittle rather than a ductile manner – losing its capacity of plastic strain. The factors that may cause these different behaviors include temperature, loading rate, plate thickness, steel quality, and corrosive environment.[2]

*Temperature.* Temperature has a significant affect on the ductility of materials. Low temperature decreases ductility, while high temperature increases it. Low temperature is the most apparent reason for the bridges in Belgium that failed in the late 1930s. When the temperature is normal and the loading rate is relatively low, steel is capable of elongating approximately 20% (ductile behavior). However, when steel is subjected to impact loading, in the presence of notches, and at cold temperatures, it can fracture in a brittle manner (i.e. without elongating).

*Loading rate.* When the loading rate increases slowly, there is enough time for microscopic movements in the material to occur and the material deforms plastically before failure occurs. Since there is not enough time for microscopic movements to take place, a rapid loading rate change often causes a ductile material to behave in a brittle manner. Both the yield strength and the tensile strength increase with an increase in loading rate and the result will be a reduced ability to elongate plastically.

Severe temperature change could also have the same negative effect as an increased loading rate on the capacity to respond in a ductile manner. When a sharp notch exists, a fast temperature drop may initiate cleavage fracture. Investigations showed that this was the case in the Rüdersdorf Bridge. Just before the brittle fracture of its bridge beams the temperature fell from 0°C to −12°C within a very short period of time. As the material is fast contracting there is not only a decrease in the capacity to respond in a ductile manner, but also a high possibility that rapid high tensile stresses can be produced because of an uneven temperature change.[2]

*Plate thickness.* The thicker the plate is, the higher the possibility of brittle behavior by a steel material. Conversely, thin plates are more likely to fail in a ductile manner. Thin parts will usually have a shear lip or fracture at an angle – this is characteristic of a ductile fracture. The shear lip becomes smaller as the thickness increases and the fracture becomes more brittle.

---

9 http://www.scoopweb.com/T2-Tanker, accessed April 2, 2013.

*Material quality.* The tensile strength increases with increased carbon content in steel. However, material ductility reduces. Sulfur and phosphorus content is also a factor affecting ductility negatively. Materials with high carbon content may not be suitable for welding.

*Corrosive environment.* High stresses combined with a corrosive environment can cause critical components to crack and fail, sometimes with little warning. While flying from Hilo to Honolulu, Aloha Airlines Flight 243 suffered extensive damage after an explosive decompression in flight, but it was still able to land safely at Kahului Airport. One flight attendant was blown out of the airplane and another 65 passengers and crew were injured. This was a significant event in the history of aviation, with far-reaching effects on aviation safety policies and procedures. Investigation showed that the accident was caused by metal fatigue by crevice corrosion. The plane was 19 years old and operated in a coastal environment, with exposure to salt and humidity.[10]

To reduce the risk of catastrophic failure by fracture mentioned in this section, designers and analysts should become familiar with fracture control requirements. Fracture criticality of structural parts and components must be documented and requires analysis. Inspection and other fracture control activities need to be implemented and monitored to ensure that the remaining life of the components will not create failure risk.

### 6.8.1 Fracture Mechanics Analysis of Fatigue Resistance

Fracture mechanics approach assumes:

a) there are preexisting flaws or cracks in a structural component; and
b) the fatigue life of the component is determined by the rate of growth of these cracks under cyclic loading.

Three possible material separation modes for crack extension under external load are shown in Figure 6.34. In most engineering problems, mode II in-plane shear and mode III out-of-plane

**Figure 6.34** Fracture modes.

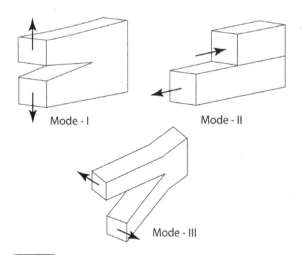

Mode - I

Mode - II

Mode - III

---

10 http://en.wikipedia.org/wiki/Aloha_Airlines_Flight_243, accessed April 3, 2013.

shear have limited application. Hence, only mode I, tensile opening mode, will be discussed in what follows.

### 6.8.1.1 Stress State in a Crack (Mode I)

LEFM can be used to provide preliminary design guidelines. As shown in Figure 6.35, a plate is subjected to a tensile stress, $\sigma_y$, at infinity with a crack length of $2a$. An element $dxdy$ of the plate at a distance $r$ from the crack tip and at an angle $\theta$ with respect to the crack plane will have the following stress field:[11]

$$\sigma_y = \frac{K_I}{\sqrt{2\pi r}} \cos \frac{\theta}{2} \left( 1 + \sin \frac{\theta}{2} \sin \frac{3\theta}{2} \right), \tag{6.48}$$

$$\sigma_y = \frac{K_I}{\sqrt{2\pi r}} \cos \frac{\theta}{2} \left( 1 + \sin \frac{\theta}{2} \sin \frac{3\theta}{2} \right), \tag{6.49}$$

$$\tau_{xy} = \frac{K_I}{\sqrt{2\pi r}} \left( \sin \frac{\theta}{2} \cos \frac{\theta}{2} \cos \frac{3\theta}{2} \right). \tag{6.50}$$

For plane stress, $\sigma_z = 0$. For plain strain, $\sigma_z = \nu(\sigma_x + \sigma_y)$, $\tau_{xy} = \tau_{yz} = 0$. The magnitude of the stress intensity factor, $K_I$, depends on load (in this case, mode I: tensile load), structural geometry, size, and the location of the crack. It is given by

$$K_I = \sigma \sqrt{\pi a}. \tag{6.51}$$

**Figure 6.35** Crack in an infinite plate subjected to a tensile stress.

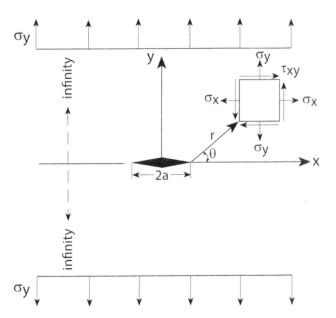

11 Irwin, G.R., "Analysis of stresses and strains near the end of a crack traversing a plate," *Transactions, ASME, Journal of Applied Mechanics*, 24, 1957.

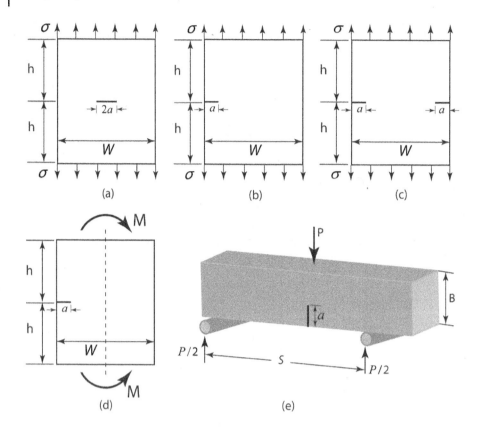

**Figure 6.36** Crack in an infinite plate subjected to a normal stress.

Using a correction factor $f(a/W)$, equation 6.51 can be modified to determine the stress intensity factor of the crack type commonly seen in crack configurations under an applied stress as shown in Figure 6.36:

$$K_I = \sigma\sqrt{\pi a}f(a/W) \tag{6.52}$$

Functions $f(a/W)$ to correct the stress intensity factor for different crack configurations are as follows:[12]

a) Central through-thickness crack (accuracy better than 0.5% for $a/W \leq 0.5$):

$$f(a/W) = \left[1 + 0.128\left(\frac{a}{W}\right) - 0.288\left(\frac{a}{W}\right)^2 + 1.525\left(\frac{a}{W}\right)^3\right]. \tag{6.53}$$

---

12 Tada, H., Paris, P.C., and Irwin, G.R., *The Stress Analysis of Cracks Handbook*, ASME Press, New York, 2000.

b) Single-edge through-thickness crack (accuracy is 0.5% for $a/W \leq 0.6$):

$$f(a/W) = \left[ 1.122 - 0.231 \left( \frac{a}{W} \right) + 10.550 \left( \frac{a}{W} \right)^2 - 21.710 \left( \frac{a}{W} \right)^3 + 30.382 \left( \frac{a}{W} \right)^4 \right].$$

(6.54)

c) Double-edge through-thickness crack (accuracy better than 0.5% for $a/W > 0.4$):

$$f(a/W) = \left[ 1.12 + 0.203 \left( \frac{a}{W} \right) - 1.197 \left( \frac{a}{W} \right)^2 + 1.930 \left( \frac{a}{W} \right)^3 \right].$$

(6.55)

d) Single-edge through-thickness crack under pure bending (accuracy is 0.2% for $a/W \leq 0.6$):

$$f(a/W) = \left[ 1.122 - 1.40 \left( \frac{a}{W} \right) + 7.33 \left( \frac{a}{W} \right)^2 - 13.08 \left( \frac{a}{W} \right)^3 + 14.0 \left( \frac{a}{W} \right)^4 \right].$$

(6.56)

e) Three-point bend specimen (accuracy is 0.5% for any $a/W$):

$$\sigma = \frac{6M}{b^2}, \quad \text{where } M = \frac{PS}{4}.$$

(6.57)

For $S/b = 4$,

$$f(a/W) = \frac{1}{\sqrt{\pi}} \times \frac{1.99 - a/W(1 - a/W)[2.15 - 3.93a/W + 2.7(a/W^2]}{(1 + 2a/W)(1 - a/W)^{3/2}}.$$

(6.58)

When the normal stress, $\sigma$, is equal to $S_y$, the material becomes unstable and plastic deformation occurs. Note that a similar analogy can be used for material failure such that when the stress-intensity factor, $K_I$, reaches the critical stress-intensity factor (also known fracture toughness), $K_{IC}$, significant crack propagation occurs. When the material's thickness is greater than some critical value, repeatable fracture toughness, $K_{IC}$, can only be obtained under the condition of plane strain rather than plane stress conditions. Thus, the designer must keep the value lower than $K_{IC}$ in the same way that the normal stress, $\sigma$, due to the applied force must be lower than the yield strength, $S_y$, for a safe design. Once the value of $K_{IC}$ for a material of a particular thickness is known, the designer can determine the crack size that can be allowed in structural members for a given stress level. Values for various material toughnesses are given in Table 6.14.[13]

Toughness is a measure of how much energy a material can absorb before rupturing. On the other hand, hardness is a measure of how much energy it takes to deform a material. If it only requires a small amount of energy to deform a material, it is called a soft material (e.g., rubber). If the material takes a lot of energy to deform, it is called a hard material (e.g., steel).

High toughness is especially important for machine components which may experience impact or for components where a fracture would be catastrophic such as pressure vessels and

---

13 http://www.sv.vt.edu/classes/MSE2094_NoteBook/97ClassProj/exper/gordon/www/fractough.html, accessed April 10, 2013.

**Table 6.14** Values for various material toughnesses, $K_{IC}$.

| Material type | Material | $K_{IC}$ (MPa $\sqrt{m}$) |
|---|---|---|
| Metal | Aluminum alloy (7075) | 24 |
| | Steel alloy (4340) | 50 |
| | Titanium alloy | 44–66 |
| | Aluminum | 14–28 |
| Ceramic | Aluminum oxide | 3–5 |
| | Silicon carbide | 3–5 |
| | Soda-lime glass | 0.7–0.8 |
| | Concrete | 0.2–1.4 |
| Polymer | Polymethyl methacrylate | 0.7–1.6 |
| | Polystyrene | 0.7–1.1 |
| Composite | Mullite-fibre composite | 1.8–3.3 |
| | Silica aerogels | 0.0008–0.0048 |

aircraft. Toughness changes with temperature; some materials change from being tough to brittle as temperature decreases (e.g., steel or rubber). As discussed earlier, a known example of this problem in steels was the ships which broke in two in cold seas during the Second World War.

## Example 6.9

Assume that an $a = 6$ mm through-thickness crack is subjected to a tensile stress as shown in Figure 6.37. The steel plate has a critical stress intensity factor, $K_{IC} = 25$ MPa $\sqrt{m}$. Determine the maximum tensile stress for failure if:

a) the through-thickness crack is at the center of the steel plate with finite width ($W = 80$ mm) and length ($h = 100$ mm), and

b) the through-thickness crack is in a long and wide steel plate.

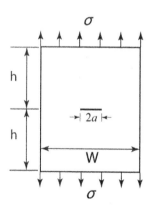

**Figure 6.37** Through-thickness crack in a plate.

## Solution

a) Since the crack is at the center of a plate with finite width and length, the functions $f(a/W)$ to correct the stress intensity factor is

$$f(a/W) = \left[1 + 0.128\left(\frac{a}{W}\right) - 0.288\left(\frac{a}{W}\right)^2 + 1.525\left(\frac{a}{W}\right)^3\right]$$

$$= \left[1 + 0.128\left(\frac{6}{80}\right) - 0.288\left(\frac{6}{80}\right)^2 + 1.525\left(\frac{6}{80}\right)^3\right]$$

$$= 1.009.$$

Equation (6.52) can be modified to calculate the maximum tensile stress for failure as

$$\sigma_{max} = \frac{K_{IC}}{\sqrt{\pi a} f(a/W)}$$

$$= \frac{25\sqrt{10^3}}{\sqrt{6 \times \pi} 1.009} = 180.5 \text{ MPa.}$$

b) Since the steel plate is long and wide, we assume that a through-thickness crack is at the center of the plate. Then the infinite plate formula as in equation (6.51) will be modified to calculate the maximum tensile stress for failure:

$$\sigma_{max} = \frac{K_{IC}}{\sqrt{\pi a}}$$

$$= \frac{25\sqrt{10^3}}{\sqrt{6 \times \pi}} = 182.13 \text{ MPa.}$$

### 6.8.1.2 Elliptical Crack in an Infinite Plate

The problem of an imbedded elliptical or semielliptical crack in an infinite solid has attracted much attention in engineering applications. In pressurized components and vessels such as pressure vessels and pipelines, a crack can easily develop from small defects and material imperfections.

The commonly used approximation for the stress-intensity factor at any point along the perimeters of internal elliptical or circular cracks within an infinite solid, loaded by a uniform uniaxial tension (see Figure 6.38) is defined by[14]

$$K_I = \frac{\sigma\sqrt{\pi a}}{\Phi}\left(\sin^2\theta + \frac{a^2}{c^2}\cos^2\theta\right)^{1/4}, \tag{6.59}$$

---

14 Irwin, G.R., "The crack extension force for a part through crack in a plate," *Transactions, ASME, Journal of Applied Mechanics*, 29(4), 651–654, 1962.

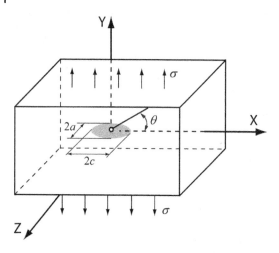

**Figure 6.38** Elliptical crack in an infinite solid (from Ertas, A., and Jones, J.C., *The Engineering Design Process*, 2nd edn, John Wiley & Sons, Inc., New York, 1996. Reproduced with permission of John Wiley & Sons, Inc.).

where $\Phi$ is the elliptical integral and is defined as

$$\Phi = \int_0^{\pi/2} \left( \sqrt{1 - k^2\sin^2\phi} \right) d\phi, \tag{6.60}$$

with

$$k^2 = 1 - \left( \frac{a}{c} \right)^2. \tag{6.61}$$

The elliptic integral has the series expansion

$$\Phi = \frac{\pi}{2} \left[ 1 - \frac{1}{4} \left( \frac{c^2 - a^2}{c^2} \right) - \frac{3}{64} \left( \frac{c^2 - a^2}{c^2} \right)^2 - \cdots \right]. \tag{6.62}$$

Equation (6.62) can be approximated by neglecting higher-order terms:

$$\Phi = \frac{3\pi}{8} + \frac{\pi}{8} \left( \frac{a}{c} \right)^2. \tag{6.63}$$

Equation (6.59) can be modified by including the back crack free-correction factor 1.1 and the effective crack length, $a^*$, which takes into account the plastic zone at the crack tip to determine the stress intensity factor for the semielliptical surface crack shown in Figure 6.39:

$$K_I = 1.1 \frac{\sigma}{\Phi} \sqrt{\pi a^*} \left( \sin^2\theta + \frac{a^2}{c^2}\cos^2\theta \right)^{1/4}, \tag{6.64}$$

where

$$a^* = a + \frac{K_I^2}{4\pi\sqrt{2}S_y^2}. \tag{6.65}$$

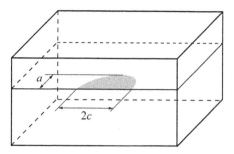

**Figure 6.39** Elliptical surface crack (from Ertas, A., and Jones, J.C., *The Engineering Design Process*, 2nd edn, John Wiley & Sons, Inc., New York, 1996. Reproduced with permission of John Wiley & Sons, Inc.).

Including the effect of the crack shape around the crack front, equation (6.64) can be further modified for the maximum value of the stress intensity factor at the minor axis ($\theta = \pi/2$) as follows:

$$K_I = 1.1\sigma\sqrt{\pi\frac{a}{Q}},\tag{6.66}$$

where $Q$ is the crack-shape parameter given by

$$Q = \Phi^2 - 0.212\left(\frac{\sigma}{S_y}\right)^2.\tag{6.67}$$

Finally, a magnification factor or deep cracks, such as those shown in Figure 6.40,[15] can be used to find the maximum stress intensity factor for a semielliptical surface flow as

$$K_I = 1.1M_k\sigma\sqrt{\pi\frac{a}{Q}}.\tag{6.68}$$

As shown in Figure 6.40, we assume that $M_k$ values vary linearly as $a/t$ varies from 0.5 to 1.0. On the other hand, $M_k$ is assumed to be unity for $a/t$ values less than 0.5.

**Figure 6.40** Magnification factor $M_k$.[15]

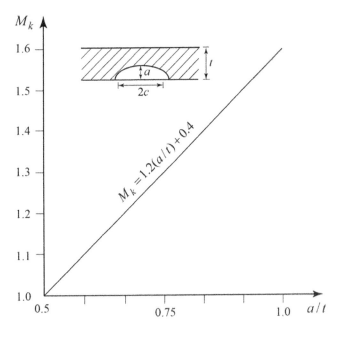

15 Adapted from Rolfe, S.T., and Barsom, J.M., *Fracture and Fatigue Control in Structures; Application of Fracture Mechanics*, Prentice Hall, Englewood Cliffs, NJ, 1977.

### Example 6.10

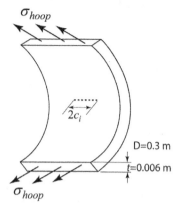

Hydrogen is a renewable energy source and an alternative to petroleum-based fuels. Because of its low energy density, hydrogen is compressed to a pressure of 20 MPa in a pressure vessel. The dimensions of the pressure vessel are shown in Figure 6.41. Assume that the hoop stress created by the internal pressure is acting perpendicular to the semielliptical surface crack depth $a$ (see Figure 6.41). The initial flow depth is $a_i = 2$ mm and the ratio $a/2c = 0.25$ remains constant. If the yield strength of the material is 1000 MPa, calculate the stress intensity factor corresponding to the hoop stress created by the internal pressure.

**Figure 6.41** Hoop stress created by internal pressure.

### Solution

From the thin-walled pressure vessel theory, stress created by the hoop stress perpendicular to the semielliptical crack depth is

$$\sigma_{\text{hoop}} = \frac{pD}{2t}$$

$$= \frac{20 \times 0.3}{2 \times 6 \times 10^{-3}} = 500 \text{ MPa}.$$

From equation (6.68), the stress intensity factor, $K_I$, for a semielliptical surface crack is

$$K_I = 1.1 M_k \sigma \sqrt{\pi \frac{a}{Q}}.$$

The shape factor, $Q$, is

$$Q = \Phi^2 - 0.212 \left(\frac{\sigma}{S_y}\right)^2,$$

where

$$\Phi = \frac{3\pi}{8} + \frac{\pi}{8}\left(\frac{a}{c}\right)^2$$

$$= \frac{3\pi}{8} + \frac{\pi}{8}(0.5)^2 = 1.276,$$

so that

$$Q = (1.276)^2 - 0.212\left(\frac{500}{1000}\right)^2 = 1.575.$$

From Figure 6.40, since $a/t$ is less than 0.5, $M_k$ is equal to 1. Then the stress intensity factor is

$$K_I = 1.1 \times 1 \times 500\sqrt{\pi \frac{2 \times 10^{-3}}{1.575}} = 34.73 \text{ MPa}$$

**Figure 6.42** Leak-before-break criteria (from Ertas, A., and Jones, J.C., *The Engineering Design Process*, 2nd edn, John Wiley & Sons, Inc., New York, 1996. Reproduced with permission of John Wiley & Sons, Inc.).

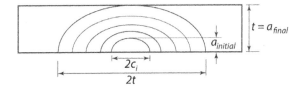

### 6.8.1.3 Critical Crack Length

For catastrophic failure, the initial crack depth, $a$, must reach the critical depth, $a_{cr}$. This critical crack depth can be determined from equation (6.69) by considering the critical stress intensity factor, $K_{IC}$, and the maximum applied stress, $\sigma_{max}$:

$$a_{cr} = \left(\frac{K_{IC}}{1.1 M_k \sigma_{max}}\right)^2 \frac{Q}{\pi}. \tag{6.69}$$

### 6.8.1.4 Leak-before-Break

Leak-before-break (LBB) is a term first proposed by Irwin et al.[16] This concept is used widely to estimate the material fracture toughness, $K_{IC}$, required for a surface flow to grow through the thickness, $t$, thus allowing the pressurized components and vessels to fail from leakage prior to a fracture occurring in service. As shown in Figure 6.42, the first mode of failure (leakage) assumes that a flaw twice the wall thickness in length should be stable at a stress equal to the design stress.[15] The critical crack size at the nominal design stress level of a material should be greater than the wall thickness.

### Example 6.11

In Example 6.10, if the critical stress intensity factor of the pressure vessel material is 45 MPa, will the pressure vessel leak?

**Solution**

From equation (6.69), the critical crack size, $a_{cr}$, is

$$a_{cr} = \left(\frac{K_{IC}}{1.1 M_k \sigma_{max}}\right)^2 \frac{Q}{\pi}$$

$$= \left(\frac{45}{1.1 \times 1 \times 500}\right)^2 \frac{1.575}{\pi} = 3.356 \times 10^{-3} \text{ m}.$$

Since the thickness $t = 6$ mm is larger than the critical crack length, $a_{cr} = 3.356$ mm, the pressure vessel will not leak.

---

16 Irwin, G. R., *Materials for Missiles and Spacecraft*, McGraw-Hill, New York, 1963.

### 6.8.1.5  Fatigue Crack Propagation

In the main, fatigue crack propagation can be divided into three stages as shown in Figure 6.43: stage I (crack formation), stage II (crack propagation) and stage III (unstable crack growth and fracture). Stage II represents the fatigue-crack propagation behavior, which permits the use of the power-law relationship to calculate fatigue life. By using the power-law relationship, the fatigue-crack growth rate, $da/dN$, of a material in terms of the range of applied stress-intensity factor, $\Delta K$, can be written as

$$\frac{da}{dN} = A(\Delta K_I)^m, \tag{6.70}$$

where

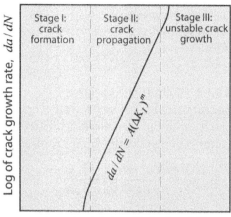

**Figure 6.43** Fatigue crack propagation stages.

$\quad a = $ crack depth,

$\quad N = $ number of cycles,

$\quad \Delta K_I = $ stress-intensity factor occurring at the crack tip,

$A$ and $m$ are the constants for a particular material, environment, and loading conditions.

### 6.8.1.6  Procedure for Analyzing Crack Growth

1) Assume an initial flaw size $a_i$.
2) Calculate the critical crack size, $a_{cr}$, that would cause catastrophic failure.
3) Assume an increment of crack growth $\Delta a$.
4) Determine $\Delta K_I$, using the proper equation for $K_I$, with $\Delta\sigma$ and $a_{ave}$, where

$$a_{ave} = a_i + \frac{\Delta a}{2}. \tag{6.71}$$

5) Determine the number of cycles at a given stress level by using equation (6.70). Direct integration should continue until $a$ reaches $a_{cr}$. If $a_{cr}$ is larger than the material thickness, integration should stop when $a$ is equal to the material thickness.

### Example 6.12

Using the theory of fracture mechanics, find the cumulative fatigue damage inflicted on the drill pipe of Example 6.8 during one pass through the riser upper ball joint.[a]

### Solution

The drill pipe is assumed to have a semielliptical (part-through) surface crack, as shown in Figure 6.44. This type of crack is typical of the starting crack shape customarily found. As shown,

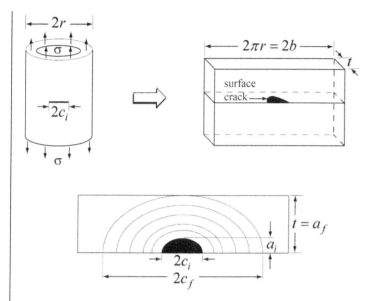

**Figure 6.44** Semielliptical surface crack in drill pipe.

the drill pipe is assumed to be cut along a longitudinal plane and opened into a plate. On the basis of quality inspection, the estimated initial flaw size $a_i$ present in the drill pipe is assumed to be 0.0312 in. The critical stress-intensity factor, $K_{IC}$, is assumed to be 56 kpsi $\sqrt{in.}$

*Given Design Data*

Modulus of elasticity of drill-pipe material, $E$ = $30 \times 10^6$ psi

Critical stress-intensity factor, $K_{IC}$ = 56 kpsi $\sqrt{in.}$

Yields strength of drill-pipe material, $\sigma_{ys}$ = 90 kpsi

Crack growth (remains constant), $\frac{a}{c}$ = 0.5

Penetration speed of drill pipe, $V$ = 120 in./h

Outside diameter of drill pipe, $D_o$ = 5.00 in.

Inside diameter of drill pipe, $D_i$ = 4.276 in.

Initial crack size, $a_i$ = 0.0312 in.

Rotational speed of drill pipe, $n$ = 100 rpm

Tension load, $T$ = 100 kips

The sinusoidal fluctuating stress-time history for a drill pipe is shown in Figure 6.45. From this figure, the stress range of the drill pipe is observed to be

$$\sigma_r = \Delta\sigma = \sigma_{max} - \sigma_{min},$$

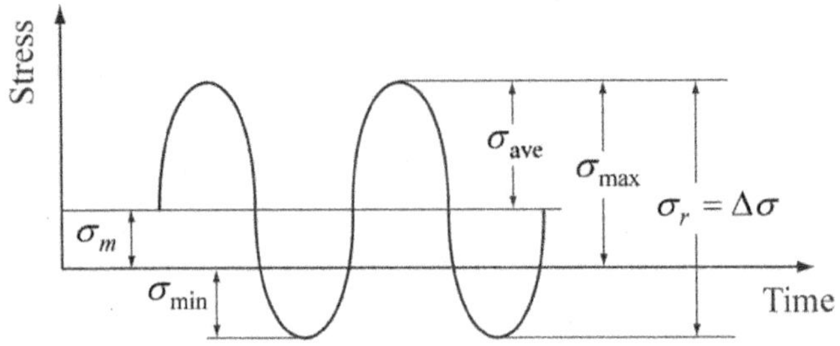

**Figure 6.45** Stress-time history of a drill pipe.

where

$$\sigma_{max} = \sigma_m + \sigma_{ave},$$
$$\Delta\sigma = \sigma_r = 2\sigma_{ave}.$$

Table 6.15 is obtained by using $\sigma_m = 18.96$ kpsi and the values of $\sigma_{ave}$ from Table 6.13 (see Example 6.8). The following steps are performed to determine the cumulative fatigue damage:

**Table 6.15** Fatigue damage calculations.

| EL No. | $\sigma_{av}$ (kpsi) | $\Delta\sigma$ (kpsi) | $\sigma_{max}$ (kpsi) |
|--------|--------|--------|--------|
| EL 1 | 19.13 | 38.26 | 38.09 |
| EL 2 | 16.73 | 33.46 | 35.69 |
| EL 3 | 13.37 | 26.74 | 32.33 |
| EL 4 | 9.10 | 18.20 | 28.06 |

*Step 1.* Divide a joint of the drill pipe into a number of finite elements as shown in Figure 6.33.

*Step 2.* Calculate the average stress amplitude, $\sigma_{ave}$, for each of these finite elements (see Example 6.8).

*Step 3.* Calculate the critical flaw size, $a_{cr}$, that would cause failure by brittle fracture for the first element as follows. As discussed previously, for catastrophic failure, the initial crack depth, $a_i$, must reach the critical crack depth calculated using

$$a_{cr} = \left(\frac{K_{IC}}{1.1M_k\sigma_{max}}\right)^2 \frac{Q}{\pi},$$

where $\sigma_{max} = 38.09$ kpsi, for element 1. Using equation (6.63),

$$\Phi = \frac{3\pi}{8} + \frac{\pi}{8}\frac{a^2}{c^2} = \frac{3\pi}{8} + \frac{\pi}{8}(0.5)^2 = 1.276.$$

For simplicity, $M_k$ is assumed to be a unity, and

$$\sqrt{Q} = \sqrt{\Phi^2 - 0.212\left(\frac{\sigma_{\max}}{S_y}\right)^2}.$$

Then

$$Q = (1.276)^2 - 0.212\left(\frac{38.09}{90}\right)^2 = 1.59.$$

Hence,

$$a_{\mathrm{cr}} = \left(\frac{56}{1.1 \times 38.09}\right)^2 \frac{1.59}{\pi} = 0.905 \text{ in.}$$

Since the critical crack size is larger than the wall thickness ($t = 0.362$ in) the drill pipe will fail by leaking and not by catastrophic propagation to failure.

*Step 4.* Assume an increment of crack growth, $\Delta a$, and determine the number of failure cycles for the crack to grow by this increment at the given stress level for each finite element by using equation (6.70) with $A = 0.614 \times 10^{-10}$ and $m = 3.16$ (specific to drill pipe material grade E):

$$\frac{da}{dN} = 0.614 \times 10^{-10}(\Delta K_I)^{3.16},$$

where $\Delta K_I$ can be obtained from equation (6.68) as

$$\Delta K_I = 1.1\Delta\sigma\sqrt{\pi\frac{a_{\mathrm{ave}}}{Q}}.$$

Note that in the above equation, $M_k$ is assumed to be 1. To determine the incremental crack growth, equation (6.70) can be modified to

$$\frac{\Delta a}{\Delta N} = 0.506 \times 10^{-9}\left[\Delta\sigma\sqrt{\frac{a_{\mathrm{ave}}}{Q}}\right]^{3.16},$$

where $\Delta N$ is the number of cycles required to propagate the incremental crack growth $\Delta a$, and

$$a_{\mathrm{ave}} = a_i + \frac{\Delta a}{2}.$$

Hence, the number of cycles, $\Delta N$, required to propagate the crack from initial depth $a_i$ to depth $a_i + \Delta a$, can be written as

$$\Delta N = \frac{1.976 \times 10^9 \Delta a}{\left(\Delta\sigma\sqrt{\dfrac{a_i + 0.5\Delta a}{Q}}\right)^{3.16}}.$$

The incremental crack growth length should be assumed to be reasonably small to obtain an accurate result; assume $\Delta a = 0.01$ in. The first iteration gives

$$\Delta N = \frac{1.976 \times 10^9 (0.01)}{\left(38.26 \sqrt{\dfrac{0.0312 + 0.5 \times 0.01}{1.59}}\right)^{3.16}} \approx 77{,}600 \text{ cycles.}$$

For the second iteration, the initial crack size will be $a_i = 0.0312 + 0.01 = 0.0412$ in. Hence, the number of cycles for the next $\Delta a = 0.01$ increment is

$$\Delta N = \frac{1.976 \times 10^9 (0.01)}{\left(38.26 \sqrt{\dfrac{0.0412 + 0.5 \times 0.01}{1.59}}\right)^{3.16}} \approx 52{,}800 \text{ cycles.}$$

Since the critical crack size was found to be larger than the wall thickness of the drill pipe, iteration for incremental crack growth should stop when it reaches a final crack size, $a_f$, which is equal to the wall thickness of 0.362 in. Thus,

$$N(\text{EL.1}) = \sum \Delta N = 77{,}600 + 52{,}800 + \cdots = 400{,}281 \text{ cycles.}$$

The damage for the first element is

$$\frac{n_1}{N_1} = \frac{300}{400{,}281} \times 100 = 0.075\%.$$

The same procedure is repeated for the remaining elements and the results are summarized in Table 6.16. Note that since the stress is changing, $Q$ and $a_{cr}$ should be recalculated for each element. The total damage to the drill pipe during one pass through the upper joint is

$$\text{Total damage} = 2 \times \left(\frac{n_1}{N_1} + \frac{n_2}{N_2} + \frac{n_3}{N_3} + \frac{n_4}{N_4}\right) \times 100$$

$$= 2 \times (0.075 + 0.098 + 0 + 0) = 0.346\%.$$

**Table 6.16** Fatigue damage calculations.

| EL No. | $\sigma_{ave}$ (kpsi) | $\Delta\sigma$ (kpsi) | $\sigma_{max}$ (kpsi) | $n$ (cycles) | $N$ (cycles) | $n/N$ % |
|---|---|---|---|---|---|---|
| EL 1 | 19.13 | 38.26 | 38.09 | 300 | 400,281 | 0.075 |
| EL 2 | 16.73 | 33.46 | 35.69 | 600 | 611,994 | 0.098 |
| EL 3 | 13.37 | 26.74 | 32.33 | 900 | $1.24 \times 10^6$ | 0 |
| EL 4 | 9.10 | 18.2 | 28.06 | 1800 | $4.18 \times 10^6$ | 0 |

The results of Examples 6.8 and 6.12 show that the fracture mechanics approach provides conservative results in determining the fatigue damage of drill pipe at a low-tension load such as $T = 100$ kips and small riser angles such as $\theta = 1$ degree.

[a]Ertas, A., Ghulam, M., and Cuvalci, O, "A comparison of fracture mechanics and S–N curve approaches in designing drill pipe," ASME Journal of Offshore Mechanics and Arctic Engineering, 114, 205–211, 1992.

### 6.8.1.7 Superposition of Stress-Intensity Factors

Machine components contain cracks that may be subjected to more than one type of loading. The principle of superposition can be applied to stress-intensity factors for different types of loading of the same mode type. In other words, the stress-intensity factors for different modes of crack growth can not be algebraically added. Also, in the summation process (superposition) the stress-intensity factors must be associated with the same structural geometry, including crack geometry. For example, stress-intensity factors associated with edge crack problems cannot be added to those of semielliptical crack problems.

### Example 6.13

Determine the stress-intensity factor of a plate subjected to combined tension and bending loadings shown in Figure 6.46.

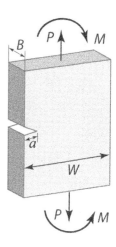

**Solution**

(a) The stress-intensity factor for tension load is

$$K_I^T = \sigma^T \sqrt{\pi a} f(a/W).$$

From equation (6.54), the correction factor for a single-edge through-thickness crack is

$$f(a/W) = \left[1.122 - 0.231\left(\frac{4.8}{40}\right) + 10.550\left(\frac{4.8}{40}\right)^2 - 21.710\left(\frac{4.8}{40}\right)^3 \right.$$
$$\left. + 30.382\left(\frac{4.8}{40}\right)^4\right] = 1.215.$$

**Figure 6.46** Plate with through thickness crack.

Then the stress-intensity factor for tension load is

$$K_I^T = 1.215\sigma^T\sqrt{\pi a},$$

where

$$\sigma^T = \frac{P}{A} = \frac{P}{B \times W}.$$

(b) The stress-intensity factor for bending is

$$K_I^M = \sigma^M \sqrt{\pi a} f(a/W).$$

From equation (6.56), the correction factor for a single-edge through-thickness crack for pure bending is

$$f(a/W) = \left[ 1.122 - 1.40 \left( \frac{4.8}{40} \right) + 7.33 \left( \frac{4.8}{40} \right)^2 - 13.08 \left( \frac{4.8}{40} \right)^3 \right.$$
$$\left. + 14.0 \left( \frac{4.8}{40} \right)^4 \right] = 1.04.$$

Then the stress-intensity factor for bending is

$$K_I^M = 1.04 \sigma^M \sqrt{\pi a},$$

where

$$\sigma^M = \frac{M}{I/c} = \frac{6M}{B \times W^2}$$

The algebraic addition of the two above stress-intensity factors will give the stress-intensity factor for combined loadings, $K_I^C$:

$$K_I^C = K^T + K^M = 1.215 \sigma^T \sqrt{\pi a} + 1.04 \sigma^M \sqrt{\pi a}$$

or

$$K_I^C = K^T + K^M = \left[ 1.215 \frac{P}{B \times W} \sqrt{\pi a} \right] + \left[ 1.04 \frac{6M}{B \times W^2} \sqrt{\pi a} \right]$$
$$= \frac{\sqrt{\pi a}}{BW} \left[ 1.215 P + \frac{6.24}{W} \right].$$

## Example 6.14

Rotating machine components, such as turbine blades, compressor disks, spacers, and cooling fan blades are exposed to cyclic stresses during engine startup, normal operation, and shutdown. Combined stresses acting in the cross-section of the turbine blade are a tensile stress component due to centrifugal forces and a bending stress component introduced by the action of the air flow pressure. Although combined loadings are steady during normal operation, fan blades are subjected to cyclic stresses during engine startup and shutdown (see Figure 6.47).

A blade of a rotating axial flow fan shown in Figure 6.47 has $N = 2400$ rpm rotational speed, $A = 0.00072$ m$^2$ cross-sectional area, and $M = 0.498$ kg mass. The distance from the center of rotation, $r$ is 0.55 m. The properties of the blade (aluminum alloy) are: $S_y = 380$ MPa, $S_{ut} = 510$ MPa, and $K_{IC} = 32$ MPa. Assume that the amplitude of the bending stress component, $\sigma_a^M$, is 6 MPa.

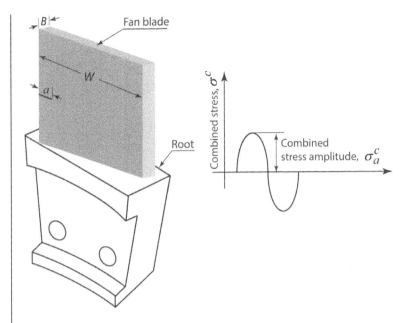

**Figure 6.47** Fan blade.

Using scaling factor $A = 9.5 \times 10^{-13}$ and $m = 4.9$ for the expression of the crack growth rate, investigate whether the design is adequate for 20,000 cycles. Assume that the blade width, $W$, is 0.08 m, thickness $B = 0.009$ m, and the initial single-edge through-thickness crack length, $a$, is 0.001 m. Use only one iteration for the fatigue life calculation.[a]

**Solution**

From equation (6.70), the fatigue life is

$$\Delta N = \frac{\Delta a}{A(\Delta K_I)^m}. \tag{a}$$

From Example 6.13, combine stress-intensity factor for tensile and bending loads is given by

$$K_I^C = K^T + K^M = 1.215\sigma^T \sqrt{\pi a} + 1.04\sigma^M \sqrt{\pi a}. \tag{b}$$

Then

$$\Delta K_I^C = 1.215\Delta\sigma^T \sqrt{\pi a_{av}} + 1.04\Delta\sigma^M \sqrt{\pi a_{\text{ave}}}, \tag{c}$$

where

$$\Delta\sigma = 2\sigma_a = \sigma_{range}. \tag{d}$$

The ensile stress amplitude, $\sigma_a^T$, due to centrifugal force can be calculated as

$$\sigma_a^T = \frac{Mr(2\pi N)^2}{\overline{A}} = \frac{0.498 \times 0.55 \times (2\pi \times 2400/60)^2}{0.00072} = 24 \times 10^6 \text{ Pa}$$

Knowing that the maximum stress is, $\sigma_{max} = \sigma_a^T + \sigma_a^M = 24 + 6 = 30$ MPa, from equation (6.52) we have the critical crack length

$$a_{cr} = \frac{1}{\pi}\left[\frac{K_{IC}}{\sigma_{max} \times f(a/W)}\right]^2 = \frac{1}{\pi}\left[\frac{32}{30 \times 1.13}\right]^2 = 0.28 \text{ m}.$$

Note that, in the above calculation for $f(a/W)$, $(1.215 + 1.04)/2 = 1.13$ is used. Since $a_{cr}$ is larger than the blade thickness, iteration will stop at the thickness value of the fan blade ($W = 0.08$ m). The average crack length, $a_{cr}$, in equation (c), is

$$a_{ave} = a_i + \frac{\Delta a}{2}.$$

For one iteration, $\Delta a = W - a_i = 0.08 - 0.001 = 0.079$ m. Then

$$a_{ave} = 0.001 + \frac{0.079}{2} = 0.0405 \text{ m}.$$

From equation (c), the combined stress-intensity factor becomes

$$\Delta K_I^C = 1.215(2 \times 24)\sqrt{\pi \times 0.0405} + 1.04(2 \times 6)\sqrt{\pi \times 0.0405} = 25.25 \text{ MPa}.$$

Using equation (a), we have the fatigue life expectation for the fan blade

$$\Delta N = \frac{0.079}{9.5 \times 10^{-13}(25.25)^4.9} = 11{,}190 \text{ cycles}.$$

Design is not adequate for 5,000 cycles.

[a]Adapted from Sameezadeh, M., and Farhangi, H., *Fracture Analysis of Generator Fan Blades*, Applied Fracture Mechanics, Intech. http://www.intechopen.com/books/applied-fracture-mechanics/fracture-analysis-of-generator-fan-blades, accessed May 12, 2013.

## 6.8.2 Design with Materials

Since material selection is an integral part of design, the designer must be aware of each class of material and its advantages, disadvantages, and design limitations. Composites offer impressive properties and can be used in many design applications. However, cost and problems related to the manufacturing of composite structures are primary barriers to large-scale use.

Most materials are subjected to environmental effects such as sunshine, temperature, rainfall, and wind. Metals show the least resistance against corrosion. Although the toughness of polymers is often low, they find a wide range of applications because of their superior corrosion resistance and low coefficient of friction. Polymers can be formed and manufactured in almost any shape, and the components can be designed to snap together, making assembly fast and cheap.

In the past decade, research has been under way on the development of high-performance engineering ceramics. These materials have an important role as tool and die materials as well as in engine components such as turbochargers and valves. Ceramics have the highest hardness of all the solid materials. For example, corundum ($Al_2O_3$), silicon carbide (SiC), and diamond (C) are used to cut, grind, and polish a wide variety of materials. Metals have very low hardness compared with ceramics. However, ceramics have very low toughness because of their brittleness. Moreover, ceramics always contain small surface flaws. Hence, the design strength of ceramic materials is determined by their fracture toughness and by the size of preexisting cracks. If the longest flaw size, $2a$, is known, referring to equation (6.51), the fracture toughness $K_{IC}$ of a ceramic can be determined by

$$K_{IC} = \sigma_{max} \sqrt{\pi a}, \tag{6.72}$$

where $\sigma_{max}$ is the tensile strength of the ceramic. As we can see from this equation, the strength of a ceramic can be improved by decreasing the crack length through careful quality control or by increasing $K_{IC}$ by making the ceramic into a composite.

## 6.9 Design Analysis for Composite Materials

Material composites are not a new concept; nature is abundant with examples of them. In fact, practically everything in the world is made of composite materials,[17] from bone to common metals. Originally, the human-made composites were laboratory curiosities. During the 1940s, with the advent of glass-reinforced plastics, the engineering application potential became evident.[18] It took two decades before composite material science became a distinct discipline. Since the 1960s, the demand for materials with high strength-to-weight ratio has steadily increased. This demand has contributed to advances in the development of composites, such as polymers and lightweight metal matrix composites.

A composite can simply be viewed as a combination of material components, usually a reinforcing agent and a binder. The regions formed by the materials involved should be large enough to be regarded as a continuum. The process of component bonding in composites is usually done to maximize the favorable properties of the components while mitigating the effects of some of their less desirable characteristics. The properties to be optimized may be physical, chemical, or mechanical.

---

17  Chawla, K.K., *Composite Materials*, Springer-Verlag, New York, 1987.
18  Luce, S., *Introduction to Composite Technology*, Society of Manufacturing Engineers, Dearborn, MI, 1988.

The elements involved in composite fabrication are fibers or reinforcing elements, matrices, and interface bonding. The fibers can be made from glass, boron, carbon, organic material, ceramic, or metal. The matrix materials are normally polymers, metals, or ceramics. The interfacial bonding can be either mechanical or chemical, but generally chemical bonding is most widely used. There are many techniques available for composite fabrication.[17] Fabrication techniques include hand and automated tape lay-up, resin injection, vacuum bag and autoclave molding, pultrusion, and filament winding. The technique used depends on the kind and quality of composite to be manufactured (e.g., for matrix composites, diffusion bonding is usually used).

Composites have found wide use in practically all engineering disciplines.[19] The applications have mainly been driven by the fact that composites can be tailor-made as per specification for an optimum design. This means that the properties can be easily modified to suit the design. From a design point of view, composites also exhibit superior material properties as compared with conventional monolithic materials. Composites have less weight, less thermal expansion, greater stiffness, greater strength, and higher fatigue resistance. In aerospace, composites are used for helicopter rotor blades, rocket nozzles, and reentry shields. In automotive engineering, applications range from doors to body moldings. In chemical engineering, composites are used for containers, pipe-work, pressure vessels, and the like. In civil engineering, composites have found important use as glass-reinforced plastics and form work for concrete. In electrical engineering, the applications range from high-strength insulators to printed circuit boards. Bio-engineering uses carbon fiber composites as prosthetic devices, and other composites have been used to manufacture heart valves.

The solid mechanics of composites is more involved than that of conventional metals (which, for simplicity, are often assumed to be isotropic). Generally, there are two types of information that determine the properties of a composite material: the internal phase geometry and the physical characteristics of the phases, that is, their stress–strain relations. Hashin[20] provides a detailed survey of the analysis of the properties of composite materials. The properties presented in this survey include static strength, fatigue failure, elasticity, thermal expansion, moisture swelling, viscoelasticity, and conductivity. Among the composite structures, the laminates have been the most thoroughly investigated. The methods used to analyze the properties of laminates are based on the well-developed laminate theory.[21,22]

### 6.9.1 Stress–Strain Relations in Composite Materials

In the elementary analysis of composite materials, the stress–strain convention is chosen such that the orthogonal axis corresponds to three mutually perpendicular planes of material symmetry (Figure 6.48). This axis is also refered to as the principal material axis. For the following

19 Harris, B., *Engineering Composite Materials*, The Institute of Metals, Brookfield, VT, 1986.

20 Hashin, Z., "Analysis of composite materials – a survey," *Transactions, ASME Journal of Applied Mechanics*, 50, 481–505, 1983.

21 Carlsson, L.A., and Pipes, R.B., *Experimental Characterization of Advanced Composite Materials*, Prentice Hall, Englewood Cliffs, NJ, 1987.

22 Phillips, L.N. (ed.), *Design of Composites in Design with Advanced Composites Materials*, Springer-Verlag, New York, 1989.

analysis, the fiber direction is assumed to be parallel to axis 1, and the direction transverse to it is axis 2. Generally in linear elasticity, the stress–strain relations are expressed in the form

$$
\begin{Bmatrix} \sigma_1 \\ \sigma_2 \\ \sigma_3 \\ \tau_{23} \\ \tau_{31} \\ \tau_{12} \end{Bmatrix} = \begin{bmatrix} C_{11} & C_{12} & C_{13} & C_{14} & C_{15} & C_{16} \\ C_{21} & C_{22} & C_{23} & C_{24} & C_{25} & C_{26} \\ C_{31} & C_{32} & C_{33} & C_{34} & C_{35} & C_{36} \\ C_{41} & C_{42} & C_{43} & C_{44} & C_{45} & C_{46} \\ C_{51} & C_{52} & C_{53} & C_{54} & C_{55} & C_{56} \\ C_{61} & C_{62} & C_{63} & C_{64} & C_{65} & C_{66} \end{bmatrix} \begin{Bmatrix} \varepsilon_1 \\ \varepsilon_2 \\ \varepsilon_3 \\ \gamma_{23} \\ \gamma_{31} \\ \gamma_{12} \end{Bmatrix}, \quad (6.73)
$$

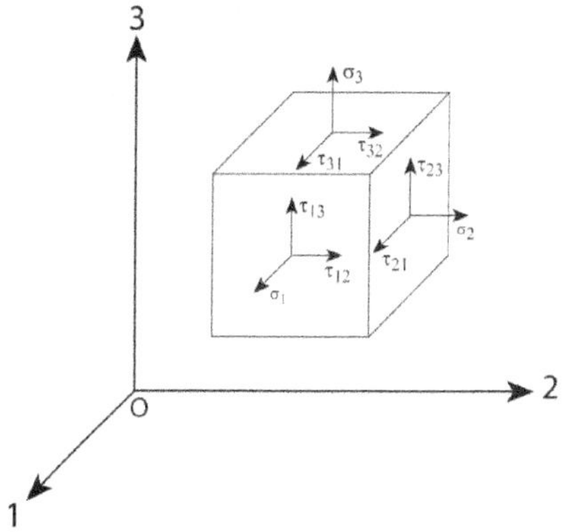

**Figure 6.48** Stresses acting on a cubic element.

where the coefficients $C_{ij}$ are called *elastic constants* or the *stiffness matrix* of the material, $\varepsilon$ is the strain, $\gamma$ is the shear strain, $\sigma$ is the normal stress, and $\tau$ is the shear stress. As can be seen from the symmetric matrix $[C]$, there are 15 off-diagonal terms and six diagonal terms, resulting in a total of 21 independent elastic constants for most anisotropic materials. For isotropic materials only two elastic constants, $C_{11}$ and $C_{12}$, are independent and the elastic properties are the same in all directions. The stiffness matrix is reduced to

$$
[C] = \begin{bmatrix} C_{11} & C_{12} & C_{12} & 0 & 0 & 0 \\ C_{12} & C_{11} & C_{12} & 0 & 0 & 0 \\ C_{12} & C_{12} & C_{11} & 0 & 0 & 0 \\ 0 & 0 & 0 & \dfrac{C_{11} - C_{12}}{2} & 0 & 0 \\ 0 & 0 & 0 & 0 & \dfrac{C_{11} - C_{12}}{2} & 0 \\ 0 & 0 & 0 & 0 & 0 & \dfrac{C_{11} - C_{12}}{2} \end{bmatrix}. \quad (6.74)
$$

Using equations 6.73 and 6.74, the stress–strain relationship for isotropic materials can be formulated as

$$
\{\varepsilon\} = [S]\{\sigma\}, \quad (6.75)
$$

where the square matrix $[S]$, called the matrix of elastic compliances, is equal to the inverse of the $[C]$ matrix, and is given by

$$
[S] = \begin{bmatrix} S_{11} & S_{12} & S_{12} & 0 & 0 & 0 \\ S_{12} & S_{11} & S_{12} & 0 & 0 & 0 \\ S_{12} & S_{12} & S_{11} & 0 & 0 & 0 \\ 0 & 0 & 0 & 2(S_{11} - S_{12}) & 0 & 0 \\ 0 & 0 & 0 & 0 & 2(S_{11} - S_{12}) & 0 \\ 0 & 0 & 0 & 0 & 0 & 2(S_{11} - S_{12}) \end{bmatrix}. \quad (6.76)
$$

For isotropic materials composed of a number of sufficiently thin laminae so that the through-thickness stresses are zero, equation (6.76) can be reduced to

$$
\left\{ \begin{array}{c} \varepsilon_1 \\ \varepsilon_2 \\ \gamma_{12} \end{array} \right\} = \left[ \begin{array}{ccc} S_{11} & S_{12} & 0 \\ S_{12} & S_{22} & 0 \\ 0 & 0 & S_{66} \end{array} \right] \left\{ \begin{array}{c} \sigma_1 \\ \sigma_2 \\ \tau_{12} \end{array} \right\}.
\tag{6.77}
$$

The relationships between compliances $S$ and elastic constants $E$, $G$, and $v$ are given as follows:

$$
S_{11} = \frac{1}{E},
$$
$$
S_{22} = S_{11},
$$
$$
S_{12} = -\frac{v}{E},
$$
$$
S_{66} = 2(S_{11} - S_{12}) = \frac{1}{G},
\tag{6.78}
$$

where the shear modulus of elasticity $G$ can be defined in terms of Young's modulus $E$ and Poisson's ratio $v$ as

$$
G = \frac{E}{2(1 + v)}.
\tag{6.79}
$$

For thin material, it is customary to replace the stiffness coefficients $C$ by the symbol $Q$. Then equation (6.73), for plane-stress isotropic composite materials, can be reduced to

$$
\left\{ \begin{array}{c} \sigma_1 \\ \sigma_2 \\ \tau_{12} \end{array} \right\} = \left[ \begin{array}{ccc} Q_{11} & Q_{12} & 0 \\ Q_{12} & Q_{22} & 0 \\ 0 & 0 & Q_{66} \end{array} \right] \left\{ \begin{array}{c} \varepsilon_1 \\ \varepsilon_2 \\ \gamma_{12} \end{array} \right\},
\tag{6.80}
$$

where the reduced stiffnesses $Q_{ij}$ are given by

$$
Q_{11} = \frac{E}{1 - v^2},
$$
$$
Q_{22} = Q_{11},
$$
$$
Q_{12} = \frac{vE}{1 - v^2},
$$
$$
Q_{66} = \frac{1}{2}(Q_{11} - Q_{12}) = G.
\tag{6.81}
$$

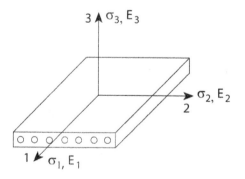

**Figure 6.49** Principal material directions for an orthotropic material.

Equations 6.77 and 6.80 describe the stress–strain relationships of the lamina assuming plane-stress conditions for isotropic materials such as thin aluminum plates stacked on top of each other. However, laminae containing fiber–matrix materials are not isotropic. They behave as orthotropic materials having three mutually perpendicular planes of material symmetry. As shown in Figure 6.49, the material properties in the three axes are different. However, for a thin lamina, it can be shown that the elastic properties are independent of direction in the plane normal to axis 1 (plane stress). Hence, elastic compliances $S_{ij}$ in equation (6.77) for such an orthotropic thin lamina are defined by independent elastic constants as

$$S_{11} = \frac{1}{E_1},$$

$$S_{22} = \frac{1}{E_2},$$

$$S_{66} = \frac{1}{G_{12}},$$

$$S_{12} = -\frac{v_{12}}{E_1} = -\frac{v_{21}}{E_2}. \tag{6.82}$$

Similarly, reduced stiffnesses $Q_{ij}$ in equation (6.80) are defined as

$$Q_{11} = \frac{E_1}{1 - v_{21}v_{12}},$$

$$Q_{22} = \frac{E_2}{1 - v_{21}v_{12}},$$

$$Q_{12} = \frac{v_{12}E_2}{1 - v_{21}v_{12}} = \frac{v_{21}E_1}{1 - v_{21}v_{12}},$$

$$Q_{66} = G_{12}. \tag{6.83}$$

In the above equation, Poisson's ratio, $v_{12}$, refers to the strain produced along axis 2 when the line of action of the load application is in axis 1 direction. It should be realized that Poission's ratio for orthotropic materials can be greater than the maximum of 0.5 that is inherent in isotropic materials.

## Example 6.15

Consider a lamina with material properties $E_1 = 20 \times 10^6$ *psi*, $E_2 = 1.5 \times 10^6$ *psi*, $G_{12} = 1.1 \times 10^6$ *psi*, $v_{12} = 0.26$, $v_{21} = 0.020$, $\theta = 0°$, and let the strains be $\varepsilon_1 = 3.2 \times 10^{-4}$, $\varepsilon_2 = 0.08 \times 10^{-4}$, and $\gamma_{12} = 0.00$. Calculate the stresses in the lamina.

## Solution

As indicated in equation (6.80) the reduced stiffness matrix $[Q]$ is

$$[Q] = \begin{bmatrix} Q_{11} & Q_{12} & 0 \\ Q_{12} & Q_{22} & 0 \\ 0 & 0 & Q_{66} \end{bmatrix}.$$

Using equation (6.83) and substituting the respective values into the above matrix yields

$$[Q] = \begin{bmatrix} 20.10 & 0.39 & 0 \\ 0.39 & 1.51 & 0 \\ 0 & 0 & 1.10 \end{bmatrix} 10^6 \ psi$$

Using equation (6.80), the stress in the lamina can be calculated as

$$
\left\{
\begin{array}{c}
\sigma_1 \\
\sigma_2 \\
\tau_{12}
\end{array}
\right\}
=
\left[
\begin{array}{ccc}
20.10 & 0.39 & 0.00 \\
0.39 & 1.51 & 0.00 \\
0.00 & 0.00 & 1.10
\end{array}
\right]
\left\{
\begin{array}{c}
3.20 \\
0.08 \\
0.0
\end{array}
\right\}^{10^2 \ \text{psi}}
$$

Solution of the above matrix yields $\sigma_1 = 64.63 \times 10^2$ psi, $\sigma_2 = 2.46 \times 10^2$ psi, $\tau_{12} = 0.0$ psi.

Equation (6.77) for orthotropic materials shows that there is no coupling between tensile and shear strains. However, this is not true when the principal material axes of a lamina are oriented at an angle $\theta$ with respect to the $(x, y)$ reference axes (Figure 6.50). In this case, both the stresses and strains need to be transformed to the reference axes. For example, the matrix relationship of the stresses and strains in the principal material axes $(1, 2)$ and the reference axes $(x, y)$ can be formulated as

$$
\left\{
\begin{array}{c}
\sigma_1 \\
\sigma_2 \\
\tau_{12}
\end{array}
\right\}
= [T]
\left\{
\begin{array}{c}
\sigma_x \\
\sigma_y \\
\tau_{xy}
\end{array}
\right\}
\tag{6.84}
$$

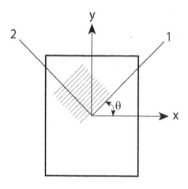

**Figure 6.50** Rotation of principal material axes $(1, 2)$ from arbitrary $(x, y)$ axes.

or

$$
\left\{
\begin{array}{c}
\sigma_x \\
\sigma_y \\
\tau_{xy}
\end{array}
\right\}
= [T]^{-1}
\left\{
\begin{array}{c}
\sigma_1 \\
\sigma_2 \\
\tau_{12}
\end{array}
\right\}
\tag{6.85}
$$

and

$$
\left\{
\begin{array}{c}
\varepsilon_1 \\
\varepsilon_2 \\
\dfrac{\gamma_{12}}{2}
\end{array}
\right\}
= [T]
\left\{
\begin{array}{c}
\varepsilon_x \\
\varepsilon_y \\
\dfrac{\gamma_{xy}}{2}
\end{array}
\right\}
\tag{6.86}
$$

or

$$
\left\{
\begin{array}{c}
\varepsilon_x \\
\varepsilon_y \\
\dfrac{\gamma_{xy}}{2}
\end{array}
\right\}
= [T]^{-1}
\left\{
\begin{array}{c}
\varepsilon_1 \\
\varepsilon_2 \\
\dfrac{\gamma_{12}}{2}
\end{array}
\right\},
\tag{6.87}
$$

where the transformation matrices are

$$[T] = \begin{bmatrix} m^2 & n^2 & 2mn \\ n^2 & m^2 & -2mn \\ -mn & mn & (m^2 - n^2) \end{bmatrix} \tag{6.88}$$

and

$$[T]^{-1} = \begin{bmatrix} m^2 & n^2 & -2mn \\ n^2 & m^2 & 2mn \\ mn & -mn & (m^2 - n^2) \end{bmatrix}, \tag{6.89}$$

with $m = \cos\theta$ and $n = \sin\theta$. The stress–strain relation becomes

$$\begin{Bmatrix} \sigma_x \\ \sigma_y \\ \tau_{xy} \end{Bmatrix} = \begin{bmatrix} \overline{Q}_{11} & \overline{Q}_{12} & \overline{Q}_{16} \\ \overline{Q}_{12} & \overline{Q}_{22} & \overline{Q}_{26} \\ \overline{Q}_{16} & \overline{Q}_{26} & \overline{Q}_{66} \end{bmatrix} \begin{Bmatrix} \varepsilon_x \\ \varepsilon_y \\ \gamma_{xy} \end{Bmatrix}. \tag{6.90}$$

The matrix $[\overline{Q}_{ij}]$ is called the transformed reduced-stiffness matrix and the stiffnesses $\overline{Q}_{ij}$ are given by

$$\overline{Q}_{11} = m^4 Q_{11} + 2m^2 n^2 (Q_{12} + 2Q_{66}) + n^4 Q_{22},$$
$$\overline{Q}_{12} = m^2 n^2 (Q_{11} + Q_{22} - 4Q_{66}) + (m^4 + n^4)Q_{12},$$
$$\overline{Q}_{22} = n^4 Q_{11} + 2m^2 n^2 (Q_{12} + 2Q_{66}) + m^4 Q_{22},$$
$$\overline{Q}_{16} = m^3 n (Q_{11} - Q_{12}) + mn^3 (Q_{12} - Q_{22}) - 2mn(m^2 - n^2)Q_{66},$$
$$\overline{Q}_{26} = mn^3 (Q_{11} - Q_{12}) + m^3 n (Q_{12} - Q_{22}) + 2mn(m^2 - n^2)Q_{66},$$
$$\overline{Q}_{66} = m^2 n^2 (Q_{11} + Q_{22} - 2Q_{12} - 2Q_{66}) + (m^4 + n^4)Q_{66}, \tag{6.91}$$

where $m = \cos\theta$ and $n = \sin\theta$. Similarly, the compliance relation becomes

$$\begin{Bmatrix} \varepsilon_x \\ \varepsilon_y \\ \gamma_{xy} \end{Bmatrix} = \begin{bmatrix} \overline{S}_{11} & \overline{S}_{12} & \overline{S}_{16} \\ \overline{S}_{12} & \overline{S}_{22} & \overline{S}_{26} \\ \overline{S}_{16} & \overline{S}_{26} & \overline{S}_{66} \end{bmatrix} \begin{Bmatrix} \sigma_x \\ \sigma_y \\ \tau_{xy} \end{Bmatrix}. \tag{6.92}$$

The transformed compliances $\overline{S}_{ij}$ have the following values ($m = \cos\theta$, $n = \sin\theta$):

$$\overline{S}_{11} = m^4 S_{11} + m^2 n^2 (2S_{12} + S_{66}) + n^4 S_{22},$$
$$\overline{S}_{12} = m^2 n^2 (S_{11} + S_{22} - S_{66}) + (m^4 + n^4)S_{12},$$
$$\overline{S}_{22} = n^4 S_{11} + m^2 n^2 (2S_{12} + S_{66}) + m^4 S_{22},$$
$$\overline{S}_{16} = 2m^3 n (S_{11} - S_{12}) + 2mn^3 (S_{12} - S_{22}) - mn(m^2 - n^2)S_{66},$$
$$\overline{S}_{26} = 2mn^3 (S_{11} - S_{12}) + 2m^3 n (S_{12} - S_{22}) + mn(m^2 - n^2)S_{66},$$
$$\overline{S}_{66} = 4m^2 n^2 (S_{11} - S_{12}) - 4m^2 n^2 (S_{12} - S_{22}) + (m^2 - n^2)^2 S_{66}. \tag{6.93}$$

## Example 6.16

Consider a thin laminate with the material properties $E_1 = 20 \times 10^6$ psi, $E_2 = 1.5 \times 10^6$ psi, $G_{12} = 1.0 \times 10^6$ psi, $v_{12} = 0.2900$, $\theta = +30°$, and let the strain in the laminate be $\varepsilon_x = 3.13 \times 10^{-4}$, $\varepsilon_y = 0.91 \times 10^{-4}$, $\gamma_{xy} = 0.0$. Determine the stresses along the principal material axes of symmetry ($\sigma_1$, $\sigma_2$, and $\tau_{12}$).

**Solution**

Poisson's ratio, $v_{21}$, can be determined from

$$v_{21} = \frac{v_{12}E_2}{E_1} = \frac{(0.2900)1.5 \times 10^6}{20 \times 10^6} = 0.02176.$$

Then

$$Q_{11} = \frac{E_1}{1 - v_{12}v_{21}} = \frac{20 \times 10^6}{1 - (0.2900)(0.02176)} = 20.13 \times 10^6 \text{ psi}.$$

Similarly, from equation (6.83) we obtain

$$Q_{22} = 1.51 \times 10^6 \text{ psi}, \quad Q_{12} = 0.44 \times 10^6 \text{ psi}, \quad Q_{66} = G_{12} = 1.0 \times 10^6 \text{ psi}.$$

The elements of the transformed reduced stiffness matrix $[\overline{Q}_{ij}]$ are

$$\begin{aligned}
\overline{Q}_{11} &= m^4 Q_{11} + 2m^2 n^2 (Q_{12} + 2Q_{66}) + n^4 Q_{22} \\
&= \{(0.866)^4(20.13) + 2(0.866)^2(0.5)^2[0.44 + 2(1.0)] + (0.5)^4(1.51)\} \times 10^6 \\
&= 12.33 \times 10^6 \text{ psi}.
\end{aligned}$$

Similarly, from equation (6.91) we have

$$\overline{Q}_{12} = 3.58 \times 10^6 \text{ psi}, \quad \overline{Q}_{22} = 3.02 \times 10^6 \text{ psi},$$
$$\overline{Q}_{16} = 5.85 \times 10^6 \text{ psi}, \quad \overline{Q}_{26} = 2.22 \times 10^6 \text{ psi},$$
$$\overline{Q}_{66} = 4.14 \times 10^6 \text{ psi}.$$

Using equations (6.84) and (6.90), the stresses along the principal material axes can be formulated as

$$\begin{Bmatrix} \sigma_1 \\ \sigma_2 \\ \tau_{12} \end{Bmatrix} = [T][\overline{Q}_{ij}] \begin{Bmatrix} \varepsilon_x \\ \varepsilon_y \\ \gamma_{xy} \end{Bmatrix},$$

where

$$[T] = \begin{bmatrix} m^2 & n^2 & 2mn \\ n^2 & m^2 & -2mn \\ -mn & mn & (m^2 - n^2) \end{bmatrix}$$

$$= \begin{bmatrix} (0.866)^2 & (0.5)^2 & 2(0.866)(0.5) \\ (0.5)^2 & (0.866)^2 & -2(0.866)(0.5) \\ -(0.866)(0.5) & (0.866)(0.5) & ((0.866)^2 - (0.5)^2) \end{bmatrix},$$

$$[\overline{Q}_{ij}] = \begin{bmatrix} \overline{Q}_{11} & \overline{Q}_{12} & \overline{Q}_{16} \\ \overline{Q}_{12} & \overline{Q}_{22} & \overline{Q}_{26} \\ \overline{Q}_{16} & \overline{Q}_{26} & \overline{Q}_{66} \end{bmatrix} = \begin{bmatrix} 12.33 & 3.58 & 5.85 \\ 3.58 & 3.02 & 2.22 \\ 5.85 & 2.22 & 4.14 \end{bmatrix},$$

and

$$\begin{Bmatrix} \varepsilon_x \\ \varepsilon_y \\ \gamma_{xy} \end{Bmatrix} = \begin{Bmatrix} 3.13 \\ 0.91 \\ 0.0 \end{Bmatrix}.$$

Substituting $[T]$, $[\overline{Q}_{ij}]$, and $\{\varepsilon_i\}$ into $\{\sigma_i\}$ yields $\sigma_1 = 52.468 \times 10^2$ psi, $\sigma_2 = 3.339 \times 10^2$ psi, and $\tau_{12} = -1.923 \times 10^2$ psi.

### 6.9.2 Failure Criteria of Composite Laminates

The analysis of failure in composite materials is more involved than that for isotropic materials. This is mainly because the anisotropic property of composite materials leads to different stiffnesses in varying loading directions. Also, the failure mode is greatly dependent on the composite materials. The difficulty of failure analysis of composite materials is further compounded by the fact that the ultimate strength behavior of composite materials is often different in tension and compression.

Most of the commonly used failure theories are set forth to predict the strength of a unidirectional lamina based on the strength in the fiber direction, the strength in the transverse fiber direction, and the strength under in-plane shear. The strength in the fiber direction is either tensile, $\sigma_1^t$, or compressive, $\sigma_1^c$; the strength in the transverse fiber direction is either tensile, $\sigma_2^t$, or compressive, $\sigma_2^c$; and the in-plane shear is in plane 1 in direction 2, $\tau_{12}^*$ (Figure 6.50). This implies that in developing the failure criteria for composite materials, the stress conditions must be considered in relation to the normal and shearing components relative to their principal axes. For simplicity, the fiber-reinforced lamina will be treated as a homogeneous, orthotropic material. Generally, three main failure criteria are used for composite materials, as set out below.[17]

### 6.9.2.1 Maximum-Stress Criterion

This criterion assumes that failure will occur when any one of the stress components is equal to or greater than its corresponding critical value. Thus, a lamina fails if

$$
\begin{aligned}
\sigma_1 &\geq \sigma_1^t, \quad \sigma_1 \leq \sigma_1^c, \\
\sigma_2 &\geq \sigma_2^t, \quad \sigma_2 \leq \sigma_2^c, \\
\tau_{12} &\geq \tau_{12}^*, \quad \tau_{12} \leq \tau_{12}^*.
\end{aligned}
\tag{6.94}
$$

where $\sigma_1^t$ and $\sigma_1^c$ are respectively the ultimate tensile and compressive strengths in direction 1; $\sigma_2^t$ and $\sigma_2^c$ are respectively the ultimate tensile and compressive strengths in direction 2; and $\tau_{12}^*$ is the ultimate planar-shear strength in plane 1 in direction 2.

### 6.9.2.2 Maximum-Strain Criterion

The maximum-strain criterion assumes that failure occurs when any one of the strain components is equal to or greater than the corresponding ultimate value. In this case the modes of failure are

$$
\begin{aligned}
\varepsilon_1 &\geq \varepsilon_1^t, \quad \varepsilon_1 \leq \varepsilon_1^c, \\
\varepsilon_2 &\geq \varepsilon_2^t, \quad \varepsilon_2 \leq \varepsilon_2^c, \\
\gamma_{12} &\geq \gamma_{12}^*, \quad \gamma_{12} \leq \gamma_{12}^*,
\end{aligned}
\tag{6.95}
$$

where $\varepsilon_1^t$ and $\varepsilon_1^c$ are respectively the ultimate tensile and compressive strains in direction 1; $\varepsilon_2^t$ and $\varepsilon_2^c$ are respectively the ultimate tensile and compressive strains in direction 2; and $\gamma_{12}^*$ is the ultimate planar-shear strain on plane 1 in direction 2.

### 6.9.2.3 Tsai–Hill Criterion

This criterion is derived from the von Mises failure criterion used for homogeneous isotropic materials. The Tsai–Hill criterion states that failure of an orthotropic lamina will occur under a general stress state when

$$
\frac{\sigma_1^2}{(\sigma_1^t)^2} - \frac{\sigma_1 \sigma_2}{(\sigma_1^t)^2} + \frac{\sigma_2^2}{(\sigma_2^t)^2} + \frac{\tau_{12}^2}{(\tau_{12}^*)^2} \leq 1,
\tag{6.96}
$$

where $\sigma_1^t$ and $\sigma_2^t$ are the respective lamina strengths in tension; and $\tau_{12}^*$ is the laminate planar shear strength. If the stresses are compressive, then the corresponding compressive failure strength ($\sigma_1^c$ and $\sigma_2^c$) is to be used in equation (6.96). The Tsai–Hill failure criterion has been shown to be more realistic than the previous two criteria.

## Example 6.17

Consider an orthotropic laminate under simple uniaxial tension $\sigma_o$ as shown in Figure 6.51 with properties $E_1 = 25 \times 10^6$ psi, $E_2 = 1.0 \times 10^6$ psi, $G_{12} = 0.5 \times 10^6$ psi, $v_{12} = 0.2500$, and $\theta = +45°$, and let the stresses in the lamina be $\sigma_1^t = 180$ kpsi, $\sigma_2^t = 12$ kpsi, and $\tau_{12}^* = 18$ kpsi. Using the Tsai–Hill criterion, find the smallest $\sigma_o$ for which laminate failure will occur.

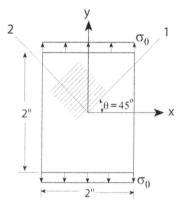

**Figure 6.51** Material axes of a lamina oriented at an angle $\theta$.

### Solution

The stresses in the principal directions are given by the transformation

$$
\begin{Bmatrix} \sigma_1 \\ \sigma_2 \\ \tau_{12} \end{Bmatrix} = \begin{bmatrix} m^2 & n^2 & 2mn \\ n^2 & m^2 & -2mn \\ -mn & mn & (m^2 - n^2) \end{bmatrix} \begin{Bmatrix} \sigma_x \\ \sigma_y \\ \tau_{xy} \end{Bmatrix},
$$

where $m = \cos\theta$ and $n = \sin\theta$. For the given plane stress problem,

$$
\begin{Bmatrix} \sigma_1 \\ \sigma_2 \\ \tau_{12} \end{Bmatrix} = \begin{bmatrix} m^2 & n^2 & 2mn \\ n^2 & m^2 & -2mn \\ -mn & mn & (m^2 - n^2) \end{bmatrix} \begin{Bmatrix} 0.0 \\ \sigma_o \\ 0.0 \end{Bmatrix}.
$$

For $\theta = 45°$, $m = n$. It then follows that

$$
\begin{Bmatrix} \sigma_1 \\ \sigma_2 \\ \tau_{12} \end{Bmatrix} = \begin{Bmatrix} 1 \\ 1 \\ 1 \end{Bmatrix} n^2 \sigma_o.
$$

Using the Tsai–Hill criterion,

$$
\frac{(n^2\sigma_o)^2}{(180)^2} - \frac{(n^2\sigma_o)^2}{(180)^2} + \frac{(n^2\sigma_o)^2}{(12)^2} + \frac{(n^2\sigma_o)^2}{(18)^2} = 1,
$$

and rearranging, we find that

$$
(n^2\sigma_o)^2 \left[ \frac{1}{12^2} + \frac{1}{18^2} \right] = 1.
$$

Hence,

$$
\sigma_o^2 = \frac{100}{n^4}
$$

$$
\sigma_o = 20 \text{ kpsi.}
$$

## 6.10 Residual (Internal) Stress Considerations[23]

Residual stresses, also referred to as internal or locked-in stresses, are the self-equilibrating stresses that exist in a body in the absence of external forces or constraints. The self-equilibrating nature of residual stresses means that the resultant force and the resultant moment produced by these stresses must be zero.

Residual stresses can be divided into two main categories: macro residual stresses and micro residual stresses. Macro residual stresses refer to a system of internal stresses that varies continuously through the volume of the body and acts over large regions. Micro residual stresses are limited to regions as small as a unit cell and may extend as far as several grains.

### 6.10.1 Sources of Residual Stress

Residual stresses arise from various steps involved in the processing and manufacturing of materials. For example, during most metal forming operations, such as rolling and extrusion, the material undergoes non-uniform plastic deformation resulting in a pattern of residual stresses throughout the material's cross-section. To illustrate this point, consider a metal sheet being rolled as shown in Figure 6.52. During the rolling process, plastic flow occurs only near the surfaces that are in direct contact with the rollers. The ensuing non-uniform deformation results in elongation of the surface fibers, while the fibers near the center of the sheet remain unchanged. However, for the sheet to remain as a continuous body, the center fibers tend to restrain the surface fibers from stretching, whereas the surface fibers tend to elongate the central fibers. This results in a pattern of residual stresses throughout the sheet with a compressive value at the surface and a tensile value at the center of the sheet (Figure 6.52(b)).

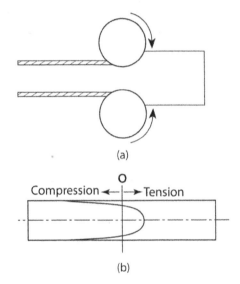

**Figure 6.52** (a) Rolling operation. (b) Residual-stress pattern in rolled metal sheet. (From Ertas, A., and Jones, J.C., *The Engineering Design Process*, 2nd edn, John Wiley & Sons, Inc., New York, 1996. Reproduced with permission of John Wiley & Sons, Inc.)

Similarly, during extrusion processes (Figure 6.53), because of the high friction values between the material and the extrusion chamber/dies, the fibers on the outer surface of the billet travel at a slower rate than the fibers near the center of the billet. The non-uniform deformation throughout the body (Figure 6.53(c)) results in tensile stresses at the outer surface and compressive stresses near the center of the extruded product.

---

23 Adapted from Ertas, A., and Jones, J.C., *The Engineering Design Process*, 2nd edn, John Wiley & Sons, Inc., New York, 1996.

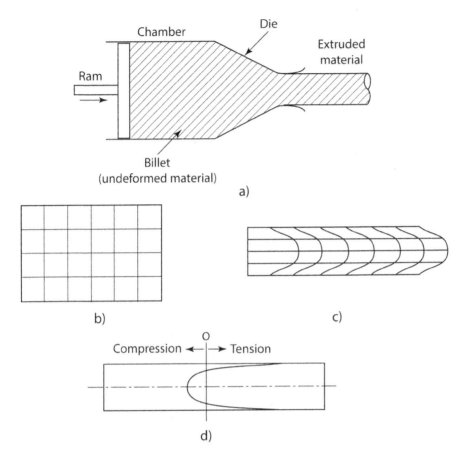

**Figure 6.53** (a) Extrusion process. (b) Deformation pattern before extrusion. (c) Material grid pattern after extrusion. (d) Residual-stress pattern in extruded metal. (From Ertas, A., and Jones, J.C., *The Engineering Design Process*, 2nd edn, John Wiley & Sons, Inc., New York, 1996. Reproduced with permission of John Wiley & Sons, Inc.)

Residual stresses are sometimes introduced during the cooling of hot products owing to the temperature differences between the surface and the center of metal. The evolution of residual stresses during the cooling of a block of metal is illustrated in Figure 6.54. Since the metal fibers on the outer surface of the block cool at a faster rate than the fibers at the center, there will be more contraction on the outer surface than at the center of the block. This non-uniform contraction induces compressive strain in the center fibers, whereas the outer fibers will be subjected to a tensile strain (Figure 6.54(c)). The hot center fibers cannot support the imposed compressive strains and yield in compression because they have a lower yield stress than the cooler outer fibers. The compressive plastic deformation of the center fibers results in the shrinkage of the center of the block in order to relieve some of the stress (Figure 6.54(c)). Upon cooling of the

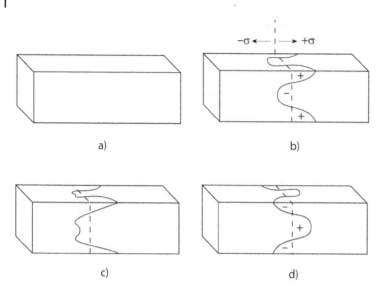

**Figure 6.54** Evolution of residual stress during quenching process (from Ertas, A., and Jones, J.C., *The Engineering Design Process*, 2nd edn, John Wiley & Sons, Inc., New York, 1996. Reproduced with permission of John Wiley & Sons, Inc.).

entire block, the total shrinkage of the center fibers will be greater than that of the outer fibers because of both cooling and compressive plastic deformation. This non-uniform deformation throughout the block produces compressive stresses at the outer surface and tensile stresses at the inner surface (Figure 6.54(d)). More extensive coverage of the formation of residual stresses can be found in the chapter bibliography.

### 6.10.2   Effect of Residual Stresses

When a mechanical component is subjected to external forces, the material behaves in a manner that is governed by the total stress acting on the material. Since the total stress experienced by a material involves the superposition (or summation) of the external stresses and the residual (internal) stresses, knowledge of residual stresses plays an important role in designing and/or predicting the behavior of materials under various loading conditions.

### Example 6.18

Consider the case of a block of A96061-T6 aluminum that has been hot-forged and quenched. Assume that because of the quenching process, a residual-stress pattern exists in the block as shown in Figure 6.55. If the block is subjected to a tensile force of 50,000 lb, determine the factor of safety guarding against yielding.

**Figure 6.55** Hot-forged and quenched aluminum block (from Ertas, A., and Jones, J.C., *The Engineering Design Process*, 2nd edn, John Wiley & Sons, Inc., New York, 1996. Reproduced with permission of John Wiley & Sons, Inc.).

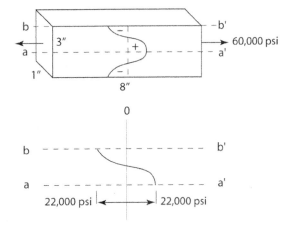

**Solution**

The yield strength of Al A96061-T6 is $S_y = 40$ kpsi. The stress due to the force is

$$\sigma = \frac{F}{A} = \frac{60,000}{3 \times 1} = 20,000 \text{ psi.}$$

The factor of safety without considering the residual stresses is

$$n = \frac{S_y}{\sigma} = \frac{40}{20} = 2.0.$$

However, since the center of the block contains tensile residual stresses, the total stress at the center of the block is

$$\sigma_{\text{center}} = 20,000 + 22,000 = 42,000 \text{ psi,}$$

which exceeds the yield stress of the material. Therefore,

$$n = \frac{40}{42} = 0.95.$$

This example illustrates the importance of accounting for residual stresses during the design process. Without considering the residual stresses, the analysis shows that the block will be capable of handling the applied load. The addition of tensile residual stresses to the total stress shows that the block will not be able to withstand the operating load.

Other effects of residual stresses include warping of components during machining and stress-corrosion cracking. When machining components contain residual stresses, the remaining residual stresses are enhanced. Since residual stresses are in self-equilibrium, removing part of the material disturbs the equilibrium of internal forces and moments. The component material responds to this by adjusting its shape (distorting) in order to establish a new equilibrium condition. Stress-corrosion cracking is a type of failure where the effect of a corrosive medium is accentuated in the presence of stresses. Residual stresses can be particularly detrimental in conjunction with corrosive environments (e.g., ammonia compounds with brass and chlorides with austenitic stainless steel).

Not all types of residual stresses are detrimental; many types are beneficial and serve as a strengthening mechanism for the material. For example, autofrettage is a process routinely used in the pressure vessel industry to induce compressive residual stresses at the bore of pressure vessels. This is done by straining the vessel beyond its elastic limit at the bore. When the vessel is unloaded, the inner surface (bore) will contain compressive residual stresses. Upon application of internal pressure, the resulting total stress (tensile) at the bore will be lower because of the presence of compressive residual stresses.

### 6.10.3 Measurement of Residual Stresses

To assess the effect of residual stresses on the intended service capability of the component, one must be able to quantify the magnitude and distribution of residual stresses. Experimental methods for measurement of residual stresses are divided into two main categories:

1) non-destructive methods;
2) destructive methods.

Examples of non-destructive techniques are:

1) acoustoelastic;
2) X-ray.

All destructive methods of residual stress measurement take advantage of the principle that when part of a residual stressed material is removed, the remaining material distorts in an attempt to reach a new equilibrium condition. This change in shape (strain) can be measured via strain gauges. By utilizing appropriate equations, the measured strains are converted to residual stresses in the removed material. By removing small consecutive layers of material, the magnitude and distribution of residual stresses throughout the entire cross-section can be determined.

Examples of destructive methods are:

1) hole-drilling methods;
2) Sach's boring-out method;
3) sectioning method;
4) deflection method.

A comprehensive description of non-destructive and destructive methods of residual-stress measurement techniques can be found in references 4 and 16 in the chapter bibliography or in handbooks on experimental stress analysis.

## 6.11 Material Standards and Specifications

Material standards and specifications are used to identify the physical and chemical characteristics and requirements for materials used in design. Specifications are normally used in the procurement process to describe accurately the essential requirements for materials, including the procedures that are used to determine that the requirements have been met. A standard is a document used by general agreement to determine whether or not something is as it should be. Standards are used to control the variability of materials and to provide uniformity in quality and performance over a wide range of applications. Many material standards fall into the category called *consensus standards*. A consensus standard is a standard developed by a body representing all of the parties interested in using and complying with the standards. Standards are implemented in procurement through specifications; these commonly refer to standards to identify material or other requirements.

### 6.11.1 Organizations Involved in Standards and Specifications Preparation[24]

There are several hundred organizations in the United States involved in the preparation of voluntary consensus standards. They include branches of the government, professional and technical societies, trade associations, public service and consumer groups, and testing and inspection bodies.[25] Two of these, the American National Standards Institute (ANSI) and the American Society for Testing and Materials (ASTM), are involved almost exclusively with the preparation of voluntary consensus standards. The Standards Development Section of the National Bureau of Standards (NBS) within the US Department of Commerce performs a similar function as do major parts of other organizationsm including the Codes and Standards Division of the American Society for Mechanical Engineers (ASME).

ANSI standards are used widely throughout industry. ANSI is the coordinator of the US national standards system and assists participants in the program in reaching agreements on standards, needs, and priorities, arranges for standards development efforts, and ensures that fair and effective procedures are used in standards development. All standards developed by this organization have an alphanumeric code that begins with the prefix ANSI.[26]

ASTM is concerned with the development of standards having to do with the characteristics and performance of materials, products, systems, and services. It is the world's largest source of voluntary consensus standards. All of its standards are identified with the prefix ASTM followed by an alphanumeric code. ASME and ASTM have cooperated for many years in the preparation of material specifications for ferrous and non-ferrous materials. In 1969, the American Welding Society (AWS) began publishing specifications for welding rods, electrodes, and filler metals. ASME now works with the AWS on these specifications as well. All identical specifications are identified by ASME/ASTM symbols or ASME/AWS symbols. When changes are required to make the specification acceptable for ASME Code usage only, the ASME symbol is used. The ASME Code, which includes a large number of material specifications, has been adopted into law by 45 states and all of the Canadian provinces.[27]

---

24 Adapted from Ertas, A., and Jones, J.C., *The Engineering Design Process*, 2nd edn, John Wiley & Sons, Inc., New York, 1996.

25 "Materials and Process Specifications and Standards," National Academy of Sciences, 1977.

26 "Guide to Materials Engineering Data and Information," American Society for Metals, 1986.

27 "ASME Boiler and Pressure Vessel Code," 1988.

The use of the ASME Codes relative to weld joint efficiency is demonstrated by the following example.[28]

## Example 6.19

As shown in Figure 6.56, a horizontal vessel that is 60 ft long is fabricated from carbon steel SA 442 using six rings 10 ft long. The vessel is supported by 120° saddles located 2 ft 6 in. from each head attachment seam. The heads are ellipsoidal and attached using type no. 2 butt joints. The shell courses have type no. 1 longitudinal joints, which are spot-radiographed in accordance with UW-52. The circumferential welds joining the courses are type no. 2 with no radiography. Given the following design parameters, determine the required shell thickness.[a]

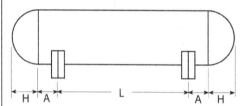

**Figure 6.56** Horizontal pressure vessel on two supports (from Ertas, A., and Jones, J.C., *The Engineering Design Process*, 2nd edn, John Wiley & Sons, Inc., New York, 1996. Reproduced with permission of John Wiley & Sons, Inc.).

*Design Parameters*

| | |
|---|---|
| Assumed joint efficiency (circumferential seams), $E$ | = 0.85 |
| Assumed joint efficiency (long seams), $E$ | = 0.65 |
| Maximum allowable stress, $S$ (see Table UCS-23 in ASME Boiler and Pressure Vessel Code) | = 13,800 psi |
| Design pressure, $P$, including static head | = 60 psi |
| Reaction at each saddle, $Q$ | = 175,000 lb |
| Saddle to tangent line, $A$ | = 30 in. |
| Vessel diameter, $OD$ | = 120 in. |
| Design temperature, $T$ | = 100°F |
| Assumed shell thickness, $t$ | = 0.3125 in. |
| Shell length, $L$ | = 720 in. |
| Weight of vessel, $W$ | = 30,000 lb |
| Weight of contents, $W_c$ | = 320,000 lb |
| Total weight, $W_T$ | = 350,000 lb |
| Head depth, $H$ | = 30 in. |

---

28 Section VIII, Division 1, Appendix L, "Examples Illustrating the Application of Code Formulas and Rules," pp. 714–716.

## Solution

The following three cases must be investigated:

*Case 1: Circumferential tensile stress due to internal pressure.* Using the equation in UG-27(c)(1) p. 23, Pressure Vessels Code, Section VIII-1, the radius $R$ is given by

$$R = \frac{OD}{2} - t = \frac{120}{2} - 0.3125 = 59.6875 \text{ in.}$$

Then

$$t = \frac{PR}{SE - 0.6P} = \frac{60(59.6875)}{13,800(0.85) - 0.6(60)} = 0.306 \text{ in.} \tag{6.97}$$

*Case 2: Longitudinal tensile stress due to bending and internal pressure.* The following equation combines the longitudinal tensile stress due to pressure with the longitudinal tensile stress due to bending at the midpoint between the saddles:

$$t = \frac{PR}{2SE + 0.4P} \mp \frac{QL}{4\pi R^2 SE} \times \left[ \frac{1 + \dfrac{2(R^2 - H^2)}{L^2}}{1 + \dfrac{4H}{3L}} - \frac{4A}{L} \right] \tag{6.98}$$

$$= \frac{60(59.6875)}{2(13,800)(0.65) + (0.4)(60)} + \frac{175,000(720)}{4\pi(59.6875)^2(13,800)(0.65)}$$

$$\times \left[ \frac{1 + 2(59.6875^2 - 30^2)/720^2}{1 + 4(30)/3(720)} - \frac{4(30)}{720} \right] = 0.199 + 0.248 = 0.447 \text{ in.}$$

This is greater than the actual thickness, so we must either thicken the shell or increase the efficiency of the welded joint by changing the weld type or the amount of radiography.

*Action.* Spot-radiograph to increase the efficiency of the welded joint from 0.65 to 0.80. Recalculating $t$ by using $E = 0.80$ from equation (6.98) yields a thickness of 0.364 in. Since the calculated thickness is still greater than the actual thickness, increase the value of $E$ further by changing the circumferential seam to a type 1 fully radiographed weld. From Table UW-12, p. 104 of Rules of Construction of Pressure Vessels, Section VIII-1, choose $E = 1.0$. Recalculation of $t$ with $tE = 1.0$ yields a thickness of 0.291 in. Since this value is less than the actual thickness, it is safe.

*Conclusion.* The circumferential joint at the center of the vessel must be type no. 1 fully radiographed. This is at the point of maximum positive moment. The maximum negative moment is at the supports, but there is no joint there. Other circumferential joints must be investigated by using the moment at the joint in calculating the combined stresses. It should be noted that many other areas of stress due to saddle loadings exist and should be investigated (see Appendix G of Section VIII-1).

*Case 3: Longitudinal compressive stress due to bending.* First determine the allowable compressive stress (see UG-23(b), p. 20 of Section VIII-1). Using the assumed value of $t$ and $R$, calculate the value of the constant $A$ using the following formula:

$$A = \frac{0.125}{R_o/t} \tag{6.99}$$

$$= \frac{0.125}{60/0.3125} = 0.000651.$$

This value of $A$ is now used to find the allowable compressive stress for a given material and design temperature from Fig. 5-UCS-28.2 (Appendix 5, p. 520 of Section VIII). For $A = 0.000651$ and temperature $100°$F, we obtain the allowable stress, $B = 9446$ psi.

The general equation for thickness is the same as for longitudinal tensile stress except that the pressure portion drops out, and the most severe condition occurs when there is no pressure in the vessel. Then the thickness calculation for this case for the allowable stress of 9446 psi is

$$t = \frac{QL}{4\pi R^2 SE} \times \left[ \frac{1 + \dfrac{2(R^2 - H^2)}{L^2}}{1 + \dfrac{4H}{3L}} - \frac{4A}{L} \right] \tag{6.100}$$

$$= \frac{175,000(720)}{4\pi(59.6875)^2(9446)(1.0)} \times \left[ \frac{1 + 2(59.6875^2 - 30^2)/720^2}{1 + 4(30)/3(720)} - \frac{4(30)}{720} \right]$$

$$= 0.236 \text{ in.}$$

The required design thickness, $t = 0.306$ in., is governed by circumferential tensile stress.

[a]from Ertas, A., and Jones, J.C., *The Engineering Design Process*, 2nd edn, John Wiley & Sons, Inc., New York, 1996.

## 6.12 Corrosion Considerations[29]

There is no argument among scientists today that corrosion is a major factor in the design, cost, and implementation of engineering applications. Corrosion costs US industries and the American public an estimated $170 billion per year.[30] Also, according to estimates from the International Association of Drilling Contractors, about 75–85 percent of all drill pipe losses in the field are caused by some type of corrosion.[31] What is in dispute, however, is the way in which the various forms of corrosion are categorized; their overlapping similarities make this a

29 Adapted from Ertas, A., and Jones, J.C., *The Engineering Design Process*, 2nd edn, John Wiley & Sons, Inc., New York, 1996.
30 *Metals Handbook*, 9th edn, Vol. 13, ASM International, 1987, p. 79.
31 Al-Marhoun, M.A., and Rahman, S.S., "Treatment of drilling fluid to combat drill pipe corrosion," *Corrosion*, 46(9), 778–782, 1990.

difficult task. A general approach must be utilized keeping in mind that each type of corrosion is not necessarily limited to its assigned major category.

The different types of corrosion are grouped according to the ASM *Handbook on Corrosion*. Among the many types of corrosion, some are encountered frequently in industry. The following is a brief explanation of these prevalent types.

### 6.12.1 Atmospheric Corrosion

Atmospheric corrosion can be defined as the deterioration of a material when exposed to air and its impurities. As one might expect, atmospheric corrosion is very widespread and accounts for more of the cost due to failures than any other form. Some important factors that influence the rate of atmospheric corrosion are temperature, climate, and relative humidity.

### 6.12.2 Galvanic Corrosion

All metals have the ability to act as either an anode or a cathode. The potential difference between two dissimilar metals causes galvanic corrosion. Current flows from the anode to the cathode, so the less resistant metal (anode) is corroded relative to the cathodic metal. There are two ways to minimize galvanic corrosion. One way is to use metals that are close together in the galvanic series. Another way is to insulate the dissimilar metals from each other.

### 6.12.3 Crevice Corrosion

This form of corrosion is highly localized and occurs within gaps or spaces between metals. Crevice corrosion may occur at cracks, seams, or other metallurgical defects. This type of corrosion is usually associated with small volumes of stagnant liquid that have become trapped, allowing corrosion to occur. Eliminating possible crevices as well as keeping parts clean helps to prevent crevice corrosion.

### 6.12.4 Pitting Corrosion

Pitting corrosion is extremely localized and forms cavities or pits. These cavities are small compared with the overall size of the surface but can cause equipment failure after only minimal weight loss of the part. The depth of pitting (pitting factor) is defined as the ratio of the deepest metal penetration to the average metal penetration. Pitting corrosion can be minimized by specifying a clean and smooth metal surface.

### 6.12.5 Intergranular Corrosion

This type of corrosion occurs at the grain boundaries of metals. The limited area of grain boundary material acts as an anode, and the larger grain area acts as a cathode. The flow of energy from the anode to the cathode results in penetrating, corrosive attack. Intergranular corrosion is common to austenitic steels that have been heated to their sensitizing range of 950– −1450°F. Reducing the occurrence of this type of corrosion is possible by annealing the steel after it has been sensitized.

### 6.12.6 Erosion Corrosion

Material degradation can be caused by relative movement between a corrosive fluid and a surface. This type of attack is called erosion corrosion. Erosion destroys the protective surface films of the material, making it easier for chemical attack to occur. Two types of erosion are cavitation and fretting. Cavitation arises from the formation and collapse of vapor bubbles near the metal surface, resulting in local deformation. Fretting occurs between two surfaces under cyclic load and results in surface pits or cracks. Reducing fluid velocity as well as eliminating instances where direct surface contact may occur are ways to control erosion corrosion.

### 6.12.7 Stress-Corrosion Cracking

Stress-corrosion cracking (SCC) is a term used to describe a type of accelerated corrosion caused by internal residual stresses or externally applied stresses coupled with a corrosive reaction. Only certain combinations of metals and corrosive environments will cause SCC to occur. Presently, more than 80 different combinations of alloys and corrosive environments are known to cause SCC. To prevent the occurrence of SCC, one should select alloys that are less susceptible to cracking in the service environment and that will maintain low stress levels.

## Bibliography

1 ALMEN, J.O., and Black, P.H., *Residual Stresses and Fatigue in Metals*, McGraw-Hill, New York, 1963.
2 AMERICAN SOCIETY FOR METALS, *Metals Handbook*, Vol. 1, ASM, Metals Park, OH, 1961.
3 ASHBY, M.F., and JONES, D.R.H., *Engineering Materials 2: An Introduction to Microstructures, Processing and Design*, Pergamon Press, Oxford, 1986.
4 BARRET, C.S., "A critical review of various methods of residual stress measurement," in *Proceedings of Society for Experimental Stress Analysis*, Vol. II, no. 1, pp. 147–156, 1945.
5 BROEK, D., *Elementary Engineering Fracture Mechanics*, Martinus Nijhoff, The Hague, 1982.
6 CRANE, F.A., and CHARLES, J.A., *Selection and Use of Engineering Material*, Butterworth & Co., London, 1984.
7 DIETER, G.E., *Engineering Design: A Materials and Processing Approach*, McGraw-Hill, New York, 1983.
8 ERTAS, A., and JONES, J.C., *The Engineering Design Process*, John Wiley & Sons, inc., New York, 1996.
9 GOEL, V.S., *Analyzing Failures, the Problems and Solutions*, American Society of Metals, Metals Park, OH, 1986.
10 GOEL, V.S., *Fatigue Life Analysis and Prediction*, American Society of Metals, Metals Park, OH, 1986.

**11** HORGER, O.J., "Residual stresses," in M. Hetenyi (ed.), *Handbook of Experimental Stress Analysis*, John Wiley & Sons, New York, 1950.

**12** KERN, R.F., and SUESS, M.E., *Steel Selection: A Guide for Improving Performance and Profits*, John Wiley & Sons, New York, 1979.

**13** MAGAD, E.L., and AMOS, J.M., *Total Materials Management*, Van Nostrand Reinhold, New York, 1989.

**14** PATTON, W.J., *Materials in Industry*, Prentice Hall, Englewood Cliffs, NJ, 1986.

**15** PSARAS, P.A., and LANGFORD, H.D., *Advancing Materials Research*, National Academy Press, Washington DC, 1987.

**16** SOCIETY FOR AUTOMOTIVE ENGINEERS, "Methods of Residual Stress Measurement," SAE report J936, SAE, New York, 1965.

**17** ROLFE, S.T., and BARSOM, J.M., *Fracture and Fatigue Control in Structures: Applications of Fracture Mechanics*, Prentice Hall, Englewood Cliffs, NJ, 1977.

**18** TAIT, R.B., and GARRETT, G.G., *Fracture and Fracture Mechanics* Pergamon Press, Elmaford, NY, 1984.

**19** TRUCKS, H.E., *Designing for Economical Production*, Society of Manufacturing Engineers, Dearborn, MI, 1987.

**20** VERINK, E.D., *Methods of Materials Selection*, Gordon and Breach, New York, 1966.

**21** WEI, R.P., and STEPHENS, R.I., *Fatigue Crack Growth under Spectrum Loads*, American Society for Testing and Materials, Philadelphia, PA, 1975.

## CHAPTER 6 Problems

**6.1** A typical stress–strain curve for rubbers and polymers is shown in Figure P6.1. Materials show strong non-linear behavior and it is difficult to identify any yield point.
   a) Estimate the yield point.
   b) Estimate the value of the elastic modulus.

**Figure P6.1** Result of tensile test.

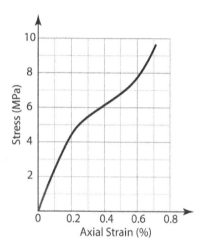

**6.2** Assume that due to design requirements, a bicycle frame should be light (maximum 3,000 Kg/m³) and stiff (minimum 1.5 GPa). Using the Young's modulus–density chart shown in Figure 6.9, find the most suitable material for this application.

**6.3** Light bulb filaments are subject to fatigue damage and eventual failure due to repetitive thermal stress when the light bulb is turned on and off (see Figure P6.3). If the maximum thermal stress is 60 kpsi when the light bulb is on, after how many cycles (switching on and off) will the light bulb fail? Calculate the fatigue damage and the remaining life of the light bulb after 600,000 cycles. Assume that $S_{ut} = 80$ kpsi and that the $S–N$ curve end coordinates are A($N_f = 10^3$, $S_f = 0.9S_{ut}$), B($N_e = 10^7$, $S_e = 0.4S_{ut}$).

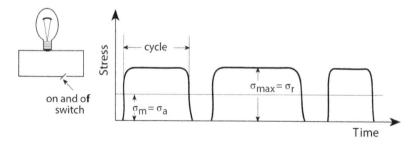

**Figure P6.3** Stress time history.

**6.4** Figure P6.4 shows an electric drive and control subsystem of an electric vehicle. A shaft-gear system mounted on a DC motor that drives the wheels is subjected to:

$$\sigma_X = +56 \text{ and} - 52 \text{kpsi from bending and trust load,}$$
$$\sigma_Y = \sigma_Z = -34 \text{ kpsi from press fit of gear,}$$
$$\tau_{XY} = +1.5 \text{ and} - 1.0 \text{ kpsi from bidirectional torque,}$$
$$\tau_{YZ} = \tau_{ZX} = 0.0.$$

**Figure P6.4** Electric drive and control subsystem of an electric vehicle.

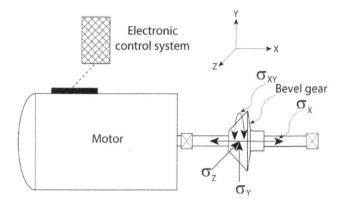

a) If the ultimate tensile strength of the shaft material is 150 kpsi, determine the fatigue life of the shaft.

b) The vehicle electronic control system is subjected to vibrations with an acceleration of 1.1G peak at a constant frequency of 380 Hz. The desired life span for the control unit is 5 hours/day of service for 9 years. If the qualification test is planned with a 2.4G peak acceleration input level, what should be the test time period?

**6.5** Derive the expression for the bending moment and deflection for the drill pipe given in Example 6.8.

**6.6** Construct an operation chart (variation of fatigue damage with respect to ball joint angle) for the drilling system used for drilling offshore wells from floating vessels by means of the $S$–$N$ curve given in Example 6.7. Use the same design data as in the example with $\theta = 1°, 2°, 3°, 4°$, $T = 100$ kips and $T = 200$ kips.

**6.7** A rotating shaft carrying an electric motor is subjected to a cyclic stress (due to pure bending) of $\sigma = 30$ kpsi for 500 cycles, then the magnitude of cyclic stress is reduced to $\sigma = 20$ kpsi for 55,000 cycles, and finally the cyclic stress level is increased to $\sigma = 40$ kpsi for 600 cycles (see Figure P6.7). Using the $S$–$N$ curve given in Figure P6-7(b), calculate the damage and the remaining life of the rotating shaft.

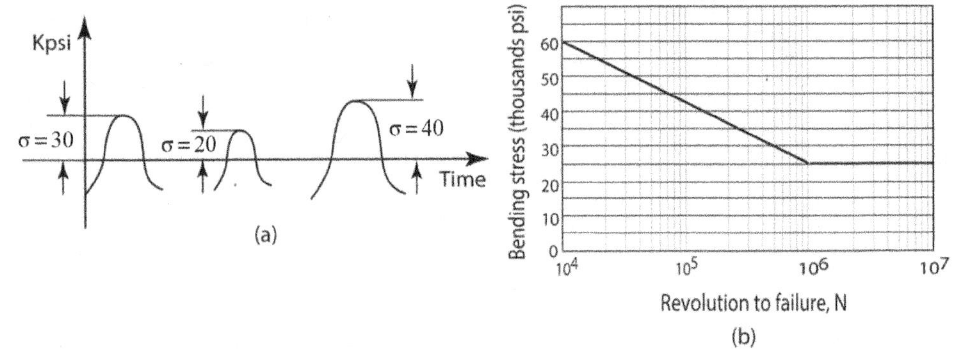

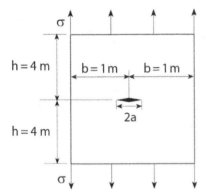

(a)

**Figure P6.7** (a) Stress time history. (b) S–N curve for rotating shaft material.

**6.8** As shown in Figure P6.8, a wide and long steel plate with small thickness is subjected to a tensile stress. If the steel has a critical stress-intensity factor $K_{IC} = 28$ MPa$\sqrt{m}$, calculate the tensile stress for failure. The steel plate has a through-thickness crack with $a = 45$ mm.

**Figure P6.8** Through-thickness crack in an infinite plate.

**6.9** Construct an operation chart (variation of fatigue damage with respect to ball joint angle) for the drilling system in Example 6.7 using the theory of fracture mechanics. Use the design data given in Problem 6.6 and compare the results of the two methods.

**6.10** The hydraulic brake system of an automobile uses a master cylinder as shown in Figure P6.10(a). The inside diameter is 0.5 in. and the cylinder thickness is 0.25 in. For an initial flaw size of 0.125 in., calculate whether the cylinder will fail or leak first and estimate the number of life cycles with respect to the stress time history shown in Figure P6.10(b). The material used is cast iron with a yield strength of 26 kpsi and $K_{IC} = 10$ kpsi$\sqrt{in}$. Assume $a/2c = 0.2$ remains constant. Assume $A = 0.66 \times 10^{-8}$,

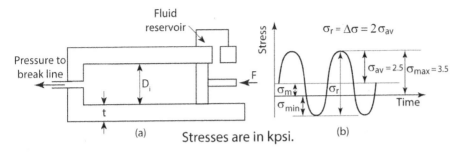

Figure P6.10 Through thickness-crack in an infinite plate.

$m = 2.25$, and $M_k = 1$. Use two iterations for the life calculation. List all your assumptions.

**6.11** The starting valves of a two-stroke marine engine are opened by means of compressed air from an air pressure vessel. For safety, assume that the pressure of the compressed air in the pressure vessel should not exceed 16 MPa and the minimum air pressure should be 12 MPa to be able to start the engine. If the inside diameter of the pressure vessel is 1.0 m and the length is 3 m, using the principles of fracture mechanics, select the most suitable materials from Table P6.11. Assume an initial surface flaw of 8 mm depth and a constant $a/2c$ ratio of 0.25. Perform the analysis based on a constant mean pressure of 14 MPa.

**Table P6.11** Materials for pressure vessel design.

| Material | $S_y$ (MPa) | $K_{IC}$ (MPa $\sqrt{m}$) | $\rho$ ton/m³ | $E$ (GPa) | Cost $/ton |
|---|---|---|---|---|---|
| Steel #1 | 1750 | 87 | 7.7 | 210 | 650 |
| Steel #2 | 1450 | 115 | 7.7 | 210 | 650 |
| Steel #3 | 1200 | 148 | 7.7 | 210 | 450 |
| Steel #4 | 1500 | 105 | 7.7 | 210 | 550 |
| Steel #5 | 1600 | 58 | 7.7 | 210 | 500 |
| Steel #6 | 1400 | 77 | 7.7 | 210 | 350 |

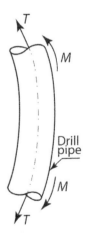

Figure P6.12 Drill pipe under bending.

**6.12** As shown in Figure P6.12, assume that a drill pipe is subjected to a constant axial tension load of 100,000 lb, and a bending stress amplitude of 22,600 psi. Assume the initial crack depth $a_i = 1/32$ in., initial length $2c_i = 1/8$ in., and a constant $a/c$ of 0.5. Select a suitable steel from Table P6.12 for this application. Consider strength, cost, service life, and leak-before-break as the desired criteria.

**Table P6.12** Drill pipe material properties.

| Steel | $S_y$ (ksi) | $K_{IC}$ (ksi $\sqrt{in.}$) | Cost $/lb |
|---|---|---|---|
| A | 90 | 56 | 1.5 |
| B | 80 | 38 | 1.00 |

**Design Data**

Density of the steel = 0.283 lb/in.$^3$

Drill pipe outside diameter = 5.0 in.

Drill pipe inside diameter = 4.276 in.

Assume that $M_k = 1$.

For both steels use $da/dN = 0.614 \times 10^{-10} \, (\Delta K)^{3.16}$.

**6.13** A car radiator fan as shown in Figure P6.13 has a through crack ($a = 0.3$ in.) introduced during the manufacturing process. The yield strength of the fan material is 90 kpsi and the critical stress intensity factor is 120 kpsi $\sqrt{in}$. Due to the change in acceleration in the radial direction, the force acting on the blade varies between 17,280 lb and 23,040 lb. Using a scaling factor $A = 0.66 \times 10^{-8}$ and slope $m = 2.25$ for the expression for the crack-growth rate, investigate whether the design is adequate for 500,000 cycles. If not, what changes would you recommend so that the design will be adequate for 500,000 cycles? Assume that the radial component of the force is causing the crack growth. Use $b = 2.4$ in. and thickness 0.12 in. For the life calculation use two iterations.

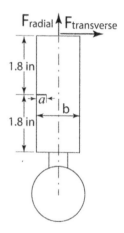

**Figure P6.13** Car radiator fan.

**6.14** The landing gear on passenger airliners is operated by a hydraulic pump. When the gear is cycled the hydraulic pressure reaches 30 kpsi. When it is locked up, the hydraulic pressure is a constant 10 kpsi. The hydraulic lines are made of 1-in. inside diameter tubing with a wall thickness of 1/16 in. The yield strength of the steel is 255 kpsi with a $K_{IC}$ of 79 kpsi $\sqrt{in}$. With an initial flaw of $a_i = \frac{1}{64}$ in. in the tubing and an $a/2c$ ratio of 0.25, find the number of cycles before failure. The density of steel is 0.283 lb/in.$^3$. Use a scaling factor $A = 0.66 \times 10^{-8}$ and the slope $m = 2.25$ for the expression for the crack growth rate. Assume $M_k = 1$.

**6.15** Consider a thin laminate with the material properties $E_1 = 20 \times 10^6$ psi, $E_2 = 1.5 \times 10^6$ psi, $G_{12} = 1.0 \times 10^6$ psi, $v_{12} = 0.2900$, $v_{21} = 0.02176$, $\theta = +45°$, and let the stresses in the laminate be $\sigma_x = 63.396 \times 10^2$ psi, $\sigma_y = 2.744 \times 10^2$ psi, $\tau_{xy} = 0.0$. Calculate the strains in the laminate ($\varepsilon_x$, $\varepsilon_y$, and $\gamma_{xy}$).

**6.16** Consider an orthotropic laminate under simple uniaxial tension as shown in Figure P6.16, with the following properties: $\sigma_1^T = 200$ kpsi, $\sigma_2^T = 100$ kpsi, $\tau_{12}^* = -20$ kpsi. For $\theta = 30°$, using the maximum stress criteria, determine the smallest stress $\sigma_o$ at which the laminate will fail.

**Figure P6.16** Material axes of a laminate oriented at an angle $\theta$.

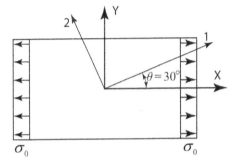

**6.17** An aircraft propeller shown in Figure P6.17 is operated at 2500 rpm with a 5 percent variance. Pick three different materials and, through analysis, select the best material to survive the possibility of introducing a through crack (single edge) during the manufacturing process with a width of $2a = 0.25$ in.

**Figure P6.17** Aircraft propeller.

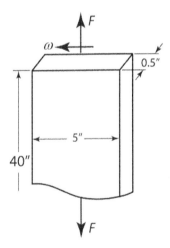

# 7

# Statistical Decisions

Statistics is the transdisciplinary science of data, employing mathematical relationships for probability and uncertainty to understand and generate a new knowledge. Statistics is a shared method which crosses disciplinary boundaries including engineering, medicine, government, education, agriculture, business, law and many others.

## 7.1  Random Variables

A function whose values cannot be predicted in a sample space is called a *random variable*. If a sample is selected in an arbitrary manner from a population, such a sample is said to be a *random sample*. It should be noted that the word *randomness* is used for the operation of selecting the sample. Rotary machines, such as turbines and compressors, are designed to operate in a vibration-free manner. However, bearing misalignment, material heterogeneity, and geometric variations, collectively or selectively, cause the rotor axial mass center to be non-coincident with the bearing axis. Such a deviation is termed *rotor mass center eccentricity*; it causes time-dependent bearing forces to occur in the rotor housing.[1,2] The mass center eccentricity is not a design condition. Instead, the eccentricity is a randomly occurring event (random variable) and differs from rotor to rotor within each machine class.

A similar argument can be made for a radial clearance of the journal bearings of a high-speed *integrally geared compressor* as shown in Figure 7.1. Surface roughness, out-of-roundness and radial clearance are the main micro-geometry parameters which influence the bearing performance.

It is crucial to note that the surface roughness and out-of-roundness of the bearing components are controlled for achieving high performance. The radial clearance of a journal bearing is

1  Ertas, A., "The response of rotating machinery to a random eccentricity distribution in the rotor," *3rd Reunión Académica de Ingeniería Mecánica*, San Luis Potosí, October 14–18, 1985.
2  Ertas, A., and Kozik, T.J., "Fatigue loads on the foundation due to turbine rotor eccentricity," *Transactions of the ASME, Journal of Energy Resources Technology*, 109, 174–179, 1987.

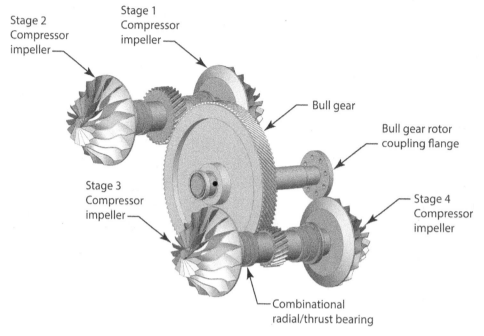

Stage 2
Compressor
impeller

Stage 1
Compressor
impeller

Stage 3
Compressor
impeller

Bull gear

Bull gear rotor
coupling flange

Stage 4
Compressor
impeller

Combinational
radial/thrust bearing

**Figure 7.1** Four-stage integrally geared compressor.

an important design parameter and bearing performance primarily depends upon this parameter. A bearing designer should take this fact into account while designing bearings.

Research studies showed that the radial clearance measurements can differ from one measuring device to another and the specified clearance may not necessarily meet the design criteria of specific oil film thickness. The radius of the bearing or the shaft also varies along the circumference, generally due to out-of-roundness.[3] The out-of-roundness contributes to the error in radial clearance measurement. In conclusion, the radial clearance of a journal bearing is a randomly occurring event and differs from bearing to bearing within each bearing class.

The bull gear rotor in Figure 7.1 is supported by two grease-packed angular contact ball bearings. These bearings can be used in applications where shaft speeds are low and minimal heating is generated. However, the low- and high-speed pinion rotors are supported by fluid film combinational journal (radial) and thrust (axial) bearings. Fluid film bearings are often used in applications where bearing loads are high in combination with high shaft surface speeds. Although requiring an ancillary lubrication system, these fluid film bearings are required for successful operation under loads and speeds representative of integrally gear compression systems.

For example, the observation obtained by measuring the radial clearance of a journal/trust bearing shown in Figure 7.1 is called a population whose sizes are countless. Suppose 18 high-speed compressors (random sampling) are selected from a larger (conceivably countless) number of compressors forming a population.

3 Sharma, S., Hargreaves, D., and Scott, W., "Journal bearing performance and metrology issues," *Journal of Achievements in Materials and Manufacturing Engineering*, 32(1), 2009.

**Table 7.1** Typical set of radial clearance measurements of 18 bearings.

| Observations Average Clearance, c, mils ($10^{-3}$ in.) | Frequency of Bearing 1 | Frequency of Bearing 2 | Frequency of Bearing 3 |
|:---:|:---:|:---:|:---:|
| 2 | 1 | 1 | 1 |
| 3 | 2 | 6 | 1 |
| 4 | 3 | 4 | 2 |
| 5 | 6 | 3 | 3 |
| 6 | 3 | 2 | 4 |
| 7 | 2 | 1 | 6 |
| 8 | 1 | 1 | 1 |

Each compressor includes four stages as shown in Figure 7.1. Table 7.1 illustrates a typical set of radial clearance measurements of bearings randomly selected out of four bearings from each compressor, which represents a sample drawn from that population.

Figure 7.2 shows the bar charts (frequency histogram) for the data given in Table 7.1. The values plotted along the *x*-axis are the bearing clearance measurements of the first, second, and third bearings of the 20 compressors, and the number of times the measurements occur (frequency) is plotted along the vertical axis. A distribution is called *symmetric*, as shown in Figure 7.2(a), if

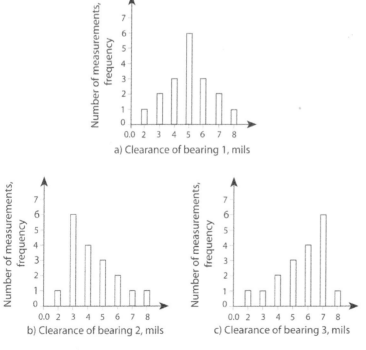

a) Clearance of bearing 1, mils

b) Clearance of bearing 2, mils

c) Clearance of bearing 3, mils

**Figure 7.2** Bearing clearance frequency distributions.

**Table 7.2** Cumulative frequency distributions of bearing clearance measurements.

| Clearance c, mils | Bearing 1 | | Bearing 2 | | Bearing 3 | |
|---|---|---|---|---|---|---|
| | Cum. freq. | Cum. freq. % | Cum. freq. | Cum. freq. % | Cum. freq. | Cum. freq. % |
| 2 | 1 | 5.6 | 1 | 5.6 | 1 | 5.6 |
| 3 | 3 | 16.7 | 7 | 38.9 | 2 | 11.1 |
| 4 | 6 | 33.3 | 11 | 61.1 | 4 | 22.2 |
| 5 | 12 | 72.2 | 14 | 77.8 | 7 | 38.9 |
| 6 | 15 | 83.3 | 16 | 88.9 | 11 | 61.1 |
| 7 | 17 | 94.4 | 17 | 94.4 | 17 | 94.4 |
| 8 | 18 | 100 | 18 | 100 | 18 | 100 |

the curve can be folded with respect to a vertical axis. A distribution that is not symmetric with respect to a vertical axis is called *skewed*. Hence, distributions like those shown in Figure 7.2(b) and Figure 7.2(c) are called *positively skewed* and *negatively skewed*, respectively. In order to understand how the distribution of the bearing clearance measurements (data) accumulates, consider another useful description called *cumulative frequency*. Table 7.1 is used to construct Table 7.2, which shows a cumulative frequency distribution.

As Figure 7.3 reveals, the cumulative frequency curve for the symmetric distribution approaches an S shape. Note that the trend of the cumulative frequency curves for the negatively and positively skewed distributions differs from that of the symmetrical distribution.

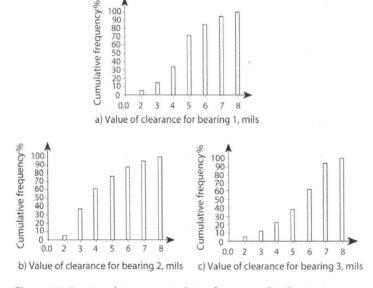

a) Value of clearance for bearing 1, mils

b) Value of clearance for bearing 2, mils    c) Value of clearance for bearing 3, mils

**Figure 7.3** Bearing clearance cumulative frequency distributions.

## 7.2 Measures of Central Tendency

There are three measures of central tendency that are generally used to represent the general trend of the eccentricity measurements shown in Table 7.1. These are described in this section.

### 7.2.1 Mean or Arithmetic Average

The mean is defined as the sum of all of the values of the random sample variables $x_i$, divided by the size of the random sample; that is,

$$\bar{x} = \frac{\sum_{i=1}^{n} x_i}{n}. \tag{7.1}$$

The mean of a random sample is denoted by $\bar{x}$, whereas the mean of a population is denoted by $\mu$. Note that the population mean $\mu$ is constant for a particular population, whereas the sample mean $\bar{x}$ can vary from sample to sample.

Suppose the overall mean of the bearing clearances given in Table 7.1 is desired. The individual means of each bearing clearance $\bar{x}_1$, $\bar{x}_2$, and $\bar{x}_3$, can be combined by the formula

$$\bar{X}_{\text{overall}} = \frac{n_1\bar{x}_1 + n_2\bar{x}_2 + n_3\bar{x}_3}{n_1 + n_2 + n_3}. \tag{7.2}$$

☐ **Example 7.1**

Using Table 7.1, estimate the mean clearance of each bearing and the overall mean of the clearance measurements of the 18 bearings.

**Solution**

The random sample size $n$ is 18. Individual means of each bearing clearance of each bearing are as follows:

$$\bar{x}_1 = \frac{\sum_{i=1}^{n} x_i}{n} = \frac{(2)(1) + (3)(2) + \cdots}{18} = 0.500 \text{ mm},$$

$$\bar{x}_2 = \frac{\sum_{i=1}^{n} x_i}{n} = \frac{(2)(1) + (3)(6) + \cdots}{18} = 4.33 \text{ mm},$$

$$\bar{x}_3 = \frac{\sum_{i=1}^{n} x_i}{n} = \frac{(2)(1) + (3)(1) + \cdots}{18} = 5.67 \text{ mm}.$$

Combine the individual means of each bearing clearance, $\bar{x}_1$, $\bar{x}_2$, and $\bar{x}_3$, to determine the overall mean of the bearing clearance distribution:

$$\bar{X}_{\text{overall}} = \frac{18(5.00) + 18(4.33) + 18(5.67)}{18 + 18 + 18} = 5.00 \text{ mm}.$$

### 7.2.2 Median

The median of a set of observations arranged in order of magnitude may be defined as the observation that divides the distribution curve into two equal parts. The value of the median depends on whether the total number of observations is odd or even. If odd, the value of the median is the $((n/2) + 1))$th observation. If even, the value of the median is the average of the $(n/2)$th and $((n/2) + 1)$th observations.

With the same data as in Example 7.1 for bearing 1, the number of observations $n$ is equal to 18. Since the total number of observations is even, we use the value corresponding to the average of the $(n/2)$th and $[(n/2) + 1]$th observations to find the median, which is 5 mm. Note that, to find the value of the median this way, the values of the random sample variable should be in increasing or decreasing order of magnitude.

### 7.2.3 Mode

The mode is the observation that occurs most frequently. If the same data are used as in Example 7.1 for bearing 1, the mode is equal to 5 mm, corresponding to the maximum frequency of 6.

If a distribution is symmetrical, the value of the mean, median, and mode will be the same. However, as shown in Figure 7.3, in the case of positively skewed distributions, the mode is less than the median, and when the mode is larger than the median, the distribution shows a negative skew.

## 7.3 Measures of Variability

This section discusses measures of variation: range, mean deviation, standard deviation, and variance.

### 7.3.1 Range

The range, which is the simplest measure of variation (spread from the center), is the absolute difference between the largest number and the smallest number in the distribution.

Suppose a set of stage 1-2 compressor bearing shafts 10 mm in diameter with an allowable tolerance of 0 to 0.034 mm is needed (see Figure 7.4). As shown in Table 7.3, there are two groups and the mean value of the tolerances in each group is $\bar{x}_1 = 0.032$ mm and $\bar{x}_2 = 0.037$ mm. Now the question is, which group should be selected? Perhaps the selection will be the first group because the mean value of the tolerance is within the acceptable range. However, when the groups are closely examined, it is noted that the number of bearing

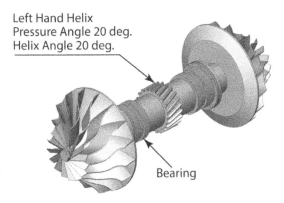

Left Hand Helix
Pressure Angle 20 deg.
Helix Angle 20 deg.

Bearing

**Figure 7.4** Stage 1-2 of compressor.

shafts that can be used is greater in the second group, even though the mean value of the tolerance in that group is not within the acceptable range. It is obvious that in this case the mean does not show the distribution of bearing shaft tolerances in each group. It follows that in some instances the range can be used to show how far the data in a distribution are from the mean.

## Example 7.2

Using the data in Table 7.3, calculate the range in each group.

### Solution

$$\text{Range for the first box} = 0.05 - 0.01 = 0.04 \text{ mm.}$$
$$\text{Range for the second box} = 0.07 - 0.02 = 0.05 \text{ mm.}$$

**Table 7.3** Clearance tolerances of the bearings in each group.

| Group 1 | Group 2 |
|---|---|
| Tolerance, mm | Tolerance, mm |
| 0.01 | 0.02 |
| 0.02 | 0.02 |
| 0.02 | 0.03 |
| 0.03 | 0.03 |
| 0.03 | 0.03 |
| 0.04 | 0.03 |
| 0.04 | 0.03 |
| 0.04 | 0.05 |
| 0.04 | 0.06 |
| 0.05 | 0.07 |
| Mean | Mean |
| $\bar{x}_1 = 0.032$ | $\bar{x}_2 = 0.037$ |

## 7.3.2 Mean Deviation

Another measure of variability is the mean deviation, which is the mean of the absolute values of the deviations from the mean. The mean deviation can be calculated using the formula

$$\text{Mean deviation} = \frac{\sum_{i=1}^{n} |x_i - \bar{x}|}{n}. \tag{7.3}$$

## 7.3.3 Standard Deviation

While the mean deviation is a useful measure of variability, it may not be practical to use because of the absolute value involved in calculating the mean deviation. Thus, the standard deviation, which is the most reliable measure of variability, is used in most cases. The standard deviation, $S$, of a random sample can be calculated using the formula

$$S = \sqrt{\frac{\sum_{i=1}^{n}(x_i - \bar{x})^2}{n - 1}}, \tag{7.4}$$

where $n$ is the number of observations. To calculate the standard deviation of a population, $\bar{x}$ and $n$ are replaced by $\mu$ and $N$ in equation (7.4):

$$\hat{\sigma} = \sqrt{\frac{\sum_{i=1}^{N}(x_i - \mu)^2}{N - 1}} \tag{7.5}$$

where $\hat{\sigma}$ is the standard deviation of a population.

### 7.3.4 Variance

The square of the standard deviation is another important measure known as the *variance* of a random variable; it shows the spread of the distribution. The variances of the random sample and the population, respectively, are given by the formulas

$$S^2 = \frac{\sum_{i=1}^{n}(x_i - \bar{x})^2}{n - 1} \tag{7.6}$$

and

$$\hat{\sigma}^2 = \frac{\sum_{i=1}^{N}(x_i - \mu)^2}{N - 1}. \tag{7.7}$$

### Example 7.3

Assume that for a good clearance fit between a pin and bushing, 1 million pin-bushings were manufactured. Suppose clearance tolerances for two sets of pin-bushing samples drawn from the 1 million population are 0.02, 0.02, 0.02, 0.03, 0.03, 0.04, 0.04, 0.04, 0.04, 0.04 (first box) and 0.01, 0.02, 0.02, 0.03, 0.03, 0.03, 0.03, 0.04, 0.05, 0.06 (second box). Calculate the sample mean, standard deviation, and variance for the data.

### Solution

As shown above, both data sets give the same mean of 0.032 mm tolerance. When both sample data are carefully checked, one would have more confidence in a mean value calculated from the first sample data set. To understand which sample truly represents the population mean, we calculate the standard deviation of the samples. We have

$$S_1 = \left[\frac{(0.02 - 0.032)^2 + (0.02 - 0.032)^2 + \cdots (0.04 - 0.032)^2}{10 - 1}\right]^{1/2}$$

$$= 8.27 \times 10^{-3} \text{ mm}$$

Hence, $S_1^2 = 6.84 \times 10^{-5}$ mm$^2$. Further,

$$S_2 = \left[ \frac{(0.01 - 0.032)^2 + (0.02 - 0.032)^2 + \cdots (0.06 - 0.032)^2}{10 - 1} \right]^{1/2}$$

$$= 1.48 \times 10^{-2} \text{ mm}.$$

Hence, $S_2^2 = 2.19 \times 10^{-4}$ mm$^2$.

As shown, calculated standard deviation from the first sample is small compared with the second sample. This means that the sample values are grouped closely together for the first sample. In other words, the range of values is small and close to the actual mean value. This is what we expect from a manufacturing process that does not have large amounts of rejections (defects). Normal probability density curves for the same mean value and different standard deviation of two samples are given in Figure 7.5.

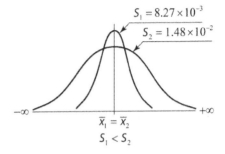

**Figure 7.5** Normal probability.

### 7.3.5 Skewness

As discussed in the preceding sections, a distribution that is not symmetric with respect to a vertical axis is said to be *skewed*. The clearance distribution of the bearings shown in Figure 7.3(b) has a *positive skew* (is skewed to the right), and in Figure 7.3(c) the distribution has a *negative skew* (is skewed to the left). The easy way to learn whether a distribution is skewed or not is to check whether the mean, mode, and median have the same value. Also, the skewness of the distribution can be easily determined by the formula

$$\text{Skewness} = \frac{3(\text{mean} - \text{median})}{\text{standard deviation}}. \tag{7.8}$$

☐ **Example 7.4**

Calculate the skewness of Figure 7.2(b, c) from the data given in Table 7.1.

**Solution**

*Skewness of Figure 7.2(b)*

From Example 7.1, the calculated sample mean for the second bearing is equal to 4.33 mm and the median is 4. The standard deviation is

$$S = \sqrt{\frac{\sum_{i=1}^{n} (x_i - \bar{x})^2}{n - 1}}$$

$$= \sqrt{\frac{1(2 - 4.33)^2 + 6(3 - 4.33)^2 + \cdots + 1(8 - 4.33)^2}{18 - 1}} = 1.61 \text{ mm}.$$

Therefore

$$\text{Skewness} = \frac{3(\text{mean} - \text{median})}{\text{standard deviation}} = \frac{3(4.33 - 4.0)}{1.61} = 0.615.$$

*Skewness of Figure 7.2(c)*

From Example 7.1, the calculated sample mean for the third bearing is equal to 5.67 mm and the median is 6. Following the same procedure, the standard deviation is found to be $S = 1.61$. Therefore,

$$\text{Skewness} = \frac{3(\text{mean} - \text{median})}{\text{standard deviation}} = \frac{3(5.67 - 6)}{1.61} = -0.615.$$

The minus sign indicates that the third bearing has negative skewness. From the results of this example, we conclude that if the mean is larger than the median, the distribution is positively skewed, and if the mean is smaller than the median, the distribution is negatively skewed. Note that the value of skewness is found to be the same for both cases because the data distribution for the third bearing is the mirror image of the second bearing (see Table 7.1).

## 7.4 Probability Distributions

The two most common continuous probability distributions used in engineering are briefly defined below.

### 7.4.1 Weibull Distribution

The Weibull distribution is used in engineering to predict the life of machine components. It is one of the most widely used life prediction methods, because it fits many different machine and electronic component failure distributions. The Weibull distribution was proposed by W. Weibull in 1951[4] and applied to ball bearing failures by Lieblein and Zelen in 1956.[5]

The Weibull distribution can take on the characteristics of other distribution types, based on the value of the shape parameter, $b$. The Weibull probability density function $f(t)$ is given by

$$f(t) = \frac{b}{\theta}\left(\frac{t - \gamma}{\theta}\right)^{b-1} \exp\left[-\left(\frac{t - \gamma}{\theta}\right)^b\right], \quad b > 0, \theta > 0, t \geq 0, \gamma \geq 0, \tag{7.9}$$

where

$b = $ shape parameter or Weibull slope,

$\theta = $ scale parameter,

$\gamma = $ location parameter (or failure-free life).

---

4  Weibull, W., "A statistical distribution function of wide application," *Journal of Applied Mechanics*, 18, 293–297, 1951.

5  Lieblein, J., and Zelen, M., "Statistical investigation of the fatigue life of deep-groove ball bearings," *Journal of Research of the National Bureau of Standards*, 57(5), 273–316, Research Paper 2719, pp. 1956.

The standard two-parameter Weibull probability distribution is obtained by setting $\gamma = 0$ in equation (7.9):

$$f(t) = \frac{b}{\theta}\left(\frac{t}{\theta}\right)^{b-1} \exp\left[-\left(\frac{t}{\theta}\right)^{b}\right]. \tag{7.10}$$

For different values of the shape parameter $b$, the Weibull distribution can be reduced to two special distributions: $b = 1$ yields the *exponential distribution,*

$$f(t) = \frac{1}{\theta} \exp\left(-\frac{t}{\theta}\right); \tag{7.11}$$

and $b = 2$ yields the Rayleigh distribution,

$$f(t) = \frac{2t}{\theta^2} \exp\left[-\left(\frac{t}{\theta}\right)^2\right]. \tag{7.12}$$

Note that in Figure 7.6, for values of $b > 2$, the curve becomes bell shaped (normal or Gaussian distribution).

The shape parameter, $b$, has also a distinct effect on the Weibull failure rate as shown in Figure 7.7. Weibull distributions with $b < 1$ show a failure rate that decreases with time, akin to infant mortality (early-life failures). Weibull distributions with $b = 1$ show a constant failure rate during the useful life period. Weibull distributions with $b > 1$ show a failure rate that increases with time during the wear-out period.

The Weibull method works with extremely small samples. This characteristic is important in development testing with small samples. The following are examples of engineering problems solved with Weibull analysis:[6]

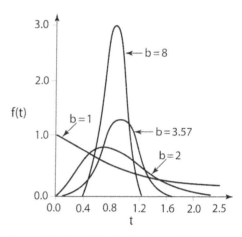

**Figure 7.6** The Weibull distribution for different values of shape parameter, *b*.

- A project engineer reports three failures of a component in service operations during a three-month period. The Program manager asks, "How many failures will we have in the next quarter, six months, and year?" What will it cost? What is the best corrective action to reduce the risk and losses?
- To order spare parts and schedule maintenance labor, how many units will be returned to depot for overhaul for each failure mode month-by-month next year?
- A state Air Resources Board requires a fleet recall when any part in the emissions system exceeds a 4 percent failure rate during the warranty period. Based on the warranty data, which parts will exceed the 4 percent rate and on what date?

---

6 Abernethy, R.B., "An overview of Weibull analysis," http://quanterion.com/Publications/WeibullHandbook/ChapterOne.pdf, accessed June 10, 2013.

- After an engineering change, how many units must be tested for how long, without any failures, to verify that the old failure mode is eliminated, or significantly improved, with 90 percent confidence?
- An electric utility is plagued with outages from boiler tube failures. Based on limited inspection data, forecast the life of the boiler based on plugging failed tubes.
- A machine tool breaks more often than the vendor promised. The vendor says the failures are random events caused by abusive operators, but you suspect premature wear-out is the cause.
- The cost of an unplanned failure for a component, subject to a wear-out failure mode, is 20 times the cost of a planned replacement. What is the optimal replacement interval?

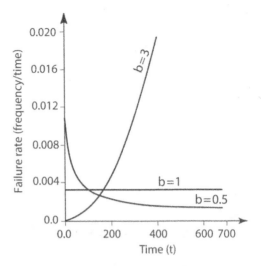

**Figure 7.7** The Weibull failure rate for different values of shape parameter, *b*.

### 7.4.2 Normal Distribution

The normal distribution (Figure 7.8), which is also called the *Gaussian distribution*, describes the distribution of many sets of data that occur in engineering. As shown in Figure 7.8, the normal distribution curve is symmetrical about the mean value and has a characteristic bell shape. Hence, the mean, mode, and median have the same value. The normal distribution is defined by the following equation:

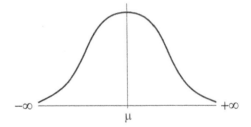

**Figure 7.8** Normal distribution.

$$f(x) = \frac{1}{\hat{\sigma}\sqrt{2\pi}} \exp\left[-\frac{(x-\mu)^2}{2\hat{\sigma}^2}\right], \quad -\infty < x < \infty, \tag{7.13}$$

where $\mu$ and $\hat{\sigma}$ are the population mean and standard deviation, respectively. The total area under the normal curve is equal to 1. That is,

$$\int_{-\infty}^{\infty} f(x)dx = 1 \tag{7.14}$$

As shown in Figure 7.9, normal distribution curve obeys the following rule:

- About 68.26% of the distribution falls within 1 standard deviation of the mean.
- About 95.46% of the distribution falls within 2 standard deviations of the mean.
- About 99.73% of the distribution falls within 3 standard deviations of the mean.

**Figure 7.9** Normal distribution ranges.

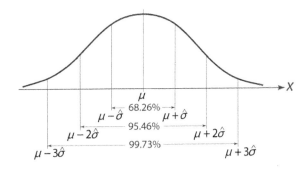

### 7.4.2.1 Standard Normal Distribution

A standard normal distribution is a normal distribution with mean equal to 0 and standard deviation equal to 1. Transformation of all the observations can be made to a new set of observations by dividing the mean deviation of each random variable $x$ by the standard deviation

$$z = \frac{x - \mu}{\hat{\sigma}}. \tag{7.15}$$

Equation (7.15) is known as the transformation formula. It transforms the $x$ value to its corresponding $z$ value. The new probability density function of the standard normal distribution becomes

$$f(z) = \frac{1}{\sqrt{2\pi}} \exp\left(-\frac{z^2}{2}\right). \tag{7.16}$$

Now Figure 7.9 can be transformed into a standard normal distribution, as shown in Figure 7.10. Note that the total area under the curve is equal to 1. In Figure 7.10, $\alpha/2$ is called a *two-sided critical region*. The values of critical regions (rejection regions) corresponding to $z$ values can be obtained from Table C.1 (see Appendix C).

The standard normal distribution is used to determine the probability of defects and the number of defects expected in a production process. For example, a set of bolts 14 mm in root diameter has been manufactured for use in a joint containing a gasket. Assume that the distribution of bolt root diameter is approximately normal with a mean of 14.05 mm and a standard deviation of 2.2 mm.

**Figure 7.10** Normal distribution ranges.

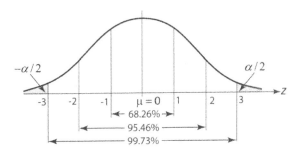

- If we require the bolts to be 14.05 ∓ 2.2 mm (1$\sigma$), the bolts will pass inspection 68.27% of the time. If 1 million bolts are manufactured, then (100% − 68.26%) × 10$^6$ = 317,400 bolts will be rejected (defective).
- If we require the bolts to be 14.05 ∓ 4.4 mm (2$\sigma$), the bolts will pass inspection 95.46% of the time. If 1 million bolts are manufactured, then (100% − 95.46%) × 10$^6$ = 45,400 bolts will be rejected (defect).
- If we require the bolts to be 14.05 ∓ 6.6 mm (3$\sigma$), the bolts will pass inspection 99.73% of the time. If 1 million bolts are manufactured, then (100% − 99.73%) × 10$^6$ = 2,700 bolts will be rejected (defect).

## Example 7.5

A set of bolts 14 mm in root diameter has been manufactured for use in a joint containing a gasket. Assume that the distribution of bolt root diameter is approximately normal with a mean of 14.05 mm and a standard deviation of 2.2 mm.

1) What percentage of pins picked at random will have diameter greater than 15.5 mm?
2) What percentage of pins picked at random will have diameter between 13.0 and 15.5 mm?

**Solution**

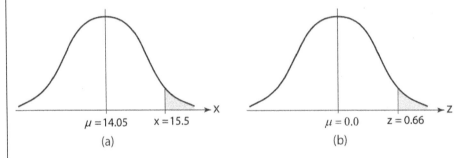

$\mu = 14.05$   $x = 15.5$    $\mu = 0.0$   $z = 0.66$

(a)    (b)

**Figure 7.11** (a) Original problem. (b) Transformed problem.

1) To find the probability of picking a bolt with a root diameter greater than 15.5 mm (area lying to the right of $x = 15.5$), transform Figure 7.11(a) into the standard normal form, as shown in Figure 7.11(b):

$$z = \frac{x - \mu}{\hat{\sigma}} = \frac{15.5 - 14.05}{2.2} = 0.66.$$

From Table C.1 in Appendix C, we obtain the area to the left of $z$:

$$\text{Area}(z < 0.66) = \frac{1}{\sqrt{2\pi}} \int_{-\infty}^{z} e^{-\frac{z^2}{2}} dz = 0.7454.$$

Then the shaded area lying to the right of $z = 0.66$ is

$$\text{Area}(z > 0.66) = 1 - 0.7454 = 0.2546.$$

Therefore, the probability of selecting pins with a diameter greater than 15.5 mm is

$$p(x > 15.5) = 0.2546.$$

2) Similarly, Figure 7.12(a) is transformed into standard normal form as shown in Figure 7.12(b):

$$z = \frac{13 - 14.05}{2.2} = -0.48.$$

The shaded area lying to the left (negative) of $z = -0.48$ is 0.3156, and from part (a) area $(z > 0.66) = 0.2546$. Thus, the probability of having a diameter between 13 and 15.5 mm is

$$p(13 < x < 15.5) = 1 - (0.2546 + 0.3156) = 0.4298.$$

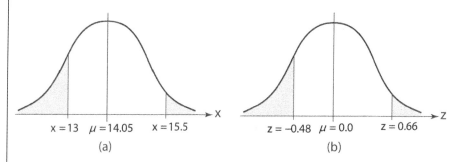

**Figure 7.12** (a) Original problem. (b) Transformed problem.

### 7.4.3 Standard Error of the Mean

The standard deviation of the sample mean is known as the *standard error of the mean* and indicates how close the sample mean is to the population mean. The standard error of the mean, $\hat{\sigma}_{\bar{x}}$, is given by

$$\hat{\sigma}_{\bar{x}} = \frac{\hat{\sigma}}{\sqrt{n}}, \tag{7.17}$$

where $\hat{\sigma}$ is the standard deviation and $n$ is the number of measurement.

### Example 7.6

Referring to Example 7.3, calculate the standard error of the mean and discuss the results for the two samples using 95% confidence (approximately between $-2$ and $+2$ times the standard error of the mean) interval. Assume that the population distribution is close to normal.

### Solution

*First Sample*

Rearranging equation (7.17) for the first sample distribution, we have

$$S_{\bar{x}_1} = \frac{S_1}{\sqrt{n}} = \frac{8.27 \times 10^{-3}}{\sqrt{10}} = 2.65 \times 10^{-3}.$$

Since true value of the mean will lie within the 95% confidence interval (meaning, approximately twice the standard error of the mean), design tolerance limits will have the following form:

$$\bar{x}_1 \mp 2 \times S_{\bar{x}_1}$$

$$0.032 \mp 2(2.65 \times 10^{-3})$$

$$\mp 5.30 \times 10^{-3},$$

that is, the confidence interval will be

$$0.027 \leq \bar{x}_1 = 0.032 \leq 0.037.$$

### Second Sample

For the second sample standard error of the mean is

$$S_{\bar{x}_2} = \frac{S_2}{\sqrt{n}} = \frac{1.48 \times 10^{-2}}{\sqrt{10}} = 4.68 \times 10^{-3}.$$

Then the design tolerance limits for the second sample are

$$\bar{x}_2 \mp 2 \times S_{\bar{x}_2}$$

$$0.032 \mp 2(4.68 \times 10^{-3})$$

$$\mp 9.36 \times 10^{-3},$$

that is, the confidence interval will be

$$0.023 \leq \bar{x}_1 = 0.032 \leq 0.041.$$

### Conclusion

Data from the first sample shows that the manufacturing process demands a high degree of precision (small tolerances and small mean variations). However, this example does not tell us if the manufacturing process is capable of producing parts (pin and bushing in this case) within the tolerances (lower and upper tolerance limits) and close to the specified target value.

## 7.5 Sampling Distributions

Frequency distributions in statistics based on samples of $n$ observations drawn from a population can be characterized by a sampling distribution. The sampling distribution plays an important role in estimating population parameters by means of confidence intervals. In the remainder of this chapter some of the important sampling distributions of statistics and their application to problems in statistical inference are discussed.

### 7.5.1 Sampling Distributions Based on Sample Means and the Central Limit Theorem (Sampling Distribution of the Mean)

As is well known, the degree of confidence in estimating parameters of a population from a sample drawn from it is dependent on the sample size $n$. It seems intuitively obvious that as the sample size $n$ becomes large, the sample mean $\bar{x}$ approaches the population mean $\mu$. A very important mathematical statistics theorem called the *central limit theorem* states that if a random variable $x$ is normally distributed in the population, then the randomly drawn samples of $n$ observations from this population will also tend toward a normal distribution as the sample size $n$ increases. Hence, the expected value $E\bar{x}$ of the sample mean can be assumed to be equal to the population mean,

$$E\bar{x} = \mu. \tag{7.18}$$

The variance of the sample mean is

$$\hat{\sigma}_{\bar{x}}^2 = \frac{\hat{\sigma}^2}{n}. \tag{7.19}$$

By taking the square root of equation (7.19) the standard deviation of the sample mean, which is known as the *standard error of the mean*, is found. This indicates how close the sample mean is to the population mean, and is expressed as

$$\hat{\sigma}_{\bar{x}} = \frac{\hat{\sigma}}{\sqrt{n}}. \tag{7.20}$$

Equation (7.20) can also be written in terms of the standard normal distribution $z_{\bar{x}}$ with a variance of 1 and a mean of 0:

$$z_{\bar{x}} = \frac{\bar{x} - \mu}{\hat{\sigma}_{\bar{x}}} = \frac{\bar{x} - \mu}{\hat{\sigma}/\sqrt{n}}. \tag{7.21}$$

Many frequency distributions experienced in engineering are approximately normal. In general, the normal approximation of the random sample mean is good if $n \geq 30$. This statement is valid even if the population distribution is not normal. If $n < 30$, the approximation is valid only if the population distribution is close to normal.

### 7.5.2 Student's *t* Distribution

In the previous section, it was assumed that the standard deviation of the population is known. Unfortunately, this is not always true. If the standard deviation of the population from which the random sample is drawn is unknown, the standard deviation of a randomly selected sample can be used in equation (7.21), provided the sample size $n \geq 30$. However, if the randomly selected sample size $n$ is less than 30, an estimate of the standard deviation of the sample mean will be less dependable. In such cases, assuming that $x$ is normally distributed, the following $t$ distribution is used:

$$t = \frac{\bar{x} - \mu}{S/\sqrt{n}}. \tag{7.22}$$

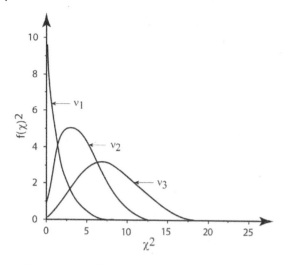

**Figure 7.13** The chi-square distribution for different degrees of freedom ($v_3 > v_2 > v_1$).

The values of the Student $t$ distribution for various values of $\alpha$ are given in Table C.2 (see Appendix C).

### 7.5.3 Chi-Square Distribution

Another important sampling distribution used to estimate the population variance $\hat{\sigma}^2$ by using the sample variance $S^2$ is the chi-square ($\chi^2$) distribution.

As shown in Figure 7.13, the form of the chi-square distribution depends on the degrees of freedom $v$ and approaches a normal distribution as the degrees of freedom increase. The chi-square distribution with $v = n - 1$ degrees of freedom is given by the formula

$$\chi^2_{\alpha,v} = \frac{(n-1)S^2}{\hat{\sigma}^2}. \tag{7.23}$$

It can be seen that because of the square terms, $\chi^2$ can never be negative. The values of the chi-square distribution for various values of $\chi^2_{\alpha,v}$ are given in Table C.3 (see Appendix C). To illustrate the use of Table C.3, assume a chi-square distribution with 18 degrees of freedom, leaving an area of 0.05 to the right, is $\chi^2_{0.05,18} = 28.869$. This is shown graphically in Figure 7.14. A chi-square distribution with a 95 percent level of confidence and 18 degrees of freedom lies between $\chi^2_{0.975,18} = 8.231$ (lower confidence limit) and $\chi^2_{0.025,18} = 31.526$ (upper confidence limit).

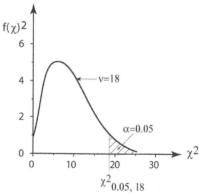

**Figure 7.14** Percentage point $\chi^2_{0.05,18}$ of the chi-square distribution.

### 7.5.4 F Distribution

The $F$ distribution is used in conjunction with statistical quality control work and is defined as the ratio of two independent sampling distributions drawn from two normal populations with variances $\hat{\sigma}_1$ and $\hat{\sigma}_2$. For sampling distributions $n_1$ and $n_2$ with variances $S_1^2$ and $S_2^2$, respectively, the $F$ distribution is

$$F_{v_1,v_2} = \frac{S_1^2/\hat{\sigma}_1^2}{S_2^2/\hat{\sigma}_2^2}, \tag{7.24}$$

with $v_1 = n_1 - 1$ and $v_2 = n_2 - 1$ degrees of freedom. The values of the $F$ distribution for various combinations of $v_1$ and $v_2$ are given in Table C.4 (see Appendix C).

## 7.6 Statistical Inference

Statistical inference can be divided into two categories: *estimation* and *hypothesis testing*. In this section some methods involving these two areas are discussed and compared.

### 7.6.1 Estimation

As we have learned, parameters like the population mean $\mu$ and the population standard deviation, $\hat{\sigma}$, can be estimated from sample data taken randomly from a population. There are two ways of estimating population parameters, namely point estimates and interval estimates. A point estimate is a single estimated value of a population parameter. For example, the sample mean $\bar{x}$ and variance $S^2$, which estimate the population mean $\mu$ and variance $\hat{\sigma}^2$, respectively, are point estimates. To estimate these parameters in terms of probabilities, an interval estimate is used that is the estimate of a population parameter within a bounded range of values rather than a single estimated value. The accuracy of these parameters depends on how large the randomly selected sample size is. The closer the size of a sample to the population size, the greater the accuracy of the estimated mean and standard deviation of the population. However, selecting a large sample size for estimating these parameters may be impractical. What, then, is the smallest sample size for accurate estimation of the population parameters, and what is the degree of confidence in them? The degree of confidence in a *confidence interval*, in which the parameters fall. The limits of this interval are called *confidence limits*.

#### 7.6.1.1 Confidence Interval Estimation of the Mean: $\hat{\sigma}$ Is Known
In the previous sections, the population mean was estimated by using the sample mean $\bar{x}$. However, $\mu$ was not assumed to be exactly equal to $\bar{x}$. Since $\bar{x}$ is a random quantity, it may be smaller or greater than the population mean $\mu$, thus introducing the error $\mu - \bar{x}$. If the population variance $\hat{\sigma}^2$ is known, this error can be defined between two boundary points (two-sided confidence interval) as

$$-z_{\alpha/2}\frac{\hat{\sigma}}{\sqrt{n}} \leq \mu - \bar{x} \leq z_{\alpha/2}\frac{\hat{\sigma}}{\sqrt{n}}. \tag{7.25}$$

Adding $\bar{x}$ to both sides of the above inequality results in

$$\bar{x} - z_{\alpha/2}\frac{\hat{\sigma}}{\sqrt{n}} \leq \mu \leq \bar{x} + z_{\alpha/2}\frac{\hat{\sigma}}{\sqrt{n}}. \tag{7.26}$$

#### 7.6.1.2 Confidence Interval Estimation of the Mean: $\hat{\sigma}$ Is Unknown but a Large Sample Size Is Selected

In most manufacturing applications, the standard deviation of the population is not known. However, if the sample size $n$ is large enough, the distribution of the sample can be assumed to be approximately normal. Consequently, the standard deviation of the population can be estimated by the sample standard deviation $S$. As discussed in Section 7.5.1, this assumption is valid if the sample size is large enough ($n \geq 30$). In this case, for equation (7.26) the standard deviation of the population $\hat{\sigma}$ is replaced by the standard deviation of the sample $S$:

$$\bar{x} - z_{\alpha/2}\frac{S}{\sqrt{n}} \leq \mu \leq \bar{x} + z_{\alpha/2}\frac{S}{\sqrt{n}}. \tag{7.27}$$

#### 7.6.1.3 Confidence Interval Estimation of the Mean: $\hat{\sigma}$ Is Unknown but a Small Sample Size is Selected

In some situations having a sample size $n \geq 30$ can be too expensive. If the distribution of the population is normal, confidence intervals can be computed when the population standard deviation $\hat{\sigma}$ is unknown, and the sample size is less than 30 by using the $t$ distribution:

$$\bar{x} - t_{\alpha/2}\frac{S}{\sqrt{n}} \leq \mu \leq \bar{x} + t_{\alpha/2}\frac{S}{\sqrt{n}}. \tag{7.28}$$

#### 7.6.1.4 Confidence Interval Estimation on the Variance of a Normal Distribution

If the standard deviation of a randomly selected sample size $n$ is drawn from a normally distributed population, then, by using a chi-square distribution, a two-sided confidence interval can be calculated using the formula

$$\frac{(n-1)S^2}{\chi^2_{\alpha/2,v}} \leq \hat{\sigma}^2 \leq \frac{(n-1)S^2}{\chi^2_{1-\alpha/2,v}}. \tag{7.29}$$

#### 7.6.1.5 Confidence Interval Estimation of the Ratio of the Variance of Two Normal Populations

Using the $F$ distribution, discussed in Section 7.5.4, the two-sided confidence interval for the ratio of the two variances $\hat{\sigma}_1^2/\hat{\sigma}_2^2$ can be calculated from

$$\left(\frac{S_1^2}{S_2^2}\right)F_{1-\alpha/2,v_1,v_2} \leq \frac{\hat{\sigma}_1^2}{\hat{\sigma}_2^2} \leq \left(\frac{S_1^2}{S_2^2}\right), F_{\alpha/2,v_1,v_2} \tag{7.30}$$

where the lower tail, $1 - \alpha/2$, of the $F$ distribution is given by

$$F_{1-\alpha/2,v_1,v_2} = \frac{1}{F_{\alpha/2,v_1,v_2}}.$$

If the confidence interval includes $\hat{\sigma}_1^2/\hat{\sigma}_2^2 = 1$, it can be concluded that there is no statistical difference in the two different events at a given level of confidence.

### 7.6.1.6 Confidence Interval Estimation Based on the Difference in Two Means (Variance Known)

Consider two independent random samples from two populations, $x_1$ of size $n_1$ with unknown mean $\mu_1$ and known variance $\hat{\sigma}_1^2$, and $x_2$ of size $n_2$ with unknown mean $\mu_2$ and known variance $\hat{\sigma}_2^2$. If the random samples $x_1$ and $x_2$ are drawn from normal populations, or the sample sizes $n_1$ and $n_2$ are both greater than 30, a two-sided $100(1 - \alpha)$ percent confidence interval on the difference in means $(\mu_1 - \mu_2)$ is computed by

$$\bar{x}_1 - \bar{x}_2 - z_{\alpha/2}\sqrt{\hat{\sigma}_1^2/n_1 + \hat{\sigma}_2^2/n_2}$$
$$\leq \mu_1 - \mu_2 \leq (\bar{x}_1 - \bar{x}_2) + z_{\alpha/2}\sqrt{\hat{\sigma}_1^2/n_1 + \hat{\sigma}_2^2/n_2}. \tag{7.31}$$

From equation (7.31) the $100(1 - \alpha)$ percent lower and upper confidence limits on $\mu_1 - \mu_2$ can be written as

$$\bar{x}_1 - \bar{x}_2 - z_{\alpha/2}\sqrt{\hat{\sigma}_1^2/n_1 + \hat{\sigma}_2^2/n_2} \leq \mu_1 - \mu_2 \tag{7.32}$$

and

$$\mu_1 - \mu_2 \leq \bar{x}_1 - \bar{x}_2 + z_{\alpha/2}\sqrt{\hat{\sigma}_1^2/n_1 + \hat{\sigma}_2^2/n_2}, \tag{7.33}$$

respectively. This method is often used to compare the effectiveness of two manufacturing processes.

### 7.6.1.7 Confidence Interval Estimation Based on the Difference in Two Means (Variance Unknown)

If the sample size drawn from the normal population is less than 30, the $t$ distribution must be used to compute the confidence interval. To find a $100(1 - \alpha)$ percent confidence interval for the difference in means $(\mu_1 - \mu_2)$, assume that $\hat{\sigma}_1^2 = \hat{\sigma}_2^2 = \hat{\sigma}^2$. This assumption is often made in comparing manufacturing processes. This unknown common variance, $\hat{\sigma}^2$, can be estimated by using a "combined" or "pooled" estimator. The "pooled" estimator equation is

$$S_p^2 = \frac{(n_1 - 1)S_1^2 + (n_2 - 1)S_2^2}{n_1 + n_2 - 2}. \tag{7.34}$$

Therefore, a $100(1 - \alpha)$ percent two-sided confidence interval for the difference in means $(\mu_1 - \mu_2)$ is given by

$$\bar{x}_1 - \bar{x}_2 - t_{\alpha/2, n_1+n_2-2}S_p\sqrt{1/n_1 + 1/n_2} \leq \mu_1 - \mu_2$$
$$\leq \bar{x}_1 - \bar{x}_2 + t_{\alpha/2, n_1+n_2-2}S_p\sqrt{1/n_1 + 1/n_2}. \tag{7.35}$$

For testing the difference in two variables, the above-mentioned test hypothesis is used to calculate the confidence interval. If the confidence interval includes $\mu_1 - \mu_2 = 0$, it is concluded that there is no statistical difference in the performance of the two manufacturing processes at a given level of confidence.

## 7.6.2 Statistical Hypothesis Testing

A decision problem in which one of two arguments must be chosen is known as a hypothesis testing problem. It is often necessary to make a decision about population parameters based on information obtained from the random sample drawn from the population, assuming that the sample and the population have the same distribution. However, if the size of the sample is small, we may not have sufficient evidence for judging that the decision is correct. Hence, the decision must be supplemented by other knowledge. This can be done by postulating a hypothesis and then checking to see whether the statistics of the sample are comparable with the observed results of the population. Statistical hypothesis testing is a procedure that leads to a decision to *reject* or *accept* the hypothesis under consideration.

In statistical testing, two hypotheses are set. A null hypothesis (initial), denoted $H_0$, is the hypothesis under test or consideration. The alternative hypothesis is denoted by $H_1$. After the observation, if the data prove correct with respect to $H_0$, $H_0$ is accepted. If the data prove false with respect to $H_0$, $H_0$ is rejected. Suppose that the null hypothesis $H_0$ is that $\mu = \mu_H$. where $\mu_H$ is the hypothetical population mean. The alternative hypothesis $H_1$ could be $\mu > \mu_H$, $\mu < \mu_H$, or $\mu \neq \mu_H$. For example, a drilling company is interested in purchasing drill pipe that, according to the material properties, has a mean yield strength of $\mu_0 = 100$ kpsi, with a standard deviation of $\hat{\sigma}_0 = 10$ kpsi. The drilling company seeks to verify the yield strength of the drill pipe. To do this, a sample of test specimens is prepared and is tested to find the mean yield strength of the sample. On the basis of this sample mean $\bar{x}$ the hypothesis that the true mean is 100 kpsi will be rejected or accepted. Suppose it is found that the sample mean $\bar{x}$ does not differ in any significant manner from the unknown mean yield strength of 100 kpsi, then we state that

$H_0 : \mu = \mu_H \Rightarrow$ the null hypothesis, and
$H_1 : \mu \neq \mu_H \Rightarrow$ the alternative hypothesis.

From the foregoing discussion, it can be understood that the approach is to make a decision about the population parameters based on the information obtained from the sample drawn from it. However, information obtained from the sample may not exactly represent the population parameters, thus, introducing two types of decision errors:

1) A *Type I* error is committed if the null hypothesis is rejected when it is true. For example, during finished product evaluation, a Type I error is committed if the quality control inspector rejects the product when it is acceptable. This will cause a risk to the producer (producer risk). In a statistical process, the probability of rejecting the null hypothesis when it is true (a Type I error) is called the *level of significance* and is denoted by $\alpha$.

2) A *Type II* error is committed if we accept the null hypothesis when it is false. If the quality control inspector accepts the product when it should be rejected, a Type II error is committed, denoted by $\beta$. This will cause a risk to the consumer (consumer risk).

Of course, an attempt should be made to minimize the above-mentioned errors. Unfortunately, this may not always be possible. In fact, for a fixed sample size, when the probability of committing a Type I error is reduced, the probability of committing a Type II error is increased.

**Figure 7.15** Critical region for testing $\mu = \mu_H$ versus $\mu \neq \mu_H$.

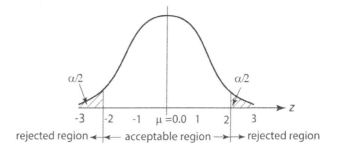

In general, designers control the probability of making a Type I error, and they set the level of $\alpha$ to a desired level, such as 0.01 or 0.05. There are two ways of minimizing both errors:

- Increase the sample size.
- Fix the level of $\alpha$ (Type I error), and specify a critical region that gives the smallest Type II error.

In testing a statistical hypothesis, the following steps are carried out:

1) State the hypothesis. The population mean should be equal to some defined hypothetical value $\mu_H$ with a given population standard deviation $\hat{\sigma}$. Set
   - null hypothesis $\Rightarrow H_0: \mu = \mu_H$,
   - alternative hypothesis $\Rightarrow H_1: \mu \neq \mu_H$.
2) Decide $\alpha$ and $\beta$.
3) Select the sample size $n$.
4) Select the statistic to be used in testing the hypothesis. For example,

$$z_{\bar{x}} = \frac{\bar{x} - \mu_H}{\hat{\sigma}/\sqrt{n}}. \tag{7.36}$$

5) Define the rejection region at the selected $\alpha$ level of significance as shown in Figure 7.15.
6) Perform the experimental test by using the sample observation, compute $z_{\bar{x}}$, and determine whether it falls in the rejected region.

☐ **Example 7.7** *Type I Error: Hypothesis Test about the Mean of a Normal Distribution when the Standard Deviation Is Known.*

Because of severe drilling conditions for a deep drilling operation, a manufacturer of drill pipes has developed a new welding process to weld the tool joints at the end of the drill pipe; hence, two drill pipe joints can be connected to each other by the use of tool joints. The quality control manager claims that the mean failure strength and the standard deviation of the drill pipe at the tool joint welding location, as shown in Figure 7.16, are 121 kpsi and 10 kpsi, respectively (normally distributed). To check the claim, a sample of 10 drill pipes is tested. The results are shown in Table 7.4. What is the decision about the hypothesis?[a]

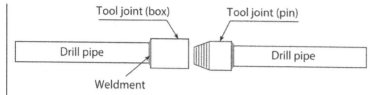

Tool joint (box)    Tool joint (pin)

Drill pipe    Drill pipe

Weldment

**Figure 7.16** Drill pipe tool joint.

**Table 7.4** Results of failure testing.

| Test number | Failure strength (kpsi) |
|:-----------:|:-----------------------:|
| 1  | 118 |
| 2  | 118 |
| 3  | 114 |
| 4  | 106 |
| 5  | 106 |
| 6  | 120 |
| 7  | 106 |
| 8  | 108 |
| 9  | 116 |
| 10 | 116 |

## Solution

Following the procedure outlined above:

1) Set the null hypothesis:
   - $H_0$: $\mu = \mu_H = 121$ kpsi with $\hat{\sigma} = 10$ kpsi (assume that the null hypothesis is, in fact, true);
   - $H_1$: $\mu \neq \mu_H \neq 121$ kpsi (alternative hypothesis).
2) Assume a level of significance, $\alpha = 0.05$ (quality control manager chooses).
3) Size of sample, $n = 10$.
4) Since the standard deviation of the population is known, the sampling distribution of the statistic can be written

$$z_{\bar{x}} = \frac{\bar{x} - \mu_H}{\hat{\sigma}/\sqrt{n}}$$

$$\bar{x} = \frac{\sum x_i}{n} = \frac{118 + 118 \cdots}{10} = 112.8$$

$$z_{\bar{x}} = \frac{\bar{x} - \mu_H}{\hat{\sigma}/\sqrt{n}} = \frac{112.8 - 121}{10/\sqrt{10}} = -2.60.$$

5) Since the critical region is two-sided, dividing the probability equally between two tails ($\alpha/2 = 0.025$), from Table C.1 in Appendix C, the values of $z$ for the critical region are found to be $z = \mp1.96$; hence,

$$-1.96 < z < 1.96.$$

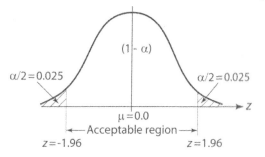

**Figure 7.17** Critical region for testing $\mu = 121$ versus $\mu \neq 121$.

*Decision*

As shown in Figure 7.17, $z_{\bar{x}} = -2.60$ is not in the acceptance region; therefore, reject the hypothesis $H_0$ and decide that the mean failure strength is not equal to $\mu_H = 120$ kpsi.

[a] Adapted from Ertas, A., and Jones, J.C., *The Engineering Design Process*, 2nd edn, John Wiley & Sons, Inc., New York, 1996.

---

☐ **Example 7.8   *Type I Error: Hypothesis Test about the Mean of a Normal Distribution When the Standard Deviation Is Unknown and the Sample Size Is Large ($n \geq 30$).***

Electronic components like the capacitors shown in Figure 7.18 are often supported by electrical lead wire through the solder terminals. If this suspended electronic component is subjected to vibration, the electrical lead wire and capacitor will act like a one-degree-of-freedom mass–spring system that develops alternating stress loads. Hence, lead wires must be good electrical conductors and must also have high fatigue life. An electronic company has used a new lead wire material and claims that the mean fatigue life of this new electrical lead wire is

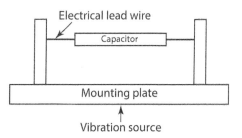

**Figure 7.18** Capacitor supported by electrical lead wire.

$5.8 \times 10^6$ cycles. To check whether the mean fatigue life of the new material is equal to the specified value, a random sample of 49 electronic components is tested. The sample mean is $\bar{x} = 5.6 \times 10^6$ cycles, and the standard deviation is $S = 4.9 \times 10^5$ cycles. Determine whether the sample data gives enough evidence to prove that the fatigue life of this new material has been improved.[a]

**Solution**

1) Set the null hypothesis:
   • $H_0 : \mu = \mu_H = 5.8 \times 10^6$ cycles (assume that the null hypothesis is, in fact, true);
   • $H_1 : \mu \neq \mu_H \neq 5.8 \times 10^6$ cycles (alternative hypothesis).
2) Assume a level of significance, $\alpha = 0.01$ (two-sided).
3) Size of sample, $n = 49$.

4) Since the sample size is large enough, the distribution of $\bar{x}$ will be approximately normal. Hence, the standard deviation of the population can be assumed to be equal to the standard deviation of the sample. Then

$$Z_{\bar{x}} = \frac{\bar{x} - \mu_H}{S/\sqrt{n}} = \frac{5.6 \times 10^6 - 5.8 \times 10^6}{4.9 \times 10^5/\sqrt{49}} = -2.86.$$

5) Assuming the two-sided critical region as shown in Figure 7.19, and $\alpha/2 = 0.005$ from Table C.1 in Appendix C, the value of $z$ for the critical region is found to be $z = \mp 2.575$.

*Decision:*

Since $z_{\bar{x}} = -2.86$ is not in the acceptance region, reject the hypothesis $H_0$ and decide that there is not sufficient evidence to indicate that the new material increases the fatigue life at a 0.01 level of significance.

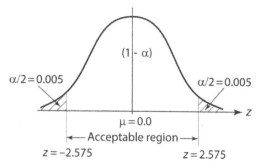

**Figure 7.19** Critical region for testing $\mu = 5.8 \times 10^6$ versus $\mu \neq 5.8 \times 10^6$.

[a]Adapted from Ertas, A., and Jones, J.C., *The Engineering Design Process*, 2nd edn, John Wiley & Sons, Inc., New York, 1996.

## Example 7.9

The temperature and color of the exhaust gases $t_{eg}$ of a diesel engine are important parameters of the operating process. A thermocouple is installed in the exhaust manifold near the exhaust valves to measure the exhaust temperature. Consider the two-stroke, eight-cylinder 4,000 horse power main diesel engine of a cargo ship. For optimum performance, the engine operator must be sure that the exhaust gas temperature of each of the cylinders is close to the temperature of the others. If, for some reason, some of the cylinder exhaust gas temperatures are higher than others, the cylinders with higher temperatures are assumed to be overloaded. The non-uniform stress distribution over the crank shaft bearing and excessive vibration may result in a possible failure of the machine parts.

Assume that, for the optimum performance of the diesel engine, the mean value of the gas temperature for each cylinder is 360°C. If the mean exhaust temperature of a cylinder reaches 370°C, that particular cylinder may be assumed to be overloaded. In both cases, assume that the standard deviation is 5°C (normally distributed). Because of thermal inertia, temperature readings of the gas flow may show changes. Suppose that during engine operation, the operator takes a reading of the exhaust gas temperature of approximately 364°C (average of four readings) from a cylinder. Which of the two following actions should the engine operator take?[a]

1) Stop the engine, and start checking the injector valve adjustment.
2) Continue running the engine.

## Solution

Several factors may cause the high exhaust temperature of a cylinder. Perhaps the most important one is improper fuel injection valve adjustment for the delivery of the required amount of fuel oil into the cylinder. If the engine is under long-duration operation, the injection valves may get out of adjustment. When this situation occurs, the engine operator must take one of the courses of action listed below.

If the operator stops the engine when the engine can still operate normally until the ship reaches port, he or she is making a Type I error. This wrong decision results in loss to the ship company because of down time during the troubleshooting period. Assume that this relative cost is 1 unit and designate it regret $r$ for making a Type I error.

If the operator does not stop the engine when the engine cannot operate normally, he or she is making a Type II error. Making this error also results in loss, in some cases replacement, of very expensive engine parts for not detecting the problem with the engine and system in time. Assume that this relative cost is 6 units and designate it regret $R$ for making a Type II error.

Assume from past experience that the probability of damage to the engine when any of the cylinders are running with relatively high exhaust gas temperature is 25 percent. Hence, 75 percent of the time engine operation is normal even though some of the cylinders may be experiencing high exhaust gas temperature. Let normal operation of the engine under high exhaust gas temperature be the *null hypothesis*. Table 7.5 shows the probabilities and the regrets of making two different decisions. The expected loss for not stopping the engine is

$$(0)(0.75) + (6)(0.25) = 1.5,$$

while the expected loss for stopping the engine is

$$(1)(0.75) + (0)(0.25) = 0.75.$$

**Table 7.5** Decision Table.

| | State | |
| --- | --- | --- |
| **Decision** | **Normal Operation $p$(normal) = 0.75** | **Abnormal Operation $p$(abnormal) = 0.25** |
| Don't stop the engine | $r = 0$ | $R = 6$ |
| Stop the engine | $r = 1$ | $R = 0$ |

Evidently the expected loss is less for stopping the engine. Hence, stopping is the correct decision. However, consider the statistical decision information as to engine operation. As Figure 7.20 shows, the two temperature distribution curves intersect at a decision-making point corresponding to a temperature of 365°C. To the left of this point, the normal operation curve is above the abnormal operation curve, whereas to the right of this point, the abnormal operation curve is

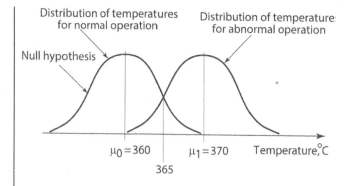

Distribution of temperatures for normal operation

Distribution of temperature for abnormal operation

**Figure 7.20** Decision-making curves.

Null hypothesis

$\mu_0 = 360$    $\mu_1 = 370$    Temperature,°C

365

above the normal operation curve. The point 365°C is called the unaffected decision-making point (UDMP), and it can be calculated by taking the average of the two mean values:

$$\text{UDMP} = \frac{\mu_0 + \mu_1}{2} = \frac{360 + 370}{2} = 365°\text{C}.$$

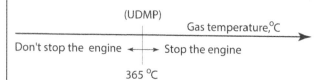

(UDMP)

Gas temperature,°C

Don't stop the engine ← → Stop the engine

365 °C

**Figure 7.21** Critical region for decision-making.

As shown in Figure 7.21, if the temperature is higher than 365°C, the engine operator should stop the engine and start troubleshooting. If the temperature is lower than 365°C, the engine operator should continue operation. Note that in the above analysis, for both the normal and abnormal cases, equal weight has been given. However, under high exhaust gas temperature operation, 75 percent of the time engine operation is normal and 25 percent of the time engine operation is abnormal. Also, the relative weighting of the regrets is not taken into account. Therefore, the following two steps will be performed to calculate the change in UDMP when the weighting of probabilities and regrets is considered.

1) The change in UDMP when probabilities are taken into consideration is[7]

$$\Delta'_{\text{UDMP}} = \frac{\hat{\sigma}^2}{\mu_1 - \mu_0} \times \ln \frac{P(\text{null hypothesis is true})}{P(\text{null hypothesis is false})}$$

$$= \frac{5^2}{370 - 360} \times \ln \frac{0.75}{0.25} = 2.75°\text{C}.$$

7  Robinson, E.A., *Statistical Reasoning and Decision Making*, Goose Pond Press, Houston, TX, 1981.

2) The change in UDMP when the regret is taken into consideration is

$$\Delta''_{UDMP} = \frac{\hat{\sigma}^2}{\mu_1 - \mu_0} \times \ln \frac{P(\text{regrets of Type I Error})}{P(\text{regrets of Type II Error})}$$

$$= \frac{5^2}{370 - 360} \times \ln \frac{1}{6} = -4.48°C.$$

3) The net change in the decision-making point is $-4.48 + 2.75 = -1.73°C$, which is a change of 1.73 to the left. Hence, the resulting decision-making point is $365 - 1.73 = 363.27°C$. Now Figure 7.20 can be modified for the new decision-making point, as shown in Figure 7.22. Since the modified decision-making point is $363.27°C$ and the observed exhaust gas temperature was $364°C$, the operator's best decision would be to stop the engine. The probability of a Type I error $\alpha$ can be determined by calculating the area under the normal curve for normal operation to the right of the decision-making point:

**Figure 7.22** Modified decision-making curves.

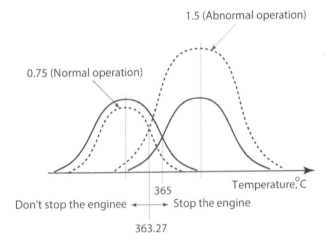

4) The probability of a Type II error $\beta$ can be determined by calculating the area under the normal curve for abnormal operation to the left of the decision-making point:

$$\alpha = 1 - p\left(z < \frac{\bar{x} - \mu_0}{\hat{\sigma}/\sqrt{n}}\right) = 1 - p\left(z < \frac{363.27 - 360}{5/\sqrt{4}}\right)$$

$$= 1 - p(z < 1.308) \approx 0.095.$$

$$\beta = p\left(z < \frac{x - \mu_1}{\hat{\sigma}/\sqrt{n}}\right) = p\left(z < \frac{363.27 - 370}{5/\sqrt{4}}\right)$$

$$= p(z < -2.69) = 0.0036.$$

5) These areas are shown in Figure 7.23.

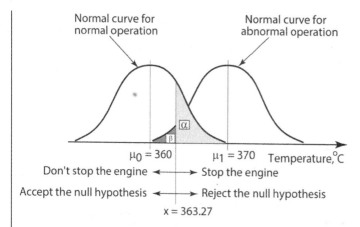

**Figure 7.23** Probability of Type I and Type II error.

Figure 7.24 shows that, without taking the temperature reading, the probability of normal operation of the engine running under high exhaust gas temperature is 75 percent. However, after the operator took the temperature reading of 364°C it was found that the observed temperature fell in the critical region to the right of the decision-making point of 363.27°C. Hence, the operator's decision should be to reject the null hypothesis $H_0$ and decide that engine operation should be stopped. From the same tree diagram we observe that the path of probability for

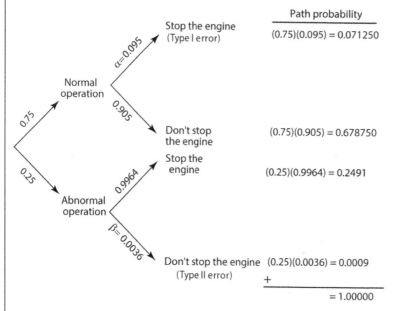

**Figure 7.24** Tree diagram.

a Type I error is 0.071250 with regret 1. The path of probability for a Type II error is 0.0009 with regret 6. Therefore, the expected loss is

$$(0.071250)(1) + (0.0009)(6) = 0.07665.$$

The previously calculated smallest loss was 0.75 if the operator did not take the temperature readings. However, by taking the temperature reading it can be seen that the expected loss is reduced by a factor of approximately 10.

[a]From Ertas, A., and Jones, J.C., *The Engineering Design Process*, 2nd edn, John Wiley & Sons, Inc., New York, 1996 (used with permission).

## 7.7 Design of Experiments

> Today's scientists have substituted mathematics for experiments, and they wander off through equation after equation, and eventually build a structure which has no relation to reality.
>
> **Nikola Tesla**

Experimental design is a transdisciplinary technique which has found broad application in many disciplines. It is a scientific process that teaches us about systems characteristics and how they work. The design of experiments is involved with the selection of sample size, the values of independent variables to be used in experimentation, and experimentation to confirm the quality of observed responses.

### 7.7.1 Experimentation

There is always a need for experimentation for process and product optimization. This is a costly and time-consuming effort; yet, by implementing statistical experimental design, cost and time can be reduced. The goal in experimentation is to develop a robust product, reliable process, and optimize the performance characteristics. Important considerations in design of experimentation are as follows:

- The exact questions that the experiment is intended to answer should be clearly identified before performing the experiment.
- Experiments should be capable of distinguishing clearly between different hypotheses.
- Experiments must be repeated enough times so that the results are consistent and statistically valid.
- The experiment must be well controlled so that the other factors will not affect the results.
- Data collection and measurements in experimentation must be accurate.
- Which method of data analysis should be used?

Consider galling as an example. Galling is a form of lubricant-related adhesive wear which occurs on connection thread load and/or metal-to-metal sealing surfaces during the makeup operation. It is sometimes severe enough to weaken the performance characteristics of the connection. The occurrence of galling is a matter of probability and statistics, similar to pitting fatigue failure of gears and rolling-element bearings. Galling depends on a number of factors, including axial pressure applied on the thread, rotational velocity, lubrication, surface roughness, and specimen material properties.

Suppose you would like to investigate the galling resistance of a tubular connection thread made of a steel pipe which has the characteristics and the properties of L-80 (yield strength is 80 kpsi). As discussed, a galling experiment involves several factors. The objectives of this experiment should be as follows:

- to identify the main factors affecting the galling;
- to identify the factors whose contribution in galling is not significant so that they can be eliminated from the experiment if necessary;
- to decide on the factor levels and range;
- to find out which factor most affects the galling of the tubular connection thread;
- to establish the percentage contributions of each factor affecting the galling; and
- to find the interaction between the factors affecting the galling.

In this section, the up-and-down and Taguchi methods of experimentation, which offer the most useful analysis for the least amount of data, are discussed.

### 7.7.2 Up-and-Down Method

The up-and-down method, the parameter estimation by sequential testing technique, was developed to obtain data from sensitivity experiments. This sensitivity experiment, a trial-by-trial tracking procedure, was originally developed for testing explosives in World War II. If the subject under experimentation detects a stimulus, the stimulus magnitude is decreased by a step for the next trial. If the subject does not detect the stimulus, the stimulus is then increased by a step. In this experiment, once a test has been carried out, the specimen is exchanged for a new specimen – a valid second statistical result cannot be obtained from the same specimen. This technique is used extensively in various disciplines, ranging from engineering to human neuropsychology. Applications include determining the strength of materials, evaluating the sensitivity of explosives to shock, and psychophysical research dealing with threshold stimuli.[8]

To use the up-and-down technique in any experiment, three major conditions must be satisfied:

1) The variant under analysis has to be normally distributed.
2) The standard deviation of the normally distributed variant has to be approximately estimated.
3) The increments (steps) between testing levels should be chosen at between 0.5 and 2.0 times the standard deviation.

---

8 Dixon, W., and Massey, F., *Introduction to Statistical Analysis*, McGraw-Hill, New York, 1957.

| Stimulus Level | Test No: | 1 | 2 | 3 | 4 | 5 | 6 | 7 | 8 | 9 | 10 | 11 |
|---|---|---|---|---|---|---|---|---|---|---|---|---|
| 90 | | | | | | | X | | | X | | |
| 80 | | | | | 0 | | X | | 0 | | X | |
| 70 | | | | 0 | | | | 0 | | | | X |
| 60 | | | | 0 | | | | | | | | |
| Initial value → 50 | | | 0 | | | | | | | | | |

Arbitrary Units

Step size { 70, 60 }

Figure 7.25 Simple example for up-and-down experiment.

The increments by which the stimulus is either increased or decreased are called *steps*. For the simple up-and-down experiment shown in Figure 7.25, a constant step size of 10 is used throughout. A series of steps in one direction only is termed as a *run*. Hence, in the example given, trials 1–5 are the first run, 5–7 the second run, 7–9 the third run, and 9–11 the fourth run. The level used (50) on the very first trial is the initial value.

The up-and-down technique will be discussed in terms of the onset of galling between two moving surfaces.[9] Since galling is stimulated by the factors mentioned in Section 7.7.1, the up-and-down technique can be applied by varying one parameter and holding the others constant. Of these factors, axial travel is probably the most important, as it is essentially a direct measurement of the interference produced in the test specimen and thus the contact pressure generated. For example, the procedure can be performed for varying the axial travel (draw) exerted on the test specimen with nominal dimensions shown in Figure 7.26, at rotational constant speed and with constant materials properties.

The first step is to choose the interval for the draw. The interval must be chosen so that the galling mean draw is enveloped. In selecting the level of interference as the so-called stimulus for the sensitivity test, the rotational velocity and other factors affecting galling will be held constant.

An initial draw $d_0$ and a succession of draws $d_{+1}, d_{+2}, d_{+3}, \ldots$ above $d_0$ and $d_{-1}, d_{-2}, d_{-3}, \ldots$ below $d_0$ are chosen. The first specimen pair is run at draw $d_0$ and constant rotational speed. If the posttest inspection reveals that galling has occurred, an "x" is marked on the data sheet and the second specimen pair is tested at $d_{-1}$. If galling does not occur, "0" is marked on the data sheet and the second specimen pair is tested at $d_{+1}$. This implies that any specimen pair would be tested at a draw immediately above or below that of the previous test, depending on whether

9 Ertas, A., et al., "Experimental investigation of galling resistance in OCTG connections," *ASME Proceedings of the Twelfth Annual Energy Resources Technology Conference and Exhibition*, PD-VOL. 29, pp. 15–20, 1990 (with permission from ASME).

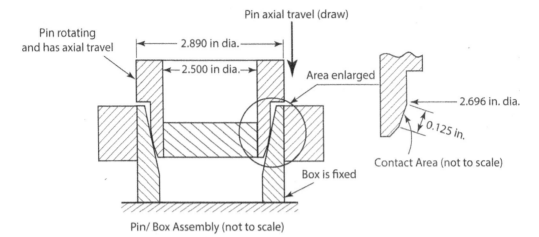

**Figure 7.26** Test specimen configuration. (From Ertas, A., and Jones, J.C., *The Engineering Design Process*, John Wiley & Sons, Inc., New York, 1996. Reproduced with permission of John Wiley & Sons, Inc.)

the specimen revealed *gall* or *no gall*. Using this technique, the mean $\mu$ and standard deviation $\hat{\sigma}$ of the population can be estimated by the sample mean $\bar{x}$ as

$$\bar{x} = D_0 + I\left(\frac{A}{N} \mp \frac{1}{2}\right),\tag{7.37}$$

where $D_0$ is the lowest test level at which the condition under evaluation was recorded, $I$ is the interval between test levels (step size), $+\frac{1}{2}$ is used if analyzing the 0s (no galling condition), $-\frac{1}{2}$ is used if analyzing the xs (galling condition), and

$$N = \sum_{i=0}^{k} n_i,\tag{7.38}$$

$$A = \sum_{i=0}^{k} i n_i,\tag{7.39}$$

where $i$ designates the number of increments above the lowest testing level at which the symbol under the analysis occurred and $n_i$ designates the frequency of occurrence of the symbols at the various levels. The standard deviation is calculated from

$$S = 1.620 \times I\left(\frac{NB - A^2}{N^2} + 0.029\right),\tag{7.40}$$

where

$$B = \sum_{i=0}^{k} i^2 n_i.\tag{7.41}$$

If $(NB - A^2)/N^2$ is less than 0.3 this formula is not valid. The standard deviation of the mean is calculated from

$$S_{\bar{x}} = \frac{6S + I}{7N^{0.5}}. \tag{7.42}$$

Equation (7.42) is valid only if $I$ is less than $3S$. For small samples, the approximate confidence limits for the population mean can be computed by

$$\bar{x} \mp t_{v,\alpha} S_{\bar{x}} \tag{7.43}$$

where $v$ is the degrees of freedom, given by

$$v = N - 1, \tag{7.44}$$

and $t$ is taken from the Student's $t$ distribution table.

## Example 7.10

Analyze the data given in Table 7.6 by the up-and-down method discussed in this section:

1) Estimate the mean.
2) Estimate the standard deviation.
3) Estimate the standard deviation of the mean.
4) Estimate the 90 percent confidence limits.

### Solution

This test series was composed of 15 tests on the L-80 tubing specimens. For this study, an initial trial at a draw of 24 mils was chosen, with subsequent trials ranging between 16 and 32 mils in steps of 4 mils. Table 7.6 presents the data as they were obtained in a sensitivity test conducted in accordance with the up-and-down test method. A "0" indicates no galling and "x" indicates that galling occurred. Since there are fewer xs than 0s, the data analysis involves only the xs in the analysis (in the event of a tie in the numbers of xs and 0s, either could be chosen). Since galling did not appear at 16 mils, data at this level were eliminated.

**Table 7.6** Up-and-down test for L-80 tubing specimens.

| Test No: → | | 1 | 2 | 3 | 4 | 5 | 6 | 7 | 8 | 9 | 10 | 11 | 12 | 13 | 14 | 15 | x | 0 |
|---|---|---|---|---|---|---|---|---|---|---|---|---|---|---|---|---|---|---|
| ↓ | 32 | | | | | | | | | | | | | | | x | 1 | 0 |
| Draw | 28 | | x | | | | | | | | | | x | | 0 | | 2 | 1 |
| (mils) | 24 | 0 | | x | | x | | | | x | | 0 | | 0 | | | 3 | 3 |
| | 20 | | | | 0 | | x | | 0 | | 0 | | | | | | 1 | 3 |
| | 16 | | | | | | | 0 | | | | | | | | | 0 | 1 |
| | | | | | | | | | | | | | | | | Total: | 7 | 8 |

**Table 7.7** Statistical calculations for L-80 tubing tests.

| Draw (mils) | $i$ | $n_i$ | $i \times n_i$ | $i^2 \times n_i$ |
|---|---|---|---|---|
| 32 | 3 | 1 | 3 | 9 |
| 28 | 2 | 2 | 4 | 8 |
| 24 | 1 | 3 | 3 | 3 |
| 20 | 0 | 1 | 0 | 0 |
| | | $N = 7$ | $A = 10$ | $B = 20$ |

The data obtained from Table 7.6 are tabulated as shown in Table 7.7. In this table, $i$ denotes the number of increments above the 16 mil level at which the symbol (x or 0) was recorded. $n_i$ denotes the frequency of occurrence of the symbols at the various levels. The column sums $N$, $A$, and $B$ are used to estimate the mean, standard deviation, and standard deviation of the mean as follows. The mean is

$$\bar{x} = D_0 + I\left(\frac{A}{N} - \frac{1}{2}\right)$$

$$= 20 + 4\left(\frac{10}{7} - \frac{1}{2}\right) = 23.71 \text{ mils.}$$

The standard deviation of the sample is

$$S = 1.620 \times I\left(\frac{NB - A^2}{N^2} + 0.029\right)$$

$$= 1.62 \times 4\left(\frac{7 \times 20 - 10^2}{7^2} + 0.029\right) = 5.48 \text{ mils.}$$

The standard deviation of the sample mean is

$$S_{\bar{x}} = \frac{6S + I}{7 \times N^{0.5}}$$

$$= \frac{6 \times 5.48 + 4}{7 \times 7^{0.5}} = 1.99 \text{ mils.}$$

Since $I = 4$ is less than $3S$, calculation of standard deviation of the sample mean using the above equation is correct. Finally, the 90 percent confidence limits are

$$\bar{x} \mp t_{v,\alpha}S_{\bar{x}} = 23.71 \mp (1.943)(1.99)$$

$$= 23.71 \mp 3.87 \Rightarrow 19.84 \text{ to } 27.58 \text{ mils.}$$

Note that for $v = 7 - 1 = 6$ and a two-sided critical region the value of the $t$ distribution is 1.943 (see Appendix C, Table C.2).

## Example 7.11

Analyze the data given in Table 7.8 by the up-and-down method.[a]

1) Estimate the mean.
2) Estimate the standard deviation.
3) Estimate the standard deviation of the mean.
4) Estimate the 90 percent confidence limits.

**Table 7.8** Up-and-down test for L-80 tubing specimens.

| | Test No: | 1 | 2 | 3 | 4 | 5 | 6 | 7 | 8 | 9 | 10 | 11 | 12 | 13 | 14 | 15 | 16 | 17 | 18 | x | 0 |
|---|---|---|---|---|---|---|---|---|---|---|---|---|---|---|---|---|---|---|---|---|---|
| | 80.5 | | | | | x | | x | | | | | | | | | | | | 2 | 0 |
| Normal | 70.5 | | | | 0 | | 0 | | x | | x | | x | | | | | | | 3 | 2 |
| contact | 60.5 | | | 0 | | | | | | 0 | | 0 | | x | | x | | | | 2 | 3 |
| stress, | 50.5 | | 0 | | | | | | | | | | | | 0 | | x | | 0 | 1 | 3 |
| kpsi | 40.5 | 0 | | | | | | | | | | | | | | | | 0 | | 0 | 2 |
| | | | | | | | | | | | | | | | | | | | Total: | 8 | 10 |

## Solution

This test series consists of 18 tests on the L-80 tubing specimens shown in Table 7.8. For this test, instead of measuring draw, normal contact strresses were measured for galling failure. A 10 kpsi stress interval with the initial trial at a nominal contact stress of 40.5 kpsi was chosen. The data obtained from Table 7.8 are tabulated as shown in Table 7.9. Table 7.8 shows that there are fewer xs than 0s, thus the data analysis involves only the xs. Since galling did not appear at the 40.5 kpsi level, data were eliminated at this level.

**Table 7.9** Statistical calculations for L-80 tubing tests.

| Nominal contact stress (kpsi) | $i$ | $n_i$ | $in_i$ | $i^2 n_i$ |
|---|---|---|---|---|
| 80.5 | 3 | 2 | 6 | 18 |
| 70.5 | 2 | 3 | 6 | 12 |
| 60.5 | 1 | 2 | 2 | 2 |
| 50.5 | 0 | 1 | 0 | 0 |
| | | $N = 8$ | $A = 14$ | $B = 32$ |

The mean is

$$\bar{x} = D_0 + I\left(\frac{A}{N} - \frac{1}{2}\right)$$

$$= 50.5 + 10\left(\frac{14}{8} - \frac{1}{2}\right) = 63 \text{ kpsi.}$$

The standard deviation of the sample is

$$S = 1.620 \times I\left(\frac{NB - A^2}{N^2} + 0.029\right)$$

$$= 1.62 \times 10\left(\frac{8 \times 32 - 14^2}{8^2} + 0.029\right) = 15.66 \text{ kpsi.}$$

The standard deviation of the sample mean is

$$S_{\bar{x}} = \frac{6S + I}{7 \times N^{0.5}}$$

$$= \frac{6 \times 15.66 + 10}{7 \times 8^{0.5}} = 5.25 \text{ kpsi.}$$

Since $I = 10$ is less than $3S$, the calculation of standard deviation of the sample mean using above equation is correct. Finally, the 90 percent confidence limits are

$$\bar{x} \mp t_{v,\alpha} S_{\bar{x}} = 63 \mp (1.895)(5.25)$$

$$= 63 \mp 9.95 \Rightarrow 53.05 \text{ to } 72.95 \text{ kpsi.}$$

Note that for $v = 8 - 1 = 7$ and the two-sided critical region the value of the $t$ distribution is 1.895 (see Appendix C, Table C.2).

Another statistical method involving a cumulative frequency distribution was also used to compare the results of the up-and-down technique.[9] Four series of tests were conducted, with 15 specimens tested in each series. These test series were conducted at nominal contact stresses of 51.5, 61.5, 71.5, and 81.5 kpsi, with an interval of 10 kpsi. Figure 7.27 shows the results of these test series. Evidently, the straight line fits the data reasonably well, indicating that the galling failure has a normal (Gaussian) distribution. As seen from the figure, 50 percent of the specimens galled at a contact stress of about 67 kpsi. We know from normal distribution theory that approximately 68 percent of the distribution lies within $\mu \mp \hat{\sigma}$. The nominal contact stress corresponding to the 16 percent failed level is about 47 kpsi. Subtracting this value from the mean stress of 67 kpsi results in a standard deviation of approximately 20 kpsi.

A comparison of the up-and-down test and frequency distribution results is presented in Table 7.10. This table indicates that the up-and-down test predicts the mean to within 6 percent of that of the frequency distribution. The standard deviation of the sample, as predicted by the up-and-down method, differs by approximately 22 percent from the frequency distribution. The up-and-down technique can thus be used in predicting the mean, although it predicts the standard deviation less accurately.

**Figure 7.27** Frequency distribution for galling failure of L-80 specimens. (From Ertas, A., and Jones, J.C., *The Engineering Design Process*, John Wiley & Sons, Inc., New York, 1996. Reproduced with permission of John Wiley & Sons, Inc.)

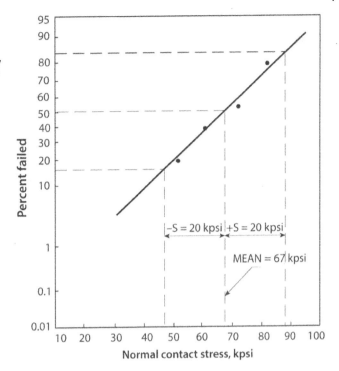

**Table 7.10** Comparison between results of the up-and-down test and frequency distribution.

| Quantity | Up-and-down prediction (kpsi) | Frequency distribution (kpsi) | Percent difference |
|---|---|---|---|
| Mean nominal contact stress | 63 | 67.0 | 6% |
| Standard deviation | 15.66 | 20.0 | 22% |

As indicated before, the interval between testing levels should be set between 0.5 and 2.0 times the standard deviation. The assumed interval (10 kpsi) is 0.5 times the calculated standard deviation of 20 kpsi; therefore, the experiment with the up-and-down method satisfies this requirement. The up-and-down method is an economical and valuable method for predicting the mean value. It allows the researchers to carry out fast inclusive parametric studies for the design of experiments.

[a]From Ertas, A., and Jones, J.C., *The Engineering Design Process*, 2nd edn, John Wiley & Sons, Inc., New York, 1996 (adapted with permission).

## 7.8 Taguchi Methods

The focus of Taguchi methods is on designing and manufacturing quality products and the emphasis is on losses or costs. In this section, Taguchi's loss function and orthogonal or factorial experiment will be discussed.

### 7.8.1 Taguchi's Loss Function

The Taguchi quality loss function is a statistical method developed by Genichi Taguchi to improve the quality of products. The use of tolerances in the specification of parameters in design and testing and in the requirement of dimensions on drawings was first initiated about 1870, with the introduction of *go, no-go* tolerance limits. There was a need for the application of tolerances because manufacturers could not make identical quality products – a phenomenon now referred to as variability. Moreover, it was not necessary that all products be exactly identical and trying to make them so was too costly. As shown in Figure 7.28, the initiation of the *go, no-go* gage about this same time offered a simple device for the production technician to check his work quickly, resulting in reduced production costs.[10] Unfortunately, this led to an outlook that

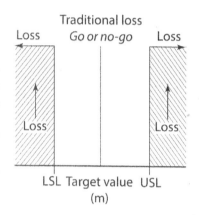

**Figure 7.28** Go or no-go (traditional loss).

has returned to trouble us during the latter part of this century: do not waste time trying to be exact.

In today's environment it is desirable to designate target values for specified parameters to which processes are controlled – to control the costs of quality variation with the products deviation on either side of the mean. Deviations from the target values result in additional costs to society (both producer and consumer).

In the early 1980s, Genichi Taguchi made the following statement on product quality: " Quality is the functional loss to society after the article is shipped." He also stated that: " A product or service has good quality if it performs its intended functions without variability, and causes little loss through harmful side effects, including the cost of using it."

The main aim of the loss function is to improve process and product design through the identification of controllable factors and their settings, which minimize the variation of a product quality around the target value. In this concept, minimizing the *loss to society* is the strategy that results in uniform product quality and reduces costs at both the production stage and at the point of consumption.[11]

---

10 Shewhart, W.A., *Statistical Method, From the Viewpoint of Quality Control*, Department of Agriculture, Washington, DC, 1939.
11 Ross, P.J., *Taguchi Techniques for Quality Engineering*, McGraw-Hill, New York, 1988.

Generally, there are three types of loss function: nominal is best; larger is better; and smaller is better. Taguchi introduced the following quadratic loss function for a nominal-is-best situation:[12]

$$L(Y) = k(Y - m)^2, \tag{7.45}$$

where $L$ is the loss associated with a particular quality characteristic $Y$, $m$ is the performance target value, and $Y - m$ is the deviation from target value.

The loss, $A_o$, to the customer when the performance characteristic is not within the customer tolerance limit, $\Delta_o$, and be found by substituting $Y = m + \Delta_o$ into equation (7.45):

$$A_o = k(y - m)^2 = k[(m + \Delta_o) - m]^2. \tag{7.46}$$

From equation (7.46), the constant $k$ can be derived as

$$k = \frac{A_o}{\Delta_o^2}. \tag{7.47}$$

If the product is rejected when it is not within the manufacturing tolerance limit, $\Delta$, this results in a cost to the manufacturer. If the manufacturer's cost, $A$, is known, the manufacturing tolerance, $\Delta$, can be determined by

$$\Delta = \sqrt{\frac{A}{A_o}} \times \Delta_o. \tag{7.48}$$

When there is more than one piece of product, the loss function takes the form

$$L(Y) = k(MSD) \tag{7.49}$$

where the mean squared deviation from the target value is given by

$$MSD = (Y - m)^2. \tag{7.50}$$

Using equation (7.45), the loss function for the nominal-is-best case can be modified for a sample population to calculate the average loss function, $\overline{L}$, as

$$\overline{L}(Y) = k\frac{\sum_{i=1}^{N}(Y - m)^2}{N}, \tag{7.51}$$

where $N$ is the total number of observations. Equation (7.51) can also be modified into the form

$$\overline{L}(Y) = kS^2 + k(\overline{Y} - m)^2, \tag{7.52}$$

where $\overline{Y}$ and $S^2$ are respectively the mean and variance of the measurements of the quality characteristic, and $m$ is the nominal value (target) of the product (or process).

Equation (7.52) is made up of two parts: the first part represents the loss due to variability and the second part represents the loss due to the off-target value. To minimize the loss to society the product characteristic needs to be centered at the nominal value ($\overline{Y} = m$). In a manufacturing operation, this is the responsibility of production personnel. The variance of the characteristic ($S^2$) must also be reduced for minimum loss. The variance should be established

12 Taguchi, G., *Introduction to Quality Engineering*, Kraus International Publications, New York, 1987.

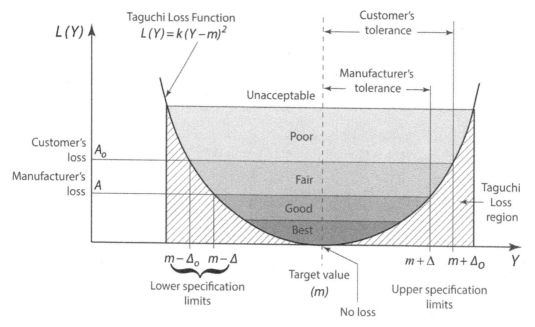

**Figure 7.29** Taguchi loss function.

before production is initiated by design and process engineering and should be reduced as production continues.[13]

Taguchi proposed that the loss is proportional to the square of the distance from the target value, *m*. As shown in Figure 7.29, the parabolic curve illustrates the cost to society (customer and manufacturer) as the product moves away from the target value, *m*. The cost of this variation increases quadratically as the quality moves farther from the target value. If there is little or no manufacturing loss then the target value of *m* coincides with the parabolic curve as shown in Figure 7.29. As the variation away from the target value increases, the customer will gradually become dissatisfied – product quality decreases.

In the smaller-is-better characteristic case, the ideal target value is defined as zero (see Figure 7.30). The loss function for this case is

$$L = L(Y) = kY^2. \tag{7.53}$$

As shown in Figure 7.31, for the larger-is-better case the most desirable value of the characteristic, *Y*, is infinity. As an example, for heat exchanger efficiency the highest value would be the best.

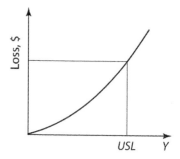

**Figure 7.30** Loss function for smaller-is-better case.

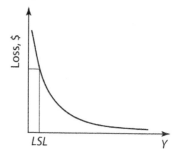

**Figure 7.31** Loss function for larger-is-better case.

13 Ertas, A., and Jones, J.C., *The Engineering Design Process*, 2nd edn, John Wiley & Sons, Inc., New York, 1996

In this case, the characteristic $Y$ has no target value except *as high as possible*. The loss function for this case is given by

$$L = L(Y) = k\frac{1}{Y^2}. \tag{7.54}$$

The constant, $k$, is determined for the above cases as for the nominal-is-best case, based on the loss associated with a particular value of $Y$. The average loss functions for the lower-is-better and higher-is-better cases, respectively, can also be written as

$$L(Y) = kS^2 + k\overline{Y}^2 \quad \text{(lower is better),} \tag{7.55}$$

$$L(Y) = k(3S^2/\overline{Y}^2) + k\frac{1}{\overline{Y}^2} \quad \text{(higher is better).} \tag{7.56}$$

## Example 7.12

A diesel engine is an internal combustion engine that uses the heat of compression to initiate ignition and burn the fuel that has been injected into the combustion chamber. The design of the diesel fuel injector nozzle is crucial to the performance of a diesel engines. Figure 7.32 shows a typical diesel engine enjector.

Table 7.11 shows the diameter of eight injector nozzle holes manufactured by a company. Assuming that the customer loss, $A_o$, is $30 when a injector nozzle is outside the customer's tolerance limit and that the tolerance limit, $\Delta_o$, equals $\mp 0.03$, determine the following:

a) Calculate the manufacturer's lower and upper tolerance limits when the average manufacturing cost due to scrapping or reworking a nozzle hole, $A$, is $14.

b) Estimate the loss parameter, $k$.

c) Calculate the average loss.

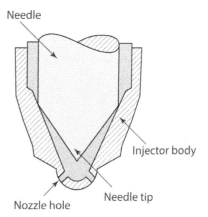

**Figure 7.32** Typical diesel engine injector.

**Table 7.11** Diesel engine injector data.

| Injector | Diameter ($Y$, mm) | Deviation from target ($m = 0.30$ mm) ($Y - m$) |
|---|---|---|
| 1 | 0.34 | 0.04 |
| 2 | 0.29 | −0.01 |
| 3 | 0.34 | 0.04 |
| 4 | 0.29 | −0.01 |
| 5 | 0.36 | 0.06 |
| 6 | 0.32 | 0.02 |
| 7 | 0.33 | 0.03 |
| 8 | 0.35 | 0.05 |

## Solution

This problem is a nominal-is-best case.

a) The manufacturer's tolerance is

$$\Delta = \sqrt{\frac{A}{A_o}} \times \Delta_o$$

$$= \sqrt{\frac{14}{30}} \times 0.03 = 0.0205.$$

The manufacturer's lower and upper limits are

$$LSL = m - \Delta = 0.30 - 0.03 = 0.27 \text{ mm},$$
$$USL = m + \Delta = 0.30 + 0.03 = 0.33 \text{ mm}.$$

b) The loss parameter, $k$, is

$$k = \frac{A_o}{\Delta_o^2} = \frac{30}{(0.03)^2} = 33{,}333.$$

c) The average loss is

$$\bar{L} = k \sum_{i=1}^{N} \frac{(y - m)^2}{N} = k(MSD),$$

where the mean squares deviation is

$$MSD = \frac{1}{N} \sum_{i=1}^{N} (y - m)^2 = \frac{1}{8}[0.04^2 + (-0.01)^2 + \cdots + 0.05^2]$$

$$= \frac{1}{8} \times 0.0108 = 0.00135.$$

Then

$$\bar{L} = k \sum_{i=1}^{N} \frac{(y - m)^2}{N} = 33{,}333 \times 0.00135 = \$45 \text{ per injector}.$$

Now assume that the upper and lower specification limits ($USL, LSL$), set by the customer, are as shown in Figure 7.33. Assume also the two distribution curves represent output from a process. As seen from this figure, as loss increases the output curve becomes wider (dashed line). This curve has its mean output on target and stays within the specification limit set by customers. The narrow (solid line) curve's mean is also on target, but is covering approximately 50 percent of the specified tolerance limits. The majority of the output is closer to the target value of $m$ – conceptual Six Sigma.

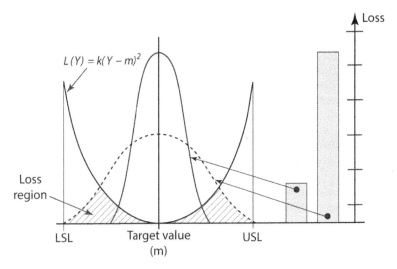

**Figure 7.33** Variation of Taguchi Loss Function.

*Six Sigma* is a data-driven methodology to measure quality of services and processes that strives for near perfection. The statistical representation of Six Sigma defines quantitatively how a process is performing. The Six Sigma process has at least six standard deviations of spread between the mean and the nearest specification limit, which predicts 3.4 defects per million opportunities. This subject is discussed in detailed in Chapter 5.

### 7.8.2 Taguchi Technique (Orthogonal or Factorial Experiment)

As discussed previously, the up-and-down technique is applied by varying one parameter and holding the others constant. However, to repeat this procedure for all the parameters involved in the experimental design is time-consuming and costly.

The Taguchi technique is a robust design method that significantly improves engineering productivity. The Taguchi method allows for the measurement of a number of different parameters and their interactions at the same time. When this method is used, a high-quality product at lower cost can be achieved.[14] In this section the analysis of variance (ANOVA) and some new terminologies commonly used in the design of experiments will be introduced.

ANOVA is the statistical method used to analyze data obtained from an experiment to see if there is any difference between groups on some variable. This important technique is called ANOVA because it takes into account the analysis of variation among the statistical measures, such as the mean. Several cases of ANOVA will be discussed in this section.[15]

---

14  Ross, P.J., *Taguchi Technique for Quality Engineering*, McGraw-Hill, New York, 1988.
15  Ryan, T.P., *Statistical Methods for Quality Measurements*, John Wiley & Sons, New York, 1989.

### 7.8.2.1   No-Way ANOVA

Suppose that the data on the weight of a 30% solution of calcium chloride at 25°C shown in Table 7.12 have been collected over a period of one week. A no-way ANOVA considers the total variation of the data in two categories, as follows:

1) The variation of the mean of the data relative to zero.
2) The variation of individual data points about the mean (experimental error).

The mean of all the observations, $Y_i$, given in Table 7.12 is

$$\overline{Y} = \frac{T}{N} = \frac{1.29 + 1.31 + \cdots + 1.27}{7} = \frac{9.77}{7} = 1.40 \text{ kg/liter,} \tag{7.57}$$

where $T$ is the sum of all the observations (grand total) and $N$ is the total number of observations. No-way ANOVA of the data shown in Table 7.12 is graphically illustrated in Figure 7.34. The total sum of the squares (*total variation*), $SS_T$, is the sum of the square of each observation $Y_i$ given by

$$SS_T = \sum_{i}^{N} Y_i^2 \tag{7.58}$$

$$= 1.29^2 + 1.31^2 + \cdots + 1.27^2 = 13.89.$$

Substituting values from Table 7.12, the sum of the squares due to the mean is

$$SS_m = N\left(\frac{T}{N}\right)^2 = \frac{T^2}{N} \tag{7.59}$$

$$= \frac{(9.77)^2}{7} \approx 13.64.$$

**Table 7.12**  Weight of CaCl$_2$ solution at 30% concentration and 25°C.

| Day | Weight (kg/liter) observations, $Y_i$ |
| --- | --- |
| M | 1.29 |
| T | 1.31 |
| W | 1.27 |
| Th | 1.67 |
| F | 1.24 |
| Sat | 1.72 |
| Sun | 1.27 |

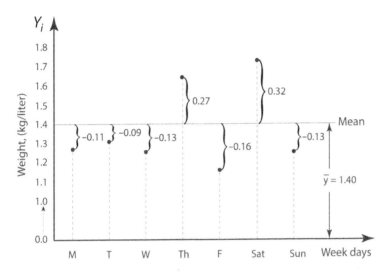

**Figure 7.34** Variation of individual data points about the mean.

In equation (7.59), $SS_m$ provides the variation due to the mean of the data points relative to zero. Variation relative to mean$(Y_i - \overline{Y})$ of each data point is squared and summed to obtain the error sum of squares:

$$SS_E = \sum_{i}^{N} (Y_i - \overline{Y})^2 \tag{7.60}$$
$$= (-0.11)^2 + (-0.09)^2 + (-0.13)^2 + \cdots + (-0.13)^2 \approx 0.25.$$

The error sum of squares can also be calculated by subtraction:

$$SS_E = SS_T - SS_m \tag{7.61}$$
$$= 13.89 - 13.64 = 0.25.$$

The total degrees of freedom $v_T$ for a no-way ANOVA is equal to the total number of observations, $N$. Note that for this case the degrees of freedom of the mean, $v_m$, are always equal to 1; then the degrees of freedom associated with the error $v_E$ can be determined from

$$v_E = v_T - v_m \tag{7.62}$$
$$= 7 - 1 = 6.$$

The mean squared variance, $V_m$, and error variance, $V_E$, can be calculated by dividing the sum of their squares by the respective degrees of freedom:

$$V_m = \frac{SS_m}{v_m} = \frac{13.64}{1} = 13.64, \tag{7.63}$$

$$V_E = \frac{SS_E}{v_E} = \frac{0.25}{6} = 0.042. \tag{7.64}$$

This variation calculation is identical to

$$S^2 = V = \frac{\sum_i^N (Y_i - \overline{Y})^2}{N - 1}. \tag{7.65}$$

Note that, in equation (7.65), $N - 1$ is equal to the degrees of freedom, $v_E$.

### 7.8.3 One-Way ANOVA

One-way (one-factor) ANOVA is used to test hypotheses regarding the equality of three or more treatments. The basis of ANOVA is the partitioning of the sum of squares, into a between-treatments sum of squares, $SS_{between}$, and a within-treatments sum of squares, $SS_{within}$. One-way ANOVA can be used to compare the observations with each other simultaneously rather than individually. In this analysis, the following assumptions are used:

1) The populations from which the samples are obtained must be approximately normally distributed.
2) The samples must be independent.
3) Groups should have approximately equal variance on the dependent variable.

ANOVA is a relatively robust procedure with respect to violations of the normality assumption. Assumptions 2 and 3 are satisfied when large sample sizes are used. Assumption 3 is satisfied when the sample sizes are roughly the same for each group. This is true when even small sample sizes are used. In this study, all the sample sizes are same and the sample sizes are reasonably large. For this case the total sum of squares, $SS_T$, is given by

$$SS_T = SS_Y - C, \tag{7.66}$$

where

$$SS_Y = \sum_{i=1}^N Y_i^2 \tag{7.67}$$

$$C = \frac{T^2}{N}, \tag{7.68}$$

in which $T$ is the sum of all observations (grand total), and $N$ is the total number of observations, $SS_Y$ is the sum of the squares of each observation $Y$, and $C$ is the correction factor equal to the variation due to the mean, $SS_m$.

**Variation between Treatments**

The sum of squares between treatments, $SS_{between}$, is the variation due to the interaction between the sample treatments and is given by

$$SS_{between} = \frac{L_1^2 + L_2^2 + L_3^2 + L_4^2}{n} - C, \tag{7.69}$$

where $L$ is the sum of observations in each treatment and $n$ is the sample size in each treatment. The between-treatment mean square, $MS_{\text{between}}$, is defined by

$$MS_{\text{between}} = \frac{SS_{\text{between}}}{v_{\text{between}}}, \tag{7.70}$$

where $v_{\text{between}}$ is the degrees of freedom for $SS_{\text{between}}$, given by

$$v_{\text{between}} = \text{number of groups} - 1. \tag{7.71}$$

**Variation within Treatments**

The sum of squares within treatments, $SS_{\text{within}}$, is the variation due to the interaction between the sample treatments and is given by

$$SS_{\text{within}} = SS_T - SS_{\text{between}}. \tag{7.72}$$

The within-treatment mean square, $MS_{\text{within}}$, is

$$MS_{\text{witin}} = \frac{SS_{\text{witin}}}{v_{\text{within}}}, \tag{7.73}$$

where $v_{\text{within}}$ is the degrees of freedom for $SS_{\text{within}}$, given by

$$v_{\text{within}} = (n - 1)k, \tag{7.74}$$

in which $k$ is the number of treatments under consideration and $n$ is the number of samples in each group. It is now possible to evaluate the null hypothesis using an $F$ test calculated by

$$F = \frac{MS_{\text{between}}}{MS_{\text{within}}}. \tag{7.75}$$

It should be noted that if $F \ll 1$ then it is likely that differences between treatment means exist. To test the hypothesis, the $F$ value calculated by the above equation is compared with the critical value of $F$. The critical value, $F_{\text{cr}}$, is determined from statistical tables using the degrees of freedom between treatments and the degrees of freedom within treatment values. If the value of $F$ is greater than the value of $F_{\text{cr}}$ then the probability of the obtained result occurring due to chance is low and we reject the null hypothesis. In the case where there are more than two treatments of independent variables, statistical analysis should be carried out in two steps:

*Step 1*. Perform the $F$ test to determine if any significant differences exist among any of the means. If the $F$ test value shows statistical significance, then we carry out a second step.

*Step 2*. In the second step, a post hoc analysis should be performed to show where the inequalities are. A post hoc test is used when we have three or more means to compare. This test provides us the critical difference between all possible pairs of means. For this study, *Fisher's protected t-test* will be used. The formula is

$$F_{\text{compare}} = \frac{(M_i - M_j)^2}{MS_{\text{within}}(1/n_i + 1/n_j)}, \tag{7.76}$$

where $i$ and $j$ are the treatments being compared, and $n_i$ and $M_i$ are the number of observations and the mean of treatment $i$, respectively. The test statistics will be performed for each pair of means by using the values of $F_{compare}$ and $F_{cr}$. Note that, for the application of the protected $t$-test, $F_{cr}$ is found by using $v = 1$ for the numerator and $v_{within}$ for the denominator.

## Example 7.13

Considering three treatments which will be explained below, evaluate the survey results of the transdisciplinary research process given in Table 7.13.

**Solution**

Assume the following objectives of transdisciplinary research:

- To improve the speed and quality of research for new discoveries.
- To collaborate and interact in new and more integrated ways to confront the engineering and social challenges of the future.

Using the first and second objectives along with three treatments A, B and C, the first and second questions can be written as follows:

*Question #1* Please rank the items from 1 to 5 according to what is most important when looking "to improve the speed and quality of research for new discoveries." Place a 1 next to the item that is least important and place a 5 next to the item that is most important.

- — Cross-communication and collaboration
- — Use of shared concepts and methods (knowledge integration from different disciplines)
- — Appropriate research team composed of researchers from diverse disciplines

*Question #2* Please rank the items from 1 to 5 according to what is most important when looking "to collaborate and interact in new and more integrated ways to confront the engineering and social changes of the future." Place a 1 next to the item that is least important and place a 5 next to the item that is most important.

- — Cross-communication and collaboration
- — Use of shared concepts and methods (knowledge integration from different disciplines)
- — Appropriate research team composed of researchers from diverse disciplines

Now consider question #1. Suppose we want to test the efficacy of treatments A, B, and C to improve the speed and quality of research for new discoveries. In other words, we are interested in whether treatment A will improve the speed and quality of research for new discoveries more than treatment B or treatment C. Our null and alternative hypotheses are as follows:

$H_0$ : There are no significant differences between the treatments' mean scores.
$H_1$ : There is a significant difference between the treatments' mean scores.

For this case, we have 28 samples for each of the three treatments with the survey results shown in Table 7.13. The mean level of *to improve the speed and quality of research for new discoveries* is the dependent variable and treatments A, B, and C are independent variables. The results of one-way ANOVA are as follows.

**Table 7.13** Survey result for three treatments A, B and C (for questions #1, #2).

| Q #1 | | | | Q #2 | | | |
|---|---|---|---|---|---|---|---|
| Samples | A | B | C | Samples | A | B | C |
| 1 | 4 | 5 | 3 | 1 | 4 | 3 | 5 |
| 2 | 4 | 5 | 4 | 2 | 5 | 5 | 4 |
| 3 | 4 | 3 | 5 | 3 | 5 | 5 | 3 |
| 4 | 3 | 3 | 4 | 4 | 4 | 4 | 4 |
| 5 | 1 | 5 | 3 | 5 | 5 | 1 | 3 |
| 6 | 5 | 4 | 5 | 6 | 5 | 4 | 5 |
| 7 | 4 | 4 | 3 | 7 | 5 | 5 | 4 |
| 8 | 4 | 4 | 4 | 8 | 5 | 4 | 5 |
| 9 | 5 | 2 | 3 | 9 | 5 | 3 | 4 |
| 10 | 4 | 3 | 5 | 10 | 4 | 3 | 5 |
| 11 | 3 | 4 | 3 | 11 | 4 | 4 | 4 |
| 12 | 5 | 4 | 4 | 12 | 5 | 3 | 4 |
| 13 | 4 | 5 | 3 | 13 | 4 | 3 | 5 |
| 14 | 5 | 4 | 5 | 14 | 5 | 3 | 5 |
| 15 | 4 | 4 | 4 | 15 | 4 | 4 | 4 |
| 16 | 4 | 4 | 3 | 16 | 5 | 5 | 4 |
| 17 | 5 | 4 | 4 | 17 | 5 | 4 | 5 |
| 18 | 5 | 3 | 5 | 18 | 5 | 4 | 5 |
| 19 | 5 | 4 | 4 | 19 | 5 | 5 | 4 |
| 20 | 3 | 3 | 3 | 20 | 5 | 5 | 5 |
| 21 | 5 | 4 | 5 | 21 | 5 | 3 | 4 |
| 22 | 4 | 3 | 5 | 22 | 3 | 4 | 5 |
| 23 | 3 | 4 | 5 | 23 | 5 | 5 | 5 |
| 24 | 5 | 3 | 5 | 24 | 5 | 3 | 5 |
| 25 | 5 | 4 | 3 | 25 | 5 | 3 | 5 |
| 26 | 4 | 3 | 5 | 26 | 4 | 3 | 5 |
| 27 | 4 | 3 | 5 | 27 | 4 | 5 | 3 |
| 28 | 2 | 4 | 3 | 28 | 4 | 5 | 3 |
| $\sum_{i=1}^{N} A_i = 113$ | $\sum_{i=1}^{N} B_i = 105$ | $\sum_{i=1}^{N} C_i = 113$ | | $\sum_{i=1}^{N} A_i = 129$ | $\sum_{i=1}^{N} B_i = 108$ | $\sum_{i=1}^{N} C_i = 122$ | |
| $\sum_{i=1}^{N} A_i^2 = 483$ | $\sum_{i=1}^{N} B_i^2 = 409$ | $\sum_{i=1}^{N} C_i^2 = 477$ | | $\sum_{i=1}^{N} A_i^2 = 603$ | $\sum_{i=1}^{N} B_i^2 = 444$ | $\sum_{i=1}^{N} C_i^2 = 546$ | |
| $\bar{x}_1 = 4.04$ | $\bar{x}_2 = 3.75$ | $\bar{x}_3 = 4.04$ | | $\bar{x}_1 = 4.61$ | $\bar{x}_2 = 3.86$ | $\bar{x}_3 = 4.36$ | |
| $S_1 = 1.0$ | $S_2 = 0.75$ | $S_3 = 0.88$ | | $S_1 = 0.57$ | $S_2 = 1.01$ | $S_3 = 0.73$ | |

We begin by calculating the total sum of squares:

$$SS_T = SS_Y - C,$$

where

$$SS_Y = \sum_{i=1}^{N} Y_i^2 = \sum_{i=1}^{N} A_i^2 + \sum_{i=1}^{N} B_i^2 + \sum_{i=1}^{N} C_i^2$$

$$= 483 + 409 + 477 = 1,369.$$

We have

$$C = \frac{T^2}{N},$$

where

$$T^2 = (A_i + B_i + C_i)^2$$

$$= (113 + 105 + 113)^2 = 109,561,$$

and the total number of observations is $N = 28 \times 3 = 84$. Thus

$$C = \frac{109,561}{84} = 1,304.30.$$

Then

$$SS_T = 1369 - 1304.30 = 64.70.$$

The sum of squares between treatments, $SS_{between}$, is

$$SS_{between} = \frac{L_1^2 + L_2^2 + L_3^2 + L_4^2}{n} - C$$

$$= \frac{113^2 + 105^2 + 113^2}{28} - 1304.30 = 1.52.$$

The sum of squares within treatments, $SS_{within}$, is

$$SS_{within} = SS_T - SS_{between}$$

$$= 64.70 - 1.52 = 63.18.$$

The degrees of freedom associated with each sum of squares are:

$$\nu_{between} = \text{number of groups} - 1 = 3 - 1 = 2,$$

$$\nu_{within} = (n-1)k,$$

where $k = 3$ is the number of treatments under consideration and $n = 28$ is the number of samples in each group. Substituting in the above equation yields

$$\nu_{within} = (28 - 1)3 = 81.$$

The total degrees of freedom are therefore

$$\nu_T = \nu_{between} + \nu_{within} = 2 + 81 = 83.$$

The mean square between treatments, $MS_{between}$, is

$$MS_{between} = \frac{SS_{between}}{\nu_{between}}$$

$$= \frac{1.52}{2} = 0.76.$$

The mean square within treatments, $MS_{within}$, is

$$MS_{witin} = \frac{SS_{witin}}{\nu_{within}}$$

$$= \frac{63.18}{81} = 0.78.$$

It is now possible to evaluate the null hypothesis using an $F$ test:

$$F = \frac{MS_{between}}{MS_{within}}$$

$$= \frac{0.76}{0.78} = 0.98.$$

**Table 7.14** Summary of ANOVA results for question #1 (group: researchers).

| SOURCE | SS | df | MS | F |
|---|---|---|---|---|
| Between | 1.52 | 2 | 0.76 | 0.98 |
| Within | 63.18 | 81 | 0.78 | |
| Total | 64.70 | 83 | | |

The analysis results for question #1 are summarized in Table 7.14. Assume that the confidence level is 95%. Then, for a two-tailed critical region with $\alpha/2$ set at 0.025, and with 2 degrees of freedom for $MS_{between}$ and 81 for $MS_{within}$, the critical value of $F_{cr}$ can be found from the $F$ table as approximately 3.88 (see Appendix C, Table C.4). Since the calculated value, $F = 0.98$, is smaller than the value $F_{cr} = 3.88$, we accept the null. This means that there are no significant differences in the three treatments *to improve the speed and quality of research for new discoveries.* They are equally important.

**Table 7.15** Summary of ANOVA results for question #2 (Group: Researchers).

| SOURCE | SS | df | MS | F |
|--------|-------|----|------|------|
| Between | 8.17 | 2 | 4.08 | 6.54 |
| Within | 50.54 | 81 | 0.62 | |
| Total | 58.70 | 83 | | |

Now let us consider question #2. The question #2 ANOVA results for research group are given in Table 7.15. Since the degrees of freedom are the same as question #1, the critical value of $F$ remains the same, $F_{cr} = 3.88$. Because the calculated value of $F$, 6.54, is greater than the critical value of $F_{cr}$, for question #2, one must reject the null and conclude that there are significant differences in the three treatments *to collaborate and interact in new and more integrated ways to confront the engineering and social changes of the future*. In summary, there are no significant differences in the three treatments for the first transdisciplinary research objective, whereas for the second objective there are significant differences in all three treatments. The calculated $F$ value obtained by ANOVA tells us that the means are not equal.

To know exactly which means are significantly different from each other, we use *post hoc* tests. There are several types of post hoc tests based on different assumptions and for different purposes. For this example problem, *Fisher's protected t test* will be used. With three treatments (treatments A, B and C) three comparisons between pairs of means are possible (A vs. B, A vs. C, and B vs. C). The calculated Fisher's protected $t$ test values are:

$$F_{A \times B} = \frac{(M_A - M_B)^2}{MS_{within}(1/n_A + 1/n_B)} = \frac{(4.61 - 3.86)^2}{0.62(1/28 + 1/28)} = 12.70,$$

$$F_{A \times C} = \frac{(M_A - M_C)^2}{MS_{within}(1/n_A + 1/n_C)} = \frac{(4.61 - 4.36)^2}{0.62(1/28 + 1/28)} = 1.41,$$

$$F_{B \times C} = \frac{(M_B - M_C)^2}{MS_{within}(1/n_B + 1/n_C)} = \frac{(3.86 - 4.36)^2}{0.62(1/28 + 1/28)} = 5.65.$$

Comparing these results with the critical value $F_{cr} = 5.24$ (note that to find the critical value for this case the degrees of freedom are 1 for the numerator and 81 for the denominator), one can conclude that significant difference lies between treatments A and B also between B and C. Because B appears in both pairs, we may further assert that it is the most dominant factor for this case.

## 7.8.4  Two-Way ANOVA

Two-way ANOVA is also known as a factorial ANOVA, with two factors. In experimental design, the parameters that may be varied from trial to trial for an experiment are generally called *factors*, and the value of the factors used in the experiment are referred to as *levels*. A two-way ANOVA is useful when we want to compare the effect of several levels of two factors and we have multiple observations at each level.

In Table 7.16, the two-way ANOVA is introduced to evaluate the effect of two factors, namely *temperature* and *concentration*, on the density of the $CaCl_2$ solution with factor concentration having two levels and temperature having three levels.

In general, the computation of two-way ANOVA problems can be summarized as shown in Table 7.17. For a two-level experiment, when sample sizes are equal, the following formulas are used.

1) The total sum of squares, $SS_T$, is

$$SS_T = SS_Y - C, \tag{7.77}$$

where $SS_Y$ is the sum of the squares of each observation $Y$, and $C$ is the correction factor given by

$$C = \frac{T^2}{N}, \tag{7.78}$$

in which $T$ is the sum of all observations (grand total), and $N$ is the total number of observations.

**Table 7.16** Density of $CaCl_2$ solution (lb/gal).

| Temperature, °F | Concentration, Weight % | |
| --- | --- | --- |
| | 25% | 30% |
| 60 | 10.35 | 11.33 |
| 80 | 10.29 | 11.27 |
| 100 | 10.24 | 11.22 |

**Table 7.17** Complete two-way ANOVA summary for two factorial experiments.

| Source | Degree of freedom ($v$) | Sum of squares ($SS_i$) | Mean squares ($MS_i$) | F test $F_i$ | Percent of variance $PC_i$ |
| --- | --- | --- | --- | --- | --- |
| Factor A | | | | | |
| Factor B | | | | | |
| Interaction between A and B | | | | | |
| Error | | | | | |

2) The error sum of squares, $SS_E$, and the sum of squares $SS_i$ for each two-level factor are calculated using

$$SS_E = SS_T - \left( \sum_{i=1}^{k} SS_i \right), \tag{7.79}$$

$$SS_i = \sum_{i}^{k} \left( \frac{L_i^2}{n_{Li}} - \frac{T^2}{N} \right) = \frac{L_1^2 + L_2^2 + L_3^2 + L_4^2}{n} - C, \tag{7.80}$$

where $k$ is the total number of factors (including interactions). The sum of squares $SS_i$ for each two-level factor can also be calculated (for linear graphs only) using

$$SS_i = \frac{N}{4} \times MEF^2, \tag{7.81}$$

where $MEF$ is the main effect of a factor and can be obtained from

$$MEF = \text{(average of observations at high level)}$$
$$- \text{(average of observations at low level).} \tag{7.82}$$

3) The mean square for each factor $MS_i$ and mean square error $MS_E$ are calculated by

$$MS_i = \frac{SS_i}{v_i}, \tag{7.83}$$

$$MS_E = \frac{SS_E}{v_{error}}, \tag{7.84}$$

where $v$ is the degrees of freedom, calculated using

$$v_T = N - 1,$$
$$v_i = \text{number of levels} - 1,$$
$$v_{error} = (N - 1) - \sum_{i=1}^{k} v_i,$$

where $v_T$, $v_e$, $v_i$, are respectively the degrees of freedom for the total, for error, and for each factor, and and $N$ is the number of rows in the orthogonal array.

4) $F$ is the ratio of sample variances that provides a decision as to whether there are significant differences in the estimates at known confidence levels. This useful tool is calculated by

$$F_i = \frac{MS_i}{MS_E}. \tag{7.85}$$

This ratio is compared with $F$ table values, and if $F_i < F_{\text{Table}}$, it is concluded that the effect of factor $i$ under consideration is not significant. In the case of no-way ANOVA, $MS_i$ and $MS_E$ in equation (7.85) are replaced by $V_m$ and $V_E$, respectively.

5) Percent of variance: The percentage contribution of factor $i$ to the total variance is determined as

$$PC_i = \frac{100(MS_i - MS_E)}{SS_T}. \tag{7.86}$$

The percent contribution to error is

$$PC_E = 100 - \left(\sum_{i=1}^{k} PC_i\right). \tag{7.87}$$

Obviously, the total percent contribution should add up to 100 percent.

## Example 7.14

Considering Example 7.12, Investigate whether the manufacturing process is on target with 95 percent confidence. If the process is not on target, calculate the savings from centering the process on target.

### Solution

Using values $k = 33,333$, $\bar{L} = \$45$, and $MSD = 0.00135$ from Example 7.12, to investigate whether the manufacturing process is on target, the sum of the squares of the mean deviation from the target $(y - m)$ should be calculated. We substitute $T = (y - m)$ into

$$SS_m = \frac{T^2}{N}.$$

Using Table 7.11,

$$T = \sum_{i=1}^{N}(y - m) = 0.04 + (-0.01) + \cdots + 0.05 = 0.22.$$

The mean sum of squares is

$$SS_m = \frac{(0.22)^2}{8} = 0.00605.$$

The total sum of squares is

$$SS_T = \sum_{i=1}^{N} Y_i^2 = \sum_{i=1}^{N} (y - m)^2$$

$$= 0.04^2 + (-0.01)^2 + \cdots + 0.05^2 = 0.0108.$$

The error sum of squares is

$$SS_E = SS_T - SS_m = 0.0108 - 0.00605 = 0.00475.$$

The mean squared variance is

$$V_m = \frac{SS_m}{v_m} = \frac{0.00605}{1} = 0.00605.$$

The error variance is

$$V_E = \frac{SS_E}{v_E} = \frac{0.00475}{7} = 0.000679.$$

Note that $v_E = N - 1 = 8 - 1 = 7$.

We now perform the $F$ test:

$$F = \frac{V_m}{V_E} = \frac{0.00605}{0.000679} = 8.91.$$

The percent contributions are

$$PC_m = \frac{100(V_m - V_E)}{SS_T} = \frac{100(0.00605 - 0.000679)}{0.00108} = 50\%.$$

Then the percent of error is

$$PC_E = 100 - 50 = 50.0\%.$$

**Table 7.18** ANOVA table.

| Source | Degrees of freedom, ($v$) | Sum of squares ($SS_i$) | Mean square ($V_m$) | F Test | Percent of variance |
|--------|---------------------------|-------------------------|---------------------|--------|---------------------|
| Mean | $v_m = 1$ | $SS_m = 0.00605$ | $V_m = 0.00605$ | $F = 8.91$ | 50.0 |
| Error | $v_E = 7$ | $SS_E = 0.00475$ | $V_E = 0.000679$ | | 50.0 |
| Total | $v_T = 8$ | $SS_T = 0.00108$ | | | |

All the information just calculated can be summarized in the ANOVA table (Table 7.18). Because the calculated $F$ value of 8.91 is larger than the tabulated $F$ value of 5.59, the data indicate that the process is not on target. The average loss $\bar{L}$ with the present process is found to be \$22.5 per unit (see Example 7.12). The average loss with the process on target would be

$$\bar{L}_T = k(MSD)PC_E$$
$$= 33{,}333(0.00135)(0.50) = \$22.5.$$

Thus the savings from centering the process are

$$\text{Savings} = \bar{L} - \bar{L}_T = \$45 - \$22.5 = \$22.5.$$

### 7.8.5 Two-Level Orthogonal Array

A two-level orthogonal array (OA)[12] involves $2^k$ data points, where $k$ is the number of factors. A two-level, two-factor OA requires $2^2 = 4$ data points. This is a so-called $L_4$ experiment and used for evaluating two factors and their interaction. A linear graph for an $L_4$ experiment is shown in Figure 7.35. Nodes 1

**Figure 7.35** Linear graph for $L_4$.

and 2 indicate major parameters that will be switched from high (2) to low (1) values, and the line indicates that an interaction is allowed between the major parameters. The $L_4$ array shown in Table 7.19 has four trials and three columns. The factors $X_1$, $X_2$, and interaction $X_1X_2$ between these factors are allocated to columns 1, 2, and 3, respectively. If columns 1 and 2 are acting independently, column 3 is allocated to residual (error). This is shown more clearly in Table 7.20. In this table, "Col." indicates the column in the orthogonal array, and "No." indicates the number of the experiment. Referring to Table 7.20 and equation (7.82), the formulas to calculate the main effect factors $MEF$ for an $L_4$ orthogonal array can be written as

$$MEF_{X_1} = \left(\frac{Y_3 + Y_4}{2}\right) - \left(\frac{Y_1 + Y_2}{2}\right), \tag{7.88}$$

$$MEF_{X_2} = \left(\frac{Y_2 + Y_4}{2}\right) - \left(\frac{Y_1 + Y_3}{2}\right), \tag{7.89}$$

$$MEF_{Error} = \left(\frac{Y_2 + Y_3}{2}\right) - \left(\frac{Y_1 + Y_4}{2}\right). \tag{7.90}$$

Other orthogonal arrays are used in experimental design. The most important one is the two-level OA with three factors, called the $L_8$ OA. A linear graph for an $L_8$ ANOVA is shown in Figure 7.36. The interactions between the factors of the $L_8$ OA are shown in Tables 7.21 and 7.22, respectively. As shown in Table 7.22, the factors $X_1$, $X_2$, and $X_3$ are assigned to columns 1, 2, and 4, respectively. This places the $X_1X_2$ interaction in column 3, the $X_1X_3$ interaction in column 5, and the $X_2X_3$

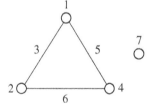

**Figure 7.36** Linear graph for $L_4$.

**Table 7.19** $L_4$ array.

| Col. → | 1 | 2 | 3 |
|---|---|---|---|
| No. ↓ | | | |
| 1 | 1 | 1 | 1 |
| 2 | 1 | 2 | 2 |
| 3 | 2 | 1 | 2 |
| 4 | 2 | 2 | 1 |

**Table 7.20** Two-factor $L_4$ orthogonal array.

| Col. → | Factors, $X_i$ 1 | 2 | Interaction 3 | Observations, $Y_i$ |
|---|---|---|---|---|
| No. ↓ | | | | |
| 1 | 1 | 1 | 1 | $Y_1$ |
| 2 | 1 | 2 | 2 | $Y_2$ |
| 3 | 2 | 1 | 2 | $Y_3$ |
| 4 | 2 | 2 | 1 | $Y_4$ |

**Table 7.21** $L_8$ Array.

| Col. → | 1 | 2 | 3 | 4 | 5 | 6 | 7 |
|---|---|---|---|---|---|---|---|
| No. ↓ | | | | | | | |
| 1 | 1 | 1 | 1 | 1 | 1 | 1 | 1 |
| 2 | 1 | 1 | 1 | 2 | 2 | 2 | 2 |
| 3 | 1 | 2 | 2 | 1 | 1 | 2 | 2 |
| 4 | 1 | 2 | 2 | 2 | 2 | 1 | 1 |
| 5 | 2 | 1 | 2 | 1 | 2 | 1 | 2 |
| 6 | 2 | 1 | 2 | 2 | 1 | 2 | 1 |
| 7 | 2 | 2 | 1 | 1 | 2 | 2 | 1 |
| 8 | 2 | 2 | 1 | 2 | 1 | 1 | 2 |

interaction in column 6. Column 7 is allocated to error accumulation, the effect of which is believed to be the smallest among the factors under study. Referring again to equation (7.82) and Table 7.22, general formulas for the main effect of factors can be written as follows:

1) The main effect factors for the major parameters are

$$MEF_{X_1} = \frac{(Y_5 + Y_6 + Y_7 + Y_8) - (Y_1 + Y_2 + Y_3 + Y_4)}{4}, \tag{7.91}$$

$$MEF_{X_2} = \frac{(Y_3 + Y_4 + Y_7 + Y_8) - (Y_1 + Y_2 + Y_5 + Y_6)}{4}, \tag{7.92}$$

$$MEF_{X_3} = \frac{(Y_2 + Y_4 + Y_6 + Y_8) - (Y_1 + Y_3 + Y_5 + Y_7)}{4}. \tag{7.93}$$

2) The main effect factors for the interactions, $X_1X_2$, $X_1X_3$, and $X_2X_3$, are

$$MEF_{X_1X_2} = \frac{(Y_3 + Y_4 + Y_5 + Y_6) - (Y_1 + Y_2 + Y_7 + Y_8)}{4}, \tag{7.94}$$

**Table 7.22** Three-Factor $L_8$ Orthogonal Array.

| Col. → | Ftr.<br>$x_1$<br>1 | Ftr.<br>$x_2$<br>2 | Int.<br>$x_1x_2$<br>3 | Ftr.<br>$x_3$<br>4 | Int.<br>$x_1x_3$<br>5 | Int.<br>$x_2x_3$<br>6 | Err.<br>7 | Obser.<br>$Y_i$ |
|---|---|---|---|---|---|---|---|---|
| No. ↓ | | | | | | | | |
| 1 | 1 | 1 | 1 | 1 | 1 | 1 | 1 | $Y_1$ |
| 2 | 1 | 1 | 1 | 2 | 2 | 2 | 2 | $Y_2$ |
| 3 | 1 | 2 | 2 | 1 | 1 | 2 | 2 | $Y_3$ |
| 4 | 1 | 2 | 2 | 2 | 2 | 1 | 1 | $Y_4$ |
| 5 | 2 | 1 | 2 | 1 | 2 | 1 | 2 | $Y_5$ |
| 6 | 2 | 1 | 2 | 2 | 1 | 2 | 1 | $Y_6$ |
| 7 | 2 | 2 | 1 | 1 | 2 | 2 | 1 | $Y_7$ |
| 8 | 2 | 2 | 1 | 2 | 1 | 1 | 2 | $Y_8$ |

$$MEF_{X_1X_3} = \frac{(Y_2 + Y_4 + Y_5 + Y_7) - (Y_1 + Y_3 + Y_6 + Y_8)}{4}, \tag{7.95}$$

$$MEF_{X_2X_3} = \frac{(Y_2 + Y_3 + Y_6 + Y_7) - (Y_1 + Y_4 + Y_5 + Y_8)}{4}. \tag{7.96}$$

3) The main effect factor for error is

$$MEF_{\text{Error}} = \frac{(Y_2 + Y_3 + Y_5 + Y_8) - (Y_1 + Y_4 + Y_6 + Y_7)}{4}. \tag{7.97}$$

The following example is a demonstration of this method.

## Example 7.15

When two optical fibers are coupled, power loss may occur mainly due to following three types of misalignment:

- *End separation misalignment* (E). As in Figure 7.37(a), due to the end separation misalignment (distance $S$) between optical fibers, power dispersion in an optical cone from one core is not coupled to the second core (less light is coupled into the core).
- *Lateral misalignment* (L). This is also called cross-sectional mismatch and designated by $L$ as shown in Figure 7.37(b). Optical fibers may be perfectly aligned but laterally displaced by a few microns.
- *Angular misalignment* (A). In this case fibers are pointed in the wrong direction as shown by $\theta$ in Figure 7.37(c).

Using the experimental data given in Table 7.23, design a two-level experiment to determine the relative effects of the end separation, lateral, and angular misalignment on power loss.

**Table 7.23** $L_8$ orthogonal array for optical fiber experiment.

| | Lateral (L) | Angular (A) | L × A | End Sep. E | L×E | A×E | Error | $Y_i$ (Loss, dB) |
|---|---|---|---|---|---|---|---|---|
| Col. → | 1 | 2 | 3 | 4 | 5 | 6 | 7 | |
| No. ↓ | | | | | | | | |
| 1 | 1 | 1 | 1 | 1 | 1 | 1 | 1 | 2.2 |
| 2 | 1 | 1 | 1 | 2 | 2 | 2 | 2 | 3.0 |
| 3 | 1 | 2 | 2 | 1 | 1 | 2 | 2 | 4.2 |
| 4 | 1 | 2 | 2 | 2 | 2 | 1 | 1 | 4.6 |
| 5 | 2 | 1 | 2 | 1 | 2 | 1 | 2 | 4.4 |
| 6 | 2 | 1 | 2 | 2 | 1 | 2 | 1 | 4.8 |
| 7 | 2 | 2 | 1 | 1 | 2 | 2 | 1 | 5.8 |
| 8 | 2 | 2 | 1 | 2 | 1 | 1 | 2 | 6.2 |

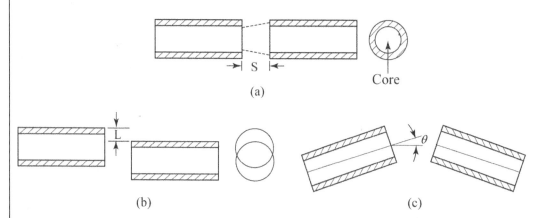

(a)

(b)                                                    (c)

**Figure 7.37** Optical fibers.

## Solution

Main effect factors are calculated as

$$MEF = \frac{\sum Y_i \text{ at high level} - \sum Y_i \text{ at low level}}{4}.$$

For $L_8$ orthogonal array, $N = 8$. Then

$$MEF_L = \frac{(4.4 + 4.8 + 5.8 + 6.2) - (2.2 + 3.0 + 4.2 + 4.6)}{4} = 1.8,$$

$$MEF_A = \frac{(4.2 + 4.6 + 5.8 + 6.2) - (2.2 + 3.0 + 4.4 + 4.8)}{4} = 1.6,$$

$$MEF_E = \frac{(3.0 + 4.6 + 4.8 + 6.2) - (2.2 + 4.2 + 4.4 + 5.8)}{4} = 0.5,$$

$$MEF_{L \times A} = \frac{(4.2 + 4.6 + 4.4 + 4.8) - (2.2 + 3.0 + 5.8 + 6.2)}{4} = 0.2,$$

$$MEF_{L \times E} = \frac{(3.0 + 4.6 + 4.4 + 5.8) - (2.2 + 4.2 + 4.8 + 6.2)}{4} = 0.1,$$

$$MEF_{A \times E} = \frac{(3.0 + 4.2 + 4.8 + 5.8) - (2.2 + 4.6 + 4.4 + 6.2)}{4} = 0.1,$$

$$MEF_{error} = \frac{(3.0 + 4.2 + 4.4 + 6.2) - (2.2 + 4.6 + 4.8 + 5.8)}{4} = 0.1.$$

## ANOVA Table Setup

1) Calculation of degrees of freedom:

$$v_T = N - 1 = 8 - 1 = 7, \quad v_L = \text{number of levels} - 1 = 2 - 1 = 1$$

Similarly,

$$v_A = 1, \quad v_E = 1, \quad v_{L \times A} = 1, \quad v_{L \times E} = 1,$$

$$v_{A \times E} = 1,$$

$$v_{error} = (N - 1) - \sum_{i=1}^{k} v_i$$

$$= (8 - 1) - (1 + 1 + 1 + 1 + 1 + 1) = 1.$$

2) Calculation of total sum of squares:

$$SS_T = SS_y - C$$

$$SS_Y = \sum_i \overset{2}{Y} = 2.2^2 + 3.0^2 + 4.2^2 + 4.6^2 + 4.4^2 + 4.8^2 + 5.8^2 + 6.2^2$$

$$= 167.12$$

$$C = \frac{T^2}{N} \quad \text{where } N = 8, \text{ and the grand total is}$$

$$T = 2.2 + 3.0 + 4.2 + 4.6 + 4.4 + 4.8 + 5.8 + 6.2 = 35.2$$

$$C = \frac{(35.2)^2}{8} = 154.88$$

$$SS_T = 167.12 - 154.88 = 12.24.$$

3) Calculation of sum of squares:

$$SS_i = 2(MEF)^2$$
$$SS_L = 2(1.8)^2 = 6.48$$
$$SS_A = 2(1.6)^2 = 5.12$$
$$SS_E = 2(0.5)^2 = 0.5$$
$$SS_{L \times A} = 2(0.2)^2 = 0.08$$
$$SS_{L \times E} = 2(0.1)^2 = 0.02$$
$$SS_{A \times E} = 2(0.1)^2 = 0.02$$

The sum of squares for error is

$$SS_{ER} = SS_T - \sum_{i=1}^{k} SS_i$$
$$= 12.24 - (6.48 + 5.12 + 0.5 + 0.08 + 0.02 + 0.02)$$
$$= 0.02.$$

4) Calculate the mean squares:

$$MS_i = \frac{SS_i}{v_i}$$
$$MS_L = \frac{6.48}{1} = 6.48$$

Similarly,

$$MS_A = 5.12, \quad MS_E = 0.5, \quad MS_{L \times A} = 0.08,$$
$$MS_{L \times E} = 0.02, \quad MS_{A \times E} = 0.02, \quad MS_{ER} = 0.02.$$

5) Perform the $F$ test:

$$F_i = \frac{MS_i}{MS_{ER}},$$
$$F_L = \frac{6.48}{0.02} = 324.$$

Similarly,

$$F_A = 256, \quad F_E = 25, \quad F_{L \times A} = 4, \quad F_{L \times E} = 1, \quad F_{A \times E} = 1.$$

6) Calculate the percentage contribution:

$$PC_i = \frac{100(MS_i - MS_E)}{SS_T}$$
$$PC_L = \frac{100(6.48 - 0.0.02)}{12.24} = 52.77\%.$$

Similarly,

$$PC_A = 41.66\%, \quad PC_E = 3.92\%, \quad PC_{L \times A} = 0.49, \quad PC_{L \times E} = 0.0\%,$$

$$PC_{A \times E} = 0,$$

and the percent contribution of error is

$$PC_E = 100 - \sum_{i=1}^{k} PC_i$$
$$= 100 - (52.77 + 41.66 + 3.92 + 0.49) = 1.16\%.$$

All the information calculated above can be summarized in the ANOVA table (Table 7.24). As seen from the table, factors that affect the power lost most are lateral and angular misalignments. The factor of end separation misalignment has relatively little effect in this experiment. The total statistical error is 1.16 percent and the interactions $L \times E$ and $A \times E$ have zero effect – no interaction.

**Table 7.24** ANOVA summary for three factorial optical fiber experiment.

| Source | Degrees of freedom $(v)$ | Sum of squares $(SS_i)$ | Mean squares $(MS_i)$ | F test | Percent of variance |
|---|---|---|---|---|---|
| Lateral (L) | 1 | 6.48 | 6.48 | 324 | 52.77 |
| Angular (A) | 1 | 5.12 | 5.12 | 256 | 41.66 |
| End. Sep. (E) | 1 | 0.5 | 0.5 | 25 | 3.92 |
| L × A | 1 | 0.08 | 0.08 | 4 | 0.49 |
| L × E | 1 | 0.02 | 0.02 | 1 | 0.0 |
| A × E | 1 | 0.02 | 0.02 | 1 | 0.0 |
| Error | 1 | 0.02 | 0.02 | | 1.16 |

## Example 7.16

Referring to Example 7.15, test whether there is a significant interaction between the lateral and angular misalignments at the 95% confidence level ($\alpha = 0.05$).

## Solution

The computed value corresponding to the $L \times A$ interaction is

$$F_i = \frac{MS_i}{MS_E}$$
$$F_{L \times A} = \frac{0.08}{0.02} = 4.$$

For $\alpha = 0.05$, the $F$ ratio to look for in the $F$ distribution table is $F_{0.05}(v_1 = 1, v_2 = 1)$. From Table C.4 (see Appendix C),

$$F_{0.05}(v_1 = 1, v_2 = 1) = 161.4.$$

Because the calculated $F$ value of 4 is less than the critical $F$ ratio from the table, the effect is not significant at the 0.05 level.

### 7.8.5.1 Lumped Model

Some factors assigned to an experiment may not be significant. For example, in Example 7.15, the effects of interactions are not significant. In such cases, an alternative approach is to assume that the interactions are minor enough to be placed in a category called residual (error). Thus, only three parameters are needed to do the calculations for the major effects, which simplifies the analysis of the problem significantly. This approach is called the *lumped model*.

### Example 7.17

To simplify the analysis of the problem in Example 7.15, assume that the effect of the interactions is insignificant. Using the lumped interaction model shown in Table 7.25, design a two-level, three-factor experiment to determine the relative effects of the end separation, lateral, and angular misalignments on power loss.

### Solution

From Example 7.15, the main effect factors are

$$MEF_L = 1.8,$$
$$MEF_A = 1.6,$$
$$MEF_E = 0.5.$$

**Table 7.25** $L_8$ Lumped model for optical fiber experiment.

| Col. → | Lateral (L) 1 | Angular (A) 2 | End-Sep. (E) 3 | $Y_i$ |
|---|---|---|---|---|
| **No. ↓** | | | | |
| 1 | 1 | 1 | 1 | 2.2 |
| 2 | 1 | 1 | 2 | 3.0 |
| 3 | 1 | 2 | 1 | 4.2 |
| 4 | 1 | 2 | 2 | 4.6 |
| 5 | 2 | 1 | 1 | 4.4 |
| 6 | 2 | 1 | 2 | 4.8 |
| 7 | 2 | 2 | 1 | 5.8 |
| 8 | 2 | 2 | 2 | 6.2 |

*Analysis of Variance*

1) Calculation of degrees of freedom:

$$v_T = N - 1 = 8 - 1 = 7, \quad v_L = \text{number of levels} - 1 = 2 - 1 = 1.$$

Similarly,

$$v_A = 1, \quad v_E = 1.$$

The degrees of freedom for error are

$$v_{\text{error}} = (N - 1) - \sum_{i=1}^{k} v_i$$
$$= (8 - 1) - (1 + 1 + 1) = 4.$$

2) Calculation of the sum of squares, from Example 7.15:

$$SS_L = 6.48,$$
$$SS_A = 5.12,$$
$$SS_E = 0.5.$$

3) The total sum of squares from Example 7.15 is

$$SS_T = 12.24.$$

The error sum of squares, $SS_{ER}$, is

$$SS_{ER} = SS_T - \sum_{i=1}^{k} SS_i$$
$$= 12.24 - (6.48 + 5.12 + 0.5) = 0.14.$$

**Table 7.26** ANOVA summary for three factorial optical fiber experiment.

| Source | Degree-of-Freedom ($v$) | Sum of Squares ($SS_i$) | Mean Squares ($MS_i$) | F Test | Percent of Variance |
|---|---|---|---|---|---|
| L | 1 | 6.48 | 6.48 | 185.1 | 52.66 |
| A | 1 | 5.12 | 5.12 | 146.3 | 41.54 |
| E | 1 | 0.5 | 0.5 | 14.3 | 3.8 |
| Error | 4 | 0.14 | 0.035 | | 2 |

4) Calculation of mean squares, from Example 7.15:

$$MS_L = 6.48,$$
$$MS_A = 5.12,$$
$$MS_E = 0.5,$$

and the error mean square is

$$MS_{ER} = \frac{SS_{ER}}{v_{error}} = \frac{0.14}{4} = 0.035.$$

5) Calculate the $F$ test:

$$F_i = \frac{MS_i}{MS_{ER}}$$

$$F_L = \frac{6.48}{0.035} = 185.1.$$

Similarly,

$$F_A = 146.3, \quad F_E = 14.3.$$

6) Calculate the percent contribution

$$PC_i = \frac{100(MS_i - MS_{ER})}{SS_T}$$

$$PC_L = \frac{100(6.48 - 0.035)}{12.24} = 52.66\%.$$

Similarly,

$$PC_A = 41.54\%, \quad PC_E = 3.8\%.$$

7) The percent contribution of error is

$$PC_E = 100 - \sum_{i=1}^{k} PC_i$$

$$= 100 - (52.66 + 41.54 + 3.8) = 2.00\%.$$

Because some of the factors were eliminated from the analysis, the error has increased to 2.0 percent as summarized in Table 7.26.

## Example 7.18

If the mean square of lateral misalignment, $MS_L$, is 4.75, determine the minimum confidence level and the risk $\alpha$ associated with the factor *lateral misalignment*.

### Solution

The computed value corresponding to lateral misalignment is

$$F_L = \frac{MS_L}{MS_{ER}}$$

$$= \frac{4.75}{0.035} = 135.7.$$

For an $\alpha$ level of confidence, the $F$ ratio to look for in the $F$ distribution table is $F_\alpha(v_1 = 1, v_2 = 1)$. From Table C.4 (see Appendix C),

$$F_{0.25}(v_1 = 1, v_2 = 1) = 5.83,$$
$$F_{0.10}(v_1 = 1, v_2 = 1) = 38.66,$$
$$F_{0.05}(v_1 = 1, v_2 = 1) = 161.4,$$

Minimum confidence level is 90% at $\alpha = 0.10$ risk.

## Bibliography

1 ERTAS, A., and JONES J.C., *The Engineering Design Process*, John Wiley & Sons, Inc., New York, 1996.

2 HICKS, C.R., *Fundamental Concepts in the Design of Experiments*, 3rd edn, CBS College Publishing, New York, 1982.

3 HINES, W.W. and MONTGOMERY, D.C., Probability and Statistics in Engineering and Management Science, 3rd ed. Wiley, New York, 1990.

4 ROBINSON, E.A., *Statistical Reasoning and Decision Making*, Goose Pond Press, Houston, TX, 1981.

5 TAGUCHI, G., *Introduction to Quality Engineering*, UNIPUB/Kraus International Publications, New York, 1987.

6 TAGUCHI, G., *System of Experimental Design*, Vol. 2, 2nd ed, UNIPUB/ Kraus International Publications, New York, 1988.

## CHAPTER 7 Problems

**7.1** Suppose 20 rotors are randomly selected from a larger (conceivably countless) number of rotors forming a population. Each rotor is discretized into three sections whose shapes are circular cylinders (disks), as shown in Figure P7.1. Table P7.1 illustrates a typical set of eccentricity measurements, representing a sample drawn from that population. Using Table P7.1, estimate the mean eccentricity of each disk and the overall mean of the eccentricity measurements of the 20 rotors.

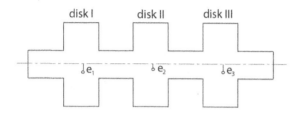

**Figure P7.1** Rotor with three discretized disks.

**Table P7.1** Typical set of eccentricity measurements of 20 rotors.

| Observations average eccentricity e, mm | Frequency of disk I | Frequency of disk II | Frequency of disk III |
|---|---|---|---|
| 0.1 | 1 | 3 | 1 |
| 0.2 | 2 | 6 | 1 |
| 0.3 | 4 | 4 | 2 |
| 0.4 | 6 | 3 | 2 |
| 0.5 | 4 | 2 | 5 |
| 0.6 | 2 | 1 | 6 |
| 0.7 | 1 | 1 | 3 |

**7.2** Find the following values for a $t$ distribution:
a) $t_{0.05}$ with $v = 16$;
b) $t_{0.01}$ with $v = 12$;
c) $t_{\alpha/2}$ so that $p(-t_{\alpha/2} < t < t_{\alpha/2}) = 0.90$. with $v = 12$

**7.3** Find the following values for a chi-square distribution:
a) $\chi^2_{0.025}$ with $v = 16$;
b) $\chi^2_{0.05}$ with $v = 24$;
c) $\chi^2_{\alpha}$ so that $p(\chi^2 < \chi^2_{\alpha}) = 0.990$ with $v = 8$.

**7.4** Find the following values for an $F$ distribution:
  a) $v_1 = 10$, $v_2 = 22$
  b) $v_1 = 12$, $v_2 = 4$
  c) $v_1 = 30$, $v_2 = 30$

**7.5** The following are the spark plug gaps, in mm, of ten spark plugs in a package (drawn from a normally distributed population):
2.20, 2.18, 2.23, 2.17, 2.19, 2.21, 2.24, 2.23, 2.25, 2.26
Find a 90 percent confidence interval for the variance of all spark plug gaps (use the chi-square distribution).

**7.6** Increasing the stiffness of a circuit board results in decreasing the deflection and, hence, decreased electrical lead wire and circuit board stresses during vibration. Assume that, for minimum deflection, it is desirable that the standard deviation $\hat{\sigma}$ of the stiffness be 3.1 kips/in. with a 90 percent confidence level. If the stiffness of a random sample of 30 circuit boards results in a sample variance $S^2 = 2.6$ kips/in., does the given data support the minimum deflection requirement? Assume that the circuit board manufacturing process for a specified stiffness is normally distributed.

**7.7** Consider manufacturing welding processes for the drill pipe described in Example 7.7. Eight drill pipes assemblies were manufactured using both old and new welding processes. The failure strength test results of eight drill pipes at the tool joint welding location are given in Table P7.7. Considering a 95 percent two-sided confidence interval, determine if there is a difference between the two manufacturing processes.

**Table P7.7** Comparison of drill pipe welding processes.

| Old method (failure strength, kpsi) | New method (failure strength, kpsi) |
| --- | --- |
| 128 | 122 |
| 126 | 126 |
| 132 | 123 |
| 118 | 116 |
| 119 | 124 |
| 129 | 128 |
| 130 | 130 |
| 127 | 127 |

**7.8** During the past two years the Federal Aviation Administration has been investigating a number of glider aircraft accidents. The FAA concluded that the majority of the accidents were due to failure occurring in the composite wing spar joints. A new method has been developed that requires vacuum bagging the composite sandwich that makes up the spar joint. The new method will increase the failure strength in the joint, therefore reducing the number of wing spar joint failures. Six composite glider wings were manufactured using both the old air dry process and the new vacuum bagging process. The strength test results of the six wings at the spar joint location are given in Table P7.8. Considering a 90 percent two-sided confidence interval, determine if there is a difference between the two manufacturing processes.

Table P7.8 Comparison of wing manufacturing processes.

| Old method (failure strength, kpsi) | New method (failure strength, kpsi) |
|---|---|
| 12 | 14 |
| 9 | 12 |
| 13 | 15 |
| 14 | 16 |
| 11 | 17 |
| 10 | 13 |

**7.9** A six-cylinder, 4,000 horse power marine diesel engine has six fuel injectors. Because of heavy fuel oil, the injectors get dirty after approximately 1,000 hours. Consequently, pressure adjustment is lost, resulting in poor combustion and possible component failure. Assume that the chance of an injector failing is 2 in 200 after the 1,000 hours of operation. Assume that the occurrences of failure of the injectors are independent of each other and that the engine can run if more than three injectors keep operating. What is the probability of the engine failing after 1,000 hours of operation? (Use the binomial distribution.)

**7.10** If the sample size of Example 7.8 is reduced to 16 electronic components, and the sample mean and the standard deviation are respectively found to be $5.1 \times 10^6$ cycles and $3.8 \times 10^5$ cycles, determine whether the sample data gives enough evidence that the fatigue life of the materials has been improved. Assume that the life of the components is normally distributed.

**7.11** The lubricating oil pressure of an engine that is coupled with a pump that delivers crude oil to shore from an offshore well should remain almost constant during the operation. Low lubricating oil pressure may cause engine failure. This is especially true for high-speed engines. However, because of vibrations, correct readings of the oil pressure using a gage are almost impossible. Suppose, during the operation of the engine, the operator takes a reading of approximately 65 psi. Assume that the engine can operate with no trouble if the mean value of normally distributed lubricating oil pressure is 68 psi and that it may fail if the mean value of the lubricating pressure is 64 psi. In both cases, assume that the standard deviation is 2.5 psi. If the lubricating oil pressure is low, close to 64 psi, two possible decisions can be made by the operator:

a) Stop the pumping operation and start troubleshooting.

b) Continue pumping.

The operator should know the relative cost of a wrong decision and the failure probability of the engine before making the above decision. Assume that, from experience, the failure probability of the engine when it is running under low pressure is 70 percent. Hence, the expected abnormal engine operation is 0.70 and the operator's null hypothesis is that the engine operation would not be normal.

If the operator does not stop the engine when it is not operating safely, he or she is making a Type I error. Making this error results in a loss and, in some instances, replacement of the engine for not detecting the problem in time. Assume that the relative cost of this is 1 unit, and it is designated regret $r$.

If the operator stops the engine for troubleshooting when the engine can still operate, at least until the pumping operation finishes, he or she is making a Type II error. Consequently, this wrong decision results in loss to the company because of the down time. Assume that the relative cost of this is 5 units, and it is designated regret $R$.

At the observed operating oil pressure of 65 psi (average of five readings), what is the best decision the operator can make? Calculate the Type I and Type II errors, show them on a tree diagram, calculate the expected loss, and discuss the results.

**7.12** To evaluate the sensitivity of an explosive, 18 specimens were tested by dropping them from a certain height. The result of the experiment is shown in Table P7.12, where the xs represent explosions (successes) and the 0s represents non-explosions (failures). By using the up-and-down method, estimate the mean and standard deviation of the height for the explosion experiment.

Table P7.12 Up-and-Down Test for Sensitivity of Explosive.

| Test No. | | 1 | 2 | 3 | 4 | 5 | 6 | 7 | 8 | 9 | 10 | 11 | 12 | 13 | 14 | 15 | 16 | 17 | 18 |
|---|---|---|---|---|---|---|---|---|---|---|---|---|---|---|---|---|---|---|---|
| | 3.0 | | | | | | x | | x | | x | | | | | | | | |
| Normalized | 2.6 | | | | | 0 | | 0 | | 0 | | | x | | | | | | |
| Height | 2.2 | | | x | | 0 | | | | | | | | | x | | x | | x | |
| | 1.8 | | 0 | | 0 | | | | | | | | | | | 0 | | 0 | | 0 |
| | 1.4 | 0 | | | | | | | | | | | | | | | | | |

x indicates explosion, 0 indicates non-explosion.

**7.13** If the result of the experiment in Problem 7.12 is as shown in Table P7.13, calculate the mean and standard deviation of the height for the explosion experiment.

**Table P7.13** Up-and-Down Test for Sensitivity of Explosive.

| Test No. | | 1 | 2 | 3 | 4 | 5 | 6 | 7 | 8 | 9 | 10 | 11 | 12 | 13 | 14 | 15 | 16 | 17 | 18 |
|---|---|---|---|---|---|---|---|---|---|---|---|---|---|---|---|---|---|---|---|
| | 3.0 | x | | | | | | | | | | | | | | | | | |
| Normalized | 2.6 | | x | | x | | | | | | | | | | x | | | | |
| Height | 2.2 | | | 0 | | x | | x | | x | | x | | 0 | | x | | | |
| | 1.8 | | | | | | 0 | | 0 | | 0 | | 0 | | | | x | | 0 |
| | 1.4 | | | | | | | | | | | | | | | | | 0 | |

x indicates explosion, 0 indicates non-explosion.

**7.14** Seventy-five explosive specimens were tested by dropping them from different heights to investigate the mean and standard deviation of the explosion height. The results showed that when the explosive specimens were:
- dropped from a height of 400 meters, out of 15 specimens 3 explosions were observed;
- dropped from a height of 500 meters, out of 15 specimens 4 explosions were observed;
- dropped from a height of 600 meters, out of 15 specimens 6 explosions were observed;
- dropped from a height of 700 meters, out of 15 specimens 9 explosions were observed;
- dropped from a height of 800 meters, out of 15 specimens 12 explosions were observed.

Use an appropriate method to determine the mean and standard deviation of the explosion height.

**7.15** Referring to Example 7.11, using the experimental data given in Table P7.15, design a two-level experiment with two factors to determine the relative effects of roughness and speed on the friction coefficient between the box and pin. From the table, it can be seen that the main factors are acting independently; thus, the error is assigned to the third column instead of interaction. Discuss the result of the ANOVA table.

**Table P7.15** Two-factor $L_4$ orthogonal array.

| | Roughness | Speed | Error | Observations, $Y_i$ |
|---|---|---|---|---|
| Col. → | 1 | 2 | 3 | coefficient of friction |
| No.↓ | | | | |
| 1 | 1 | 1 | 1 | 0.050 |
| 2 | 1 | 2 | 2 | 0.054 |
| 3 | 2 | 1 | 2 | 0.061 |
| 4 | 2 | 2 | 1 | 0.069 |

**7.16** Design a two-level experiment if the main factors are dependent in Problem 7.15. Discuss the result of the ANOVA table.

**7.17** Referring to Example 7.15, determine the minimum confidence level and risk $\alpha$ associated with factor lateral misalignment.

**7.18** Compressed natural gas vehicles use methane stored at 3,000 psi to achieve a density of 10 lb/ft$^3$. A typical vehicle with 10 cubic feet of storage has a range of 100 miles. An adsorbed natural gas (ANG) storage tank uses activated carbon as an adsorbent and can achieve storage densities as high as 24 lb/ft$^3$ at 500 psi. The disadvantage of ANG storage comes from the heat generated during the adsorption process. When natural gas is adsorbed on the activated carbon, the gas changes phase from a "three-dimensional" gas to a "two-dimensional" liquid on the surface of the carbon. The heat of adsorption is equal to or greater than the heat of vaporization. The changing process heats the tank and limits the amount of gas that can be absorbed. The bulk thermal properties of the activated carbon and methane gas under charging conditions need to be known so that optimal ANG tanks can be designed. These properties are a function of temperature and pressure.

To evaluate the effects of pressure and temperature on the bulk thermal conductivity, $k$, of activated carbon and adsorbed methane gas, design an experiment using:
(a) the $L_4$ orthogonal array model with interaction;
(b) the $L_8$ lamped model;
(c) the full $L_8$ orthogonal array model.

The effect of pressure and temperature on bulk thermal conductivity, $k$, were tested for the following parameters:
• low-level pressure, $p_1 = 20$ psi;
• high-level pressure, $p_2 = 500$ psi;
• low-level temperature, $T_1 = 25°C$;
• high-level temperature, $T_2 = 50°C$.

The results are illustrated in Table P7.18. Note that two test results are used in Table P7.18. For example, at $P_1$ and $T_1$, for the first test, $k = 0.143$ W/°C, and for the second test $k = 0.145$ W/°C.
(a) Construct ANOVA table and discuss the results.
(b) Calculate the confidence level for pressure.
(c) For $\alpha = 0.01$, test if there is a significant interaction between pressure and temperature.

**Table P7.18** Experimental results of bulk thermal conductivity, $k$.

| Old method parameters | New method observations of $k$ |
| --- | --- |
| at $P_1$ and $T_1$ | 0.143, 0.145 |
| at $P_1$ and $T_2$ | 0.156, 0.158 |
| at $P_2$ and $T_1$ | 0.211, 0.215 |
| at $P_2$ and $T_2$ | 0.229, 0.231 |

**7.19** Electrical components have changed dramatically in the last 20 years. Instead of vacuum tubes, microcircuits are now used. These microcircuits incorporate many hundreds or thousands of electrical components; they are referred to as very-large-scale integrated circuits (VLSICs). These electrical devices, because of their size, are very susceptible to contamination; a particle of dust can bridge the gap between very small wires (called leads) and cause reliability problems. Because contamination is such a serious problem, engineers spend much time and money on limiting the number of particles that can get on a wafer. The higher the contamination, the larger the yield and, subsequently, the lower the profit.

This problem deals with a wafer undergoing one of two annealing processes, whether or not inspection under a microscope introduces particles, and whether the wafer cleaning itself introduces particles. Different manufacturing processes generate different contaminants; thus, it is necessary to use clean processes. Inspection during the various processing steps is accomplished under microscopes but, since operators are looking down at the wafer under a scope, skin particles can fall on the wafer. Cleaning the wafer is usually done in a bath, and then the wafers are placed in a machine that rotates the wafers to spin off the solvent. This is referred to as spin rinse drying. This can also introduce particles onto the surface from loosening of the photoresist.

**Table P7.19** Orthogonal array $L_8$ for spin rinse drying experiment.

| Col. → | 1 | 2 | 3 | 4 | 5 | 6 | 7 | $Y_i$ |
|---|---|---|---|---|---|---|---|---|
| No. ↓ | | | | | | | | |
| 1 | 1 | 1 | 1 | 1 | 1 | 1 | 1 | 40.0 |
| 2 | 1 | 1 | 1 | 2 | 2 | 2 | 2 | 335.0 |
| 3 | 1 | 2 | 2 | 1 | 1 | 2 | 2 | 25.0 |
| 4 | 1 | 2 | 2 | 2 | 2 | 1 | 1 | 80.0 |
| 5 | 2 | 1 | 2 | 1 | 2 | 1 | 2 | 23.0 |
| 6 | 2 | 1 | 2 | 2 | 1 | 2 | 1 | 248.0 |
| 7 | 2 | 2 | 1 | 1 | 2 | 2 | 1 | 45.0 |
| 8 | 2 | 2 | 1 | 2 | 1 | 1 | 2 | 196.0 |

Column 1 = process, column 2 = inspection, column 3 = process × inspection, column 4 = spin rinse dry, column 5 = process × spin rinse dry, column 6 = inspection × spin, column 7 = error.

The task is to find the largest generator of profit-robbing particles so that the problem can be quickly and adequately addressed. Using the experimental data given in Table P7.19, for particle count $Y_i$, design a two-level experiment to determine the relative effects of process, inspection, spin rinse drying, and the interaction of the major components on contamination of VLSIC wafers. (This problem is adapted from Experimental Design and Taguchi Method, course notes, Jones Reilly & Associates, Inc., 1988.)

**7.20** To simplify the analysis given in Problem P7.19, assume that the interactions are minor and can be placed in a catagory called residual (error). Design a two-level lumped experiment to determine the relative effects of process, inspection, and spin rinse drying on contamination of VLSI wafers.

**7.21** A machine known as an ion implanter has made possible the fabrication of tiny microcircuits. This machine is a plasma generator and extractor that dopes wafers preferentially. The plasma field from which the dopants are extracted depends mainly on three parameters: arc current, extraction current, and pressure inside the plasma. The arc current is the current from the filament to the arc chamber, and the extraction current is the electromotive force applied to pull dopant ions from the arc chamber, as shown in Figure P7.21. Using the experimental data given in Table P7.21, design a three-parameter experiment to find the relationship between the arc current, extraction current, pressure, and the output, which is beam current. Optimizing the beam current is essential for the most efficient use of the ion implanter. Assume that the interactions are very small.

**Table P7.21** Orthogonal array $L_8$ for arc chamber experiment.

| Col. → | Arc current | Extraction current | pressure | Beam current $Y_i$ |
|--------|---------|---------|---------|---------|
| No.↓ | | | | |
| 1 | 1 | 1 | 1 | 0.80 |
| 2 | 1 | 1 | 2 | 0.88 |
| 3 | 1 | 2 | 1 | 0.95 |
| 4 | 1 | 2 | 2 | 1.045 |
| 5 | 2 | 1 | 1 | 1.050 |
| 6 | 2 | 1 | 2 | 1.155 |
| 7 | 2 | 2 | 1 | 1.400 |
| 8 | 2 | 2 | 2 | 1.540 |

**Figure P7.21** Arc chamber.

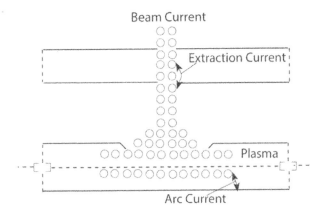

Beam Current

Extraction Current

Plasma

Arc Current

# 8

# Risk, Reliability, and Safety

Often the difference between a successful man and a failure is not one's better abilities or ideas, but the courage that one has to bet on his ideas, to take a calculated risk, and to act.

**Maxwell Maltz**

## 8.1 Introduction

This chapter introduces the concept of risk management, reliability, and safety as applied to engineering practice. Risk, reliability, and safety are three main components of a system design. There is a natural conflict between reliability and safety. High system reliability and performance are often achieved through the proper balance between performance and reliability so that an adequate safety factor is provided.

Customers expect to obtain reliable and maintainable systems that are of high quality, readily available and dependable, and able to satisfy their needs with some fair and reasonable cost parameters. A focus on maintainability in the design process results in a system that can be maintained realistically within given time constraints. Availability is also an important parameter in complex systems. Having high reliability does not ensure that the system will be operational (available) when needed. Conversely, a system can be available but not reliable. System dependability is an important design parameter that provides a measure of the system condition combining its reliability and maintainability. This chapter examines the mathematical relationships involved to determine the importance of the above-mentioned parameters in system design.

## 8.2 What Is Risk?

The EPA considers risk to be the chance of harmful events to human health or to ecological systems resulting from exposure to an environmental stressor. A stressor is any physical, chemical, or biological object that can generate a hostile response.

For example, security risk can be formulated mathematically as a function of threats ($T$), vulnerabilities ($V$), and potential consequences ($C$): that is,

$$R = f(T, V, C). \tag{8.1}$$

To take a specific instance, the immune system is the body's defense against disease-causing viruses (threat). When the immune system is weak, it may not be able to destroy virus-infected cells; the success of a virus causing disease depends on the immune system's vulnerability. A human body with a strong immune system may not be as vulnerable as a body with a weak immune system. Thus, in equation (8.1), vulnerability is coupled with threat and target.

Equation (8.1) defining risk can be written as the product of $T$, $V$, and $C$.

$$R = T \times V \times C \tag{8.2}$$

In equation (8.2), $T$ is the probability that an attack will be attempted by a virus, $V$ is the (conditional) probability that an attempt by viruses will lead to disease, and $C$ is the consequence.

Equation (8.2) assumes that $T$ and $V$ are independent – it assumes that the virus causing the disease has nothing to do with the vulnerability of the target (immune system of the body). This assumption may not hold in all cases. It is difficult, if not impossible, to determine $T$ and $V$ independently. Therefore, as shown in Figure 8.1, we will combine $T$ and $V$ in equation (8.2) as the probability of the event, $P$(event).

Thus, risk, expected loss due to a failure, can be expressed as the product of probability of a failure, $P$(event), and its consequence, $C$: risk is a combination of consequences and probability of occurrence associated with an event:

$$R(\text{Risk}) = P(\text{event}) \times C. \tag{8.3}$$

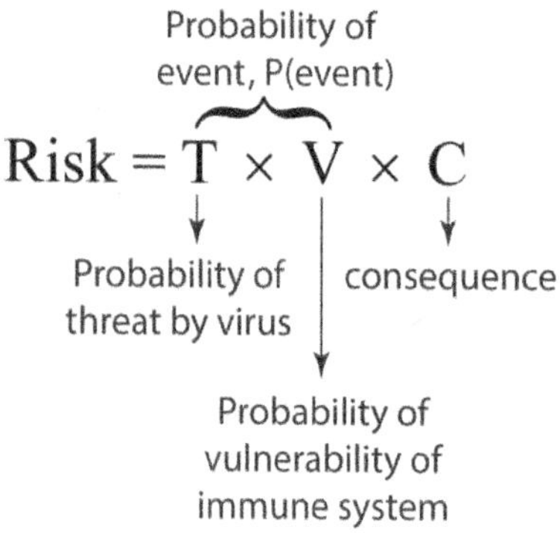

**Figure 8.1** Combined risk.

In terms of the cost, risk can be defined as

Risk (\$/yr) = frequency of occurrence ($n$/yr) $\times$ consequence of occurrence (\$).

## Example 8.1

Researchers claimed that the main sweetener in diet soft drinks may cause cancer. For example, studies showed that 3,000 people had cancer because of diet soft drinks every year in the United States, and approximately 1 in 100 people suffering cancer results in a fatality. Calculate the annual fatality risk.

### Solution

Probability, $P = 1/100 = 1\% = 0.01$.

Consequence, $C$, due to diet soft drinks is that 3,000 people have cancer each year. Then the annual fatality risk is

$$R = P \times C = 0.01 \times 3,000 = 30.$$

Thus, 30 people may die because of the diet soft drink every year.

### 8.2.1 Relative Risk

Relative risk, $RR$, is the ratio of the probability of an event occurring (e.g., developing a disease) in an exposed group to the probability of the event occurring in a comparison, non-exposed group. Namely,

$$RR = \frac{P_{exposed}}{P_{non\text{-}exposed}} = \frac{P_1}{P_2}. \tag{8.4}$$

### Example 8.2

Suppose that a group (cohort) study of 100 diet soda drinkers and 100 diet soda non-drinkers are followed for 15 years for the development of a cancer. Assume that 20 of the diet soda drinkers and 10 of the non-drinkers develop a form of cancer. Calculate the relative risk, $RR$.

### Solution

Table 8.1 shows the calculations of the probabilities.

$$RR = \frac{P_1}{P_2} = \frac{0.2\%}{0.1\%} = 2.$$

Thus, in this experiment people who drink diet soda have twice the risk of cancer compared to the people who do not drink diet soda. If the risk ratio, $RR$, is less than 1, then the event being considered shows a reduction in risk.

**Table 8.1** Probability calculations.

| Diet soda drinkers | Cancer | No cancer | Total | Probability of event, $P$ |
|:---:|:---:|:---:|:---:|:---|
| Yes | 20 | 80 | 100 | $P_1 = 20/100 = 0.2\%$ |
| No | 10 | 90 | 100 | $P_2 = 10/100 = 0.1\%$ |

### 8.2.2 Risk Categories

It is important to identify the types of risk involved in a typical project for the purpose of risk management. Risks can be classified into the following four main categories:

- Strategic risks
- Operational risks

- Financial risks
- Safety and hazard risks.

Potential risk expectancy is highest with strategic and operational risks, and lowest with financial and safety and hazard risks. Each of these risk categories has unique factors that require different risk management techniques. Factors involved with each risk category are given in Figure 8.2.

### 8.2.2.1 Strategic Risks

Strategic risks allow the project management to look at external risks which may affect the final project outcome. Strategic risks are those that arise from the fundamental decisions that project management take concerning a project's objectives. Strategic risks include following risk factors:

- *Technology risks.* A bad technology decision can disrupt or destroy an otherwise compelling project. It is also important to mention that bad project management is a much more likely cause of technology risk. Frequently changing product development or testing environment can also lead to this risk. Technology risks must be evaluated as a part of risk management.

**Figure 8.2** Risk categories.

- *Customer risks.* A customer proposing new requirements, with their consequences for the project schedule, and insisting on not accepting the product even though it meets all specifications is an example of customer risk.
- *Project management (PM) risks.* Unclear success measures to verify the successful completion of each project phase lead to PM risks.
- *Requirement risks.* Poorly defined, inconsistent, incomplete, or frequently changing requirements lead to requirement risks.
- *Technical risks.* Technical risk is the possible impact changes could have on a project when an implementation does not work as anticipated.

#### 8.2.2.2   Operational Risks

Operational risk is the potential loss resulting from insufficient or failed procedures, systems or policies. Operational risks include following risk factors:

- *Quality and process risks.* Incorrect application of process tailoring, unskilled or incompetent people to perform the task at hand will lead to this risk.
- *Scheduling risks.* An unrealistic schedule, missing tasks, or delays in completion of tasks lead to schedule risk.
- *Personnel risks.* Lack of input of key personnel to a project or personnel deliberately acting against the project success will lead to personnel risk. Also, accidents and illness, outdated professional skills, personnel disputes, information leaks, and theft are included in personnel risks.
- *Legal risks.* Legal risks include regulatory, litigation, and contractual issues. Personnel misconduct, accidents, and product liability are examples of litigation legal risk. Prior to litigation, areas of uncertainty need to be identified that affect the project objectives and narrow the possible outcomes from particular events which will lead to litigation. Contract risk is the most difficult to track among legal risks. One contract may create significant legal risk such as loss incurred due to the buyer canceling the contract. Changes in laws and regulations may considerably impact a project's success.
- *Liability risks.* Liability risks involve the threat of the company or individual having to accept the consequences of damage or of violating standards due to operations, a product, an act, or neglect.

#### 8.2.2.3   Financial Risks

Financial risks include following risk factors:

- *Credit risks.* Credit risks refer to the probability of loss due to a borrower's or counterparty's failure to perform on an obligation – making payments on any type of debt. For many banks, loans are the largest and most apparent source of credit risk.
- *Budget risks.* Incorrect budget assessment or project goal change leads to budget risk. This risk may cause a delay in the delivery of the project.
- *Liquidity risks.* Liquidity risk is a financial risk that for some period of time a given financial asset, security, or commodity cannot be bought or sold quickly enough to avoid or minimize a loss. Liquidity risk management is a crucial banking activity and an integral part of the asset and liability management process.

#### 8.2.2.4 Hazard and Safety Risks

- *Hazard risks.* A hazard is any source of possible damage, harm or undesirable health effects on something or someone at work. "Hazard risk is the likelihood that a person may be harmed or suffers adverse health effects if exposed to a hazard."[1] Hazard risks include chemical and substance hazards, biological hazards and infectious diseases, physical hazards, environmental hazards, psychosocial hazards such as fatigue, and work-related stress.
- *Safety risks.* Safety risk is a quantitative representation of the possible accidents resulting from the operation and maintenance of machinery, equipment, and any other physical objects.
- *Security risks.* Securing company interests against operational risks is an important issue for any management team. Loss of life, intellectual property, resources, and reputation can have a damaging impact on a company's business. Protecting classified information from unauthorized access, use, or disclosure is also included in security risks.

### 8.2.3 Risk Management

Risk appetite is the amount and type of risk that an organization wishes to take in order to meet its strategic objectives. Organizations will have different risk appetites that may change over time depending on their sector, culture, and objectives. All complex projects have risks. Managing these risks and preventing them from damaging the project is a crucial activity of project management. The risk management process depicted in Figure 8.3 begins with the practice of identifying events that may have a negative effect on the project's ability to achieve its performance.

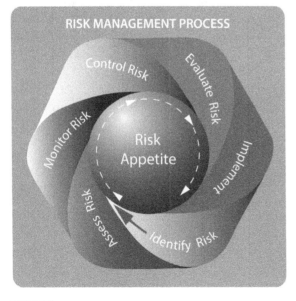

**Figure 8.3** Risk management.

---

1 Occupational Safety and Health.

### 8.2.3.1 Identifying Risks

There is no simple way of identifying risks. The project manager should help to identify risks that may threaten the aim of the project. The full project team and other relevant stakeholders should also be actively involved in this process. Risk identification should begin early in the design process when uncertainty is highest; action should be taken when the risks are easier to address. There are a number of methods in use for risk identification. Surveys, brainstorming, interviews, working groups, experimental knowledge, risk trigger questions, direct observations, lessons learned, checklists, cause effect diagrams, the Delphi technique, and decision tree analysis can be used for the identification of project risks. Risk identification methods vary in complexity and each method has advantages and disadvantages.

Any possible risk identified by the team member should be recorded, regardless of whether other members of the group consider it to be significant. With a large number of potential risk events, it might be difficult to address each and every setting. Hence, It may be necessary to prioritize risks. All possible risks identified by brainstorming should be recorded in a consistent form using the following sentence structure: cause ⇒ risk ⇒ effect. For example:

because of [cause], the [risk] may or may not occur, which results in [effect].

Table 8.2 shows how to record identified risk consistently. This list of risks is documented in the form of a register called a risk register. It is important to note that at the risk identification stage, the impacts on cost and time are not analyzed; that occurs in the qualitative risk assessment or quantitative risk assessment processes.

### 8.2.3.2 Risk Assessment

The risk identification process precedes risk assessment, as shown in Figure 8.4. The first activity in the risk assessment process is to define risk criteria that will be used for evaluation of risks to be used in design and project development. Risk criteria for evaluating the significance of risks need to be identified by the project developer.

The risk criteria should reflect the objectives and framework for the risk assessment. The risk criteria selected should be constantly reviewed. Before we detail the risk criteria, the risk categories shown in Figure 8.2 for which risks will be evaluated should be defined.

In general, risk assessment considers two types of risk analysis: quantitative and qualitative risk analysis. Quantitative risk analysis is the process of providing numerical values (such as profit and loss, investment returns, environmental consequences) of the overall effect of risks on the project objectives. Qualitative analysis estimates the magnitude and likelihood of possible

**Table 8.2** Consistent record of identified risks (risk register).

| Cause | Risk | Effect |
|---|---|---|
| If we use an unknown contractor | then unexpected scheduling problems may occur <br> (*an uncertain risk*) | which would result in overspending on the project <br> (*an effect on the budget objective*) |

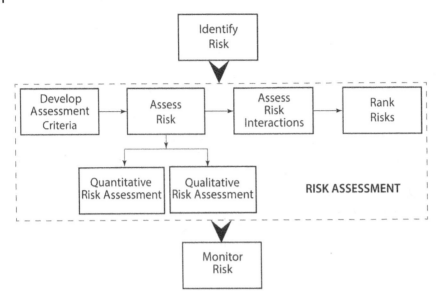

**Figure 8.4** Risk assessment.

consequences that are defined and described in detail. Risk analysis is qualitative and includes qualifying and ranking the potential risks according to their potential effect on the project objectives that have been identified in terms of probability and impact on the project. Qualitative analysis is used as an initial assessment to identify risks which will be the subject of further detailed analysis. A comprehensive analysis of a project management should include looking at both the qualitative and quantitative factors that would impact a project decision-making.

The potential interdependence of risks should not be overlooked. Some of the risks identified may create or enhance other risks – the activation of one risk may initiate the activation of another risk. Although understanding how different risks relate to one another is a difficult task, risks interactions should be comprehensively analyzed during the life cycle of design project.

### 8.2.3.3 Risk Monitoring and Control

Risk monitoring and control is the process of discovering, analyzing, and planning for newly identified risks. It is important to note that risk monitoring is expected to be a regular, ongoing process across the entire project life cycle. Newly recognized risks and symptoms of earlier identified risks should be tracked and communicated right away for evaluation and action.

### 8.2.3.4 Risk Evaluation and Implementation

Risk evaluation is the process of defining what the estimated risk really means to project management; it compares the estimated risk against the given risk criteria so as to determine the significance of the risk. A main part of this evaluation will be the understanding of how people perceive risks. During the risk evaluation process decisions have to be made regarding which risks need to be considered and which do not, as well as the risk priorities.

**Figure 8.5** Risk probability/impact chart.

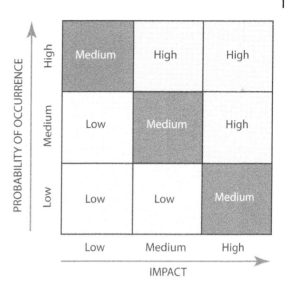

As shown Figure 8.5 two measures (probability occurrence and the impact of the risk) help us to understand the priority of risks and then decide what resources will be allocated to manage each particular risk. When evaluating risks, we would like to know how likely it is that the risk will occur. A risk that is highly unlikely can be discounted from the evaluation process. The research project team may accept risks with low magnitude of impact and low probability of occurrence. However, the project team can transfer the risks with high magnitude of impact and high probability of occurrence to third-party organizations such as insurance providers.

A number of standards have been developed worldwide for implementing risk management systematically. These standards pursue the establishment of a common view on related problems, and are generally set by professional international standards bodies, societies, or industry groups. Some of the commonly used standards are: ISO 31000 2009 – Risk Management Principles and Guidelines; ISO/IEC 31010:2009 - Risk Management - Risk Assessment Techniques; A Risk Management Standard – IRM/Alarm/AIRMIC 2002.

---

**CASE STUDY 8.1**

Tank cleaning and gas freeing is an extremely hazardous activity. When working in an enclosed space workers are exposed to a number of hazards that in some cases have led to injury or even death. The additional risk from the toxic effect of petroleum gas during the tank cleaning cannot be overestimated. It is also important to note that oil tanks do not have natural ventilation and they are a place which is difficult to enter and escape from in case of an emergency. Therefore, it is important that the greatest possible care is exercised in all operations related to tank cleaning and gas freeing.

*(Continued)*

**CASE STUDY 8.1 (Continued)**

**Figure 8.6** Tank cleaning.

**Scenario**

Suppose that a tank in an old crude oil tanker (Figure 8.6) is to be cleaned. The chief officer of the ship supervises all crude oil tank cleaning operations. The following items should be checked before the cleaning operation:[2]

- Identify all the operational risks related to cleaning and reduce them as much as possible.
- Various precautions and start of work must be notified to crew members.
- Inform the crew to wear proper personal protective equipment.
- The ship's fire hoses and fire-fighting equipment must be ready for immediate use.
- Combustible gas detectors and oxygen content detectors must be ready for use and in good working order.
- Flammable mixture checks must be made 1 meter below deck and middle level of the tank(s) to confirm $O_2$ level below 8%.
- An effective deck watch in attendance on board and adequate supervision of operations on the ship must be provided.

---

2 Summary from ShipsBusiness.com, http://shipsbusiness.com/tank-cleaning-purging-gas-free-check-items.html, accessed November 1, 2016.

The following is a list of items to be checked during tank cleaning:

- Is oxygen content of delivered inert gas below 5%?
- Do the cleaning tanks have positive pressure?
- Is the cleaning line pressure proper?
- Are cleaning machines working properly?
- Are stripping systems working properly?
- Is the surrounding sea surface in good condition?

Identify all the operational risks related to cleaning and reduce them to an acceptable level as much as possible. To illustrate the risk management with a simplified hypothetical scenario, the following four risks are considered and presented in Tables 8.3 and 8.4. The probability values are indicative and used for the purpose of this example. Since risks R-3 and R-4 are interdependent, the occurrence of these two risks may double the risk effects.

**Table 8.3** Qualitative analysis of risk scenario.

| Risk | Risk name | Risk effect | Risk category | Risk type | Risk owner | P | Impact | Risk score |
|------|-----------|-------------|---------------|-----------|------------|---|--------|------------|
| R-1 | Fire/explosion | If a tank contains flammable mixture of vapor and air exists inside a tank, then any source of ignition may cause a fire and/or explosion. | Technical | Threat | Mark T. | 3 | 3 | 9 |
| R-2 | $O_2$ level | If $O_2$ level inside the tank is above 8%, then any source of ignition may cause a fire and/or explosion. | Technical | Threat | John K. | 3 | 3 | 9 |
| R-3 | Untrained workers | Unsafe work environment | Technical | Threat | Adam J. | 2 | 2 | 4 |
| R-4 | Mishandled operation/human error | Risk with high loss | Technical | Threat | Kent A. | 1 | 2 | 2 |
| R-5 | Implementing non-human cleaning | Reusable robotics | Budget | Opportunity | Eric N. | 1 | 4 | 4 |

*(Continued)*

**CASE STUDY 8.1 (Continued)**

Table 8.4 Quantitative analysis of risk scenario.

| Risk# | Risk name | Risk effect | Risk owner | P% | Impact | Expected value |
|---|---|---|---|---|---|---|
| R-1 | Fire/ explosion | If a tank contains flammable mixture of vapor and air exists inside a tank, then any source of ignition may cause a fire and/or explosion. | Mark, T | 30% | $5 million | $1.5 million |
| R-2 | $O_2$ level | If $O_2$ level inside the tank is above 8%, then any source of ignition may cause a fire and/or explosion. | John K. | 20% | $5 million | $1.0 million |
| R-3 | Untrained workers | Unsafe work environment | Stephen J. | 1% | $2 million | $20K |
| R-4 | Mishandled operation/ human error | Risk with high loss | Kent A. | 1% | $2 million | $20K |

*R-1: fire and explosion.* If a tank contains flammable mixture of vapor and air exists inside a tank, then any source of ignition may cause a fire and/or explosion.

*R-1: assess and control.* Flammable mixture checks must be made 1 meter below deck and middle level of the tank(s) to confirm flammable mixture of vapor level is below the lower flammable limit (LFL).

*R-2: source of ignition.* If the oxygen content of the tank is above 8% and the flammable mixture of vapor level is above the LFL, then any source of ignition may cause a fire and/or explosion.

*R-2: assess and control.* Check the level of $O_2$ by measuring at fore and aft, high, middle and lower sections of the tank. Make sure the $O_2$ level is below 8%.

*R-3: untrained workers.* If untrained workers are assigned to clean tanks, the work environment will be unsafe. During cleaning, operation health and safety should be a top priority.

*R-3: assess and control.* Before the cleaning work assignment starts, the employer must provide proper training for all workers who are required to work in permit spaces. After the training, employers must ensure that the employees have the understanding, knowledge and skills needed to safely carry out their duties.

*R-4: mishandled operation/human error.* If the cleaning operation is mishandled by the personnel there will be high loss, potential for injury, immediate danger to health, even to life.

*R-4: assess and control.* Appropriate personnel must be supervised and their daily activities controlled to prevent mishandling.

As shown in Table 8.3, in qualitative risk analysis, the probability and impact values are assigned and the risk score is decided – probability values are recorded as 1 for low risk and 5 for high risk. Each risk is assigned a probability value and an impact value and the risk score is calculated. Based on the risk scores, we can now perform the prioritization of risks. In this example, risks R-1 and R-2 are the most important risks and deserve more attention. These values should be derived from different sources. Of course, the correctness of the analysis depends on the accuracy of the assigned values. As mentioned earlier, quantitative analysis is performed after the qualitative analysis.

### 8.2.3.5  Threats and Opportunities in Project Management

Although risk events are threats and reflect a negative or bad situation in some cases we can consider opportunities as risks as well. In other words, risk events could have negative or positive effects on the objective of the project. In risk management, project teams should try to balance the possible negative consequences of risk against the possible benefits of the opportunity.

Robots are ideal for use in hazardous environments by removing people from direct contact with hostile conditions, as in the example scenario above. However, development of a system which uses robots requires substantial investment which may create budget risks. On the other hand, if such a system is developed it can be reused in the tank cleaning operation over and over. In short, there is no doubt that when implementing a system which uses robots many hazards are dramatically reduced and some even can be eliminated.

## 8.3  Basic Mathematical Concepts in Reliability Engineering

Reliability engineering, as a technical discipline, developed along with the growth of commercial aviation following World War II. The cost of aviation-related accidents motivated the aviation industry to contribute seriously to the advancement of the reliability engineering discipline.

The term *reliability* is usually defined as the probability that a system, device, or component will successfully perform a required function for:

- a given range of operating conditions;
- a specific environmental condition;
- a stated survival time.

### 8.3.1  Basic Reliability Equation

Discrete or continuous density functions, distribution functions, and probabilistic parameters such as random variables are utilized in the development of reliability theory.

The failure cumulative density function of a component or system at time $t$ is defined by

$$F(t) = \int_0^t f(t)dt, \tag{8.5}$$

where $f(t)$ is the failure probability density function of the random variable. $F(t)$ is the *unreliability function* when considering a failure. If the random variable is discrete, the integral in equation (8.5) is replaced by a summation. The reliability, $R(t)$, or the probability that the component or system experiences no failures during the time interval from zero to $t$, is given by

$$R(t) = 1 - F(t) = 1 - \int_0^t f(t)dt. \tag{8.6}$$

The properties of the probability density function are such that

$$R(t = 0) = 1, \tag{8.7}$$

$$R(\infty) = 0. \tag{8.8}$$

After differentiating, equation (8.6) can be rearranged to yield the failure density function in terms of the reliability:

$$f(t) = -\frac{d}{dt}R(t). \tag{8.9}$$

The *hazard rate or instantaneous failure rate* is the conditional probability that the component has not failed by time $t$:

$$h(t) = \lim_{\Delta t \to 0} \frac{F(t + \Delta t) - F(t)}{\Delta t} \frac{1}{R(t)} = \frac{f(t)}{R(t)}. \tag{8.10}$$

Equation (8.10) is the fundamental relationship in reliability analysis. For example, if one knows the density function of the time to failure, $f(t)$, and the reliability function, $R(t)$, then the hazard rate function, $h(t)$, for any time $t$ can be obtained. Substituting equation (8.10) into equation (8.9) yields

$$h(t) = -\frac{1}{R(t)}\frac{dR(t)}{dt}. \tag{8.11}$$

Equation (8.11) can also be written as

$$h(t) = -\frac{d}{dt}\ln R(t). \tag{8.12}$$

Integrating both sides of equation (8.12) yields

$$\ln R(t) = -\int_0^t h(t)dt \tag{8.13}$$

and, finally, reliability can be expressed as

$$R(t) = \exp\left[-\int_0^t h(t)dt\right]. \tag{8.14}$$

Equation (8.14) is the general expression for the reliability function. Note that $h(t)$ is the time-dependent failure rate. If a constant failure rate is required for a component, as is often the case for electronic equipment, $h(t)$ is set to $\lambda$:

$$h(t) = \lambda. \tag{8.15}$$

Substituting equation (8.15) into equation (8.14) gives

$$R(t) = e^{-\lambda t}. \tag{8.16}$$

In equation (8.16), $\lambda$ is the *failure rate*, also referred to as the *hazard* or *instantaneous failure rate*, which forecasts the number of failures of a system or component that has occurred over a particular length of time. Equation (8.16) is simple and can be applied at any time other than after component failures. Only an exponential distribution has this unique feature. Note that $e$ is the base of natural logarithms (2.718281828).

### 8.3.2 Mean Time to Failure

The *mean time to failure* (MTTF) is the average time that a non-repairable system, device, or component will operate before experiencing a failure. From the basic probability definitions, the mean value of a random continuous variable $x$ is given by

$$E(x) = \int_{-\infty}^{\infty} xf(t)dx. \tag{8.17}$$

Hence, for time to failure, at $t \geq 0$,

$$E(t) = \text{MTTF} = \int_{0}^{\infty} tf(t)dt, \tag{8.18}$$

and it can be shown that

$$\text{MTTF} = \int_{0}^{\infty} R(t)dt. \tag{8.19}$$

## 8.4 Probability Distribution Functions Used in Reliability Analysis

There are many statistical distributions which can be used in reliability analysis. However, a relatively small number of statistical distributions satisfy most needs in reliability analysis. This section briefly discusses selected probability distributions for finding the failure rate of a single component.

### 8.4.1 Normal Distribution

There are two main applications of the normal distribution to reliability. One application deals with the failure due to wear, and the other application is the analysis of manufactured parts variability and their ability to meet specifications.

The normal distribution probability density function is given by

$$f(t) = \frac{1}{\hat{\sigma}\sqrt{2\pi}} \exp\left[-\frac{1}{2}\left(\frac{t-\mu}{\hat{\sigma}}\right)^2\right]. \tag{8.20}$$

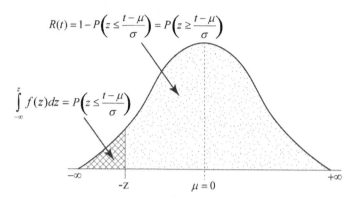

$$R(t) = 1 - P\left(z \le \frac{t-\mu}{\sigma}\right) = P\left(z \ge \frac{t-\mu}{\sigma}\right)$$

$$\int_{-\infty}^{z} f(z)dz = P\left(z \le \frac{t-\mu}{\sigma}\right)$$

**Figure 8.7** Normal distribution.

The standard normal distribution density function is

$$f(z) = \frac{1}{\sqrt{2\pi}} \exp\left(-\frac{z^2}{2}\right), \tag{8.21}$$

where $\mu = 0$ and $\sigma^2 = 1$. In Figure 8.7, $\int_{-\infty}^{z} f(z)dz$ is the area of failure, which is probability of failure, $P(z \le \frac{t-\mu}{\sigma})$. By definition, reliability refers to no failure, thus

$$R(t) = 1 - \int_{-\infty}^{z} f(z)dz = 1 - P\left(z \le \frac{t-\mu}{\sigma}\right). \tag{8.22}$$

Note that $z$ converts from the normal to standard normal distribution by using the transformation

$$z = \frac{t-\mu}{\sigma}. \tag{8.23}$$

The hazard rate, $h(t)$, for the normal distribution is

$$h(t) = \frac{f(t)}{R(t)} \tag{8.24}$$

where $f(t)$ is the failure distribution function given by

$$f(t) = \frac{\int_{-\infty}^{z} f(z)dz}{\hat{\sigma}}. \tag{8.25}$$

## Example 8.3

A main turbine engine powering a cargo ship has been observed to follow a normal distribution with $\mu = 400$ hours and $\hat{\sigma} = 50$ hours. What is the reliability of the turbine for a mission time (or time before maintenance) of 350 hours? What is the hazard rate at 300 hours?

## Solution

For the reliability at 350 hours we require

$$z = \frac{t - \mu}{\hat{\sigma}} = \frac{350 - 400}{50} = -1.0.$$

From Table C.1 in Appendix C, we obtain the area (probability of failure) to the left of $z$: $\int_{-\infty}^{z} f(z) = \text{area}(z < -1.0) = 0.1587$ (see Figure 8.7). Then the reliability is

$$R(t = 350) = 1 - \int_{-\infty}^{z} f(z)dz = 1 - P\left(z \leq \frac{t - \mu}{\sigma}\right) = 1 - 0.1587 \approx 0.84.$$

The hazard rate is

$$h(t) = \frac{f(t)}{R(t)},$$

and for the rate at 300 hours we require

$$z = \frac{t - \mu}{\hat{\sigma}} = \frac{300 - 400}{50} = -2.0.$$

From Table C.1 in Appendix C, we obtain the area (probability of failure) to the left of $z$: area $(z < -2.0) = 0.0228$. Then the reliability is

$$R(t = 300) = 1 - \int_{-\infty}^{z} f(z)dz = 1 - P\left(z \leq \frac{t - \mu}{\sigma}\right) = 1 - 0.0228 \approx 0.98,$$

and we have

$$f(t) = \frac{\int_{-\infty}^{z} f(z)dz}{\hat{\sigma}} = \frac{0.0228}{50} \approx 0.00046.$$

Then

$$h(t) = \frac{f(t)}{R(t)} = \frac{0.00046}{0.98} \approx 0.00047 \text{ failures/hour.}$$

### 8.4.2 Lognormal Distribution

The lognormal distribution is a continuous probability distribution of a random variable whose logarithm is normally distributed: it is the normal distribution with $\ln(t)$ as the variate. The lognormal distribution is mainly used in reliability analysis of semiconductors, fatigue life of mechanical components, and in maintainability analysis. The probability density function of the lognormal distribution is of the form

$$f(t) = \frac{1}{\hat{\sigma} t \sqrt{2\pi}} \exp\left[-\frac{1}{2}\left(\frac{\ln(t) - \mu}{\hat{\sigma}}\right)^2\right]. \tag{8.26}$$

Equation (8.26) can be related to the standard normal variant $z$ by

$$f(z) = \frac{1}{\sqrt{2\pi}} \exp\left(-\frac{z^2}{2}\right).$$

(8.27)

For the lognormal distribution, the $z$ equation becomes

$$z = \frac{\ln t - \mu}{\hat{\sigma}},$$

(8.28)

and the reliability equation is

$$R(t) = 1 - \int_{-\infty}^{z} f(z)dz = 1 - P\left(z \le \frac{\ln t - \mu}{\hat{\sigma}}\right).$$

(8.29)

The hazard rate, $h(t)$, is

$$h(t) = \frac{f(t)}{R(t)},$$

(8.30)

where the failure distribution function $f(t)$ is

$$f(t) = \frac{\int_{-\infty}^{z} f(z)dz}{t\hat{\sigma}}.$$

(8.31)

### Example 8.4

Repeated firing will produce fatigue cracks in a gun barrel. When a crack size reaches a critical length, the barrel is no longer useful. Assume that gun barrel failure due to fatigue occurs according to the lognormal distribution with $\mu = 9$ and $\hat{\sigma} = 3$.

a) Determine the reliability for a 1200 firing mission.
b) What is the hazard rate when the gun has been fired 900 times?

### Solution

a) The $z$ value required is

$$z = \frac{\ln t - \mu}{\hat{\sigma}} = \frac{\ln(1200) - 9}{3} = -0.64.$$

From Table C.1 in Appendix C, we obtain the area (probability of failure) to the left of $z$: $\int_{-\infty}^{z} f(z) = \text{area}(z < -0.64) = 0.2611$. Then the reliability is

$$R(t = 1200) = 1 - \int_{-\infty}^{z} f(z)dz = 1 - P\left(z \le \frac{\ln t - \mu}{\sigma}\right) = 1 - 0.2611 \approx 0.74.$$

b) The hazard rate when the gun has been fired 900 times is

$$h(t = 900) = \frac{f(t)}{R(t = 900)}$$

and we require

$$z = \frac{\ln t - \mu}{\hat{\sigma}} = \frac{\ln(900) - 9}{3} = -0.73.$$

From Table C.1 in Appendix C, we obtain the area (probability of failure) to the left of $z$: area $(z < -0.73) = 0.2327$. Then the reliability is

$$R(t = 900) = 1 - \int_{-\infty}^{z} f(z)dz = 1 - P\left(z \le \frac{\ln t - \mu}{\sigma}\right) = 1 - 0.2327 \approx 0.77$$

and we have

$$f(t) = \frac{\int_{-\infty}^{z} f(z)dz}{t\hat{\sigma}} = \frac{0.77}{900 \times 3} \approx 0.000285.$$

Then

$$h(t) = \frac{f(t)}{R(t)} = \frac{0.000285}{0.77} = 0.000370 \text{ failures/firing.}$$

### 8.4.3 Exponential Distribution

This is the simplest and most widely used distribution in reliability engineering and is used for reliability prediction of electronic equipment. The failure probability density function $f(t)$ of a component is given by

$$f(t) = \frac{1}{\theta} \exp\left(-\frac{t}{\theta}\right), \quad t > 0, \tag{8.32}$$

where $\theta$ is the mean life of a component. Let $\lambda = 1/\theta$. Then

$$f(t) = \lambda e^{-\lambda t}. \tag{8.33}$$

Substituting equation (8.33) into equation (8.6) and integrating yields the reliability function

$$R(t) = e^{-\lambda t}, \tag{8.34}$$

where $\lambda$ is the constant failure rate of a component. Substituting equation (8.34) into equation (8.19) and integrating leads a component's *mean time to failure*:

$$\text{MTTF} = \int_{0}^{\infty} e^{-\lambda t} dt = \frac{1}{\lambda}. \tag{8.35}$$

The hazard rate is

$$h(t) = \frac{f(t)}{R(t)}. \tag{8.36}$$

### Example 8.5

A copy machine has a constant breakdown rate of one failure every 30 days of continuous operation. What is the reliability associated with the copy machine for non-stop operation that requires 15 hours? Find the hazard rate after 10 hours of operation.

## Solution

The MTTF is $24 \times 30 = 720$ hours. So

$$\lambda = \frac{1}{\text{MTTF}} = \frac{1}{720} = 0.00139 \text{ failure/hour}$$

So

$$R(t = 15) = e^{-\lambda t} = e^{-(0.00139)15} = 0.98$$

and

$$h(t) = \frac{f(t)}{R(t)} = \frac{\lambda e^{-\lambda t}}{e^{-\lambda t}} = \lambda = 0.00139 \text{ failures/hour}.$$

### 8.4.4 Weibull Distribution

Another life-predicting distribution is the Weibull distribution. This is a general distribution which, by adjusting the shape parameter, $b$, and scale parameter, $\theta$, can model a wide range of life distribution characteristics of different classes of engineered products. The failure probability density function $f(t)$ of a component is given by

$$f(t) = \frac{b}{\theta}\left(\frac{t-\gamma}{\theta}\right)^{b-1} \exp\left[-\left(\frac{t-\gamma}{\theta}\right)^{b}\right], \quad \text{for} \quad b > 0, \quad \theta > 0, \quad t \geq 0. \tag{8.37}$$

Since the minimum life, $\gamma$, is often zero (failure assumed to start at $t = 0$), the failure density function becomes

$$f(t) = \frac{b}{\theta}\left(\frac{t}{\theta}\right)^{b-1} \exp\left[-\left(\frac{t}{\theta}\right)^{b}\right]. \tag{8.38}$$

Substituting equation (8.38) into equation (8.6) and integrating yields the component reliability function

$$R(t) = \exp\left[-\left(\frac{t}{\theta}\right)^{b}\right]. \tag{8.39}$$

By utilizing equation (8.10) with equations (8.38) and (8.39), the hazard rate can be determined:

$$h(t) = \frac{bt^{b-1}}{\theta^{b}}. \tag{8.40}$$

At $\theta = 1$, the curves of equation (8.40) for various values of $b$ are shown in Figure 8.8. Note that when $b = 1$ the Weibull hazard funwhich is a linearly increasing hazard rate function. Substituting equation (8.39) into equation (8.19) and integrating yields the MTTF of a component,

$$\text{MTTF} = \theta\Gamma\left(\frac{1}{b} + 1\right), \tag{8.41}$$

**Figure 8.8** Weibull distribution.

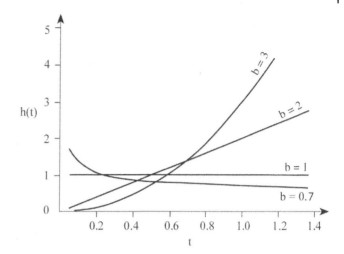

where $\Gamma$ is the gamma function given by

$$\Gamma(\beta) = \int_0^\infty t^{\beta-1}e^{-t}dt, \quad \text{for } \beta > 0. \tag{8.42}$$

Values of $\Gamma$ corresponding to $\beta$ can be obtained from Figure 8.9.

**Figure 8.9** The $\Gamma$ function.

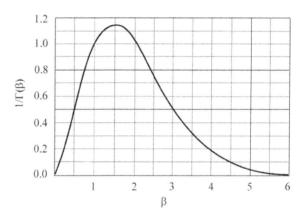

### ☐ Example 8.6

Boiler reliability is crucial for power plants. Fired boilers are liable to reheated tube failures due to oxidation pitting. Reheated tube failure times are found to be Weibull distributed with $b = 0.8$ and $\theta = 1500$ hours. Calculate:

1) The reliability for 600 hours of operation.
2) The mean time to failure.
3) The hazard rate.

**Solution**

1. Using equation (8.39), we obtain

$$R(t = 200) = \exp\left[-\left(\frac{t}{\theta}\right)^b\right] = \exp\left[-\left(\frac{600}{1500}\right)^{0.8}\right] = 0.62.$$

2. The time to failure is

$$\text{MTTF} = \theta\Gamma\left(\frac{1}{b} + 1\right) = 1500\Gamma\left(\frac{1}{0.8} + 1\right)$$
$$= 1500\Gamma(2.25) = 1500 \times 1.16 = 1{,}740 \text{ hours.}$$

3. The hazard rate is

$$h(t = 600) = \frac{bt^{b-1}}{\theta^b} = \frac{0.8 \times (600)^{0.8-1}}{(1500)^{0.8}} = 0.00064 \text{ failures/hour.}$$

Note that from Figure 8.9, we obtain $1/\Gamma(2.25) = 0.86$, from which $\Gamma(2.25) = 1.16$.

## 8.5 Failure Modeling

Failure modeling is crucial for reliability engineering. Validated failure rate models are important for the development of prediction techniques, design, and analysis methodologies. Over the years, the "bathtub" curve has become widely accepted by reliability engineers. It is accepted as especially appropriate for electronic components and systems. Figure 8.10 displays a typical time versus failure rate curve for a component – typical hazard rate curves for different values of $b$ at the constant value of $\theta = 1$. The characteristics of a hazard function are often associated

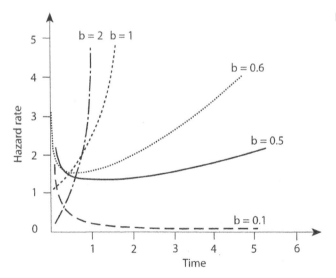

**Figure 8.10** Hazard rate curves for $\theta = 1$.

with certain components and applications. Different hazard functions are modeled with different failure distribution models.

Experience indicates that the failure behavior of many components follows the bathtub hazard rate function shown in Figure 8.11. The entire life cycle of a component is plotted on the *x*-axis and the number or percentage of failures is plotted on the *y*-axis. This curve represents the failure of components that have three different life zones.

In zone I, the earliest period of operation, the failure rate will be high but will decrease with time. This zone is known as the *infant mortality* (decreasing failure rate) period and is attributed to design or manufacturing defects due to the lack of adequate controls in the manufacturing process. When these mistakes are not discovered by quality control inspections, an early failure is likely to result. At the end of zone I, the defective components are eliminated.

Zone II (the flat part) of the curve depicts the lowest constant failure rate over an extended period, and is known as *useful life*. The cause of failure during this stage is not readily evident, and failures are assumed to be random. This is the period controlled by *chance failures*. Components are designed to operate under certain conditions and at certain stress levels. When these stress levels are exceeded due to random unexpected events, a chance failure will occur. Since failure at this stage is constant, the exponential distribution is used as the basis for engineering design and failure prediction – it is widely applicable for complex components and systems reliability design.

Zone III, the *wear-out* period, is characterized by an increasing failure rate as a result of equipment deterioration due to fatigue or weakening of materials in time. For example, mechanical components such as bearings have a certain period of survival and will finally wear out and fail. Wear-out can be prevented by replacing or repairing the deteriorating component before it fails.

Different statistical distributions can be used to describe each zone shown in Figure 8.11. The infant mortality period can be modeled by the Weibull or gamma, the useful life period by the exponential, and the wear-out period by the gamma or normal distribution.

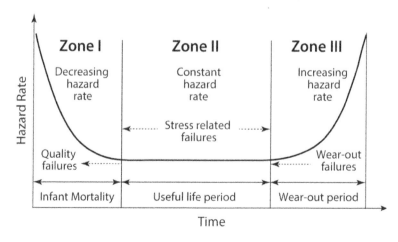

**Figure 8.11** The bathtub hazard rate curve.

## 8.6 Probability Plotting

Reliability data can be presented in two ways: using the mathematical equations described previously or by plotting the data on graphs. The use of probability plots enables us to estimate important parameters accurately and with simplicity. For example, Weibull distribution plotting can be illustrated with the times to failure of a component for a set of 16 samples of manufactured components as shown in Table 8.5. The following steps are used for probability plotting:

1) Data should be ranked in order from the smallest to the largest value.
2) Estimate $E_i$, for the cumulative percent for each data point. Different equations have been used for this calculation. We will use the *midpoint plotting positions*,

$$E_i = \frac{100(i - 0.3)}{n + 0.4}, \tag{8.43}$$

As an example, as shown in Table 8.5, suppose that eight out of sixteen newly manufactured components failed at the following time periods: 85, 160, 190, 250, 300, 380, 420, 480 hours. Using the *median plotting position method*, estimate the reliability of newly manufactured components:

1) for a mission of 260 hours;
2) for a mission of 280 hours, given that the component has survived up to 600 hours; and
3) calculate the hazard rate at $t = 600$ hours.

**Table 8.5** Cumulative percent failure of manufactured components.

| Rank order $i$ | Time to failure (hours) | Cumulative percent failure $E_i = \dfrac{100(i - 0.3)}{n + 0.4}$ |
|:---:|:---:|:---:|
| 1 | 85 | 4.27 |
| 2 | 160 | 10.37 |
| 3 | 210 | 16.46 |
| 4 | 250 | 22.56 |
| 5 | 300 | 28.66 |
| 6 | 380 | 34.76 |
| 7 | 420 | 40.85 |
| 8 | 480 | 49.85 |
| ... | ... | ... |
| ... | ... | ... |
| | | $n = 16$ |

Estimated $E_i$ results are given in Table 8.5.

1) The data in Table 8.5 are used to develop Weibull plot shown in Figure 8.12. From Figure 8.12, the Weibull slope $b = 1.4$ is obtained. Figure 8.12 shows the percent failure versus time. For example, 5% of the components will fail before 82 hours. From the figure the cumulative

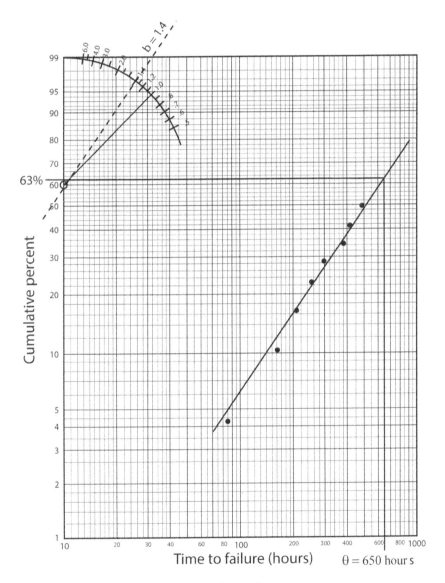

**Figure 8.12** Weibull plot for newly manufactured components.

percent failure at 260 hours starting at time zero is approximately $F(t) = 22\%$. Hence, the reliability for a mission of 260 hours is

$$R(t = 260) = 1 - F(t = 260)$$

$$= 1 - 0.22$$

$$= 0.78.$$

2) The probability that the component will fail at time $t_o + \Delta t$, given that the component has survived up to $t_o$, is called the *conditional reliability*. For a mission of $\Delta t$ hours this is given by

$$R(t_o, t_o + \Delta) = \frac{R(t_o + \Delta t)}{R(t_o)}, \tag{8.44}$$

where $t_o$ is the time at the beginning of the mission and $\Delta t$ is the length of the mission. From Figure 8.12, $F(t_o = 600)$ and $F(t_o + \Delta = 880)$ are estimated to be 58% and 78%, respectively. Hence, the corresponding reliabilities are

$$R(t_o = 500) = 1 - F(t = 600) = 1 - 0.58 = 0.42,$$

$$R(t_o + \Delta = 750) = 1 - F(t = 880) = 1 - 0.78 = 0.22.$$

Thus, from equation (8.44),

$$R(t_o = 600, t_o + \Delta = 880) = \frac{0.22}{0.42} = 0.52.$$

Note that the reliability for a mission of 280 hours starting at 600 hours ($R = 0.22$) is less than the reliability for a mission of 280 hours starting at zero hours ($R = 0.78$).

3) The hazard rate at $t = 600$ hours is

$$h(t = 600) = \frac{bt^{b-1}}{\theta^b}$$

$$= \frac{1.4(600)^{1.4-1}}{650^{1.4}}$$

$$= 0.0021 \text{ failures/hour.}$$

Note that $\theta = 650$ hours is obtained by reading the value of the time to failure, $t$, at $F(t) = 63\%$ from the Weibull plot.

## 8.7 Basic System Reliability

An important problem in system design is to make sure that a system will function with acceptable performance and reliability. This becomes an important design criterion, especially when the system involves many components and a number of design constraints. The performance

**Figure 8.13** Component block diagram of a series system.

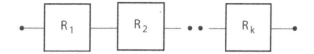

of a system with a collection of $n$ components depends on the performance of each and every component. In this section the reliability of multicomponent systems that are connected in series, parallel, and in more complex configurations is discussed.

### 8.7.1 Series Systems

The simplest and most commonly occurring configuration in reliability engineering is the series system. A series reliability configuration is represented by the block diagram shown in Figure 8.13 with $k$ components. Let $R_1$ be the reliability of component 1, $R_2$ the reliability of component 2, and so on. The arrangement shown in Figure 8.13 implies that all of the components must work to ensure system success. The system fails if just one element fails. The reliability of the system, $R_s$, is given by the product of the component reliabilities,

$$R_s = \prod_{k=1}^{n} R_k = R_1 R_2 R_3 \dots R_n, \tag{8.45}$$

where $R$ is the reliability of the component. Assuming an exponential distribution for the reliability function and a constant failure rate, $\lambda$, then the system reliability becomes

$$R_s = e^{-\lambda_1 t} \cdot e^{-\lambda_2 t} \cdot \dots \cdot e^{-\lambda_n t} = \exp\left[-\sum_{i=1}^{n} \lambda_i t\right] = \exp[-\lambda t], \tag{8.46}$$

where

$$\lambda = \lambda_1 + \lambda_2 + \dots + \lambda_n = \frac{1}{\theta}. \tag{8.47}$$

The system failure rate, $\lambda$, is the sum of the individual component failure rates and $\theta$ is the system mean.

### Example 8.7

A high school student has been dreaming of running his own radio station. He has a good idea of what he will broadcast but he has a limited knowledge of electronics and radio. He searched through the internet to learn exactly what equipment he will need. He learned that the transmitter is the main component of his radio broadcast. It takes his broadcast signal, encodes it, and transmits as radio waves that can be picked up by any receiver. Figure 8.14 shows a breakdown of the basic equipment for a transmitter having reliabilities of 0.65, 0.70, 0.80 and 0.88. For proper system operation he learned that the four components must be connected in series, as shown in Figure 8.14. What is the reliability of the radio station?[a]

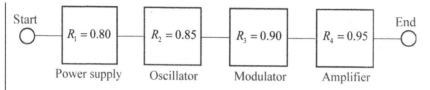

**Figure 8.14** Component block diagram of a series system.

**Solution**

Using equation (8.45), the reliability of the radio station is

$$R_s = \prod_{k=1}^{n} R_k = R_1 R_2 R_3 R_4 = (0.80)(0.85)(0.90)(0.95) = 0.58.$$

[a] Adapted from Ertas, A., and Jones, J.C., *The Engineering Design Process*, John Wiley & Sons, Inc., New York, 1996.

### 8.7.2 Parallel Systems

When components of a system can be connected in parallel, the system will only fail if all of its components fail. The block diagram of a $k$-component parallel system is shown in Figure 8.15. The parallel system reliability, $R_p$, can be obtained from

$$R_p = 1 - \prod_{k=1}^{n} (1 - R_k). \tag{8.48}$$

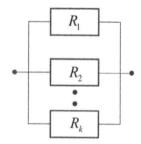

**Figure 8.15** Component block diagram of a parallel system.

For components with the same reliability, equation (8.48) becomes

$$R_p = 1 - (1 - R_k)^k. \tag{8.49}$$

For parallel systems, the system reliability increases as the number of components increases. However, increasing the number of parallel components increases the initial cost, weight, and volume of the system as well as the maintenance requirements. Therefore, the number of components must be carefully optimized.

### Example 8.8

To increase the reliability of the system, the student decides to buy a complete backup system. Determine the increase in system reliability.

## Solution

The student used a simple approach to interconnect the two systems as shown in Figure 8.16. Because this is a parallel configuration, the total system reliability is

$$R_p = 1 - (1 - R_k)^k = 1 - (1 - 0.58)^2 = 0.82.$$

The increase in reliability $\Delta R$ is

$$\Delta R = 0.82 - 0.58 = 0.24.$$

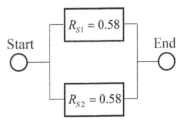

**Figure 8.16** Component block diagram of a parallel system.

### 8.7.3 Multistage Systems with Parallel Redundancy

Figure 8.17 shows a multistage system with parallel redundancy. Many industrial processes, such as power plants, can be represented as multistage systems. As shown, the system consists of $n$ stages in series where the components have identical reliabilities and are connected in parallel at each stage. The system reliability is the product of reliabilities of each stage. Using equations (8.45) and (8.49), the reliability of a multistage system with parallel redundancy is determined as

$$R_{sp} = \prod_{k=1}^{n} [1 - (1 - R_k)^{Y_k}], \tag{8.50}$$

where $Y_k$ is the number of components in each stage, and $R_k$ is the common reliability of the components.

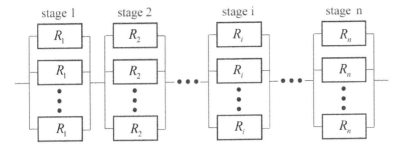

**Figure 8.17** Block diagram of a multistage system with parallel redundancy.

## Example 8.9

In Example 8.8, the student increased the reliability of the system. However, it needs to be increased further. The student has a limited budget and cannot afford to buy another system to increase the system reliability further. He searched the internet for other solutions to increase the reliability of the radio station, and learned about a multistage system with parallel redundancy. Determine the reliability of the system shown in Figure 8.18.

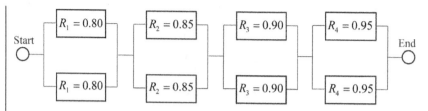

**Figure 8.18** Block diagram for radio station.

## Solution

The reliability of the new system is

$$R_{sp} = \prod_{k=1}^{n} [1 - (1 - R_k)^{Y_k}]$$

$$= [1 - (1 - 0.80)^2][1 - (1 - 0.85)^2][1 - (1 - 0.90)^2][1 - (1 - 0.95)^2]$$

$$= 0.93.$$

The result shows that proper combination of the components can improve the system reliability.

### 8.7.4 Parametric Method in System Reliability Evaluation

In 1972 Banerjee and Rajamani proposed a parametric method to evaluate the reliability of complex systems.[3] Calculation of complex system reliability using the classical approach becomes very difficult and requires a long computational time. However, by using the parametric method, the reliability of complex systems can be easily evaluated. This method introduces a new parameter, $\phi$, defined by

$$\phi = \frac{1 - R}{R}, \tag{8.51}$$

where $R$ is the reliability of a component.

*Series systems.* Substituting equation (8.51) into equation (8.45) yields

$$1 + \phi_s = \prod_{k=1}^{n} (1 + \phi_k). \tag{8.52}$$

Consider a two-element system, $n = 2$, where the system parameter is

$$\phi_s = \phi_1 + \phi_2 + \phi_1 \phi_2. \tag{8.53}$$

Assuming $\phi \ll 1$, the higher-order terms of equation (8.53) can be neglected, and the corresponding parameter for the series system $\phi_s$ can be approximated as

$$\phi_s = \sum_{k=1}^{n} \phi_k. \tag{8.54}$$

---

3 © 1972 IEEE. Banerjee, S.K., and Rajamani, K., "Parametric representation of probability in two-dimensions – a new approach in system reliability evaluation," *IEEE Transactions on Reliability*, R-21, 56–60, 1972.

Once the parameter $\phi_s$ is known, the system reliability is given by

$$R_s = \frac{1}{1 + \phi_s}. \tag{8.55}$$

*Parallel systems.* Substitution of equation (8.51) into equation (8.48) yields

$$\frac{\phi_p}{1 + \phi_p} = \prod_{k=1}^{n} \frac{\phi_k}{1 + \phi_k}. \tag{8.56}$$

For small values of $\phi$, the $1 + \phi$ term can be approximated by unity. Hence, the parameter, $\phi_p$ for a parallel system becomes

$$\phi_p = \prod_{k=1}^{n} \phi_k. \tag{8.57}$$

Similarly, after calculating $\phi_p$, equation (8.55) can be used to determine the system reliability.

*Multistage systems with parallel redundancy.* Because the value of the parameter is the same for all components in a stage, equation (8.57) can be written as

$$\phi_p = \phi_k^{Y_k}, \tag{8.58}$$

where $Y_k$ is the number of components in the $k$th stage. The parameter for the system shown in Figure 8.17 can be determined by modifying equation (8.54):

$$\phi_{sp} = \sum_{k=1}^{n} \phi_k^{Y_k}. \tag{8.59}$$

Complex system reliability can be determined by using the parametric method with the simpler equations. Note that the parametric method assumes that $\phi \ll 1$; however, when $R$ is less than 0.5, $\phi \approx 1$. Hence, the range of $\phi$ is an important criterion to consider in determining the error of calculations using the parametric method. The parametric method can be used to determine the system reliability for systems having components with high reliability.

## Example 8.10

Repeat Example 8.7 using the parametric approach.

**Solution**

The parameters, $\phi_1$, $\phi_2$, $\phi_3$, and $\phi_4$, for the first, second, third, and fourth components, respectively, are

$$\phi_1 = \frac{1 - R_1}{R_1} = \frac{1 - 0.80}{0.80} = 0.25,$$

$$\phi_2 = \frac{1 - R_2}{R_2} = \frac{1 - 0.85}{0.85} = 0.18,$$

$$\phi_3 = \frac{1 - R_3}{R_3} = \frac{1 - 0.90}{0.90} = 0.11,$$

$$\phi_4 = \frac{1 - R_4}{R_4} = \frac{1 - 0.95}{0.95} = 0.053.$$

The system parameter, $\phi_s$, is

$$\phi_s = \phi_1 + \phi_2 + \phi_3 + \phi_4 = 0.25 + 0.18 + 0.11 + 0.053 = 0.593.$$

The system reliability can now be obtained:

$$R_s = \frac{1}{1 + \phi_s} = \frac{1}{1 + 0.593} = 0.62.$$

The error between the method used in Example 8.7 and this example is $0.62 - 0.58 = 0.04$. If the reliability of system components is relatively low, this error will be noticeable.

### 8.7.5  Quantitative System Reliability Parameters

*Mean time to failure.* One of the the most common reliability parameters is the mean time to failure for a non-repairable component or system. The MTTF is the total number of hours of operation of all components divided by the number of components. That is,

$$\text{MTTF} = \frac{\text{Total operating time}}{\text{Number of units under test}}. \tag{8.60}$$

Mathematical equations for when the failures are random samples were presented in Section 8.3.2.

*Mean time between failures.* The mean time between failures (MTBF) is widely used as the measurement of a product's reliability and performance. It is the total number of hours of operation of all components divided by the number of failures:

$$\text{MTBF} = \frac{\text{Total operating time}}{\text{Number of failures}}. \tag{8.61}$$

For example, 14 components are tested for 400 hours and during the test 4 failures occur. The mean time between failures is

$$\text{MTBF} = \frac{14 \times 400}{4} = 1{,}400 \text{ hours/failure.}$$

The mean time to failure is

$$\text{MTTF} = \frac{14 \times 400}{14} = 400 \text{ hours/failure.}$$

Note that until the steady state is reached, the MTBF may change as a function of time – before reaching the steady state, the calculated MTBF changes as the system ages. Therefore, the MTBF equation should be used cautiously.

*Mean time to repair.* Another important metric is the amount of time required to repair a system and bring it back to normal operating conditions, the *mean time to repair* (MTTR). It is given by

$$\text{MTTR} = \frac{1}{\mu}, \tag{8.62}$$

where $\mu$ is the constant repair rate of a component or system. MTTR provides important information that can help reliability engineers to make decisions such as repair or replace, store parts onsite or a change parts strategy.

### 8.7.5.1 Maintainability

An emphasis on maintainability requirements during the design and development phase for designing and developing maintainable products and systems is crucial. Maintainability is defined as the probability that a component or system will be restored to operational effectiveness within a given period of time when the maintenance action is performed in accordance with prescribed procedures.[4] That is, maintainability measures the easiness and speed with which a system can be restored to optimum operational status after a failure occurs. As shown in Figure 8.19, maintainability analysis requires time to repair rather than time to failure as in the case of reliability. As shown in this figure, if a component or a system has an 80% probability of repair for one hour, this means that there is an 80% chance that the component will be repaired within an hour.

The designer must be careful not to improve maintainability by introducing maintenance requirements that reduce reliability. To achieve satisfactory maintainability, the following maintenance design guidelines should be considered by the designer:[5]

1) Minimize the number and complexity of maintenance tasks by using a simple design that includes optimum interchangeability and standardized equipment.
2) Permit rapid and positive identification of multifunctions, defective parts, or assembly.
3) Minimize the need for high degrees of personnel skill and training.
4) Minimize the special tools and test equipment necessary to perform maintenance tasks.

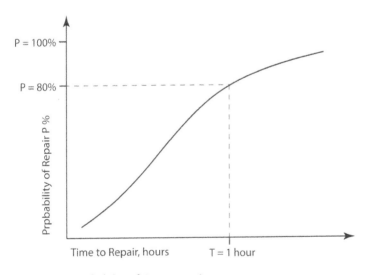

**Figure 8.19** Probability of time to repair.

4 Goldman, A.S., and Slattery, T.B., *Maintainability: A Major Element in Systems Effectiveness*, John Wiley & Sons, Inc., New York, 1964.
5 Taylor, V.J., "Weapon system design and supportability, a function of failure prediction," *IRE Transactions on Reliability and Quality Control*, RQC-11, 13–17, 1962.

5) Provide optimum accessibility to all systems, equipment, and components so that they can be reached easily for maintenance.
6) Provide maximum safety for both equipment and personnel.
7) Minimize the required maintenance time to ensure the availability of the system to satisfy operational demands.
8) Maintain the lowest life cycle cost possible.

When system maintainability and reliability analysis are combined together, some insightful results can be obtained about the overall performance of the system.

Maintenance tasks can be classified into two main categories:[6]

1) *Preventive maintenance* is regularly performed on a piece of equipment to keep it in a satisfactory operational condition. Preventative maintenance is performed while the equipment is still working, so that it does not fail unexpectedly. In other words, preventive maintenance will keep equipment failure rates from increasing above the design levels. Periodically scheduled maintenance of equipment will help to improve equipment life and prevent any unplanned maintenance activity. For example, consider a boiler used for numerous purposes in a cargo ship. Suppose that the main job of a boiler is to make high-pressure steam to power the steam turbine. All sub-components included in the operating system, such as the feed water pump supplying water to the boiler drum, fuel oil atomizers providing fuel oil in a combustion chamber inside the boiler, and the condenser condensing the water vapor to liquid water, should be regularly checked in accordance with the maintenance schedule. If preventive maintenance is not done in accordance with the maintenance schedule, the company may have to bear the cost of an unexpected shutdown of the main steam turbine due to sub-system component failure. Preventive maintenance is extremely important in this circumstance because if the main steam engine of the ship is not operational at sea during bad weather conditions, the safety of neither the ship nor the crew can be assured.
2) *Corrective maintenance* is a maintenance task performed to identify, isolate and restore faulty equipment as soon as possible to proper working order. This maintenance can be achieved by replacing, repairing, adjusting, or calibrating the failure source that has caused interruption or breakdown of the system.

It is impossible to design components or system with perfect reliability so that failures are not experienced. In almost every system, components are designed for limited life and hence the consideration of corrective maintenance is essential for any program. However, if equipment is unusually costly and difficult to repair, it may complete its entire life cycle without maintenance (e.g., satellites).

The maintainability function $M(t)$ for any given distribution can be expressed as

$$M(t) = \int_0^t f(t)dt, \tag{8.63}$$

---

6 From Ertas A., and Jones, J.C., *The Engineering Design Process*, John Wiley & Sons, Inc., 1996.

where $f(t)$ is the repair time probability density function and $t$ is the allowable repair time constraint. Assuming the exponential probability density function and substituting it in equation (8.63), the maintainability function $M(t)$ is found to be

$$M(t) = 1 - \exp(-\mu t), \tag{8.64}$$

where $\mu$ is the constant repair rate of a component, given by

$$\mu = \frac{1}{\text{MTTR}}, \tag{8.65}$$

MTTR being the mean time to repair of a particular component after a malfunction has occurred.

## Example 8.11

Any critical component on a military tank that is knocked out (e.g., track or turret drive) will be repairable in a combat situation in a very short time. Assume that during battle, a maximum 0.55-hour delay can be allowed for repair in the event of turret failure. If the component constant repair rate $\mu$ is 1.2 repairs/hour, what is the probability of achieving maintenance success?

### Solution

For a repair time constraint of 0.55 hours, the probability of maintenance success is

$$M(t) = 1 - \exp(-\mu t) = 1 - \exp[-(0.55)(1.2)] = 0.483,$$

and the MTTR becomes

$$\text{MTTR} = \frac{1}{\mu} = \frac{1}{1.2} = 0.833.$$

The result shows that the design presents only a 48.3 percent probability of meeting the specified time constraint for repair in the event of turret failure.

With a design change, if the repair rate is increased to 2.5 repairs/hour the MTTR becomes

$$\text{MTTR} = \frac{1}{\mu} = \frac{1}{2.5} = 0.4.$$

The new maintenance success probability within the time constraint of 0.55 hours is

$$M(t) = 1 - \exp(-\mu t) = 1 - \exp[-(0.55)(2.5)] = 0.7472.$$

Now with the new design we are able to offer a 74.72 percent probability of maintenance success. For any repair time constraint, the probability of achieving maintenance success depends on ensuring a low MTTR. To the customer, a high $M(t)$ represents a greater probability of reducing maintenance expenditure and a substantial step toward establishing system effectiveness.

### 8.7.5.2  Maintenance of Complex Systems
Complex high-technology devices are increasing in use in industry, service sectors, and daily life. Their reliability and maintenance are of extreme importance in view of their cost and critical

functions. Proper use of past experience and data available to predict failure of these devices may lead to significant cost saving, reducing the downtime to an absolute minimum, maximizing the availability, and increasing the overall level of safety. There are many different kinds of maintenance management framework which describe the concepts, methods, and processes of modern maintenance management of complex systems – focusing specifically on modern modeling tools for maintenance planning and scheduling.

There are many factors to consider when designing a new complex system and putting it into production. They are even more critical if the new system is replacing or upgrading an old one. In maintenance management, one of the most important factors is to understand the relationships (dependencies) between the various system components. Any expectations regarding the operation and maintenance of the system components or their relationships, any complex and unusual design and programming techniques used must be analyzed and documented carefully. It is important to note that maintenance management also involves identifying which attributes of a system contribute to maintenance complexity.

In a discussion of the maintainability of commercial off-the-shelf (COTS) systems, Gansler and Luxyshyn say: "Greater use of [COTS] systems and components is one strategy that can enable achieving the required DoD transformation, and help to ensure American military success in the twenty-first century. [COS] is a term for software or hardware that is commercially made and available for sale, lease, or license to the general public and that requires little or no unique government modifications to meet the needs of the procuring agency. Because of their rapid availability, lower costs, and low risk, COTS products must be considered as alternatives to in-house, government-funded developments."[7] Identifying which attributes of COTS-based system (CBS) contribute to maintenance complexity by analyzing the complexity of the deployment and investigating the relationships of the components is an important operational maintenance issue.

---

### CASE STUDY 8.2  Maintenance Complexity of a COTS-Based System

This case study outlines a diagnostic approach to quantifying the maintainability of a CBS by analyzing the complexity of the deployment of the system components. We show how interpretive structural modeling (ISM) is used to support identification and understanding of interdependencies among COTS components and how they affect the complexity of the maintenance of the CBS.[a]

#### Background

It is important to understand the deployment dependencies between all of the COTS products (components) in a system's architecture. Identifying the dependencies allows the integration team to define the complexity of a system deployment and allows proper decisions to be made to either

---

7 Gansler, J.S., and Lucyshyn, W., "Commercial-off-the-shelf (cots): doing it right," http://www.dtic.mil/dtic/tr/fulltext/u2/a494143.pdf, accessed November 2016.

accept the complexity or reduce the complexity through deployment or design changes. Once the complexity is identified, accurate costing can be determined which leads to more accurate project costing which, in turn, leads to fewer cost overruns.

Current measures of the maintainability of CBSs are focused on the number of COTS components that are part of a system. However, as with biological systems, it is not the number of components that creates complexity, but instead the interactions between the components.[8] The measure of maintainability described in this case study is a metric that is based on the interactions of the components.

### Case Study Statement

The system involved in this case study includes eight COTS components and allows users to query, view, and manipulate images (see Figure 8.20). The system has two operating systems (Windows XP in component 8 and Red Hat Linux in component 5) which oftentimes leads to complexity in system design. The system also has MySQL server and client (components 1 and 2, respectively) which always have dependencies on one another, and finally the Tungsten client (component 7) and server (component 6) software products.

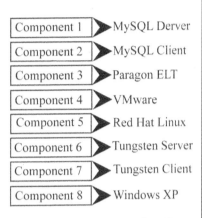

**Figure 8.20** Components of COTS-based system.

The Tungsten products recover and store data in the MySQL databas, creating a relationship between the two. Otherwise, the system only has a few other miscellaneous products on which the major components rely.

The Tungsten products are installed on a virtualized host; therefore, VMware (component 4) is also necessary. Component 3 is Paragon electronic light table, which is a tool for viewing and manipulating images.

The concept of operations for the system compel data storage within the database (MySQL) and the mix of operating systems is driven by need to interact with the user on a Windows XP host while hosting the core mission applications on Linux.

### Case Study Analysis

The ISM process shown in Figure 8.21 was used to understand the interdependencies among COTS components and how they affect the complexity of the maintenance of the CBS. The first step in the ISM process was to organize a group of people with relevant knowledge, skills, and background into a team to identify and define the components (factors) affecting the complexity of the maintenance of the CBS. For a CBS, a group of experts in the COTS products comprising

*(Continued)*

---

8  Mitchell, M., *Complexity: A Guided Tour*, Oxford University Press, Oxford, 2009.

**CASE STUDY 8.2 (Continued)**

the system, experts in the system requirements, as well as system administrators and software developers gather and identify factors that influence the COTS installation of the program. By brainstorming and an iterative process the group identified the eight most important components (see Figure 8.21) impacting the system maintainability, and these are used for the model development.

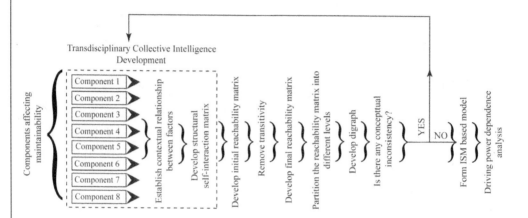

**Figure 8.21** Sequence of activities of ISM model for CBS.

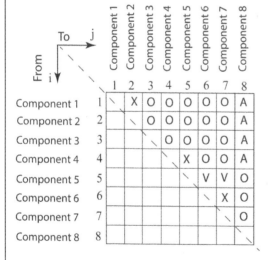

**Figure 8.22** Structural self-interaction matrix.

Following the ISM steps, a structural self-interaction matrix (SSIM) was created (see Figure 8.22). The matrix shows the direction of the contextual relationships of the factors influencing the system. The direction of the influence is important in the next steps to identify which components in the system create complexity in the system. It is important to note that influence of one component over another is subjective. The consensus of the group determines the size of influence one component has over another.

The next steps in the analysis were the computation of the adjacency matrix, the reachability matrix with transitivity, and the final reachability matrix which includes driving power and dependence of factors. For the decomposition of the CBS into levels, the final reachability matrix along with antecedents of each element of prospects were used. Finally, the digraph is converted to an ISM to see a broad representation of the interrelationship between the COTS components (see Figure 8.23). For detail information on this kind of analysis see Chapter 1.

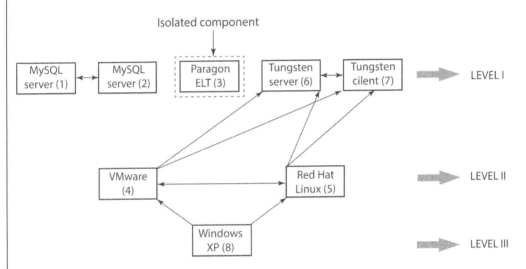

**Figure 8.23** ISM-based model of CBS.

## Discussion

Some of the COTS components that have weak relationships in the CBS shown in Figure 8.20 are eliminated from the ISM analysis. After removing the transitivities based on the reachability matrix as described in the ISM approach, the structural hierarchy of the performance measure factors of CBS is developed as shown in Figure 8.23. This figure depicts visually the direct and indirect relationships between the factors affecting the complexity of the maintenance of the CBS. Components 3 is isolated since it is not adjacent to any component – this COTS component does not influence the maintainability of the CBS. Component 8 is the source component since it has only outgoing paths.

*(Continued)*

**CASE STUDY 8.2 (Continued)**

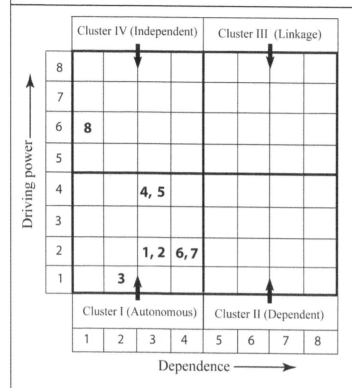

Figure 8.24 MICMAC Analysis.

The final step is a MIMAC analysis as shown in Figure 8.24. The performance measures of components affecting CBS maintainability have been classified into four categories. Cluster I includes autonomous components (1, 2, 3, 4, 5, 6, 7). They have low driving power and low dependence. The components in this cluster have little influence on the other components in the system; neither are they influenced by other components. In a CBS, COTS components in this cluster would require little effort to maintain because they are not linked to other components in the system. They could be changed without impacting any other components.

Cluster II identifies dependent factors that have low driving power and high dependence. In a CBS, there is no component in this cluster. COTS products in this cluster would have been end-user products such an internet browser.

In a CBS, changes to components in cluster III will require significant effort because of the potential impact to other components in the system. To determine the maintainability of a COTS deployment, the maintainers of a CBS must take special care in maintenance of these components. An action on any of the components in cluster III will affect the entire system. A component like a database typically fits into this cluster because a database installation is impacted by underlying components like the operating system and is likely to impact application servers that can be used to preserve the state of application processing. Evidently, no component falls in this cluster for the CBS under study.

The Windows XP operation system is the only component falls in cluster IV. This component stands independently, that is, it has few if any dependencies, but because every other application builds on its existence, the impact of a change to a component in cluster IV has the potential to cause significant effort.

### Conclusion

The MICMAC analysis demonstrates which components in the CBS contribute most significantly to the complexity of the system maintainability. With the ISM, architects, system integrators, and system maintainers can isolate the COTS products that cause the most complexity, and therefore the most effort to maintain, and take precautions to only change those products when necessary or during major maintenance efforts. The analysis also clearly shows the components that can be easily replaced or upgraded with very little impact on the rest of the system. The MICMAC analysis provides valuable information for minimizing the effort associated with the CBS and also reducing risk associated with product reconfiguration and upgrades.

[a]Adapted from Ertas, A., et al., "Complexity of system maintainability analysis based on the interpretive structural modeling methodology: transdisciplinary approach," *Journal of Systems Science and Systems Engineering*, 25(2), 254–268, 2016.

#### 8.7.5.3 Availability

Considering both the reliability and maintainability properties, availability is an important design parameter used to assess the performance of repairable systems. In system design, it is crucial to ensure that the system can be maintained easily and inexpensively. Preventive maintenance of repairable components or systems can be very useful in reducing repair and replacement costs, and improving system availability.

Availability can be defined as the ability of a product/equipment to perform a required function under given conditions for a predicted time duration $t$. In other words, availability is a measure of the proportion of time that a product/equipment will be able to perform its intended function when needed.

In reliability and maintainability studies, the terms *steady-state availability* and *instantaneous availability* are commonly used. The steady-state availability, $A$, is defined as the proportion of total operation time that a component is available for use. It is also called *inherent availability*, $A_I$, when considering only the corrective maintenance downtime of the system.

For a single component, the formula for calculating inherent availability is

$$A_I = \frac{\text{MTTF}}{\text{MTTF} + \text{MTTR}}. \tag{8.66}$$

For a system, equation (8.66) becomes

$$A_I = \frac{\text{MTBF}}{\text{MTBF} + \text{MTTR}}. \tag{8.67}$$

If the MTBF is known, one can calculate the constant failure rate, $\lambda$, as the inverse of the MTBF,

$$\lambda = \frac{1}{\text{MTBF}}. \tag{8.68}$$

Similarly, the constant repair rate, $\mu$, is

$$\mu = \frac{1}{\text{MTTR}}. \tag{8.69}$$

Substituting equations (8.68) and (8.69) into equation (8.67) gives

$$A_I = \frac{\mu}{\lambda + \mu}. \tag{8.70}$$

The time-dependent instantaneous availability is defined as the probability that at time $t$ the system will be available. A single-system instantaneous availability $A(t)$ at a given time $t$ is given by

$$A(t) = \frac{\mu}{\lambda + \mu} + \frac{\lambda}{\lambda + \mu} \exp[-(\lambda + \mu)t]. \tag{8.71}$$

Note that, as $t$ goes to infinity, equation (8.71) approaches the stationary steady-state availability. Equation (8.71) reveals that, as the repair rate $\mu$ increases, the probability of maintenance success increases and availability increases.

The product law can be used with the individual component availabilities to calculate the steady-state system availability for repairable component installation in series as

$$A_{ss} = \prod_{k=1}^{n} A_k. \tag{8.72}$$

Similarly, steady-state availability for a repairable parallel component installation is given by

$$A_{ps} = 1 - \prod_{k=1}^{n} (1 - A_k). \tag{8.73}$$

### Example 8.12

A boiler feed pump feeds condensed steam from condenser into a steam boiler, as on board a marine vessel. Suppose a feed pump has constant failure and repair rates of 0.003 failure/hour and 0.02 repair/hour, respectively. Calculate the inherent availability, $A_I$, of the feed pump.

### Solution

For $\mu = 0.02$ and $\lambda = 0.003$, the inherent availability is

$$A_I = \frac{\mu}{\lambda + \mu} = \frac{0.02}{0.003 + 0.02} = 0.870.$$

In this example, if the repaired feed pump is as good as new, we can compute the availability of the feed pump at time $t = 80$ hours. The time-dependent instantaneous availability is

$$A(t = 80) = \frac{\mu}{\lambda + \mu} + \frac{\lambda}{\lambda + \mu} \exp[-(\lambda + \mu)t]$$

$$= \frac{0.02}{0.003 + 0.02} + \frac{0.003}{0.003 + 0.02} \exp[-(0.003 + 0.02)(80)]$$

$$= 0.89.$$

In other words, 89 percent of the time, the feed pump will be available for operation after 80 hours of usage.

#### 8.7.5.4 Relationships of Reliability, Availability, and Maintainability

Reliability, availability, and maintainability (RAM) are system design attributes that collectively affect both the utility and the life cycle costs of a system. RAM analysis identifies components of systems whose failure affects the expected system performance based on system design, operations, and maintenance.

There is an intimate relation between cost and availability that affects design performance. Figure 8.25 shows how the design performance changes with respect to constant cost profiles and availability.[9] As can be seen from this figure, there are numerous design solutions for different levels of performance. For example, for constant availability, $A_c$, there are three possible solutions, varying from low cost with low performance to high cost with high performance. Cost profile 2 may be preferred because it gives fair performance with reasonable cost. The decision-maker should be aware of the tradeoff between performance and availability in selecting

**Figure 8.25** Important design parameters.

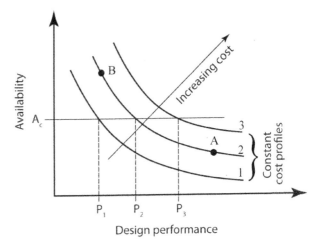

9 © 1976 IEEE, David, H., "Availability – concept and definitions," *Proceedings, Annual Reliability and Maintainability*, pp. 482–490, 1976.

systems between $A$ and $B$ on curve 2. If the choice is a system with high availability, the system will have low performance. It is clear from the figure that a system with high design performance and excellent availability is expensive. In some instances it is very difficult to make a decision on which situation is best. Before the designer makes a choice, he/she has to study the information and the resulting alternative solutions from the tradeoff analysis.

Figure 8.26 shows the relationship between reliability, maintainability, and availability. As stated earlier, availability is the ability of a product or piece of equipment to perform a required function under given conditions for a predicted time duration $t$ – availability is not only a function of reliability, but also a function of maintainability. As seen from the Figure 8.26(a), if the reliability is held constant, even at a high value of reliability, it does not ensure a high availability. As the time to repair increases (an increase in repair means a decrease in the time it takes to perform maintenance), the availability decreases. Even a system with a low reliability could have a high availability if the time to repair is short. It is clear from Figure 8.26(b) that when availability increases, the reliability increases. Conversely, when availability decreases, reliability decreases.

### 8.7.5.5 Dependability

The dependability of a component is another important design parameter that provides a single measure of system condition(s), combining reliability and maintainability. Dependability can be defined as the probability that a component either does not fail or fails and will be repaired within an allowable time interval $t$. The concept of dependability is significant because it permits cost tradeoffs between reliability and maintainablity. Dependability can be determined as follows:[10]

$$D(t) = 1 - \left(\frac{1}{d-1}\right)(e^{-\lambda t})[1 - e^{-(d-1)\lambda t}] \tag{8.74}$$

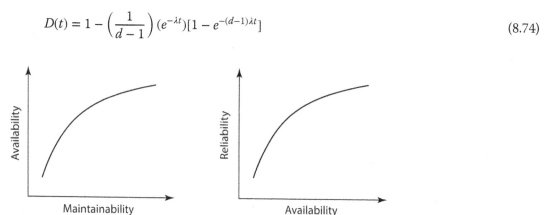

**(a) Constant reliability**  **(b) Constant maintainability**

**Figure 8.26** Relationships of reliability, maintainability, and availability.

---

10 © 1966 IEEE, Wohl, J.G., "System operational readiness and equipment dependability," *IEEE Transactions on Reliability*, vol. R-15(1), 1–6, 1966.

where $d$ is the component dependability ratio

$$d = \frac{\mu}{\lambda}.$$ (8.75)

Equation (8.70) can be modified to establish the relationship between availability and dependability as

$$A = \frac{\mu}{\mu + \lambda} = \frac{\mu/\lambda}{1 + \mu/\lambda} = \frac{d}{1 + d}.$$ (8.76)

Component availability as a function of the dependability ratio is shown in Figure 8.27.[10] As shown in the figure, the component dependability ratio, $d$, increases rapidly above the availability level of 0.9 (point A) and decreases rapidly for values of availability less than 0.1 (point B). These extremes define the regions of maximum sensitivity to change of the dependability ratio.

For a given value of $d$, the minimum value of $D(t)$ in relation to $\lambda t$ can be obtained by taking the derivative of equation (8.74) with respect to $\lambda t$:

$$\frac{\partial [D(t)]}{\partial (\lambda t)} = 0 = \left( \frac{e^{-\lambda t} - e^{-d\lambda t}}{d - 1} - e^{-d\lambda t} \right).$$ (8.77)

Solving for $\lambda t$ yields

$$\lambda t = \frac{\ln d}{d - 1}.$$ (8.78)

Substituting $\lambda t$ in equation (8.74), the minimum obtainable $D(t)$ is

$$D(t)_{min} = 1 - \left( \frac{1}{d - 1} \right) \left( e^{-\ln d/(d-1)} - e^{-d \ln d/(d-1)} \right).$$

**Figure 8.27** Component availability versus dependability ratio.

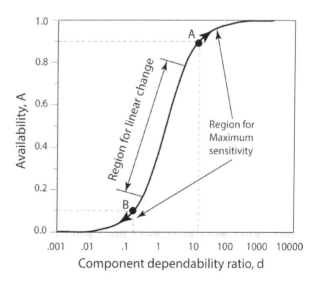

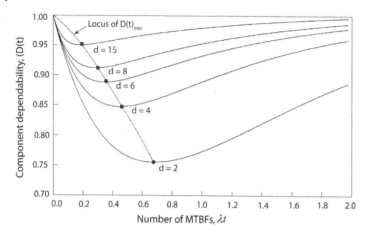

**Figure 8.28** Component dependability $D(t)$ versus dependability ratio $d$.

Figure 8.28 shows the component dependability $D(t)$ versus $\lambda t$ for different values of the component dependability ratio $d$.[10] As shown in the figure, dependability $D(t)$ increases with increasing dependability ratio $d$. This figure also shows that minimum dependability, $D(t)_{min}$ increases with increasing $d$. The concepts of dependability and dependability ratio are important design parameters. However, it is important to remember that reliability, maintainability, availability, and dependability are very closely interrelated and their effects on design performance and effectiveness should be considered as a whole.

### Example 8.13

The radio station in Example 8.7 has components connected in series. The high school student changed his mind and wants to sell his radio station. A buyer is willing to buy the radio station for the asking price provided that the dependability of the station for 1200 hours of operation is more than 0.95. The data for the four main components are available from the manufacturers as shown in Table 8.6. Can the high school student sell the radio station for the asking price?

**Table 8.6** Component data for $\lambda$ and $\mu$.

| Component | $\lambda$ | $\mu$ |
| --- | --- | --- |
| 1 | 0.0010 | 0.020 |
| 2 | 0.0025 | 0.025 |
| 3 | 0.0020 | 0.030 |
| 4 | 0.0020 | 0.040 |

## Solution

The dependability ratios for the components are

$$d_1 = \frac{\mu_1}{\lambda_1} = \frac{0.020}{0.0010} = 20,$$

$$d_2 = \frac{\mu_2}{\lambda_2} = \frac{0.025}{0.0025} = 10,$$

$$d_3 = \frac{\mu_3}{\lambda_3} = \frac{0.030}{0.0020} = 15,$$

$$d_4 = \frac{\mu_4}{\lambda_4} = \frac{0.040}{0.0020} = 20.$$

The dependabilities of the components are

$$D(t) = 1 - \left( \frac{1}{d-1} \right) (e^{-\lambda t}) \left[ 1 - e^{-(d-1)\lambda t} \right],$$

$$D(t_1) = 1 - \left( \frac{1}{20-1} \right) e^{-(0.001)(1200)} \left[ 1 - e^{-(20-1)(0.001)(1200)} \right] = 0.984,$$

$$D(t_2) = 1 - \left( \frac{1}{10-1} \right) e^{-(0.0025)(1200)} \left[ 1 - e^{-(10-1)(0.0025)(1200)} \right] = 0.994,$$

$$D(t_3) = 1 - \left( \frac{1}{15-1} \right) e^{-(0.002)(1200)} \left[ 1 - e^{-(15-1)(0.002)(1200)} \right] = 0.994,$$

$$D(t_4) = 1 - \left( \frac{1}{20-1} \right) e^{-(0.002)(1200)} \left[ 1 - e^{-(20-1)(0.002)(1200)} \right] = 0.955.$$

The system dependability in series is

$$D(t)_s = \prod_{k=1}^{n=4} D(t)_k = D(t_1)D(t_2)D(t_3)D(t_4)$$

$$= (0.984)(0.994)(0.994)(0.955)$$

$$= 0.928$$

The high school student will not be able to sell his radio station for the asking price since the system dependability is 0.928, which is lower than 0.95. He needs to reconsider the values of design parameters, $\lambda$ and $\mu$, for the components and optimize the system.

### 8.7.5.6  Attributes of Dependability

Dependability was first proposed as a generic term including attributes such as reliability, availability, maintainability, and safety. It was defined in terms of "task accomplishment" and "delivery

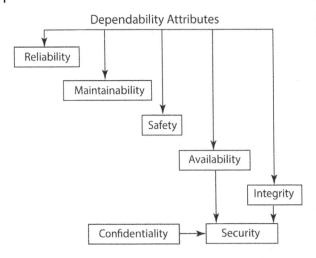

Dependability Attributes

**Figure 8.29** Integration of dependability and security.

of expected service".[11] The attributes of dependability state the performance parameters which are anticipated from a system to perform acceptable service that could be justifiable and trusted.

Figure 8.29 explains the relationship between dependability and security in terms of their main attributes. In order to accept the trust in dependability the following basic interrelated attributes can be considered:[12]

- Availability: the ability of readiness of a system to deliver desired services when requested
- Reliability: the ability of a system to deliver correct services continuously for a period of time
- Safety: the ability of a system to operate without catastrophic failures on the user(s) and the environment
- Maintainability: the ability to restore to operational effectiveness when the maintenance action is performed
- Integrity: preventing improper system state variations.

When speaking of security, an additional attribute, *confidentiality* (preventing unauthorized disclosure of information) will be considered. As shown in Figure 8.29, two attributes of dependability (availability, integrity) and confidentiality are combined as the security concept. The security concept is integrated with dependability which describes not only the system behavior, as the other attributes do, but also the system's ability to resist external attacks (e.g., integrity). A system must be both dependable and secure to be considered *trustworthy*. The concept of trustworthiness is proposed as an extension of dependability as a property of

11 Laprie, J.C., and Costes, A., "Dependability: A unifying concept for reliable computing," in *Proc. 12th IEEE International Symposium on Fault-Tolerant Computing* (FTCS-12), pp. 18–21, 1982.

12 Avizienis, A., Laprie, J.C., Randell, R., and Landwehr, C., "Basic concepts and taxonomy of dependable and secure computing," *IEEE Transactions on Dependable and Secure Computing*, 1(1), 1–23, 2004.

a system.[13] This concept is especially appropriate for large and complex systems with intense human interaction for which the requirements are not completely and clearly defined.[14]

## 8.8 Failure Mode and Defects Analysis

Failure mode and defects analysis (FMEA) is a qualitative methodology to help designers to discover potential problems that may exist within the design of a system, product or process.[15] FMEA is widely used in manufacturing, banking, service centers, and almost all business industries to analyze potential reliability problems early in the product and process development cycle to mitigate failure – to design out failures and produce reliable, safe, and customer-satisfying product/process. Although the FMEA method is effective when analyzing an existing process, service or a product for improvement, it can also be used when developing a new process or a process in producing a product or service. The three main classifications of FMEAs are as follows:

1) *System FMEA* is used to discover potential system failures modes related to system safety, system integration, sub-systems interfaces, and interactions.
2) *Design FMEA* explores the possibility of failures associated with the product that could cause product malfunctions, shortened product life, and safety hazards, and regulatory concerns that might be caused by material properties and irregularities, improper tolerances, interfaces with other components, etc.
3) *Process FMEA* is used to discover potential process failure modes that can cause product quality, process reliability, customer dissatisfaction, safety issues, and environmental hazards.

Following the FMEA basic principles and steps shown in Figure 8.30, they are used to reduce risk associated with the system, sub-system, and component or manufacturing processes to an acceptable level:

1) *Team development.* The first important step is to organize an interactive management (CIM) workshop (see Chapter 1) to make a decision on potential failure modes through cross-functional team. The involvement of engineers (design, manufacturing/assembly, process, quality, test, reliability, materials, field service engineers), project manager, subject-matter experts, practitioners, and other stakeholders in the CIM workshop will create alternative viewpoints during the brainstorming to define the effects of the failure and possible causes. At the workshop, team members will be organized for each FMEA (sytem, design or product) and they will agree on a ranking system.

13 Dobson, J., McDermid, J., and Randell, B., "On the trustworthiness of computer systems," ESPRIT/BRA Project 3092 Technical Report Series No. 14, 1990.
14 Jonsson, E., "An integrated framework for security and dependability," in *NSPW '98 Proceedings of the 1998 Workshop on New Security Paradigms*, pp. 22–29.
15 McDermott, R.E., Mikulak, R.J., and Beauregard, M.R., *The Basics of FMEA*, 2nd edn, CRC Press, New York, 2009.

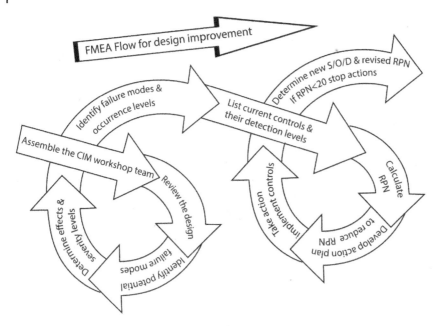

**Figure 8.30** FMEA flow for design improvement.

2) *Design review.* During the review process, every components of the product under study will be identified with respect to their functional requirements (FRs). For each FR, the team suggests all possible failure modes that would prevent the design from failing to satisfy each FR. A review of any similar products or processes of FMEAs can be a starting point for making suggestions. Coupling, interfaces, and interactions among the components and sub-components will be documented at all levels. For each failure mode, the team identifies causes and effects. Information obtained from the design review is used as the input to the FMEA worksheet shown in Table 8.7.

**Table 8.7** Generic FMEA worksheet.

| FMEA Objective and Goal(s): | | | | | | | | | FMEA Type: Process | | |
|---|---|---|---|---|---|---|---|---|---|---|---|
| System: | | | | | | | | | Prepared By: Process Engineer | | |
| Sub-system: | | | | | | | | | Date: xxxxx | | |
| Component: | | | | | | | | | Team members: xxxxx | | |

| Function/ Process | Potential failure mode | Effect(s) of failure | Severity | Cause(s) of failure | Occurrence | Current Control (design or prevention) | Detection | RPN | Recomended action(s) to reduce (O) & to improve D | Actions taken | Revised rating S \| O \| D \| RPN |
|---|---|---|---|---|---|---|---|---|---|---|---|
| | | | | | | | | | | | |
| | | | | | | | | | | | |

3) *Potential failure modes.* During this step, the team will list the design requirements for tolerances, materials, finishes, tests, and verifications, etc., and describe the potential failure modes for each requirement. The impact of each potential failure mode on process performance, product functionality, and system performance will be determined. Possible causes in the system, design or process for the associated failure modes will also be identified.

4) *Severity level (S).* Severity is a numerical subjective ranking that shows the importance of the effect on customer requirements. Suggested severity ratings to be used in FMEAs are shown in Table 8.8. The severity of each potential failure will be evaluated using a 1–10 scale, where 1 indicates no impact and 10 indicates hazardous impact.

5) *Occurrence level (O).* Occurrence is a numerical rating of the frequency of a cause that occurs and creates failure modes. Table 8.9 gives the probability of occurrence. The appropriate occurrence value as shown in Table 8.9 will be assigned.

6) *Detection level (D).* The probability that the failure would be detected before product reaches the customer will be estimated using a 1–10 scale. The appropriate detection value as shown in Table 8.10 will be assigned for the failure mode.

7) *Current controls.* Define the current design controls, if they exist, that are in place to limit the failure mode causes.

8) *Calculate the risk priority number (RPN).* The RPN is a measure used to assess critical failure modes related to a design. The RPN is as a function of the *severity* of the effect, frequency of the *occurrence* of the cause, and the ability to *detect* the failure:

$$RPN = S \times O \times D. \tag{8.79}$$

**Table 8.8** Rating for severity of failure (**S**).

| Effect | Description of the Severity of the Effect | Rating |
| --- | --- | --- |
| None | There is no effect on product/process performance. | 1 |
| Very slight | Negligible effect on product/process performance. Customer may not notice failure. | 2 |
| Slight | Slight effect noticed by customer on product/process performance. | 3 |
| Minor | Minor effect noticed by customer on product/process performance. | 4 |
| Moderate | Moderate effect noticed by customer on product/process performance. Customer dissatisfied. | 5 |
| Severe | Severe effect noticed by customer on product/process performance. Customer dissatisfied. | 6 |
| High severity | Very poor product/process performance noticed by customer. Very dissatisfied customer. | 7 |
| Very high severity | Product/process is non-functional. Very dissatisfied customer. | 8 |
| Extreme severity | Safety-related potential failure. | 9 |
| Hazardous | Safety related sudden failure with hazardous effects. | 10 |

**Table 8.9** Rating for probability of occurrence (**O**).

| Likelihood of failure | Possible failure rate | Rating |
|---|---|---|
| Very high | $\geq 1$ per 10 | 10 |
| High | 1 in 20 | 9 |
| | 1 in 50 | 8 |
| | 1 in 100 | 7 |
| Moderate | 1 in 500 | 6 |
| | 1 in 2,000 | 5 |
| | 1 in 10,000 | 4 |
| Low | 1 in 100,000 | 3 |
| | 1 in 1,000,000 | 2 |
| Very low | Failure is eliminated by prevention control | 1 |

**Table 8.10** Rating for detection (**D**).

| Likelihood of Detection | Detection by Design Control | Rating |
|---|---|---|
| Absolutely uncertain | Failures are not detectable: no inspection | 10 |
| Very remote | Very remote chance of detection by inspection | 9 |
| Remote | Remote chance of detection by inspection | 8 |
| Very low | Very low chance of detection | 7 |
| Low | Low chance of detection | 6 |
| Moderate | Moderate chance of detection | 5 |
| Moderately high | Moderately high chance of detection | 4 |
| High | High chance of detection | 3 |
| Very high | Very high chance of detection | 2 |
| Almost certain | Almost certain detection | 1 |

The RPN is a measure of the overall impact of the failure on a company. Although a critical value of RPN is usually optional because of its subjectivity, if the value of RPN is less than 20 we may consider stopping the action plan.[16]

9) *Action plan.* Using the calculated RPN and the results of the potential failure mode verification studies, decide on which failure modes require action. Sort design requirements by RPN number a using a scale of 1–10 to prioritize the action plan for maximum impact. Record a brief description of the action on the FMEA form.

---

16 "Design failure modes and effects analysis," http://www.raytheon.com/connections /rtnwcm/groups/public/ documents/content/rtn_connect_dfmea_pdf.pdf; accessed January 30, 2017.

10) *Implement control.* If the controls for the FMEA analysis are inadequate, define how the failure can be better delimited and/or eliminated. Consider implementing new or more effective design controls such as:[16]
    - Specifying a requirement as a "critical characteristic"
    - Reliability tests/design verification tests
    - Design reviews
    - Worst-case stress analysis
    - Robust/parameter design
    - Environmental stress testing
    - Designed experiments
    - Computational modeling
    - Variation simulation and statistical tolerance analysis
    - Component de-rating
    - Fault tree analysis.
11) *Calculate revised RPN.* If the design control is in place and the value of RPN is approximately 20, then no further action plan is required. Otherwise, repeat the cycle.

---

**CASE STUDY 8.3**

Bulgur (cracked wheat) is a whole grain that can be used to make pilafs, soups, bakery goods, or as stuffing. Its high nutritional value makes it a good substitute for rice. Whole-grain, high-fiber bulgur can be found in natural food stores. It has a light, nutty flavor. For more consistent cooking time and results the particle sizes of bulgur must be uniform (see Figure 8.31). For this hypothetical case study, let us use the process of producing bulgur as an example.[a]

**Figure 8.31** Bulgur.

---

*(Continued)*

**CASE STUDY 8.3 (Continued)**

**Bulgur Production Process**

Grain processing starts with washing the grains to remove stones and light impurities. The next step is a cooking process – grains are boiled in water in large cauldrons. During cooking, grain absorbs water to obtain uniform gelatinization of starch.

After the cooking process, the product undergoes drying by spreading the grain under the sun (traditional drying method). The dried grains are then softened with moisture to remove the outer layers. Then the bulgur undergoes a second drying process. After this stage, the grains are cracked into coarse particles. Grinding (cracking) is a critical step as it determines either the cooking time or the yield of the end product. All these operations are traditionally done by hand.

**Case Study Analysis**

This family-owned small business producing bulgur was beginning to pose a serious competitive challenge to other companies. Since they were using traditional methods, the family knew the business faced critical strategic challenges.

FMEA was used to discover potential process failure modes that caused poor product quality, customer dissatisfaction, and possible revenue loss. The FMEA team included subject-matter experts who made decisions during an imaginary brainstorming session – they came up with the initial FMEA worksheet shown in Table 8.11. Under the potential failure modes, the team listed three possible failures in bulgur making and its respective effects and potential causes.

Potential failure modes were: (a) inconsistent bulgur particles, (b) deformation of wheat kernels, (c) bulgur particles were sticky. The initial RPNs for the three failure modes were found to be 128, 112, and 80 (see Table 8.11). Based on the FMEA worksheet, priority was given to all the failure modes because they all have very high RPN. The FMEA team came up with following action plans to counter these failures:

1) Instead of traditional method, a roller mill was used to improve cracking wheat grains to produce uniform bulgur particle sizes.
2) 100% gelatinization of starch must be achieved to produce wheat kernels without any deformation. This was accomplished by cooking wheat grains at 75°C for 60 minutes.
3) Complete gelatinization of starch was obtained to prevent stickness.

Due to the implementation of the courses of action and process control improvements, the severity, occurrence and detection scorings changed and the new calculated RPNs were 72, 24, and 12, respectively (see Table 8.11).

Given the results of the first cycle FMEA, the failure on inconsistent bulgur particles (first failure mode) was still ranking as the top failure in the bulgur-making process. However, after one cycle of implementation of the action plans, the second and third failure modes were significantly improved – the RPN scores dropped to 24 and 12, respectively.

**Table 8.11** FMEA worksheet.

FMEA Objective and Goal(s): Producing good quality bulgur

System:
Sub-system:
Component:
Process: **Bulgur production process**

FMEA Type: PFMEA
Prepared By: Process Engineer
Team members: xxxxx
Team Leader: xxxxx
Date: xxxx

| Function/ Process | Potential failure mode | Effect(s) of failure | Severity | Cause(s) of failure | Occurrence | Current Control (design or prevention) | Detection | RPN | Recomended action(s) to reduce (O) & to improve D | Actions taken | Revised rating S \| O \| D \| = RPN |
|---|---|---|---|---|---|---|---|---|---|---|---|
| Bulgur making | Inconsistent coarse particles | Displeased customer | 8 | Cracking was done by hand | 4 | None | 4 | 128 | Use industrial equipment | Roller mill used | 6 × 3 × 4 = 72 |
| | Deformation of wheat kernel | Company losses | 7 | Cooking time and temperature | 4 | None | 4 | 112 | Use proper time and temperature | Cooked at 75°C for 60 min. | 2 × 3 × 4 = 24 |
| | Sticky | Upset customer | 10 | Gelatinization of starch was not obtained | 4 | None | 2 | 80 | Improve precooking process | Complete gelatinization of starch obtained | 3 × 2 × 2 = 12 |

*(Continued)*

---

**CASE STUDY 8.3 (Continued)**

The team again assessed the process and its first failure mode and decided that the design modification was not effective at eliminating the inconsistent bulgur particles of the failure – the RPN was not reduced to an acceptable value. After the team's research effort, the roller mill was improved by using four successive rollers with three gaps. This new process provided a high production yield as well as uniform particle sizes due to multiple milling stages – severity and occurrence scores dropped to 4 and 2, respectively. The new revised RPN found to be $4 \times 2 \times 4 = 32$.

Knowing the second cycle FMEA results, the failure on inconsistent bulgur particles still ranked as the top failure in the bulgur-making process. The team recommended conducting FMEA analysis periodically to check on the performance of the process. The problem of bulgur particles stickiness has been addressed significantly as it has the biggest drop in RPN due to the drop in its severity. Note that the ratings for the detection remained the same. This is because no action plan was implemented to solve the problem – no inspection process was implemented.

For all FMEA's benefits, it is a team effort – it is only as good as the team behind it. Any problem beyond their knowledge will not likely to be resolved.

[a]Sissons, M., et al. (eds), *Durum Wheat: Chemistry and Technology*, Chapter 10, AACC International, St. Paul, MN, 2012.

---

## 8.9 Fault-Tree Analysis[17]

Fault-tree analysis (FTA) was first introduced by Bell Laboratories and is a commonly used technique for describing system reliability, maintainability, and safety analysis. FTA is a systematic deductive process that gives a quantitative interpretation to resolve the causes of system failure.

The following basic steps are generally used in an FTA:

- Define an undesired event.
- Resolve the event into its immediate causes.
- Continue with resolution of events until basic causes are identified.
- Construct a fault tree showing the logical event relationships.
- Evaluate the fault tree.
- Interpret the result and recommend corrective action.

FTA is carried out to identify effects of human error, the causes of failure and weaknesses in a system. As shown in Figure 8.32, the construction of a basic fault-tree structure starts with the identification of an undesired event (system failure) leading to the *top event*. The *top event* is linked to the *intermediate event* by logic gates and event statements. This resolution continues

---

17 Adapted from Ertas, A., and Jones, J.C., *The Engineering Design Process*, John Wiley & Sons, Inc., New York, 1996.

**Figure 8.32** Basic fault-tree structure.

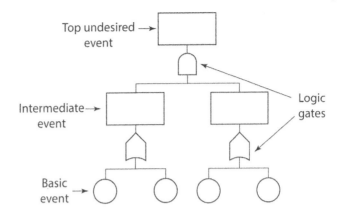

until the basic causes are identified. As shown in Table 8.12, event statements and gates are represented by specific logic symbols to illustrate the event relationships. The following example[18] demonstrates fault tree construction by using some of the symbols included in Table 8.12.

As shown in Figure 8.33, a condenser is used to convert the exhaust steam of an engine into water so that it can be reused in the boiler. It also creates a partial vacuum at the engine exhaust through the use of a vacuum pump and thereby increases the efficiency of the engine. To condense the steam, cold water is circulated through the condensing tubes by the use of a circulating pump. Finally, condensed exhaust steam is pumped by a feed-water pump to the boiler.

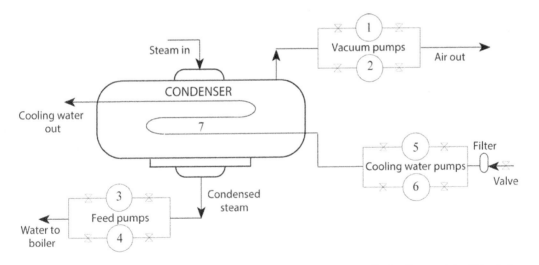

**Figure 8.33** A condenser system. (Ertas, A., and Jones, J.C., *The Engineering Design Process*, John Wiley & Sons, Inc., New York, 1996. Reproduced with permission of John Wiley & Sons, Inc.)

18 Ertas, A., and Jones, J.C., *The Engineering Design Process*, John Wiley & Sons, Inc., New York, 1996.

**Table 8.12** Fault-tree symbols.

| Symbols | Definitions |
|---|---|
| | *Resultant event*: represents the fault event above the gates, which is a result of the combination of other fault events. |
| | *Basic fault event*: represents a fault where the failure probability can be driven from empirical data. This is the limit of resolution of the fault tree. |
| | *House event*: revents a basic event that is expected to occur during the system operation. |
| in / out | *Transfer symbol*: transfer-in and transfer-out triangles used to transfer fault tree from one part to another. |
| | *AND gate*: output fault event occurs only when all the input fault events occur simultaneously. |
| | *Priority AND gate*: The output event occurs if all input events occur in a specific sequence. |
| | *Inhibit gate*: The input event occurs if all input events occur and an additional conditional event occurs. |
| | *OR gate*: output fault event occur when one or more input faults occur. |
| | *XOR gate*: The output event occurs if exactly one input event occurs. |

**Figure 8.34** A block diagram. (Ertas, A., and Jones, J.C., *The Engineering Design Process*, John Wiley & Sons, Inc., New York, 1996. Reproduced with permission of John Wiley & Sons, Inc.)

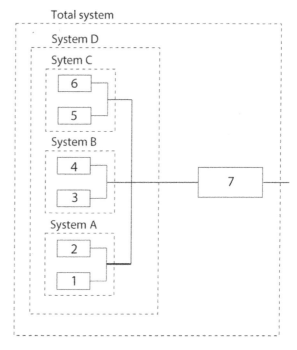

The system, as shown in the figure, is composed of series and parallel sub-systems. As the first step, a block diagram can be constructed by breaking the system down into series and parallel subsystem configurations, as shown in Figure 8.34. It is now relatively easy to develop a fault-tree network representing the condenser system.

Several conditions can cause the failure of a condenser system. The most important one is the loss of the vacuum in the condenser. Other problems include the failure of pumps, dirty filters, leaking gaskets, and high boiler pressure. Consider the fault-tree network for the condenser system shown in Figure 8.35. Each block in the figure represents an independent mechanical component and is assigned a number. The *top event* is the condenser system failure that may result from several prior faults (events) represented by the numbered circles. An OR gate is used because a failure may occur when any one of two events, 7, or system D failure, occur. Note that three AND gates are used because failure occurs only if pumps 1 and 2, 3 and 4, or 5 and 6 fail simultaneously.

Although the concepts of fault-tree network development appear to be simple, FTA for complex systems requires a thorough understanding of the system being analyzed and the techniques of fault-tree development.

## 8.9.1 Root Cause Analysis

Root cause analysis (RCA) is a common and widely used technique that helps people answer the question why the problem occurred in the first place. It tries to identify the root of a problem using a set of steps, with appropriate tools to find the main cause. The *five whys* or *fishbone* tools

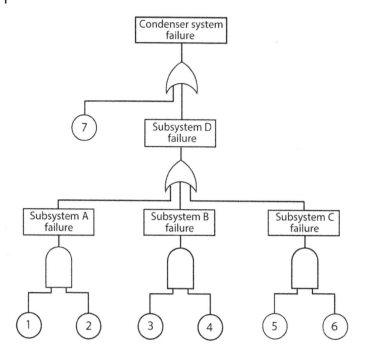

**Figure 8.35** A fault tree network. (Ertas, A., and Jones, J.C., *The Engineering Design Process*, John Wiley & Sons, Inc., New York, 1996. Reproduced with permission of John Wiley & Sons, Inc.)

can be used to identify the contributing root causes likely to be causing a problem. Both tools are used in the "analyze" phase of Six Sigma's DMAIC approach to problem solving.

### 8.9.1.1 Root Cause Analysis Using Fishbone Diagram

The *cause and effect* or *fishbone diagram* was first used by Dr. Kaoru Ishikawa of the University of Tokyo in 1943. It is a visual way to look at cause and effect and can help in brainstorming to identify the possible causes of a problem and in grouping ideas into useful categories. The fishbone tool is most effective when used in a team or group setting.

The team using the fishbone diagram should carry out the steps shown in Figure 8.36. After identifying the problem, a team should have collective understanding of the problem. Through CIM, the team has brainstormed all the potential causes. The facilitator helps the team to rate the potential causes according to their level of importance and create a diagram of a hierarchy. The design of the diagram has a fishbone shape as shown in Figure 8.37. The customer problem or effect is shown at the head or mouth of the fish.

Agree on the main categories of causes of the problem. The main categories usually include people/staff, process/methods, equipment/supply, environment, rules/policy/procedure, management, materials, service and requirements factors.

Then brainstorm around each cause of the problem. Ask "why does this happen?" As each idea is given, the facilitator of the CIM writes the underlying factor as a branch from the related category as shown in Figure 8.37. Causes can be written in different places if they relate to different categories. Repeat asking the question "Why does this happen?" about each cause and identify sub-causes branching off the cause branches until the root causes have been identified. The five whys or another questioning process can be used to develop the fishbone diagram.

Use the fishbone diagram tool to keep the team focused on the causes of the problem, rather than the symptoms. For example, if the spine becomes overly strained or compressed, a disc may rupture or bulge outward and create back pain. If you take painkillers, that will only take away the symptoms; you will need a different treatment to help your back heal properly.

As a part of the CIM, during the brainstorming, consider having team members write each idea of cause on sticky notes. Idea writing is an efficient process for producing many ideas relevant to the stated problem – it is self-documenting. This process will continue until all ideas are exhausted. No more than six people will take part in a team in each instance of the process. But any number of these instances can be carried out simultaneously.

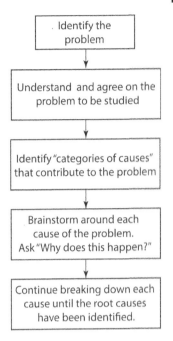

**Figure 8.36** Flow chart for fishbone diagram.

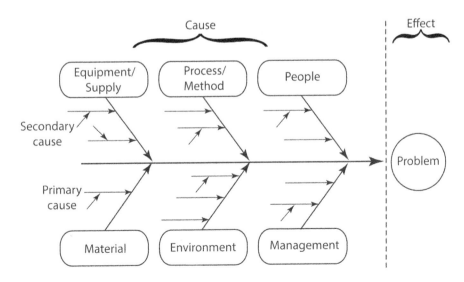

**Figure 8.37** Fishbone diagram.

## Example 8.14

Using fishbone diagram identify all the possible reasons for bad wine in order to determine the root cause.

**Solution**

*Step 1:* Before starting a fish bone diagram the problem statement must be well defined. The problem statement might be: *I had a truly bad wine at somebody's home last night.*

*Step 2:* Once the problem statement has been defined, to better identify possible causes, data related to the production of wine and how the wine was stored may be collected. This will make the process of checking the possible causes and theories a lot easier.

*Step 3:* Draw the backbone of the fishbone diagram, a long arrow pointing right toward the problem as shown in Figure 8.38. Then include the possible main categories of causes of the bad wine. They are:
1) Process
2) Equipment
3) Material
4) Environment.

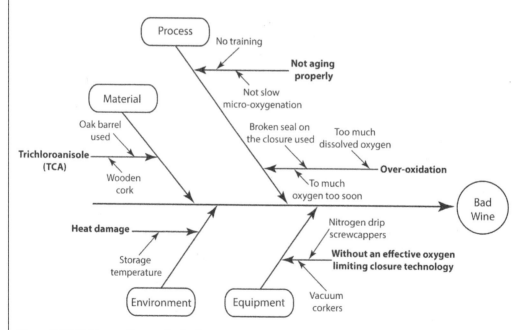

**Figure 8.38** Fishbone diagram for bad wine.

*Step 4:* For each category of cause, brainstorm the possible causes and hypotheses related to that category of cause. The production and storage data collected in step 2 can be very helpful in this step.

*Step 5:* For each of the causes identified in step 4 the same thought process is repeated to identify the root of each cause. The result is shown in Figure 8.38.

### 8.9.1.2 Root Cause Analysis Using the Five Whys Technique

Sakichi Toyoda, one of the fathers of the Japanese industrial revolution, developed the *five whys* technique in the 1930s. He was an inventor and founder of Toyota Industries.

The five whys is a technique used in the analyze phase of the Six Sigma DMAIC methodology. It is a good Six Sigma tool that does not require data segmentation, hypothesis testing, or other advanced statistical tools. In many cases the root cause analysis can be completed without a data collection plan.

The five whys technique is really effective when the answers come from people who have hands-on experience of the problem being examined. It is really simple: when a problem arises, you uncover its nature and source by repeatedly asking the question "why" until root cause(s) is/are found. You may ask the question fewer or more times than five before you find the causes related to a problem.

Table 8.13 can be utilized to identify the root cause of a problem when the five whys technique is used.

**Table 8.13** Five whys technique.

| Problem ⇒ Statement | One sentence description of issue or problem |
| --- | --- |
| **Five Whys** ⇓ | |
| **Why?** ⇒ | |
| **Why?** ⇒ | |
| **Why?** ⇒ | |
| **Why?** ⇒ | |
| **Why?** ⇒ | |
| **Record the root cause(s)** | 1. <br> 2. <br> 3. <br> 4. <br> If the issue or problem is solved by removing the identified root causes, then STOP. |
| **Action** | Provide solution. |

◻ **Example 8.15**

Yesterday on my drive home, I ended up behind an older vehicle with a carburetor. A massive amount of black smoke was coming from its tailpipe. Using the five whys technique, investigate the root cause of black smoke.

**Solution**

This simple technique can quickly direct you to the root of the problem – that is air filter was clogged (see Table 8.14). Note that the five whys technique may not always help to identify the root cause. For more complex problems, it can lead you to pursue a single track of inquiry when there could be more than one cause. In these cases, cause and effect analysis (the fishbone diagram) can be more effective.

**Table 8.14** Example of five whys technique.

| Problem ⇒<br>Statement | *Black smoke is coming out of the exhaust.* |
| --- | --- |
| **Five Whys** ⇓ | |
| Why was black smoke coming out of the tailpipe? | Too much fuel and not enough air going into the engine |
| Why was too much fuel and not enough air going into the engine? | Air/fuel ratio was not correct. |
| Why was the air/fuel ratio not correct? | The air filter was dirty. |
| Why was the air filter dirty? | The air filter was well beyond its useful service life and not replaced. |
| Why was the air filter not changed? | Forgot to replace it. |
| **Root Cause is** | Dirty air filter. |
| **Action** | Get a new air filter. |

## 8.10  Probabilistic Design[19]

Probabilistic design takes into account uncertainties/variabilities in the behavior of a structure that is under consideration. Over the years significant progress has been made in the development of probabilistic design.

Safety is an important responsibility of a designer. Because of the uncertainties or variabilities involved in design parameters, in many cases, it is difficult to make a decision on how safe a design is. The variability of design parameters may include material properties, geometry,

---

19 Adapted from Ertas, A., and J. Jones, J.C., *The Engineering Design Process*, John Wiley & Sons, Inc., New York, 1996.

tolerances, and environmental and time-dependent effects, such as external loads and their distributions. All these uncertainties make it difficult to ensure that the structure is safe. Therefore, it is important to understand uncertainty as it relates to design activity – the designer has to make sure that there is no risk of failure because of uncertainty.

When a probabilistic approach is used in design, each design variable (parameter) is viewed as a probability distribution. To minimize the risk of failure, a designer can reduce the random variability in design parameters to improve the performance characteristics of a structure. For example, the risk of failure of a wind turbine blade depends on the external loads to which it will be subjected during its useful lifetime. In this example, the parameter is the external load, which depends on certain uncontrollable factors such as weather conditions (variation of wind speed), large deflections, coupling of bending and torsional vibration in the blades, and the like.

Suppose a wind turbine blade will be designed to resist certain external loadings. As shown in Figure 8.39, the stress created by random loading is viewed as a probability distribution. Due to the variability in the blade material, the strength of the blade is also viewed as a probability distribution. Overlapping of the two probability distribution curves creates a region for possible risk of failure – if a weak structural component is exposed to heavy load, which exceeds its strength, the structural element will fail. Therefore, safety factors are used to minimize the effect of uncertainties in loading conditions and material strength.

### 8.10.1 Concept of Limit States

During the design stage, first the functions of the system components are defined and then possible failure modes are stated. Limit states are conditions of potential failure – system components no longer serve the purpose for which they were designed and built. Structural limit states fall into two major categories: (a) limit state of strength and (b) limit state of serviceability.

A general limit states equation, $g$, takes the form

$$g = \text{performance capacity} - \text{required demand.} \tag{8.80}$$

**Figure 8.39** Risk of failure.

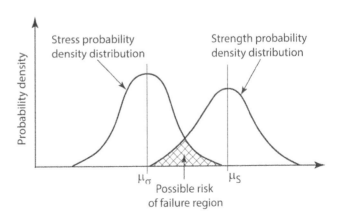

Stress probability density distribution

Strength probability density distribution

Probability density

$\mu_\sigma$    $\mu_S$

Possible risk of failure region

### 8.10.1.1 Limit State of Strength

Strength-based limit states are potential modes of structural failure – loss of stability of the structure – and can be written as

$$g(\mathbf{X}) = R(\mathbf{X}) - S(\mathbf{X}), \tag{8.81}$$

where $\mathbf{X}$ is a vector of random variables defining the geometry of the structure, the external loads that are applied, the strength of material (yielding, buckling, rupture), etc. $R(\mathbf{X})$ is the strength (resistance) of a system component as a function of $\mathbf{X}$ and $S(\mathbf{X})$ is the response (e.g. internal stress due to applied load) of a system component as a function of $\mathbf{X}$.

In equation (8.81), negative values of $g(\mathbf{X})$ indicate failure and positive values indicate safe states. Thus, the probability of failure, $P_f$ can be defined as

$$P_f = P[g(\mathbf{X}) \le 0]. \tag{8.82}$$

In other words, the probability of failure equals the probability that a combination of values of the random variables, $\mathbf{X}$, lies within the failure domain as shown in Figure 8.40. That is,

$$P_f = P(\mathbf{X} \in F). \tag{8.83}$$

### 8.10.1.2 Limit State of Serviceability

The limit state of serviceability refers to conditions that are not strength-based but still make the system components' performance unacceptable. Examples are deflection, vibration, corrosion, repairable damage due to fatigue, slenderness, and clearance. A general serviceability limit states equation can written as:

$$\text{Actual behavior} \le \text{allowable behavior}. \tag{8.84}$$

If the actual behavior of the structure is larger than the allowable behavior, structure will not be in the safe state.

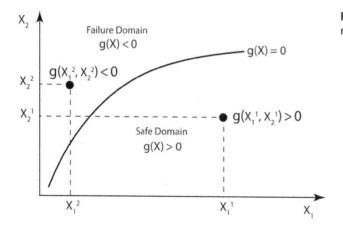

**Figure 8.40** Limit state, graphical representation.

### 8.10.1.3 Safety Factor

It is common practice to size the machine components, so that the maximum design stress created by external loadings is below the strength of the material, $S$ (ultimate tensile strength or yield strength). Then the safety factor, $n$, is defined as the ratio of strength to stress due to external loadings,

$$n = \frac{S}{\sigma}, \tag{8.85}$$

or the ratio of yield strength to actual working stress,

$$n = \frac{S_y}{\sigma_{\text{working}}}. \tag{8.86}$$

However, uncertainties will cause a variation of strength, $\Delta S$, and of stress, $\Delta\sigma$; therefore, the lowest possible strength can be written as[20]

$$S_{\text{min}} = S - \Delta S, \tag{8.87}$$

and the highest possible stress as

$$\sigma_{\text{max}} = \sigma + \Delta\sigma. \tag{8.88}$$

For no failure the following inequality must be satisfied:

$$S_{\text{min}} \geq \sigma_{\text{max}} \tag{8.89}$$

or

$$S\left[1 - \frac{\Delta S}{S}\right] \geq \sigma\left[1 + \frac{\Delta\sigma}{\sigma}\right]. \tag{8.90}$$

For conservative design, the minimum safety factor can be obtained from the above equation as

$$n = \frac{S}{\sigma} = \frac{1 + \frac{\Delta\sigma}{\sigma}}{1 - \frac{\Delta S}{S}}. \tag{8.91}$$

For example, if the maximum variations of stress and strength are 15 percent and 20 percent, respectively, the minimum safety factor is

$$n = \frac{1 + 0.15}{1 - 0.20} = 1.44.$$

### 8.10.1.4 Interference Model

The failure probability of a component can be determined by plotting the stress and strength probability density curves as shown in Figure 8.41. Whenever there is an overlap between two distributions and a mean difference $(\mu_S - \mu_\sigma)$ in probability less than zero a component failure results. The mean difference $\mu_D$ is given by

$$\mu_D = \mu_S - \mu_\sigma < 0, \tag{8.92}$$

---

20 Dimitri, K., and David, C., "Designing a specified reliability directly into a component," in *SAE Third Annual Aerospace Reliability and Maintainability Conference*, paper 640617, pp. 546–565, 1964.

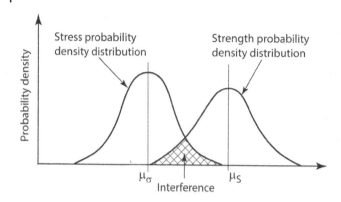

**Figure 8.41** Stress and strength density functions.

where $\mu_S$ is the mean strength and $\mu_\sigma$ is the mean stress due to external loadings. The difference standard deviation, $\hat{\sigma}_D$, can be expressed as

$$\hat{\sigma}_D = \sqrt{\hat{\sigma}_S^2 + \hat{\sigma}_\sigma^2}, \tag{8.93}$$

where $\hat{\sigma}_S$ and $\hat{\sigma}_\sigma$ are the standard deviations for the strength and stress, respectively.

When the stress and strength are represented by normal distributions, the probability of failure is the probability of having a negative difference between the stress and strength. This probability can be shown to be the negative tail end of a difference distribution, as shown in Figure 8.42. The probability of failure $P_f(D)$ is equal to the area of the negative tail and is given by[21]

$$P_f(D) = \int_{-\infty}^{0} f_f(D)d\mu_D, \tag{8.94}$$

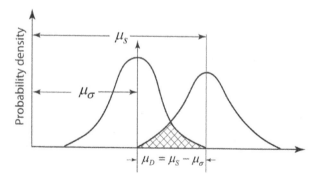

**Figure 8.42** Difference diagram for normal distribution.

---

21 Bratt, M.J., Reethof, G., and Weber, G.W., "A model for time varying and interfering stress-strength probability density distributions with consideration for failure incidence and property degradation," in *SAE Third Annual Aerospace Reliability and Maintainability Conference*, paper 640616, pp. 566–575, 1964.

where $f_f(D)$ is the probability density of the difference. For a normal distribution, the negative tail area can be determined using

$$z = \frac{\mu_D}{\hat{\sigma}_D}.$$ 
(8.95)

After calculating the value of $z$, the probability of failure (area), $P_f(D)$, is obtained from Table C.1 (Appendix C). Once the probability of failure is known, the component reliability can be determined from

$$R = 1 - P_f(D).$$ 
(8.96)

## Example 8.16

The data collected from an airplane engine structural member are as follows:[a]

1) The mean estimated stress at flight conditions as measured by simulated conditions in the test cell and flight test:
   a) mean thermal stress, $\mu_{\sigma t} = +30,000$ psi
   b) mean maneuver stress, $\mu_{\sigma m} = +1,800$ psi
   c) mean residual stress, $\mu_{\sigma r} = -10,000$ psi.
2) The standard deviations determined from test data, manufacturing tolerances, and technical requirements are:
   a) $\hat{\sigma}_{thermal} = +2,600$ psi
   b) $\hat{\sigma}_{maneuver} = +400$ psi
   c) $\hat{\sigma}_{residual} = 700$ psi.
3) The mean material strength (from material data) is $\mu_S = 38,000$ psi.
4) The standard deviation of the strength (from material data) is $\hat{\sigma}_S = 3,600$ psi.

Using the above given data, calculate the probability of failure of the engine structural member.

**Solution**

The net mean stress $\mu_\sigma$ in the structural member is

$$\mu_\sigma = 30,000 + 1,800 - 10,000 = 21,800 \text{ psi.}$$

The mean difference is

$$\mu_D = \mu_S - \mu_\sigma = 36,000 - 21,800 = 14,200 \text{ psi.}$$

The combined standard deviation of the average stress is

$$\hat{\sigma}_\sigma = \sqrt{2,600^2 + 400^2 + 700^2} = 2,722 \text{ psi.}$$

The difference standard deviation is

$$\hat{\sigma}_D = \sqrt{2,722^2 + 3,600^2} = 4,513 \text{ psi.}$$

Then the $z$ value is

$$z = \frac{\mu_D}{\hat{\sigma}_D} = \frac{14{,}200}{4{,}513} = 3.15.$$

From Table C.1 of Appendix C, the failure probability $P_f(D) = 0.0008$ can be obtained. The reliability of the structure can be calculated using equation (8.96):

$$R = 1 - 0.0008 = 0.9992.$$

[a]This example has been reprinted with permission from SAE Technical paper 640579, Society of Automotive Engineers, Inc.

## 8.11 Worst-Case Design[22]

As discussed previously, because of both external loads and material property variabilities the choice of a meaningful safety factor to increase the reliability of a design can be complicated. To overcome this difficulty, one may chose an unrealistically high safety factor to exceed the desirable confidence level. However, such a decision will result in expensive and over-designed components and the presence of excessive material that may actually weaken the component and cause poor performance.

Parameters that control the design performance are related in a manner that causes variations in one to cancel or increase the effect of variations in the other. Manufacturing tolerances or material property variations are examples. The poorest combination of parameters results in the most reliable component design, called the *worst-case design*.

Based on the theory of variances, the worst-case design can be defined as

$$d\phi = \left(\frac{\partial \phi}{\partial x}\right) dx + \left(\frac{\partial \phi}{\partial y}\right) dy, \tag{8.97}$$

where $\phi$ is a function of two independent variables $x$ and $y$. The following example illustrates the use of equation (8.97).

### Example 8.17

The maximum axial stress $\sigma$ in a pressure vessel (shown in Figure 8.43) due to bending can be approximated as[a]

$$\sigma = \mp \frac{138KQL}{R^2 t}, \tag{8.98}$$

---

22 Adapted from Ertas, A., and Jones, J.C., *The Engineering Design Process*, John Wiley & Sons, Inc., New York, 1996.

where

$K$ = constant corresponding to the ratio $A/L$

$L$ = pressure vessel length, ft

$Q$ = load on the supports, lb

$R$ = radius of the pressure vessel, ft

$t$ = thickness of the pressure vessel, in.

If the allowable stress is 10,000 psi, $A/L = 0.1$, $L = 40$ ft ($\mp0.2$ ft), and $R = 4$ ft ($\mp0.1$ ft), determine the thickness of the pressure vessel for the worst-case design.

**Figure 8.43** A fault-tree network(Ertas, A., and Jones, J.C., *The Engineering Design Process*, John Wiley & Sons, Inc., New York, 1996. Reproduced with permission of John Wiley & Sons, Inc.).

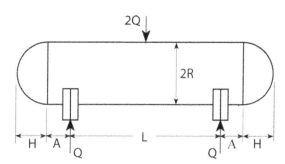

## Solution

If $A/L = 0.1$, $K$ can be approximated as 0.6. To simplify equation (8.98) further, assume $Q = 1$ lb. Then the thickness is

$$t = \frac{138KL}{\sigma R^2} = \frac{(138)(0.6)(L)}{(10,000)(R^2)} = \frac{0.00828L}{R^2}. \tag{8.99}$$

To calculate the nominal thickness, substitute $L = 40$ and $R = 4$ ft:

$$t = \frac{(0.00828)(40)}{4^2} = 0.0207 \quad \text{in.}$$

As can be seen from equation (8.99), thickness corresponds to the function $\phi$, and length and radius are the parameters under consideration. Taking the partial derivative of equation (8.99) with respect to $L$ yields

$$\frac{\partial t}{\partial L} = \frac{0.00828}{R^2} = \frac{0.00828}{4^2} = 0.00052;$$

and the partial derivative with respect to $R$ is

$$\frac{\partial t}{\partial R} = \frac{-0.01656L}{R^3} = \frac{(-0.0248)(40)}{4^3} = -0.01035.$$

The total variance equation is

$$dt = \left(\frac{\partial t}{\partial L}\right) dL + \left(\frac{\partial t}{\partial R}\right) dR$$
$$= (0.00052)(\mp 0.2) + (-0.01035)(\mp 0.1)$$

The possible thickness calculations are shown in Table 8.15. The results show that case 3 yields the greatest wall thickness. Hence, $t = 0.02184$ in. would be the thickness for the worst-case design (adapted from reference [8] in the bibliography below, pp. 101–102).

**Table 8.15** Cases for thickness calculations.

| Case | dL | dR | dt | t(0.0207 $\mp$ dt) |
|------|------|------|-----------|---------|
| 1 | 0.2 | 0.1 | −0.000931 | 0.01977 |
| 2 | −0.2 | 0.1 | −0.001139 | 0.01956 |
| 3 | 0.2 | −0.1 | 0.001139 | 0.02184 |
| 4 | −0.2 | −0.1 | 0.000931 | 0.02163 |

[a]Zick, L.P., "Stresses in large horizontal cylinder pressure vessels on two saddle supports," *ASME Pressure Vessel and Piping Design and Analysis*, 2, 959–970, 1972.

## Bibliography

1 BARBER, C.F., "Expanding maintainability concept and techniques," *IEEE Transactions on Reliability*, R-16(1), 5–9, 1967.

2 BILLINTON, R., and ALLAN, N. R., *Reliability Evaluation of Engineering Systems*, Plenum Press, New York, 1983.

3 CONDRA, L.W., *Reliability Improvement with Design of Experiments*, Marcel Dekker, New York, 1993.

4 DHILLON, B.S., *Quality Control, Reliability, and Engineering Design*, Marcel Dekker, New York, 1985.

5 DHILLON, B.S., *Mechanical Reliability: Theory, Models and Applications*, American Institute of Aeronautics and Astronautics, Washington, DC, 1988.

**6** FREBERG, D.D., and SPECTOR, R.B., "Reliability analysis and prediction for turbojet engines–results versus needs," in *SAE Third Annual Aerospace Reliability and Maintainability Conference*, paper 640579, pp. 253–262, 1964.

**7** HENLY, J.E., *Reliability Engineering and Risk Assessment*, Prentice Hall, Englewood Cliffs, NJ, 1981.

**8** KIVENSON, G., *Durability and Reliability in Engineering Design*, Hayden Book Co., New York, 1971.

**9** LEWIS, E. E., *Introduction to Reliability Engineering*, John Wiley & Sons, Inc., New York, 1987.

**10** PETERSON, E. L., "Maintainability application to system effectiveness quantification," *IEEE Transactions on Reliability*, R-20(1), 3–7, 1971.

**11** WILLIAM, J.M.F., "Bayes' equation, reliability, and multiple hypothesis testing," *IEEE Transactions on Reliability*, R-21(3), 136–139, 1972.

**12** WOHL, J. G., "System operational readiness and equipment dependability," *IEEE Transactions on Reliability*, R-15(1), 1–6, 1966.

## CHAPTER 8 Problems

**8.1** Research suggests that regularly eating even small amounts of cold cuts, bacon, sausage, and hot dogs increases colorectal cancer risk. For example, studies show that 4,200 people had colorectal cancer because of eating 3.0 ounces of such foods every day, and approximately 1 out of 100 people who have cancer go on to die. Calculate the annual fatality risk.

**8.2** Assuming that a group (cohort) study of 100 hot dog eaters and 100 hot dog non-eaters are followed for 20 years for the development of a cancer. Assume that 30 of the hot dog eaters and 15 of the non-eaters develop a form of cancer. Calculate the relative risk, $RR$.

**8.3** Assume that the failure of rolling bearings is described by a Weibull distribution. If the estimated values are $\theta = 1400$ hours and $b = 1.25$, determine:
a) The reliability for a 600-hour operational period.
b) The mean time to failure (MTTF) of the rolling bearing.
c) The hazard rate.

**8.4** The purpose of a condenser is to convert the exhaust steam of an engine into water so that it can be reused in the boiler. Both the vacuum pump, which removes the water–air mixture from the condenser, and the cooling water circulating pump must work simultaneously for system success. If the overall reliability must not be less than 0.96, what should be the minimum reliability of each pump?

**8.5** Assume a system consists of two components having reliabilities 0.82 and 0.88. For mission success, at least one of the components must be in operation. What is the reliability of the system?

**8.6** A water cooling tower operates with two independent and identical circulating pumps, both of which must work for cooling to be successful. What is the cooling tower reliability if each pump has reliability of 0.98?

**8.7** Assume that 200 psi of compressed air is provided by two identical compressors to a pressure vessel. This system will fail if both compressors fail. If the constant failure rate of both compressors is 0.00015 failures/hour, calculate the system reliability and MTTF for 1800 hours of operation.

**8.8** Solve Problem 8.5 by using the parametric approach.

**8.9** Develop a maintainability prediction chart as a function of time and $M(t)$ for the two constant repair rates $\mu_1 = 0.5$ repairs/hour, and $\mu_2 = 1.0$ repairs/hour. Discuss the design charts in detail.

**8.10** Obtain the maintainability function $M(t)$ expression when a Weibull probability density function is used.

**8.11** Suppose the failure of a system fits the Weibull distribution

$$R(t) = \exp\left[-\left(\frac{t}{1800}\right)^{0.29}\right].$$

Calculate:
a) The reliability for 1500 hours of operation
b) The mean time to failure
c) The hazard rate
d) The maintenance success probability for the time constraint of 2 hours.

**8.12** You are in charge of evaluating components that may be used at a waste water treatment facility. The criteria for selection include an availability of no less than 0.86 and a minimum obtainable dependability of 0.92. Given the values of $\mu = 0.022$ and $\lambda = 0.0018$ for a particular component, determine if it satisfies the criteria.

**8.13** Calculate the dependability assuming that in Example 8.13 the components are connected in parallel with a complete backup system.

**8.14** In Problem 8.13, if the system is connected as multistage with parallel redundancy, will the dependability increase or decrease?

**8.15** In turbocharged two-stroke diesel engines, the axial blower is connected directly to a gas turbine driven by the engine exhaust gases. In this instance the turbine and blower are designed as a single turboblower unit. Because of the high speed, wear of the turboblower's thrust bearings is the main concern. Assume a constant failure rate of $\lambda = 0.018$ failures/hour and a constant repair rate of $\mu = 0.06$ repairs/hour. For 18 hours of operation, determine the steady-state and instantaneous availability.

**8.16** A pressurized water tank as shown in Figure P8.16 distributes water to stations A, B, C, ..., N. Water is first transferred from the storage tank to the pressurized tank by means of a transfer pump. The pressurized tank provides enough head to transfer the water to the stations. The maximum allowable water level and air pressure are 5 ft and 100 psi, respectively. Two control schemes are used with the system:
- The water transfer pump stops when the maximum height is reached and starts when the minimum water level is reached.
- The water transfer pump stops when the maximum pressure is reached and starts when the minimum pressure is reached.

Develop a fault-tree network for this system if no water is provided to the stations.

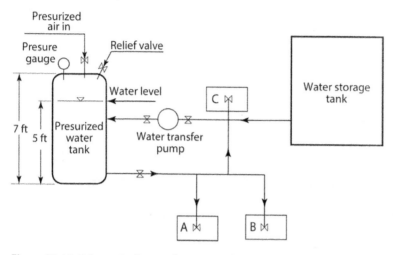

**Figure P8.16** Schematic diagram for a water distribution system.

# 9

# Optimization in Design

The designer must consider the optimum solution to a design problem so that a better, more efficient, less expensive solution that is in harmony with the laws of man and nature can be achieved.

## 9.1 Introduction

Problems associated with optimum design have been the subject of considerable attention for a number of years. During the last several decades the field of optimum design has made remarkable progress in systems design, control of dynamics, and engineering analysis. With recent advances in the field of computer technology, many modern optimization methods have been developed, and designing complex systems with the optimum configuration has become possible within a reasonable computation time.

There are two kinds of *effects* inherently associated with any mechanical element or system:

1) Undesirable effects, such as high cost, excessive weight, large deflections, and vibrations.
2) Desirable effects, such as long useful life, efficient energy output, good power transmission capability, and high cooling capacity.

*Optimum design* can be defined as the best possible design from the standpoint of the most significant effects, that is, minimizing the most significant undesirable effects and/or maximizing the most significant desirable effects.

The application of optimization to a design problem requires formulation of an *objective function* such as weight, cost, or shape, and the expression of design *constraints* as equalities or inequalities. The objective function, $U$, in terms of the *independent variables* (design variables) from which an optimum solution is sought can be written as

$$U = U(x_1, x_2, \ldots, x_n), \tag{9.1}$$

where $x$ is the design variable and $n$ is the number of design variables.

In an optimum design solution there are certain restrictions that are satisfied by the design variables for an acceptable design. These restrictions or limitations are called *constraints*. One seeks the optimum solution for a given objective function that fulfills the following equality and inequality constraints:

$$h_i(x) = h_i(x_1, x_2, \dots, x_n) = 0, \quad i = 1, 2, \dots, m, \tag{9.2}$$

$$g_j(x) = g_j(x_1, x_2, \dots, x_n) \geq 0, \quad j = 1, 2, \dots, p. \tag{9.3}$$

The design variable can be a material property, structural dimension (thickness, width, length), or geometric data (coordinates of node points of a structure).

## Example 9.1

Show the equation for the objective function for the minimum volume of a fixed supported beam subjected to a concentrated load $F$ as shown in Figure 9.1.

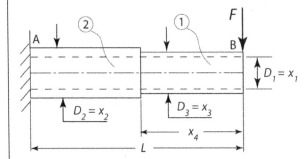

**Figure 9.1** Fixed supported beam.

### Solution

Since a minimum-volume design is sought, the objective function is the volume of the beam, which can be defined in terms of the four design variables, $x_1$, $x_2$, $x_3$, and $x_4$, as

$$U(x) = \left[ \frac{\pi(x_3^2 - x_1^2)}{4} \right] x_4 + \left[ \frac{\pi(x_2^2 - x_1^2)}{4} (L - x_4) \right].$$

Similarly, the objective functions can be written for factors like those for static deflection and cost.

If the design requires the beam to resist a maximum load of 8,000 N, the equality constraint can be written thus:

$$h_1 = F = 8{,}000 \text{ N}.$$

If a permissible stress $\sigma_p$ of 300 MPa is used, the beam will be safe against bending failure. The inequality constraint is

$$g_1 = \sigma_p = 500 \text{ MPa} \geq \sigma_b,$$

where $\sigma_b$ is the bending stress. The appropriate optimization method can now be applied to the above objective function (with the given constraints) to define the volume region that includes the optimum-design point.

Optimum-design problems may involve the following:

1) Equality constraints only
2) Inequality constraints only
3) Both equality and inequality constraints
4) No constraint.

For case (4), which is called an *unconstrained-optimization* problem, there are no restrictions imposed on the design variables. In general, there are two types of optimization problems:

1) *Single-criterion optimization problems.* Problems in which the designer's goal is to minimize or maximize one objective function. In this situation the optimum point is simply the minimum or maximum.
2) *Multicriterion optimization problems.* Problems in which the designer's goal is to minimize or maximize more than one objective function simultaneously. In this situation, all the objective functions should be considered to find the optimum solution.

## 9.2 Mathematical Models and Optimization Methods

The mathematical methods for optimum design can be divided into two categories:

1) *Analytical methods.* This method includes differentiation, variational methods, and the use of Lagrange multipliers.
2) *Numerical methods.* This method includes linear (simplex method) and nonlinear programming methods such as those given in Figure 9.2.[1]

This chapter discuses some of the common analytical and numerical optimization methods used in design. Additional information on these methods can be found in references 1, 3, and 5 in the chapter bibliography.

### 9.2.1 The Differential Calculus Method

The differential calculus method uses the first and second derivatives to find maximum and minimum values of a given differentiable function $U(x)$. To find a maximum or minimum using this method, take the first derivative of the objective function $U(x)$ and set it equal to zero,

$$\frac{dU}{dx} = 0,$$ 
(9.4)

---

1 Farkas, J., *Optimum Design of Metal Structures*, Halsted Press, John Wiley & Sons Inc., New York, 1984.

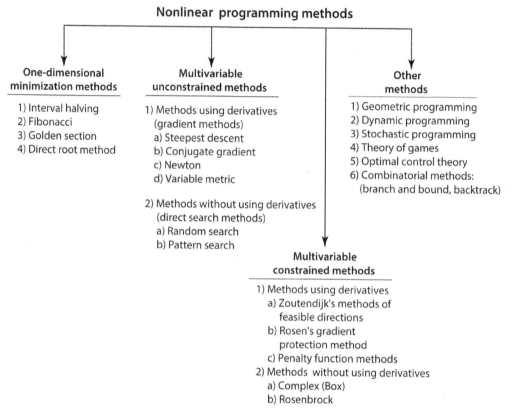

**Figure 9.2** Selected nonlinear programming methods. (Ertas, A., and Jones, J.C., *The Engineering Design Process*, John Wiley & Sons, Inc., New York, 1996. Reproduced with permission of John Wiley & Sons, Inc.)

then solve for the independent variable $x$. Equation (9.4) gives the value of the critical points (optimum points). If the second derivative evaluated at the critical point is less (greater) than zero, then the point is a maximum (minimum):

$$\frac{d^2U}{dx^2} < 0 \quad \text{implies a local maximum,} \tag{9.5}$$

$$\frac{d^2U}{dx^2} > 0 \quad \text{implies a local minimum.} \tag{9.6}$$

As shown in Figure 9.3, if the slope and the second derivative at a point are zero (as at point $x_2$), this is an inflection point. The function must be continuous to have a maximum or minimum. Thus, point $x_6$ in Figure 9.3 is not a maximum.

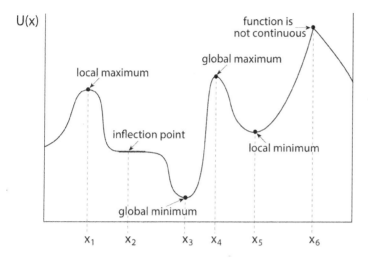

**Figure 9.3** Global and local optima. (Ertas, A., and Jones, J.C., *The Engineering Design Process*, John Wiley & Sons, Inc., New York, 1996. Reproduced with permission of John Wiley & Sons, Inc.)

### Example 9.2

Priority mail charges by size – a large flat-rate box is \$15.45 unless the sum of the dimensions as shown in Figure 9.4 is greater than 29.5 in. Determine the dimensions that will maximize the volume without exceeding the 29.5 in. requirement.

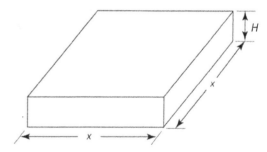

**Figure 9.4** Priority mail flat rate box size.

### Solution

The objective function is the volume; hence,

$$U = x^2 H.$$

The equality constraint is as follows:

$$h(x) = H + 4x - 29.5 = 0.$$

By eliminating the free variable $H$ from the above equations,

$$U = -4x^3 + 29.5x^2.$$

Consider the first derivative:

$$\frac{dU}{dx} = -12x^2 + 59x = 0.$$

The solution is given by $x_1 = 0$ and $x_2 = 4.9$. To ensure that $x_2 = 4.9$ is a maximum, take the second derivative,

$$\frac{d^2U}{dx^2} = -24x + 59,$$

and evaluate at $x_2 = 4.9$,

$$\frac{d^2U}{dx^2}\bigg|_{x_2=4.9} = -58.6.$$

This indicates a maximum for $U$ at $x_2 = 4.9$, where $dU/dx = 0$. Now use $x_2 = 4.9$ to solve for $H$ from the equality constraint:

$$H = 29.5 - 4x$$
$$= 29.5 - 4(4.9) = 9.9 \text{ in.}$$

Then maximum volume is

$$V = x^2 H = (4.9)^2(9.9) = 237.7 \text{ in.}^3.$$

### 9.2.2 The Lagrange Multiplier Method

The Lagrange multiplier method of optimization is named for its developer, Joseph Louis Lagrange (1736–1814), a French mathematician and astronomer. This method is important when dealing with nonlinear optimization problems. It uses a function called the Lagrange expression, LE, which consists of the objective function, $U(x, y, z)$, and constraint functions $h_i(x, y, z)$, multiplied by Lagrange multipliers, $\lambda_i$:

$$LE = U(x, y, z) + \lambda_1 h_1(x, y, z) + \cdots + \lambda_i h_i(x, y, z). \tag{9.7}$$

The additional unknown, $\lambda_i$, is introduced into the Lagrange expression so that in determining the optimum values of $x$, $y$, and $z$, the problem can be treated as though it were unconstrained. The conditions that must be satisfied for the optimum points are as follows:

$$\frac{\partial LE}{\partial x} = \frac{\partial LE}{\partial y} = \frac{\partial LE}{\partial z} = 0, \tag{9.8}$$

$$\frac{\partial LE}{\partial \lambda_1} = \frac{\partial LE}{\partial \lambda_2} = \cdots = \frac{\partial LE}{\partial \lambda_i} = 0, \tag{9.9}$$

where $i$ is the number of Lagrange multipliers.

## Example 9.3

Solve Example 9.2 using the Lagrange multiplier method.

**Solution**

From Example 9.2 the objective function is

$$U = x^2 H.$$

The equality constraint is

$$h(x) = H + 4x - 29.5 = 0.$$

Applying the Lagrange expression,

$$\text{LE} = x^2 H + \lambda(H + 4x - 295),$$

and differentiating with respect to $x$, $H$, and $\lambda$ yields

$$\frac{\partial \text{LE}}{\partial x} = 2xH + 4\lambda = 0,$$

$$\frac{\partial \text{LE}}{\partial H} = x^2 + \lambda = 0,$$

$$\frac{\partial \text{LE}}{\partial \lambda} = H + 4x - 29.5 = 0.$$

The solution of three equations gives $x = 4.9$ in. and $H = 9.9$ in.; hence the optimum volume is

$$V = x^2 H = (4.9)^2 (9.9) = 237.7 \text{ in.}^3$$

## 9.2.3 Search Methods

Search methods are based on examining simultaneous or sequential trial solutions over the entire domain of feasible designs to determine which point is optimal. These methods provide information about a region in which an optimum point is located. However, since these methods are not exact, a rapid rate of convergence is not possible. If the function has a single variable to locate the optimal point in a given interval, single-variable search methods are used. For functions having more than one variable, multiple-variable search methods are used.

A simultaneous search examination provides the approximate value of a minimum of a function having several minima in a given interval. However, because of the large number of functions evaluated, considerable computer time is required. For a function having one minimum or maximum (unimodal function), it is more efficient to use either the golden section search method or the Fibonacci search method.

*Equal interval search.* Assume that an objective function $U = f(x)$ is a function of a single variable $x$ and has a single minimum value of $U_{\min}$ at $x^*$ as shown in Figure 9.5. In using the equal interval search method, the function $U(x)$ is evaluated at points with equal $\Delta x$ increments. The calculated values of the function at two successive points, (say, $x_i$ and $x_{i+1} = x_i + \Delta x$), are then compared. When an increase in function value at any final point, $x_f = x_{i+1}$, is sensed, then

$$U(x_f - \Delta x) < U(x_f), \qquad (9.10)$$

and the minimum value of the function has passed. Thus, lower and upper limits for the interval of uncertainty can be written as

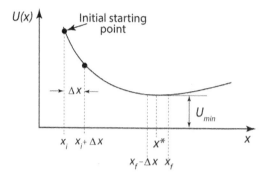

**Figure 9.5** Function $U(x)$ with a single variable. (Ertas, A., and Jones, J.C., *The Engineering Design Process*, John Wiley & Sons, Inc., New York, 1996. Reproduced with permission of John Wiley & Sons, Inc.)

$$x_f - \Delta x < x^* < x_f, \qquad (9.11)$$

where $x^*$ is the point at which the function $U(x)$ has a minimum. To reduce the interval of uncertainty to an acceptable value, the search process is reversed by a sign change with the increment $\Delta x$ cut in half. Now the search process is restarted by evaluating the function value at $x_f - \Delta x/2$, and the process is repeated for the next smaller interval of uncertainty.[2] This search process is repeated until the final interval of uncertainty is reduced to a small convergence criterion $\varepsilon$.

### Example 9.4

Find a minimum point of the function given below using the equal interval search method.

$$U(x) = 4x^2 + \frac{1310}{x}.$$

**Solution**

Following the flowchart given in Figure 9.6 and assuming that $\Delta x = 0.5$ and an initial value of $x_i$ is equal to 1, the value of the function $U$ is

$$U(x) = 4(1)^2 + \frac{1310}{1} = 1314.00.$$

---

2 Johnson, C.R., *Optimum Design of Mechanical Elements*, John Wiley & Sons, New York, 1980.

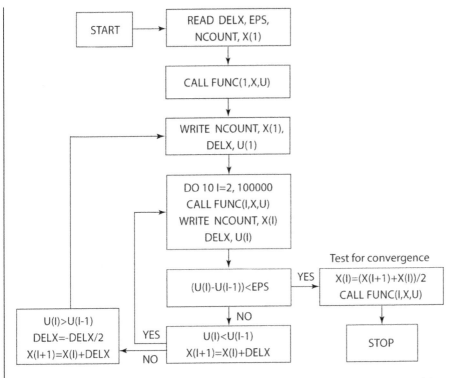

**Figure 9.6** The solution of $U_{min}$ using the equal interval search method. (Ertas, A., and Jones, J.C., *The Engineering Design Process*, John Wiley & Sons, Inc., New York, 1996. Reproduced with permission of John Wiley & Sons, Inc.)

The initial value of $x$ for the next iteration is

$$x_i + \Delta x = 1.0 + 0.5 = 1.5.$$

Then

$$U_{i+1} = 4(1.5)^2 + \frac{1310}{1.5} = 882.33.$$

Note from Table 9.1 that an increase in $U$ is sensed at the 11th iteration. Thus, the increment $\Delta x$ is reduced to one half and the sign is changed. This is shown in Table 9.1 beginning with the 12th iteration. The iteration process is repeated until the solution converges to a minimum value of the function. Based on a convergence criterion of $\varepsilon = 0.001$, the solution to the problem is (see Appendix A.2 for MATLAB code for calculations)

$$U_{min} = 359.17209 \text{ at } x = 5.47266.$$

**Table 9.1** Results of iterative calculation for equal interval search method.

| i | $x_i$ | $\Delta x$ | $U(x)$ |
|---|---|---|---|
| 1 | 1.00000 | 0.50000 | 1314.00000 |
| 2 | 1.50000 | 0.50000 | 882.33333 |
| 3 | 2.00000 | 0.50000 | 671.00000 |
| 4 | 2.50000 | 0.50000 | 549.00000 |
| 5 | 3.00000 | 0.50000 | 472.66667 |
| 6 | 3.50000 | 0.50000 | 423.28571 |
| 7 | 4.00000 | 0.50000 | 391.50000 |
| 8 | 4.50000 | 0.50000 | 372.11111 |
| 9 | 5.00000 | 0.50000 | 362.00000 |
| 10 | 5.50000 | 0.50000 | 359.18182 |
| 11 | 6.00000 | 0.50000 | 362.33333 |
| 12 | 5.75000 | −0.25000 | 360.07609 |
| 13 | 5.50000 | −0.25000 | 359.18182 |
| 14 | 5.25000 | −0.25000 | 359.77381 |
| 15 | 5.37500 | 0.12500 | 359.28343 |
| — | — | — | — |
| — | — | — | — |
| 26 | 5.46094 | 0.00781 | 359.17290 |
| 27 | 5.46875 | 0.00781 | 359.17176 |
| 28 | 5.47656 | 0.00781 | 359.17209 |
| | $x = 5.47266$ | | $U(x) = 359.17174$ |

*Golden section search.* In the equal interval search method the increment $\Delta x$ is kept constant until an increase in the function is sensed. The golden section search method provides an alternate procedure in which the increment varies at each search step. This search method finds the minimum value of a given function over a specified interval of length $L$. The first step of this method is to find the two interior points of the interval at which the function is to be calculated (see Figure 9.7).

Consider the interval to be from the lower limit $XL$ to the upper limit $XU$. The length of the interval is given by

$$L = XU - XL \tag{9.12}$$

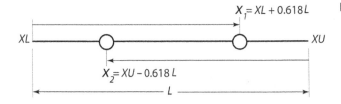

$X_1 = XL + 0.618L$

$X_2 = XU - 0.618L$

**Figure 9.7** Golden section intervals.

The two interior points $x_1$ and $x_2$ are found by using the equations

$$x_1 = XL + 0.618L, \tag{9.13}$$
$$x_2 = XU - 0.618L. \tag{9.14}$$

The interior points at which the function is evaluated are not selected arbitrarily; they are based on the ratio 0.618, known as the *golden ratio*.[3]

Suppose that the goal is to find the minimum value of a given function. If the value of the function evaluated at $x_1$ is larger than the value at $x_2$, the region to the right of $x_1$ is eliminated. The new interval is from $XL$ to $XU = x_1$, and the calculations are repeated. If the value calculated at $x_2$ is larger than the value calculated at $x_1$, the region to the left of $x_2$ is eliminated. In this case the new interval is from $XL = x_2$ to $XU$. This process is continued until the desired accuracy is obtained. The accuracy is determined by the remaining length of the interval. When the interval length becomes sufficiently small, say $\varepsilon = 0.005$, then the value is at least that close to the actual value of the minimum and the iterative process is stopped. The reverse applies to find the maximum of the function. The flow chart for this method is shown in Figure 9.8.

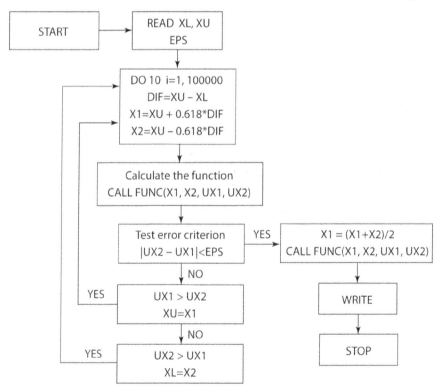

**Figure 9.8** The solution of $U_{min}$ using the golden section search method. (Ertas, A., and Jones, J.C., *The Engineering Design Process*, John Wiley & Sons, Inc., New York, 1996. Reproduced with permission of John Wiley & Sons, Inc.)

---

3 Reklaitis, G.V., Ravindran, A., and Ragsdell, K.M., *Engineering Optimization Methods and Applications*, John Wiley & Sons, Inc., New York, 1983.

## Example 9.5

Based on the convergence criterion $\varepsilon = 0.001$, use the golden section search method to find the minimum point of the function

$$U(x) = F(x) = 4x^2 + \frac{1310}{x}$$

in the interval of $2 \le x \le 10$.

### Solution

The length of interval $L$ is

$$L = XU - XL$$
$$= 10 - 2 = 8.$$

The interior points $x_1$ and $x_2$ are

$$x_1 = XL + 0.618L$$
$$= 2 + 0.618(8) = 6.944,$$
$$x_2 = XU - 0.618L$$
$$= 10 - 0.618(8) = 5.056.$$

Evaluating the function $U$ at interior points $x_1$ and $x_2$ yields

$$U(x_1 = 6.94) = 381.528$$
$$U(x_2 = 5.06) = 361.350$$

Compare the function values evaluated at $x_1$ and $x_2$:

$$U(x_1) > U(x_2).$$

Therefore, as shown in Figure 9.9, the region to the right of $x_1$ is eliminated and $XU = x_1 = 6.944$. The new interval for the next calculation is from $XU = 6.944$ to $XL = 2$. The same iteration process is repeated until the solution converges to a minimum value of the function. As shown in Table 9.2, based on a convergence criterion $\varepsilon = 0.001$, the solution to the problem is (see Appendix A.3 for MATLAB code for calculations)

$$U_{min} = 359.17175 \quad \text{at } x = 5.46897.$$

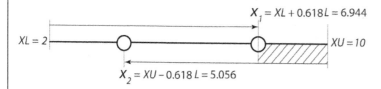

**Figure 9.9** Golden section intervals.

**Table 9.2** Results of iterative calculation for the golden section search method.

| $i$ | XL | XU | $x_1$ | $x_2$ | UX1 | UX2 |
|---|---|---|---|---|---|---|
| 1 | 2.00000 | 10.00000 | 6.94400 | 5.05600 | 381.52862 | 361.35065 |
| 2 | 2.00000 | 6.94400 | 5.05539 | 3.88861 | 361.35722 | 397.36656 |
| 3 | 3.88861 | 6.94400 | 5.77684 | 5.05577 | 360.25510 | 361.35315 |
| 4 | 5.05577 | 6.94400 | 6.22270 | 5.77707 | 365.40746 | 360.25672 |
| 5 | 5.05577 | 6.22270 | 5.77693 | 5.50153 | 360.25572 | 359.18291 |
| 6 | 5.05577 | 5.77693 | 5.50145 | 5.33125 | 359.18285 | 359.40989 |
| 7 | 5.33125 | 5.77693 | 5.60668 | 5.50150 | 359.38930 | 359.18289 |
| 8 | 5.33125 | 5.60668 | 5.50147 | 5.43647 | 359.18286 | 359.18601 |
| 9 | 5.43647 | 5.60668 | 5.54166 | 5.50149 | 359.23124 | 359.18288 |
| 10 | 5.43647 | 5.54166 | 5.50147 | 5.47665 | 359.18287 | 359.17210 |
| 11 | 5.43647 | 5.50147 | 5.47664 | 5.46130 | 359.17210 | 359.17282 |
| | $x = 5.46897$ | | | | | $U(x) = 359.17175$ |

## 9.2.4 Multivariable Search Method

As discussed previously, in a single-variable unconstrained optimization problem a one-dimensional search is required. In most engineering problems the coordinate of the optimum design point is necessary. Hence, to optimize a given multivariable function, a multidimensional search is required. Although several methods can be used to find the extremes of multivariable functions, the *steepest descent*, one of the simplest methods, is discussed here.[2]

*Steepest descent method.* Suppose an objective function with two variables $x_1$ and $x_2$ is given:

$$U = U(x_1, x_2). \tag{9.15}$$

For the given function, $U$, contour curves of constant $U$ can be obtained by changing values of $x_1$ and $x_2$ as shown in Figure 9.10. The point $O$, shown in the figure, is the optimum point of function $U$. $A_1$ is the arbitrary starting point, called the *base point*. Consider the tangent and normal vectors with the unit vectors $\bar{t}$ and $\bar{n}$ at point $A_1$:

$$\overline{T} = T \cdot \bar{t}, \tag{9.16}$$
$$\overline{N} = N \cdot \bar{n}. \tag{9.17}$$

The normal vector at point $A_1$ is called the gradient vector, defined as

$$\overline{N} = \nabla U = \left(\frac{\partial U}{\partial x_1}\right)\bar{I}_1 + \left(\frac{\partial U}{\partial x_2}\right)\bar{I}_2. \tag{9.18}$$

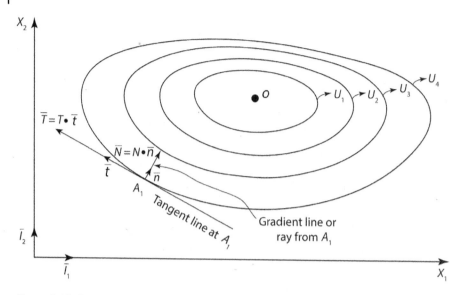

**Figure 9.10** Contour curves of constant *U*.

Thus, the unit vector can be written as

$$\bar{n} = \frac{\left(\dfrac{\partial U}{\partial x_1}\right)\bar{I}_1 + \left(\dfrac{\partial U}{\partial x_2}\right)\bar{I}_2}{\sqrt{\left(\dfrac{\partial U}{\partial x_1}\right)^2 + \left(\dfrac{\partial U}{\partial x_2}\right)^2}}, \tag{9.19}$$

which shows the direction of the gradient as the directional vector of the next step. The steepest search method can be used for multivariable problems with the increment in the gradient direction as shown in Figure 9.11. To move in the direction of the gradient vector, $\overline{N}$, the increment $\Delta$ is multiplied by the unit vector $\bar{n}$:

$$\overline{\Delta} = \Delta\bar{n} \tag{9.20}$$

or

$$\overline{\Delta} = \left[\frac{\left(\dfrac{\partial U}{\partial x_1}\right)\Delta\bar{I}_1 + \left(\dfrac{\partial U}{\partial x_2}\right)\Delta\bar{I}_2}{\sqrt{\left(\dfrac{\partial U}{\partial x_1}\right)^2 + \left(\dfrac{\partial U}{\partial x_2}\right)^2}}\right]. \tag{9.21}$$

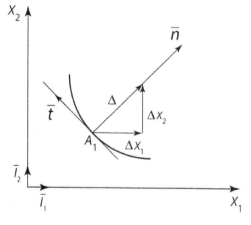

**Figure 9.11** Components of step size Δ.

The above equation can be written as

$$\bar{\Delta} = \Delta x_1 \bar{I}_1 + \Delta x_2 \bar{I}_2,$$
(9.22)

where

$$\Delta x_1 = \frac{-\left(\dfrac{\partial U}{\partial x_1}\right)_{A_1} \Delta}{\sqrt{\left(\dfrac{\partial U}{\partial x_1}\right)_{A_1}^2 + \left(\dfrac{\partial U}{\partial x_2}\right)_{A_1}^2}},$$
(9.23)

$$\Delta x_2 = \frac{-\left(\dfrac{\partial U}{\partial x_2}\right)_{A_1} \Delta}{\sqrt{\left(\dfrac{\partial U}{\partial x_1}\right)_{A_1}^2 + \left(\dfrac{\partial U}{\partial x_2}\right)_{A_1}^2}}.$$
(9.24)

The same $\Delta$ and gradient direction are used until an increase in $U$ is sensed. The direction of the gradient is then changed and $\Delta$ is reduced to $\Delta/2$. This procedure is repeated until the $\Delta$ is reduced to a value corresponding to that of the convergence criterion, $\varepsilon_1$. The next gradient direction (ray) is then determined, and the same steps are repeated until $\partial U/\partial x_1$ or $\partial U/\partial x_2$ is reduced to a value corresponding to that of the convergence criterion $\varepsilon_2$. When $x_1$ and $x_2$ reach a minimum or maximum, their values will be equal to the roots of $\partial U/\partial x_1$ and $\partial U/\partial x_2$. Hence, the smaller the partial derivative, $\partial U/\partial x_i$ the closer the $x_i$ values are to the roots. The following example will illustrate the application of this method.

## Example 9.6

To illustrate the steepest descent method, consider the function[a]

$$U = F(x) = 3(x_1^2 + x_2^2) + 1{,}296 \left( \frac{1}{x_1} + \frac{1}{x_2} \right).$$

**Solution**

Following the flowchart given in Figure 9.12, choose an arbitrary starting point and increment as

$$x_1 = 8, \quad x_2 = 1, \quad \Delta = 1.0.$$

Assume a convergence criterion for $\Delta$ of $\varepsilon = 0.001$, and a convergence criterion for the ray (gradient line) of $\varepsilon = 0.01$. Calculate the function $U$ for $x_1$ and $x_2$:

$$U = 3(8^2 + 1^2) + 1{,}296 \left( \frac{1}{8} + \frac{1}{1} \right) = 1{,}653.00.$$

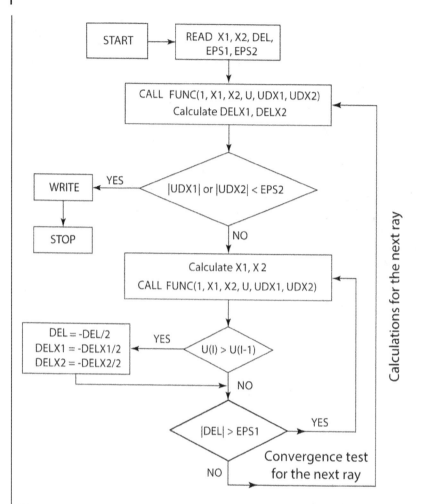

**Figure 9.12** The solution of $U_{\min}$ using the steepest descent method. (Ertas, A., and Jones, J.C., *The Engineering Design Process*, John Wiley & Sons, Inc., New York, 1996. Reproduced with permission of John Wiley & Sons, Inc.)

To find the components of the increment, $\Delta$, determine

$$\left(\frac{\partial U}{\partial x_1}\right) = 6x_1 - \frac{1,296}{x_1^2}$$

$$= 6(8) - \frac{1,296}{8^2} = 27.75,$$

$$\left(\frac{\partial U}{\partial x_2}\right) = 6x_2 - \frac{1,296}{x_2^2}$$

$$= 6(1) - \frac{1,296}{1^2} = -1,290.$$

From equations (9.23) and (9.24),

$$\Delta x_1 = \frac{-\left(\dfrac{\partial U}{\partial x_1}\right)\Delta}{\sqrt{\left(\dfrac{\partial U}{\partial x_1}\right)^2 + \left(\dfrac{\partial U}{\partial x_2}\right)^2}}$$

$$= \frac{-(27.75)(1.0)}{\sqrt{27.75^2 + 1290^2}} = -0.021506$$

$$\Delta x_2 = \frac{-\left(\dfrac{\partial U}{\partial x_2}\right)\Delta}{\sqrt{\left(\dfrac{\partial U}{\partial x_2}\right)^2 + \left(\dfrac{\partial U}{\partial x_2}\right)^2}}$$

$$= \frac{-(-1{,}290)(1.0)}{\sqrt{27.75^2 + 1{,}290^2}} = 0.99977.$$

Calculate $x_1$ and $x_2$ for the first iteration of the first ray:

$$x_1 = x_1(\text{initial}) + \Delta x_1$$
$$= 8 - 0.021506 = 7.97849,$$
$$x_2 = x_2(\text{initial}) + \Delta x_2$$
$$= 1 + 0.99977 = 1.99977.$$

Calculate the value of $U$ at the new points $x_1$ and $x_2$:

$$U = 3(7.97849^2 + 1.99977^2) + 1{,}296 \left(\frac{1}{7.97849} + \frac{1}{1.99977}\right) = 1{,}013.47800.$$

Calculations for the next iteration are

$$x_1 = 7.97849 - 0.021506 = 7.95699,$$
$$x_2 = 1.99977 + 0.99977 = 2.99954,$$

and the new value of the function $U$ is 811.875.

As shown in Table 9.3, at the sixth iteration the values for $x_1$ and $x_2$ are 7.87096 and 6.99861, respectively. Note that at this iteration, the value of the function is increased. Thus, for the next iteration, the values of $\Delta$, $\Delta x_1$, and $\Delta x_2$ are reduced to one half and the gradient direction sign is changed:

$$\Delta = -\frac{1.0}{2} = -0.500000,$$
$$\Delta x_1 = -\left(\frac{-0.021506}{2}\right) = 0.01075,$$
$$\Delta x_2 = -\left(\frac{0.99977}{2}\right) = -0.49988.$$

Hence,

$$x_1 = 7.87096 + 0.01075 = 7.88171,$$

$$x_2 = 6.99861 - 0.49988 = 6.49873,$$

and the function $U$ is

$$U = 3(7.88171^2 + 6.49873^2) + 1296 \left( \frac{1}{7.88171} + \frac{1}{6.49873} \right) = 676.91956.$$

The same iteration procedure is repeated until the increment $\Delta$ is equal to the convergence criterion $\varepsilon = 0.001$. As shown in Table 9.3, this occurs at the 37th iteration. The values calculated at this iteration are used as initial values for the next ray's first iteration.

The *second ray calculations* start from the coordinates $x_1 = 7.89179$ and $x_2 = 6.03009$ with $\Delta = 1.00000$. Note that the value of the increment, $\Delta = 1.00000$, remains the same. Calculate the new ray direction as follows:

$$\left( \frac{\partial U}{\partial x_1} \right) = 6x_1 - \frac{1{,}296}{x_1^2}$$

$$= 6(7.89179) - \frac{1{,}296}{7.89179^2} = 26.54160,$$

$$\left( \frac{\partial U}{\partial x_2} \right) = 6x_2 - \frac{1{,}296}{x_2^2}$$

$$= 6(6.03009) - \frac{1{,}296}{6.03009^2} = 0.53892.$$

Hence,

$$\Delta x_1 = \frac{-\left( \dfrac{\partial U}{\partial x_1} \right) \Delta}{\sqrt{\left( \dfrac{\partial U}{\partial x_1} \right)^2 + \left( \dfrac{\partial U}{\partial x_2} \right)^2}}$$

$$= \frac{-(26.54160)(1.0)}{\sqrt{26.54160^2 + 0.53892^2}} = -0.99979,$$

$$\Delta x_2 = \frac{-\left( \dfrac{\partial U}{\partial x_2} \right) \Delta}{\sqrt{\left( \dfrac{\partial U}{\partial x_2} \right)^2 + \left( \dfrac{\partial U}{\partial x_2} \right)^2}}$$

$$= \frac{-(0.53892)(1.0)}{\sqrt{26{,}54160^2 + 0.53892^2}} = -0.020300.$$

**Table 9.3** Results of iterative calculation for steepest descent.

| Ray | $i$ | $\Delta$ | $x_1$ | $x_2$ | $U$ |
|---|---|---|---|---|---|
| | | 1.00000 | 8.00000 | 1.00000 | 1653.00000 |
| | 1 | 1.00000 | 7.97849 | 1.99977 | 1013.47800 |
| | 2 | 1.00000 | 7.95699 | 2.99954 | 811.87500 |
| | 3 | 1.00000 | 7.93548 | 3.99931 | 724.27228 |
| | 4 | 1.00000 | 7.91397 | 4.99907 | 685.87415 |
| | 5 | 1.00000 | 7.89247 | 5.99884 | 675.08038 |
| | 6 | 1.00000 | 7.87096 | 6.99861 | 682.63324 |
| | 7 | −0.50000 | 7.88171 | 6.49873 | 676.91956 |
| 1 | 8 | −0.50000 | 7.89247 | 5.99884 | 675.08032 |
| | 9 | −0.50000 | 7.90322 | 5.49896 | 677.76300 |
| | 10 | 0.25000 | 7.89784 | 5.74890 | 675.80719 |
| | 11 | 0.25000 | 7.89247 | 5.99884 | 675.08032 |
| | 12 | 0.25000 | 7.88709 | 6.24879 | 675.47998 |
| | 13 | −0.12500 | 7.88978 | 6.12382 | 675.14508 |
| | — | — | — | — | — |
| | — | — | — | — | — |
| | 36 | −0.00195 | 7.89175 | 6.03204 | 675.07050 |
| | 37 | 0.00098 | 7.89179 | 6.03009 | 675.07056 |
| | | 1.00000 | 7.89179 | 6.03009 | 675.07056 |
| | 1 | 1.00000 | 6.89200 | 6.00979 | 654.54395 |
| | 2 | 1.00000 | 5.89220 | 5.98949 | 648.10687 |
| 2 | 3 | 1.00000 | 4.89241 | 5.96919 | 660.71576 |
| | 4 | −0.50000 | 5.39231 | 5.97934 | 651.57721 |
| | — | — | — | — | — |
| | — | — | — | — | — |
| | 31 | −0.00195 | 6.00156 | 5.99171 | 648.00067 |
| | 32 | 0.00098 | 6.00351 | 5.99175 | 648.00073 |
| | | 1.00000 | 6.00351 | 5.99175 | 648.00073 |
| | 1 | −0.50000 | 5.61280 | 6.91226 | 656.24243 |
| 3 | — | — | — | — | — |
| | — | — | — | — | — |
| | 29 | −0.00195 | 6.00046 | 5.99894 | 648.00000 |
| | 30 | 0.00098 | 6.00122 | 5.99714 | 648.00012 |
| | | 1.00000 | 6.00122 | 5.99714 | 648.00012 |
| | 1 | −0.50000 | 5.60919 | 6.91710 | 656.33893 |
| 4 | — | — | — | — | — |
| | — | — | — | — | — |
| | 28 | −0.00195 | 5.99969 | 6.00074 | 648.00000 |
| | 29 | 0.00098 | 6.00045 | 5.99894 | 648.00006 |
| | NCOUNT= 128 | | $x_1 = 6.000455$ | $x_2 = 5.998939$ | $U = 648.0001$ |

Calculate $x_1$ and $x_2$ for the first iteration:

$$x_1 = 7.89179 - 0.99979 = 6.89200,$$
$$x_2 = 6.03009 - 0.020300 = 6.00979,$$

and

$$U = 3(6.89200^2 + 6.00979^2) + 1{,}296 \left( \frac{1}{6.89200} + \frac{1}{6.00979} \right) = 654.54395.$$

Iterations for the second ray are repeated until the increment is reduced to the assumed convergence criterion, $\varepsilon_1 = 0.001$.

The iteration procedure for each ray is repeated until $\partial U / \partial x_1$ or $\partial U / \partial x_2$ is reduced to a convergence criterion $\varepsilon_2 = 0.01$. As shown in Table 6.3, based on the convergence criterions $\varepsilon_1 = 0.001$ and $\varepsilon_2 = 0.01$, the minimum solution is

$$U_{min} = 648.00006 \quad \text{at } x_1 = 6.00045, \quad x_2 = 5.99894.$$

Using elementary calculus, an exact solution of the given problem can also be found as

$$\left( \frac{\partial U}{\partial x_1} \right) = 6x_1 - \frac{1{,}296}{x_1^2} = 0 \Rightarrow x_1 = 6.00000,$$

$$\left( \frac{\partial U}{\partial x_2} \right) = 6x_2 - \frac{1{,}296}{x_2^2} = 0 \Rightarrow x_2 = 6.00000.$$

Substituting the above determined roots $x_1$ and $x_2$ into the equation yields the minimum value of the function

$$U_{min} = 3(6^2 + 6^2) + 1{,}296 \left( \frac{1}{6} + \frac{1}{6} \right) = 648.00000.$$

The result of the exact solution agrees with the approximate solution found by the steepest descent method.

[a]From Ertas, A., and Jones, J.C., *The Engineering Design Process*, John Wiley & Sons, Inc., New York, 1996.

## 9.2.5 Linear Programming

The method of linear programming is applicable to a linear objective function subject to a number of linear constraints. As shown in Figure 9.13, the linear objective function $U(x) = a + bx$, without constraints, has a minimum and maximum at $x \to -\infty$ and $x \to \infty$, respectively.

The maximization problem for a linear objective function subject to linear constraints can be formulated as follows:

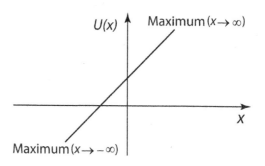

**Figure 9.13** The maximum and minimum for an unconstrained obective function.

1) Optimize the objective function

$$U(x) = a_1 x_1 + a_2 x_2 + \cdots + a_n x_n \tag{9.25}$$

2) Set the linear constraints as

$$h_i(x) = b_{i1} x_1 + b_{i2} x_2 + \cdots + b_{in} x_n \leq, =, \geq C_i \tag{9.26}$$

$$x_1, x_2, \ldots, x_n \geq 0, \quad i = 1, 2, \ldots, m,$$

where $x$ is the design variable, $n$ is the number of design variables, and $a$, $b$, and $C$ are given constants.

Linear programming problems can be solved either analytically or graphically. If there are few unknowns associated with the optimum design, a graphical method can be used to find the optimum points of a given objective function. If the number of unknowns is relatively high, the most common numerical method of solution, the *simplex method*, is used. The graphical method uses the following steps:

1) Identify design variables, objective function, and constraints.
2) Identify the boundaries of the feasible region by using the given constraints. If any inequality constraints are given, use them as equality constraints to find the boundaries of the constraint region so that the feasible region will simultaneously satisfy all the constraints.
3) Plot the objective function to identify the best design point that optimizes the objective function.

## Example 9.7

Using the graphical method, maximize

$$U(x) = 4x_1 + 2x_2 \tag{9.27}$$

subjected to the following constraints:

$$2x_1 + x_2 \leq 40,$$
$$x_1 + x_2 \leq 30,$$
$$36x_1 + 94x_2 \leq 2{,}520,$$
$$x_1 \geq 0,$$
$$x_2 \geq 0.$$

### Solution

The two constraints $x_1 \geq 0$ and $x_2 \geq 0$ form the boundaries $x_1 = 0$ and $x_2 = 0$ of the feasible region. After converting the constraint equations to equality constraints,

$$2x_1 + x_2 = 40,$$
$$x_1 + x_2 = 30,$$
$$36x_1 + 94x_2 = 2{,}520,$$

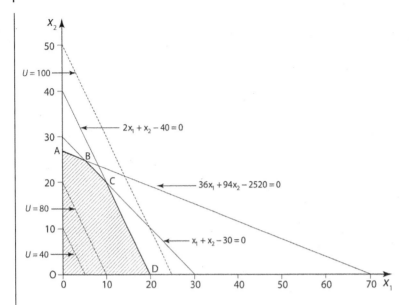

**Figure 9.14** Feasible region for the optimal solution.

the feasible region (shaded area) for the optimum solution can be drawn as shown in Figure 9.14. This feasible region simultaneously satisfies all five constraints and represents the area of possible optimum solutions to the design problem. The objective function $U(x) = 4x_1 + 2x_2$ is also shown on the figure for $U(x) = 40$, $U(x) = 80$, and $U(x) = 100$. As can be observed from the figure, when parallel lines that define the objective function move away from the point of origin, the value of the objective function is increased. The objective function for $U(x) = 100$ does not intersect the feasible region boundaries; hence, it should not be included in the optimum design analysis. The set of feasible points lies on the boundaries of the feasible region. The point that maximizes the objective function occurs at the intersection of two or more constraints. Thus, the first step in finding the maximum point is to find the intersection points of pairs of constraints. Next, reduce the intersection points to one point that maximizes the objective function. Table 9.4 shows the coordinates of points $A$, $B$, $C$, $D$, and $O$ and the corresponding objective function values. The optimum solutions are obtained at points $C$ and $D$, where the value of the objective function is 80.

**Table 9.4** Set of solutions for $U(x)$.

| Points | $U(x) = 4x_1 + 2x_2$ |
| --- | --- |
| $A(0, 26.8)$ | 53.6 |
| $B(5.2, 24.8)$ | 70.4 |
| $C(10, 20)$ | 80.0 |
| $D(20, 0)$ | 80.0 |
| $O(0, 0)$ | 0.0 |

## 9.2.6 Nonlinear Programming Problems

If the objective function or any other constraints that define the optimization problem are nonlinear, the design problem is called a *nonlinear programming* problem. As an example of an optimum design problem that can be solved by using nonlinear programming, consider the beam problem given in Example 9.1. The problem is to find the dimensions of the beam that satisfy strength and geometric constraints and minimize the volume of the beam.

To simplify the problem, assume predetermined values of

$$x_2 = D_2 = 50 \text{ mm},$$
$$x_3 = D_3 = 40 \text{ mm},$$
$$L = 600 \text{ mm}.$$

The goal is to determine the dimensions $x_1$ and $x_4$ for the minimum volume. The objective function is

$$U(x) = \left[ \frac{\pi(40^2 - x_1^2)}{4} \right] x_4 + \left[ \frac{\pi(50^2 - x_1^2)}{4}(600 - x_4) \right].$$

Simplifying,

$$U(x) = -707x_4 - 471x_1^2 + 1{,}178{,}097.$$

The problem may be further formulated to find the dimensions, $x_1$ and $x_4$, which minimize the objective function, $U$, and satisfy the following inequality constraints:

1) The beam should resist the maximum load $F = 8{,}000$ N and the allowable bending stress $\sigma_a = 500$ MPa. Hence, the bending strength inequality constraint for part 1 of the beam is

$$\sigma_b \le \sigma_a,$$

where $\sigma_b$, the bending stress due to $F$, is

$$\sigma_b = \frac{32Fx_4D_3}{\pi(D_3^4 - x_1^4)}.$$

Substituting into the inequality constraint yields

$$\frac{3{,}259{,}493x_4}{2.56 \times 10^6 - x_1^4} \le 500.$$

Similarly, the bending strength inequality constraint for part 2 of the beam is

$$\frac{32FLD_2}{\pi(D_2^4 - x_1^4)} \le \sigma_a$$

$$\frac{768 \times 10^7}{\pi(50^4 - x_1^4)} \le 500.$$

The above equation yields

$$x_1 \le 34 \text{ mm}$$

2) The inside diameter of the beam $x_1$ should not be less than 20 mm, and the length $x_4$ must be equal to or larger than zero. Thus, the additional inequality constraints are

$$x_1 \geq 20 \text{ mm},$$

$$x_4 \geq 0.$$

In summary, the strength and geometric inequality constraints are

$$g_1 \equiv 500 - \frac{3{,}259{,}493x_4}{2.56 \times 10^6 - x_1^4} \geq 0,$$

$$g_2 \equiv 34 - x_1 \geq 0,$$

$$g_3 \equiv x_1 - 20 \geq 0,$$

$$g_4 \equiv x_4 \geq 0.$$

A graphical solution for the feasible region of the beam is shown in Figure 9.15. In this example, since $g_1$ is a nonlinear function of the design variable, $x_1$, the problem is called a nonlinear programming optimization problem. The designer may also consider an equality constraint such as the length of part 1 of the beam being equal to five times the inside diameter $x_1$. Thus, the equality constraint is

$$h_1 = x_4 - 5x_1 = 0.$$

Figure 9.15 is duplicated as shown in Figure 9.16. Now the feasible region is represented by the dark line. It is interesting to observe that the area of the feasible region is reduced to a line when the equality constraint is included.

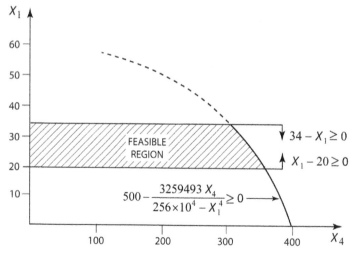

**Figure 9.15** The constraint on design variables $x_1$ and $x_4$.

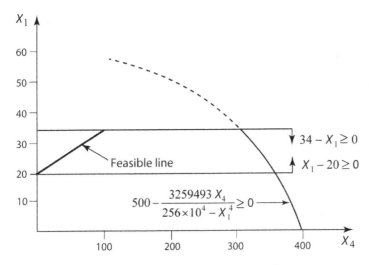

**Figure 9.16** The constraint on design variables $x_1$ and $x_4$.

### 9.2.7 Multicriterion Optimization

Almost all real-world problems involve more than one objective. In a multicriterion optimization the aim is to maximize or minimize more than one objective function simultaneously. The advantage of this method compared with that of single-criterion optimization is that a set of alternative solutions can be obtained rather than a single solution. In optimizing design problems that involve more than one objective function, the designer may have to:

1) maximize all the objective functions;
2) minimize all the objective functions;
3) maximize some and minimize others.

There are many different methods that can be used for multicriterion optimization problems.[4] A simple beam problem can be solved to illustrate a multicriterion optimization. The previously discussed problem can be reformulated as follows. Find the dimensions $x_1$ and $x_2$ that satisfy the geometric and strength constraints and minimize the following criteria:

1) the volume of the beam;
2) the static deflection of the beam under the load $2F$.

Both functions are to be minimized. Note that these are contrasting criteria; that is, the best solution for the first criterion yields the worst solution for the second. More information on multicriterion optimization problems is given by Osyczka and Steuer (see the chapter bibliography).

---

4 Cohon, J.L., *Multiobjective Programming and Planning*, Academic Press, New York, 1978.

## 9.3 Optimization of System Reliability[5]

As shown in Figure 9.17, the overall reliability of a series system can be increased by providing redundancy at each stage. As mentioned before, increasing the number of components at each stage increases the cost and weight of the system. To design a system with high performance requires compromises between reliability and cost or weight. The problem is to optimize the reliability with respect to cost or weight by a reasonable tradeoff among the various components at each stage. This section discusses three cases, each using a parametric approach, of reliability optimization of a multistage parallel redundant system with different linear constraints:[6]

1) Maximum reliability for a given cost constraint
2) Minimum cost for a given reliability constraint
3) Maximum reliability for given cost and weight constraints.

To optimize the redundant system mentioned above, the equation

$$\phi_{sp} = \sum_{k=1}^{n} \phi_k^{Y_k} \tag{9.28}$$

is used. This equation is simple in structure but requires the minimization of $\phi$ instead of the maximization of $R$, as in the classical method. Hence, the use of the parametric method to optimize the reliability of a complex system provides considerable simplification and time reduction in computations.

### 9.3.1 Reliability Optimization for a Given Cost Constraint

Consider a system where the number of stages and component reliabilities are given. It is desirable that the total cost of the system not exceed a given cost limit $C_L$. Thus

$$C_L \geq \sum_{k=1}^{n} C_k Y_k, \tag{9.29}$$

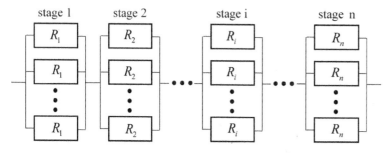

**Figure 9.17** Block diagram of a multistage system with parallel redundancy.

5 From Ertas, A., and Jones, J.C., *The Engineering Design Process*, John Wiley & Sons, Inc., New York, 1996.
6 © 1973 IEEE, Banerjee, S.K., and Rajamani, K., "Optimization of system reliability using a parametric approach," *IEEE Transactions on Reliability*, R-22, 35–39, 1973.

where $C_k$ is the cost of one component. The constraint function can be written as

$$\psi = \sum_{k=1}^{n} C_k Y_k - C_L = 0. \tag{9.30}$$

Since reliability is subject to optimization, the objective function $U$ is defined by

$$U = \sum_{k=1}^{n} \phi_k^{Y_k}. \tag{9.31}$$

The Lagrange expression, $\text{LE} = U + \lambda\psi$, for reliability optimization is

$$\text{LE} = \sum_{k=1}^{n} \phi_k^{Y_k} + \lambda \left( \sum_{k=1}^{n} C_k Y_k - C_L \right) \tag{9.32}$$

To optimize the number of components $Y_i$ in the $i$th stage, as shown in Figure 9.17, differentiate the Lagrange equation with respect to $Y_i$:

$$\frac{\partial \text{LE}}{\partial Y_i} = \ln \phi_i \phi_i^{Y_i} + \lambda C_i = 0. \tag{9.33}$$

Because only one stage is considered, the summation in equation (9.33) is deleted. The above equation can be written in the form

$$Y_i = a_i \ln \lambda + b_i, \tag{9.34}$$

where

$$a_i = \frac{1}{\ln\phi_i}, \tag{9.35}$$

$$b_i = \frac{\ln K_i}{\ln\phi_i}, \tag{9.36}$$

$$K_i = -\frac{C_i}{\ln\phi_i}. \tag{9.37}$$

Using equations (9.29) and (9.34), the expression for the Lagrange multiplier $\lambda$ can be obtained in the form

$$\lambda = e^s, \tag{9.38}$$

where

$$s = \frac{C_L - \sum_{k=1}^{n} C_k b_k}{\sum_{k=1}^{n} C_k a_k}. \tag{9.39}$$

To obtain a sequence of numbers that will optimize the reliability for the given cost limit, it is assumed that a sequence $\bar{n} = (n_1, n_2, \ldots, n_i, \ldots, n_n)$ is produced for the system reliability of $R(\bar{n})$ and cost $C(\bar{n})$. Let there be another series $\bar{m} = (m_1, m_2, m_3, \ldots, m_i, \ldots, m_n)$, where $R(\bar{m}) > R(\bar{n})$ and $C(\bar{m}) > C(\bar{n})$. The sequence $\bar{m}$ dominates sequence $\bar{n}$, provided that the cost constraint is satisfied. A sequence of undominated numbers, $\bar{n}$, can be obtained from the smallest $n_i$, which will satisfy the inequality[7]

$$\frac{\phi_i^{Y_i}}{[1 + \phi_i]^{(1+Y_i)}} < \lambda C_i. \tag{9.40}$$

By calculating $\lambda$ from equation (9.38), the smallest numbers that will satisfy the above inequality can be obtained. By changing the value of $\lambda$ in decrements, a different sequence that corresponds to each $\lambda$ can be obtained. By choosing proper decrements for $\lambda$, optimum component numbers for each stage that satisfy the cost constraint can be obtained by using the iteration procedure shown in the flowchart of Figure 9.18. As can be seen from the flowchart, the value of $\lambda$ changes with the cost and the reliability of the element. The decrements of $\lambda$ should be chosen for different sets of data (see Appendix A.4 for MATLAB code for calculations).

### 9.3.2 Cost Minimization for a Given Reliability Constraint

Consider the case where the specific system should achieve a certain reliability requirement while keeping the lowest possible price. In this situation, the objective function that should be minimized is

$$U = \sum_{k=1}^{n} C_k Y_k. \tag{9.41}$$

The constraint equation to be satisfied is

$$\phi_L \le \sum_{k=1}^{n} \phi_k^{Y_k}, \tag{9.42}$$

where the limiting value of $\phi_L$ is given by

$$\phi_L = \frac{1 - R_L}{R_L}, \tag{9.43}$$

$R_L$ being the reliability limit. From equation (9.42) the reliability constraint function can be written as

$$\psi = \sum_{k=1}^{n} \phi_k^{Y_k} - \phi_L. \tag{9.44}$$

The Lagrangian equation can be written as

$$LE = \sum_{k=1}^{n} C_k Y_k + \lambda \left( \sum_{k=1}^{n} \phi_k^{Y_k} - \phi_L \right) \tag{9.45}$$

To optimize the number of components $Y_i$ in the $i$th stage, differentiate the Lagrangian equation with respect to $Y_i$,

$$\frac{\partial LE}{\partial Y_i} = C_i + \lambda \ln \phi_i \phi_i^{Y_i} = 0. \tag{9.46}$$

---

7 Barlow, R.E., and Proschan, F., *Mathematical Theory of Reliability*, John Wiley & Sons, Inc., New York, 1965.

**Figure 9.18** Flowchart for reliability maximization (©1973 IEEE, Banerjee, S.K., and Rajamani, K., "Optimization of reliability using a parametric approach," *IEEE Transactions on Reliability*, R-22, 35–39, 1973).

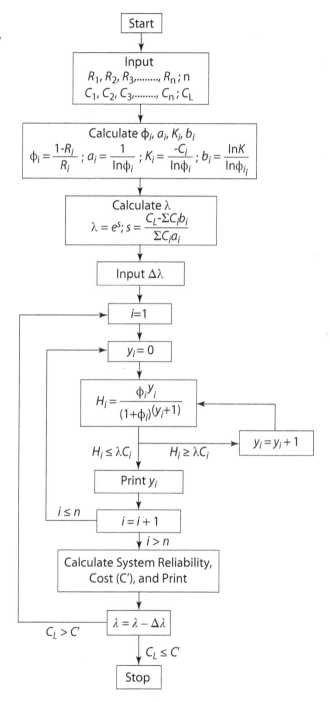

For $i = 1$, equation (9.46) becomes

$$\lambda \ln\phi_1 \phi_1^{Y_1} = -C_1.$$
(9.47)

Eliminating $\lambda$ from equations (9.46) and (9.47), the resulting equation is

$$\phi_i^{Y_i} = K_i \phi_1^{Y_1},$$
(9.48)

where

$$K_i = \frac{\ln\phi_1 C_i}{\ln\phi_i C_1}.$$
(9.49)

From equations (9.47) and (9.48), $\lambda$ can be calculated as

$$\lambda = -\frac{C_1}{S \ln\phi_1},$$
(9.50)

where

$$S = \frac{\phi_L}{\kappa},$$
(9.51)

$$\kappa = \sum_{k=1}^{n} K_k.$$
(9.52)

From the theory of undominated modes, a sequence of numbers can be generated by finding the smallest number that will satisfy the inequality[6]

$$\lambda \frac{\phi_i^{Y_i}}{(1 + \phi_i)^{(1+Y_i)}} < C_i.$$
(9.53)

To determine the optimum cost for a given reliability constraint, the MATLAB code given in Appendix A.5 is used. The algorithm of the program is shown in Figure 9.19.

### 9.3.3 Reliability Optimization for a Given Cost and Weight Constraint

Consider a system similar to that of Section 9.3.2 but subject to two linear constraints. The objective function that should be minimized is

$$U = \sum_{k=1}^{n} \phi_k^{Y_k}.$$
(9.54)

The weight and the cost of the system are both linear functions of the number of elements and are used as constraint functions. The total weight and cost of the system should not exceed a given weight limit, $W_L$, and a cost limit, $C_L$:

$$C_L \geq \sum_{k=1}^{n} C_k Y_k,$$
(9.55)

$$W_L \geq \sum_{k=1}^{n} W_k Y_k,$$
(9.56)

**Figure 9.19** Flowchart for reliability maximization (©1973 IEEE, Banerjee, S.K., and Rajamani, K., "Optimization of Reliability using a parametric approach," *IEEE Transactions on Reliability*, R-22, 35–39, 1973).

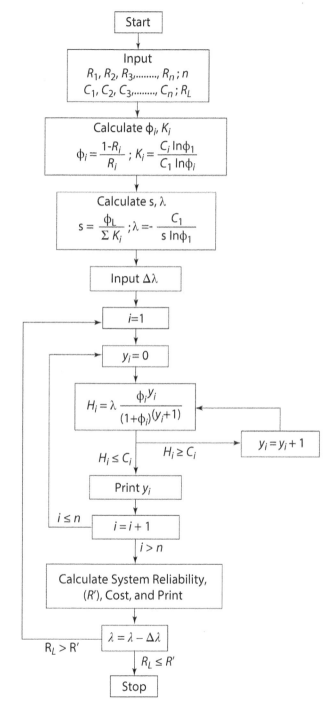

where $W_k$ is the weight of a component at the $k$th stage. The constraint functions can be written as

$$\psi_1 = \sum_{k=1}^{n} C_k Y_k - C_L = 0, \tag{9.57}$$

$$\psi_2 = \sum_{k=1}^{n} W_k Y_k - W_L = 0. \tag{9.58}$$

Then the Lagrangian equation can be expressed as

$$LE = \sum_{k=1}^{n} \phi_k^{Y_k} + \lambda_1 \left( \sum_{k=1}^{n} C_k Y_k - C_L \right) + \lambda_2 \left( \sum_{k=1}^{n} W_k Y_k - W_L \right). \tag{9.59}$$

Differentiating with respect to $Y_i$ gives

$$\frac{\partial LE}{\partial Y_i} = \ln\phi_i \phi_i^{Y_i} + \lambda_1 C_i + \lambda_2 W_i = 0. \tag{9.60}$$

Rearranging equation (9.60) yields

$$Y_i = \frac{1}{\ln\phi_i}[\ln(a_i\lambda_1 + b_i\lambda_2)], \tag{9.61}$$

where

$$a_i = -\frac{C_i}{\ln\phi_i}, \tag{9.62}$$

$$b_i = -\frac{W_i}{\ln\phi_i}. \tag{9.63}$$

Differentiating the Lagrange equation with respect to $\lambda_1$ and $\lambda_2$, respectively,

$$\frac{\partial LE}{\partial \lambda_1} = \sum_{k=1}^{n} C_k Y_k - C_L = 0 \tag{9.64}$$

$$\frac{\partial LE}{\partial \lambda_2} = \sum_{k=1}^{n} W_k Y_k - W_L = 0 \tag{9.65}$$

Substituting equation (9.61) into equations (9.64) and (9.65),

$$C_L = -\sum_{k=1}^{n} a_k[\ln(a_k\lambda_1 + b_k\lambda_2)], \tag{9.66}$$

$$W_L = -\sum_{k=1}^{n} b_k[\ln(a_k\lambda_1 + b_k\lambda_2)]. \tag{9.67}$$

MATLAB code to solve this problem is given in Appendix A.6. The algorithm of the program is shown in Figure 9.20. The nonlinear simultaneous equations are solved by the Newton–Raphson method. The above equations can be easily evaluated for systems with three or more linear constraints by simply introducing a $\lambda_i$ for each constraint.

**Figure 9.20** Flowchart for reliability maximization (©197-IEEE, S.K. Banerjee and K. Rajamani, "Optimization of Reliability Using a Parametric Approach," *IEEE trans. on Reliability,* vol. R-22, pp. 35-39, 1973).

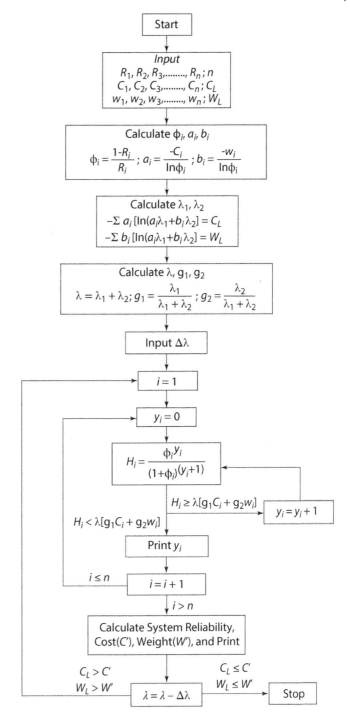

To determine $\lambda_1$ and $\lambda_2$, the above two nonlinear equations should be solved simultaneously. Approximate values of $\lambda$ are sufficient to start the iterations. Again, a sequence of undominated modes can be generated from the smallest numbers that will satisfy the inequality[6]

$$\frac{\phi_i^{Y_i}}{[1+\phi_i]^{(1+Y_i)}} < \lambda(g_1 C_i + g_2 W_i), \tag{9.68}$$

where

$$\lambda = \lambda_1 + \lambda_2, \tag{9.69}$$

$$g_1 = \frac{\lambda_1}{\lambda}, \tag{9.70}$$

$$g_2 = \frac{\lambda_2}{\lambda}. \tag{9.71}$$

### ☐ Example 9.8

Find the number of components to optimize the reliability of the four-stage redundant system given in Table 9.5. Assume the cost constraint $C_L \leq \$40$.[a]

**Table 9.5** Model for cost constraint.

| Stage, $i$ | Cost, $C_i$ | Reliability, $R_i$ |
|---|---|---|
| 1 | 2 | 0.63 |
| 2 | 5 | 0.72 |
| 3 | 7 | 0.86 |
| Constraint | $C_L = 40$ | |

### Solution

Referring to the flowchart in Figure 9.18, the calculations are as follows:

*Step 1.* Calculate $\phi_i$, $a_i$, $b_i$, and $K_i$. The results are shown in Table 9.5.
*Step 2.* Calculate the summations

$$\sum_{k=1}^{n} C_k b_k = -19.0022,$$

$$\sum_{k=1}^{n} C_k a_k = -12.9080,$$

and determine $S$ for the cost limit $C_L = 40$,

$$S = \frac{C_L - \sum_{k=1}^{3} C_k b_k}{\sum_{k=1}^{3} C_k a_k} = \frac{40 - (-19.0022)}{-12.9080} = -4.57099.$$

Then $\lambda$ is

$$\lambda = e^S = e^{-4.57099} = 1.034786 \times 10^{-2}.$$

*Step 3.* Assume the decrement $\Delta\lambda = 0.001$ and start the iteration with $\lambda_1 = 1.034786 \times 10^{-2}$. The number of components for the first stage is determined as follows:

a) For $i = 1$, the inequality equation can be written as

$$\frac{\phi_1^{Y_1}}{[1 + \phi_1]^{(1+Y_1)}} < \lambda_1 C_1. \tag{9.72}$$

**Table 9.6** Results of parameters.

| Stage $i$ | $\phi_i$ | $a_i$ | $b_i$ | $K_i$ |
|---|---|---|---|---|
| 1 | 0.5873 | −1.8789 | −2.4874 | 3.7578 |
| 2 | 0.3888 | −1.0588 | −1.7646 | 5.2940 |
| 3 | 0.1628 | −0.5508 | −0.7435 | 3.8561 |

The right-hand side (RHS) of equation (9.72) is

$$\lambda_1 C_1 = (1.034786 \times 10^{-2})(2) = 2.069572 \times 10^{-2}.$$

For the inequality to be satisfied the left-hand side (LHS) of equation (9.72) must be less than the RHS. The LHS of the equation for the initial iteration $Y_1 = 0$ is

$$\frac{\phi_1^0}{(1 + \phi_1)^{(1+0)}} = \frac{1}{1 + \phi_1} = \frac{1}{1 + 0.5873} = 0.6300.$$

Because this is larger than the LHS, increment $Y_1$ by 1 and recalculate the LHS:

$$\frac{\phi_1^1}{(1 + \phi_1)^{(1+1)}} = \frac{\phi_1^1}{(1 + \phi_1)^2} = \frac{0.5873}{(1 + 0.5873)^2} = 0.2331.$$

When $Y_1 = 4$, the RHS is equal to $1.18807 \times 10^{-2}$, which satisfies the inequality equation. Thus, the number of components in the first stage is 4. Similarly, this same calculation can be performed for stages 2 and 3. Using $\lambda = 1.034786 \times 10^{-2}$, the number of components for stages 2 and 3 can be found to be 3 and 2, respectively.

b) Calculate the system reliability

$$R_{sp} = \prod_{k=1}^{n} [1 - (1 - R_k)^{Y_k}]$$

$$= [1 - (1 - 0.63)^4][1 - (1 - 0.72)^3][1 - (1 - 0.86)^2] = 0.9409,$$

and the system cost $C_s$ is

$$C_s = \sum_{k=1}^{n} C_k Y_k = (2 \times 4) + (5 \times 3) + (7 \times 2) = \$37.$$

Because the calculated systems cost is less than the cost limit, the same iteration must be repeated for a new $\lambda$:

$$\lambda_2 = \lambda_1 - \Delta\lambda = 1.034786 \times 10^{-2} - 0.001 = 0.0093478.$$

When the same calculation procedure is repeated with $\lambda_2 = 0.0093478$, the same solution is found as with $\lambda_1 = 1.034786 \times 10^{-2}$. If $\Delta\lambda$ is selected to be very small, repeating solutions may occur as in this example. It is safe to select $\Delta\lambda$ small in order not to miss the correct optimum solution. The only drawback is computation time. A good rule of thumb to select $\Delta\lambda = 0.2\lambda_1$. In this example, for $\lambda_2 = 0.0093478$, $\lambda_3 = 0.0083478$, $\lambda_4 = 0.0073478$, and $\lambda_5 = 0.0063478$, the same solutions are obtained. For $\lambda_6 = 0.0063478 - 0.001 = 0.0053478$, the RHS of the inequality is

$$\lambda_6 C_1 = (0.0053478)(2) = 0.0106956.$$

Now start the iteration with $Y_1 = 0$, and repeat the foregoing solution procedure. The correct optimum solution with the cost constraint $C_L = 40$ is then obtained. The number of components is 5, 3, and 2 for the first, second, and third stages, respectively. The reliability of this system is

$$R_{sp} = \prod_{k=1}^{n} [1 - (1 - R_k)^{Y_k}]$$
$$= [1 - (1 - 0.63)^5][1 - (1 - 0.72)^3][1 - (1 - 0.86)^2] = 0.9522,$$

and the system cost $C_s$ is

$$C_s = \sum_{k=1}^{n} C_k Y_k = (2 \times 5) + (5 \times 3) + (7 \times 2) = \$39.$$

[a]From Ertas, A., and Jones, J.C., *The Engineering Design Process*, John Wiley & Sons, Inc., New York 1996.

# Bibliography

1 GALLAGHER, R.H., and ZIENKIEWICZ, O.C., *Optimum Structural Design: Thgeory and Applications*, John Wiley & Sons, Inc., New York, 1973.

2 GERO, J.S., *Design Optimization*, Academic Press, New York, 1985.

3 HAUG, E.J., and ARORA, J.S., *Applied Optimal Design*, John Wiley & Sons, Inc., New York, 1979.

**4** JOHNSON, C.R., *Optimum Design of Mechanical Elements*, John Wiley & Sons, Inc., New York, 1980.

**5** KIRSCH, U., *Optimum Structural Design*, McGraw-Hill, New York, 1982.

**6** MORRIS, A.J., *Foundations of Structural Optimization: A Unified Approach*, John Wiley & Sons, Inc., New York, 1982.

**7** OSYCZKA, A., *Multicriterion Optimization in Engineering*, Ellis Horwood Ltd., Chichester, 1984.

**8** REKLAITIS, G.V., RAVINDRAN, A., and RAGSDELL, K.M., *Engineering Optimization Methods and Applications*, John Wiley & Sons, Inc., New York, 1983.

**9** SPUNT, L., *Optimum Structural Design*, Prentice Hall, Englewood Cliffs, NJ, 1971.

**10** STEUER, R.E., *Multiple Criteria Optimization: Theory, Computation, and Application*, John Wiley & Sons, Inc., New York, 1986.

**11** WILDE, D.J., *Optimum Seeking Methods*, Prentice Hall, Englewood Cliffs, NJ, 1964.

## CHAPTER 9 Problems

**9.1** Show the equation for the objective function for the minimum volume of a simply supported beam subject to a concentrated load $F$ as shown in Figure P9.1.

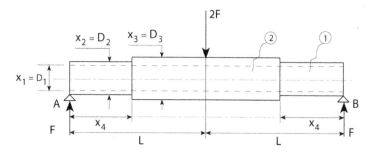

**Figure P9.1** Pressure vessel.

**9.2** If the permissible stress $\sigma_p$ is 400 MPa and predetermined values are

$$x_2 = D_2 = 50 \text{ mm},$$
$$x_3 = D_3 = 60 \text{ mm},$$
$$L = 800 \text{ mm},$$

find the permissible bending stress equations in part 1 and part 2 of the beam.

**9.3** As additional inequality constrains, if the inside diameter of the beam $x_1$ should not be
less than 30 mm, and the length $x_4$ must be equal to or larger than zero, define the feasible
region and discuss the result.

**9.4** United Parcel Service (UPS) charges by weight to ship
a parcel unless the sum of the dimensions as shown in
Figure P9.4 is greater than 112.5 in.; otherwise, there is an
extra charge per unit volume. Determine the dimensions that
will maximize the volume without exceeding the 112.5 in.
requirement.

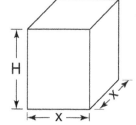

**Figure P9.4** UPS Required
parcel.

**9.5** Solve Problem 9.4 using the Lagrange multiplier method.

**9.6** Using differential calculus, maximize the volume of a box made of cardboard as shown
in Figure P9.6, subject to the following constraints:
a) The allowable area of the cardboard is equal to 40 in.²
b) The length of the box $L$ is equal to its width $W$.

**Figure P9.6** Carboard box.

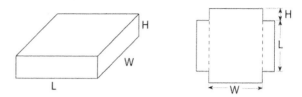

**9.7** Solve Problem 9.6 using the Lagrange multiplier method for the allowable area of 80 in.$^2$.

**9.8** Using the graphical method, find the maximum of the objective function $U = x_1 + 4x_2$ subject to

$$x_1 \geq 0,$$
$$x_2 \geq 0,$$
$$3x_1 + 2x_2 \leq 6,$$
$$x_1 + 2x_2 \leq 4.$$

**9.9** Find a minimum point of the function given below using equal interval search method

$$U(x) = 3x^2 + \frac{1,296}{x}.$$

**9.10** Based on the convergence criterion $\varepsilon = 0.001$, use the golden section search method to find the minimum point of the function

$$U(x) = F(x) = 3x^2 + \frac{1,296}{x}$$

in the interval of $1 \leq x \leq 11$.

**9.11** Recall from Chapter 8 the high school student who has been dreaming of running his own radio station. Suppose he wants to increase the radio station's reliability to at least 0.95. Find the number of components required for the three-stage redundant system given in Table P9.11.

**Table P9.11** Three-stage redundant system.

| Stage | Cost per unit | Reliability |
|-------|---------------|-------------|
| 1 | 2 | 0.63 |
| 2 | 5 | 0.72 |
| 3 | 7 | 0.86 |

**9.12** Consider Problem 9.11. If the student has limited space of $V_L \leq 40$ unit volume in the station and a cost constraint $C_L \leq \$40$, find the number of components for optimum reliability (refer to Table P9.12).

**Table P9.12** Three-stage, two-constraint redundant system.

| Stage | Cost per unit | Volume | Reliability |
|-------|---------------|--------|-------------|
| 1 | 2 | 3 | 0.63 |
| 2 | 5 | 2 | 0.72 |
| 3 | 7 | 4 | 0.86 |

**9.13** Find the number of components to minimize the cost of the four-stage redundant system given in Table P9.13. Assume a reliability constraint of $R_L \geq 0.99$.

**Table P9.13** Four-stage redundant system.

| Stage | Cost per unit | Reliability |
|-------|---------------|-------------|
| 1 | 4 | 0.95 |
| 2 | 2 | 0.92 |
| 3 | 1.5 | 0.80 |
| 4 | 1.0 | 0.70 |

**9.14** Find the number of components to optimize the reliability of the four-stage redundant system given in Table P9.14. Assume the cost constraint $C_L \leq 35$.

**Table P9.14** Four-stage redundant system.

| Stage | Cost per unit | Reliability |
|-------|---------------|-------------|
| 1 | 4.5 | 0.90 |
| 2 | 3.0 | 0.86 |
| 3 | 2.0 | 0.80 |
| 4 | 1.5 | 0.70 |

**9.15** Find the number of components to optimize the reliability of the five-stage redundant system given in Table P9.15. Assume a cost constraint $C_L \leq 90$ and weight constraint $W_L \leq 100$.

**Table P9.15** Five-stage, two-constraint redundant system.

| Stage | Cost per unit | Weight per Unit | Reliability |
|-------|---------------|-----------------|-------------|
| 1 | 4 | 7 | 0.90 |
| 2 | 4 | 9 | 0.75 |
| 3 | 8 | 7 | 0.60 |
| 4 | 7 | 7 | 0.80 |
| 5 | 5 | 8 | 0.80 |

**9.16** A schematic of a power supply system is given in Figure P9.16, which is made up of a steam turbine, condenser, boiler, and diesel generator. Fuel, lubrication, water pumps, and valves

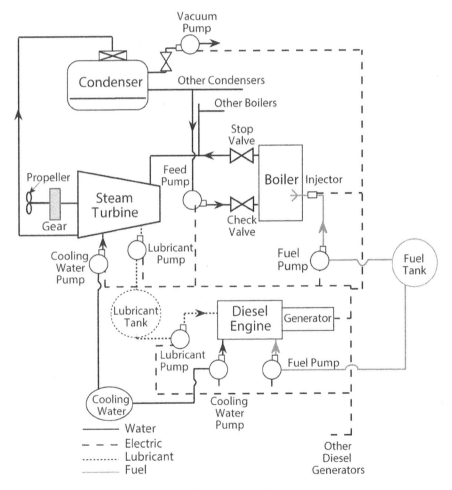

**Figure P9.16** Representative picture of a power supply system.

are also used in the system. The objective is to calculate the optimum number of devices that should be included in the system to increase the system reliability. The system can be analyzed in four stages. Stage 1 is formed by the diesel engine and devices that support its operation (cooling water pump, fuel pump, generator, and lubrication pump). Stage 2 includes the boiler and other devices (feed pump, fuel pump, boiler, burner, stop valve, and check valve). Stage 3 includes the steam turbine and its components (lubrication pump, cooling pump, reduction gear, and pinion gear). Stage 4 includes the condenser and its devices (vacuum pump, vacuum gauges, and main control valve). The cost and reliability of these devices are given in Table P9.16.

1. Draw block diagrams of each stage and the whole system.
2. Optimize the system reliability for the cost limit of $C_L = 2,400,000$. How many diesel generators, boilers, steam turbines, and condensers should a ship have for maximum reliability?
3. Minimize the cost for a given reliability limit of $R_L = 0.99$.

**Table P9.16** Cost and reliability of the stages.

| Stage | Component | Cost $ | Reliability |
| --- | --- | --- | --- |
| 1. Diesel generator | Cooling pump | 500 | 0.985 |
| | Fuel pump | 700 | 0.980 |
| | Diesel engine | 10,000 | 0.995 |
| | Generator | 4,000 | 0.995 |
| | Lubrication pump | 800 | 0.980 |
| 2. Boiler generator | Feed pump | 500 | 0.980 |
| | Fuel pump | 500 | 0.980 |
| | Boiler | 20,000 | 0.999 |
| | Burner | 400 | 0.999 |
| | Stop valve | 350 | 0.980 |
| | Check valve | 250 | 0.980 |
| 3. Steam turbine | Lubrication pump | 600 | 0.980 |
| | Cooling pump | 500 | 0.980 |
| | Reduction gear | 50,000 | 0.999 |
| | Turbine | 1,000,000 | 0.999 |
| | Pinion gear | 900 | 0.999 |
| 4. Condenser | Condenser | 6,000 | 0.990 |
| | Vacuum pump | 3,500 | 0.985 |
| | Vacuum gauge | 250 | 0.980 |
| | Main control valve | 750 | 0.999 |

# 10

# Modeling and Simulation

Through modeling engineers expand and enrich their vision, exercise their sensibilities, formulate unique and personal interpretations, process and optimize a large number of alternative solutions, originate innovations, enhance their understanding of physical problems – in short, they develop the qualities which we marvel at in children.

## 10.1 Modeling in Engineering

One of the fundamental activities in which engineers are involved is model building. The process of model building is a way to present knowledge and to explore alternative solutions. A scale model of a physical system can be used to predict accurately the performance of the prototype. Model studies have proved useful in design and development for many years, especially when experimental testing of a full-size prototype is either impossible or prohibitively expensive. The advantages of model testing in experimental design are as follows:

1) When the problem is too complex for an analytical solution, an empirical solution can be developed.
2) Analytical techniques can be substantiated by correlating the predicted model behavior with the actual behavior of the model.
3) Prototypes with non-attainable characteristics can be studied, such as those with:
   a) Large structures
   b) Molecular structures
   c) An environment that cannot be simulated
   d) High-speed reactions
   e) Dangerous situations.

The selection of a proper model and its implementation require a priori knowledge about the system to be modeled. Figure 10.1 shows a general modeling process for model development. As shown in the figure, the system goal must be defined and studied in the early stage of the modeling effort. At this preliminary model development stage, the system analyst must determine the need for the model, what kinds of analysis to perform, and an appropriate measure of performance. The second stage is the system analysis required for model development. During this stage, salient components, interactions, relationships, and dynamic behavior mechanisms of a system are isolated. System synthesis, which is the next stage in the modeling process, deals with structuring and implementing the various models in accordance with the findings from the system analysis stage. During this stage, the cost-effectiveness and method of implementation of the various models are determined. Selection of the optimum model is then made by comparing model accuracies, implementation approaches, maintainability, and projected costs of the experimentation in subsequent stages of the modeling process. Once the model has been selected, verification of the model response is required. As shown in Figure 10.1, the verification stage of a modeling process serves as a check on the system

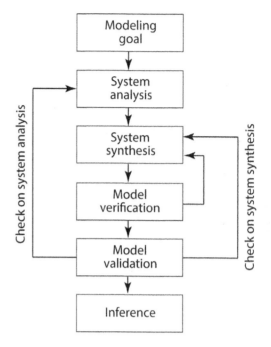

**Figure 10.1** Modeling process. (Ertas, A., and Jones, J.C., *The Engineering Design Process*, John Wiley & Sons, Inc., New York, 1996. Reproduced with permission of John Wiley & Sons, Inc.)

synthesis stage. Before the model can be considered acceptable, similar checks on the system analysis stage are necessary. This is done during the validation stage by comparing responses from the model with corresponding responses recorded from the actual (modeled) system. Experimentation should be conducted with both the model and with the actual system. If the actual system is not available for experimentation, validation tests will not be dependable; thus, the analyst takes a risk in drawing conclusions regarding the actual system. In the inference stage of a modeling process, experiments are conducted solely with the verified and validated model.

## 10.2 Heuristic Modeling

This phrase is used to describe what might be called *common-sense* or *minimum-cost* physical modeling. The word heuristic relates to discovery learning and, by implication, it embodies the concept of learning by comprehending the total problem. It is used here to emphasize the value of simple, inexpensive models in helping to grasp the relative size and interrelationship of individual elements in a design and in resolving problems associated with interfaces and interferences.

The product of design is normally a material entity that did not exist previously. Most inexperienced and many experienced designers have difficulty envisioning the interrelationships of the various elements in a design, and when the total project involves elements from different disciplines, the potential problems are compounded. Fortunately, with some of the analytical features incorporated in computer-aided design programs, this difficulty can be minimized. However, many design organizations do not have this capability. Furthermore, many interface and interrelationship problems are not recognized until after the design is completed and construction begins. An example of what is meant by heuristic modeling is given in the following description:

The Atlas F missile emplacement program was a large effort in involving many different contractors with contracts managed by both the US Air Force (USAF) and the US Army Corps of Engineers (COE). The purpose of this effort was to install the Atlas missile and all supporting equipment in a vertical underground concrete silo. The silo was first poured and then the structure surrounding the missile was constructed and supporting equipment was installed. Finally, the missile was lowered through open doors at the ground surface and interconnections to the ground support equipment were made. A significant number of propellant and pressurization lines provided interconnection to ground support equipment within the silo and supply vehicles at the surface. These lines were located in an appendage (propellant systems shaft) to the silo that entered the silo wall 30 to 40 feet below the ground surface. Propellant lines were delivered to the construction site already fabricated, cleaned for propellant service and sealed for protection from contamination. Shortly after the installation of these lines was initiated the COE, the agency responsible for overseeing the construction contractor's effort, contacted the USAF, who had the overall responsibility for completing the effort, about a problem in installing one of the propellant lines. One of the prefabricated piping sections could not be maneuvered into the propellant systems shaft. The design of this piping section was complicated by several bends at various angles and the construction contractor maintained that the overall configuration was such that it could not be passed into the propellant systems shaft. The contractor that manufactured the piping sections had been contacted, remanufacturing the piping sections for the 72 missile sites was going to cost over $300,000. Thus, the incentive to find some other solution was high. The USAF had a small contingent of consulting engineers on their staff and the problem was passed on to two of them. After considerable discussion and analysis of the drawings, the two engineers decided that the piping section configuration was too complicated for analysis from the drawings and they decided to make a small model of the silo, propellant shaft, and piping section to get a better understanding of the interference problem. A crude model of the silo/propellant shaft was constructed out of cardboard and a pipe cleaner was used to construct a model of the piping section. As had been indicated by the construction contractor, there was definitely a problem working the piping section into the propellant shaft. However, using the model one method for manipulating the piping section was devised that indicated that it could be passed into the shaft, barely. The next question was, how closely did the model actually reflect the dimensional configuration of the piping section and silo? To verify this, a demonstration was arranged at one of the missile sites. When the two engineers arrived for the demonstration, the construction contractor already had the piping section rigged, hanging from a crane, and was demonstrating how the piping section could not be passed into the propellant shaft. The USAF Colonel in charge of the operation was glum, whereas the COE Colonel had an *I told you so* expression on his face. Upon inspection by the two engineers it was noted that the piping section was rigged in a different manner than that required to duplicate the method devised using the model. The piping section was rerigged accordingly, and another attempt was made to pass the piping section into the shaft. To almost everyone's surprise (including the two engineers) the piping section passed into the shaft with essentially no room to spare. The USAF Colonel was so delighted that he almost danced a jig. The COE Colonel was now glum.

Although the use of this modeling technique is not likely to result in any scenario as dramatic (or as accompanied by the theatrics) as this example, nevertheless there are several lessons that can be learned by using heuristic modeling:

1) Physical modeling of the object in question provides a grasp of the problem that cannot be achieved by any other technique, even using sophisticated computer analytical tools.
2) Crude models can be developed from basic materials (wood, fiberglass, sheet metal, paper, cardboard, glue, rubber bands, paper clips, and even pipe cleaners) that are available almost everywhere. The cost of these models is minimal.
3) A good bit of caution needs to be applied to any conclusions reached on the basis of using such models, but for gross indications relative to appropriate overall proportion, one structural member being too large relative to another, interferences, interfaces, and installation difficulties, as in the example, techniques of this kind can prove to be invaluable.
4) It is almost always helpful to develop a crude model first before spending time and effort on more sophisticated models. Development of the crude model will provide insight as to what should be included in the sophisticated model.

## 10.3  Mathematical Modeling

As the name implies, mathematical modeling is a process of writing mathematical expressions describing the behavior of a physical system. The physical system may range from a spacecraft in orbit under the mutual gravitation of Earth and Moon, concentrations of reactants in a chemical reactor, air flow around an object in a wind tunnel, behavior of structures, or machine elements when acted on by external forces. Mathematical modeling begins by assuming that the system obeys certain constitutive laws – these are the basic laws of physics. In the case of the orbiting spacecraft, the laws governing the dynamics are Newton's laws: the total external forces are equated to the inertial forces and the equations of motion are written in terms of acceleration of the spacecraft. Other laws such as the conservation of mass, energy, and momentum may also be needed. For instance, to describe the reactions in a chemical reactor, one starting point might be to write the equations of the reactions, and supplement them with the equations for mass and energy conservation; the second law of thermodynamics determines the conditions under which the reactions are possible. In addition to the constitutive laws, simplifying assumptions can be made. These assumptions can make the problem under consideration mathematically tractable. Natural processes are very complex; however, interest is normally confined to a particular behavior observed, for example, in an experiment. It is therefore important that the simplifying assumptions are such that, on the one hand, they make the problem mathematically simple while, on the other, they include all the physical behavior to be investigated. From the example of the orbiting spacecraft, it might be assumed that the motion takes place in an elliptical orbit, and the problem could be formulated in such a way as to investigate small variations in the original elliptical motion. For the chemical reactor, a similar simplifying assumption could be postulated by stating that the concentrations vary about the original chemical equilibrium point.

Once a reliable mathematical model has been formulated, the next step is to obtain the solution. For a large class of problems, numerical solutions obtained by computer are acceptable. There are a variety of numerical schemes available to deal with all types of problems. To obtain the solution for the orbiting spacecraft, a step-by-step time integration algorithm, such as the Runge–Kutta scheme, can be used to obtain its trajectory. For large structural problems, finite-element and finite-difference formulations are often used. For very large and complex problems, the computer is the only reliable tool available.

In recent years, another tool called symbolic manipulation has become available. Compared to numerical simulation, which manipulates numbers to obtain solutions, symbolic manipulators use symbols, and all of the mathematical operations are performed on symbols. The advantage of using symbolic manipulators is that they can be incorporated in the modeling process itself to derive the equations. In fact, it is desirable to use symbolic manipulators to derive the equations because this eliminates a great deal of tedious algebra and the accompanying algebraic mistakes.

As an example, suppose that the natural frequencies and mode shapes of a cantilever beam as shown in Figure 10.2 are taken from any elementary book on vibration. The equation of motion for the beam is

$$\frac{\partial^4 y}{\partial x^4} - \beta^4 y = 0, \tag{10.1}$$

where

$$\beta^4 = \frac{\rho A \omega^2}{EI},$$

$\rho A$ = mass of the beam per unit length,

$\omega$ = natural frequency.

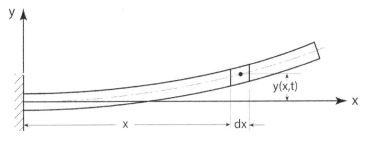

**Figure 10.2** Cantilever beam model.

In addition to the differential equatio,n the boundary conditions are needed. For the cantilever beam, the deflection and the slope at $x = 0$ are zero, that is,

$$y(x = 0) = y'(x = 0) = 0, \tag{10.2}$$

where the prime denotes the derivative with respect to $x$. At the free end, the shear force and the bending moment are zero, that is,

$$y'''(x = l) = y''(x = l) = 0. \tag{10.3}$$

Equation (10.1), with the boundary conditions from equations (10.2) and (10.3), completely model the vibration of the cantilever beam. At this stage, the symbolic manipulation software packages such as Maple, Macsyma, or Mathematica can be used. For this problem, Maple was used. The program was written to mimic the solution procedure described below.

The general solution of the fourth-order beam equation is given by

$$y(x) = A \cosh \beta x + B \cos \beta x + C \sinh \beta x + D \sin \beta x. \tag{10.4}$$

By applying the boundary conditions at $x = 0$, we get

$$B = -A, \quad D = -C. \tag{10.5}$$

Substituting in equation (10.4),

$$y(x) = A(\cosh \beta x - \cos \beta x) + C(\sinh \beta x - \sin \beta x). \tag{10.6}$$

To apply the bending moment and the shear force conditions at $x = l$, differentiate equation (10.6) twice and three times with respect to $x$ and substitute $x = l$ to obtain

$$y''(x = l) = Ac_{11} + Cc_{12} = 0,$$
$$y'''(x = l) = Ac_{21} + Cc_{22} = 0, \tag{10.7}$$

or in matrix form

$$\begin{bmatrix} c_{11} & c_{12} \\ c_{12} & c_{22} \end{bmatrix} \begin{bmatrix} A \\ C \end{bmatrix} = \begin{bmatrix} 0 \\ 0 \end{bmatrix}, \tag{10.8}$$

where

$$c_{11} = \cosh \beta l + \cos \beta l,$$
$$c_{12} = \sinh \beta l + \sin \beta l,$$
$$c_{21} = (\sinh \beta l - \sin \beta l), \beta$$
$$c_{22} = (\cosh \beta l + \cos \beta l, )\beta$$

and the determinant of the $2 \times 2$ matrix is the characteristic polynomial

$$p(\lambda l) = \cosh \beta l \, \cos \beta l + 1. \tag{10.9}$$

The roots of the characteristic polynomial are the eigenvalues $\beta$ from which the natural frequency $\omega$ of the beam can be obtained. To find the roots of $p(\beta l)$ requires considerable numerical effort; however, with Maple, the roots can easily be evaluated. Equation (10.9) has infinite roots; however, only the few lower roots are of interest. The following steps are used to find the roots:

1) Plot the polynomial as shown in Figure 10.3.
2) Find the approximate location of the intersection with the $\beta$ axis.
3) To find the roots accurately, solve the polynomial in the neighborhood of the approximate root locations using Maple.

   Once the roots are known, they can be substituted in equation (10.6) to obtain the corresponding mode shape of the beam as shown in Figure 10.4. The complete symbolic program along with the output is given in Appendix A.7.

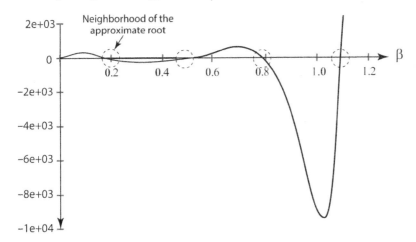

**Figure 10.3** Roots of the characteristic polynomial. (Ertas, A., and Jones, J.C., *The Engineering Design Process*, John Wiley & Sons, Inc., New York, 1996. Reproduced with permission of John Wiley & Sons, Inc.)

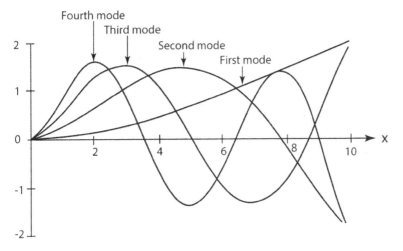

**Figure 10.4** Mode shape of the beam. (Ertas, A., and Jones, J.C., *The Engineering Design Process*, John Wiley & Sons, Inc., New York, 1996. Reproduced with permission of John Wiley & Sons, Inc.)

## 10.4   Dimensional Analysis

When the functional relation between parameters or the governing equations of the system are unknown, dimensional analysis is used to obtain a valid scaling law. The Buckingham $\pi$ theorem can be used to develop a dimensionless parameter, which indicates the state of a given physical system. To obtain a correct relationship the correct physical characteristics must be identified. The following are considered physical characteristics:

1) Characteristic length
   - Length
   - Cross-sectional area
   - Volume
   - Moment of inertia
2) Characteristic motion
   - Velocity
   - Acceleration
   - Angular velocity
   - Moment of inertia
3) Characteristic material properties
   - Young's modulus
   - Density
   - Viscosity
   - Strength
4) Characteristic phenomenon involved
   - Surface tension coefficient
   - Heat transfer coefficient
   - Temperature difference.

   To determine the dimensionless ratio $\pi$ terms, the following steps can be used. Any dimensionless physical characteristic, such as the width–length ratio, can be selected as a $\pi$ term.

1) Select basic dimensions such as mass ($M$), length ($L$), time ($T$), and the like, for use as repeating variables from the physical characteristics. As a rule of thumb, select the repeating variable so that one is a characteristic length (length, cross-sectional area, volume, moment of inertia, etc.), one is a characteristic motion (velocity, acceleration, angular velocity, etc.), one is a material characteristic (Young's modulus, density, viscosity, tensile strength, etc.), and one is a characteristic of the phenomenon involved (surface tension coefficient, heat transfer coefficient, mass transfer coefficient, etc.).
2) Use the repeating variables together with one of the remaining physical characteristics for each $\pi$ term. Equate the exponents of the basic units using unknowns for the repeating variables, unity for the remaining characteristics, and zero for the dimensionless $\pi$ term. Solve for the unknowns and form the $\pi$ terms.

The number of independent dimensionless parameters required to obtain $\pi$ terms can be determined by

$$S = n - b, \tag{10.10}$$

where $S$ is the number of $\pi$ terms, $n$ is the total number of physical characteristics, and $b$ is the number of basic units involved. The following example illustrates how to calculate the necessary $\pi$ terms.

### Example 10.1

Suppose that a sphere of diameter $d$ falls through a fluid in time $t$ with initial velocity $v$ (see Figure 10.5). Considering the resistance of the fluid through which the sphere falls, do the following:

1) Determine the number of $\pi$ terms necessary.
2) Form the $\pi$ terms.

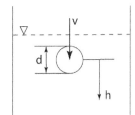

**Figure 10.5** Sphere falling in fluid media.

### Solution

a) The first step is to select the important physical characteristics involved in the problem. These characteristics and the corresponding basic units are shown in Table 10.1.

From Table 10.1, the number of selected physical characteristics is $n = 8$, and the repeating basic dimensions are $M$, $L$, and time $T$; therefore, $b = 3$. Hence, the number of necessary $\pi$ terms $S$ is

$$S = n - b$$
$$= 8 - 3 = 5$$

b) The physical characteristics can be written in a functional equation as

$$F(h, t, v, g, \rho, \mu, m, d) = 0, \tag{10.11}$$

**Table 10.1** Physical characteristics.

| Physical characteristic | Symbol | Basic dimension |
| --- | --- | --- |
| Distance | $h$ | $L$ |
| Time | $f$ | $T$ |
| Velocity | $v$ | $LT^{-1}$ |
| Gravity | $g$ | $LT^{-2}$ |
| Fluid density | $\rho$ | $ML^{-3}$ |
| Fluid viscosity | $\mu$ | $ML^{-1}T^{-1}$ |
| Sphere mass | $m$ | $M$ |
| Sphere diameter | $d$ | $L$ |

and the functional equation can be written in dimensional form as

$$F(L,\ T,\ LT^{-1},\ LT^{-2},\ ML^{-3},\ ML^{-1}T^{-1},\ M,\ L) = 0. \tag{10.12}$$

The repeating variables are $d\ (L), v\ (LT^{-1})$, and $\mu\ (ML^{-1}T^{-1})$. As mentioned previously, repeating variables are selected, one from each physical characteristic. Following step 2, the $\pi$ terms can be formed as shown in Table 10.2. Referring to Table 10.2, the dimensionless form for the $\pi_4$ term can be written in the form of an equation as

$$\pi_4 = \mu^{x_4} d^{y_4} v^{z_4} \rho. \tag{10.13}$$

**Table 10.2** Forming $\pi$ Terms.

| $\pi$ Terms | Repeating variables | Remaining physical characteristics |
|---|---|---|
| $\pi_1 =$ | $\mu^{x_1} d^{y_1} v^{z_1}$ | $h$ |
| $\pi_2 =$ | $\mu^{x_2} d^{y_2} v^{z_2}$ | $t$ |
| $\pi_3 =$ | $\mu^{x_3} d^{y_3} v^{z_3}$ | $g$ |
| $\pi_4 =$ | $\mu^{x_4} d^{y_4} v^{z_4}$ | $\rho$ |
| $\pi_5 =$ | $\mu^{x_5} d^{y_5} v^{z_5}$ | $m$ |

Following step 3, equation (10.13) can be written in basic units as

$$M^0 L^0 T^0 = (ML^{-1}T^{-1})^{x_4}(L)^{y_4}(LT^{-1})^{z_4}(ML^{-3})^1. \tag{10.14}$$

Because the $\pi_4$ term has no unit, the left-hand side of the above equation should be dimensionless; thus, the exponents of $M$, $L$, and $T$ are assumed to be zero. Equating exponents for $M$, $L$, and $T$ yields

$$\text{for } M, \quad 0 = x_4 + 1,$$
$$\text{for } L, \quad 0 = -x_4 + y_4 + z_4 - 3,$$
$$\text{for } T, \quad 0 = -x_4 - z_4.$$

Solution of the above equations yields $x_4 = -1$, $y_4 = 1$, and $z_4 = 1$. Hence, from equation (10.13), the $\pi_4$ term is written as

$$\pi_4 = \mu^{-1} d^1 v^1 \rho^1 = \frac{\rho v d}{\mu} \tag{10.15}$$

The $\pi_4$ term is known as Reynolds number. Following the same procedure, other $\pi$ terms can be determined.

The Reynolds number found in the above example is the ratio of inertia forces to viscous forces. The critical value of Reynolds number can be used to distinguish between laminar and turbulent flow. Other dimensionless parameters that frequently occur in fluid flow studies are as follows:

1) The Mach number $M$ is used as a parameter to characterize the compressibility effects in fluid flow and is defined by

$$M = \frac{v}{c}. \tag{10.16}$$

2) The Froude number $F_R$ is used for flows with free surface effects. This is a key parameter in the design of ship and hydraulic structures and is defined by

$$F_R = \frac{v^2}{lg}. \tag{10.17}$$

3) The Weber number $W_E$ is another important dimensionless parameter that is used in studies involved with gas–liquid or liquid–liquid interfaces. It is given by

$$W_E = \frac{\rho l v^2}{\sigma}. \tag{10.18}$$

In equations (10.16)–(10.18), the physical characteristics $v, c, l, g, \rho$, and $\sigma$ represent velocity, speed of sound, length, acceleration of gravity, mass density, and surface tension, respectively. These equations can be obtained by the same procedure as used to find the Reynolds number equation (10.15).

## 10.5 Similarity Laws in Model Testing

Dimensional analysis is an important tool often used to increase the accuracy and efficiency of experimental design. To predict prototype behavior from measurements on the model, there must be similitude between the model and the prototype. The laws that quantify the scale model behavior are called the laws of similitude. Similitude is used to establish a set of scaling factors to obtain the relationship between the model and the prototype. To obtain the correct relationship, the correct or important parameters that affect the experimental design must be identified. To select the important parameters requires some experience. If an incorrect parameter is selected or an important one is left out, no subsequent analysis of the basic units can correct the error. The most common types of similarity are geometric, kinematic, and dynamic.

### 10.5.1 Geometric Similarity

The model and prototype are assumed to be geometrically similar if, and only if, all pairs of points on the model and prototype have the same ratio of distances in all three coordinates. Hence, ratios of model lengths to the corresponding prototype lengths must be the same. For example,

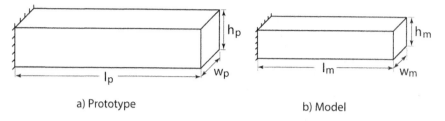

a) Prototype             b) Model

**Figure 10.6** Cantilever beams.

consider the two cantilever beams depicted in Figure 10.6. In order for the cantilever beams be geometrically similar, the following equality must be satisfied:

$$\lambda = \frac{l_p}{l_m} = \frac{w_p}{w_m} = \frac{h_p}{h_m}, \tag{10.19}$$

where $\lambda$ is called the scale factor and the subscript $m$ refers to the model and $p$ to the prototype.

### 10.5.2 Kinematic Similarity

The motions of the model and prototype are kinematically similar if the ratio of the corresponding velocities $v/u$ in the flow fields are constant. For example, kinematic similarity of a prototype and a model of a wind turbine (Figure 10.7) requires the following velocity relation:

$$\lambda = \frac{v_p}{u_p} = \frac{v_m}{u_m}, \tag{10.20}$$

where $v = r\omega$, $r$ being the radius and $\omega$ the angular velocity.

### 10.5.3 Dynamic Similarity

Dynamic similarity requires geometric similarity and ensures kinematic similarity between the model and the prototype. Dynamic similarity exists if prototype and model force and pressure

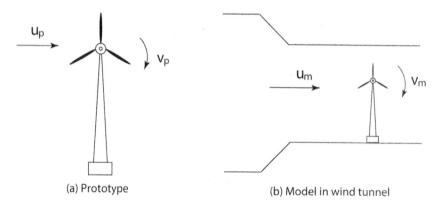

(a) Prototype             (b) Model in wind tunnel

**Figure 10.7** Wind turbine model. (Ertas, A., and Jones, J.C., *The Engineering Design Process*, John Wiley & Sons, Inc., New York, 1996. Reproduced with permission of John Wiley & Sons, Inc.)

coefficients are identical. In the wind turbine example, for dynamic similarity to exist, the Reynolds numbers of the prototype and model must be equal, that is,

$$R_{e(m)} = R_{e(p)}. \tag{10.21}$$

## 10.6 Wind and Water Tunnels

Aerodynamics plays a vital role in many engineering fields such as aerospace, architectural, automotive, and marine. Aerodynamic testing can be conducted in either a wind tunnel or a water tunnel, depending on the facility available and the information required. These tunnels are used to examine the streamlines and to determine the behavior patterns of aerodynamic forces acting on test objects. These facilities allow designers to model new body shapes without going to the expense of constructing full-scale units for the test. Scale models are usually necessary due to the constraint of tunnel size. For example, full-scale testing of a Boeing 747 would be impractical. In the aerospace industry, the use of scale models has resulted in the saving of countless lives and enormous expense.

A water tunnel uses the same principle as a wind tunnel but uses water as the fluid instead of air. Any closed testing system or tunnel has the following four major components, regardless of the fluid employed:

1) A contoured duct to control and direct fluid flow
2) A drive system to move the fluid through the duct
3) A model of the object to be tested
4) Instrumentation to measure forces exerted on the model.

Because the kinematic viscosity of water is 16 times greater than that of air, model studies at high Reynolds numbers should be conducted in water tunnels rather than wind tunnels. Linear tow tanks are another means of performing aerodynamic testing in water. The major disadvantage of this type of testing is that the model can only be subjected to intermittent, limited-duration testing. Another common water testing method uses a thin layer of water flowing over a flat table on which a model is placed. The disadvantage of water table testing is that it can only be used for two-dimensional analysis and cannot be used when accurate Reynolds number testing is required. A water tunnel overcomes these problems by providing a three-dimensional flow field.

## 10.7 Numerical Modeling

The solution of the partial differential equations (PDEs) that govern a structure's response may not always be possible; thus, numerical techniques must be utilized. In general, there are two alternative methods of approximating the solution of PDEs: the finite-difference method and the finite-element method.

### 10.7.1 The Finite-Difference Method

The familiar finite-difference model of a problem gives a pointwise approximation to the exact solution of a PDE. To find the approximate solution of a PDE using the finite-difference technique, a network of *grid points* is established through the domain of interest defined by the independent variables. As shown in Figure 10.8, $x$ and $y$ are the two independent variables, and $\Delta x$ and $\Delta y$ are the respective grid spacings. Subscripts $i$ and $j$ are used to denote space points having coordinates $i\,\Delta y$, $j\,\Delta x$.

Let the exact solution to the PDE be $U(x, y)$, and its approximation determined at each grid point be $\phi_{ij}$. Each partial derivative of the original PDE is replaced by an appropriate finite-difference expression involving $\Delta x$, $\Delta y$, and $\phi_{ij}$, which leads to a set of algebraic equations in $\phi_{ij}$. By solving this system of equations, the values of $\phi_{ij}$ can be determined at each grid point $(i, j)$. To obtain a close approximation, sufficiently small grid spacings should be selected. Note that improvement in accuracy by using a small grid spacing is achieved at a price, because it increases the computational time. Moreover, a very small grid size may increase the roundoff error, leading to a less accurate solution.

There are basically three finite-difference expressions used in the finite-difference method: the forward, backward, and central difference expressions. The central difference expressions that approximate the first- and second-order derivatives of the original PDE are formulated as

$$\phi_1' = \frac{\partial \phi}{\partial x} = \frac{\phi_{i+1} - \phi_{i-1}}{2\,\Delta x}, \tag{10.22}$$

$$\phi_1'' = \frac{\partial^2 \phi}{\partial x^2} = \frac{\phi_{i+1} - 2\phi_i + \phi_{i-1}}{2\,\Delta x^2}. \tag{10.23}$$

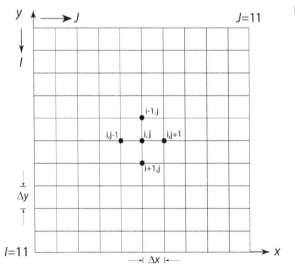

**Figure 10.8** Arrangement of grid points.

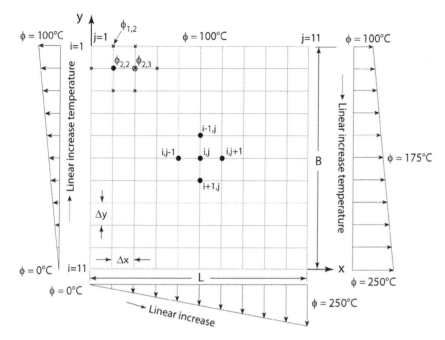

**Figure 10.9** Thin plate with the boundary conditions. (Ertas, A., and Jones, J.C., *The Engineering Design Process*, John Wiley & Sons, Inc., New York, 1996. Reproduced with permission of John Wiley & Sons, Inc.)

To illustrate this method, consider the problem of determining the temperature distribution in a thin homogeneous rectangular plate with prescribed boundary conditions (BCs) along the edges of the plate as shown in Figure 10.9. The top BC is held at 100°C, and the temperature of the other three BCs changes linearly as shown in the figure. The problem can be mathematically formulated as

$$\frac{\partial^2 \phi}{\partial x^2} + \frac{\partial^2 \phi}{\partial y^2} = \frac{C\rho}{k} \frac{\partial \phi}{\partial t}. \tag{10.24}$$

It is convenient to start by non-dimensionalizing the independent variables $x$ and $y$ as

$$X = \frac{x}{L}, \quad \text{for } 0 \le X \le 1, \tag{10.25}$$

$$Y = \frac{y}{B}, \quad \text{for } 0 \le Y \le 1. \tag{10.26}$$

Equation (10.24) is then rearranged as

$$\frac{\partial^2 \phi}{\partial X^2} + \frac{L^2}{B^2} \frac{\partial^2 \phi}{\partial Y^2} = \frac{L^2 C\rho}{k} \frac{\partial \phi}{\partial t}. \tag{10.27}$$

Non-dimensionalizing time $t$,

$$t = \frac{k}{L^2 C \rho} T, \tag{10.28}$$

yields

$$\frac{\partial^2 \phi}{\partial X^2} + \frac{L^2}{B} \frac{\partial^2 \phi}{\partial Y^2} = \frac{\partial \Phi}{\partial T}, \tag{10.29}$$

where

$$\frac{\partial \Phi}{\partial T} = \frac{L^2 C \rho}{k} \frac{\partial \phi}{\partial t}. \tag{10.30}$$

$L/B$ is called the *aspect ratio* and must be given for a specific problem. For the steady-state solution, the right-hand side (time-dependent term) drops out. The problem can then be described by the differential equation

$$\frac{\partial^2 \phi}{\partial X^2} + \left(\frac{L}{B}\right)^2 \frac{\partial^2 \phi}{\partial Y^2} = 0, \quad \text{for } 0 \leq X \leq 1 \quad \text{and} \quad 0 \leq Y \leq 1. \tag{10.31}$$

Using equations (10.22) and (10.23), the derivatives of equation (10.31) are now replaced by the central difference expressions

$$\frac{\partial^2 \phi}{\partial X^2} = \frac{\phi_{i,j+1} - 2\phi_{i,j} + \phi_{i,j-1}}{\Delta X^2}, \tag{10.32}$$

$$\frac{\partial^2 \phi}{\partial Y^2} = \frac{\phi_{i+1,j} - 2\phi_{i,j} + \phi_{i-1,j}}{\Delta Y^2}. \tag{10.33}$$

Substituting in equation (10.31) yields

$$\frac{\phi_{i,j+1} - 2\phi_{i,j} + \phi_{i,j-1}}{\Delta X^2} + \left(\frac{L}{B}\right)^2 \frac{\phi_{i+1,j} - 2\phi_{i,j} + \phi_{i-1,j}}{\Delta Y^2} = 0 \tag{10.34}$$

or

$$\phi_{i,j+1} - 2\phi_{i,j} + \phi_{i,j-1} + \left(\frac{L}{B}\right)^2 \left(\frac{\Delta X}{\Delta Y}\right)^2 \phi_{i+1,j} - 2\phi_{i,j} + \phi_{i-1,j} = 0. \tag{10.35}$$

Let

$$R = \left(\frac{L}{B}\right)^2 \left(\frac{\Delta X}{\Delta Y}\right)^2 \tag{10.36}$$

and solve for $\phi_{i,j}$:

$$\phi_{i,j} = \frac{\phi_{i,j+1} + \phi_{i,j-1} + R\phi_{i-1,j} + R\phi_{i+1,j}}{2(1+R)}. \tag{10.37}$$

For simplicity, assume that $\Delta X = \Delta Y$ and $L/B = 1$. Then equation (10.37) reduces to

$$\phi_{i,j} = \frac{\phi_{i,j+1} + \phi_{i,j-1} + \phi_{i-1,j} + \phi_{i+1,j}}{4}. \qquad (10.38)$$

Equation (10.38) indicates that the temperature at each grid point is the average of temperatures at surrounding grid points; hence, no interior grid point can have a temperature greater than the hottest boundary temperature. The first step in determining the grid point temperature by using equation (10.38) is to initialize all of the interior grid point temperatures to some value and start iteration. For this example, assume that all the interior grid point temperatures are equal to zero. Then the temperature at $\phi_{2,2}$ is

$$\phi_{2,2} = \frac{\phi_{1,2} + \phi_{2,3} + \phi_{3,2} + \phi_{2,1}}{4}$$

$$= \frac{100 + 0 + 0 + 90}{4} = 47.5°C.$$

Knowing the temperature at $\phi_{2,2}$, the temperature at $\phi_{2,3}$ can be calculated:

$$\phi_{2,3} = \frac{\phi_{1,3} + \phi_{2,4} + \phi_{3,3} + \phi_{2,2}}{4}$$

$$= \frac{100 + 0 + 0 + 47.5}{4} = 36.875°C.$$

After all the values of grid points are calculated, use these values as initial values and start the second iteration. This iteration process is repeated until the largest change in the $\phi_{i,j}$ component is less than the assumed convergence criterion $\varepsilon$. It is obvious that the number of iterations required for convergence depends on how good the initial estimate was and on the assumed convergence criterion $\varepsilon$. In most instances, setting all of the interior grid points equal to the mean of all of the boundary temperatures gives a good initial estimate for starting the iteration.

## 10.7.2   The Finite-Element Method

A finite-element model of a problem gives a piecewise approximation to the governing equations of a structure response. The basic premise of the finite-element method is that the domain can be analytically modeled or approximated by replacing it with an assemblage of discrete elements. Because these elements can be put together in several ways, they can be used to represent extremely complex shapes. Although both the finite-difference method and the finite-element method can give accurate results if used properly, one advantage of the latter is the relative ease with which the boundary conditions of the problem are handled. Many physical problems have irregularly shaped boundaries. Boundaries of this type are relatively difficult to handle using the finite-difference method.

The first step in the finite-element method is to divide the region into a finite number of elements and to label the elements and the node numbers, as shown in Figure 10.10. The points at which the elements are connected are called the grid points. They are identified by grid point numbers to define their location in space. The accuracy of the approximation to a problem solution depends on the characteristics and number of elements that are used in the domain. The most common finite elements used in modeling are shown in Figure 10.11(a,b), one-dimensional elements with two and three nodes, respectively. They have a single degree of freedom at each grid point

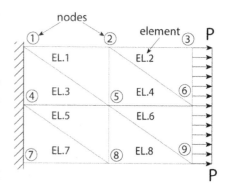

**Figure 10.10** Subdivision of a plate into triangular elements.

and are used for one-dimensional problems involving two force (truss) members. There are two general element families used in modeling the two-dimensional domain: triangular and quadrilateral elements. Figure 10.11(c,d) shows linear elements that have straight sides. Higher-order two-dimensional elements (quadratic or cubic) have either straight or curved sides or both (see Figure 10.11(e,f). Two-dimensional elements are used to model membranes, plates, and shells. Because of the greater accuracy, quadrilateral elements are preferred over triangular elements. Figure 10.11(g–i) depicts three-dimensional elements, also called *solid elements*, which are used to model thick plates, thick shells, and three-dimensional solid media, respectively.

Monte Carlo simulation belongs to the family of Monte Carlo methods. Monte Carlo methods constitute a branch of experimental mathematics that is concerned with experiments on random numbers. In essence, this is an elegant approach to approximating physical and mathematical problem solutions by simulation of random quantities.[1]

During the mid-1940s, John von Neumann and his colleagues introduced the term "Monte Carlo" as a code name for their classified research on neutron diffusion problems.[2] The name Monte Carlo was probably chosen because of the method's random number basis, which can supposedly be duplicated by roulette wheels. Such roulette wheels are often found in gambling houses, of which, in the mid-1940s, the most famous were in the principality of Monte Carlo. In some literature, Monte Carlo methods are also referred to as methods of statistical trials, since they involve a scheme of producing random events where each trial is independent of the rest.

There are various forms of Monte Carlo methods that are used, depending on the problem to be solved. The simplest form is that used for the direct simulation of a probabilistic

1  Hammersley J.M., and Handscomb, D.C., *Monte Carlo Methods*, John Wiley & Sons, New York, 1964.
2  Spanos P.D., and Lutes, L.D., "A primer of random vibration techniques in structural engineering," *Shock and Vibration Digest*, 18(4), 3–9, 1986.

**Figure 10.11** Some common finite elements.

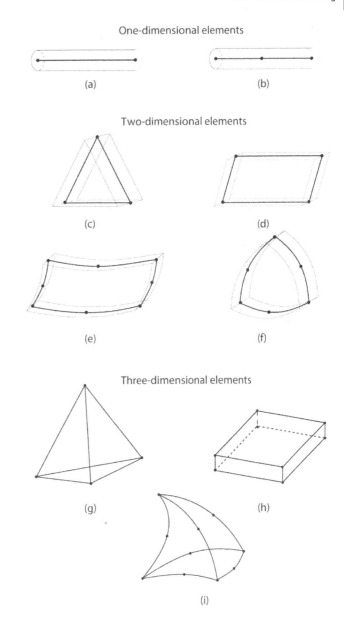

One-dimensional elements

(a)

(b)

Two-dimensional elements

(c)

(d)

(e)

(f)

Three-dimensional elements

(g)

(h)

(i)

problem. In its most general form, Monte Carlo simulation can be characterized by three steps:[3]

1) Application of random numbers to simulate representative samples of random variables with prescribed distributions

3 Kareem, A., "Wind effects on structures: A probabilistic viewpoint," *Probabilistic Engineering Mechanics*, 20(4), 166–200, 1987.

2) Solution of the problem on the basis of large realizations
3) Statistical analysis of results (by derivation of means, mean squares, variances, etc.).

Because Monte Carlo simulation involves statistical trials, its accuracy is proportional to the number of trials. Increasing the number of trials or observations is not always an economic course of action, since at some point the computer time cost becomes prohibitive. For example, if $n$ is the number of trials, it has been shown that the computer cost increases in proportion to $n$ while the statistical uncertainty decreases in proportion to $1/\sqrt{n}$.[4] Due to the nature of Monte Carlo simulation, it is most effective in solving problems in which the required accuracy of the statistics sought is within 5–10 percent of the actual values.[5] In some instances, variance-reduction techniques can be used to reduce the uncertainty of the solutions. In essence, these techniques improve the accuracy of the results without increasing the number of tests (trials) and without sacrificing reliability.

Monte Carlo simulation can be used to simulate most processes influenced by random factors. Some of the problems that can be solved by this method include the following:

1) Solution of linear algebraic equations
2) Engineering design problems
3) Structural dynamics problems
4) Operations research problems (e.g., queueing systems)
5) Control of floodwaters and construction of dams
6) Ecological competition among species
7) Chemical kinetics
8) Diffraction of waves on random surfaces.

Consider a simple example of the application of the Monte Carlo method.[5] Assume that a unit square circuit board has been designed. The available area $P$ on the circuit board is to be estimated (Figure 10.12). Choose $n = 40$ points at random within the square. Note that the selected coordinates of the points are random and uniformly distributed over the entire unit square. The random coordinates can be generated by using a computer as shown in Table 10.3. In order to assign the numbers 0–9 to the coordinates of the points, the unit square is divided into 10 parts along both the $x$- and $y$-axes. Only the first digit of the respective two-digit random numbers is used for the location of the points within the unit square. In the table, let columns 1, 3, 5, and 7 and columns 2, 4, 6, and 8 be the $x$ and $y$ coordinates of the points, respectively. In Figure 10.12, the number of points enclosed in the shape is $n' = 17$. From geometrical considerations, the area $P$ can be approximated by the ratio $n'/n = 17/40 = 0.43$. The latter value is close to the true area of $P$ of 0.38. Because this experiment involves trials, the accuracy will be increased by increasing

---

4 Branstetter, L.J., Jeong, G., Yao, J.T.P., Wen, Y.K., and Lin, Y.K., "Mathematical modelling of structural behaviour during earthquakes," *Probabilistic Engineering Mechanics*, 3(3), 130–145, 1988.
5 Sobol, I.M., *The Monte Carlo Method*, University of Chicago Press, Chicago, 1974.

**Figure 10.12** Shape of available area on circuit board. (Ertas, A., and Jones, J.C., *The Engineering Design Process*, John Wiley & Sons, Inc., New York, 1996. Reproduced with permission of John Wiley & Sons, Inc.)

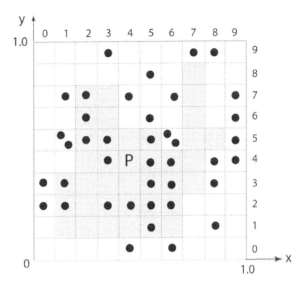

**Table 10.3** Generated random numbers.

| 1 | 2 | 3 | 4 | 5 | 6 | 7 | 8 |
|---|---|---|---|---|---|---|---|
| x | y | x | y | x | y | x | y |
| 03 | 27 | 43 | 73 | 56 | 16 | 96 | 47 |
| 97 | 74 | 24 | 67 | 62 | 42 | 81 | 14 |
| 16 | 72 | 62 | 27 | 66 | 56 | 50 | 26 |
| 12 | 56 | 85 | 99 | 36 | 96 | 96 | 68 |
| 55 | 59 | 56 | 35 | 64 | 38 | 54 | 82 |
| 16 | 22 | 77 | 94 | 39 | 49 | 54 | 43 |
| 84 | 42 | 17 | 53 | 31 | 57 | 24 | 55 |
| 63 | 01 | 63 | 78 | 49 | 26 | 95 | 55 |
| 33 | 21 | 12 | 34 | 29 | 78 | 64 | 56 |
| 57 | 60 | 87 | 32 | 44 | 09 | 07 | 37 |

the number of times the experiment is performed and also by increasing the number $n$ of random points. Hence, if the experiment is repeated a number of times, the estimated areas can be shown to closely cluster around the true area. As a word of caution, the Monte Carlo method is certainly not the most efficient way to estimate the area of a plane surface, but as the space dimensions of the body increase, so does the efficiency of this method. Therefore, this method is well suited to the solution of multidimensional problems.

## 10.8  Discrete Event Simulation

Discrete event simulation, or simply simulation, is a form of descriptive modeling. Simulation experiments explore the behavior of a given system. A discrete event simulation model is basically a computer program that can be used to simulate system behavior. An engineer developing a simulation model has at his or her disposal various special purpose simulation languages such as Simscript, Simula, or the General Purpose Simulation System (GPSS). Alternatively, a simulation program can be developed by using general purpose programming languages such as Fortran, Pascal, or C. A simulation model can be used for describing new systems, predicting the behavior of systems, demonstrating a system's behavior, and training. The advantages of simulation modeling are:[6]

1) Ease in performing controlled experiments
2) Time compression in the sense that a simulation experiment takes a small fraction of time compared with actual system operation time
3) Sensitivity analysis for observation of the behavior limits of the system
4) Experimentation without disturbing the real system
5) Use of the simulation model as an effective training tool.

Consider the single-server oil change station shown in Figure 10.13 as an example to simulate. In this model, cars arrive for service according to some distribution such as Poisson, and they are serviced according to some service time distribution. If the service station is busy, the arriving car waits in the queue for service. If the service station is idle, the car moves into the station, the oil is changed, and the vehicle departs. In both of these cases, systems statistics are collected. The flow of the simulation model of this simple oil change service station is shown in Figure 10.14.

The above simulation example can be used to support a variety of studies such as:

1) Evaluating different tasks to achieve balanced workloads and improve system efficiency
2) Performing sensitivity analyses to determine which parameter has the most impact on the system efficiency
3) Determining which parameter combination will provide the most feasible operation of the system
4) Analyzing the performance impact of different degrees of automation.

**Figure 10.13** Oil change service station.

6 Graybeal, W.T., and Pooch, U.W., *Simulation: Principles and Methods*, Little, Brown, Boston, 1980.

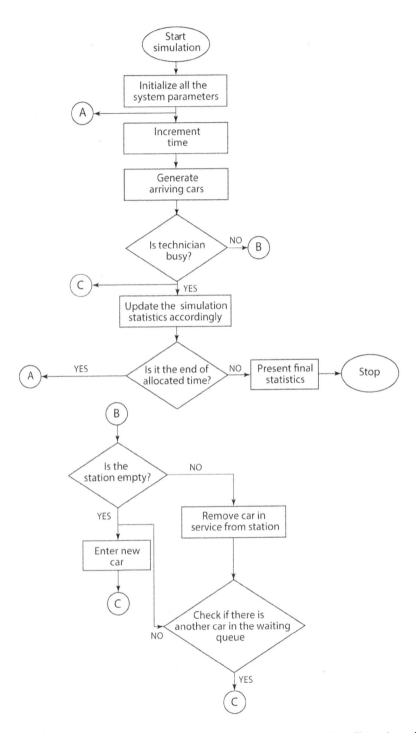

**Figure 10.14** Simulation model flow for an oil change service station. (Ertas, A., and Jones, J.C., *The Engineering Design Process*, John Wiley & Sons, Inc., New York, 1996. Reproduced with permission of John Wiley & Sons, Inc.)

## 10.9 Knowledge-Based Systems in the Design Process

The idea of capturing the knowledge of an expert in an automated medium and then using this knowledge when needed, has been one of the most productive areas in the application of artificial intelligence. The class of technical activities representing the knowledge of an expert in an electronic medium for later use are collectively called expert systems technology. The systems produced with the aid of this technology are collectively called expert systems, knowledge-based systems, or decision systems.

Because of their flexibility in handling incomplete and inconsistent information as human experts would, the use of knowledge-based systems is gaining importance in the engineering design process.[7,8,9] The following capabilities characteristic of expert systems typically separate them from conventional systems:

1) Solving problems normally solved by experts in the field
2) Producing solutions for problems that are not suitable for traditional techniques
3) Functioning with incomplete and inconsistent data
4) Producing quick prototypes
5) Providing explanations of decisions reached.

Production of an expert system with the above capabilities requires that knowledge be gathered (knowledge acquisition), knowledge needs to be represented in an electronic medium (knowledge representation), knowledge needs to be used to derive inferences (knowledge manipulation), and interfacing with a user of the systems needs to be continuous throughout the process (user interface).[10]

These necessities for knowledge acquisition, knowledge representation, knowledge manipulation, and user interface also dictate the overall components of expert systems. As shown in Figure 10.15, the components of a typical expert system are user interface, knowledge base (rule base), interpreter, and working memory. The user interface is the obvious component to be able to achieve input to the system and to present the output to the user. The knowledge base is the location holding the collective set of rules capturing the expertise obtained from experts. The interpreter is a program that uses logical inferences in arriving at conclusions dictated by logic. Working memory performs logical inference operations on selected rules from the knowledge base.

7 Tanik, M.M., and Chan, E.C., *Fundamentals of Computing for Software Engineers*, Van Nostrand Reinhold, New York, 1991.

8 Flemming, J., Elghadamsi, E., and Tanik, M.M., "A knowledge-based approach to preliminary design of structures," *Journal of Energy Resources Technology* 112, 213–219, 1990.

9 Tanik, M.M. and Ertas, A. (eds), *Expert Systems and Applications*, Proceedings of ASME, New York, 1991.

10 Tanik, M.M., and Yun, Y.Y.D., "Guest Editor's Introduction: Interactions between expert systems and software engineering," *IEEE Experts: Intelligent Systems and Their Applications*, 3(4), 5–6, 1988.

**Figure 10.15** Components of a typical expert system.

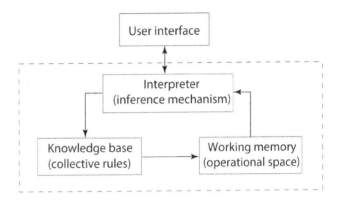

### 10.9.1 An Expert System for Preliminary Design of Structures

Designing a structure, such as the tension leg platform (TLP) shown in Figure 10.16, is an engineering process that requires creativity, experience, and knowledge of engineering principles.[11] As in any engineering activity, there is usually more than one solution that is considered appropriate. Typically, the speed and success of a structural designer in selecting a suitable structural scheme depend on his or her experience.

The structural design process follows the standard stages of design, including preliminary design, detailed design, and drawing production. Currently, computers are extensively used in most of the phases involved in the production of structures. In this example, the use of computers with expert systems technology in the preliminary design stage will be demonstrated. The analysis, detailed design, and drawing production stages present problems suitable for conventional solutions described elsewhere in this book.

The preliminary design stage has two main goals. The first is the conceptual design in which the framing scheme is selected. The second is preliminary member sizing in which the sizes

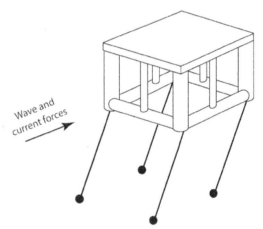

**Figure 10.16** Tension leg platform.

of the various components of the selected design are determined. Typically, the engineer may have to perform an analysis to select preliminary member sizes. Experienced engineers are normally able to make reasonable estimates of the required sizes. An expert system, as shown in Figure 10.15, can be used by novice engineers in the preliminary design phase. In this example, the prototype of an expert system used for selecting initial structural member sizes is discussed.

---

11 Flemming, J., Eighadamsi, E., and Tanik, M.M.,"A knowledge-based approach to preliminary design of structures," *Journal of Energy Resources Technology*, 112, 213–219, 1990, with permission from ASME.

For the design of an expert system for structural member selection, three key issues need to be decided: representation of experience; finding matching structures; and selection of member sizes.

There are various methods for representing experience. One of the most widely used techniques is the rule-based approach. In this approach, attributes related to the problem are stored in the knowledge base. In case of the platform problem, examples of attributes are as follows:

1) Geographical location of platform
2) Purpose of platform
3) Water depth
4) Soil type
5) Wave and current loading descriptions
6) Wind loading descriptions
7) Seismic zone
8) Design specification
9) Structural framing cost
10) Foundation cost.

To be able to utilize previous experience, designs of previously built structures that have features matching the intended new design need to be found. Table 10.4 shows a set of criteria for the TLP example that can be used to find matching structures. For example, water depth is a required criterion. If the new TLP is to be designed for a water depth of 1,000 feet, structures designed for depths considerably less than 1,000 feet or considerably more than 1,000 feet should not be considered. The expert system should not find (through the matching process) these overly conservative or unreliable structures for the knowledge base for the new structure design.

The last phase of the preliminary design process is to select the initial member sizes. Member attributes such as type, length, location, and special loading conditions can be used to identify matching members. Once the members are identified, they should be ranked in a manner similar to the ranking of the matching structures. To determine the required section properties for desired members, the rankings of the similar members, their sizes, and the ranking of the structure are used.

The design of the expert system is completed at this point. The process involves the selection of an expert system implementation language (tool, shell) and the encoding of the rules developed above using the selected implementation language and computer.

## 10.9.2 Selection of Expert System Tools

To be able to actually implement the expert system in question, a suitable implementation tool must be selected. This is comparable to selecting a programming language and a compiler for

**Table 10.4** Example criteria.

| Criteria | Comments |
| --- | --- |
| Geographical location of platform | Desirable criterion |
| Purpose of platform | Essential criterion |
| Water depth | Essential criterion (the tolerance for a match will vary within the water depth) |
| Soil type | Desired criterion |
| Wind speed | Essential criterion |
| Wave loads | Essential criterion |
| Current loads | Essential criterion |
| Seismic zone | Essential criterion |
| Design specification | Desirable criterion |
| Structural framing cost | Desirable criterion |
| Foundation cost | Desirable criterion |

conventional programming purposes. Selection of implementation tools is by itself a complicated process.[12] Some of the features of tools which seem promising for designers are:

1) Ability to query the user during the inference process
2) Existence of an explanation support mechanism
3) Graphical display support system
4) Ease of use
5) Ability to detect incomplete and inconsistent data.

### 10.9.3 Determining the Suitability of the Candidate Problem

The first step in selecting an expert system for a particular problem is to determine whether the problem is likely to be suitable for solution by an expert system method. This determination is based on *essential* and *desirable* features of the problem. Each of these features is assigned a weight indicating its relative importance in assessing the candidate problem.[13] These weights are not problem dependent, but are related to the capabilities of expert system technology. The *essential* and desirable features and associated weights used to evaluate the candidate problem and associated weights are listed in Tables 10.5 and 10.6, respectively.

---

12 Mills, S., and Tanik, M.M., "Selection of expert system tools for engineering design applications," *Proc. ASME Expert Systems and Applications*, 35, 41–45, 1990.
13 Mills, S., Ertas, A., Hurmuzlu, Y., and Tanik, M.M., "Selection of expert system development tools for engineering applications," *Journal of Energy Resources Technology*, 114(1), 38–46, 1992.

**Table 10.5** Essential features of candidate problem.

| Description | Weight (1–10) |
| --- | --- |
| The problem is not natural language dependent | 10 |
| The problem is not knowledge intensive | 7 |
| Solution of the problem requires use of heuristics | 8 |
| Test cases for the problem are available | 10 |
| The problem is decomposable into subproblems | 7 |
| "Common sense" is not needed to solve the problem | 10 |
| An optimal solution is not required | 8 |
| Problem solution will take place in the future | 10 |
| A domain expert exists and is available and cooperative | 10 |
| Transfer of expertise for the problem domain is difficult | 7 |
| Solution should not depend upon physical senses | 10 |

**Table 10.6** Desirable features of candidate problem.

| Description | Weight (1–10) |
| --- | --- |
| The candidate problem should previously have been identified | 4 |
| Solutions should be explainable and require interaction | 5 |
| Similar applications have been successfully implemented | 8 |
| The problem should be solved at many different locations | 5 |
| The environment in which the system will operate is hostile | 3 |
| Solution of the problem requires subjective judgment | 4 |
| The expert will not be available in the future | 3 |

### 10.9.4 Knowledge Representation Schemes

Separating domain-specific knowledge from the program is the fundamental idea of expert systems. The term *knowledge base* refers to the collection of domain knowledge, and the term *inference engine* refers to the program reasoning through that knowledge. The specification of knowledge representation comprises both of these components.

The inference engine works on the knowledge base and generates solutions to the problems of the user. A knowledge representation supported by a powerful inference scheme allows the user to solve a problem by simply storing the facts in the knowledge base. There are several techniques being used for knowledge representation. In this section, *production systems*, which is a commonly used scheme, will be discussed.[14]

The terms *production-rule systems* and *rule-based systems* are used to refer to the knowledge representation scheme. In this scheme, knowledge is expressed as a set of rules. A production system consists of three parts:[15]

1) *Rule base.* The rule base is composed of a set of production rules. A production rule is an expression in the form

> IF <condition-part> THEN <action-part>

The <condition-part> states the conditions that must be satisfied for the production to be applicable, and the <action-part> defines the actions to be executed.
2) *Context.* The context is the short-term memory buffer that contains the initial inputs by the user and the deductions that have been made by the inference engine. The actions of the production rules that are invoked can expand the context with the addition of current deductions.
3) *Interpreter.* The interpreter decides the order of rules to be invoked. The production systems operate according to the cycles of the interpreter. A basic cycle of the interpreter is given in Figure 10.17.

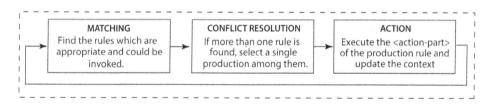

**Figure 10.17** Basic cycle of the rule interpreter. (Ertas, A., and Jones, J.C., *The Engineering Design Process*, John Wiley & Sons, Inc., New York, 1996. Reproduced with permission of John Wiley & Sons, Inc.)

---

14 Demirors, E., et al., "Rule based expert systems: A PC + experience," Technical Report 91-cse-19, Department of Computer Science and Engineering, Southern Methodist University, May 1991.
15 Barr, A., and Feigenbaum, E., *The Handbook of Artificial Intelligence*, Addison-Wesley, Reading, MA, © 1989, by Addison-Wesley Publishing Company, Inc. Reprinted with permission of the publisher.

In the following example "On–CL X" refers to the fact that the symbol X is currently in the context list CL and "Put–On–CL X" is used to refer to X being added to the current context list.

---

PRODUCTION

---

**Rule 1** ⇒ IF On–CL green

THEN Put–On–CL produce

**Rule 2** ⇒ IF On–CL packed in small container

THEN Put–On–CL delicacy

**Rule 3** ⇒ IF On–CL refrigerated OR On–CL produce

THEN Put–On–CL perishable

**Rule 4** ⇒ IF On–CL weighs 15 lb AND On–CL inexpensive AND NOT

On–CL perishable

THEN Put–On–CL staple

**Rule 5** ⇒ IF On–CL weighs 15 lb each AND On–CL produce

THEN Put–On–CL watermelon

---

In the above example, the following steps are performed to differentiate the staple food from produce:

1) If more than one production is applicable, then deactivate any production whose action adds duplicate deductions to the CL.
2) If more than one production rule is applicable, then execute the action with the lowest (arbitrarily chosen) numbered application production.

The above production system can be used to identify food items. Let the existing knowledge about the unknown food be "it is green" and "weighs 15 lb." Therefore, at the beginning the context list is

CL = (green, weighs 15 lb).


1) Cycle 1 ⇒ since rule 1 is applicable, the action part of rule 1 is executed. This adds "produce" to the context representing a new fact about the unknown food item. Therefore, the updated context list is

CL = (produce, green, weighs 15 lb).

2) Cycle 2 ⇒ rule 1, rule 2, and rule 3 are applicable. Since rule 1 adds "produce," which is duplication, rule 1 is eliminated from the execution. Hence, rule 3 is executed. Therefore, the undated context list is

CL = (perishable, produce, green, weighs 15 lb).

3) Cycle 3 ⇒ rule 1, rule 2, and rule 3 are applicable. Rule 5 is executed. Therefore, the updated context list is

$$CL = (\text{watermelon, perishable, produce, green, weighs 15 lb}).$$

4) Cycle 4 ⇒ No redundant productions to activate.


## Bibliography

**1** BARBER, C.F., "Expanding maintainability concept and techniques," *IEEE Transactions on Reliability*, R-16(1), 5–9, 1967.

**2** DOEBELIN, E.O., *System Modeling and Response*, John Wiley & Sons, Inc., New York, 1980.

**3** FOX, R.W., and MCDONALD, A.T., *Introduction to Fluid Mechanics*, John Wiley & Sons, Inc., New York, 1973.

**4** GIORDANO, R.F., and WEIR M., *A First Course in Mathematical Modeling*, Brooks/Cole, Monterey, CA, 1985.

**5** GORDON, G., *System Simulation*, Prentice Hall, Englewood Cliffs, NJ, 1969.

**6** MARTIN, F.F., *Computer Modeling and Simulation*, John Wiley & Sons, Inc., New York, 1968.

**7** WELLSTEAD, P.E., *Introduction to Physical System Modelling*, Academic Press, London, 1979.

## CHAPTER 10 Problems

**10.1** Develop a 1/10 scale model of your room with all the furniture, using cardboard, wood, glue, and the like. Furniture should be capable of being relocated so that various configurations can be considered. Decide on the best layout and justify why you think it is optimum. Compare the optimum to the present furniture arrangement.

**10.2** If an automobile shown in Figure P10.2 is excited by a harmonic force $F_1 \sin \omega t$, model the system using (1) a single degree of freedom and (2) two degrees of freedom, write the mathematical expressions that describe the behavior of the automobile.

**Figure P10.2** Automobile on the road.

**10.3** Determine the $\pi_1$, $\pi_2$, $\pi_3$, and $\pi_5$ terms for Example 10.1.

**10.4** Find an expression for the distance traveled by a freely falling mass $M$ in time $t$ under the influence of gravity $g$ if the initial velocity is zero.

**10.5** A simply supported aluminum beam of rectangular cross section ($E = 10.5 \times 10^6$ psi) is 8 in. wide and 14 in. in height. A concentrated load of 6000 lb is applied 6 ft from the left end of the 14-ft beam. Design a model with a 10-in. span using a steel beam to accurately predict deflection at any location along the aluminum prototype. Ignore the weight and deflection of the beam.

**10.6** A bending moment of 3000 $lb_f$-in. is required to bend a model aluminum bar into an $L$ shape. A prototype bar is made of the same material as the model with a 50 percent larger diameter. Determine the bending moment required to bend the prototype into an L shape.

**10.7** Write a Matlab Code for the temperature distribution in the thin plate shown by Figure 10.7 which evaluates the temperatures at the grid points by iteration.

**10.8** Write a Matlab Code to evaluate the temperatures at the grid points for the steady-state heat conduction in the thin plate shown by Figure P10.8. Compare the result with the analytical solution at $a = 5$ in. and $b = 5$ in. The heat transfer for this problem follows the equation

$$\frac{\partial^2 T}{\partial x^2} + \frac{\partial^2 T}{\partial y^2} = 0$$

The boundary conditions are:
a) $T(0, y) = 0°$
b) $T(x, b) = 0°$
c) $T(a, y) = 0°$
d) $T(x, 0) = f(x)$

**Figure P10.8** Thin plate with the boundary conditions.

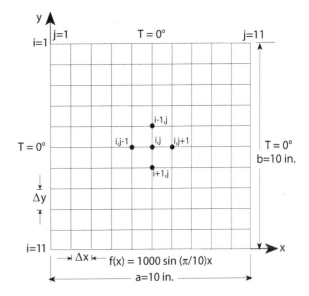

**10.9** Consider the available area on a circuit board shown in Figure P10.9. Using the second digits of the respective random number pair in Table 10.3 as the coordinates of the random points, estimate the available area $P$.

**Figure P10.9** Shape of available area on circuit board.

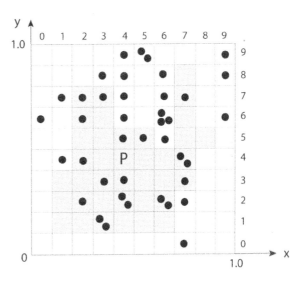

# 11

## Engineering Economics

> It would be well if engineering were less generally thought of, and even defined, as the art of constructing. In a certain important sense it is rather the art of not constructing; or, to define it rudely but not inaptly, it is the art of doing that well with one dollar which any bungler can do with two after a fashion.

> **A. M. Wellington**

## 11.1   Project/Product Cost and the Engineer

During the second half of the twentieth century significant changes occurred around in the United States and the rest of the world in regard to product cost and the meaning of competition. Through the end of World War II and into the 1950s US industry operated with a philosophy whereby new, modified, and unique products were developed within a particular product line with pricing based on development, manufacturing, and marketing costs, plus a reasonable profit. Competitive product costs were considered, of course, but this comparison had more to do with product cost–utility ratios than just lower cost. Industry took pride in the *product* of its efforts and little of the *copycat* mentality that seems so prevalent today was present. Patents for unique product ideas usually meant monetary returns for the inventor and long-term profit for the company with the manufacturing rights. Today, many companies with innovative ideas for products or processes deliberately do not make patent applications because they do not want to disclose their idea publicly. They know that as soon as the innovation is made public it will be copied, manufactured in a cheaper labor location or using cheaper materials or methods, and sold at lower cost than the original. Unfortunately, the copycat seldom has any real commitment to the product, and after the initial market demand has been met will often go on to copy some other product or, possibly worse, close the manufacturing plant down and declare bankruptcy for protection against creditors. Fortunately, this is not a fair representation of all of the consumer product industry nor of industry involved in developing highly technical products. There are many reputable manufacturers concerned with producing a quality product and standing behind the product after the sale. Automobile manufacturers now emphasize the quality of their products and their commitment to the customer after the sale in their advertisements. Unfortunately, reputable manufacturers operate in the same economy as the less reputable ones and are thus driven to compromises and cost-cutting practices they might normally shun. Who gains from all this pencil sharpening? The buyer gains in the sense that he or she is paying less (initially) for the product, but with lower cost (in general) comes lower quality, decreased reliability, and poorer product support, and the ultimate cost to the consumer is, more often than not, equal to or greater than the higher-quality product.

Another significant influence on the cost of products and the potential market came about as a result of US participation in the global economy. When US industry was more or less isolated from industry worldwide, producers competed on a so-called *level playing field* in that labor costs were pretty well controlled by management, labor unions, and legislation, and government subsidies and support were applied quite equitably. In competing in a global economy, no such even-handedness prevails and many US industries have found themselves in difficulty as a result. To maintain their competitiveness, many companies have located manufacturing facilities in low-cost labor regions of the world to take advantage of lower manufacturing costs. There are several disadvantages that should be considered in taking such action, however, including capricious foreign governments, loss of effective control, availability of management capable of directing the operation while maintaining loyalty to the company, availability of an adequately trained work force, product quality, transportation, availability of supplies, and bad publicity in the United States. Another factor to consider in locating manufacturing facilities outside the USA is the trend for third-world countries (where labor may be cheap) to become more like the major industrialized countries as their economy grows. Labor cost differentials may diminish or totally disappear over a period of a few years.

Product cost and facility investment decisions in today's environment are certainly more complex and fraught with more uncertainty than those of forty or fifty years ago. Since much of the data input to these decisions comes from engineers it is especially important that engineering graduates have some grasp of the factors affecting such decisions. For years almost all engineering curricula in the USA included a requirement for three semester hours (equivalent) of economics, usually at the sophomore level. Although the courses offered to satisfy this requirement (more often than not) presented a broad view of economics, with no intentional emphasis on engineering considerations, they did provide engineering students with a general understanding of economic principles and awakened a vague awareness that the engineering profession was governed by these principles just as was the rest of the business community. With increasing emphasis on other subjects that were considered too important to be left out of the curriculum, many universities subsequently eliminated such courses, often substituting an engineering economics course as an elective. In recognition of the increasing importance of economics and cost in engineering, most universities have more recently made an engineering economics course a requirement.

The information provided in this chapter is intended to augment the engineering economic analysis material presented in such courses and to introduce the student to cost analysis scenarios thought to be representative of those in which a typical working engineer might be involved.

Although it can be argued that few new graduate engineers will have the opportunity to apply economic analysis and decision-making during the early years of their employment, it is nevertheless essential that they have some understanding of the significance of cost to the success or failure of the projects to which they are assigned. To this end, some of the many and varied cost and economic analyses to which beginning, as well as journeyman, engineers are exposed are outlined in this chapter. Typical design situations and job assignments in which these analyses are necessary are listed below:

1) *Obtaining an accurate cost for a significant element in the design of a system to which the new engineer is assigned.* This is a simple task but it requires a good understanding of the supplier's operation so that the validity and long-term reliability of the estimate can be ascertained. For significant elements it is especially important that confidence in the integrity of the supplier's management be ensured. Establishing a reasonable delivery schedule is also important in providing an environment in which the supplier can effectively comply with the contract requirements.

2) *Estimating the fabrication costs for relatively simple design elements and components.* This is a straightforward task of gathering information on the cost of materials and supplies used in the design (usually from outside sources) and in determining the cost of the various production tasks (usually in-house). Manufacturing firms will have a breakdown of the various machining, assembly, and inspection costs that can be utilized in compiling this estimate.

3) *Estimating the cost of modifications to previously designed components and systems accomplished by the company's work force either in-house, in the field, or by an outside contractor in-plant or at the field location.* This task is considerably more complex, even when the modification effort is accomplished by the in-house work force, since it usually involves making material and component takeoffs from drawings prepared by other organizations and working with unfamiliar detail design and assembly drawings. This type of estimating is further complicated by the fact that in-house labor task cost data is not usable for tasks accomplished in another contractor plant or in the field. Estimating is also complicated by general unfamiliarity with the field situation: what tools and equipment, labor skills, supervision, materials, supplies, and other resources are available and at what cost?

   Estimates for contract changes can have significant ramifications, resulting in bad feelings between the customer and the contractor as well as other parties involved in the contract arrangement. The difficulties usually arise from assessments as to who is at fault. The customer may feel that the change was required due to inadequate initial design, whereas the contractor will usually attribute the change to a modification of the customer's requirements. It is especially important in this environment to have a thorough and valid estimate of the cost and a willingness to negotiate a change cost and schedule that is mutually agreeable to both parties.

4) *Evaluating the estimated costs on proposals and bids received from contractors.* Engineers usually do a comparative evaluation of the various estimates to determine which constitutes the best offer. If it is a response to a request for bid (RFB), cost is the principal evaluation criterion and the analysis consists of evaluating the cost elements and how they combine to

make up the total cost. It is important to ensure that the bid complies with the RFB and, if so, that the contract is awarded to the qualified bidder with the lowest bid price. For a request for proposal (RFP) the analysis is more complicated. In this case both the cost and the proposed solutions of the various proposers must be evaluated. An RFP competition usually involves additional negotiation with the two to three top proposers to establish the comparability of cost elements in the proposals and to bring the proposals to a state of acceptability to the contracting agency. In-house cost estimates are often made to establish a basis for evaluation.

5) *Preparing proposals, either as a result of an RFP or through effort on an unsolicited proposal.* These situations require more than a valid estimate of the cost of the effort or product. Answers are needed to questions such as: How important is the project to the company? Is this effort likely to lead to further contracts? Will this effort place the firm in a unique competitive position? Does this work integrate well with the company's long-term marketing and technology goals? Will this effort have a serious impact on staffing and other company resources? Whether the company has the financial resources and cash flow to handle the contract is all-important and must be addressed. Large and well-respected companies have been forced out of business for failure to ensure adequate cash flow, even when a large backlog of work existed. During the mid-1960s, Douglas Aircraft Company, Santa Monica, CA, a firm with a significant work backlog, was forced to sell out to become part of McDonnell Aircraft Company, St. Louis, MO, to secure adequate cash flow to meet its aircraft delivery commitments.

6) *Internal company requests or assignments may also precipitate proposal effort involving engineers.* Because these situations involve internal company operations, the data developed are primarily technical, although departmental goals and capabilities have to be considered. The costs associated with these efforts are more often than not reflected in types of labor, man-hours, quantities of materials, and other special requirements.

7) *Managing outside contracts and studies.* Feasibility studies constitute a special situation in which cost is usually one of the primary factors on which a decision is made to proceed with a project. In the role as one of the team members helping to manage a feasibility study, the engineer is required to use experience and knowledge in directing the design philosophy and approach adopted, in part, to ensure that cost limitations and/or goals are met.

8) *Participating in (or making) decisions as to whether the firm should make certain capital investments.* Investment decisions involve not only the time value of money but also the long-term future of the company. This usually involves a decision as to where the company wants to be a number of years in the future and what other competing investment options are available for the funds. This is the case even though the funds are not readily available and will have to be raised by borrowing or other means. Investments always involve projections of what may happen in the future and are thus fraught with uncertainty. Shorter investment time periods generally mean safer and more predictable outcomes. Interest rates vary, and the longer the investment period the greater the uncertainty. Until the early 1970s it was common for capital investment periods in the USA to be 10–15 years or longer. As interest rates rose from 6–8 percent at that time to 18–20 percent in the early 1980s, firms reduced the period

over which they were willing to make investments. Although interest rates have moderated considerably in today's environment, firms continue to require recovery of invested funds in a period of two to three years. This, of course, significantly reduces the number and variety of investment options open to most firms, and makes the capital investment task considerably more difficult.

## 11.2  Cost Analysis and Control

In today's world of rapidly advancing technology and intensifying competition, both worldwide and at the local, state, and national levels, it is essential that industrial organizations have viable and responsive cost analysis and control systems. Effective analysis of product and project costs and the ability to implement cost control measures are two sides of the same coin. The term *management* includes management of the product or project cost, and this requires a knowledge and understanding of the cost elements and their sensitivity to various control parameters. Cost analysis forms the basis for cost control, and without accurate and timely cost data, effective cost control is impossible. Management must have the authority to implement cost control procedures and policies as well. The most accurate and timely cost data are useless unless coupled with an effective cost control mechanism.

In the early stages of a project, cost control is usually accomplished at the work package level. Work packages are commonly identified based on the organizational breakdown, and cost control is accomplished through the appropriate organizational entity when this is true. This may result in some difficulty in controlling cost at the project level, however, since the project manager does not have direct management control over the organizational entity responsible for the work package. An additional problem that occurs at this stage of the project is that cost estimates made up to this time are projections based largely on experience with similar projects. Thus, budget allocations to the various organizational entities for direct labor based on these cost estimates have no relation to the actual work accomplished since they were made before the existing level of effort was initiated. In addition, the contribution of indirect labor to project output is very difficult to measure. For example, the costs of labor associated with management, supervision, personnel, procurement, labor relations, and the like, can be identified accurately from the payrolls of the various departments, but relating the output of this effort to completion of the project defies measurement. Thus, data essential to effective cost control are often lacking at this stage of the project. Because of the lack of definition at this early stage, it is very difficult to determine what percentage of the work has been completed and whether the effort is behind or ahead of schedule. Fortunately, the rate at which costs are incurred early in the project is usually slow, but increasing, and is primarily associated with the cost of labor charged against a particular work package.

As the project evolves, cost data, and the ability to use these data for analysis, continue to improve and to be better defined. Unfortunately (in a cost control sense), the activity and number of personnel involved in the project also increase and, even though better cost data are available, cost control will be even more difficult than at the earlier stages of the project. For large projects, cost control is closely integrated with project planning and the tools of management. When a planning network is utilized to manage the project, all the cost-incurring activities should be included in the network, and changes in the network must provide for adjustments in the project cost. Planning is an activity that looks to the future and tries to predict the most effective way to accomplish the required activities. Control considers what has been accomplished relative to the plan, including the costs that have been incurred, and implements any changes required to complete the effort within cost and schedule.[1] Regardless of the size of the project, some degree of planning and control must be implemented. For small projects, or projects of relatively short life, sophisticated planning and cost control procedures are usually not required (or desired). Nevertheless, basic cost tracking and control principles should be understood and applied as appropriate.

### 11.2.1 Cost Categories

Several cost categories are available to serve as a basis for economic and cost analysis. Classification into these categories provides an understanding of the effect of the various types of costs on the project and helps to ensure that all project costs are accounted for.[2]

*First cost (investment cost).* First cost is the cost of initiating an activity or project. It is usually limited to one-time costs only: those that occur only once for any given undertaking. First cost includes elements such as the cost of equipment, shipping and installation costs, and any required training costs. For an item that is not off-the-shelf, it includes design and development costs and construction or production costs as well as shipping, installation, and training costs. First cost is an important element in the decision-making process since it quantifies the amount of capital required to get the project started and often determines whether a project or activity can be undertaken. Projects that appear to be profitable over a period of time may entail such a high first cost and concomitant level of investment that they are beyond the financial capabilities of the organization.

*Operations and maintenance costs.* Operations and maintenance (O&M) costs are the costs incurred in operating and maintaining entire plants, systems, subsystems, items of equipment, or individual components. This cost category includes labor, fuel, and power costs, materials and supplies, spare parts, repairs, insurance, taxes, and an allocated portion of indirect costs called overhead or burden. O&M costs occur over the life of the item being operated and maintained and usually increase with time. A very common evaluation method for determining when an item should be replaced uses a plot of average maintenance and capital costs versus the economic life of the asset. O&M costs generally increase with the age of the equipment, whereas the cost of investment decreases. The ideal time to replace the asset using this type of analysis is at the minimum life cycle cost.

---

1 Operational Research Society and Institute of Cost and Works Accountants, *Project Cost Control Using Networks* (C. Staffurth, ed.), Heinemann, London, 1969.
2 Fabrycky, W.J., and Blanchard, B.S., *Life-Cycle Cost and Economic Analysis*, Prentice Hall, Englewood Cliffs, NJ, 1991.

*Fixed and variable costs.* Fixed costs are costs that remain relatively constant over the operational life of the facility, system, subsystem, equipment, or component. Fixed costs are independent of the production volume or output and include elements such as depreciation, taxes, insurance, interest on invested capital, sales, and some administrative expenses. Variable costs vary with production volume and can change rapidly, affecting labor costs, materials and supplies, utilities, etc. Fixed costs vary at a much slower rate and thus must be established based on a production volume and mode of operation considered appropriate by management. Fixed costs arise from predictions about where the company wants to be in the future. Equipment purchased now may allow reduction of labor costs in the future or may provide for product improvement or diversification. Research is conducted that is not related to ongoing production in the hope that some future payoff may justify the expenditure. Fixed cost investments are made in anticipation that profit will be increased in the future by reduced variable costs or by increased sales and income.

Variable costs are a function of the level of production or activity. As production levels or activity increase, certain (variable) costs will increase in response. Direct labor, materials, utilities, and other costs that can be allocated on a per production unit basis increase with increased production levels. Materials costs would normally vary directly with the number of units produced, but increased purchases due to the increased production may result in lower costs due to discounts and reduced transportation costs. Labor costs vary based on how much the production line efficiency can be increased with greater volume, use of overtime, additional staffing, etc.

*Incremental or marginal cost.* Incremental and marginal cost both refer to increases in cost. Marginal cost is a term used to describe the situation in which the cost of an incremental level of output is just covered by the income generated. Incremental costs are normally applied to unit costs and are referred to as incremental cost per pound, incremental cost per unit of production, etc. Incremental costs are applied in manufacturing operations to project the added cost of increased production. In this case it is necessary to estimate the cost per unit for increasing the output by some increment.

*Direct and indirect costs.* The costs incurred in all business and industrial operations are assigned to two principal categories – direct and indirect costs. Direct costs include material, labor, and any subcontracts that provide elements used in the product. The quantity and type of direct materials used in manufacturing processes can be determined from the engineering drawing bill of materials. The cost of these materials is obtained from suppliers. Direct labor for manufacturing processes usually includes only the hourly workers involved and excludes salaried personnel, even though they may be supervising and managing the production process. Thus, the indirect labor cost includes supervision and management as well as other support operations labor necessary to produce the product. Indirect material costs include materials used in the overall production operation but not used in the product itself. Other indirect expenses include the cost of utilities, building rentals, depreciation, taxes, insurance, maintenance, etc.

*Recurring and non-recurring costs.* Recurring costs are those that continue to occur over the life of the project and include categories such as manufacturing costs, engineering support required during production or construction, customer support over the life of the product, and ongoing project management. Non-recurring costs are costs that occur only once during the project and include cost categories such as product applied research, design and development, testing (other than acceptance testing), new construction, and manufacturing tools and equipment. When evaluating the costs associated with product and project changes, schedule modifications, and increased or decreased production levels, the evaluation of recurring and non-recurring costs is helpful in providing insight into the actual costs and their source.

When evaluating the life cycle cost of the product, all recurring and non-recurring costs are included. During design and development the costs incurred are primarily non-recurring. As the product moves into the production and utilization phases, recurring costs are incurred. Life cycle cost analysis considers all of these costs, in addition to the investment costs.

*Sunk (past) cost.* Sunk or past costs are project expenditures that have occurred in the past and cannot be recovered by any future action. Since these sunk costs are not relevant to cost analyses concerned with future events, they must be ignored when project decisions are made based on a projection of future events.

### 11.2.2 Cost Estimating

The ability to estimate costs accurately is essential if a firm is to stay in business. Unfortunately, cost estimating is not an exact science and in the best of circumstances will only provide an approximation of the costs that will actually be incurred. It is thus essential that the various methods and tools available for accomplishing this task be understood and be applied so that the approximation will be as accurate as possible. Various methods for estimating costs are described in the following discussion.[2]

*The engineering estimate.* An engineering estimate is made up of the individual costs of many elements of the design at a low level of detail. The estimate includes costs for every element of the design and fabrication process, including the cost of production tools, jigs, and fixtures. Components, lengths of piping and tubing, the number and size of welds, the quantities and types of materials, surface treatments, test requirements, and the like, are identified from the engineering drawings and are multiplied by the element unit cost that may come from vendors, or from production and in-house cost standards. Element costs are determined for each of these elements at the lowest reasonable level of detail. Some elements of the overall cost, such as inspection and production control, are estimated by taking a percentage of direct labor, which invariably results in uncertainty in the final estimate. The overall estimate is made by summing all these individual element costs.

There are two basic drawbacks to the use of engineering estimates – the time required to prepare the estimate and the difficulty in *seeing the forest and not just the trees*. When a cost estimate is made up of many individual cost elements, it is difficult to keep an overall perspective and not get lost in all the detail. When this cost estimating technique is used it is often augmented by the use of other (more gross) techniques to ensure that the overall estimate is valid. Engineering estimates are often used in construction projects, especially when field changes are being negotiated.

*Estimating by use of analogy.* Gross estimates are often made by using cost data from similar programs. When little detailed information is available on which to base the cost, using data from similar programs may be the only way to develop a reasonable estimate. This is the type of estimating that companies use when initiating new product lines or projects. Cost data from the analogous program are adjusted to correspond to changes in performance, numbers of major elements, differences in size, and so forth, for the new program. An example of this type of estimating is documented in the experience of the aircraft companies during the missile development era of the mid-1950s. Since no aircraft company had any experience in the development and production of missiles the extensive cost data that were available from aircraft programs were adapted for use in estimating the cost of missile development and production.

This type of estimating is also used at lower levels of detail such as basing the direct labor hours on the number of hours required to fabricate a similar part or on the cost of direct material.

*Statistical estimating methods.* Statistical cost estimating is based on determining the relationship between cost and the factor(s) on which it depends. Factors such as power rating, flowrate, mass, volume, and production quantities are used to make statistical cost estimates. For example, rough estimates of the fabrication costs for certain materials are estimated based on the mass of the direct material used. Thus, if a product made of steel weighs 1,000 lb, its cost of fabrication might be estimated as $1,000 or $1.00/lb.

This estimating technique is often used for long-range planning. When little information is available, as at the beginning of a project, it will normally be necessary to include a cost for contingencies in the estimate to provide for unforeseen events that result in changes to the design. At this point in the effort, the cost estimate is based on the total cost of the end-product with little or no cost detail below the system or end product level. Cost estimates can be made by using statistical data and can be validated by using cost data from analogous programs or products. As the project evolves and the end product becomes better defined, estimates can be made with less contingency cost, since there is less uncertainty in the overall design and corresponding cost. At this stage, estimates can be based on the costs of the elements included in the overall system by making an engineering estimate. As in the earlier program phase, the engineering estimate can be validated by making an estimate using cost data from analogous programs.

### 11.2.3   Life Cycle Costing

Life cycle costing refers to the cost estimating approach that encompasses the costs of all phases in the life of the project or product, including research and development costs, production costs, operation and support costs, and retirement and disposal costs.[2] In the early stages of a design effort firm cost targets should be established and used to manage the various project phases. One technique that has been used successfully in this regard is the design-to-cost approach whereby cost is established as a project constraint similar to performance or schedule. As with other design and development project performance or schedule requirements, design- to-cost requires the assignment of cost targets to significant elements of the project. The primary difference in design-to-cost management is in the aggressive manner in which these elements are held to these cost targets. Tradeoff studies are used extensively in an attempt to find alternative ways in which effort can be accomplished with reduced cost. System performance requirements are constantly reviewed to identify areas of potential cost reduction. Test requirements are scrutinized to ensure that only essential development testing is accomplished. This often places the program manager at odds with the engineering staff since some testing thought to be essential by engineering may be eliminated by the program manager to reduce costs. When this occurs, the program manager must be especially sensitive to engineering input and try to gain a consensus decision based on the criticality of cost relative to the need for information derived from the test. The development of the NASA Space Shuttle is a notable example of a design-to-cost effort wherein the balance between cost and development testing was managed effectively. Unfortunately, this same effectiveness did not extend to the operations phase of this program. The result was a serious and widely publicized accident that caused a significant launch schedule delay and program cost increase. When the cost of the program is paramount, the use of this management/costing technique is essential, but excellent judgment must be applied in making decisions regarding test and performance tradeoffs, especially when personnel safety is involved.

One method of managing cost in a major program is to use a cost breakdown structure. This ties project activities to available resources by subdividing the total cost into logical categories such as the functional areas and major tasks shown by Figure 11.1. For a program to be managed effectively, the categories shown in Figure 11.1 must be expanded to include greater detail and to a lower (subsystems and component) level, such as that provided by the work breakdown structure shown in Figure 3.5.

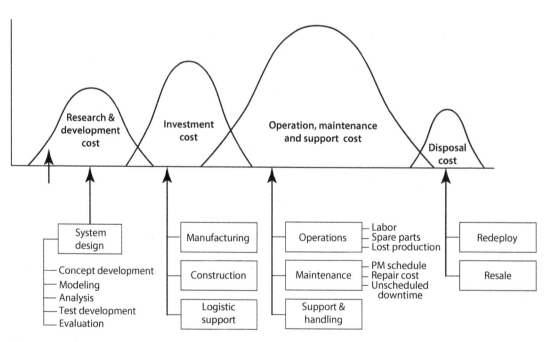

**Figure 11.1** Typical life cycle cost.

The level of definition used in the cost breakdown structure is tailored to the manner in which cost is managed. It must provide the necessary cost data to the organizational level charged with managing each element of the effort. In accomplishing this, the cost breakdown structure should have the following characteristics:[2]

1) All costs associated with the project must be included.
2) Cost categories must be well defined and clearly understood by all project personnel.
3) The cost breakdown structure must be broken down to the level necessary for management to identify areas of high cost and cause-and-effect relationships.
4) The cost breakdown structure must include a cost coding scheme that allows the collection of cost data for analysis of specific areas of interest, such as distribution costs as a function of manufacturing.
5) The cost coding scheme must allow for separation of producer, supplier, and consumer costs expeditiously.
6) The cost breakdown structure must be compatible (through the use of the cost coding scheme and other techniques) with planning documentation, the work breakdown structure, the organizational structure, scheduling techniques, and management information systems used.

Neglecting the time value of money, the life cycle cost of an asset can be determined by

$$\text{ALCC} = \frac{P}{n} + O + (n-1)\frac{M}{2}, \tag{11.1}$$

where

$\text{ALCC}$ = average annual life cycle cost,

$O$ = constant annual operating cost (equal to first-year operating

cost including a portion of maintenance),

$M$ = annual increase in maintenance costs,

$n$ = life of the asset in years,

$P$ = first cost of asset.

The minimum-cost life can be found by differentiating equation (11.1) and setting it equal to zero:

$$\frac{d\text{ALCC}}{dn} = -\frac{P}{n^2} + \frac{M}{2} = 0. \tag{11.2}$$

Hence,

$$n^* = \sqrt{\frac{2P}{M}}. \tag{11.3}$$

Substituting equation (11.3) into equation (11.1) yields the minimum annual life cycle cost as

$$\text{ALCC}^* = \sqrt{2PM} + O - \frac{M}{2}. \tag{11.4}$$

As an example of the use of equation (11.3), consider the purchase of a new automobile with a first cost of $12,150 and a first-year operating and maintenance cost of $1,000, with maintenance costs increasing by $300/year. The minimum-cost life from equation (11.3) is

$$n^* = \sqrt{\frac{2(12,150)}{300}} = 9 \text{ years.}$$

The minimum annual life cycle cost for this period of time using equation (11.4) is

$$ALCC^* = \sqrt{2(12,150)(300)} + 1,000 - 150$$
$$= \$3,550.$$

Unfortunately, annual maintenance cost increases do not usually change uniformly as required when using equations (11.3) and (11.4). For situations in which the annual maintenance cost increases vary and to account for the salvage value of the investment the tabular approach *must* be used.

## 11.3 Important Economic Concepts

When the consequences of economic decisions are immediate, or occur over a short period of time, the change in the value of money with time does not have to be accounted for. The positive and negative aspects of the decision can be evaluated and the decision can be made and implemented. Most business economic decisions involve the element of time, however, and it is thus necessary to consider factors such as interest, the time value of money, and depreciation of equipment.

### 11.3.1 Interest

Interest can be thought of as the rental charge levied by financial institutions on the use of money. Like other rental charges, the level or rate of interest is determined by supply and demand. If individuals as a whole increase the portion of their earnings put into savings the supply of money will increase and rates will generally go down. If savings are reduced, the amount of money available for financial institutions to provide for loans will decrease and rates will normally rise. Governments usually believe that it is in the best interest of their citizens to influence the money market, however. Thus, if the US government believes that the economy is sliding into recession, it may, through the Federal Reserve Board, reduce the interest rate charged to member banks (referred to as the *prime rate*) and thus strongly influence the rate to remain low to stimulate the economy. In this sense the money market is unlike other free market commodities that operate unencumbered by government influence.

The interest rate is the ratio (expressed as a percentage) of the rental charge on the borrowed money to the total amount of money borrowed over a period of time, usually one year. For example, if $100 is charged for the use of $1,000 for one year the interest rate is 10 percent per annum. For simple interest, the amount of interest paid on repayment of a loan is proportional to the length of time that the principal sum has been borrowed. Simple interest can be calculated using

$$I = Pni, \tag{11.5}$$

where

$I =$ interest earned,

$P =$ amount of money loaned (principal),

$n =$ interest time period, usually years,

$i =$ interest rate.

**Simple interest.** A simple interest loan can be negotiated for any period of time. The principal and interest become due only at the end of the loan period, which can be for less than one year or for multiple years. For loans of less than one year the interest would be calculated using equation (11.5) with $n$ equal to the loan period in days divided by the number of days in a year.

### Example 11.1

Assume that a college student borrows $1,000 at a simple interest rate of 10% per year for 5 years. Calculate the interest charge that will be paid at the end of 5 years.

#### Solution

From equation (11.5) the interest charge can be calculated as

$$I = Pni$$
$$= (\$1000)(0.10)(5).$$
$$= \$500$$

At the end of the fifth year the total amount required to be paid is $1,500.

**Compound interest.** When a loan is made that covers several interest periods, interest is determined at the end of each period. This interest could be paid at the end of each period or could be accumulated until the loan is due for repayment. If the interest is accumulated and the borrower is charged interest on the total amount owed, including principal and interest, the interest is said to be compounded.

### Example 11.2

Assume that, in Example 11.1, the student wants to pay the interest at the end of fifth year. Calculate the total amount that has to be paid at the end of the fifth year.

#### Solution

Table 11.1 shows the total amount that has to be paid is $1,610.51 when the loan is terminated. The difference of $110.51 is because of the compounding of interest over the five years.

**Table 11.1** Calculation of interest when interest is paid annually.

| Year | Amount owed at beginning of year A | Interest charge for year B | Amount owed at end of year A + B |
|---|---|---|---|
| 1 | $1,000 | $1,000 × 0.10 = 100 | $1,100.00 |
| 2 | $1,100 | $1,100 × 0.10 = 110 | $1,210.00 |
| 3 | $1,210 | $1,210 × 0.10 = 121 | $1,331.00 |
| 4 | $1,331 | $1,331 × 0.10 = 133.10 | $1,464.10 |
| 5 | $1,464.10 | $1,464.10 × 0.10 = 146.41 | $1,610.51 |

### 11.3.2 Cash Flow

A cash flow diagram provides a graphical description of an alternative's receipts and disbursements over a selected period of time. Receipts are reflected by upward-pointing arrows (an increase in cash) and disbursements are shown by downward-pointing arrows (a decrease in cash). Figure 11.2 shows cash flow diagrams of Example 11.2 for both the borrower and the lender for a loan of $1,000. The borrower receives $1,000 initially at time $t = 0$ and pays $100, $110, $121, $133.10 for 4 years. The $1,000 principal is also repaid at the end of the loan period with the last interest payment of $146.41. The lender experiences a negative cash flow initially as shown by the downward-pointing arrow and receives the $100, $110, $121, $133.10 in each of the first four years. The lender also receives the fifth year's interest in addition to the $1000 principal at the end of the loan agreement.

If both negative and positive cash flows are experienced during an interest period, the net value is shown and the length and direction of the arrow for that interest period are determined by the algebraic sum of the positive and negative cash flows. Disbursements made to initiate an investment are shown at the beginning of the investment period. Receipts and disbursements that occur during the life of the investment are considered to have transpired at the end of the period in which they occur.

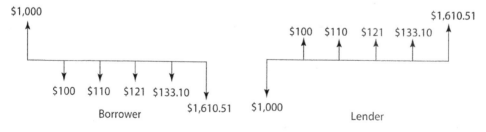

**Figure 11.2** Cash flow diagrams.

### 11.3.3  Present Worth Analysis

Present worth is the amount of money that a future sum of money is worth today. The present worth amount at an interest rate $i$ over a period of $n$ years is found from

$$P(i) = \sum_{t=0}^{n} F_t(1 + i)^{-t} \tag{11.6}$$

where

$P(i)$ = present worth (value) at interest rate $i$, and

$F_t$ = future amount at time $t$, years.

The present worth amount includes the time value of money through the value of $i$ selected. It also provides an equivalent value of the net cash flow at any point in time from $t = 0$ to $t = n$. A single unique value for the present worth is associated with each interest rate used no matter what the cash flow pattern may be. Present worth is thus a valuable tool in determining the viability of alternatives in the decision-making process.

For a single payment, to calculate present value $P(i)$ for a given future amount $F_t$, equation (11.6) can be modified to

$$P(i) = F_t \left[\frac{1}{(1 + i)^t}\right] = F(P/F, i\%, t) \tag{11.7}$$

where $(P/F, i\%, t)$ is called present worth factor.

### ☐ Example 11.3

Calculate the present value of \$3,000 to be received 5 years later based on the market interest rate of 10%.

#### Solution

From equation (11.7), the present value is

$$P(i) = F(P/F, i\%, t)$$

where

$$(P/F, i\%, t) = \left[\frac{1}{(1 + i)^t}\right] = \left[\frac{1}{(1 + 0.10)^5}\right] = 0.62.$$

Then

$$P(i) = \$3000(0.62) = \$1,862.20.$$

When the present value, $P$, is given, the future value, $F$, can be calculated from

$$F_t = P(i)(1 + i)^t = P(i)(F/P, i\%, t), \tag{11.8}$$

where $(F/P, i\%, t)$ is called the compound amount factor.

### Example 11.4

Calculate the future value of $1,000 in 5 years.

### Solution

From equation (11.8), the future value is

$$F_t = P(i)(F/P, i\%, t).$$

where

$$(F/P, i\%, t) = (1 + i)^t = (1 + 0.10)^5 = 1.6105.$$

Then

$$F_t = \$1000(1.6105) = \$1,610.50.$$

If both $P$ and $F$ are given for a given interest rate, the number of years can be calculated from equation (11.8):

$$\frac{F_t}{P(i)} = (1 + i)^t. \tag{11.9}$$

Taking natural logarithms of equation (11.9) gives

$$\ln\left[\frac{F_t}{P(i)}\right] = t \ln(1 + i) \tag{11.10}$$

or

$$t = \frac{\ln[F_t/P(i)]}{\ln(1 + i)}. \tag{11.11}$$

### Example 11.5

In Example 11.4, how many years are required for a $1,000 investment to be worth $2,000?

### Solution

From equation (11.11), the the number of years is

$$t = \frac{\ln[F_t/P(i)]}{\ln(1 + i)}$$

$$= \frac{\ln[\$2,000/\$1,000]}{\ln(1 + 0.10)} = \frac{\ln 2}{\ln 1.1}$$

$$= \frac{0.6931}{0.0953} = 7.3 \text{ years.}$$

The interest rate $i$ can also be calculated using equation (11.8) as

$$i = \sqrt[t]{F_t/P(i)} - 1. \tag{11.12}$$

For example, at what interest rate does a $2,000 investment yield $3,000 in 8 years? Equation (11.12) yields 5.2%.

### 11.3.4 Payback Period

When the required rate of return on an investment is known, the length of time needed for the investment to pay for itself can be determined. Investments that tend to pay for themselves in short time periods are more desirable than long-period investments since the return of capital investment is quicker and there is less uncertainty with the shorter time period. If the net cash flows (revenues) are the same each period, then the payback period is given by

$$\text{Payback period} = \frac{\text{Investment}}{\text{Net cash flow}}. \tag{11.13}$$

For example, a carburetor compressed natural gas (CNG) conversion kit for eight-cylinder engines costs $600 and saves $40 per month. Then the payback period is $600/$40 per month, or 15 months.

Payback periods are often quoted without considering the time value of money ($i = 0$ percent). This approach may be warranted in situations where uncertainty as to the appropriate interest rate is high, but one should be aware of the limitations of this method of analysis. For this case, the following equation can be used to calculate the minimum number of periods, or years, required to recover the initial investment:

$$\sum_{t=0}^{n} F_t \geq 0. \tag{11.14}$$

Table 11.2 shows the cash flow for two investment alternatives for which the payback period is 3 years. Although the payback period is identical, the present worth of these investments varies considerably. The use of payback analysis with zero interest has serious drawbacks.

To include consideration of the time value of money, a method known as the discounted payback period can be used. This method allows the determination of the time required for an investment's present equivalent receipts to equal or exceed the present equivalent disbursements. Using $F_t$ and $t$ as defined previously, the discounted payback period is the smallest value of $n^*$ that satisfies the expression

$$\sum_{t=0}^{n^*} F_t \frac{1}{(1+i)^t} \geq 0. \tag{11.15}$$

Using an interest rate of 10 percent for alternative A in Table 11.2,

$$-\$800 + \$400 \left[\frac{1}{(1+0.1)^1}\right] + \$200 \left[\frac{1}{(1+0.1)^2}\right] + \$200 \left[\frac{1}{(1+0.1)^3}\right]$$

$$+ \$200 \left[\frac{1}{(1+0.1)^4}\right] \geq 0.$$

$$\$15.79 \geq 0$$

**Table 11.2** Two alternatives with a three-year payback period.

| End of year | A | B |
|---|---|---|
| 0 | −$800 | −$1,400 |
| 1 | 400 | 600 |
| 2 | 200 | 400 |
| 3 | 200 | 400 |
| 4 | 200 | 600 |
| 5 | 200 | 700 |
| Payback period | 3 years | 3 years |
| Interest rate, $i = 0$ | | |

Thus, the shortest time period that will satisfy the inequality is $n^* = 4$ years. This compares with a period of 3 years when the time value of money is not accounted for. At an interest rate of 10 percent, the discounted payback period for investment B in Table 11.2 can also be determined to be 4 years. The importance of using the discounted payback period method is thus apparent whenever the interest rate can be estimated, even approximately.

### 11.3.5 Depreciation and Taxes

Depreciation can be defined as the reduction in the value of an asset with the passage of time. Thus, an income-producing asset can be depreciated over its expected life to reflect the reduced ability to perform its intended service. The primary causes of physical depreciation are deterioration because of the action of the elements and wear and tear from use. Functional depreciation is the reduction in value of an asset as a result of changed demand for the services the asset performs. This can come about as a result of technology improvement that causes the asset to become obsolete, as a result of a change in the type of work required, or as a result of the need to increase capacity beyond the capability of the asset.

If an asset is used to produce income, depreciation can be taken into account for tax purposes. Assets can be composed of either tangible or intangible property. Intangible property includes such things as designs, licenses, patents, and copyrights. A deduction for depreciation of such assets can be taken whether or not income was actually derived from the asset. Thus, tax law in the United States is structured to encourage investment. Unfortunately, changes in the allowable depreciation schedules in 1987 seriously reduced the flexibility in use of the various depreciation options and seriously reduced the advantage of investment in some types of assets. At present the only options allowable by the federal government under the current Modified Accelerated Cost Recovery System are to use the specified depreciation schedule provided by the Internal Revenue Service.

### 11.3.6 Inflation and Deflation

Inflation and deflation are terms used to describe the change in price levels of goods and services with time. Inflation has been a much more common occurrence in world economies during the past half century, but deflation was a major malady during the 1920s and 1930s. The cost of goods and services is driven by various factors. When the supply of goods and services increases without a corresponding increase in demand, prices go down. In general, the opposite is true when supply decreases, but in recent years government policies also have had a significant and overriding inflationary effect. Government policies that tend to reduce the value of the dollar, such as price supports, deficit financing, and increasing the money supply, all result in inflationary pressure on the economy. The embargo placed on the supply of oil by the oil-producing nations in the early 1970s is a good example of how reduced supply can affect the price of goods. The price of oil experienced a tenfold increase in price in a matter of a few months as a result of this action, which was strictly arbitrary and not a result of any systemic cost increases.

A price index is used to provide a way to account for these price changes over time. A price index is the ratio of the price of a commodity at a point in time relative to the price at some earlier time. The federal government prepares various composite price indexes including the Consumer Price Index, the Producer Price Index, and the Implicit Price Index for the Gross National Product, which measure historical price level changes in the economy. Figure 11.3 depicts the Consumer Price Index for the United States for the years 1950–2015.

The price index is usually expressed as a number corresponding to the ratio of prices for the years of interest multiplied by 100. For example, if the price of a loaf of bread was $0.25 in 1967 and in 1991 the price was $1.19, the price index would be ($1.19/$0.25)(100) = 476, which indicates that the 1991 price is 476 percent times the price in 1967.

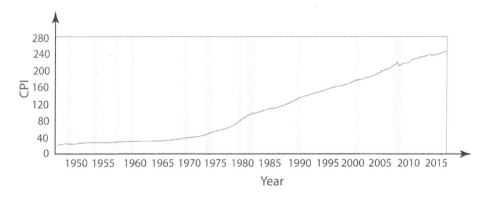

**Figure 11.3** Consumer Price Index (from Economic Research, Federal Reserve Bank of St. Louis).

The effect of inflation (or deflation) on the rate of return for an investment depends on how the future returns respond. If the return is in constant dollars that are not increased to reflect inflation, the effect will be to reduce the before-tax rate of return. If the amount of the return increases to keep up with inflation, the before-tax rate of return will not be affected by inflation. This will not be the case when an after-tax analysis is made. Even when the future returns increase to reflect inflation, the allowable depreciation schedule will not change, and the result will be increased taxable income. This obviously reduces the after-tax return on the investment and points out the importance of accounting for inflation when evaluating capital expenditure proposals.[3]

## 11.4   Selecting an Appropriate Rate of Return

Economic alternatives available to a firm must be compared with other available investment opportunities so that the maximum return is ensured. To allow a meaningful comparison, a minimum attractive rate of return (MARR) must be established that should be equal to the greatest of (1) the cost of borrowing money, (2) the cost of capital, and (3) the rate of return that the firm can earn from other investments. It is obvious that money should not be borrowed at 10 percent to fund a project with only an 8 percent return on investment. Also, the return on investment of a project should not be less than the cost of raising funds by capital restructuring. The ratio between capital and debt is significant to a firm's ability to borrow money and therefore must be actively managed so that banks and other financial entities will make favorable evaluations of the firm's solvency. Finally, business organizations usually have several options or alternatives for investment. One alternative is to not invest in any project but to place surplus funds in financial instruments totally unrelated to the firm's primary business and withdraw them when more attractive investment opportunities present themselves. This is known by economists as the *do-nothing alternative.* When money is invested in this manner it is not considered to be idle. Rather, the assumption is made that the funds yield a rate of return equal to the MARR and would be withdrawn and invested in a project only if the opportunity offered a higher rate of return.

Investments are always concerned with future events that cannot be predicted accurately, therefore probability, risk, and uncertainty should always be considered. If reasonable probabilities can be established for the future outcomes of an investment opportunity, then expected values for each outcome can be determined and the investment can be evaluated accordingly. This is known as risk analysis, where there are two or more possible outcomes and the probability of each outcome is known. The expected value for an investment can be determined using[3]

$$\text{Expected value} = \text{outcome } A \times P(A) + \text{outcome } B \times P(B) + \dots, \tag{11.16}$$

where outcome $A$ ($B$) is the value of outcome $A$ ($B$), and $P(A)$ ($P(B)$) is the probability of $A$ ($B$).

---

3 Newnan, D.G., *Engineering Economic Analysis*, Engineering Press, San Jose, CA, 1988.

## Example 11.6

A friend wants to bet $10 on which football team will win the Southwest Conference in 2018. The probabilities of the likely contenders winning are considered to be as shown in Table 11.3. As in horse racing, the outcome represents the $10 bet plus the amount won. What is the expected value of a bet on Texas A&M to win the Conference?

### Solution

$$\text{Expected value} = \text{Outcome if A\&M wins} \times P(\text{A\&M winning})$$
$$+ \text{Outcome if A\&M loses} \times P(\text{A\&M losing})$$
$$= \$15.00(0.40) + \$0(0.60) = \$6.00.$$

The expected value of the $10 bet on Texas A&M is thus equal to $6.00.

**Table 11.3** Probabilities of the likely contenders winning.

| Team | Probability of winning | Outcome of bet if you win |
|------|------------------------|---------------------------|
| Texas Tech | 0.25 | 25.00 |
| Texas | 0.20 | 30.00 |
| Texas A&M | 0.40 | 15.00 |
| Houston | 0.15 | 40.00 |

The lowest rate at which firms can borrow money is known as the prime interest rate. This is the rate that banks and other lending institutions charge their best customers for the use of money. It varies over time and is widely reported in the news media. It is primarily a function of the interest rate that the Federal Reserve charges member banks and is one of the principal ways in which the federal government manages and controls the US economy.

The cost of capital for a corporation is conventionally higher than the cost of borrowed money. The cost of capital must include consideration of the market value of the common and other stock of the firm, which varies widely. The return on total capital earned by the corporation has an impact on the market value of the stock as well as the cost of borrowed money. *Fortune* magazine reports that the after-tax rate of return on investment for individual firms averages 8 percent. The after tax rate of return on common stock and retained earnings as reported by *Business Week* magazine averages 14 percent.[3] Firms that are struggling for survival cannot invest funds in any but the most critical projects with high rates of return and for short periods of time, often with payback periods of 1 year or less. For a project with a life of several years a very high rate of return would be required. Struggling firms also cannot usually borrow money at the prime rate because of their weak financial condition. These conditions result in the need for such firms to establish a high MARR, which limits the number of investment opportunities and usually results in riskier investments. The more stable companies, which make up the bulk of all industrial enterprises, take a longer-range view of their capital investments. With a greater supply of funds to work with they can invest in projects that the struggling firm cannot consider. With a solid financial foundation these firms can also qualify for lower interest rates when borrowing money, allowing them to operate with a lower MARR. For projects requiring only a small investment, payback periods are usually short, normally 1–2 years. Larger investments are analyzed by evaluating the rate of return. For projects with average risk, the after-tax MARR for these firms typically varies from 12 percent to 15 percent. For projects with greater risk, the MARR will be considerably higher, depending on the degree of risk. These values for the MARR are obviously opportunity costs rather than the cost of borrowed money or the cost of capital, which indicates that these firms have adequate high rate of return opportunities and have not been required to consider projects with lower rates of return nearer to the cost of borrowed money or the cost of capital. Although this may be interpreted to mean that good projects are going unfunded, it reflects the fact that most firms are reluctant to invest in projects expected to earn only slightly more than the cost of borrowing money or the cost of capital because of the inherent risk and uncertainty about the future.

## 11.5   Evaluation of Economic Alternatives

To evaluate engineering proposals, decision criteria must be established so that the proposal that best satisfies the desired objectives can be selected. The option of rejecting all proposals under consideration, or *doing nothing*, should also be retained as a possibility. The decision criterion adopted will drive the selection but will not ensure that company objectives will be realized unless they are thoroughly understood. Decision criteria can be based on the following:

1) The economical equivalence of a present, annual, or future investment cash flow
2) Optimization of selection decision variables that affect the investment cost, periodic costs, or project life
3) The rate of return on incremental investment.

In all of these approaches, a decision must be made as to whether to consider risk and uncertainty or to make the simplifying assumption that the future is known with certainty. Like other decision-making processes, the decision matrix is a valuable tool to assist in evaluating the various alternatives. By using a decision matrix (see Chapters 2 and 3), the interaction between a number of alternatives can be evaluated relative to a wide variety of future outcomes over which the decision-maker has no control.

### 11.5.1   Evaluation by Economic Equivalence

Evaluation by economic equivalence is represented by the function[2]

$$PE, AE, \text{ or } FE = f(F_t, i, n). \tag{11.17}$$

**Table 11.4** Differences between mutually exclusive alternatives.

| End of year | Alternatives $A_1$ | Alternatives $A_2$ | Differences $(A_2 - A_1)$ |
|:---:|:---:|:---:|:---:|
| 0 | −$2400 | −$3200 | −$800 |
| 1 | $800 | $400 | −$400 |
| 2 | $800 | $1400 | $600 |
| 3 | $800 | $1400 | $600 |
| 4 | $800 | $1400 | $600 |
| 5 | $800 | $1400 | $600 |

The present equivalent, annual equivalent, and future equivalent amounts provide adequate bases for evaluation of a single alternative or for the comparison of mutually exclusive alternatives. Although few proposals are totally independent of each other, they can usually be arranged into mutually exclusive alternatives for evaluation purposes. Mutually exclusive proposals can be compared by evaluating their differences. This is demonstrated in Table 11.4.[4]

To compare the two alternatives, $A_1$ and $A_2$, the cash flow difference is determined. The cash flows representing alternatives $A_1$ and $A_2$ are shown in Figure 11.4(a,b). The cash flow shown in Figure 11.4(c) represents the difference between $A_1$ and $A_2$. The decision to undertake alternative $A_2$ instead of $A_1$ requires an additional investment of $800 now and $400 one year

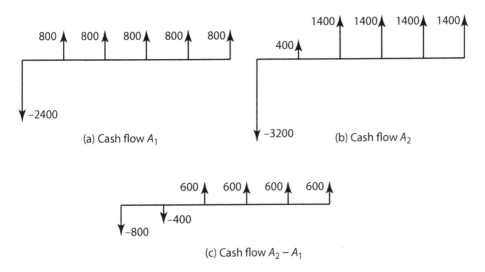

(a) Cash flow $A_1$

(b) Cash flow $A_2$

(c) Cash flow $A_2 - A_1$

**Figure 11.4** Cash flow differences between two alternatives. (Adapted from Thuesen, G.J., and Fabrycky, W.J., *Engineering Economy*, Prentice Hall, Englewood Cliffs, NJ, 1989.)

4 Adapted from Thuesen, G.J., and Fabrycky, W.J., *Engineering Economy*, Prentice Hall, Englewood Cliffs, NJ, 1989.

hence. The receipts from this extra investment amount to $600 at the end of years 2, 3, 4, and 5. The question that must be answered is whether the extra receipts justify the extra investment.

## 11.5.2 The Economic Optimization Function

The use of this method allows evaluation of both economic equivalence and economic optimization, which is appropriate when investment cost, periodic costs, and/or project life are functions of the decision variables. An economic optimization function is a mathematical model that links an evaluation measure $E$ with controllable decision variables $X$ and uncontrollable system parameters $Y$. It thus allows for evaluating decision variables in the presence of system parameters by using a mathematical test that results in an optimized value for $E$. This relationship is expressed as

$$E = f(X, Y). \tag{11.18}$$

An example of this method of evaluation is the determination of an optimal procurement quantity for inventory. The evaluation measure in this case is cost, and the objective is to select a procurement quantity based on demand, procurement cost and warehouse operational cost so that the total cost is minimized. Procurement quantity is directly under the control of the decision maker, but demand, procurement cost, and warehousing costs are not. By applying the optimization function method, the decision maker can determine the procurement quantity that trades off the other cost elements and results in a minimum total cost.

This approach can also be extended to operational and design decisions involving alternatives. In this application, decision-dependent system parameters $Y_d$ must be isolated from decision-independent parameters $Y_i$. Equation (11.18) can then be rewritten as

$$E = f(X, Y_d, Y_i). \tag{11.19}$$

An application of this version of the decision evaluation function is the establishment of a procurement and inventory system for an item that is available from several sources. The procurement level and quantity would be the decision variables in this case. For each separate source there would be a set of source-dependent parameters including the unit cost, procurement cost per procurement, replenishment rate, and procurement lead time. Uncontrollable system parameters would include the demand rate, warehousing cost per unit per period, and shortage penalty cost. The objective is to minimize the total system cost by proper selection of procurement level, quantity, and source.[2]

## 11.5.3 Evaluation Using the Rate of Return on Incremental Investment

In evaluating alternatives using this method, the rate of return on the increment or difference cash flows between alternatives is compared with the MARR. Alternatives are usually arranged in order of increasing first-year cost and are analyzed in pairs, beginning with the first two alternatives. The most favorable alternative from this initial analysis is then compared with the next alternative, and the process is continued until all alternatives have been evaluated. The analysis is accomplished as follows:[4]

1) Compute the rate of return (ROR) for each alternative. Eliminate any alternatives with ROR less than the MARR.
2) Arrange the remaining alternatives in increasing order of first-year cost.
3) Make a two-alternative analysis of the first two alternatives by using

$$\begin{bmatrix} \text{Higher-cost} \\ \text{Alternative } Y \end{bmatrix} = \begin{bmatrix} \text{Lower-cost} \\ \text{Alternative } X \end{bmatrix} + \begin{bmatrix} \text{Difference} \\ (Y - X) \end{bmatrix}. \tag{11.20}$$

Compute the incremental rate of return ($\Delta$ROR) on the increment of investment $(Y - X)$ and apply the following test:
- If $\Delta$ROR $\geq$ MARR, retain the higher cost alternative $Y$.
- If $\Delta$ROR $<$ MARR, retain the lower cost alternative $X$.
- Reject the other alternative used in the analysis.
4) Using the preferred alternative from step 3, and the next alternative in the list from step 2, proceed with another two-alternative analysis.
5) Continue this process until all alternatives have been evaluated and the best alternative has been identified.

## ☐ Example 11.7

The five mutually exclusive alternatives listed below have 10-year useful lives. Which alternative should be selected if the MARR is 8 percent?

- Alternative 1 requires an initial expenditure of $5,000 and has an annual benefit (receipts less disbursements) of $885.
- Alternative 2 requires an initial expenditure of $3,000 and has an annual benefit of $488.
- Alternative 3 requires an initial expenditure of $6,000 and has an annual benefit of $1,018.
- Alternative 4 requires an initial expenditure of $1,000 and has an annual benefit of $149.
- Alternative 5 requires an initial expenditure of $10,000 and has an annual benefit of $1,424.

## Solution

Step 1 Calculate the rate of return for each alternative and eliminate any that have ROR less than the MARR. To calculate the ROR for each alternative, use the following equation and solve for the interest rate $i$ using a trial-and-error method:

$$\frac{P}{A} = \left[ \frac{(1 + i)^n - 1}{i(1 + i)^n} \right] = \frac{\$5,000}{\$885} = 5.65$$

or

$$5.65i(1 + i)^{10} = (1 + i)^{10} - 1.$$

Assume that $i = 12$ percent:

$$5.65(0.12)(1 + 0.12)^{10} = (1 + 0.12)^{10} - 1$$
$$2.1058 = 2.1058;$$

therefore,

$$\text{ROR} = 12\%.$$

Using the same procedure the ROR for the other alternatives can be determined.

|  | 4 | 2 | 1 | 3 |
|---|---|---|---|---|
| Cost | $1000 | $3000 | $5000 | $6000 |
| Uniform annual benefits | 149 | 488 | 885 | 1018 |
| Rate of return | 8% | 10% | 12% | 12% |

Step 2  Arrange the remaining alternatives in increasing order of first-year cost. Since the ROR for alternative 5 (7 percent) is less than the MARR it can be eliminated from further consideration.

Step 3  Make a two-alternative analysis beginning with the first two alternatives. From analysis of increment 2 − 4, a ΔROR of 11 percent is computed; thus alternative 2 is preferred over alternative 4. When increment 1 − 2 is evaluated, a ΔROR of 14.9 percent is realized; thus alternative 1 is preferred over alternative 2. Since the 3 − 1 increment has a rate of return less than the MARR, alternative 3 can be discarded, which leaves alternative 1 as the best alternative and the one that would be selected for investment. Note that if adequate resources were available, there were no other mitigating circumstances, and these were the only investment options, alternatives 2, 3, and 4 would be undertaken as well.

|  | Increment 2 − 4 | Increment 1 − 2 | Increment 3 − 1 |
|---|---|---|---|
| Δ Cost | $2000 | $2000 | $1000 |
| Δ Annual benefit | 339 | 397 | 133 |
| Δ Rate of return | 11% | 14.9% | 5.6% |

## Example 11.8   *Cost Analysis for the Conversion of a School Bus Fleet to Compressed Natural Gas*[a]

This example was developed for a fleet of 80 gasoline-fueled busses that are being considered for conversion to CNG. The busses average 16,000 miles per year. The cost analysis is based on a facility life of 30 years with the compressor and conversion kits requiring overhauls at 10-year intervals. To evaluate this investment the payback period and percent worth of the investment for an interest rate of 10 percent is determined.

## Solution

The initial costs were estimated and are shown in Table 11.5. Due to the fact that interest accumulates on all monetary transactions over the life of the project all receipts and expenditures must be valued over the same period of time. Transactions are usually evaluated at their present worth. Since the compressor requires overhaul at 10-year intervals, this is a cost ($6,000/10 year) that must be included in the evaluation. The cost of overhaul for conversion is $128,000/10 year (see Table 11.5). The present value of this future expenditure is determined using the equation for present worth for a single payment as

$$P(i) = F_t(1 + i)^{-t}$$
$$PW(10\%) = \$6,000(1 + 0.10)^{-10}$$
$$= \$6,000(0.3855)$$
$$= \$2,313.$$

The present worth of the compressor overhaul at 20 years is calculated in the same manner with $t = 20$. Conversion kits are essentially replaced every 10 years and are assumed to have a salvage value of $80,000 at that time.

Residual values are estimated and the present worth is calculated in the same manner as capitalized costs since they are also considered to be single receipts that occur at the time of the present worth of the facility life (see Tables 11.6 and 11.7). For the compressor the salvage value of $20,000 is determined to be

$$P(10\%) = \$20,000(1 + 0.10)^{-30}$$
$$= \$20,000(0.05731)$$
$$= \$1,146.$$

**Table 11.5** Initial costs.

| Item | Life (years) | Initial cost |
|---|---|---|
| Land | — | $1,200 |
| Compressor | 30 | $150,000 |
| Storage tank(s) | 30 | $30,000 |
| Supply facilities | 30 | $80,000 |
| Elect. power to facility | — | $1,200 |
| Site preparation | — | $4,000 |
| Pipeline installation | — | $4,000 |
| Conversion kits (including labor) $1,600 ea | 10 | $128,000 |
| Vehicle fuel tanks $1,200 ea | 30 | $96,000 |
| **Total initial cost** | | $494,400 |

**Table 11.6** Capitalized costs.

| Item | Years | Cost of overhaul | Present worth |
|------|-------|------------------|---------------|
| Compressor: 1st o'haul | 10 | $5,000 | $2,313 |
| Compressor: 2nd o'haul | 20 | $5,000 | $892 |
| Conversions: 1st o'haul | 10 | $128,000 | $49,344 |
| Conversions: 2nd o'haul | 20 | $128,000 | $19,021 |
| **Total capitalized cost** | | | $71,570 |

**Table 11.7** Residual values.

| Component or system | Estimated salvage value | Present worth |
|---------------------|-------------------------|---------------|
| Compressor (30 years) | $20,000 | $1,146 |
| Main storage tanks(30 years) | $5,000 | $287 |
| Refueling facilities (30 years) | $5,000 | $287 |
| Vehicle kits (10 years) | $80,000 | $30,840 |
| Other (kits at 20 years) | $80,000 | $11,891 |
| **Total present worth of residual values** | | $44,451 |

The total capitalized cost (TCC) is the initial cost plus the capitalized cost minus the present worth of the residual values and land:

$$TCC = \text{Initial cost} + \text{Capitalized cost} - \text{Residual values} - \text{Land cost}$$

$$= \$494,400 + \$71,570 - \$44,451 - \$1,200$$

$$= \$520,319.$$

The total annual savings, $A$, is the sum of these annual fuel savings (Table 11.9) and the additional annual savings (Table 11.10) less the annual operating costs (Table 11.8):

$$A = \text{Annual fuel savings} + \text{Other annual savings} - \text{Annual operating costs}$$

$$= \$131,556 + \$9,480 - \$6,320$$

$$= \$134,716.$$

**Table 11.8** Annual operating costs.

| Item | Annual costs |
|---|---|
| Interest on land | $120 |
| Compressor maintenance and repair (not electrical) | $200 |
| Refueling station maintenance and repair | $500 |
| Additional electrical power costs | $5,500 |
| **Total annual operating cost** | **$6,320** |

**Table 11.9** Annual fuel savings for 80 busses.

| Fuel | Miles/year | Avg MPG | Cost/gal | Cost/year |
|---|---|---|---|---|
| Gasoline | 16,000 | 4.0 | $2.30 | $736,000 |
| CNG | 16,000 | 3.6 | $1.70 | $604,444 |
| **Total annual fuel savings** | | | | **$131,556** |

**Table 11.10** Additional annual savings for 80 busses.

| Saving per vehicle | No vehicle saving | Savings cost |
|---|---|---|
| Increased savings, engine life | $0.00 | $0 |
| Increased savings, engine Oil | $30.00 | $2,400 |
| Increased savings, oil filters | $25.00 | $2,000 |
| Increased savings, spark plugs | $3.50 | $280 |
| Increased savings, tune ups | $0.00 | $0 |
| Increased savings, exhaust system life | $60.00 | $4,800 |
| **Total additional annual savings** | | **$9,480** |

To determine the payback period the equation for present worth for an equal payment series[5] can be solved by trial and error for $n$ using the known ratio

$$\frac{P(i)}{A} = \left[\frac{(1+i)^n - 1}{i(1+i)^n}\right].$$

(11.21)

___

5 Thuesen, G.J., and Fabrycky, W.J., *Engineering Economy*, Prentice Hall, Englewood Cliffs, NJ, 1989.

Rearranging yields

$$(1+i)^n = \frac{1}{k},$$ (11.22)

where

$$k = \left[1 - \frac{P(i)i}{A}\right]$$ (11.23)

$$= \left[1 - \frac{(611,621)(0.10)}{(134,716)}\right] = 0.614.$$

Taking the natural logarithm of each side of equation (11.22) and solving for $n$ gives the relationship

$$n = \frac{\ln(1/k)}{\ln(1+i)}$$ (11.24)

$$= \frac{\ln(1/0.614)}{\ln(1+0.10)} = 5.12 \text{ years.}$$

Note that the annual savings must be smaller than the present worth, otherwise the natural logarithm of the numerator cannot be computed. Also the interest rate must be less than 100 percent.

The present value of the total investment can also be determined to evaluate the cost-effectiveness of the conversion project. To determine the present value of an annual receipt of $134,716 use the equation for the present value of an equal payment series:

$$P(i) = A\left[\frac{(1+i)^n - 1}{i(1+i)^n}\right]$$

$$= \$134,716\left[\frac{(1+0.10)^{30} - 1}{0.10(1+0.10)^{30}}\right]$$

$$= \$1,269,954.$$

The present worth, $PW$, of the investment is now be determined to be

$$PW = \$1,268,954 - \$520,319$$

$$= \$749,635,$$

which makes the investment attractive if other, more profitable investments are not available.

---

[a]Adapted from Maxwell, T.T., and Jones, J.C., *Alternative Fuels: Emissions, Economics, and Performance*, Society of Automotive Engineers, Warrendale, PA, 1995.

### 11.5.4  Cost Model for Decision-Making

For many proposed new designs, engineers must prepare a cost model and may be expected to demonstrate a *proof of concept*. A typical application of the use of cost models is the use of this technique for evaluating energy savings projects. Energy savings cost models require that certain conditions be estimated to determine the validity of the project. These estimates concern the identity of targeted end users of the technology, degree of awareness by the public, likelihood of adoption, appropriate timing, and technical feasibility. The following equation can be used to determine the expected energy savings for a proposed new design:[6]

$$
\begin{aligned}
\text{Energy savings} = &\ (\text{Targeted energy consumption}) \\
&\times (\text{Savings factor}) \\
&\times (\text{Feasibility factor}) \\
&\times (\text{Penetration factor}) \\
&\times (\text{Adoption factor}).
\end{aligned}
\tag{11.25}
$$

*Savings factor.* The savings factor, SF, represents the percentage of the energy that will be saved by the installation of a proposed system or design. The savings factor can be calculated as

$$
\text{SF} = 1 - \frac{E_{\text{new}}}{E_{\text{old}}},
\tag{11.26}
$$

where $E_{\text{new}}$ is the energy consumption for the proposed new design and $E_{\text{old}}$ is the energy consumption for the old design. For example, the savings factor may be used to represent an expected reduction in fuel consumption due to using an alternative fuel.

*Feasibility factor.* The feasibility factor deals with the probability that the desired technical goals for a project can be met within an allotted time period. As the time for demonstration and proof of concept draws near, the numerical value for the feasibility factor should increase. A timetable for completion can be set up on a yearly basis so that progress can be tracked. The feasibility factor ranges from 0 to 1, with 1 corresponding to project completion.

By carefully examining progress reports, reviewers may be able to make an educated guess as to the probability of success. An indicator of whether the project is on schedule is the time lag between projected and achieved accomplishments. The time lag provides a clue as to the likelihood that the project goals will be met within the specified time frame.

*Penetration factor.* The penetration factor represents the portion of the targeted population that is being reached annually by a program. It can be calculated as

$$
\text{PF} = \frac{R_{\text{pop}}}{T_{\text{pop}}},
\tag{11.27}
$$

---

6 "A Guide for Estimating Energy Savings, Energy Research in Applications Program," Texas Higher Education Coordinating Board, 1990.

where $R_{pop}$ is the population reached, and $T_{pop}$ is the population targeted. In general, the penetration factor varies from year to year depending on type of product and market position. The penetration factor reflects the success of marketing in advertising and publicizing the product. For example, the targeted population could be the total number of trucks on the road. The manufacturers of trucks that use CNG could advertise, hoping to convince owners to change to trucks that run on CNG. For this case, the penetration factor would be those reached by the marketing efforts for a year divided by the number of trucks on the road for that same year.

*Adoption factor.* The adoption factor is the fraction of the penetrated population who will use the design to save energy. It can be determined as the number of adopters, $A_{pop}$, divided by the population reached, $R_{pop}$:

$$AF = \frac{A_{pop}}{R_{pop}}. \tag{11.28}$$

*Targeted energy consumption.* This term refers to the amount of energy that a targeted population will consume in a year. Using this type of analysis requires that a population be identified and that an estimate be made of how energy consumption will vary for future years. Yearly energy consumption should be estimated in millions ($10^6$) or billions ($10^9$) of British thermal units (BTUs). BTUs can be readily determined for all forms of energy. including electricity, natural gas, or barrels of oil consumed.

The targeted population of energy users may consist of either a part of a sector of the economy or an entire sector. The more accurately the target population is defined, the better the projection of cost savings. The following example will demonstrate how to develop a model and how to make a decision by the use of the above methodology.[7]

## Example 11.9

Peanuts, after having been dug up, have a high moisture content, which must be reduced for marketing and safe storage. Conventional drying is usually accomplished by increasing the dry-bulb temperature of the drying air by using gas heaters as shown in Figure 11.5. An alternative drying concept is to dehumidify the air by using a liquid desiccant solution (lithium chloride mixed with water). Desiccants are materials that have a high affinity for water vapor due to the vapor pressure difference between the desiccant and the moisture in the air. The lower the vapor pressure of the desiccant, the better the dehumidification process is. A proposed peanut drying system using a liquid desiccant is shown in Figure 11.6. This system operates by using forced air obtained from the dehumidifier. When the humid air and liquid desiccant are brought into contact in the dehumidifier, moisture in the air will move from the humid air to the liquid desiccant. As shown in Figure 11.6, a regeneration tower is needed to restore

---

7 Ertas, A., "Economical models for hybrid cooling and drying systems by using desiccant technology," Report ERAP-2-#309, submitted to the Texas Higher Education Coordinating Board, December 2, 1991.

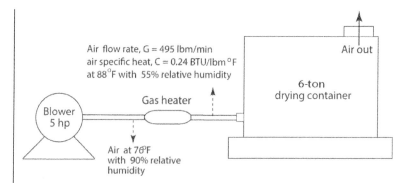

**Figure 11.5** Conventional peanut drying system. (Ertas, A., and Jones, J.C., *The Engineering Design Process*, John Wiley & Sons, Inc., New York, 1996. Reproduced with permission of John Wiley & Sons, Inc.)

the diluted liquid desiccant to the desired concentration. Regeneration is achieved by heating the diluted liquid desiccant in a heater and removing the moisture by passing air over the desiccant in the regeneration tower. Since the liquid desiccant is heated, its vapor pressure is higher than that of the moisture in the air; hence, the moisture in the desiccant will move from the desiccant to the air. The hot liquid desiccant must then be cooled down to the dehumidifier operating temperature (usually 90°F) in the cooling tower. Figure 11.6 shows the energy consumption of all the components for this system.[a]

An economical model for the system is to be developed and the feasibility of the system is to be compared with that of the conventional system. The initial investment for a 6-ton conventional system including blower and gas heater is $2,070 and the initial investment for a liquid desiccant system is $14,709. The cost of the drying containers and the maintenance costs for both systems are assumed to be the same.

### Solution

*Energy and cost analysis of the conventional system.* The time required to dry 6 tons of peanuts by circulating $G = 495$ lbm/min air through the container is estimated to be 38 hours. The blower work $W_{bl}$ during this time is

$$W_{bl} = (5 \text{ hp})(0.746 \text{ kW/hp})(38 \text{ h}) = 141.74 \text{ kWh/container}$$

$$= 483{,}759 \text{ BTU/container}.$$

The heat required to for the temperature increase, $\Delta T = 88 - 76 = 12°F$, at the gas heater is

$$Q_{gas} = G \times C_p \times \Delta T$$

$$= (495)(0.24)(12) = 1{,}425.6 \text{ BTU/min}.$$

$$= 85{,}536 \text{ BTU/h}$$

$$= (38)(85{,}536) = 3{,}250{,}368 \text{ BTU/container}.$$

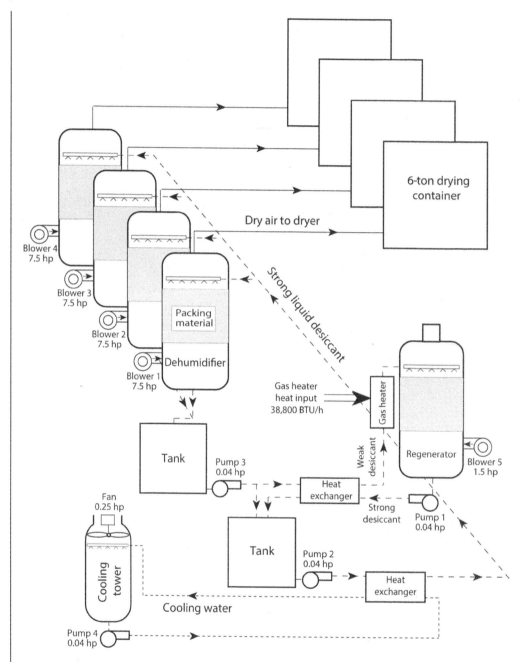

**Figure 11.6** Proposed new peanut drying system. (Ertas, A., and Jones, J.C., *The Engineering Design Process*, John Wiley & Sons, Inc., New York, 1996. Reproduced with permission of John Wiley & Sons, Inc.)

The overall energy requirement, $E_{conv}$, of the conventional system for a complete drying process is

$$E_{conv} = W_{bl} + Q_{gas}$$
$$= 483{,}759 + 3{,}250{,}368$$
$$= 3{,}734{,}127 \text{ BTU/container.}$$

The cost of natural gas per million BTU is assumed to be \$3.84, and the cost of electricity per kWh is \$0.04. The drying operation cost $DC_{conv}$ for each container can be calculated as

$$DC_{conv} = (3.25)(3.84) + (141.74)(0.04)$$
$$= \$18.15/\text{container.}$$

In a drying season, which is three months, each drying unit can dry 58 containers, hence, the annual operating cost, $AC_{conv}$, of the conventional system is

$$AC_{conv} = (18.15)(58) = \$1{,}052.7.$$

*Energy and cost analysis of the new system.* Table 11.11 shows the total power requirement of the new system, Assuming all the pumps and fans work at maximum power. The total energy $W_T$ required for the overall drying process is

$$W_T = (32.25)(0.746)(38) = 914.2 \text{ kWh/run}$$
$$= 3{,}120{,}164.6 \text{ BTU/run.}$$

From Figure 11.6, the total heat energy input to the gas heater to obtain the proper regeneration temperature is

$$Q_T = (38{,}800)(38) = 1{,}474{,}400 \text{ BTU/run.}$$

The overall energy requirement, $E_{new}$, of the new system for a complete drying process is

$$W_T + Q_T = 3{,}120{,}164.6 + 1{,}474{,}400$$
$$= 4{,}594{,}564.6 \text{ BTU/run.}$$

**Table 11.11** Power requirement of the new system.

| Component | Each (hp) | Number | Total (hp) |
| --- | --- | --- | --- |
| Pumps | 0.125 | 4 | 0.5 |
| Blower (5) | 1.5 | 1 | 1.5 |
| Blower (1–4) | 7.5 | 4 | 30 |
| Fan | 0.25 | 1 | 0.25 |
| **Total** | | | 32.25 |

The drying operation cost, $DC_{new}$, for the new system that has four containers is

$$DC_{new} = (914.2)(0.04) + (1.474)(3.84)$$
$$= \$42.23/\text{run}.$$

Assuming that the new drying system can also dry 58 times in a drying season, the annual operational cost of this system is

$$AC_{new} = (42.23)(58) = \$2{,}449.30/\text{yr}$$

*Payback calculation.* Initial investment for the conventional system is $4 \times \$2{,}070 = \$8{,}280$. Note that since the new system has four containers, the cost of the conventional system is multiplied by 4. The initial cost investment difference, $P$, between the two systems is

$$P = 14{,}709 - 8{,}280 = \$6{,}429.$$

The annual saving $F_t$, is the difference in the cost of operations between the two systems:

$$F_t = 4(AC_{conv}) - AC_{new}$$
$$= 4(1{,}052.7) - 2{,}449.3 = \$1{,}761.5.$$

Using an annual interest rate $i = 11$ percent and substituting the value of $F_t$ in equation (11.6),

$$\sum_{t=1}^{n} F_t(1+i)^{-t} \geq PW(i), \tag{11.29}$$

the payback period $n$ is found to be approximately 5 years. Therefore, the proposed liquid desiccant drying system employing packed tower regeneration and a natural gas liquid heater is a feasible option for the conventional natural gas system.

*Targeted energy.* A total of 1.872 million tons of peanuts are harvested annually in the USA, of which about 0.187 million tons are grown in Texas. From the energy calculation for the conventional system the targeted energy, $E_{targ}$, in Texas is

$$E_{targ} = \left( \frac{3{,}734{,}127 \text{ BTU/container}}{6 \text{ tons/container}} \right) (0.187 \times 10^6 \text{ tons})$$
$$= 0.11638 \times 10^{12} \text{ BTU}.$$

*Savings factor.* From equation (11.26), the savings factor is

$$SF = 1 - \frac{E_{new}}{E_{old}}$$
$$= 1 - \frac{4{,}594{,}564.6}{3{,}734{,}127 \times 4}$$
$$= 0.69.$$

Thus, the new system promises up to 69 percent energy savings.

*Feasibility factor.* Suppose that a contract to develop this new drying concept is awarded as a project in January 1992. The concept is to be proved by January 1994. If the project stays on schedule, then feasibility factor values of 0.35, 0.6, and 0.8 are considered reasonable for 1992, 1993, and 1994, respectively. The feasibility factor for the following years is assumed to be 1.

*Penetration factor.* Assume that the number of peanut drying facilities in Texas is $T_{pop}$. After the proof of concept, 5 percent of the peanut facilities are assumed to be reached by the advertisement. The penetration factor for 1994 is

$$PF_{94} = \frac{R_{pop}}{T_{pop}} = \frac{0.05T_{pop}}{T_{pop}} = 0.05.$$

During the following years, assume that the peanut facilities reached will grow linearly. Hence, for year $n$, the penetration factor will be

$$PF_n = 0.05 \times (n - 1993).$$

Using this equation, the penetration factors presented in Table 11.12 can be determined.

*Adoption factor.* A typical adoption factor versus year curve is shown in Figure 11.7. Based on Figure 11.7, adoption factors for 10 subsequent years are given in Table 11.13.

**Table 11.12** Penetration factors for a drying system.

| Year | 1992 | 1993 | 1994 | 1995 | 1996 | 1997 | 1998 | 1999 | 2000 | 2001 |
|------|------|------|------|------|------|------|------|------|------|------|
| PF   | 0.0  | 0.0  | 0.05 | 0.10 | 0.15 | 0.20 | 0.25 | 0.30 | 0.35 | 0.40 |

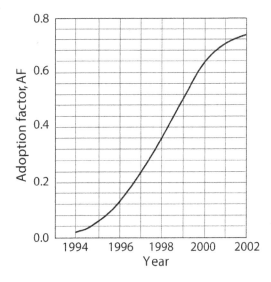

**Figure 11.7** Adoption factor versus year. (Ertas, A., and Jones, J.C., *The EngineeringDesign Process*, John Wiley & Sons, Inc., New York, 1996. Reproduced with permission ofJohn Wiley & Sons, Inc.)

**Table 11.13** Adoption factors for a drying system.

| Year | 1992 | 1993 | 1994 | 1995 | 1996 | 1997 | 1998 | 1999 | 2000 | 2001 |
|------|------|------|------|------|------|------|------|------|------|------|
| AF   | 0.0  | 0.0  | 0.02 | 0.06 | 0.14 | 0.24 | 0.34 | 0.50 | 0.64 | 0.70 |

As is apparent in this table, marketing and the acceptance of the new drying system start from the proof of concept date. The energy savings ES for 1994 (the proof of concept year) can be determined by using equation (11.25):

$$ES_{94} = (0.11638 \times 10^{12})(0.69)(0.8)(0.05)(0.02)$$

$$= 0.0642 \times 10^9. \text{ BTU}$$

Similarly, the energy savings for subsequent years are calculated and presented in Table 11.14. The summation of all the energy savings will give the total BTUs saved. Note that savings for 1992 and 1993 are equal to zero. Introducing a new drying concept employing a liquid desiccant will save $65.43 \times 10^9$ BTU in 10 years.

**Table 11.14** Adoption factors for a drying system.

| Year | 1994 | 1995 | 1996 | 1997 | 1998 | 1999 | 2000 | 2001 |
|------|------|------|------|------|------|------|------|------|
| ES   | 0.0642 | 0.4818 | 1.6863 | 3.8545 | 6.8257 | 12.0453 | 17.9877 | 22.4846 |

[a]From Ertas, A., and Jones, J.C., *The Engineering Design Process*, 2nd edn, John Wiley & Sons, Inc., New York, 1996.

# Bibliography

1 BROWN, R.J., and RUDOLPH, R.Y., *Introduction to Life-Cycle Costing*. Prentice Hall, Englewood Cliffs, NJ, 1985.
2 ESCHENBACH, T.G., *Engineering Economy Applying Theory to Practice*, Oxford University Press, New York, 2002.
3 FABRYCKY, W.J., and BLANCHARD, B.S., *Life-Cycle Cost and Economic Analysis*. Prentice Hall, Englewood Cliffs, NJ 1991.
4 GONEN, T., *Engineering Economy for Engineering Managers*, John Wiley & Sons, Inc., New York, 1990.
5 NEWNAN, D.G., *Engineering Economic Analysis*, Engineering Press, San Jose, CA, 1988.
6 RIGGS, J.L. *Engineering Economics*, McGraw-Hill, New York, 1977.
7 THUESEN, G.J., and FABRYCKY, W.J., *Engineering Economy*, Prentice Hall, Englewood Cliffs, NJ, 1989.
8 WHITE, J.A., AGEE, M.H., and CASE, K.E., *Principles of Engineering Economic Analysis*, John Wiley & Sons, Inc., New York, 1984.

# CHAPTER 11 Problems

**11.1** The parents of a first-year university student are in the process of buying a car for the student to use. Two used cars have been found. One car can be purchased for $7,500 with a first-year projected maintenance cost of $300. Annual maintenance costs are expected to increase by $200 thereafter. Operational costs for this vehicle are anticipated to amount to $1,000 per year, whereas the salvage value decreases at a rate of 15 percent per year. The second car can be purchased for $12,000 and the first-year maintenance cost is expected to be $200. Annual maintenance costs for this vehicle are expected to increase by $200 thereafter. Operational costs are also expected to be $1,000 per year and the salvage value for this second vehicle is projected to decrease at a rate of 18 percent per year.

    a) Find the minimum-cost life and minimum annual life cycle cost of the two vehicles if the time value of money is ignored.

    b) Prepare cash flow diagrams for both vehicles. Assume a simple interest rate of 12 percent that is paid annually for the minimum-cost life period calculated above.

    c) Find the minimum-cost life and minimum life cycle cost for both vehicles for an annual interest rate of 12 percent that is paid in 36 equal monthly payments. Since the vehicle belongs to the financial organization until the end of the third year, no salvage cost should be included for years 1 and 2.

**11.2** How much would the owner of a pickup truck be justified in paying for the vehicle to be converted to operate on compressed natural gas if the vehicle accumulates 20,000 miles per annum with an average 12 mpg, the savings in fuel cost amounts to $0.50/gallon, the life of the conversion system is 10 years, and the salvage value of the conversion system is 15 percent of its initial cost? Assume that the time value of money is 10 percent.

**11.3** The annual rental income from a duplex is $9,600 and annual expenses are $2,500. If the potential buyer anticipates that the duplex could be sold for $75,000 at the end of 10 years, what purchasing price could be justified? Assume that the interest rate is 8 percent.

**11.4** Calculate the ROR for the cash flow shown in Table P11.4.

Table P11.4 Cash flow with respect to years.

| Year | Cash flow, $ |
| --- | --- |
| 1 | −15,000 |
| 2 | −10,000 |
| 3 | 10,000 |
| 4 | 12,000 |
| 5 | 14,000 |
| 6 | 15,000 |

**11.5**   For the two alternatives shown in Table 11.2, determine the payback period for an interest rate of 12 percent.

**11.6**   A fleet manager is considering the construction of a compressed natural gas refueling facility. The cost of site preparation, dispensing equipment, storage tanks, and the like has been determined, but the decision as to the compressor selection, which is the greatest single expense item, has yet to be made. The two options are (a) purchasing a 50,000-SCFM compressor at a cost of $60,000 at an interest rate of 12 percent, or (b) renting a compressor at a cost of $1,000 per month, including maintenance. If a compressor is purchased, it is anticipated that annual maintenance costs will be $1000. The life of the compressor is estimated to be 10 years. Assume that the compressor is depreciated using the straight-line method and determine which is the better option for the fleet manager if the income tax rate is 25 percent.

**11.7**   Automobile dealers are increasingly advertising the leasing of vehicles instead of purchasing. In one case, a $20,000 automobile can be leased for $375 per month for 36 months, after which it is returned to the dealer. If the automobile is purchased, it could be financed for 3 years at a 10 percent annual rate with a down payment of 5 percent and 36 equal monthly payments. If at the end of the 36-month period the vehicle is estimated to be worth $8,000, which would be the preferred alternative? Assume that the value of money to the buyer is also 10 percent per annum.

**11.8**   Find the RORs for alternatives A1, A2, and A1 - A2, in Table 11.4.

**11.9**   The five mutually exclusive alternatives listed below have 15-year useful lives. If the MARR is 12 percent, which alternative should be selected?
- Alternative 1 requires an initial expenditure of $10,000 and has an annual benefit (receipts less disbursements) of $1,424.
- Alternative 2 requires an initial expenditure of $6,000 and has an annual benefit of $1,026.
- Alternative 3 requires an initial expenditure of $4,000 and has an annual benefit of $742.
- Alternative 4 requires an initial expenditure of $2,000 and has an annual benefit of $356.
- Alternative 5 requires an initial expenditure of $1,000 and has an annual benefit of $256.

**11.10**   Determine the expected value of bets on Texas Tech, Texas University, Texas A&M, and Houston winning the 2018 Southwest Conference basketball championship based on the data provided in Table P11.10.

**Table P11.10** Data for Texas universities.

| Team | Probability of winning | Outcome of bet if you win |
|------|------------------------|---------------------------|
| Texas Tech | 0.35 | 20.00 |
| Texas | 0.30 | 25.00 |
| Texas A&M | 0.20 | 40.00 |
| Houston | 0.15 | 50.00 |

**11.11** A company has several projects under consideration and is trying to decide which should be pursued. It has been decided that the projects will be ranked according to their net present worth divided by their cost. If the company's MARR is 18 percent, rank the projects in accordance with this criterion and determine which projects from Table P11.11 should be funded.

**Table P11.11** List of Properties.

| Project | Cost ($) | Uniform annual benefit,($) | Useful life (years) | Salvage ($) (%) | ROR |
|---------|----------|----------------------------|---------------------|-----------------|-----|
| 1 | 1200 | 267 | 10 | 0 | 18 |
| 2 | 2000 | 521.20 | 8 | 0 | 20 |
| 3 | 500 | 185.90 | 5 | 0 | 25 |
| 4 | 2000 | 477 | 10 | 0 | 20 |
| 5 | 2000 | 438.5 | 10 | 1000 | 20 |
| 6 | 1500 | 225 | 6 | 1500 | 15 |

**11.12** An investment of $100,000 for a new air conditioning system is being considered by a company. The expected life of the new system is estimated to be 18 years, and the salvage value of system equipment at this point is thought to be $15,000. The new system is expected to increase business such that annual gross receipts will grow by $10,000. Expenses are expected to increase by $2,500 per annum. If the time value of money is 15 percent, determine the following measures of investment worth and recommend the appropriate action:
(a) Internal rate of return
(b) Present worth
(c) Annual worth
(d) Future worth
(e) Net present worth/cost.

**11.13** For the peanut drying system discussed in Example 11.9, liquid desiccant is to be regenerated by using an open solar collector in conjunction with an auxiliary heater as shown

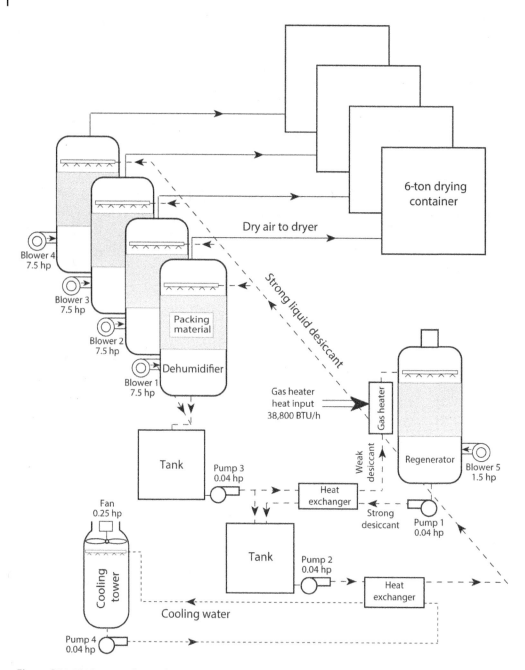

**Figure P11.13** Proposed new drying system with solar regeneration.

in Figure P11.13. If the total heat energy that will be required from the auxiliary heater is $Q_T = 429{,}748$ BTU/container, develop an economical model and determine whether this system is feasible. Use the following assumptions:

a) 58 drying times in a drying season.
b) Cost of drying system with solar regenerator is $6,340.
c) Annual interest rate, $i = 11$ percent.
d) Cost of natural gas is $3.84 per million BTU.
e) Cost of electricity is $0.04/kWh.

**11.14** A sum of money, $40,000, is to received 7 years hence but a payment of $75,000 is required 12 years from the present. If the average annual inflation rate is 6 percent and the average annual interest rate is 10 percent over this time period, compare the present worth of these payments using actual dollars and constant dollars.

**11.15** An employee has a retirement fund that earns 10 percent per annum compounded monthly. He has just turned 35 years old and plans to retire at age 65. If he makes equal monthly deposits of $200 until his 65th birthday, what equal annual withdrawals in actual dollars could be made beginning on his 65th birthday and continuing until his 80th birthday? If the annual inflation rate is 5 percent compounded monthly, find the constant dollar equal annual series over this same period of time that is equivalent to these withdrawals.

**11.16** A firm is considering the purchase of new equipment with a first cost of $800,000 and a salvage value of $80,000 at the end of 12 years of service. If the firm's MARR is 18 percent, how much must be earned on an annual equivalent basis over the life of the equipment to make the investment worthwhile?

**11.17** Interest in an oil well can be purchased for $200,000, and net annual income generated will amount to $40,000 for the life of the well, which is forecast to be 15 years. What is the annual rate of return on this investment if the well is worthless after the 15 years?

**11.18** Air conditioning for a building costs $60,000 per year. By adding 2 inches of insulation at a cost of $15,000, this cost can be reduced to $50,000 per year. Another option is to add 4 inches of insulation at a cost of $18,000, which will reduce the cost of air conditioning to $45,000 per year. Which option should be pursued?

# 12

## Engineering Ethics

I do not believe in immortality of the individual, and I consider ethics to be an exclusively human concern with no superhuman authority behind it.

**Albert Einstein**

## 12.1   Ethics in Industry[1]

In US industry, interest in the development and teaching of corporate ethics has experienced a sizeable increase. According to the Ethics Resource Center in Washington, DC, a 1979 study of Fortune 500 companies and the top 150 service firms in the United States revealed that 73 percent had written standards of ethics, half of which were 5 years old or less. A decade later, better than 9 out of every 10 companies from 2,000 surveyed reported having standards of ethical conduct. About 11 percent of these companies provide ombudsmen, an individual with whom employees can discuss ethical concerns and other issues.[2] Only since the mid-1970s have articles on ethics been commonly found in popular engineering journals. Today, it is increasingly common to find articles on ethics in journals published by professional societies such as the American Society of Mechanical Engineers, the American Society of Chemical Engineers, the American Society of Civil Engineers, and the Institute of Electrical and Electronics Engineers.[3] This is significantly different from 1959 when the writer, who was then a novice first-level supervisor for a large aerospace company, urged the company to establish a policy pertaining to employees accepting vendor lunches. During this time, very few companies had any published ethical standards, and most were unwilling to establish standards even when urged to do so by their own employees. These various factors caused the company to refuse the establishment of a policy regarding vendor lunches, choosing not to respond to the request in any form.

---

1 Adapted from Ertas, A., and Jones J.C., *The Engineering Design Process*, John Wiley & Sons, Inc., New York, 1996.
2 "Teaching ethics takes off in U. S. firms," *Engineering Times*, June 1989.
3 Martin, M.W., and Schinzinger, R., *Ethics in Engineering*, 2nd edn, McGraw-Hill, New York, 1989.

Currently, ethics is plainly recognized by leading companies as a critical element in business and policy decisions and as a valuable tool in resolving difficult management problems. However, training in corporate ethics is not restricted to just those employees who have an influential position within a company. Increasingly, industry is recognizing that all employees, regardless of their position or assignment, should be concerned with maintaining high standards of conduct on the job to sustain and increase the company's overall performance. A company's performance is dependent on the actions of its employees, acting solo or in concert with other employees, and these actions are strongly impacted by the employee's perception of the company and knowledge and understanding of the company's policies and goals. There are various techniques used to convey instructions to employees on corporate ethics. A common method that seems to be effective is based on the use of case studies concerning problems the firm has faced in the past and how to anticipate, recognize, and avoid situations involving questionable ethics in like situations. Training programs can also include specific issues and company policies having to do with ethical questions or concerns. If the company has a policy regarding the acceptance of lunches and other gratuities from vendors, this can be specifically spelled out and discussed during the training sessions.

Effective and ongoing communication with employees is the key to a successful corporate ethics program. General Dynamics Corporation has instituted a Standards of Business Ethics and Conduct booklet as part of its corporate ethics program. These standards alert employees to situations where misconduct might happen due to ignorance, misunderstanding, or a mistaken notion of what is expected. They provide simple, explicit guidelines to follow to avoid the fact, and sometimes even the appearance, of misconduct. Additional methods of employee communication used in this program include the use of videotapes featuring top management and other employees, stories on the program in internal employee news publications, posters advertising particular features of the program, and annual (and other) reports that include informative articles on the program. Twenty-nine ethics hotlines are provided throughout the company with a toll-free 800 number to the corporate office. This allows employees to have access to a readily available channel to ask questions, seek advice, voice concerns, or raise allegations. In 1987, the director of the ethics program at General Dynamics received 5,760 communications from employees, 56 percent of these coming over the hotlines. The goals of this program are to support individual employees in their daily business conduct, to improve the administrative performance of the company in basic business relationships, and to help build the bond of trust between the company and its customers, suppliers, employees, shareholders, and the communities in which it functions.[4]

## 12.2 Ethics and the University[5]

In spite of the growth of industrial ethics programs, there remains considerable controversy about the effectiveness of teaching ethics in universities. Many people believe that a person's

---

4 "Ethics Program Update," General Dynamics Corporation, 1988.
5 Adapted from Ertas, A., and Jones J.C., *The Engineering Design Process*, John Wiley & Sons, Inc., New York, 1996.

ethical outlook is established at a young age due to the combination of various factors, such as family, church, friends, and elementary and secondary school teachers, and that this outlook cannot be changed or influenced significantly at the young adult stage, the university entry age level. It is obvious from the emphasis being placed on ethics in universities that people involved in education believe that a person's basic value system can be influenced and reinforced, however slightly, by providing an appropriate environment, one that encourages strong beliefs about honesty, personal integrity, fair play, and wholesome competition. Exposing university students to ethical dilemmas in a classroom environment might cause them to be able to deal with certain situations more effectively. Case studies involving situations typical of those encountered by professional engineers in industry, government, and private practice can also be utilized to demonstrate ethical principles and applicable codes. Through this form of training the student should be able to learn how to identify situations in which ethical difficulties often arise at an early stage so that an alternative course of action can be formed. The student will also become familiar with the particular code of ethics applicable to his or her engineering discipline and how this code is enforced and applied.

The Accreditation Board for Engineering and Technology (ABET), which is the officially recognized organization for accrediting university programs in the USA, makes the following statement in reference to curriculum requirements: "An understanding of the ethical, social, economic, and safety considerations in engineering practice is essential for a successful engineering career. Course work may be provided for this purpose, but as a minimum it should be the responsibility of the engineering faculty to infuse professional concepts into all engineering course work." Thus, universities that want to keep and improve their accreditation standing should include material on ethics in their program. This can be accomplished by integrating ethical concepts into lectures, problems, and design projects, including a study of ethics as a separate topic within an upper-level course and/or developing a formal course in professionalism with specific attention to the role of ethics. ABET has formed guidelines to assist engineering institutions with the incorporation of ethics and professionalism in engineering curricula. These guidelines suggest that the following areas of instruction be included in any formal course on ethics and professionalism: an introduction, which should include justification for the course; some material on philosophical ethics, including several of the major ethical theories; concepts of professionalism, which should emphasize the importance of registration; codes of ethics, including their purpose and form; and pertinent case studies, which should emphasize real-life ethical dilemmas.

## 12.3   Ethics in Engineering[6]

On October 10, 1973, Spiro T. Agnew resigned as Vice-President of the United States amid charges of bribery and tax evasion related to his previous position as County Executive of Baltimore County, Maryland. Agnew was a civil engineer and lawyer; as County Executive

---

6 Adapted from Ertas, A., and Jones J.C., *The Engineering Design Process*, John Wiley & Sons, Inc., New York, 1996.

during the years 1962–1966 he had authority to award contracts for public works projects to architect/engineering firms. By exercising that authority, he benefited from a profitable kickback scheme whereby certain firms were given special consideration in receiving contracts for public works projects if they made private payments in return.[3] This kickback scheme was complex, with several architect/engineering firms involved, so engineers cannot justify Agnew's conduct on the basis of his position. Although this may be the most notable incident in recent times of illegal and unethical conduct by engineers, it is certainly not unique. This incident does present a problem as to what can be done to prevent intentional ethical misconduct in the engineering profession. Several lectures or even a course on ethics will prevent a person with the wrong goal from being tempted to bend or break the rules of acceptable moral and legal conduct. If the goal in an individual's life is to become rich, gain great power, or hold enormous prestige, one will probably be susceptible to shortcuts and get-rich schemes. If, on the other hand, a person's goal is to build a career and reputation as a good engineer and honest individual, such schemes will have little or no appeal. In all probability, the only deterrent to people who intentionally act in an unethical and illegal manner is to increase the likelihood of apprehension and to increase penalty severity. When the act is unethical, but not illegal, the responsibility for meting out the appropriate punishment falls on the profession itself, which is often poorly equipped to deal with such occurrences.

Breach of ethical conduct can also occur due to ignorance. We live in a complex and often confusing world in which the rules governing professional conduct are constantly evolving and being reinterpreted. Although industry is rapidly correcting this problem, there are many working engineers who are not conscious of company ethical policies and guidelines, nor familiar with professional codes of ethics that are applicable to their specific engineering discipline. This is a problem that can be solved simply by providing adequate training at the university level, augmented by further, and more specific, training as the graduating engineer enters industry. Fundamentals of the codes as well as techniques and strategies for handling ethical decisions and challenges can be taught in the university environment. Case studies can be used to clarify the principles covered in the codes and to analyze ethical dilemmas in which moral principles are in conflict or are vague. Needs, such as finding a match between the graduating student's ethical value system and that of potential employers, can be discussed. The university classroom is an appropriate place for discussing and analyzing the situation in which one's values are tested by peers, subordinates, or superiors.

ABET has defined engineering as "the profession in which a knowledge of the mathematical and natural sciences gained by study, experience and practice is applied with judgment to develop ways to utilize, economically, the materials and forces of nature for the benefit of mankind."[7] It is with the words "with judgment" that engineers have the greatest ethical difficulty. This is not a matter of knowing or unwitting violation of ethical codes and moral values, but one in which a choice must be made between options, all of which have a varying degree of ethical value and liability. Judgment is one of those attributes that varies from individual to individual and, on a grander scale, from company to company. It is self-evident that an

---

7 Accreditation Board for Engineering and Technology, 1988 Annual Report, p. 108.

individual's judgment is strongly affected by his or her ethical value system, as well as education and personal experience. Corporations, as well as individual engineers, are increasingly being called on to validate and defend their judgment in regard to technical decisions as a result of structural/equipment failures and accidents. The Space Shuttle *Challenger* accident, the DC10 cargo door failures, and the Ford Pinto gasoline tank fires are all notable examples of situations in which judgment and ethical practices have been investigated. The ethical dilemma presented by each of these situations is that in all cases the decision-maker was advised as to the possible outcome of the decision but considered the tremendously low likelihood of failure, and the attendant consequences, to be favorable to the alternative. The corporation's financial loss associated with making the recommended change(s) was undoubtedly present in the mind of the decision-maker in these examples, but other factors invariably impact decisions of this size, whether they are aware of it or not. The personal cost to the decision-maker, which could include a loss of prestige and advancement potential, could well have been involved, as well as the ethical tradeoff between responsibility to the using public for safety and the financial management responsibility to the employer.

It is in this area of "judgment" that a lot of second guessing occurs, especially after an incident has happened; nothing is as clear as hindsight. This is not to excuse faulty judgment but is pointed out to emphasize the importance of full and open discussion of all the pertinent facts and ethical consequences, and their appropriate weighting, when making significant decisions. It is extremely important that decision-makers make choices on the basis of a full and complete discussion of the facts and that they not be influenced by extraneous matters, such as ego, interpersonal relationships, customer/contractor relationships and authority, history of the program, or how hard the decision may be to sell. In the case of the *Challenger* accident, the decision-making process failed, not due to the process itself, but because of the outside factors that hindered decision-making being based on the facts. The customer/contractor relationship obviously interfered with suitable consideration of the facts. Success in past launches led the customer, NASA, to believe that the launch could be successfully accomplished in spite of the engineering data that indicated otherwise. NASA management ego was probably one of the factors that led to the decision to overrule the recommendations of the engineers close to the problem. The NASA project managers at the Marshall Space Flight Center did not accept the contractor's initial technical decision as being valid and, using pressure, persuaded the contractor to consider the "management" consequences. Contractor management was overwhelmed by the threatening conduct of the customer and subsequently agreed with NASA over the recommendations of their own engineers, this is a sad commentary on engineering decision-making and ethical conduct. Technical decisions must be made on their own merit – personal feelings should have no role in this process and, when they are present, will generally result in sub-optimal decisions.

What can be done in the university environment to enhance the ethical judgment of students? Is it possible to install values that have not been developed earlier by the student's family, religious institutions, friends, and teachers? Probably not, but the university education experience will influence the shape and structure of the student's value system, and it is the university's duty to guarantee that this influence is positive. The student should experience strong and positive value reinforcement in all interactions with faculty and university administration. A value-reinforcing climate should be nurtured in all campus facilities and functions, including dormitories, recreation centers, student newspapers, and student organizations. Finally, classroom techniques that urge students to test their own value systems can be used to strengthen and modify. This can be achieved by using case studies that highlight ethical dilemmas that demonstrate credible engineering issues.

## 12.4   Legal Responsibilities of Engineers[8]

Law and the legal field control modern American society to an unparalleled extent. Engineers are, at least, held to the same legal standards as other professions and occupations.

First, one may not practice the profession of engineering, at least, in most states, without satisfying certain legal requirements of training, education, examination, and experience. Second, there are various federal, state, and local laws, regulations, and ordinances (such as building codes) that control the work that engineers do. Third, the work of engineers is generally conducted, and disputes resolved, pursuant to some contractual agreement, which is itself governed by various requirements of the law. Lastly, engineers are not able to practice their profession without being alert at all times that there is a potential risk of becoming liable for mistakes in judgment, intentional wrongdoing, or even inadvertent mistakes.

Each individual state in the United States requires that an engineer be registered before being able to practice professional engineering in that state. Typically, although state laws vary on the requirements, a person must satisfy minimum education and experience requirements and pass an examination to be registered as a professional engineer. A person who practices professional engineering without satisfying state registration requirements is subject to various penalties and fines. Registration laws and requirements differ from state to state, and registration in one state does not give the engineer the privilege of acting as a professional engineer in another state. However, many states have reciprocal agreements which allow registration to be simplified if the engineer is already registered in another state.

Like other professions, engineering is considered to be self-regulating. This generally means that members of this profession work collaboratively to ensure that other members of the profession act ethically and competently. Professions typically organize associations and societies that, in turn, adopt codes of ethics to oversee professional conduct. Peer pressure and one's reputation within engineering are some of the informal ways that self-regulation is enforced. However, formal control does exist; many states have incorporated professional codes of ethics into the law governing the registration of engineers. In those states, violation of the code of ethics can result in one receiving legal penalties or losing one's license to practice.

Apart from direct regulation from the state, the law impacts engineers greatly through the fear of civil liability. There has been a tremendous increase in the number of lawsuits filed against

---

8 Adapted from Ertas, A., and Jones J.C., *The Engineering Design Process*, John Wiley & Sons, Inc., New York, 1996.

engineers in the last 20–30 years. There are two basic ways in which an engineer may become liable to another party or parties. English common law, which is the historical basis for the American legal system, differentiated between duties owed as result of mutual consent, that is, by contract, and duties owed by all persons to act with reasonable care, skill, and diligence so as not to injure other persons or property. The law governing liability imposed in the absence of a contract is known as torts.

Contracts are a very important aspect of the legal rights of engineers. Generally, all engineering work is done pursuant to some contract. Most often that contract is in writing and includes detailed language and many provisions. Whenever a party, including an engineer, enters into a contract, his or her rights and responsibilities are defined with reference to the language of the contract. A contract should not be signed unless one has read it in its entirety and understood the terms used. Courts do not sympathize with claims that one has not read a contract fully and/or understood the language of the contract.

Ordinarily, a breach of a provision of a contract entitles another party to the contract to recover money damages by filing a civil lawsuit for breach of contract. Most construction contracts to which engineers are party, however, contain a provision that all disputes arising under the contract will be submitted to arbitration. This is a process in which a dispute is submitted to an arbitrator or panel of arbitrators, selected by the parties, and whose decision on the matter is final and binding. Arbitration has several advantages over litigation: it usually saves significant time and cost, can be kept private and confidential if the parties so desire, and the decision can be made by a person or persons who have more knowledge and experience in the field than do judges and juries. In most states, and in all contracts that involve interstate commerce, the courts will enforce an agreement to arbitrate and will impose the decision of the arbitrator.

Under the law of negligence, persons may be held liable even without a prior contractual agreement with another party. In general, persons owe a duty to society to exercise reasonable care in going about their business so as not to injure other persons. If someone acts negligently or carelessly, that person will be held liable for anyone who is injured as a probable consequence of the negligent act. The term "malpractice" is used to refer to negligence by professionals. Malpractice is commonly defined as a dereliction of professional duty or a failure of professional skill or learning that results in injury, loss, or damage. All professionals owe a duty not only to their clients but also to the public to exercise skill and ability when performing their professional services.

However, undertaking professional services does not always result in a satisfactory result. Just because there is an error of judgement does not mean that one lacks skill or care. The question of liability in malpractice is ultimately a question of performing as a rational professional would perform under the same or similar circumstances. Often, in addition, the locality of the defendant is considered when making allowances. Thus, the question of liability hinges on the credibility of expert witnesses on the issue of how a reasonable engineer would have performed under the same circumstances.

A relatively recent development in the law of torts is strict liability, which is the imposition of liability even in the absence of negligence or other fault. Manufacturers of products can be held liable if a defective product causes injury, even in the absence of proof of negligence in the design or manufacture of the product. While this law and its impacts vary from state, engineers should be aware of the possibility of strict liability as they perform their professional duties.

## 12.5   Codes of Ethics[9]

All of the major professions in the USA have adopted codes of ethics to provide guidance and support for their profession and its membership. These codes deal with common problems such as competency, confidentiality, and conflicts of interest and, generally, are more beneficial in the guidance that they provide than they are in a negative, or disciplinary, sense. Engineering codes of ethics encourage ethical conduct and provide guidance concerning the obligations of engineers. Codes provide support to engineers seeking to act ethically in situations that involve conflict with their superiors and serve as a deterrent to unethical conduct. Codes also enhance the profession's public image and generally encourage professionalism.

There are many codes of ethics within the profession of engineering; generally each separate discipline has adopted its own code. However, all of these codes include the basic principles incorporated in the code of ethics published by the National Society of Professional Engineers (NSPE), which is a voluntary organization of engineering professionals dedicated to serving the public and the profession. The NSPE is the umbrella organization for the state professional engineering societies and, as such, represents all of the individual engineers registered to practice under the various state registration laws:

> NSPE strives to insure the application of engineering knowledge and skills in the public interest and to foster public understanding of the role of engineering in society. Further, the NSPE promotes the professional, social and economic interests of the engineering profession and its individual members.

One of the goals of NSPE is to provide "professionwide leadership on selected professional issues of importance to all segments of engineering." One of the objectives under this goal is "to seek common acceptance, understanding and enforcement of ethical standards."[10]

It is pursuant to this objective that the NSPE has adopted and promulgated a Code of Ethics for Engineers. This establishes standards of conduct for all engineers. It is enforceable only as applied to members through the action of the state societies. Each state society has adopted the Code as the state society's code, with minor changes in some states. The NSPE issues interpretations of the Code in particular circumstances through its Board of Ethical Review. As indicated previously, in some states the principal elements of the NSPE Code have become state law through the enactment of laws concerning professional engineering registration within the state. In these cases, disciplinary action can be considerably more severe.

---

9  Adapted from Ertas, A., and Jones J.C., *The Engineering Design Process*, John Wiley & Sons, Inc., New York, 1996.
10  *Texas Society of Professional Engineers Chapter/State Reference Handbook*, p. III-2, 1986.

---

**The Leaning Tower: A Timely Dilemma (Case 1001)***[a]*

---

The mission of the National Institute for Engineering Ethics (NIEE) is to promote ethics in engineering practice and education. One component of NIEE is the Applied Ethics in Professional Practice (AEPP) program, providing free engineering ethics cases for educational purposes. The following case may be reprinted if it is provided free of charge to the engineer or student. Written permission is required if the case is reprinted for resale. For more cases and other NIEE Products & Services, contact the National Institute for Engineering Ethics, Texas Tech University, www.niee.org

**The case**

A medium-sized town in the Northeast derived the bulk of the local income years ago from shoes manufactured in an extensive mill facility on the banks of the river running through the town. The mill had originally been located on this site in order to use water power as the primary energy source for running the mill equipment through a vast array of belts, pulleys and reduction gears. However, because of the site location adjacent to the river, the soils tended to be loose and/or moderately compressible, requiring deep foundations for heavier portions of the structures.

One of the most prominent mill structures was the original tower which was in excess of 15 stories in height, and over the years the large clocks on all four sides of the tower became the standard reference for correct time in the local community, even after the mill became defunct and lay idle for more than 35 years. Recently, however, the region has realized an increased growth due to a demand for computer software development for industrial, medical and personal uses. MegaBite Unlimited, a nationally known computer software development firm, surveyed a number of sites in the state and decided that because of the semi-rural atmosphere of the town, the above-average educational background of most of the residents and the attractive tax incentives offered by the local improvement board, they would design and build a new corporate facility in the town.

The prime site for the new facility was the abandoned mill and clock tower. Since the mill buildings had been left without maintenance for such a long period, it was decided to demolish them and build an efficiently designed complex of structures which would complement the colonial decor of the area.

The local planning commission concurred with this decision, but insisted that the original 15-plus story clock tower be retained as a symbol of the prosperity the town once enjoyed, and would now experience again.

During the design of the new facility, MegaBite Unlimited's architect retained I. B. Stout, a structural engineer from a nearby city, to do the structural engineering and design such systems as might be necessary during demolition of the old structures in order to protect the old clock tower, if necessary. Stout's review of the available records showed that the heavy clock tower was supported on a 48-foot square mat at a depth of about 18 feet below ground surface, and the mat in turn was supported on wooden piles driven to some depth below the water table to pick up additional support in the underlying soils. Based on this information, Stout designed a shallow

*(Continued)*

---

**The Leaning Tower: A Timely Dilemma (Continued)**

retaining system for support of the sides of the excavation which would occur during demolition of the adjacent mill buildings approximately 50 feet away.

The plans and specifications for the demolition were completed, along with the design documents for the proposed new facility, advertised to contractors for bid, and the job was let to Colonial Construction Company, a general construction contractor in business for more than 60 years in the local area. Since Colonial had the necessary heavy equipment, they proceeded with the demolition work as one of the first stages of the project. About the time that the demolition excavation had come to approximately 100 feet from the old clock tower, one of the construction crew noted that the tower seemed a bit out of plumb, but thought little of it, since the tower was so old. However, as the excavation progressed closer to the tower, it became apparent that the tower was tilting at an increasing rate, and toward the demolition excavation.

Colonial stopped work over a weekend to consider what steps, if any, should be taken with regard to the clock tower. Upon returning to the site on Monday morning, new survey measurements indicated that the tilt at the top of the tower had increased to six inches. At that point Colonial called W. E. Holdem, Inc., a specialty ground modification subcontractor, and asked for their assistance in correcting the situation so that the demolition could be completed to the extent originally planned. Realizing the seriousness of the problem, Holdem in turn called in Jonathan Turnbuckle, an engineering consultant in the Midwest who had an excellent reputation for coming up with innovative, cost-effective solutions to construction problems requiring ground modification. Holdem knew of Turnbuckle because they had worked together on a number of projects over the past several years in other areas of the country.

Turnbuckle was quick to respond; visited the site; assessed the problem; devised a solution with a reasonable chance of success; conferred with the architect, I. B. Stout, the architect's geotechnical engineer, MegaBite's representatives and the town's building officials and the mayor, and explained the solution. He also made recommendations for standby cranes and other safety precautions should it not be possible to implement the proposed ground modification scheme in time to save the clock tower and avert potential property damage and possible personal injury.

The entire group cooperated in expediting the standby equipment and procedures. Holdem mobilized on the site within two days and initiated the remediation procedure devised by Turnbuckle, successfully continuing the process and averting the collapse of the clock tower. It took a period of about 70 days to continue the process until the adjacent demolition was completed and the excavation backfilled, with constant monitoring on the site and analyses being required for at least the first 45 days. No one, including I. B. Stout whose original design had allowed the problem to occur, was sued.

About three weeks or so after the remediation procedure was initiated by Holdem, Inc., Jonathan Turnbuckle realized that since the solution he had devised was an engineering design, he should have permission to practice in the state, if only for a temporary period. He then contacted the state's Board of Registration for Engineers and Surveyors, requesting a temporary engineering license. He was told by the Board that although many states do have a provision for such a temporary or short-term permit for engineers licensed in other states, this particular state had no such provision and he would have to make a full, formal application for registration as a professional engineer in the state.

Turnbuckle obtained the necessary forms, filled them out (including references to the professional engineering registrations he held in 17 other states), and turned them in to the Board within a couple of days. Approximately three months after the remedial construction had been completed and the clock tower saved, he received notice of his acceptance by the Board as a registered professional engineer by reciprocity. Since the project was complete as far as Turnbuckle and Holdem were concerned, they each went on to other projects in other areas.

Recently, Turnbuckle has received a registered letter from the state Board of Registration for Professional Engineers and Land Surveyors notifying him that I. B. Stout, the structural engineer for the MegaBite project, has filed a formal complaint against him for practicing as an engineer during the time of the clock tower incident without a license in the state. Furthermore, Turnbuckle is advised that Stout is prepared to carry the matter to court, where Stout intends to sue Turnbuckle for a substantial sum of money, claiming that too many out-of-state engineers do designs for projects within the state without being licensed, and that practice is financially detrimental to Stout and the survival of his practice.

Is Turnbuckle at fault for not having a professional engineer's license in the state during the design and implementation of the clock tower remediation? What should he do, or have done?

### Alternate Approaches and Survey Results for "The Leaning Tower, A Timely Dilemma" (Case 1001)

1) Realizing that he was not registered in the state, Turnbuckle should have declined the request to devise a remedy for the potential imminent collapse of the clock tower right away. *Percentage of votes agreeing:* 6%

2) Before accepting the assignment, Turnbuckle should have contacted the Board of Registration for Engineers and Surveyors to request a temporary license. Upon learning that the state had no provisions for such a temporary license, and realizing that he would in fact be providing an engineering design without a license should he continue, he should have declined the assignment. Colonial Construction and W. E. Holdem could have found someone else who was a registered engineer in the state to come up with a design to save the clock tower. In any event, the matter would be out of Turnbuckle's hands, since he would be complying with the legal requirements of the state. *Percentage of votes agreeing:* 30%

3) By placing the health, safety and welfare of the public (NSPE Code of Ethics for Engineers) above the strict legal requirements of the Board of Registration, Turnbuckle should be supported for the action he took. In addition, he did make application for a temporary license in good faith, and should not be held responsible for the fact that the Board had no official mechanism through which to grant him a license, even though he was registered as a professional engineer in 17 other states. *Percentage of votes agreeing:* 64%

### Forum Comments from Respondents

1) I firmly believe that all current registration programs need reevaluating. In my opinion, all disciplines should have a national registration, much like AIPG. Once registered through a state board, individuals should be allowed to practice in all states in the US. At present, state boards

*(Continued)*

---

**The Leaning Tower: A Timely Dilemma (Continued)**

have too much power and in some instances abuse that power for their own gain. The actions taken against Turnbuckle are an example of protectionism. I. B. Stout did not adequately plan out his portion of the project and should have been investigated to determine if he was negligent in his appraisal of the situation. In such a case as this, a national certification with reciprocity would have eliminated the protectionism and in-fighting.

2) The owner and the architect should have brought I. B. Stout into the process as soon as the problem was noted. As a consequence, Turnbuckle could have been brought in by Stout as an advisor, with Stout being ultimately responsible for the design.

3) While solution #3 is ethically the correct answer, Turnbuckle could have worked closely as a consultant with a qualified engineer registered in the state, who would in turn provide the final design recommendations, although the final design could therefore be somewhat different than that proposed by Turnbuckle. Turnbuckle's change in role from designer to consultant is (1) not ethically necessary; (2) not necessary to provide an adequate design; and (3) more costly. However, it does allow Turnbuckle's vast successful experience to be utilized, is legal, and is ethical.

4) When Turnbuckle was first contacted by the contractor, he should have referred to the chart in the new book produced by NSPE entitled, "Engineering Licensure Laws: A State-by-State Summary and Analysis" (NSPE Pub. No. 2015), which would have told him that he cannot apply for a temporary license in that state. As a result, he could have told the prospective client that other arrangements would have to be made due to the registration laws in that state.

**Epilogue**

In fact, none of the design consultants associated with the project was willing to become involved with the remediation of the problem, even though it was shown that it was not physically possible to construct the original design. Despite his timely response and recommendations which averted a real disaster, Turnbuckle was subsequently fined $500 by the State Board of Registration. In addition, he was required by law to notify each of the other 17 states in which he holds professional engineering licenses of the action take against him. Having done that, only two state boards acknowledged receipt of the information. In the interim, I. B. Stout has discontinued his pursuit of monetary relief through court action against Turnbuckle.

[a]Case 1001 is used with permission from National Institute for Engineering Ethics.

---

## 12.6 Ethical Dilemmas

The following group discussion activity can be used to introduce case studies given below. Similar group discussion can also be used for any ethics discussion in the classroom. Through this discussion students will be able to differentiate between a choice based on facts and a choice based on values.[11]

---

11 "Ethical Dilemmas," Adapted from Anthropology Outreach Office, Smithsonian Institution, http://anthropology .si.edu/outreach/Teaching_Activities/edethica.html, accessed August 21, 2017.

## Process of the discussion

1) Choose four or five students to be a part of a group to discuss case studies. Present this group in front of the class so that your leadership role and the group discussion can be observed.

2) Read the case aloud and then ask the group to examine the given situation.

   a) Ask the group to describe exactly what happened. What information is missing from the case?

   b) What issues and problems does the case raise and why?

   c) How would you have acted in this same situation?

3) As a group leader demonstrate your role by primarily (a) asking questions, (b) clarifying students' answers, (c) linking together different responses, and (d) summarizing the understandings of the situation. Make sure you state clearly to the students that there are no right or wrong answers.

4) Split the class up into four or five groups and have each person from the initial demonstration group act as a group leader. Assign each group a case to analyze for 15 minutes, using the same guidelines outlined above.

5) One person from each group presents a summary of the case and the group's conclusions. As a whole class, the teacher and the students might consider:

   • What are the relevant ethical considerations in this case?

   • What are the interests of the various players?

   • Were there conflicts of interest?

   • What are your options?

---

### Case No. 08-2
### Quality of Products – Defective Chips[12]

*Facts.* Engineer A is an electrical engineer working in quality control at a computer chip plant. Engineer A's staff generally identifies defects in manufactured chips at a rate of 1 in 150. The general industry practice is for defective chips to be repaired or destroyed. Engineer B, Engineer A's supervisor, recently announced that defective chips are to be destroyed, because it is more expensive to repair a defective chip than it is to make a new chip. Engineer A proceeds on the basis of Engineer B's instructions. A few months later, Engineer B informs Engineer A that Engineer A's quality control staff is rejecting too many chips, which is having an effect on overall plant output and, ultimately, company profitability. Engineer B advises Engineer A's staff to allow a higher percentage of chips to pass through quality control. Engineer B notes that in the end, these issues can be best handled under the company's warranty policy under which the company agrees to replace defective chips based upon customer complaints. Engineer A has concerns as to whether this approach is in the best interest of the company or its clients.

*Question.* What are Engineer A's ethical obligations under the circumstances?

---

12 Cases 08-2, 08-4, and 08-12 are reprinted by permission of the National Society of Professional Engineers (NSPE), www.nspe.org.

*References*

1) Section I.6, NSPE Code of Ethics: Engineers, in the fulfillment of their professional duties, shall conduct themselves honorably, responsibly, ethically, and lawfully so as to enhance the honor, reputation, and usefulness of the profession.
2) Section II.4, NSPE Code of Ethics: Engineers shall act for each employer or client as faithful agents or trustees.
3) Section II.3.a, NSPE Code of Ethics: Engineers shall be objective and truthful in professional reports, statements, or testimony. They shall include all relevant and pertinent information in such reports, statements, or testimony, which should bear the date indicating when it was current.
4) Section II.1.b, NSPE Code of Ethics: Engineers shall approve only those engineering documents that are in conformity with applicable standards.

---

## Case No. 08-4
## Recommendation Regarding Mitigation of Electromagnetic Field (EMF) Exposure

*Facts.* A developer retains a contractor to design and build a residential subdivision near several high-voltage power lines. Engineer A, an electrical engineer employed by the contractor, recommends to the contractor and developer to include a protective steel mesh in the homes to be built to mitigate occupants' exposure to interior levels of low-frequency electromagnetic fields (EMF). While Engineer A understands that in the USA there are no widely accepted health and safety standards limiting occupational or residential exposure to 60-Hz EMF, he is aware of and concerned about certain scientific research concerning possible causal links between childhood leukemia and exposure to low-frequency EMF from power lines. Because of the added cost associated with the recommendation, the developer refuses to approve the recommendation. Contractor directs Engineer A to proceed in accordance with the developer's decision.

*Question.* What are Engineer A's ethical obligations under the circumstances?

*References*

1) Section I.1, NSPE Code of Ethics: Engineers, in the fulfillment of their professional duties, shall hold paramount the safety, health, and welfare of the public.
2) Section II.1, NSPE Code of Ethics: Engineers shall hold paramount the safety, health, and welfare of the public.
3) Section II.4, NSPE Code of Ethics: Engineers shall act for each employer or client as faithful agents or trustees.

**Case No. 08-12**
**Public Health, Safety and Welfare – Compliance with Fire Code**

*Facts.* Engineer A, a licensed electrical engineer, works for a state university on construction and renovation projects. Engineer A's immediate manager is an architect, and next in the chain of command is an administrator (Administrator), a man with no technical background. Administrator, without talking to the engineers, often produces project cost estimates that Administrator passes on to higher university officials. In cases where it becomes evident that actual costs are going to exceed these estimates, Administrator pressures the engineers to reduce design features. One such occasion involves the renovation of a warehouse to convert storage space into office space. Among the specifications detailed by Engineer A is the installation of emergency exit lights. These are mandated by the building code. As part of his effort to bring down actual costs, Administrator insists that the specification for emergency lights be deleted. Engineer A strongly objects and when Engineer A refuses to yield, Administrator accuses Engineer A of being a disruptive influence in the workplace.

*Question.* What are Engineer A's ethical obligations under the circumstances?

*References*

1) Section II.1, NSPE Code of Ethics: Engineers shall hold paramount the safety, health, and welfare of the public.
2) Section II.1.a, NSPE Code of Ethics: If engineers' judgment is overruled under circumstances that endanger life or property, they shall notify their employer or client and such other authority as may be appropriate.
3) Section II.1.b, NSPE Code of Ethics: Engineers shall approve only those engineering documents that are in conformity with applicable standards.
4) Section II.1.e, NSPE Code of Ethics: Engineers shall not aid or abet the unlawful practice of engineering by a person or firm.
5) Section II.4, NSPE Code of Ethics: Engineers shall act for each employer or client as faithful agents or trustees.

## 12.7 The NSPE Code of Ethics for Engineers[13]

Engineering is an important and learned profession. As members of this profession, engineers are expected to exhibit the highest standards of honesty and integrity. Engineering has a direct

13 Reprinted by Permission of the National Society of Professional Engineers (NSPE) www.nspe.org.

and vital impact on the quality of life for all people. Accordingly, the services provided by engineers require honesty, impartiality, fairness, and equity, and must be dedicated to the protection of the public health, safety, and welfare. Engineers must perform under a standard of professional behavior that requires adherence to the highest principles of ethical conduct.

*I. Fundamental canons.* Engineers, in the fulfillment of their professional duties, shall:

1) Hold paramount the safety, health, and welfare of the public.
2) Perform services only in areas of their competence.
3) Issue public statements only in an objective and truthful manner.
4) Act for each employer or client as faithful agents or trustees.
5) Avoid deceptive acts.
6) Conduct themselves honorably, responsibly, ethically, and lawfully so as to enhance the honor, reputation, and usefulness of the profession.

*II. Rules of practice*

1) Engineers shall hold paramount the safety, health, and welfare of the public.
   a) If engineers' judgment is overruled under circumstances that endanger life or property, they shall notify their employer or client and such other authority as may be appropriate.
   b) Engineers shall approve only those engineering documents that are in conformity with applicable standards.
   c) Engineers shall not reveal facts, data, or information without the prior consent of the client or employer except as authorized or required by law or this Code.
   d) Engineers shall not permit the use of their name or associate in business ventures with any person or firm that they believe is engaged in fraudulent or dishonest enterprise.
   e) Engineers shall not aid or abet the unlawful practice of engineering by a person or firm.
   f) Engineers having knowledge of any alleged violation of this Code shall report thereon to appropriate professional bodies and, when relevant, also to public authorities, and cooperate with the proper authorities in furnishing such information or assistance as may be required.
2) Engineers shall perform services only in the areas of their competence.
   a) Engineers shall undertake assignments only when qualified by education or experience in the specific technical fields involved.
   b) Engineers shall not affix their signatures to any plans or documents dealing with subject matter in which they lack competence, nor to any plan or document not prepared under their direction and control.
   c) Engineers may accept assignments and assume responsibility for coordination of an entire project and sign and seal the engineering documents for the entire project, provided that each technical segment is signed and sealed only by the qualified engineers who prepared the segment.
3) Engineers shall issue public statements only in an objective and truthful manner.
   a) Engineers shall be objective and truthful in professional reports, statements, or testimony. They shall include all relevant and pertinent information in such reports, statements, or testimony, which should bear the date indicating when it was current.

b) Engineers may express publicly technical opinions that are founded upon knowledge of the facts and competence in the subject matter.

c) Engineers shall issue no statements, criticisms, or arguments on technical matters that are inspired or paid for by interested parties, unless they have prefaced their comments by explicitly identifying the interested parties on whose behalf they are speaking, and by revealing the existence of any interest the engineers may have in the matters.

4) Engineers shall act for each employer or client as faithful agents or trustees.

a) Engineers shall disclose all known or potential conflicts of interest that could influence or appear to influence their judgment or the quality of their services.

b) Engineers shall not accept compensation, financial or otherwise, from more than one party for services on the same project, or for services pertaining to the same project, unless the circumstances are fully disclosed and agreed to by all interested parties.

c) Engineers shall not solicit or accept financial or other valuable consideration, directly or indirectly, from outside agents in connection with the work for which they are responsible.

d) Engineers in public service as members, advisors, or employees of a governmental or quasi-governmental body or department shall not participate in decisions with respect to services solicited or provided by them or their organizations in private or public engineering practice.

e) Engineers shall not solicit or accept a contract from a governmental body on which a principal or officer of their organization serves as a member.

5) Engineers shall avoid deceptive acts.

a) Engineers shall not falsify their qualifications or permit misrepresentation of their or their associates' qualifications. They shall not misrepresent or exaggerate their responsibility in or for the subject matter of prior assignments. Brochures or other presentations incident to the solicitation of employment shall not misrepresent pertinent facts concerning employers, employees, associates, joint venturers, or past accomplishments.

b) Engineers shall not offer, give, solicit, or receive, either directly or indirectly, any contribution to influence the award of a contract by public authority, or which may be reasonably construed by the public as having the effect or intent of influencing the awarding of a contract. They shall not offer any gift or other valuable consideration in order to secure work. They shall not pay a commission, percentage, or brokerage fee in order to secure work, except to a bona fide employee or bona fide established commercial or marketing agencies retained by them.

### III. *Professional obligations*

1) Engineers shall be guided in all their relations by the highest standards of honesty and integrity.

a) Engineers shall acknowledge their errors and shall not distort or alter the facts.

b) Engineers shall advise their clients or employers when they believe a project will not be successful.

c) Engineers shall not accept outside employment to the detriment of their regular work or interest. Before accepting any outside engineering employment, they will notify their employers.

d) Engineers shall not attempt to attract an engineer from another employer by false or misleading pretenses.

e) Engineers shall not promote their own interest at the expense of the dignity and integrity of the profession.

2) Engineers shall at all times strive to serve the public interest.

a) Engineers are encouraged to participate in civic affairs; career guidance for youths; and work for the advancement of the safety, health, and well-being of their community.

b) Engineers shall not complete, sign, or seal plans and/or specifications that are not in conformity with applicable engineering standards. If the client or employer insists on such unprofessional conduct, they shall notify the proper authorities and withdraw from further service on the project.

c) Engineers are encouraged to extend public knowledge and appreciation of engineering and its achievements.

d) Engineers are encouraged to adhere to the principles of sustainable development[14] in order to protect the environment for future generations.

3) Engineers shall avoid all conduct or practice that deceives the public.

a) Engineers shall avoid the use of statements containing a material misrepresentation of fact or omitting a material fact.

b) Consistent with the foregoing, engineers may advertise for recruitment of personnel.

c) Consistent with the foregoing, engineers may prepare articles for the lay or technical press, but such articles shall not imply credit to the author for work performed by others.

4) Engineers shall not disclose, without consent, confidential information concerning the business affairs or technical processes of any present or former client or employer, or public body on which they serve.

a) Engineers shall not, without the consent of all interested parties, promote or arrange for new employment or practice in connection with a specific project for which the engineer has gained particular and specialized knowledge.

b) Engineers shall not, without the consent of all interested parties, participate in or represent an adversary interest in connection with a specific project or proceeding in which the engineer has gained particular specialized knowledge on behalf of a former client or employer.

5) Engineers shall not be influenced in their professional duties by conflicting interests.

a) Engineers shall not accept financial or other considerations, including free engineering designs, from material or equipment suppliers for specifying their product.

---

14 "Sustainable development" is the challenge of meeting human needs for natural resources, industrial products, energy, food, transportation, shelter, and effective waste management while conserving and protecting environmental quality and the natural resource base essential for future development.

b) Engineers shall not accept commissions or allowances, directly or indirectly, from contractors or other parties dealing with clients or employers of the engineer in connection with work for which the engineer is responsible.

6) Engineers shall not attempt to obtain employment or advancement or professional engagements by untruthfully criticizing other engineers, or by other improper or questionable methods.

   a) Engineers shall not request, propose, or accept a commission on a contingent basis under circumstances in which their judgment may be compromised.

   b) Engineers in salaried positions shall accept part-time engineering work only to the extent consistent with policies of the employer and in accordance with ethical considerations.

   c) Engineers shall not, without consent, use equipment, supplies, laboratory, or office facilities of an employer to carry on outside private practice.

7) Engineers shall not attempt to injure, maliciously or falsely, directly or indirectly, the professional reputation, prospects, practice, or employment of other engineers. Engineers who believe others are guilty of unethical or illegal practice shall present such information to the proper authority for action.

   a) Engineers in private practice shall not review the work of another engineer for the same client, except with the knowledge of such engineer, or unless the connection of such engineer with the work has been terminated.

   b) Engineers in governmental, industrial, or educational employ are entitled to review and evaluate the work of other engineers when so required by their employment duties.

   c) Engineers in sales or industrial employ are entitled to make engineering comparisons of represented products with products of other suppliers.

8) Engineers shall accept personal responsibility for their professional activities, provided, however, that engineers may seek indemnification for services arising out of their practice for other than gross negligence, where the engineer's interests cannot otherwise be protected.

   a) Engineers shall conform with state registration laws in the practice of engineering.

   b) Engineers shall not use association with a nonengineer, a corporation, or partnership as a "cloak" for unethical acts.

9) Engineers shall give credit for engineering work to those to whom credit is due, and will recognize the proprietary interests of others.

   a) Engineers shall, whenever possible, name the person or persons who may be individually responsible for designs, inventions, writings, or other accomplishments.

   b) Engineers using designs supplied by a client recognize that the designs remain the property of the client and may not be duplicated by the engineer for others without express permission.

   c) Engineers, before undertaking work for others in connection with which the engineer may make improvements, plans, designs, inventions, or other records that may justify copyrights or patents, should enter into a positive agreement regarding ownership.

d) Engineers' designs, data, records, and notes referring exclusively to an employer's work are the employer's property. The employer should indemnify the engineer for use of the information for any purpose other than the original purpose.

e) Engineers shall continue their professional development throughout their careers and should keep current in their specialty fields by engaging in professional practice, participating in continuing education courses, reading in the technical literature, and attending professional meetings and seminars.

### 12.7.1 As Revised July 2007

By order of the United States District Court for the District of Columbia, former Section 11(c) of the NSPE Code of Ethics prohibiting competitive bidding, and all policy statements, opinions, rulings or other guidelines interpreting its scope, have been rescinded as unlawfully interfering with the legal right of engineers, protected under the antitrust laws, to provide price information to prospective clients; accordingly, nothing contained in the NSPE Code of Ethics, policy statements, opinions, rulings or other guidelines prohibits the submission of price quotations or competitive bids for engineering services at any time or in any amount.

## Bibliography

1 AMERICAN ANTHROPOLOGICAL ASSOCIATION, *Anthropology Newsletter*, American Anthropological Association, Arlington, VA.

2 1981 issues contain a series on ethical dilemmas prepared under the auspices of the Committee on Ethics (cases 3, 4, and 5).

3 October 1995 issue has an article by Bernard Gert, "Universal values and professional codes of ethics."

4 APPELL, G.N., *Ethical Dilemmas in Anthropological Inquiry: A Case Book*, Crossroads Press, Waltham, MA, 1978.

5 FLUEHR-LOBBAN, C. (ed.), *Ethics and the Profession of Anthropology: Dialogue for a New Era*, University of Pennsylvania Press, Philadelphia, 1991.

# 13

# Communications in Engineering

Good writing does not succeed or fail on the strength of its ability to persuade. It succeeds or fails on the strength of its ability to engage you, to make you think, to give you a glimpse into someone else's head.

**Malcolm Gladwell**

## 13.1 Introduction

Industry has pushed hard for increased emphasis on written and oral communications in the engineering curriculum to better prepare graduates in this area because engineers are often accused of lacking these skills. Universities have responded to this concern by adding courses in engineering report writing and oral communications to the curriculum; they have also increased the writing and oral presentation requirements. In spite of all this effort to improve the communications capabilities of graduates, it has not changed the perception of engineers as ineffective communicators. This could be due to a lack of interest by students in English and language courses in secondary and tertiary education. Many, if not most, engineering students have a natural inclination for, and interest in, math and science and little of either for writing and oral communications. Without a serious interest in developing communication skills, it is doubtful that a student's capabilities in this area can be significantly improved.

A comment often heard that has some applicability here is that to be an effective writer, a person needs to be an avid reader. The US education system does not effectively motivate students to read, and this could contribute to the poor writing skills and vocabulary of its students. If engineering students were required to have a good dictionary and thesaurus at their desk and would get in the habit of using them, the quality of their writing could be improved immensely. There is an inherent relationship between certain specific words and knowledge; until this meaning is grasped, the related thought cannot be experienced. Thus, improvement in writing ability, to some extent, implies increased knowledge, which would be of great value for all engineers, beginners and journeymen alike. The quality of the communication courses that are offered should be addressed. Many secondary school and university English courses emphasize other aspects of writing and not writing itself. This causes engineering students to often have little understanding of grammar and its mechanics. This results in them having considerable difficulty in writing clearly on technical subjects. To assist these students, some universities have established communication centers within their colleges of engineering to provide guidance for and critique of written reports, theses, and other manuscripts.

Information flow is critical to business and industry. Engineers in industry often comment on the large portion of their time that is committed to writing and other forms of communication. Most business and industry communications are verbal; this includes face-to-face discussions, meetings, and telephone conversations. Important communications are transmitted in writing so that others do not misunderstand the meaning. Employees who are not capable of producing clear written communications are limited in their work and advancement in their career. To a large extent, an employee's value is measured by the ability to communicate clearly and effectively. In effect, engineers market their skills through their ability to communicate. It is much better to be dealing with a bull market than a bear market in this context.

## 13.2   The Formal Engineering Report

The engineering report is likely the greatest writing challenge a typical engineer has to face. Engineering reports document a significant portion of work and, as such, represent an important commitment on the part of the firm in which the engineer is employed. The engineering report is commonly the only document that describes how and why the work was accomplished and what the results, recommendations, and conclusions were made by the engineer. It is usually the only document that is retained for future reference. The report may be the principal or only measure of the quality of the work for customers or other persons outside the organization.

Due to the variety of purposes for which engineering reports are prepared, the format must be flexible. A report on a test program will undoubtedly require a different format than that on a failure analysis. A report on an engine test will probably require a different format than that on a computer component. Some organizations will have a basic report format that can be adapted to meet their particular need. The general format outlined below includes all of the elements that should normally be considered for inclusion in a typical report, though only a small number of reports prepared by engineers will actually require all of the elements described. Engineering reports are prepared for many different purposes. A few of these are listed below:

1) Test programs
2) Finding solutions to technical problems
3) Experiments
4) Purchase of equipment
5) Research investigations
6) Failure analyses
7) To make recommendations to technical supervisors.

One must bear in mind that individuals who are not intimately familiar with the work or its author may read the engineering report. Their impressions are thus formulated by the content of the report and their ability to understand what was accomplished as well as its meaning. In some cases, engineering reports are prepared for non-technical customers, which only exacerbates this problem. The challenge is to discuss the work accomplished in terms that are understandable to someone not directly involved in the project, possibly someone who is not an engineer or technical person. It is paramount that the author knows the audience and prepares the report so that they can understand it.

Many inexperienced writers struggle to establish a logical content flow in their reports. If the reader is to understand the report, the presentation of information must follow some logical thread. The material presented should be introduced in chronological sequence within each division of the report, and periodic signposts should be provided so that the reader always knows where they have been and where they are going on their journey through the report. This goal can be achieved by previewing what is coming next in the report in each major sub-division and by beginning each paragraph with a clear topic statement.[1]

Formal engineering reports are generally written in the third person, past tense, so personal pronouns should be avoided. The report is prepared in a purely objective, impersonal manner, reflecting the writer's relationship to the material presented. Results are judged on the criteria of existing, applicable theory, and previous experiments. Opinions are introduced only when existing knowledge is not accurate.

Reports should be presented in the following order: title of the report, author's name, abstract, introduction, state of the art (literature review), technical approach (theory), experimental setup and procedure, results and discussion, conclusions, acknowledgments, references, and appendix.

## 13.2.1  The Abstract

The abstract is simply a short (normally 300 words or less), standalone summary of the essential contents of the report. The abstract often exists apart from the report in library collections. Most article databases in the online catalog of the library enable you to search abstracts so that you can review them and decide whether or not to read the main body of the document. A short summary (often called an executive summary) is sometimes used in place of an abstract, though this will never exist separately from the report. An abstract should in most cases include some of the following:

*Motivation.* Explain why readers should care about the problem and the results, why the problem is important.
*Problem statement.* Explain what problem are you trying to solve.
*Objective(s).* Set out the major objectives/hypotheses of the report.

---

1  Harbinger, S.A., Whitmer, A.B., and Price, R., *English for Engineers*, McGraw-Hill, New York, 1951.

*Procedure.* Explain how you solved the problem – method, simulation, analytic models, etc., used to satisfy the objective(s).

*Results.* Explain, as a result of completing the above procedure, what answers you cam up with.

*Conclusions.* Explain how you met your objectives. State whether you met the design requirement. Include the main contributions of the work.

An example of a typical abstract is given below.

### Abstract

Structures, due to their light weight and low stiffness, have shown a potential problem of wind-induced vibrations, a direct outcome of which is fatigue failure. In particular, if the structure is long and flexible, failure by fatigue will be inevitable if not designed properly. The main objective of this report is to perform a theoretical analysis of a novel free pendulum device as a passive vibration absorber. In this report, the beam-tip mass-free pendulum structure is treated as a flexible multibody dynamic system and the Absolute Nodal Coordinate Formulation (ANCF) is used to demonstrate the coupled nonlinear dynamics of a large deflection of a beam with an appendage consisting of a mass-ball system. The results show that the complete energy transfer between two modes occurred when the beam frequency is twice the ball frequency. Results are discussed and compared with the findings of MSC ADAMS simulation software. It is shown that this novel free pendulum device is a practical and feasible passive vibration absorber in the mitigation of large amplitude wind-induced vibrations in traffic signal structures.

### 13.2.2 The Introduction

The introduction is one of the key and the most read section of a report. The main purpose of the introduction is to provide the necessary background to the field the report is about, to describe the objective(s) of the problem, and to define the method and scope of the investigation. Figures, tables, equations and calculations should not be used in the introduction unless they are absolutely necessary. The introduction should be approximately one page, but can be either longer or shorter depending on the subject being introduced. An example introduction is given below. Note that Introduction is numbered Section 1.

### 1. Introduction

Acoustics is the science that studies the emission, transmission, and reception of sound waves. It touches on disciplines as different as psychology and meteorology, and includes many sub-disciplines such as architectural acoustics, structural acoustics, bio-acoustics, environmental acoustics, and musical acoustics. The physical indication of sound is a time-dependent pressure variation around a static pressure in a compressible fluid, such as air or water. Noise or acoustic noise, on the other hand, is defined as unwanted sound and therefore it is a sound that annoys, disturbs, bothers, irritates, perturbs, agitates, interferes with, distracts, or harms. Naturally, when talking about sound, especially noise, the audible frequency range – that is, the range between about 16 Hz and 16 kHz – is of primary interest. Vibrations and waves at lower frequencies (infrasound) generally belong to the fields of mechanical vibrations or seismics, whereas those at higher frequencies belong to the field of ultrasonics.

Since noise is an unwanted sound, there may be cases in which it should be eliminated or at least minimized. Therefore the quantification of sound radiated from structures is an important issue from a noise control point of view. It is also important for the engineer to have general expressions for acoustic properties of structural elements. In acoustics, generally speaking, finding closed-form or even approximate expressions for these properties may be difficult unless some practical restrictions on the physical shapes of the structures are included. The most important structural elements are, in the applied mechanics sense, beams, plates, or shells. Therefore, there seems to be adequate motivation for developing analytical tools for the sound radiated from structural elements to take effective noise control measures.

The main objective of this study is to develop general analytical expressions for far-field acoustic pressure distribution, acoustic intensity, and acoustic power radiation due to harmonically vibrating planar sources, namely pistons and thin plates. Along with the far-field integration approach, the following steps are used to obtain the necessary acoustic quantities:

- assumption of a displacement function for the piston, or solution of the plate equation with the boundary conditions to obtain a displacement function for the plate;
- derivation of the surface velocity distribution of the source from the displacement function.

Triangular, rectangular, circular, and elliptical sources are studied. Clamped and simply supported boundary conditions are considered for plates. The dependence of the results obtained on various design parameters, such as frequency of vibration and source geometry, is investigated.

*Background.* Acoustics is the science that studies the emission, … unwanted sound and therefore it is a sound that annoys, disturbs, bothers, irritates, perturbs, agitates, interferes with, distracts, or harms.

*Objectives.* We will develop general analytical expressions for far-field acoustic pressure distribution, acoustic intensity, and acoustic power radiation due to harmonically vibrating planar sources, namely pistons and thin plates.

*Method.* Along with the far-field integration approach, the following steps are used to obtain the necessary acoustic quantities …

> *Scope.* Triangular, rectangular, circular, and elliptical sources are studied. Clamped and simply supported boundary conditions are considered for plates. The dependence of the results obtained on various design parameters, such as frequency of vibration and source geometry, is investigated.

### 13.2.3  Literature Review

The goal of a literature review is to show that you have read, and have a good grasp of, the main published work concerning a particular topic or problem. The literature review can be a separate section, or may be included in the introductory sections of a report. This part of the report demonstrates that your work has not been previously done. An example of a literature review is given below.

> **2. Literature Review**
>
> In the present study, the analysis is limited to thin plates having three different geometric shapes, namely, rectangular, circular, and elliptical plates. Therefore, this literature survey is also limited to these cases.
> Leissa [1] investigated the vibrations of plates in detail. In this comprehensive study, he compiled the available knowledge for the frequencies and mode shapes of circular, elliptical, triangular, rectangular, parallelogram, and several other quadrilateral plates. He also analyzed anisotropic plates, plates with inplane forces, and plates with variable thickness.
> Warburton [2] studied the free transverse vibrations of rectangular plates. He considered all possible combinations of the three different boundary conditions (free, clamped, and simply supported) and used assumed deflection functions satisfying the plate equation and the boundary conditions to find the frequencies. The free vibration analysis and the assumed deflection functions presented in this paper constitute the basis for the analysis of thin rectangular plates in this study. ...

Note that references should be numbered consecutively in accordance with their appearance in the text. All publications cited in the text should be included in a numbered references list placed at the end of the report.

### 13.2.4  Technical Approach (Theory)

The technical approach/theory should include an identification of the theoretical principles involved and the equations used in making calculations from the experimental data. The manner in which the various principles and equations were used in accomplishing the work should be described. Equations should be presented in the order in which they were used in making the calculations included in the report. The use of supporting sketches, diagrams, and curves is recommended to assist in clarifying and amplifying the text in this section. Whenever figures, tables, and the like are used in a report they must always be referred to in the text ahead of the figure or table, and they must be in sequential order.

The technical approach should also include any discussion of experimental uncertainties that are appropriate. The required levels of accuracy and precision vary with the nature and purposes of experimental work, but the report on any project should always examine the accuracy of the measurements involved as they relate to the overall objectives and include the appropriate discussion and/or theory.

Portions of a typical technical approach section are given below:

### 3. Technical Approach (Theory)

The acoustic pressure produced at any far-field point by the vibration of a planar source is the sum of the pressures that would be produced by an equivalent assembly of simple sources; in other words, infinitesimal elements on the surface of the source. Each infinitesimal element of area $dS$ of a harmonically vibrating source mounted in an infinite baffle, as shown in Figure 13.1, contributes an element of pressure $dp$ given by

$$dp = -i\frac{\rho c k}{2\pi d} u dS \exp[i(-\omega t + kd)] \tag{13.1}$$

at a far-field point P. In equation (13.1), $\rho$ is the equilibrium density of the acoustic medium, $c$ is the speed of sound in the same medium, $d$ is the distance between the far-field point and the infinitesimal element, $u$ is the transverse velocity distribution on the surface of the source; and $\omega$ is the frequency of vibration.

An equation for the acoustic pressure at this far-field point can be obtained by integrating $dp$ given by equation (13.1) over the surface of the vibrating source.

## 13.2.5 Experimental Setup and Procedures

A complete identification of the component or system to be tested and of all significant test equipment should be provided in a test setup section. A neat sketch/schematic of the test setup should be prepared with the components and instruments used during the test or experiment

**Figure 13.1** A planar source and the coordinate system.

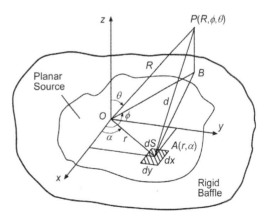

identified. Identification can be provided directly on the sketch or schematic or by including a list with numerical identification corresponding to that on the schematic. A narrative should be provided that describes the function of the major components in the test setup. The reader should be able to interrelate the narrative and the schematic to gain a full understanding of how the equipment was used in the test or experiment.

In performing critical tests wherein data may be questioned by other contractors or the customer after the test, it is important to identify all instrumentation completely, including serial numbers. With this information available, the status of the equipment during the test can be ascertained, including instrument calibration data. For this type of report, it is important to include adequate information on the test setup so that the test can be repeated, if required. An example test setup report section is given below:[2]

### 4. Experimental Setup

The test apparatus used in this study is the experimental simulation of the dynamics of an impacting spherical pendulum with large angle and parametric forcing. The pendulum system was studied with eight different bobs and two different base configurations with external frequency 25.6–25.9 Hz. Comparative analysis was performed at low and high coulomb damping values for the inverted, impacting pendulum model shown in Figure 13.2.

Figure 13.2 shows the spherical coordinate axes where the length of the pendulum is $L$, the latitudinal angular displacement is $\phi$, the longitudinal angular displacement is $\theta$, and the motion of the base is in the vertical direction. The pendulum ball is initially at rest condition on the constraint. The connecting rod in an ideal system is massless, perfectly rigid and straight. The rod for this system was made of a graphite composite arrow shaft of length 167.2 mm. To have an ideal pendulum model the bearing and sensor system had a zero mass moment of inertia. Movement of the pendulum was allowed in a 45 degree cone from the vertical for all values

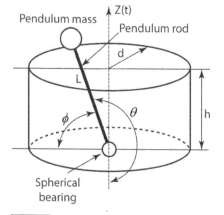

**Figure 13.2** Model of a three-dimensional spherical pendulum.

2 Adapted from Ertas, A., and Garza, G.S., "Experimental investigation of dynamics and bifurcations of an impacting spherical pendulum," *Experimental Mechanics*, 49, 653–662, 2009.

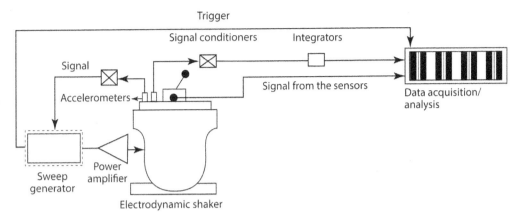

**Figure 13.3** Experimental setup.

of latitude. Two types of sensors were used. In the low coulomb damping model a linear magnetic sensor was used, and in the relatively high coulomb damping model two perpendicular potentiometers were used. In the magnetic sensor design it was assumed that the sensor does not attract, repel or damp the pendulum. In the potentiometer setup, damping was measured in both axes and found to be nearly identical.

The experimental setup is shown in Figure 13.3. The model was excited by the shaker table (the amplitude and frequency of which are controlled by the vibration control system). The electrical voltage output from the analog sensors was converted to a digital signal by using A/D converter. LabVIEW software was used to allow filtering, storage, and manipulation of the collected data. In this study, the sampling rate was 300 samples per second. The data was filtered by a fifth-order Butterworth filter above 50 Hz and stored on disk. The voltages were converted to angles for use in the analysis.

During the experiments, the frequency of the excitation remained between 25.6 Hz and 25.9 Hz. The amplitude varied from 0 to 12 mm peak to peak. The parameters used to investigate the pendulum system response for both low and relatively high coulomb damping are given in Table 13.1.

## 13.2.6 Results and Discussion

An objective discussion of the results and analysis to show that the conclusions are warranted is the most important requirement of any engineering report. Each major conclusion must be clearly substantiated, and any contradictory theories or results must be explained. All statements should be clear to readers not as well acquainted with the subject matter as the author. All discussion in the text should be supported with graphs and tables to the extent possible. Graphs supported by tables is the preferred approach for showing trends and comparisons, substantiated by the tabulated data and calculated values. All statements about the results

**Table 13.1** Multiple bob setup characteristics, $L = 167.2$ mm, $r_r = 20$ mm, $r_b = 12.69$ mm.

| Setup number | $\Omega$ (HZ) | $r_s$ (mm) | $R$ | $C$ (g-mm/s²) | $M_d$ (g) | System mass moment of inertia (g-mm²) |
|---|---|---|---|---|---|---|
| 1 | 25.6 | 17.78 | 0.945 | 82348 | 22.039 | 604997 |
| 2 | 25.9 | 12.69 | 0.931 | 82348 | 8.419 | 226076 |
| 3 | 25.7 | 15.82 | 0.896 | 82348 | 15.875 | 433267 |
| — | — | — | — | — | — | — |
| — | — | — | — | — | — | — |
| — | — | — | — | — | — | — |
| 15 | 25.6 | 18.71 | 0.87 | 315209 | 24.955 | 688113 |
| 16 | 25.9 | 15.49 | 0.844 | 315209 | 15.162 | 414649 |

and any numerical values cited must agree precisely with information in the tables and on the figures.

One way to build a good discussion is to compare each result with its predicted or expected value. Some discussion as to why the experimental data should agree with the predicted or expected data should also be included. If the data do not agree with accepted theory, an explanation should be offered, if possible. If the reason for the discrepancy cannot be determined, state that this is the case. If an error or discrepancy has occurred in the data or during the conduct of the experiment, recognize it briefly and continue with the discussion. Do not dwell on events that add little or nothing to the main thrust of the effort being described. This is the most important section of the report, so be sure to develop an adequate discussion and analysis that will convince the reader that the conclusions are valid. Example paragraphs from a typical results and discussion report section are given below.

### 5. Results and Discussion

A standard configuration was chosen for in depth study of the transitions in vibrational response. Multiple configurations were studied to determine what affected these transitions and how they could be detected. A single bob was chosen as the standard setup. This bob had a relatively high coefficient of restitution and a relatively high mass moment of inertia. During the experiments, power levels, $a\Omega$, varied from 0 to 125 mm-Hz. Each experiment was run three times in order to verify data.

As shown in Figure 13.4, experimental evidence indicated that Type I responses occur in this system. A Type I response was period 1 impacting the restraint within a localized range of the latitudinal angle, $\phi$ and with the longitudinal angle $\theta$. Type I responses for the standard bob at high coulomb damping are shown in Figure 13.4. The data shows a spike at 25 mm-Hz; an FFT at this point indicates phase-locking

**Figure 13.4** Type I response at high coulomb damping.

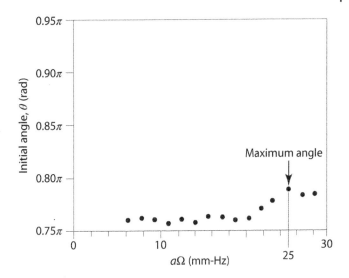

**Figure 13.5** Type I position plot at high coulomb damping (parameters specified in Table 13.1, setup 10, 25 mm-Hz).

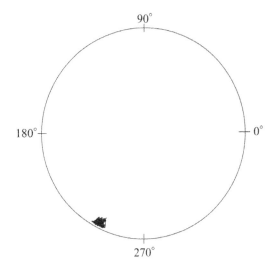

of the response. Response restricted to a Type I variety occurs in the lower range of the shaker power (less than 30 mm-Hz output).

In Figure 13.5, a position plot is shown for a Type I response of the high coulomb damping model. The position plot shows a line where the pendulum traveled. The view is from above the pendulum, and shows that the bob stayed in a tightly confined region and did not move along the periphery of the restraint.

### 13.2.7 Conclusions

Conclusions reached in the results and discussion section should be restated here in a more general and less specific manner. Conclusions are normally listed in the order of their importance, preferably in itemized form, although a narrative presentation is completely satisfactory. Already known facts should not be included in this section, and conclusions should not be confused with factual test or experimental results. If conclusions are qualified, give the assumptions and limitations that apply.

No new information should be introduced in this section. The conclusions provide a quick reference for the reader with limited time. Often a reader will only read the abstract and the conclusions sections of the report. Thus, it is imperative that conclusions be stated concisely and accurately, with whatever qualifications apply. Do not discuss the conclusion but use the discussion in the previous section to extract the significant thought(s). An example conclusions section is included below:

### 6. Conclusions

The experimental results presented in this report indicate that coulomb damping can make a difference in restraint separation. The data showed that extra energy is required to overcome the additional loss due to coulomb damping. Several bobs were evaluated that would not exhibit a Type II response regardless of the power input. The one common factor was their low coefficient of restitution (less than 0.80). Restraint separation criteria showed for experimental data that there is no correspondence to either the mass moment of inertia or the coefficient of restitution.

### 13.2.8 References

There are several acceptable systems for listing references in an engineering report. The *Publication Manual of the American Psychological Association* (APA) and *The Chicago Manual of Style* are commonly used citation listing styles. In APA style, all publications cited in the report text should be included in a numbered references list placed at the end of the report. Citations in the main text are indicated with numbers in square brackets, in line with the text; for example, [1] or Doak [1]. Inclusion of the author(s) is optional, but the reference number(s) must always be given. Number references consecutively in accordance with their appearance in the text. Number the list of references in the order in which they are cited in the text. Include the last names and first initials of each author in your references as shown in APA examples below.

### 7. References

*For Journal Articles*
[1] Mann, B. P. (2007). Experimental study of an impact oscillator with viscoelastic and hertzian contact. *International Journal of Nonlinear Mechanics, 50*(2), 587–596.

*For Books*
[2] Ertas, A., and Jones, J.C. (1996). *The Engineering Design Process.* New York: John Wiley & Sons.

*For Proceedings*
[3] Khalilolahi, A., and Torab, H. (1994). Opposing mixed convention near discretely heated boards. In Faghri, A., and Yahoubi, M. A., *Advances in Heat Transfer* (pp. 7–12). New York: American Society of Mechanical Engineers.

*Web References*
As a minimum, the full URL should be given and the date when the reference was last accessed. Any further information, if known (author names, dates, reference to a source publication, etc.), should also be given. Examples are:

[4] http://www.sfwt.org/files/education/90-75.pdf (accessed September 15, 2004).
[5] Tress, B. and Fry, G. Defining concept and the process of knowledge production in integrative research. http://library.wur.nl/frontis/landscape-research/02-tress .pdf (accessed August 18, 2012).

## 13.2.9   Appendix

Appendices are used to provide supporting data that are not significant enough to be included in the body of the report. In this manner, the body of the report can focus on material essential to the discussion and conclusions, and readers will not be distracted by supporting information of little importance in the main body of the report. Less important information can be included in the appendix without cluttering up the more significant portions of the report. Appendices normally include items such as original data sheets, sample calculations, calibration data, uncertainty analyses, instrument charts, detailed test procedures, test codes, and bulletins. Using data in the appendix, the seriously interested reader should be able to validate most, if not all, the graphs, plots, results, etc., included in the report.

## 13.2.10   Acknowledgements

The acknowledgements tell you who helped the author in carrying out the research. A page of acknowledgements is usually included at the beginning of a report, immediately after the table of contents. Acknowledgments should be brief, in a professional style, and should not exceed two pages. A simple example of acknowledgements is included below:

### Acknowledgements

Financial support for this study has been provided by the State of Texas under the Energy Research in Application Program (ERAP). The authors also gratefully acknowledge the support of [company name] and wish to thank Mr. Mike Sullivan.

Note that the acknowledgements do not have a section number.

## 13.3  Proposal Preparation

The preparation of proposals is a task that most engineers encounter at one time or another during their professional careers. Preparation of an engineering proposal can result from a formal request for proposal (RFP), from an informal suggestion from a potential customer that a proposal in an area of interest would be favorably received, in response to a request from in-house management, or from an idea or need that is thought to exist but for which there has been no formal or informal request.

The following suggestions should be kept in mind while preparing a proposal:

1) In responding to a formal RFP the format specified in the RFP should be followed verbatim and should be deviated from only in the most extreme situations. This procedure will at times cause some frustration in trying to fit the information considered essential to describing the idea into the format suggested, but it is the prudent thing to do. Proposal evaluators are inclined to use the format specified in the RFP to check and evaluate the material provided in the proposal. If it is difficult to find or to relate certain elements required by the RFP to ones included in the proposal, evaluators are likely to reflect this difficulty in their evaluation. Page limitations and other restrictions should also be complied with or the proposal may be rejected without further consideration.

2) For an unsolicited proposal, or one in which the format is not specified, organization and document format are extremely important and should be given commensurate attention. The goal should be to make the evaluator's job as easy as possible. Thus, the material submitted should be organized so that all essential information is included within the proposal in a logical and ordered fashion. Information provided in the proposal should flow from a logical beginning to a suitable conclusion and be structured so that the evaluator does not have to hunt throughout the document to find all of the information on any subject.

3) It is important to be specific in regard to the product of the proposed effort. If the product is a final report, it should be stated clearly what will be included in this document. If the product is some item of hardware, it should be described adequately. The product is what the customer is going to pay for, and it thus plays an important role in determining whether or not the project will be funded.

As with formal engineering reports, the format for proposals needs to be adapted to the effort being described. However, the sections outlined below are typically included in most proposals in one form or another. Below you will find a sample proposal for a new approach to hyperspectral characterization of the life cycle of *Escherichia coli*.

### 13.3.1  The Project Summary

The proposal project summary is certainly the most important component of any proposal. This section is not about summarizing the proposed work. The purpose of the proposal project summary is to sell your idea for solving the customer's problem. It should be persuasive, outlining why the customer should choose your proposal. An example proposal project summary is given below:

**Project Summary**

The goal of this proposal is to develop novel tools and validation methods for potential hyperspectral techniques to detect, quantify, and qualify a selected target such as *Escherichia coli*, or *E. coli*. Conventional and reliable laboratory methods remain the mainstay to isolate and identify suspected sessile and planktonic biomaterial. However, the FDA Food Safety Modernization Act, Jan. 4, 2011, is requiring additional proactive identification and monitoring. With recent outbreaks, reported by the CDC, the need for faster identification is again emphasized.

Advancement in hyperspectral data mining techniques will be used to ensure and improve the safety of our food supply, animal feed quality control, faster and more consistent biomedical analysis. A statistical method will be develop to generate specific signatures and then prove them in an independent study. With a selected target identified as *E. coli*, the signature features of *E. coli* will be characterized using Markov chain Monte Carlo implementations. The spectral signatures will lead us to understand the colonies' quality of life. Then we can develop a specific test to discover effective means to counter the defensive mechanisms of the biofilm.

Imagery will be collected for analysis and characterization such that a noninvasive method can be provided for microbiologists to determine the health, thus effectiveness of treatment, of an *E.coli* community being studied. This solves the probing problem as biologist do not really know the effective time of death nor the causal actions of death. Outcome of this research will provide microbiologist researchers an additional noninvasive tool to quantify and qualify target species and their state of life.

## 13.3.2 Background/Problem Statement

In general, all proposals include information that establishes the background for the research problem. Including background material addresses why the proposed work is important in the field and ensures that the potential customer recognizes that the proposer understands the problem and is capable of developing a solution. Sufficient details should be given in this section to address questions such as: What previous work has been done in this area (background or literature review), and what was discovered? What is the problem? What are the motivational factors? Why is this problem of significance?

The various parts of a typical background/problem statement section are as follows:

**Background**

An admittedly narrow search seems to indicate the demand for sensor-based decision support in agriculture and biomedical system is rapidly growing. A fast and precise identification of fungal, bacterial, and other pathogenic agents is sorely needed. The following literature review helps to establish the need and the current state of the art. Hyperspectral imaging systems are assuming a greater importance for a wide variety of commercial and military systems [1]. ...

## Problem Statement

This research effort will concentrate on two distinct tasks that will yield a contribution to the field of hyperspectral (HS) remote sensing and potentially to microbiology. The first goal is the development of a method to statistically create HS signatures that will allow the interested user community to scan for the existence of certain or selected biological specimens. The second goal is to create an environment or model that will allow for testing of the signature with a variable number of HS data sets (known as cubes) representing variable concentrations of heterogeneous communities. The aim of the model is to help predict signature performance against variable backgrounds. ...

## Motivation

Applicable areas of interest are quite wide. A general search on the internet demonstrates that several industrial areas are contributing to the 10 billion dollar hyperspectral industry. Military applications allow the spotting of vehicles or other objects under camouflage, and the identification and geolocation of drug crops from space-based telescopes benefit law enforcement agencies. Agriculture applications are increasing yields that help feed the world, medical imaging, environmental and weather monitoring are all benefiting from the growing technology base. Many commercial sources indicate a deep penetration and increased needs in the mineral mining industries. The news continues to announce occurrences of food supply contaminations, by *E. coli* or other bacteria that produce harmful effects in humans and animals. It is clear that hyperspectral imaging and analysis will contribute to the body of knowledge concerning passive remote sensing of the food supply. ...

## Significance

How do bacteria get into my food? Microorganisms are present on food products when you purchase them. For example, plastic-wrapped boneless chicken breasts and ground meat were once part of live chickens or cattle. Raw meat, poultry, seafood, and eggs are not sterile. Foods, including safely cooked and ready-to-eat foods (prepared salad bags), can become cross-contaminated with pathogens transferred from raw egg products and raw meat, poultry, and seafood products and their juices, other contaminated products, or from food handlers with poor personal hygiene.

What is needed? Hyperspectral imaging could be used at critical control points of food processing to inspect for potential contaminants of the supply of poultry, produce, and grains. HSI technology could contribute significantly to risk reduction goals in food safety by effectively detecting microbial pathogens such as *Salmonella* and *E. coli* O157:H7. ...

### 13.3.3  Objectives

After the problem has been described, a clear and concise statement of the objective(s) of the proposal should be given. The objective of the effort is sometimes included in the background/problem statement portion of the proposal, but if this is the case it should be located toward the end of this section. If the objective(s) is included as a separate section of the proposal, it can be brief, often only a few sentences. If the effort is to be accomplished in several phases, it is sometimes helpful to identify objectives for each separate phase. For example:

#### Objective

The primary objective of this research problem is to discover and develop a means to create hyperspectral signatures by employing a Markov chain Monte Carlo (MCMC) state machine to identify bands (supporting dimensional reduction), and feature signatures with a reasonably low $P_{fa}$. The research will focus on an *E. coli* species, specifically the K-12 strain. ...

### 13.3.4  Technical Approach

This section of the proposal describes how the effort will be conducted. In many ways this section is the most important section of the proposal, since it spells out how the proposer will go about finding a solution to the problem that has been identified. The technical approach portion of the proposal must convince the potential customer that the proposer understands the problem and that an approach has been devised for finding a solution that has a high probability of success. The technical approach should address each of the objectives and should provide an adequate level of detail so that the potential customer can understand exactly how each of the objectives would be satisfied. Individual tasks should be identified and discussed and should be shown on an easily understood schedule. Use of a bar chart/milestone schedule is recommended for this purpose, since it provides for clear delineation of the tasks and shows the associated schedule and important milestones. Typical tasks that should be discussed and scheduled include literature search activities, feasibility assessment, preliminary design, detail design, preparation of component specifications, component delivery dates, development test activity, fabrication, final assembly, qualification testing, acceptance testing, end item delivery, and final report preparation.

  For projects involving delivery of an item of hardware, the technical approach section of the proposal should also include a discussion as to how the final product will be evaluated to determine whether it meets the goals of the project. This is important to the customer, since the product evaluation plan assures the customer that the end item will perform the functions intended.

### 13.3.5 Budget

The total proposed cost is the most important element in the proposal. It is the total cost that will determine whether the customer will fund the effort. Thus, the budget must follow logically from the tasks identified, and the allocations made must agree with the manner in which the effort is to be accomplished, as described in the technical approach. For proposals that respond to RFPs, the budget breakdown will often be required to follow a set format that may specify salaries, fringe benefits, overhead, materials and supplies, equipment, and travel.

### 13.3.6 Organization and Capabilities[3]

This section of the proposal describes how the proposer will organize to accomplish the effort. The responsibilities of key personnel should be described here, and qualifications, as related to the tasks for which they are responsible, should be identified. The organizational support to be provided and the resources available for accomplishing the work also should be included in this section.

*Key personnel.* The responsibilities of key personnel should be described and the qualifications, as related to these tasks, should be briefly outlined. The narrative description of qualifications provided in this section should be supported by details on each key individual located in the appendix. It is very helpful for at least some of the key personnel to be recognized performers in the research area and to have reputations as productive and innovative workers.

*Organizational and other support.* This section of the proposal should substantiate the enthusiastic support of company management and, when appropriate, any labor, materials, or supplies that will be provided to the project but not charged to the customer should be identified. If organizations other than the proposing entity will be involved in the project, they should be identified here, and their responsibilities should be clearly described. If letters of support from leaders in the research area outside the proposer's company can be obtained, they should be referred to here and be included in the appendix.

*Facility capabilities.* Describe the facilities (laboratories, test facilities, office space, shops, computer capabilities, etc.) available to support the project. Emphasize the characteristics of the facilities that make them especially useful in accomplishing the work.

An example of a brief proposal that includes some of the features described in this section is included in Appendix B.

## 13.4 Oral Communications[4]

Webster's dictionary defines the word "communicate" as meaning to impart, share, or *to make common*. This definition implies that communication is a sharing of ideas and thoughts in a way

---

3 Adapted from Ertas, A., and Jones J.C., *The Engineering Design Process*, John Wiley & Sons, Inc., New York, 1996.
4 Adapted from Ertas, A., and Jones J.C., *The Engineering Design Process*, John Wiley & Sons, Inc., New York, 1996.

that is understood in the same way by all the participants in the discussion. This definition places considerable responsibility on the person who is doing the speaking to not only *send* his or her message, but to ensure that it is *received* and *understood.* To accomplish this goal a speaker must know his or her audience, use language that is common to them, and elicit feedback to ensure that the message is received and understood. Although many settings do not allow all of these objectives to be satisfied, these goals are ones that speakers should strive to achieve.

There are four basic modes of delivery for oral communications that are of interest to engineers. These are described as follows:[5]

*Manuscript.* A manuscript is meant to be read word for word to ensure that what is communicated is precisely what is intended. This is a seldom used mode of delivery for an engineer, but is appropriate for certain occasions when it is imperative that the information presented be thoroughly thought out and reviewed before presentation and not be embellished upon when given. This is the type of presentation that is appropriate when the information presented will go into some significant permanent record (e.g., congressional testimony, or trial depositions). It is also appropriate when the engineer is called on to present an important technical paper in substitution for another person who did the work. The disadvantages of this type of presentation are as follows: (1) most engineers are not skilled in reading in a natural and interesting fashion; (2) the speaker is forced to focus attention on the manuscript and thus is isolated from the audience; and (3) in reading a manuscript the reader sometimes appears to be detached from and uninterested in the material being presented.

*Memorization.* There appears to be of little or no use for this mode of delivery for engineers. Only the best speakers can make a presentation by memory without coming across as dull, inflexible, and distant. For technical material, such as that presented by engineers, memorization has little or no application.

*Extemporaneous speaking.* One meaning for the word *extemporaneous* is "with preparation but not read or memorized." This is the mode of delivery that most engineers use when giving oral reports, technical papers, and other less formal presentations. The degree of preparation varies widely depending on the material being presented and the audience, but should normally consist of either committing the entire presentation to writing or preparing an outline. Practicing by using audio or video recording or having someone listen and critique the presentation is helpful. For important occasions, it may be advantageous for the inexperienced speaker to write the presentation material completely and practice using the script initially, relying on it less and less with each practice session, and finally giving the presentation with little or no reference to the written version. Committing the presentation to writing sharpens the focus on the ideas presented, helps to improve the wording used in the presentation, and assists in the elimination of poor enunciation and time-stalling phrases. Using an outline assists in organizing the presentation, and in cases when the technical material is well understood by an experienced speaker,

---

5  Arthur, R.H., *The Engineer's Guide to Better Communication,* Scott, Foresman, Glenview, IL, 1984.

this may be quite adequate. The important point to remember when using this mode of delivery is to practice, several times if possible.

*Impromptu speaking.* Engineers are often required to deliver remarks, provide input to a discussion, answer a question, or give a brief summary of a project they are working on without any advanced warning or preparation. This is what is known as impromptu speaking. In this mode of delivery it is important (given the limited amount of time) that thought be given to how the remarks will be organized and structured:

1) Organize remarks to include a beginning, middle, and conclusion.
2) Clarify the purpose of the discussion at the outset for both yourself and the audience.
3) Think of the points that need to be made and start with the most important. Start the discussion on each of these points with a single thesis statement that summarizes it.
4) Conclude remarks with a concise summary and seek responses to determine whether the information presented has been understood.

## 13.5   Oral Presentations[6]

Giving an oral presentation is possibly one of the most trying experiences a newly graduated engineer has to face. Fortunately, as experience is gained and confidence increases, giving an oral presentation becomes less formidable. However, the apprehension associated with standing before an audience never totally disappears for most people and thus some means of controlling fear and turning it to advantage is needed. Possibly the most effective way to control fear is preparation. If the speaker is well prepared and thoroughly understands the material to be presented, the likelihood of experiencing uncontrolled fear is greatly diminished. Familiarity with the environment in which the presentation is to be given also helps to build confidence. Becoming familiar with the size and layout of the room, whether there is an audio system and/or podium, the location of light switches, position of projectors, and the like, before giving the presentation is very helpful in reducing anxiety and minimizing difficulties during the actual event. Wearing the proper attire also helps to establish confidence and to minimize feelings of inadequacy. Some feeling of fear can actually be an asset in that it generates concern for the importance of the event, which often results in diligent preparation and practice. Fear can also enhance alertness and may contribute to developing the "up" or "on" feeling so essential for outstanding presentations.

Uncontrollable fear and outright panic are difficult to control, since they are totally due to emotional factors not related to any correctable situation. One technique that seems to be beneficial in combating uncontrollable fear is to penetrate the barrier between the speaker and the audience by getting some sort of dialog going. If the speaker has a co-worker in the audience, the co-worker can ask a question or make some comment that requires an answer or action from the speaker. When there is no one in the audience to perform this service the speaker can

---

6 Adapted from Ertas, A., and Jones J.C., *The Engineering Design Process*, John Wiley & Sons, Inc., New York, 1996.

ask some question of the audience that requires an answer. The question or comment can be as simple as "Can you focus that projector better?" or "Can everyone see that?" Another ice breaker is to get everyone in the audience to introduce themselves, to tell where they are from or what they hope to get out of the presentation. The point is to break down the wall built up in the speaker's mind between him or herself and the audience and, in so doing, relieve the tension.

A speaker should always come across to the audience as being sincere and enthusiastic. The speaker needs to be responsive to the audience's interests by anticipating why they are there and what they will derive from the presentation. Injecting a little humor into the presentation is helpful in keeping the audience's attention, but the speaker should make sure that he or she has the knack of telling a joke successfully. Relating personal experiences that support the presentation theme is also a good technique. Tying the various parts of the presentation together so that there is a logical flow is essential. The audience should at all times know where the speaker is and where he or she is going in relation to the subject of the talk. Most importantly, the speaker must always stay in character and be him- or herself. If the audience detects any falseness on the part of the speaker, any credibility that has been established will be destroyed. Finally, a strong closing should always be made. If the presentation ends by just trailing off with "That's all I have" or "Thank you," the audience will probably not be persuaded by the speaker and will certainly not feel a strong impetus to support the thesis of the presentation. The presentation should end with a strong summary of what has been discussed, emphasizing the points of greatest importance.

## 13.5.1 Organizing the Oral Presentation

Oral presentations should be organized into three parts: an introduction, body, and conclusion. Before the presentation is actually started, there are several things to be taken care of. If visual aids are being used, which is strongly encouraged, a viewgraph or slide identifying the subject of the presentation, to whom the presentation is being given, the presenter(s) name and affiliation, and the date should be projected on the screen. This makes it clear for the audience exactly what is going to be discussed and who the presenters are. It also sets the stage for the presentation and eliminates the problem of people being in the wrong room. It provides an appropriate cover sheet for the copy of the presentation that is maintained in the speaker's files for future reference. As the speaker begins the presentation, appropriate introductions should be made. The speaker should introduce the other personnel in his/her party by having each individual stand up as their name is mentioned. The presentation is now ready to begin.

*Introduction.* An outline of what is to be discussed by what individual (if several people are involved) should be initially presented. This outline should be accompanied by an appropriate visual aid. This gives the audience an overview as to what is to be discussed and follows the military's recommendation for presentations *to tell them what you are going to tell them, tell them, and then tell them what you told them.* The background for the subject should then be presented, supported by appropriate visual aids. It is important to capture the interest and attention of the audience at this point. Fortunately, the background for most subjects is usually rich enough to provide ample material for this. The introduction should also include a clear statement and explanation of the problem as well as the objective(s) of the work being reported on. Finally, the introduction should outline the methods used to accomplish the work. The use of visual schematics and sketches of the apparatus is usually very helpful in giving the audience an understanding of the experimental method.

If theory must be included in the presentation, it can be given as section of the introduction or integrated in with the discussion in the body of the presentation. The decision as to where the theory is included depends on the degree of difficulty, the magnitude, and whether it can logically be separated from the detailed discussion in the body of the presentation. Wherever it is located, theory requires a special emphasis to make it understandable to an audience that has little time in which to absorb difficult concepts.

*Body.* The body is the *meat* of the presentation and corresponds generally to the results and discussion section of a written report. As is true of written reports, the body of the presentation should be supported with an adequate number of visual aids. Figures showing curves, drawings, or sketches are preferred over tables, which often contain too much information for the audience to grasp during the 30–60 seconds that a visual aid is normally projected. The body of the presentation should include the information necessary to substantiate the conclusions of the effort being reported on. The flow of this part of the presentation should be such that the audience can easily follow the evolution of the effort from the first steps and findings to the final conclusion.

For proposal presentations individual tasks should be described and a bar chart showing the planned schedule for completion of each task should be used. A budget depicting an appropriate cost breakdown such as that shown in Appendix E for the written proposal should also be presented.

*Conclusion.* The conclusion of an oral report or proposal should include a summary of what has been presented. Conclusions can be listed individually and discussed briefly, emphasizing the importance of major findings. If recommendations are appropriate, this is the place to include them. The important thing to keep in mind here is that statements made at the conclusion of the presentation are the ones most likely to be remembered by the audience. The presentation should be concluded dynamically, leaving the audience with the most important conclusions, suggestions, and recommendations.

## 13.6   A Final Word on Communications

Like it or not, communication is an essential and significant element in the professional life of an engineer. For those engineers who refuse to become good communicators and depend on others to do their communicating, the future probably holds an outcome similar to that of Miles Standish in Longfellow's famous poem, *The Courtship of Miles Standish.* Standish sent his friend, John Alden, to carry his proposal of marriage to the maiden he loved. The maiden refused, feeling that Standish should have spoken for himself if she were worth wooing. Furthermore, she asked John why he did not speak for himself. In the same manner, engineers cannot depend on others to do their talking unless they are prepared to have others get the credit for their work. Engineers must speak for themselves and be prepared to present their work in written or oral form, as appropriate, whenever the opportunity arises. The best opportunity that an engineer has to improve his or her communications skills is while still a student, surrounded by people interested in assisting in this endeavor. The principal requirement for making significant improvement in communication skills is having a strong desire to do so. Those who have this drive will have little difficulty in becoming effective communicators.

# Appendix A

## A.1   Matlab Code for Interpretive Structural Modeling (ISM-Case Study 2.3)[1]

```
clc;
clear;
close all;
syms X V O A %this is for symbolic variables
%INPUT matrix
Self_interaction_Matrix= [1 X O V V O V X O;
0 1 V X X O A V A;

0 0 1 A A A V A A;
0 0 0 1 A O O V O;
0 0 0 0 1 V O V O;
0 0 0 0 0 1 O X A;
0 0 0 0 0 0 1 V O;
0 0 0 0 0 0 0 1 A;
0 0 0 0 0 0 0 0 1]
%Manuplation of the Self-Interaction Matrix to Reachability Matrix
for i = 1:size(Self_interaction_Matrix,1)
for j = 1:size(Self_interaction_Matrix,2)
if Self_interaction_Matrix(i,j)==V
Self_interaction_Matrix(i,j)=1;
Self_interaction_Matrix(j,i)=0;
end
if Self_interaction_Matrix(i,j)==A
Self_interaction_Matrix(i,j)=0;
Self_interaction_Matrix(j,i)=1;
end
if Self_interaction_Matrix(i,j)==X
    Self_interaction_Matrix(i,j)=1;
```

1 All MATLAB Codes in Appendix A were developed by Mr. Utku Gulbulak and Dr. T. Batuhan Baturalp for The Academy of Transdisciplinary Learning & Advanced Studies (ATLAS). Reprinted by permission of ATLAS. Students can create MICMAC quadrant by using the final reachability matrix. For interactive ISM calculations to plot Digraph and MICMAC, refer to http://www.theatlas.org/ISM/.

```
Self_interaction_Matrix(j,i)=1;
end
if Self_interaction_Matrix(i,j)==0
Self_interaction_Matrix(i,j)=0;
Self_interaction_Matrix(j,i)=0;
end
end
end
Initial_Reachability_Matrix=double(Self_interaction_Matrix)
 %first output
%transitivity
Reachability_Matrix_w_transitivity=Initial_Reachability_Matrix;
for i = 1:size(Initial_Reachability_Matrix,1)
for j = 1:size(Initial_Reachability_Matrix,2)
if Initial_Reachability_Matrix(i,j)==0
B=transpose(Initial_Reachability_Matrix(:,j))&Initial_Reachability_
 Matrix(i,:);
if sum(B)>0
Reachability_Matrix_w_transitivity(i,j)=1;
end
end
end
end
Reachability_Matrix_w_transitivity %second output MATRIX AKA FINAL
 REACHABILITY
%Column and Row Sums
for i=1:size(Reachability_Matrix_w_transitivity,1)
Driving_Power(i) = sum(Reachability_Matrix_w_transitivity(i,:));
end
for j=1:size(Reachability_Matrix_w_transitivity,2)
Dependence(j) = sum(Reachability_Matrix_w_transitivity(:,j));
end
Driving_Power %ALSO USED IN SECOND OUTPUT TABLE
Dependence
d=1;
%THESE BELOW CODES ARE 3RD OUTPUT,
%Reachability Set
for i=1:size(Reachability_Matrix_w_transitivity,1)
k=1;
for j=1:size(Reachability_Matrix_w_transitivity,2)
if Reachability_Matrix_w_transitivity(i,j)==1
Reachability(i,k,d)=1;
end
k=k+1;
end
end
```

```
%Antecedent Set
for j=1:size(Reachability_Matrix_w_transitivity,2)
k=1;
for i=1:size(Reachability_Matrix_w_transitivity,1)
if Reachability_Matrix_w_transitivity(i,j)==1
Antecedent(j,k,d)=1;
end
k=k+1;
end
end
%Intersection Set
for i=1:size(Reachability,1)
for j=1:size(Reachability,2)
if Reachability(i,j,d)==Antecedent(i,j,d)
Intersection(i,j,d)=Antecedent(i,j,d);
else
Intersection(i,j,d)=0;
end

end
end
level_check=1;
while level_check~=0
%Check the Levels (TO FIND OUT LEVELS)
pp=1;
for i=1:size(Reachability,1)
if isequal(Reachability(i,:,d),Intersection(i,:,d))==1
if Reachability(i,:,d)==zeros
Level(d,pp)=0;
else
Level(d,pp)=i;
pp=pp+1;
end
end
end
%next level
d=d+1;
%Substract Levels
Reachability(:,:,d)=Reachability(:,:,d-1);
Antecedent(:,:,d)=Antecedent(:,:,d-1);
Intersection(:,:,d)=Intersection(:,:,d-1);
for i=nnz(Level(d-1,:)):-1:1
Reachability(Level(d-1,i),:,d)=zeros;
Reachability(:,Level(d-1,i),d)=zeros;
Antecedent(Level(d-1,i),:,d)=zeros;
Antecedent(:,Level(d-1,i),d)=zeros;
```

```
Intersection(Level(d-1,i),:,d)=zeros;
Intersection(:,Level(d-1,i),d)=zeros;
end
if Reachability(:,:,d)==zeros;
level_check=0;
end
end
Reachability;
Antecedent;
Intersection;
Level
```

Self_interaction_Matrix =

```
[ 1, X, O, V, V, O, V, X, O]
[ 0, 1, V, X, X, O, A, V, A]
[ 0, 0, 1, A, A, A, V, A, A]
[ 0, 0, 0, 1, A, O, O, V, O]
[ 0, 0, 0, 0, 1, V, O, V, O]
[ 0, 0, 0, 0, 0, 1, O, X, A]
[ 0, 0, 0, 0, 0, 0, 1, V, O]
[ 0, 0, 0, 0, 0, 0, 0, 1, A]
[ 0, 0, 0, 0, 0, 0, 0, 0, 1]
```

Initial_Reachability_Matrix =

| | | | | | | | | |
|---|---|---|---|---|---|---|---|---|
| 1 | 1 | 0 | 1 | 1 | 0 | 1 | 1 | 0 |
| 1 | 1 | 1 | 1 | 1 | 0 | 0 | 1 | 0 |
| 0 | 0 | 1 | 0 | 0 | 0 | 1 | 0 | 0 |
| 0 | 1 | 1 | 1 | 0 | 0 | 0 | 1 | 0 |
| 0 | 1 | 1 | 1 | 1 | 1 | 0 | 1 | 0 |
| 0 | 0 | 1 | 0 | 0 | 1 | 0 | 1 | 0 |
| 0 | 1 | 0 | 0 | 0 | 0 | 1 | 1 | 0 |
| 1 | 0 | 1 | 0 | 0 | 1 | 0 | 1 | 0 |
| 0 | 1 | 1 | 0 | 0 | 1 | 0 | 1 | 1 |

Reachability_Matrix_w_transitivity =

| | | | | | | | | |
|---|---|---|---|---|---|---|---|---|
| 1 | 1 | 1 | 1 | 1 | 1 | 1 | 1 | 0 |
| 1 | 1 | 1 | 1 | 1 | 1 | 1 | 1 | 0 |
| 0 | 1 | 1 | 0 | 0 | 0 | 1 | 1 | 0 |
| 1 | 1 | 1 | 1 | 1 | 1 | 1 | 1 | 0 |
| 1 | 1 | 1 | 1 | 1 | 1 | 1 | 1 | 0 |
| 1 | 0 | 1 | 0 | 0 | 1 | 1 | 1 | 0 |
| 1 | 1 | 1 | 1 | 1 | 1 | 1 | 1 | 0 |
| 1 | 1 | 1 | 1 | 1 | 1 | 1 | 1 | 0 |
| 1 | 1 | 1 | 1 | 1 | 1 | 1 | 1 | 1 |

```
Driving_Power =

     8     8     4     8     8     5     8     8     9

Dependence =

     8     8     9     7     7     8     9     9     1

Level =

     3     7     8     0
     1     6     0     0
     2     4     5     0
     9     0     0     0
```

## A.2   Matlab Code for Search Method Calculations

```
clc
clear all
close all

fprintf('Optimization in Design - Equal Interval Search Example
 \n \n')

%Input function to be minimized
fun = input(['Enter Function to be Minimized (Use x for unknown
 variable, Example:' ... '"4*x^2+1310/x"):'], 's');
%Input increment size
delx = input('Enter Delta x Value: ');
%Input convergence tolerance
eps = input('Enter Convergence Criterion (Epsilon) Value: ');

  %Input initial x value
xi = input('Enter Initial x Value: ');

f = inline(fun,'x');

%x value for the first and second iterations
x(1) = xi;
x(2) = xi+delx;

%function value for the first iteration
y(1) = feval(f, x(1,1));
```

```
%Counter lines for printing
dellx(1) = delx;
nint(1) = 1;

%Algorithm is given in the textbook
for i = 2: 100000

    y(i) = feval(f, x(i));
    dellx(i) = delx;

    if abs(y(i) - y(i-1)) < eps

        x(i+1) = (x(i-1)+x(i))*0.5;

        y(i+1) = feval(f , x(i+1));

        break

    elseif y(i) < y(i-1)
        x(i+1) = x(1,i) + delx;

    else

        delx = -delx/2;

        x(i+1) = x(1,i) + delx;

    end
end

%Printing
root = x(end);
root2 = y(end);
fprintf(' \nIteration #       x(i)        Delta x        U(x)\n')
fprintf('_____\n')
for j = 1:length(x)-1

fprintf(' %4d      %15.5f  %15.5f  %15.5f  \n',j, x(j), dellx(j),
 y(j))

    end

fprintf('\nUmin is equal to %15.5f', root2)
fprintf('\nAt x is equal to %15.5f \n\n', root)
```

**BELOW IS THE OUTPUT**

```
Enter Function to be Minimized (Use x for unknown variable):
 4*x*x + 1310/x
Enter Delta x Value: 0.5
Enter Convergence Criterion (Epsilon) Value: 0.001
Enter Initial x Value: 1
```

| Iteration # | x(i) | Delta x | U(x) |
|---|---|---|---|
| 1 | 1.00000 | 0.50000 | 1314.00000 |
| 2 | 1.50000 | 0.50000 | 882.33333 |
| 3 | 2.00000 | 0.50000 | 671.00000 |
| 4 | 2.50000 | 0.50000 | 549.00000 |
| 5 | 3.00000 | 0.50000 | 472.66667 |
| 6 | 3.50000 | 0.50000 | 423.28571 |
| 7 | 4.00000 | 0.50000 | 391.50000 |
| . | . | . | . |
| . | . | . | . |
| . | . | . | . |
| 23 | 5.48438 | -0.01563 | 359.17388 |
| 24 | 5.46875 | -0.01563 | 359.17176 |
| 25 | 5.45313 | -0.01563 | 359.17552 |
| 26 | 5.46094 | 0.00781 | 359.17290 |
| 27 | 5.46875 | 0.00781 | 359.17176 |
| 28 | 5.47656 | 0.00781 | 359.17209 |

```
Umin is equal to   359.17174
At x is equal to   5.47266
```

## A.3   Matlab Code for Golden Section Search Method Calculations

```
clc
clear
close all

fprintf('Optimization in Design - Golden Section Search Example
 \n \n')

%Input function to be minimized
fun = input(['Enter Function to be Minimized (Use x for unknown
 variable, Example:' ... '"4*x^2+1310/x"): '], 's');
```

```
%Input upper limit
U = input('Enter Upper Bound XU: ');
%Input lower limit
L = input('Enter Lower Bound XL: ');
%Input convergence tolerance
eps = input('Enter Convergence Criterion (Epsilon) Value: ');
f = inline(fun,'x');
%Initial Upper and Lower Limit Values
XU(1) = U;
XL(1) = L;
%Algorithm is given in the textbook
for i = 1:100000
    dif = XU(i) - XL(i);
    x1(i) = XL(i) + 0.618*dif;
    x2(i) = XU(i) - 0.618*dif;
    U1(i) = feval(f, x1(i));
    U2(i) = feval(f, x2(i));
    if abs(U2(i) - U1(i)) < eps
        x1f = (x1(i) + x2(i))/2;
        U1f = feval(f, x1f);
        break
    elseif U1(i) > U2(i)
        XU(i+1) = x1(i);
        XL(i+1) = XL(i);
    else
        XL(i+1) = x2(i);
        XU(i+1) = XU(i);
    end
end
root = x1f;
root2 = U1f;
fprintf(' \nIteration #    XL     XU     X1     X2     UX1     UX2\n')
fprintf('_____\n')

for j = 1:length(XU)

    fprintf(' %4d      %15.5f %15.5f %15.5f %15.5f %15.5f
  %15.5f\n',j, ... XL(j), XU(j), x1(j), x2(j), U1(j), U2(j))

end
fprintf('\nUmin is equal to %15.5f ', root2)
fprintf('\nAt x is equal to %15.5f \n\n' , root)
```

**BELOW IS THE OUTPUT**

```
Enter Function to be Minimized (Use x for unknown variable):
  4*x*x+1310/x
Enter Upper Bound XU: 10
Enter Lower Bound XL: 2
Enter Convergence Criterion (Epsilon) Value: 0.001
```

| Iteration # | XL | XU | X1 | X2 | UX1 | UX2 |
|---|---|---|---|---|---|---|
| 1 | 2.00000 | 10.00000 | 6.94400 | 5.05600 | 381.52862 | 361.35065 |
| 2 | 2.00000 | 6.94400 | 5.05539 | 3.88861 | 361.35722 | 397.36656 |
| 3 | 3.88861 | 6.94400 | 5.77684 | 5.05577 | 360.25510 | 361.35315 |
| 4 | 5.05577 | 6.94400 | 6.22270 | 5.77707 | 365.40746 | 360.25672 |
| 5 | 5.05577 | 6.22270 | 5.77693 | 5.50153 | 360.25572 | 359.18291 |
| 6 | 5.05577 | 5.77693 | 5.50145 | 5.33125 | 359.18285 | 359.40989 |
| 7 | 5.33125 | 5.77693 | 5.60668 | 5.50150 | 359.38930 | 359.18289 |
| 8 | 5.33125 | 5.60668 | 5.50147 | 5.43647 | 359.18286 | 359.18601 |
| 9 | 5.43647 | 5.60668 | 5.54166 | 5.50149 | 359.23124 | 359.18288 |
| 10 | 5.43647 | 5.54166 | 5.50147 | 5.47665 | 359.18287 | 359.17210 |
| 11 | 5.43647 | 5.50147 | 5.47664 | 5.46130 | 359.17210 | 359.17282 |

```
Umin is equal to    359.17175
At x is equal to    5.46897
```

## A.4   Maximization of Reliability for a Given Cost Constraint

```
% THIS PROGRAM CALCULATES THE OPTIMUM NUMBER OF REDUNDANCIES IN
% AN N STAGE SERIES SYSTEM FOR A GIVEN COST LIMIT AND CONSTANT
% RELIABILITY. THE LAGRANGIAN METHOD IS USED FOR THE OPTIMIZATION.
% PROGRAM CALCULATES THE SYSTEM RELIABILITY, SYSTEM TOTAL COST AND
% ALSO PRINTS OUT THE LAGRANGE COEFFICIENT LAMBDA. DECREMENTS OF
% LAMBDA SHOULD BE CAREFULLY CHOSEN TO BE ABLE TO OBTAIN SUFFICIENT
% NUMBER OF STEPS BEFORE PROGRAM QUITS.

% VARIABLES:
% A : COEFFICIENT FOR LAMBDA
% B : COEFFICIENT FOR LAMBDA
% C : COST OF EACH COMPONENT
% CL : COST CONSTRAINT
% DELLAM : CHANGE IN THE LAMBDA
% N : TOTAL NUMBER OF STAGES
```

```
% NK : COEFFICIENT OF LAMBDA
% PHI : PARAMETER
% R : RELIABILITY OF EACH COMPONENT
% REL : RELIABILITY OF THE SYSTEM
% SUM1 : DUMMY VARIABLES FOR THE CALCULATION OF LAMBDA
% SUM2 : DUMMY VARIABLES FOR THE CALCULATION OF LAMBDA
% SUMC : TOTAL COST OF THE ELEMENT
% XLAM : LAGRANGE MULTIPLIER
clc
clear all
close all
fprintf(['Design for Reliablity - Maximization of Reliability for a
 Given' ... 'Cost Constraint \n \n'])
N = input('What is the number of stages? ');
CL = input('What is the cost limit? ($) ');

for i=1:N
  C(i) = input(['Enter the cost of each stage # (Enter only one value
 at a time,' ... 'starting from the first one): ']);

  end

for i=1:N

  R(i) = input(['Enter the reliability value of each stage # (Enter
 only one value at a' ... 'time, starting from the first one): ']);
  end
% CALCULATE PARAMETERS AND CONSTANTS FOR THE CALCULATION OF LAMBDA
for I = 1 : N
      PHI(I)=(1-R(I))/R(I);
      A(I)=1/log(PHI(I));
      NK(I)=-C(I)/log(PHI(I));
      B(I)=log(NK(I))/log(PHI(I));
end
% CALCULATE LAMBDA
SUM1 = 0;
SUM2 = 0;
for I = 1 : N
      SUM1 = SUM1 + C(I)*B(I);
      SUM2 = SUM2 + C(I)*A(I);
end
S = (CL-SUM1)/SUM2;
XLAM = exp(S);
disp(['LAMBDA IS CALCULATED AS = ',num2str(XLAM)]);
```

```
DELLAM = input('\n Enter Delta Lambda Value: ');
% CALCULATE THE SMALLEST NUMBER OF ELEMENTS THAT WILL SATISFY
% THE UNDOMINATED SEQUENCE INEQUALITY FOR EVERY STAGE.
SUMC= 0;
j = 1;
while CL > SUMC
for I = 1 : N
    NY(I) = 0;
    H(I) = (PHI(I)^NY(I))/((1+PHI(I))^(NY(I)+1));
    AL = XLAM * C(I);
    while H(I) >= AL
    NY(I) = NY(I)+1;
    H(I) = (PHI(I)^NY(I))/((1+PHI(I))^(NY(I)+1));
    AL = XLAM * C(I);
    end
end
% CALCULATE THE COST AND RELIABILITY OF THE SYSTEM
SUMC = 0;
SUMP = 1;
sumpp = 0;
for NI = 1 : N
        SUMC = SUMC + C(NI) * NY(NI);
        RELL (NI) = 1 / (1+PHI(NI));
        RELLL(NI) = 1-(1-RELL(NI))^NY(NI);
        SUMP = SUMP * RELLL(NI);
end
XLAM = XLAM - DELLAM;
if SUMC <= CL
sumc(j) = SUMC;
REL(j) = SUMP;
NC(:,:,j) = NY;
j = j + 1;
end
end
stage = i;
%Printing
fprintf('\n  Stage      PHI(i)       a(i)         b(i)         K(i)\n')
fprintf('_____\n')

for j = 1:stage
    fprintf(' %4d   %15.5f  %15.5f  %15.5f  %15.5f \n' , j , PHI(j),
 A(j) , B(j), NK(j))
end
```

```
fprintf(['\nThe number of components is %4d, %4d, %4d and %4d for the
  first, second,' ... 'third and fourth stages\n'], NC(:,1,end),
  NC(:,2,end),NC(:,3,end),NC(:,4,end))
fprintf('\nSystem reliability is %4.5f\n', REL(end))
fprintf('\nSystem cost is $%4.2f\n', sumc(end))
```

**BELOW IS THE OUTPUT**

Design for Reliability - Maximization of Reliability for a Given Cost
  Constraint
What is the number of stages? 4
What is the cost limit? ($) 35
Enter the cost of each stage # (Enter only one value at a time,
  starting from the first one): 3
Enter the cost of each stage # (Enter only one value at a time,
  starting from the first one): 2
Enter the cost of each stage # (Enter only one value at a time,
  starting from the first one): 4
Enter the cost of each stage # (Enter only one value at a time,
  starting from the first one): 5
Enter the reliability value of each stage # (Enter only one value at
  a time, starting from the first one): .9
Enter the reliability value of each stage # (Enter only one value at
  a time, starting from the first one): .85
Enter the reliability value of each stage # (Enter only one value at
  a time, starting from the first one): .7
Enter the reliability value of each stage # (Enter only one value at
  a time, starting from the first one): .92
LAMBDA IS CALCULATED AS = 0.0084017
Enter Delta Lambda Value: 0.001

| Stage | PHI(i) | a(i) | b(i) | K(i) |
|-------|---------|----------|----------|---------|
| 1 | 0.11111 | -0.45512 | -0.14173 | 1.36536 |
| 2 | 0.17647 | -0.57650 | -0.08208 | 1.15300 |
| 3 | 0.42857 | -1.18022 | -1.83170 | 4.72089 |
| 4 | 0.08696 | -0.40944 | -0.29336 | 2.04721 |

The number of components is 2, 3, 3 and 2 for the first, second,
  third and fourth stages
System reliability is 0.95387
System cost is $34.00

## A.5    Minimization of Cost for a Given Reliability Constraint

```
% THIS PROGRAM CALCULATES THE OPTIMUM NUMBER OF REDUNDANCIES IN AN
% N STAGE SERIES SYSTEM FOR A GIVEN RELIABILITY LIMIT AND COST
% CONSTRAINT. THE LAGRANGIAN METHOD IS USED FOR THE OPTIMIZATION.
% PROGRAM CALCULATES THE SYSTEM RELIABILITY, SYSTEM TOTAL COST AND
% ALSO PRINTS OUT THE LAGRANGE COEFFICIENT LAMBDA. DECREMENTS OF
% LAMBDA SHOULD BE CAREFULLY CHOSEN TO BE ABLE TO OBTAIN
% SUFFICIENT NUMBER OF STEPS BEFORE PROGRAM QUITS

% VARIABLES:
% A : COEFFICIENT FOR LAMBDA
% B : COEFFICIENT FOR LAMBDA
% C : COST OF EACH COMPONENT
% DELLAM : CHANGE IN THE LAMBDA
% N : TOTAL NUMBER OF STAGES
% NK : COEFFICIENT OF LAMBDA
% PHI : PARAMETER
% R : RELIABILITY OF EACH COMPONENT
% RL: RELIABILITY CONSTRAINT
% REL : RELIABILITY OF THE SYSTEM
% SUM1 : DUMMY VARIABLES FOR THE CALCULATION OF LAMBDA
% SUM2 : DUMMY VARIABLES FOR THE CALCULATION OF LAMBDA
% SUMC : TOTAL COST OF THE ELEMENT
% SUMP : TOTAL RELIABILITY OF THE SYSTEM
% XLAM : LAGRANGE MULTIPLIER

clc
clear all
close all

fprintf(['Design for Reliability - Minimization of Cost for a
Given' ... 'Reliability Constraint \n \n'])
N = input('What is the number of stages? ');
RL = input('What is the reliability constraint? ');
RS = RL-0.0001*RL;

for i=1:N

C(i) = input(['Enter the cost of each stage # (Enter only one value
at a time,' ... 'starting from the first one): ']);
end
for i=1:N
```

```
R(i) = input(['Enter the reliability value of each stage # (Enter
only one value at' ... 'a time, starting from the first one): ']);

end

    % CALCULATE PARAMETERS AND CONSTANTS FOR THE CALCULATION OF LAMBDA
    for I = 1 : N
    PHI(I) = (1-R(I))/R(I);
    K(I) = (C(I)*log(PHI(1)))/(C(1)*log(PHI(I)));
    end

    PH = (1-RL)/RL;
    SUM1 = 0;
    for I = 1 : N
    SUM1 = SUM1 + K(I);
    end
    % CALCULATE LAMBDA
    S = PH/SUM1;
    XLAM = -C(1)/(S*log(PHI(1)));
    disp(['LAMBDA IS CALCULATED AS = ',num2str(XLAM)]);

DELLAM = input('\n Enter Delta Lambda Value: ');

    % CALCULATE THE SMALLEST NUMBER OF ELEMENTS THAT WILL SATISFY
    % THE UNDOMINATED SEQUENCE INEQUALITY FOR EVERY STAGE.
    SUMC= 0;
    j = 1;
    REL = 100;
    while REL > RS
    for I = 1 : N
        Y(I) = 0;
        H(I)=(PHI(I)^Y(I))/((1+PHI(I))^(Y(I)+1))*XLAM;
        while H(I) >= C(I)
            Y(I) = Y(I)+1;
            H(I)=(PHI(I)^Y(I))/((1+PHI(I))^(Y(I)+1))*XLAM;
        end

    end

    % CALCULATE THE COST AND RELIABILITY OF THE SYSTEM
    SUMC = 0;
    SUMP = 1;
    for NI = 1 : N
      SUMC = SUMC+C(NI)*Y(NI);
      RELL(NI) = 1/(1+PHI(NI));
      RELLL(NI) = 1-(1-RELL(NI))^Y(NI);
```

```
    SUMP = SUMP*RELLL(NI);
  end
  XLAM = XLAM - DELLAM;
  if RS <= REL
      sumc(j) = SUMC;
      REL(j) = SUMP;
      NC(:,:,j) = Y;
      j = j + 1;
  end
  end
fprintf(['\nThe number of components is %4d, %4d, %4d and %4d for the
  first, second,' ... 'third and fourth stages\n'], NC(:,1,end),
  NC(:,2,end),NC(:,3,end),NC(:,4,end))
fprintf('\nSystem reliability is %4.5f\n', REL(end))
fprintf('\nSystem cost is $%4.2f\n', sumc(end))
```

## BELOW IS THE OUTPUT

Design for Reliability - Minimization of Cost for a Given
Reliability Constraint

What is the number of stages? 4

What is the reliability constraint? 0.95

Enter the cost of each stage # (Enter only one value at a time,
starting from the first one): 2

Enter the cost of each stage # (Enter only one value at a time,
starting from the first one): 3

Enter the cost of each stage # (Enter only one value at a time,
starting from the first one): 1

Enter the cost of each stage # (Enter only one value at a time,
starting from the first one): 1.5

Enter the reliability value of each stage # (Enter only one value at
a time, starting from the first one): .93

Enter the reliability value of each stage # (Enter only one value at
a time, starting from the first one): .9

Enter the reliability value of each stage # (Enter only one value at
a time, starting from the first one): .85

Enter the reliability value of each stage # (Enter only one value at
a time, starting from the first one): .8

LAMBDA IS CALCULATED AS = 72.1443

Enter Delta Lambda Value: 0.001

The number of components is 2, 2, 2 and 2 for the first, second, third and fourth stages

System reliability is 0.92446

System cost is $15.00

## A.6   Reliability Maximization with Multiple Constraints

```
clc
clear all
close all
% THIS PROGRAM CALCULATES THE OPTIMUM NUMBER OF REDUNDANCIES IN AN N
% STAGE SERIES SYSTEM FOR A GIVEN COST AND RELIABILITY CONSTRAINT.
% THE LANGRANGIAN METHOD IS USED FOR THE OPTIMIZATION. PROGRAM
% CALCULATES THE SYSTEM RELIABILITY, SYSTEM TOTAL COST AND ALSO
% PRINTS OUT THE LANGRANGE COEFFICIENT LAMBDA. FOR THE CALCULATIONS
% IT IS NECESSARY TO GUESS TWO LAMBDA VALUES. DECREMENTS OF LAMBDA
% SHOULD BE CAREFULLY CHOSEN TO BE ABLE TO OBTAIN SUFFICIENT NUMBER
% OF STEPS BEFORE PROGRAM QUITS

% VARIABLES:
% A : COEFFICIENT FOR LAMBDA
% B : COEFFICIENT FOR LAMBDA
% C : COST OF EACH COMPONENT
% CL : COST LIMIT
% DELLAM : CHANGE IN THE LAMBDA
% N : TOTAL NUMBER OF STAGES
% NK : COEFFICIENT OF LAMBDA
% PHI : PARAMETER
% R : RELIABILITY OF EACH COMPONENT
% REL : RELIABILITY OF THE SYSTEM
% SUM1 : DUMMY VARIABLES FOR THE CALCULATION OF LAMBDA
% SUM2 : DUMMY VARIABLES FOR THE CALCULATION OF LAMBDA
% SUMC : TOTAL COST OF THE ELEMENT
% SUMP : TOTAL RELIABILITY OF THE SYSTEM
% SUMW : TOTAL WEIGHT OF THE SYSTEM
% XLAM : LAGRANGE MULTIPLIER
% XLAM1 : LAGRANGE MULTIPLIER GUESS
% XLAM2 : LAGRANGE MULTIPLIER GUESS
% G1 : FRACTION OF LAMBDA1
% G2 : FRACTION OF LAMBDA2
% W : WEIGHT OF EACH COMPONENT
% WL : WEIGHT CONSTRAINT
```

```
global R C W PHI
fprintf(['Design for Reliability - Maximization of Reliability for
 Given Cost & Weight' ... 'Constraint \n \n'])

 N = input('What is the number of stages? ');
CL = input('What is the cost limit ($)? ');
WL = input('What is the weight limit ? ');
for i=1:N
C(i) = input(['Enter the cost of each stage # (Enter only one value
 at a time, starting' ... 'from the first one): ']);
end
for i=1:N
R(i) = input(['Enter the reliability value of each stage # (Enter
 only one value at a' ... 'time, starting from the first one): ']);
end

for i=1:N
W(i) = input(['Enter the weight value of each stage # (Enter only
 one value at a time,' ... 'starting from the first one): ']);
end
%
%
XLAM1 = input('Enter your first guess for lambda: ');
XLAM2 = input('Enter your second guess for lambda: ');
x0 = [XLAM1, XLAM2];
lb = [0.001,0.001];
fun = @myfun2;
[LAM] = lsqnonlin(fun,x0,lb);
XLAM1 = LAM(1);
XLAM2= LAM(2);
XLAM = XLAM1 + XLAM2;
G1 = XLAM1/XLAM;
G2 = XLAM2/XLAM;
fprintf('Lambda is calculated as %10.4f\n', XLAM)
DELLAM = input('Choose Delta Lambda: ');
NY = 0;
j = 1;
SUMW = 0;
SUMC = 0;
while CL > SUMC && WL > SUMW
for I = 1 : N
    NY(I) = 0;
    H(I) = (PHI(I)^NY(I))/((1+PHI(I))^(NY(I)+1));
    AL = XLAM * C(I);
```

```
      while H(I) >= AL
      NY(I) = NY(I)+1;
      H(I) = (PHI(I)^NY(I))/((1+PHI(I))^(NY(I)+1));
      AL = XLAM * (G1*C(I) + G2*W(I));
      end
end
% CALCULATE THE COST AND RELIABILITY OF THE SYSTEM
SUMC = 0;
SUMP = 1;
sumpp = 0;
SUMW = 0;
for NI = 1 : N
      SUMC = SUMC + C(NI) * NY(NI);
      SUMW = SUMW + W(NI) * NY(NI);
      RELL (NI) = 1 / (1+PHI(NI));
      RELLL(NI) = 1-(1-RELL(NI))^NY(NI);
      SUMP = SUMP * RELLL(NI);
end
XLAM = XLAM - DELLAM;
if SUMC <= CL && SUMW <= WL
sumc(j) = SUMC;
sumw(j) = SUMW;
REL(j) = SUMP;
NC(:,:,j) = NY;
j = j + 1;
end

end
fprintf(['\nThe number of components is %4d, %4d, %4d, %4d and %4d
  for the first,' ... 'second, third, fourth and fifth stages\n'],
  NC(:,1,end),NC(:,2,end),NC(:,3,end), ... NC(:,4,end), NC(:,5,end))
fprintf('\nSystem reliability is %4.5f\n', REL(end))
fprintf('\nSystem cost is $%4.2f\n', sumc(end))
fprintf('\nSystem weight is %4.5f\n', sumw(end))
function F = myfun2(x)
global R C W PHI
for i = 1 : length(R)
PHI(i) = (1-R(i))/R(i);
a(i) = -C(i)/log(PHI(i));
b(i) = -W(i)/log(PHI(i));
end
F(1) = sum(a.*log(a.*x(1)+b.*x(2)))+100;
F(2) = sum(b.*log(a.*x(1)+b.*x(2)))+104;
end
```

**BELOW IS THE OUTPUT**

Design for Reliability - Maximization of Reliability for Given Cost &
 Weight Constraint

What is the number of stages? 5
What is the cost limit ($)? 90
What is the weight limit ? 104
Enter the cost of each stage # (Enter only one value at a time,
 starting from the first one): 5
Enter the cost of each stage # (Enter only one value at a time,
 starting from the first one): 4
Enter the cost of each stage # (Enter only one value at a time,
 starting from the first one): 8
Enter the cost of each stage # (Enter only one value at a time,
 starting from the first one): 7
Enter the cost of each stage # (Enter only one value at a time,
 starting from the first one): 7
Enter the reliability value of each stage # (Enter only one value at
 a time, starting from the first one): .9
Enter the reliability value of each stage # (Enter only one value at
 a time, starting from the first one): .75
Enter the reliability value of each stage # (Enter only one value at
 a time, starting from the first one): .7
Enter the reliability value of each stage # (Enter only one value at
 a time, starting from the first one): .8
Enter the reliability value of each stage # (Enter only one value at
 a time, starting from the first one): .85
Enter the weight value of each stage # (Enter only one value at a
 time, starting from the first one): 8
Enter the weight value of each stage # (Enter only one value at a
 time, starting from the first one): 8
Enter the weight value of each stage # (Enter only one value at a
 time, starting from the first one): 6
Enter the weight value of each stage # (Enter only one value at a
 time, starting from the first one): 7
Enter the weight value of each stage # (Enter only one value at a
 time, starting from the first one): 8
Enter your first guess for lambda: .1
Enter your second guess for lambda: .2
Lambda is calculated as 0.0036
Choose Delta Lambda: .0001
The number of components is 2, 3, 4, 3 and 2 for the first, second,
 third, fourth and fifth stages
System reliability is 0.93733
System cost is $89.00
System weight is 101.00000

## A.7   Program Listing of MAPLE for Evaluating the Eigenvalues

```
# Define the solution of the beam problem as a MAPLE
  function

y:=
   proc(x)
   a*cosh(beta*x) + b*cos(beta*x) + c*sinh(beta*x)
   + d*sin(beta*x) end;

y := proc(x) a*cosh(beta*x)+b*cos(beta*x)+c*sinh(beta*x)
     +d*sin(beta*x) end

# Check to see if we have the correct  function
 y(x);
        a cosh(beta x) + b cos(beta x) + c sinh(beta x)
        + d sin(beta x)

# Apply the displacement condition at x=0
# Solve for b, and substitute b=-a in y(x)
 solve(y(0),b);
                                          - a
 subs(b=-a,y(x));
        a cosh(beta x) - a cos(beta x) + c sinh(beta x)
        + d sin(beta x)

# Now apply the slope condition at x=0
# Solve for d, and substitute d=-c in y(x)
 diff(y(x),x);
        a sinh(beta x) beta - b sin(beta x) beta
        + c cosh(beta x) beta + d cos(beta x) beta
 subs(x=0,");
        a sinh(0) beta - b sin(0) beta + c cosh(0) beta
        + d cos(0) beta
 evalf(");
                             c beta + d beta
 solve(",d);

# Now define another MAPLE function v(x), which does not
# contain b and d
 v:=proc(x) subs(b=-a,d=-c,y(x)) end;
 v:= proc(x) subs(b = -a,d = -c,y(x)) end

# Define the 2nd and 3rd derivatives as functions
 d2v:=proc(x) diff(v(x),x\$2) end;
```

```
 d2v:= proc(x) diff(v(x),x \$ 2) end
 d3v:=proc(x) diff(v(x),x\$3) end;
# Now apply the bending moment and the shear force conditions
# at x=L
# Determine the coefficients of "a" and "c" in the bending
# moment and the shear force boundary conditions
#c11:= coefficient of "a" in the bending moment condition at
 x=L
c11:=coeff(d2v(L),a);
```
$$c11 := \cosh(\beta L)\,\beta^2 + \cos(\beta L)\,\beta^2$$
```
#c12:= coefficient of "c" in the bending moment condition at
 x=L
c12:=coeff(d2v(L),c);
```
$$c12 := \sinh(\beta L)\,\beta^2 + \sin(\beta L)\,\beta^2$$
```
#c21:= coefficient of "a" in the shear force condition at
 x=L
c21:=coeff(d3v(L),a);
```
$$c21 := \sinh(\beta L)\,\beta^3 - \sin(\beta L)\,\beta^3$$
```
#c22:= coefficient of "c" in the shear force condition at
 x=L
c22:=coeff(d3v(L),c);
```
$$c22 := \cosh(\beta L)\,\beta^3 + \cos(\beta L)\,\beta^3$$
```
# Now define a Wronskian Matrix, W=c(i,j) of the above
  calculated    coefficients as
MAPLE array
 W:=array([[c11,c12],[c21,c22]]);
 W:=
```
$$
[\cosh(\beta L)\,\beta^2 + \cos(\beta L)\,\beta^2,
$$
$$
\sinh(\beta L)\,\beta^2 + \sin(\beta L)\,\beta^2]
$$
$$
[\sinh(\beta L)\,\beta^3 - \sin(\beta L)\,\beta^3,
$$
$$
\cosh(\beta L)\,\beta^3 + \cos(\beta L)\,\beta^3]
$$
```
# Load the Linear Algebra Package
 with(linalg);
# Define the Characteristic polynomial as the determinant of
# Wranskian
 p:=proc(beta) simplify(det(W)) end;
 p:= proc(beta) simplify(det(W)) end
```

```
# Check the characteristic polynomial
  p(beta);
```

$$2 \cosh(\text{beta } L) \text{ beta}^5 \cos(\text{beta } L) + 2 \text{ beta}^5$$

```
# Factor out  p(beta) and simplify
  factor(p(beta));
```

$$2 \text{ beta}^5 (\cosh(\text{beta } L) \cos(\text{beta } L) + 1)$$

```
#Assign p(beta)->q is the simplified characteristic polynomial
  q:= "/(2*beta^5);
```

$$q := \cosh(\text{beta } L) \cos(\text{beta } L) + 1$$

```
# For the length L=10, plot q,for 0<beta<1.2,-10000<q<2000,
# the roots are the eigenvalues
  L:=10;
```

$$L := 10$$

```
  plot(q,beta=0.0..1.2,-10000..2000);

# Now solve for the roots in the range
# beta=[0,0.25],[0.2,0.5],[0.5,0.8],[0.8,1.2]
  beta1:=fsolve(q,beta,0.0..0.25);
```

$$\text{beta1} := .1875104069$$

```
  beta2:=fsolve(q,beta,0.2..0.50);
```

$$\text{beta2} := .469409113$$

```
  beta3:=fsolve(q,beta,0.5..0.80);
```

$$\text{beta3} := .7854757438$$

```
  beta4:=fsolve(q,beta,0.8..1.20);
```

$$\text{beta4} := 1.099554073$$

```
# Now substitute a=1,c=-c12/c11 ,beta=beta1,beta2,beta3,
  beta4 into v(x) and plot the mode-shapes
  z1:=proc(x) subs(a=1,c=-c11/c12,beta=beta1,v(x)) end;
  z1:= proc(x) subs(a = 1,c = -c11/c12,beta = beta1,v(x))
end
  z2:=proc(x) subs(a=1,c=-c11/c12,beta=beta2,v(x)) end;
  z2:= proc(x) subs(a = 1,c = -c11/c12,beta = beta2,v(x))
end
  z3:=proc(x) subs(a=1,c=-c11/c12,beta=beta3,v(x)) end;
  z3:= proc(x) subs(a = 1,c = -c11/c12,beta = beta3,v(x)) end
  z4:=proc(x) subs(a=1,c=-c11/c12,beta=beta4,v(x)) end;
  z4:= proc(x) subs(a = 1,c = -c11/c12,beta = beta4,v(x)) end

  plot({z1(x),z2(x),z3(x),z4(x)},x=0..L);
```

NOTE: Please type each command as one line in the MATLAB script file when typing the code in MATLAB.

# Appendix B

**A Proposal: The Effect of the Long-Term Use of Methanol**

## B.1 Background

Methanol, one of the leading alternatives to gasoline as a motor vehicle fuel, has been highlighted in national competitions such as the SAE Methanol Marathon in 1989 and the SAE Methanol Challenge in 1990, but little has been done in the area of long-term testing of methanol as a motor vehicle fuel. The 1988 Chevrolet Corsica modified by Texas Tech University, which finished fifth in the 1989 competition and second in the 1990 competition, is an ideal test bed to determine the long-term effects of methanol on engine and emission systems performance.

The Texas Tech Corsica was optimized to operate on M85 for the SAE competitions; however, it has recently been modified to use M100. A methanol compatible fuel system was installed for the SAE competitions. The engine has been modified by increasing the stroke to take advantage of methanol's increased energy availability, and decreasing the bore to maintain economy. The resulting displacement is 2.8 liters, which is the same as the original gasoline engine. Because methanol has a higher octane rating than gasoline, the compression ratio was increased to 11.7 : 1 by installing custom flat-top pistons. A custom camshaft was employed to compensate for the slow burn characteristics of methanol. Allied Signal, Inc., Tulsa, Oklahoma, provided the specially designed catalysts to control exhaust emissions.

## B.2 Objective

The objective of this project is to determine the effects of methanol fuel on engine performance and exhaust emissions during long-term use. Engine wear and tear, gasket performance, fuel system performance, emissions level, and overall vehicle performance will be monitored over 25,000 to 30,000 miles of vehicle operation.

## B.3 Technical Approach

A vehicle performance baseline will be established initially and used for comparative purposes during the program. The engine will be removed from the vehicle and disassembled to record all bearing and ring clearances and cam profiles to determine any pre-existing wear. Any needed repairs will be made to the engine at this time. After reassembling the engine, a Super-Flow dynamometer will be used to determine the engine performance at peak and road loads. Performance parameters to be measured will include brake torque, brake power, brake specific fuel consumption, and engine out levels of hydrocarbons, carbon monoxide, carbon dioxide, and oxygen. The engine will then be installed in the vehicle.

Once the engine is installed in the Corsica, chassis dynamometer testing will be accomplished for engine/vehicle final calibration and performance evaluation. Track testing will determine 0–60 mph and quarter-mile elapsed times. On-road tests will measure both city and highway fuel economy. The vehicle will then be driven to Southwest Research Institute in San Antonio where it will undergo a full Environmental Protection Agency (EPA) Federal Test Procedure (FTP) emissions test to determine a basis for future emission comparisons.

The vehicle will then be driven approximately 25,000 miles. To accumulate this mileage, the vehicle will be displayed at various automobile, emissions, and energy seminars in Texas and across the nation and the vehicle will be driven daily in the Lubbock area. The mileage amassed on the vehicle will consist of approximately one-third highway and two-thirds city miles. Two oil samples will be taken every 3,000 miles when the oil is changed. One sample will be analyzed locally and the other sample will be sent to Lubrizol Corp for analysis. Emissions testing will be accomplished periodically during the program using a four-gas analyzer.

When the required mileage has been accumulated on the Corsica, the vehicle will be driven back to Southwest Research Institute for a second round of emissions tests. These results will indicate any emission degradation caused by the long-term use of methanol in the Corsica. The engine will then be removed from the vehicle and retested on the dynamometer to identify any performance loss. After the dynamometer test, the engine will be disassembled and all fits and clearances will be remeasured. The gaskets will be removed and sent to FelPro Inc. for analysis. The data from the initial tests and measurements will be compared to the data from the final tests, and a detailed report will be prepared and disseminated to all participants in the project. It is also anticipated that a paper describing the project and the results will be submitted to the Society of Automotive Engineers (SAE).

## B.4   Project Personnel

The principal investigators will be Dr. Timothy T. Maxwell and Mr. Jesse C. Jones. Dr. Maxwell's interests include automotive engineering, internal combustion engines, aerodynamics, computational fluid dynamics, combustion, heat transfer and energy utilization. Dr. Maxwell is the Technical Program Manager for a Ford Motor Company funded aerodynamics research program that is currently in its 5th year. Mr. Jones has 25 years experience in the aerospace industry. He was site manager of the NASA White Sands Test Facility with responsibility for management, administration, engineering, technical support, and operation. Mr. Jones' interests include aerodynamics, automotive engineering, and design. Mr. Jones is coordinator for the cooperative industry/university design project program.

Both Dr. Maxwell and Mr. Jones are extensively involved in alternative fuels research programs. They are co-principal investigators for a project funded through the Texas Advanced Technology

Program to investigate the starting of methanol fueled engines in cold weather and they were coadvisors for the 1989–90 Methanol Marathon/Challenge and the 1991 Natural Gas Vehicle Challenge. Mr. Jones and Dr. Maxwell have several other alternate fuels programs in progress. One of these programs is a survey of the Lubbock, Texas, area to determine what vehicle fleets could be converted to burn alternate fuels. Another project is concerned with the development of a direct injection capability for natural gas fueled engines. A series of workshops was recently completed across the State of Texas on Managing Vehicle Fleets for Fuel Economy.

Mr. Jones and Dr. Maxwell will work together to manage the project. Mr. Jones will be responsible for all engine and vehicle testing and Dr. Maxwell will direct the engine disassembly, characterization, and reassembly activities. They will be assisted in these tasks by three undergraduate students who will compile the data, demonstrate the vehicle, and assist in the test and engine disassembly efforts.

## B.5   Schedule

The project will start in mid June 1992 and will continue over a two- year period. Table B.1 presents the work schedule by quarters.

**Table B.1** Project schedule.

| Tasks | Quarters Starting in June 1992 | | | | | | | |
|---|---|---|---|---|---|---|---|---|
| | 1 | 2 | 3 | 4 | 5 | 6 | 7 | 8 |
| Remove engine from car disassemble, inspect, measure clearance, reassemble | ■ | | | | | | | |
| Engine & chassis dynamometer tests | ■ | | | | | | | |
| Performance & fuel economy tests | ■ ■ | | | | | | | |
| Emissions tests | ■ | | | | | | | |
| Mileage accumulation | | ■■■ | ■■■ | ■■■ | ■■■ | ■■■ | | |
| Emissions retest | | | | | | | ■ | |
| Performance and fuel economy retest | | | | | | | ■ ■ | |
| Engine and chassis dynamometer retests | | | | | | | | ■ |
| Remove engine from car disassemble, inspect, measure clearance, and reassemble | | | | | | | | ■ |
| Prepare report | | | | | | | | ■ |

## B.6 Budget

Several organizations have agreed to provide either funding or in-kind support. FelPro Inc., will provide $10,000 and all gaskets and seals needed for the project. Air Products and Chemicals, Inc. will provide 2,000 gallons of methanol fuel and The Lubrizol Corp. will provide all the oil required for the project. Allied Signal will provide the exhaust system catalysts. The Texas Tech University Center for Energy Research has agreed to provide $10,000 for the project, and the University will charge only 10 percent indirect costs as partial cost sharing.

Table B.2 presents the details of the budget request for this project. The total budget is $43,263.

**Table B.2** Project Budget.

| | |
|---|---:|
| A. Salaries and Wages | |
|    1. Undergraduate students (500 hours @ $5/h) | $ 2,500 |
| **Total Salaries & Wages** | **$ 2,500** |
| B. Fringe Benefits (18% of A.1) | $ 450 |
| C. Parts and Supplies | |
|    1. Six Fuel Injectors | 300 |
|    2. Fuel Pump | 100 |
|    3. Custom Piston Rings | 120 |
|    4. Engine Bearings | 40 |
|    5. Gaskets | 150 |
|    6. Filters (oil and air) | 80 |
|    7. Spark Plugs | 15 |
| **Total Parts & Supplies** | **$ 805** |
| D. Engine Work | |
|    1. Remove engine from vehicle and replace (2) | 900 |
|    2. Disassemble and clean engine (2) | 200 |
|    3. Inspect, measure, record data and reassemble engine (2) | 3,000 |
| **Total Engine Work** | **$ 4,100** |
| E. Engine and Vehicle Testing | |
|    1. Dynamometer testing of engine (2) | 3,400 |
|    2. FTP emissions tests (2) | 5,000 |
| **Total Engine and Vehicle Testing** | **$ 8,500** |
| F. Fuel & Oil | |
|    1. 2,000 gallons of methanol | 3,000 |
|    2. 60 quarts of oil | 75 |
| **Total Fuel and Oil** | **$ 3,075** |
| G. Travel (20 trips) | $ 20,000 |
| H. Total Direct | $ 39,330 |
| I. Indirect Costs (10% of H) | $ 3,933 |
| J. **Total Project Costs** | **$ 43,263** |

# Appendix C

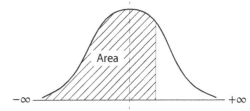

**Table C.1** Area under the normal curve.

| z | 0.00 | 0.01 | 0.02 | 0.03 | 0.04 | 0.05 | 0.06 | 0.07 | 0.08 | 0.09 |
|------|--------|--------|--------|--------|--------|--------|--------|--------|--------|--------|
| −3.4 | 0.0003 | 0.0003 | 0.0003 | 0.0003 | 0.0003 | 0.0003 | 0.0003 | 0.0003 | 0.0003 | 0.0002 |
| −3.3 | 0.0005 | 0.0005 | 0.0005 | 0.0004 | 0.0004 | 0.0004 | 0.0004 | 0.0004 | 0.0004 | 0.0003 |
| −3.2 | 0.0007 | 0.0007 | 0.0006 | 0.0006 | 0.0006 | 0.0006 | 0.0006 | 0.0005 | 0.0005 | 0.0005 |
| −3.1 | 0.0010 | 0.0009 | 0.0009 | 0.0009 | 0.0008 | 0.0008 | 0.0008 | 0.0008 | 0.0007 | 0.0007 |
| −3.0 | 0.0013 | 0.0013 | 0.0013 | 0.0012 | 0.0012 | 0.0011 | 0.0011 | 0.0011 | 0.0010 | 0.0010 |
| −2.9 | 0.0019 | 0.0018 | 0.0017 | 0.0017 | 0.0016 | 0.0016 | 0.0015 | 0.0015 | 0.0014 | 0.0014 |
| −2.8 | 0.0026 | 0.0025 | 0.0024 | 0.0023 | 0.0023 | 0.0022 | 0.0021 | 0.0021 | 0.0020 | 0.0019 |
| −2.7 | 0.0035 | 0.0034 | 0.0033 | 0.0032 | 0.0031 | 0.0030 | 0.0029 | 0.0028 | 0.0027 | 0.0026 |
| −2.6 | 0.0047 | 0.0045 | 0.0044 | 0.0043 | 0.0041 | 0.0040 | 0.0039 | 0.0038 | 0.0037 | 0.0036 |
| −2.5 | 0.0062 | 0.0060 | 0.0059 | 0.0057 | 0.0055 | 0.0054 | 0.0052 | 0.0051 | 0.0049 | 0.0048 |
| −2.4 | 0.0082 | 0.0080 | 0.0078 | 0.0075 | 0.0073 | 0.0071 | 0.0069 | 0.0068 | 0.0066 | 0.0064 |
| −2.3 | 0.0107 | 0.0104 | 0.0102 | 0.0099 | 0.0096 | 0.0094 | 0.0091 | 0.0089 | 0.0087 | 0.0084 |
| −2.2 | 0.0139 | 0.0136 | 0.0132 | 0.0129 | 0.0125 | 0.0122 | 0.0119 | 0.0116 | 0.0113 | 0.0110 |
| −2.1 | 0.0179 | 0.0174 | 0.0170 | 0.0166 | 0.0162 | 0.0158 | 0.0154 | 0.0150 | 0.0146 | 0.0143 |
| −2.0 | 0.0228 | 0.0222 | 0.0217 | 0.0212 | 0.0207 | 0.0202 | 0.0197 | 0.0192 | 0.0188 | 0.0183 |
| −1.9 | 0.0287 | 0.0281 | 0.0274 | 0.0268 | 0.0262 | 0.0256 | 0.0250 | 0.0244 | 0.0239 | 0.0233 |
| −1.8 | 0.0359 | 0.0352 | 0.0344 | 0.0336 | 0.0329 | 0.0322 | 0.0314 | 0.0307 | 0.0301 | 0.0294 |
| −1.7 | 0.0446 | 0.0436 | 0.0427 | 0.0418 | 0.0409 | 0.0401 | 0.0392 | 0.0384 | 0.0375 | 0.0367 |

*(Continued)*

**Table C.1** (Continued)

| z | 0.00 | 0.01 | 0.02 | 0.03 | 0.04 | 0.05 | 0.06 | 0.07 | 0.08 | 0.09 |
|---|------|------|------|------|------|------|------|------|------|------|
| −1.6 | 0.0548 | 0.0537 | 0.0526 | 0.0516 | 0.0505 | 0.0495 | 0.0485 | 0.0475 | 0.0465 | 0.0455 |
| −1.5 | 0.0668 | 0.0655 | 0.0643 | 0.0630 | 0.0618 | 0.0606 | 0.0594 | 0.0582 | 0.0571 | 0.0559 |
| −1.4 | 0.0808 | 0.0793 | 0.0778 | 0.0764 | 0.0749 | 0.0735 | 0.0722 | 0.0708 | 0.0694 | 0.0681 |
| −1.3 | 0.0968 | 0.0951 | 0.0934 | 0.0918 | 0.0901 | 0.0885 | 0.0869 | 0.0853 | 0.0838 | 0.0823 |
| −1.2 | 0.1151 | 0.1131 | 0.1112 | 0.1093 | 0.1075 | 0.1056 | 0.1038 | 0.1020 | 0.1003 | 0.0985 |
| −1.1 | 0.1357 | 0.1335 | 0.1314 | 0.1292 | 0.1271 | 0.1251 | 0.1230 | 0.1210 | 0.1190 | 0.1170 |
| −1.0 | 0.1587 | 0.1562 | 0.1539 | 0.1515 | 0.1492 | 0.1469 | 0.1446 | 0.1423 | 0.1401 | 0.1379 |
| −0.9 | 0.1841 | 0.1814 | 0.1788 | 0.1762 | 0.1736 | 0.1711 | 0.1685 | 0.1660 | 0.1635 | 0.1611 |
| −0.8 | 0.2119 | 0.2090 | 0.2061 | 0.2033 | 0.2005 | 0.1977 | 0.1949 | 0.1922 | 0.1894 | 0.1867 |
| −0.7 | 0.2420 | 0.2389 | 0.2358 | 0.2327 | 0.2296 | 0.2266 | 0.2236 | 0.2206 | 0.2177 | 0.2148 |
| −0.6 | 0.2743 | 0.2709 | 0.2676 | 0.2643 | 0.2611 | 0.2578 | 0.2546 | 0.2514 | 0.2483 | 0.2451 |
| −0.5 | 0.3085 | 0.3050 | 0.3015 | 0.2981 | 0.2946 | 0.2912 | 0.2877 | 0.2843 | 0.2810 | 0.2776 |
| −0.4 | 0.3446 | 0.3409 | 0.3372 | 0.3336 | 0.3300 | 0.3264 | 0.3228 | 0.3192 | 0.3156 | 0.3121 |
| −0.3 | 0.3821 | 0.3783 | 0.3745 | 0.3707 | 0.3669 | 0.3632 | 0.3594 | 0.3557 | 0.3520 | 0.3483 |
| −0.2 | 0.4207 | 0.4168 | 0.4129 | 0.4090 | 0.4052 | 0.4013 | 0.3974 | 0.3936 | 0.3897 | 0.3859 |
| −0.1 | 0.4602 | 0.4562 | 0.4522 | 0.4483 | 0.4443 | 0.4404 | 0.4364 | 0.4325 | 0.4286 | 0.4247 |
| −0.0 | 0.5000 | 0.4960 | 0.4920 | 0.4880 | 0.4840 | 0.4801 | 0.4761 | 0.4721 | 0.4681 | 0.4641 |
| 0.0 | 0.5000 | 0.5040 | 0.5080 | 0.5120 | 0.5160 | 0.5199 | 0.5239 | 0.5279 | 0.5319 | 0.5359 |
| 0.1 | 0.5398 | 0.5438 | 0.5478 | 0.5517 | 0.5557 | 0.5596 | 0.5636 | 0.5675 | 0.5714 | 0.5753 |
| 0.2 | 0.5793 | 0.5832 | 0.5871 | 0.5910 | 0.5948 | 0.5987 | 0.6026 | 0.6064 | 0.6103 | 0.6141 |
| 0.3 | 0.6179 | 0.6217 | 0.6255 | 0.6293 | 0.6331 | 0.6368 | 0.6406 | 0.6443 | 0.6480 | 0.6517 |
| 0.4 | 0.6554 | 0.6591 | 0.6628 | 0.6664 | 0.6700 | 0.6736 | 0.6772 | 0.6808 | 0.6844 | 0.6879 |
| 0.5 | 0.6915 | 0.6950 | 0.6985 | 0.7019 | 0.7054 | 0.7088 | 0.7123 | 0.7157 | 0.7190 | 0.7224 |
| 0.6 | 0.7257 | 0.7291 | 0.7324 | 0.7357 | 0.7389 | 0.7422 | 0.7454 | 0.7486 | 0.7517 | 0.7549 |
| 0.7 | 0.7580 | 0.7611 | 0.7642 | 0.7637 | 0.7704 | 0.7734 | 0.7764 | 0.7794 | 0.7823 | 0.7852 |
| 0.8 | 0.7881 | 0.7910 | 0.7939 | 0.7967 | 0.7995 | 0.8023 | 0.8051 | 0.8078 | 0.8106 | 0.8133 |

**Table C.1** (Continued)

| z | 0.00 | 0.01 | 0.02 | 0.03 | 0.04 | 0.05 | 0.06 | 0.07 | 0.08 | 0.09 |
|---|---|---|---|---|---|---|---|---|---|---|
| 0.9 | 0.8159 | 0.8186 | 0.8212 | 0.8238 | 0.8264 | 0.8289 | 0.8315 | 0.8340 | 0.8365 | 0.8389 |
| 1.0 | 0.8413 | 0.8438 | 0.8461 | 0.8485 | 0.8508 | 0.8531 | 0.8554 | 0.8577 | 0.8599 | 0.8621 |
| 1.1 | 0.8643 | 0.8665 | 0.8686 | 0.8708 | 0.8729 | 0.8749 | 0.8770 | 0.8790 | 0.8810 | 0.8830 |
| 1.2 | 0.8849 | 0.8869 | 0.8888 | 0.8907 | 0.8925 | 0.8944 | 0.8962 | 0.8980 | 0.8997 | 0.9015 |
| 1.3 | 0.9032 | 0.9049 | 0.9066 | 0.9082 | 0.9099 | 0.9115 | 0.9131 | 0.9147 | 0.9162 | 0.9177 |
| 1.4 | 0.9192 | 0.9207 | 0.9222 | 0.9236 | 0.9251 | 0.9265 | 0.9278 | 0.9292 | 0.9306 | 0.9319 |
| 1.5 | 0.9332 | 0.9345 | 0.9357 | 0.9370 | 0.9382 | 0.9394 | 0.9406 | 0.9418 | 0.9429 | 0.9441 |
| 1.6 | 0.9452 | 0.9463 | 0.9474 | 0.9484 | 0.9495 | 0.9505 | 0.9515 | 0.9525 | 0.9535 | 0.9545 |
| 1.7 | 0.9554 | 0.9564 | 0.9573 | 0.9582 | 0.9591 | 0.9599 | 0.9608 | 0.9616 | 0.9625 | 0.9633 |
| 1.8 | 0.9641 | 0.9649 | 0.9656 | 0.9664 | 0.9671 | 0.9678 | 0.9686 | 0.9693 | 0.9699 | 0.9706 |
| 1.9 | 0.9713 | 0.9719 | 0.9726 | 0.9732 | 0.9738 | 0.9744 | 0.9750 | 0.9756 | 0.9761 | 0.9767 |
| 2.0 | 0.9772 | 0.9778 | 0.9783 | 0.9788 | 0.9793 | 0.9798 | 0.9803 | 0.9808 | 0.9812 | 0.9817 |
| 2.1 | 0.9821 | 0.9826 | 0.9830 | 0.9834 | 0.9838 | 0.9842 | 0.9846 | 0.9850 | 0.9854 | 0.9857 |
| 2.2 | 0.9861 | 0.9864 | 0.9868 | 0.9871 | 0.9875 | 0.9878 | 0.9881 | 0.9884 | 0.9887 | 0.9890 |
| 2.3 | 0.9893 | 0.9896 | 0.9898 | 0.9901 | 0.9904 | 0.9906 | 0.9909 | 0.9911 | 0.9913 | 0.9916 |
| 2.4 | 0.9918 | 0.9920 | 0.9922 | 0.9925 | 0.9927 | 0.9929 | 0.9931 | 0.9932 | 0.9934 | 0.9936 |
| 2.5 | 0.9938 | 0.9940 | 0.9941 | 0.9943 | 0.9945 | 0.9946 | 0.9948 | 0.9949 | 0.9951 | 0.9952 |
| 2.6 | 0.9953 | 0.9955 | 0.9956 | 0.9957 | 0.9959 | 0.9960 | 0.9961 | 0.9962 | 0.9963 | 0.9964 |
| 2.7 | 0.9965 | 0.9966 | 0.9967 | 0.9968 | 0.9969 | 0.9970 | 0.9971 | 0.9972 | 0.9973 | 0.9974 |
| 2.8 | 0.9974 | 0.9975 | 0.9976 | 0.9977 | 0.9977 | 0.9978 | 0.9979 | 0.9979 | 0.9980 | 0.9981 |
| 2.9 | 0.9981 | 0.9982 | 0.9982 | 0.9983 | 0.9984 | 0.9984 | 0.9985 | 0.9985 | 0.9986 | 0.9986 |
| 3.0 | 0.9987 | 0.9987 | 0.9987 | 0.9988 | 0.9988 | 0.9989 | 0.9989 | 0.9989 | 0.9990 | 0.9990 |
| 3.1 | 0.9990 | 0.9991 | 0.9991 | 0.9991 | 0.9992 | 0.9992 | 0.9992 | 0.9992 | 0.9993 | 0.9993 |
| 3.2 | 0.9993 | 0.9993 | 0.9994 | 0.9994 | 0.9994 | 0.9994 | 0.9994 | 0.9995 | 0.9995 | 0.9995 |
| 3.3 | 0.9995 | 0.9995 | 0.9995 | 0.9996 | 0.9996 | 0.9996 | 0.9996 | 0.9996 | 0.9996 | 0.9997 |
| 3.4 | 0.9997 | 0.9997 | 0.9997 | 0.9997 | 0.9997 | 0.9997 | 0.9997 | 0.9997 | 0.9997 | 0.9998 |

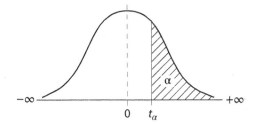

**Table C.2** Critical values of the *t* distribution.

| $\nu =$ $n-1$ | $\alpha$ | | | | | | | | |
|---|---|---|---|---|---|---|---|---|---|
| | 0.20 | 0.15 | 0.10 | 0.05 | 0.025 | 0.020 | 0.015 | 0.010 | 0.005 |
| 1 | 1.3764 | 1.9626 | 3.0777 | 6.3138 | 12.7062 | 15.8945 | 21.2049 | 31.8205 | 63.6567 |
| 2 | 1.0607 | 1.3862 | 1.8856 | 2.9200 | 4.3027 | 4.8487 | 5.6428 | 6.9646 | 9.9248 |
| 3 | 0.9785 | 1.2498 | 1.6377 | 2.3534 | 3.1824 | 3.4819 | 3.8960 | 4.5407 | 5.8409 |
| 4 | 0.9410 | 1.1896 | 1.5332 | 2.1318 | 2.7764 | 2.9985 | 3.2976 | 3.7469 | 4.6041 |
| 5 | 0.9195 | 1.1558 | 1.4759 | 2.0150 | 2.5706 | 2.7565 | 3.0029 | 3.3649 | 4.0321 |
| 6 | 0.9057 | 1.1342 | 1.4398 | 1.9432 | 2.4469 | 2.6122 | 2.8289 | 3.1427 | 3.7074 |
| 7 | 0.8960 | 1.1192 | 1.4149 | 1.8946 | 2.3646 | 2.5168 | 2.7146 | 2.9980 | 3.4995 |
| 8 | 0.8889 | 1.1081 | 1.3968 | 1.8595 | 2.3060 | 2.4490 | 2.6338 | 2.8965 | 3.3554 |
| 9 | 0.8834 | 1.0997 | 1.3830 | 1.8331 | 2.2622 | 2.3984 | 2.5738 | 2.8214 | 3.2498 |
| 10 | 0.8791 | 1.0931 | 1.3722 | 1.8125 | 2.2281 | 2.3593 | 2.5275 | 2.7638 | 3.1693 |
| 11 | 0.8755 | 1.0877 | 1.3634 | 1.7959 | 2.2010 | 2.3281 | 2.4907 | 2.7181 | 3.1058 |
| 12 | 0.8726 | 1.0832 | 1.3562 | 1.7823 | 2.1788 | 2.3027 | 2.4607 | 2.6810 | 3.0545 |
| 13 | 0.8702 | 1.0795 | 1.3502 | 1.7709 | 2.1604 | 2.2816 | 2.4358 | 2.6503 | 3.0123 |
| 14 | 0.8681 | 1.0763 | 1.3450 | 1.7613 | 2.1448 | 2.2638 | 2.4149 | 2.6245 | 2.9768 |
| 15 | 0.8662 | 1.0735 | 1.3406 | 1.7531 | 2.1314 | 2.2485 | 2.3970 | 2.6025 | 2.9467 |
| 16 | 0.8647 | 1.0711 | 1.3368 | 1.7459 | 2.1199 | 2.2354 | 2.3815 | 2.5835 | 2.9208 |
| 17 | 0.8633 | 1.0690 | 1.3334 | 1.7396 | 2.1098 | 2.2238 | 2.3681 | 2.5669 | 2.8982 |
| 18 | 0.8620 | 1.0672 | 1.3304 | 1.7341 | 2.1009 | 2.2137 | 2.3562 | 2.5524 | 2.8784 |
| 19 | 0.8610 | 1.0655 | 1.3277 | 1.7291 | 2.0930 | 2.2047 | 2.3456 | 2.5395 | 2.8609 |
| 20 | 0.8600 | 1.0640 | 1.3253 | 1.7247 | 2.0860 | 2.1967 | 2.3362 | 2.5280 | 2.8453 |
| 21 | 0.8591 | 1.0627 | 1.3232 | 1.7207 | 2.0796 | 2.1894 | 2.3278 | 2.5176 | 2.8314 |
| 22 | 0.8583 | 1.0614 | 1.3212 | 1.7171 | 2.0739 | 2.1829 | 2.3202 | 2.5083 | 2.8188 |
| 23 | 0.8575 | 1.0603 | 1.3195 | 1.7139 | 2.0687 | 2.1770 | 2.3132 | 2.4999 | 2.8073 |
| 24 | 0.8569 | 1.0593 | 1.3178 | 1.7109 | 2.0639 | 2.1715 | 2.3069 | 2.4922 | 2.7969 |
| 25 | 0.8562 | 1.0584 | 1.3163 | 1.7081 | 2.0595 | 2.1666 | 2.3011 | 2.4851 | 2.7874 |
| 26 | 0.8557 | 1.0575 | 1.3150 | 1.7056 | 2.0555 | 2.1620 | 2.2958 | 2.4786 | 2.7787 |
| 27 | 0.8551 | 1.0567 | 1.3137 | 1.7033 | 2.0518 | 2.1578 | 2.2909 | 2.4727 | 2.7707 |
| 28 | 0.8546 | 1.0560 | 1.3125 | 1.7011 | 2.0484 | 2.1539 | 2.2864 | 2.4671 | 2.7633 |

**Table C.2** (Continued)

| $v =$ | $\alpha$ | | | | | | | | |
|---|---|---|---|---|---|---|---|---|---|
| $n-1$ | 0.20 | 0.15 | 0.10 | 0.05 | 0.025 | 0.020 | 0.015 | 0.010 | 0.005 |
| 29 | 0.8542 | 1.0553 | 1.3114 | 1.6991 | 2.0452 | 2.1503 | 2.2822 | 2.4620 | 2.7564 |
| 30 | 0.8538 | 1.0547 | 1.3104 | 1.6973 | 2.0423 | 2.1470 | 2.2783 | 2.4573 | 2.7500 |
| 31 | 0.8534 | 1.0541 | 1.3095 | 1.6955 | 2.0395 | 2.1438 | 2.2746 | 2.4528 | 2.7440 |
| 32 | 0.8530 | 1.0535 | 1.3086 | 1.6939 | 2.0369 | 2.1409 | 2.2712 | 2.4487 | 2.7385 |
| 33 | 0.8526 | 1.0530 | 1.3077 | 1.6924 | 2.0345 | 2.1382 | 2.2680 | 2.4448 | 2.7333 |
| 34 | 0.8523 | 1.0525 | 1.3070 | 1.6909 | 2.0322 | 2.1356 | 2.2650 | 2.4411 | 2.7284 |
| 35 | 0.8520 | 1.0520 | 1.3062 | 1.6896 | 2.0301 | 2.1332 | 2.2622 | 2.4377 | 2.7238 |

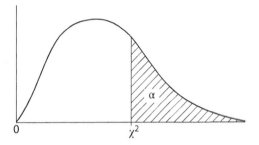

**Table C.3** Critical Values of the chi-square distribution.

| $v =$ | $\alpha$ | | | | | | | |
|---|---|---|---|---|---|---|---|---|
| $n-1$ | 0.995 | 0.99 | 0.975 | 0.95 | 0.05 | 0.025 | 0.01 | 0.005 |
| 1 | 0.0000 | 0.0002 | 0.0010 | 0.0039 | 3.8415 | 5.0239 | 6.6349 | 7.8794 |
| 2 | 0.0100 | 0.0201 | 0.0506 | 0.1026 | 5.9915 | 7.3778 | 9.2103 | 10.5966 |
| 3 | 0.0717 | 0.1148 | 0.2158 | 0.3518 | 7.8147 | 9.3484 | 11.3449 | 12.8382 |
| 4 | 0.2070 | 0.2971 | 0.4844 | 0.7107 | 9.4877 | 11.1433 | 13.2767 | 14.8603 |
| 5 | 0.4117 | 0.5543 | 0.8312 | 1.1455 | 11.0705 | 12.8325 | 15.0863 | 16.7496 |
| 6 | 0.6757 | 0.8721 | 1.2373 | 1.6354 | 12.5916 | 14.4494 | 16.8119 | 18.5476 |
| 7 | 0.9893 | 1.2390 | 1.6899 | 2.1673 | 14.0671 | 16.0128 | 18.4753 | 20.2777 |
| 8 | 1.3444 | 1.6465 | 2.1797 | 2.7326 | 15.5073 | 17.5345 | 20.0902 | 21.9550 |
| 9 | 1.7349 | 2.0879 | 2.7004 | 3.3251 | 16.9190 | 19.0228 | 21.6660 | 23.5894 |
| 10 | 2.1559 | 2.5582 | 3.2470 | 3.9403 | 18.3070 | 20.4832 | 23.2093 | 25.1882 |
| 11 | 2.6032 | 3.0535 | 3.8157 | 4.5748 | 19.6751 | 21.9200 | 24.7250 | 26.7568 |
| 12 | 3.0738 | 3.5706 | 4.4038 | 5.2260 | 21.0261 | 23.3367 | 26.2170 | 28.2995 |
| 13 | 3.5650 | 4.1069 | 5.0088 | 5.8919 | 22.3620 | 24.7356 | 27.6882 | 29.8195 |
| 14 | 4.0747 | 4.6604 | 5.6287 | 6.5706 | 23.6848 | 26.1189 | 29.1412 | 31.3193 |

*(Continued)*

**Table C.3** (Continued)

| $v =$ $n - 1$ | $\alpha$ | | | | | | | |
|---|---|---|---|---|---|---|---|---|
| | 0.995 | 0.99 | 0.975 | 0.95 | 0.05 | 0.025 | 0.01 | 0.005 |
| 15 | 4.6009 | 5.2293 | 6.2621 | 7.2609 | 24.9958 | 27.4884 | 30.5779 | 32.8013 |
| 16 | 5.1422 | 5.8122 | 6.9077 | 7.9616 | 26.2962 | 28.8454 | 31.9999 | 34.2672 |
| 17 | 5.6972 | 6.4078 | 7.5642 | 8.6718 | 27.5871 | 30.1910 | 33.4087 | 35.7185 |
| 18 | 6.2648 | 7.0149 | 8.2307 | 9.3905 | 28.8693 | 31.5264 | 34.8053 | 37.1565 |
| 19 | 6.8440 | 7.6327 | 8.9065 | 10.1170 | 30.1435 | 32.8523 | 36.1909 | 38.5823 |
| 20 | 7.4338 | 8.2604 | 9.5908 | 10.8508 | 31.4104 | 34.1696 | 37.5662 | 39.9968 |
| 21 | 8.0337 | 8.8972 | 10.2829 | 11.5913 | 32.6706 | 35.4789 | 38.9322 | 41.4011 |
| 22 | 8.6427 | 9.5425 | 10.9823 | 12.3380 | 33.9244 | 36.7807 | 40.2894 | 42.7957 |
| 23 | 9.2604 | 10.1957 | 11.6886 | 13.0905 | 35.1725 | 38.0756 | 41.6384 | 44.1813 |
| 24 | 9.8862 | 10.8564 | 12.4012 | 13.8484 | 36.4150 | 39.3641 | 42.9798 | 45.5585 |
| 25 | 10.5197 | 11.5240 | 13.1197 | 14.6114 | 37.6525 | 40.6465 | 44.3141 | 46.9279 |
| 26 | 11.1602 | 12.1981 | 13.8439 | 15.3792 | 38.8851 | 41.9232 | 45.6417 | 48.2899 |
| 27 | 11.8076 | 12.8785 | 14.5734 | 16.1514 | 40.1133 | 43.1945 | 46.9629 | 49.6449 |
| 28 | 12.4613 | 13.5647 | 15.3079 | 16.9279 | 41.3371 | 44.4608 | 48.2782 | 50.9934 |
| 29 | 13.1211 | 14.2565 | 16.0471 | 17.7084 | 42.5570 | 45.7223 | 49.5879 | 52.3356 |
| 30 | 13.7867 | 14.9535 | 16.7908 | 18.4927 | 43.7730 | 46.9792 | 50.8922 | 53.6720 |
| 31 | 14.4578 | 15.6555 | 17.5387 | 19.2806 | 44.9853 | 48.2319 | 52.1914 | 55.0027 |
| 32 | 15.1340 | 16.3622 | 18.2908 | 20.0719 | 46.1943 | 49.4804 | 53.4858 | 56.3281 |
| 33 | 15.8153 | 17.0735 | 19.0467 | 20.8665 | 47.3999 | 50.7251 | 54.7755 | 57.6484 |
| 34 | 16.5013 | 17.7891 | 19.8063 | 21.6643 | 48.6024 | 51.9660 | 56.0609 | 58.9639 |
| 35 | 17.1918 | 18.5089 | 20.5694 | 22.4650 | 49.8018 | 53.2033 | 57.3421 | 60.2748 |
| 50 | 27.9907 | 29.7067 | 32.3574 | 34.7643 | 67.5048 | 71.4202 | 76.1539 | 79.4900 |
| 60 | 35.5345 | 37.4849 | 40.4817 | 43.1880 | 79.0819 | 83.2977 | 88.3794 | 91.9517 |
| 80 | 51.1719 | 53.5401 | 57.1532 | 60.3915 | 101.880 | 106.629 | 112.329 | 116.321 |
| 110 | 75.5500 | 78.4583 | 82.8671 | 86.7916 | 135.480 | 140.917 | 147.414 | 151.949 |
| 130 | 92.2225 | 95.4510 | 100.331 | 104.662 | 157.610 | 163.453 | 170.423 | 175.278 |
| 200 | 152.241 | 156.432 | 162.728 | 168.279 | 233.994 | 241.058 | 249.445 | 255.264 |

**Table C.4** Percentage points of the $F$ distribution.

$$F_{0.25,\nu_1,\nu_2}$$

| | Degrees of freedom for the numerator ($\nu_1$) | | | | | | | | | | | | | | | | | | |
| $\nu_2$ | 1 | 2 | 3 | 4 | 5 | 6 | 7 | 8 | 9 | 10 | 12 | 15 | 20 | 24 | 30 | 40 | 60 | 120 | ∞ |
|---|---|---|---|---|---|---|---|---|---|---|---|---|---|---|---|---|---|---|---|
| 1 | 5.83 | 7.50 | 8.20 | 8.58 | 8.82 | 8.98 | 9.10 | 9.19 | 9.26 | 9.32 | 9.41 | 9.49 | 9.58 | 9.63 | 9.67 | 9.71 | 9.76 | 9.80 | 9.85 |
| 2 | 2.57 | 3.00 | 3.15 | 3.23 | 3.28 | 3.31 | 3.34 | 3.35 | 3.37 | 3.38 | 3.39 | 3.41 | 3.43 | 3.43 | 3.44 | 3.45 | 3.46 | 3.47 | 3.48 |
| 3 | 2.02 | 2.28 | 2.36 | 2.39 | 2.41 | 2.42 | 2.43 | 2.44 | 2.44 | 2.44 | 2.45 | 2.46 | 2.46 | 2.46 | 2.47 | 2.47 | 2.47 | 2.47 | 2.47 |
| 4 | 1.81 | 2.00 | 2.05 | 2.06 | 2.07 | 2.08 | 2.08 | 2.08 | 2.08 | 2.08 | 2.08 | 2.08 | 2.08 | 2.08 | 2.08 | 2.08 | 2.08 | 2.08 | 2.08 |
| 5 | 1.69 | 1.85 | 1.88 | 1.89 | 1.89 | 1.89 | 1.89 | 1.89 | 1.89 | 1.89 | 1.89 | 1.89 | 1.88 | 1.88 | 1.88 | 1.88 | 1.87 | 1.87 | 1.87 |
| 6 | 1.62 | 1.76 | 1.78 | 1.79 | 1.79 | 1.78 | 1.78 | 1.78 | 1.77 | 1.77 | 1.77 | 1.76 | 1.76 | 1.75 | 1.75 | 1.75 | 1.74 | 1.74 | 1.74 |
| 7 | 1.57 | 1.70 | 1.72 | 1.72 | 1.71 | 1.71 | 1.70 | 1.70 | 1.70 | 1.69 | 1.68 | 1.68 | 1.67 | 1.67 | 1.66 | 1.66 | 1.65 | 1.65 | 1.65 |
| 8 | 1.54 | 1.66 | 1.67 | 1.66 | 1.66 | 1.65 | 1.64 | 1.64 | 1.63 | 1.63 | 1.62 | 1.62 | 1.61 | 1.60 | 1.60 | 1.59 | 1.59 | 1.58 | 1.58 |
| 9 | 1.51 | 1.62 | 1.63 | 1.63 | 1.62 | 1.61 | 1.60 | 1.60 | 1.59 | 1.59 | 1.58 | 1.57 | 1.56 | 1.56 | 1.55 | 1.54 | 1.54 | 1.53 | 1.53 |
| 10 | 1.49 | 1.60 | 1.60 | 1.59 | 1.59 | 1.58 | 1.57 | 1.56 | 1.56 | 1.55 | 1.54 | 1.53 | 1.52 | 1.52 | 1.51 | 1.51 | 1.50 | 1.49 | 1.48 |
| 11 | 1.47 | 1.58 | 1.58 | 1.57 | 1.56 | 1.55 | 1.54 | 1.53 | 1.53 | 1.52 | 1.51 | 1.50 | 1.49 | 1.49 | 1.48 | 1.47 | 1.47 | 1.46 | 1.45 |
| 12 | 1.46 | 1.56 | 1.56 | 1.55 | 1.54 | 1.53 | 1.52 | 1.51 | 1.51 | 1.50 | 1.49 | 1.48 | 1.47 | 1.46 | 1.45 | 1.45 | 1.44 | 1.43 | 1.42 |
| 13 | 1.45 | 1.55 | 1.55 | 1.53 | 1.52 | 1.51 | 1.50 | 1.49 | 1.49 | 1.48 | 1.47 | 1.46 | 1.45 | 1.44 | 1.43 | 1.42 | 1.42 | 1.41 | 1.40 |
| 14 | 1.44 | 1.53 | 1.53 | 1.52 | 1.51 | 1.50 | 1.49 | 1.48 | 1.47 | 1.46 | 1.45 | 1.44 | 1.43 | 1.42 | 1.41 | 1.41 | 1.40 | 1.39 | 1.38 |
| 15 | 1.43 | 1.52 | 1.52 | 1.51 | 1.49 | 1.48 | 1.47 | 1.46 | 1.46 | 1.45 | 1.44 | 1.43 | 1.41 | 1.41 | 1.40 | 1.39 | 1.38 | 1.37 | 1.36 |
| 16 | 1.42 | 1.51 | 1.51 | 1.50 | 1.48 | 1.47 | 1.46 | 1.45 | 1.44 | 1.44 | 1.43 | 1.41 | 1.40 | 1.39 | 1.38 | 1.37 | 1.36 | 1.35 | 1.34 |
| 17 | 1.42 | 1.51 | 1.50 | 1.49 | 1.47 | 1.46 | 1.45 | 1.44 | 1.43 | 1.43 | 1.41 | 1.40 | 1.39 | 1.38 | 1.37 | 1.36 | 1.35 | 1.34 | 1.33 |
| 18 | 1.41 | 1.50 | 1.49 | 1.48 | 1.46 | 1.45 | 1.44 | 1.43 | 1.42 | 1.42 | 1.40 | 1.39 | 1.38 | 1.37 | 1.36 | 1.35 | 1.34 | 1.33 | 1.32 |
| 19 | 1.41 | 1.49 | 1.49 | 1.47 | 1.46 | 1.44 | 1.43 | 1.42 | 1.41 | 1.41 | 1.40 | 1.38 | 1.37 | 1.36 | 1.35 | 1.34 | 1.33 | 1.32 | 1.30 |

*(Continued)*

Table C.4 (Continued)

$$F_{0.25,v_1,v_2}$$

| $v_2$ | Degrees of freedom for the numerator ($v_1$) | | | | | | | | | | | | | | | | | | |
|---|---|---|---|---|---|---|---|---|---|---|---|---|---|---|---|---|---|---|---|
| | 1 | 2 | 3 | 4 | 5 | 6 | 7 | 8 | 9 | 10 | 12 | 15 | 20 | 24 | 30 | 40 | 60 | 120 | ∞ |
| 20 | 1.40 | 1.49 | 1.48 | 1.47 | 1.45 | 1.44 | 1.43 | 1.42 | 1.41 | 1.40 | 1.39 | 1.37 | 1.36 | 1.35 | 1.34 | 1.33 | 1.32 | 1.31 | 1.29 |
| 21 | 1.40 | 1.48 | 1.48 | 1.46 | 1.44 | 1.43 | 1.42 | 1.41 | 1.40 | 1.39 | 1.38 | 1.37 | 1.35 | 1.34 | 1.33 | 1.32 | 1.31 | 1.30 | 1.28 |
| 22 | 1.40 | 1.48 | 1.47 | 1.45 | 1.44 | 1.42 | 1.41 | 1.40 | 1.39 | 1.39 | 1.37 | 1.36 | 1.34 | 1.33 | 1.32 | 1.31 | 1.30 | 1.29 | 1.28 |
| 23 | 1.39 | 1.47 | 1.47 | 1.45 | 1.43 | 1.42 | 1.41 | 1.40 | 1.39 | 1.38 | 1.37 | 1.35 | 1.34 | 1.33 | 1.32 | 1.31 | 1.30 | 1.28 | 1.27 |
| 24 | 1.39 | 1.47 | 1.46 | 1.44 | 1.43 | 1.41 | 1.40 | 1.39 | 1.38 | 1.38 | 1.36 | 1.35 | 1.33 | 1.32 | 1.31 | 1.30 | 1.29 | 1.28 | 1.26 |
| 25 | 1.39 | 1.47 | 1.46 | 1.44 | 1.42 | 1.41 | 1.40 | 1.39 | 1.38 | 1.37 | 1.36 | 1.34 | 1.33 | 1.32 | 1.31 | 1.29 | 1.28 | 1.27 | 1.25 |
| 26 | 1.38 | 1.46 | 1.45 | 1.44 | 1.42 | 1.41 | 1.39 | 1.38 | 1.37 | 1.37 | 1.35 | 1.34 | 1.32 | 1.31 | 1.30 | 1.29 | 1.28 | 1.26 | 1.25 |
| 27 | 1.38 | 1.46 | 1.45 | 1.43 | 1.42 | 1.40 | 1.39 | 1.38 | 1.37 | 1.36 | 1.35 | 1.33 | 1.32 | 1.31 | 1.30 | 1.28 | 1.27 | 1.26 | 1.24 |
| 28 | 1.38 | 1.46 | 1.45 | 1.43 | 1.41 | 1.40 | 1.39 | 1.38 | 1.37 | 1.36 | 1.34 | 1.33 | 1.31 | 1.30 | 1.29 | 1.28 | 1.27 | 1.25 | 1.24 |
| 29 | 1.38 | 1.45 | 1.45 | 1.43 | 1.41 | 1.40 | 1.38 | 1.37 | 1.36 | 1.35 | 1.34 | 1.32 | 1.31 | 1.30 | 1.29 | 1.27 | 1.26 | 1.25 | 1.23 |
| 30 | 1.38 | 1.45 | 1.44 | 1.42 | 1.41 | 1.39 | 1.38 | 1.37 | 1.36 | 1.35 | 1.34 | 1.32 | 1.30 | 1.29 | 1.28 | 1.27 | 1.26 | 1.24 | 1.23 |
| 40 | 1.36 | 1.44 | 1.42 | 1.40 | 1.39 | 1.37 | 1.36 | 1.35 | 1.34 | 1.33 | 1.31 | 1.30 | 1.28 | 1.26 | 1.25 | 1.24 | 1.22 | 1.21 | 1.19 |
| 60 | 1.35 | 1.42 | 1.41 | 1.38 | 1.37 | 1.35 | 1.33 | 1.32 | 1.31 | 1.30 | 1.29 | 1.27 | 1.25 | 1.24 | 1.22 | 1.21 | 1.19 | 1.17 | 1.15 |
| 120 | 1.34 | 1.40 | 1.39 | 1.37 | 1.35 | 1.33 | 1.31 | 1.30 | 1.29 | 1.28 | 1.26 | 1.24 | 1.22 | 1.21 | 1.19 | 1.18 | 1.16 | 1.13 | 1.10 |
| ∞ | 1.32 | 1.39 | 1.37 | 1.35 | 1.33 | 1.31 | 1.29 | 1.28 | 1.27 | 1.25 | 1.24 | 1.22 | 1.19 | 1.18 | 1.16 | 1.14 | 1.12 | 1.08 | 1.00 |

Table C.4 (Continued)

$$F_{0.10,v_1,v_2}$$

| $v_2$ \ $v_1$ | 1 | 2 | 3 | 4 | 5 | 6 | 7 | 8 | 9 | 10 | 12 | 15 | 20 | 24 | 30 | 40 | 60 | 120 | ∞ |
|---|---|---|---|---|---|---|---|---|---|---|---|---|---|---|---|---|---|---|---|
| 1 | 38.86 | 49.50 | 53.59 | 55.83 | 57.24 | 58.20 | 58.91 | 59.44 | 59.86 | 60.19 | 60.71 | 61.22 | 61.74 | 62.00 | 62.26 | 62.53 | 62.79 | 63.06 | 63.33 |
| 2 | 8.53 | 9.00 | 9.16 | 9.24 | 9.29 | 9.33 | 9.35 | 9.37 | 9.38 | 9.39 | 9.41 | 9.42 | 9.44 | 9.45 | 9.46 | 9.47 | 9.47 | 9.48 | 9.49 |
| 3 | 5.54 | 5.46 | 5.39 | 5.34 | 5.31 | 5.28 | 5.27 | 5.25 | 5.24 | 5.23 | 5.22 | 5.20 | 5.18 | 5.18 | 5.17 | 5.16 | 5.15 | 5.14 | 5.13 |
| 4 | 4.54 | 4.32 | 4.19 | 4.11 | 4.05 | 4.01 | 3.98 | 3.95 | 3.94 | 3.92 | 3.90 | 3.87 | 3.84 | 3.83 | 3.82 | 3.80 | 3.79 | 3.78 | 3.76 |
| 5 | 4.06 | 3.78 | 3.62 | 3.52 | 3.45 | 3.40 | 3.37 | 3.34 | 3.32 | 3.30 | 3.27 | 3.24 | 3.21 | 3.19 | 3.17 | 3.16 | 3.14 | 3.12 | 3.10 |
| 6 | 3.78 | 3.46 | 3.29 | 3.18 | 3.11 | 3.05 | 3.01 | 2.98 | 2.96 | 2.94 | 2.90 | 2.87 | 2.84 | 2.82 | 2.80 | 2.78 | 2.76 | 2.74 | 2.72 |
| 7 | 3.59 | 3.26 | 3.07 | 2.96 | 2.88 | 2.83 | 2.78 | 2.75 | 2.72 | 2.70 | 2.67 | 2.63 | 2.59 | 2.58 | 2.56 | 2.54 | 2.51 | 2.49 | 2.47 |
| 8 | 3.46 | 3.11 | 2.92 | 2.81 | 2.73 | 2.67 | 2.62 | 2.59 | 2.56 | 2.54 | 2.50 | 2.46 | 2.42 | 2.40 | 2.38 | 2.36 | 2.34 | 2.32 | 2.29 |
| 9 | 3.36 | 3.01 | 2.81 | 2.69 | 2.61 | 2.55 | 2.51 | 2.47 | 2.44 | 2.42 | 2.38 | 2.34 | 2.30 | 2.28 | 2.25 | 2.23 | 2.21 | 2.18 | 2.16 |
| 10 | 3.29 | 2.92 | 2.73 | 2.61 | 2.52 | 2.46 | 2.41 | 2.38 | 2.35 | 2.32 | 2.28 | 2.24 | 2.20 | 2.18 | 2.16 | 2.13 | 2.11 | 2.08 | 2.06 |
| 11 | 3.23 | 2.86 | 2.66 | 2.54 | 2.45 | 2.39 | 2.34 | 2.30 | 2.27 | 2.25 | 2.21 | 2.17 | 2.12 | 2.10 | 2.08 | 2.05 | 2.03 | 2.00 | 1.97 |
| 12 | 3.18 | 2.81 | 2.61 | 2.48 | 2.39 | 2.33 | 2.28 | 2.24 | 2.21 | 2.19 | 2.15 | 2.10 | 2.06 | 2.04 | 2.01 | 1.99 | 1.96 | 1.93 | 1.90 |
| 13 | 3.14 | 2.76 | 2.56 | 2.43 | 2.35 | 2.28 | 2.23 | 2.20 | 2.16 | 2.14 | 2.10 | 2.05 | 2.01 | 1.98 | 1.96 | 1.93 | 1.90 | 1.88 | 1.85 |
| 14 | 3.10 | 2.73 | 2.52 | 2.39 | 2.31 | 2.24 | 2.19 | 2.15 | 2.12 | 2.10 | 2.05 | 2.01 | 1.96 | 1.94 | 1.91 | 1.89 | 1.86 | 1.83 | 1.80 |
| 15 | 3.07 | 2.70 | 2.49 | 2.36 | 2.27 | 2.21 | 2.16 | 2.12 | 2.09 | 2.06 | 2.02 | 1.97 | 1.92 | 1.90 | 1.87 | 1.85 | 1.82 | 1.79 | 1.76 |
| 16 | 3.05 | 2.67 | 2.46 | 2.33 | 2.24 | 2.18 | 2.13 | 2.09 | 2.06 | 2.03 | 1.99 | 1.94 | 1.89 | 1.87 | 1.84 | 1.81 | 1.78 | 1.75 | 1.72 |
| 17 | 3.03 | 2.64 | 2.44 | 2.31 | 2.22 | 2.15 | 2.10 | 2.06 | 2.03 | 2.00 | 1.96 | 1.91 | 1.86 | 1.84 | 1.81 | 1.78 | 1.75 | 1.72 | 1.69 |
| 18 | 3.01 | 2.62 | 2.42 | 2.29 | 2.20 | 2.13 | 2.08 | 2.04 | 2.00 | 1.98 | 1.93 | 1.89 | 1.84 | 1.81 | 1.78 | 1.75 | 1.72 | 1.69 | 1.66 |
| 19 | 2.99 | 2.61 | 2.40 | 2.27 | 2.18 | 2.11 | 2.06 | 2.02 | 1.98 | 1.96 | 1.91 | 1.86 | 1.81 | 1.79 | 1.76 | 1.73 | 1.70 | 1.67 | 1.63 |

Degrees of freedom for the numerator ($v_1$)

(Continued)

**Table C.4** (*Continued*)

$$F_{0.10, \nu_1, \nu_2}$$

| $\nu_2$ \ $\nu_1$ | 1 | 2 | 3 | 4 | 5 | 6 | 7 | 8 | 9 | 10 | 12 | 15 | 20 | 24 | 30 | 40 | 60 | 120 | ∞ |
|---|---|---|---|---|---|---|---|---|---|---|---|---|---|---|---|---|---|---|---|
| 20 | 2.97 | 2.59 | 2.38 | 2.25 | 2.16 | 2.09 | 2.04 | 2.00 | 1.96 | 1.94 | 1.89 | 1.84 | 1.79 | 1.77 | 1.74 | 1.71 | 1.68 | 1.64 | 1.61 |
| 21 | 2.96 | 2.57 | 2.36 | 2.23 | 2.14 | 2.08 | 2.02 | 1.98 | 1.95 | 1.92 | 1.87 | 1.83 | 1.78 | 1.75 | 1.72 | 1.69 | 1.66 | 1.62 | 1.59 |
| 22 | 2.95 | 2.56 | 2.35 | 2.22 | 2.13 | 2.06 | 2.01 | 1.97 | 1.93 | 1.90 | 1.86 | 1.81 | 1.76 | 1.73 | 1.70 | 1.67 | 1.64 | 1.60 | 1.57 |
| 23 | 2.94 | 2.55 | 2.34 | 2.21 | 2.11 | 2.05 | 1.99 | 1.95 | 1.92 | 1.89 | 1.84 | 1.80 | 1.74 | 1.72 | 1.69 | 1.66 | 1.62 | 1.59 | 1.55 |
| 24 | 2.93 | 2.54 | 2.33 | 2.19 | 2.10 | 2.04 | 1.98 | 1.94 | 1.91 | 1.88 | 1.83 | 1.78 | 1.73 | 1.70 | 1.67 | 1.64 | 1.61 | 1.57 | 1.53 |
| 25 | 2.92 | 2.53 | 2.32 | 2.18 | 2.09 | 2.02 | 1.97 | 1.93 | 1.89 | 1.87 | 1.82 | 1.77 | 1.72 | 1.69 | 1.66 | 1.63 | 1.59 | 1.56 | 1.52 |
| 26 | 2.91 | 2.52 | 2.31 | 2.17 | 2.08 | 2.01 | 1.96 | 1.92 | 1.88 | 1.86 | 1.81 | 1.76 | 1.71 | 1.68 | 1.65 | 1.61 | 1.58 | 1.54 | 1.50 |
| 27 | 2.90 | 2.51 | 2.30 | 2.17 | 2.07 | 2.00 | 1.95 | 1.91 | 1.87 | 1.85 | 1.80 | 1.75 | 1.70 | 1.67 | 1.64 | 1.60 | 1.57 | 1.53 | 1.49 |
| 28 | 2.89 | 2.50 | 2.29 | 2.16 | 2.06 | 2.00 | 1.94 | 1.90 | 1.87 | 1.84 | 1.79 | 1.74 | 1.69 | 1.66 | 1.63 | 1.59 | 1.56 | 1.52 | 1.48 |
| 29 | 2.89 | 2.50 | 2.28 | 2.15 | 2.06 | 1.99 | 1.93 | 1.89 | 1.86 | 1.83 | 1.78 | 1.73 | 1.68 | 1.65 | 1.62 | 1.58 | 1.55 | 1.51 | 1.47 |
| 30 | 2.88 | 2.49 | 2.28 | 2.14 | 2.03 | 1.98 | 1.93 | 1.88 | 1.85 | 1.82 | 1.77 | 1.72 | 1.67 | 1.64 | 1.61 | 1.57 | 1.54 | 1.50 | 1.46 |
| 40 | 2.84 | 2.44 | 2.23 | 2.09 | 2.00 | 1.93 | 1.87 | 1.83 | 1.79 | 1.76 | 1.71 | 1.66 | 1.61 | 1.57 | 1.54 | 1.51 | 1.47 | 1.42 | 1.38 |
| 60 | 2.79 | 2.39 | 2.18 | 2.04 | 1.95 | 1.87 | 1.82 | 1.77 | 1.74 | 1.71 | 1.66 | 1.60 | 1.54 | 1.51 | 1.48 | 1.44 | 1.40 | 1.35 | 1.29 |
| 120 | 2.75 | 2.35 | 2.13 | 1.99 | 1.90 | 1.82 | 1.77 | 1.72 | 1.68 | 1.65 | 1.60 | 1.55 | 1.48 | 1.45 | 1.41 | 1.37 | 1.32 | 1.26 | 1.19 |
| ∞ | 2.71 | 2.30 | 2.08 | 1.94 | 1.85 | 1.77 | 1.72 | 1.67 | 1.63 | 1.60 | 1.55 | 1.49 | 1.42 | 1.38 | 1.34 | 1.30 | 1.24 | 1.17 | 1.00 |

Degrees of freedom for the numerator ($\nu_1$)

Table C.4 (Continued)

$F_{0.05,\nu_1,\nu_2}$

| $\nu_2$ \ $\nu_1$ | 1 | 2 | 3 | 4 | 5 | 6 | 7 | 8 | 9 | 10 | 12 | 15 | 20 | 24 | 30 | 40 | 60 | 120 | ∞ |
|---|---|---|---|---|---|---|---|---|---|---|---|---|---|---|---|---|---|---|---|
| | | | | | | | | Degrees of freedom for the numerator ($\nu_1$) | | | | | | | | | | | |
| 1 | 161.4 | 199.5 | 215.7 | 224.6 | 230.2 | 234.0 | 236.8 | 238.9 | 240.5 | 241.9 | 243.9 | 245.9 | 248.0 | 249.1 | 250.1 | 251.1 | 252.2 | 253.3 | 254.3 |
| 2 | 18.51 | 19.00 | 19.16 | 19.25 | 19.30 | 19.33 | 19.35 | 19.37 | 19.38 | 19.40 | 19.41 | 19.43 | 19.45 | 19.45 | 19.46 | 19.47 | 19.48 | 19.49 | 19.50 |
| 3 | 10.13 | 9.55 | 9.28 | 9.12 | 9.01 | 8.94 | 8.89 | 8.85 | 8.81 | 8.79 | 8.74 | 8.70 | 8.66 | 8.64 | 8.62 | 8.59 | 8.57 | 8.55 | 8.53 |
| 4 | 7.71 | 6.94 | 6.59 | 6.39 | 6.26 | 6.16 | 6.09 | 6.04 | 6.00 | 5.96 | 5.91 | 5.86 | 5.80 | 5.77 | 5.75 | 5.72 | 5.69 | 5.66 | 5.63 |
| 5 | 6.61 | 5.79 | 5.41 | 5.19 | 5.05 | 4.95 | 4.88 | 4.82 | 4.77 | 4.74 | 4.68 | 4.62 | 4.56 | 4.53 | 4.50 | 4.46 | 4.43 | 4.40 | 4.36 |
| 6 | 5.99 | 5.14 | 4.76 | 4.53 | 4.39 | 4.28 | 4.21 | 4.15 | 4.10 | 4.06 | 4.00 | 3.94 | 3.87 | 3.84 | 3.81 | 3.77 | 3.74 | 3.70 | 3.67 |
| 7 | 5.59 | 4.74 | 4.35 | 4.12 | 3.97 | 3.87 | 3.79 | 3.73 | 3.68 | 3.64 | 3.57 | 3.51 | 3.44 | 3.41 | 3.38 | 3.34 | 3.30 | 3.27 | 3.23 |
| 8 | 5.32 | 4.46 | 4.07 | 3.84 | 3.69 | 3.58 | 3.50 | 3.44 | 3.39 | 3.35 | 3.28 | 3.22 | 3.15 | 3.12 | 3.08 | 3.04 | 3.01 | 2.97 | 2.93 |
| 9 | 5.12 | 4.26 | 3.86 | 3.63 | 3.48 | 3.37 | 3.29 | 3.23 | 3.18 | 3.14 | 3.07 | 3.01 | 2.94 | 2.90 | 2.86 | 2.83 | 2.79 | 2.75 | 2.71 |
| 10 | 4.96 | 4.10 | 3.71 | 3.48 | 3.33 | 3.22 | 3.14 | 3.07 | 3.02 | 2.98 | 2.91 | 2.85 | 2.77 | 2.74 | 2.70 | 2.66 | 2.62 | 2.58 | 2.54 |
| 11 | 4.84 | 3.98 | 3.59 | 3.36 | 3.20 | 3.09 | 3.01 | 2.95 | 2.90 | 2.85 | 2.79 | 2.72 | 2.65 | 2.61 | 2.57 | 2.53 | 2.49 | 2.45 | 2.40 |
| 12 | 4.75 | 3.89 | 3.49 | 3.26 | 3.11 | 3.00 | 2.91 | 2.85 | 2.80 | 2.75 | 2.69 | 2.62 | 2.54 | 2.51 | 2.47 | 2.43 | 2.38 | 2.34 | 2.30 |
| 13 | 4.67 | 3.81 | 3.41 | 3.18 | 3.03 | 2.92 | 2.83 | 2.77 | 2.71 | 2.67 | 2.60 | 2.53 | 2.46 | 2.42 | 2.38 | 2.34 | 2.30 | 2.25 | 2.21 |
| 14 | 4.60 | 3.74 | 3.34 | 3.11 | 2.96 | 2.85 | 2.76 | 2.70 | 2.65 | 2.60 | 2.53 | 2.46 | 2.39 | 2.35 | 2.31 | 2.27 | 2.22 | 2.18 | 2.13 |
| 15 | 4.54 | 3.68 | 3.29 | 3.06 | 2.90 | 2.79 | 2.71 | 2.64 | 2.59 | 2.54 | 2.48 | 2.40 | 2.33 | 2.29 | 2.25 | 2.20 | 2.16 | 2.11 | 2.07 |
| 16 | 4.49 | 3.63 | 3.24 | 3.01 | 2.85 | 2.74 | 2.66 | 2.59 | 2.54 | 2.49 | 2.42 | 2.35 | 2.28 | 2.24 | 2.19 | 2.15 | 2.11 | 2.06 | 2.01 |
| 17 | 4.45 | 3.59 | 3.20 | 2.96 | 2.81 | 2.70 | 2.61 | 2.55 | 2.49 | 2.45 | 2.38 | 2.31 | 2.23 | 2.19 | 2.15 | 2.10 | 2.06 | 2.01 | 1.96 |
| 18 | 4.41 | 3.55 | 3.16 | 2.93 | 2.77 | 2.66 | 2.58 | 2.51 | 2.46 | 2.41 | 2.34 | 2.27 | 2.19 | 2.15 | 2.11 | 2.06 | 2.02 | 1.97 | 1.92 |
| 19 | 4.38 | 3.52 | 3.13 | 2.90 | 2.74 | 2.63 | 2.54 | 2.48 | 2.42 | 2.38 | 2.31 | 2.23 | 2.16 | 2.11 | 2.07 | 2.03 | 1.98 | 1.93 | 1.88 |

(Continued)

**Table C.4** (Continued)

$$F_{0.05, v_1, v_2}$$

| $v_2$ \ $v_1$ | Degrees of freedom for the numerator ($v_1$) | | | | | | | | | | | | | | | | | | |
|---|---|---|---|---|---|---|---|---|---|---|---|---|---|---|---|---|---|---|---|
| | 1 | 2 | 3 | 4 | 5 | 6 | 7 | 8 | 9 | 10 | 12 | 15 | 20 | 24 | 30 | 40 | 60 | 120 | ∞ |
| 20 | 4.35 | 3.49 | 3.10 | 2.87 | 2.71 | 2.60 | 2.51 | 2.45 | 2.39 | 2.35 | 2.28 | 2.20 | 2.12 | 2.08 | 2.04 | 1.99 | 1.95 | 1.90 | 1.84 |
| 21 | 4.32 | 3.47 | 3.07 | 2.84 | 2.68 | 2.57 | 2.49 | 2.42 | 2.37 | 2.32 | 2.25 | 2.18 | 2.10 | 2.05 | 2.01 | 1.96 | 1.92 | 1.87 | 1.81 |
| 22 | 4.30 | 3.44 | 3.05 | 2.82 | 2.66 | 2.55 | 2.46 | 2.40 | 2.34 | 2.30 | 2.23 | 2.15 | 2.07 | 2.03 | 1.98 | 1.94 | 1.89 | 1.84 | 1.78 |
| 23 | 4.28 | 3.42 | 3.03 | 2.80 | 2.64 | 2.53 | 2.44 | 2.37 | 2.32 | 2.27 | 2.20 | 2.13 | 2.05 | 2.01 | 1.96 | 1.91 | 1.86 | 1.81 | 1.76 |
| 24 | 4.26 | 3.40 | 3.01 | 2.78 | 2.62 | 2.51 | 2.42 | 2.36 | 2.30 | 2.25 | 2.18 | 2.11 | 2.03 | 1.98 | 1.94 | 1.89 | 1.84 | 1.79 | 1.73 |
| 25 | 4.24 | 3.39 | 2.99 | 2.76 | 2.60 | 2.49 | 2.40 | 2.34 | 2.28 | 2.24 | 2.16 | 2.09 | 2.01 | 1.96 | 1.92 | 1.87 | 1.82 | 1.77 | 1.71 |
| 26 | 4.23 | 3.37 | 2.98 | 2.74 | 2.59 | 2.47 | 2.39 | 2.32 | 2.27 | 2.22 | 2.15 | 2.07 | 1.99 | 1.95 | 1.90 | 1.85 | 1.80 | 1.75 | 1.69 |
| 27 | 4.21 | 3.35 | 2.96 | 2.73 | 2.57 | 2.46 | 2.37 | 2.31 | 2.25 | 2.20 | 2.13 | 2.06 | 1.97 | 1.93 | 1.88 | 1.84 | 1.79 | 1.73 | 1.67 |
| 28 | 4.20 | 3.34 | 2.95 | 2.71 | 2.56 | 2.45 | 2.36 | 2.29 | 2.24 | 2.19 | 2.12 | 2.04 | 1.96 | 1.91 | 1.87 | 1.82 | 1.77 | 1.71 | 1.65 |
| 29 | 4.18 | 3.33 | 2.93 | 2.70 | 2.55 | 2.43 | 2.35 | 2.28 | 2.22 | 2.18 | 2.10 | 2.03 | 1.94 | 1.90 | 1.85 | 1.81 | 1.75 | 1.70 | 1.64 |
| 30 | 4.17 | 3.32 | 2.92 | 2.69 | 2.53 | 2.42 | 2.33 | 2.27 | 2.21 | 2.16 | 2.09 | 2.01 | 1.93 | 1.89 | 1.84 | 1.79 | 1.74 | 1.68 | 1.62 |
| 40 | 4.08 | 3.23 | 2.84 | 2.61 | 2.45 | 2.34 | 2.25 | 2.18 | 2.12 | 2.08 | 2.00 | 1.92 | 1.84 | 1.79 | 1.74 | 1.69 | 1.64 | 1.58 | 1.51 |
| 60 | 4.00 | 3.15 | 2.76 | 2.53 | 2.37 | 2.25 | 2.17 | 2.10 | 2.04 | 1.99 | 1.92 | 1.84 | 1.75 | 1.70 | 1.65 | 1.59 | 1.53 | 1.47 | 1.39 |
| 120 | 3.92 | 3.07 | 2.68 | 2.45 | 2.29 | 2.17 | 2.09 | 2.02 | 1.96 | 1.91 | 1.83 | 1.75 | 1.66 | 1.61 | 1.55 | 1.50 | 1.43 | 1.35 | 1.25 |
| ∞ | 3.84 | 3.00 | 2.60 | 2.37 | 2.21 | 2.10 | 2.01 | 1.94 | 1.88 | 1.83 | 1.75 | 1.67 | 1.57 | 1.52 | 1.46 | 1.39 | 1.32 | 1.22 | 1.00 |

**Table C.4** *(Continued)*

$$F_{0.025, v_1, v_2}$$

| $v_2$ \ $v_1$ | 1 | 2 | 3 | 4 | 5 | 6 | 7 | 8 | 9 | 10 | 12 | 15 | 20 | 24 | 30 | 40 | 60 | 120 | ∞ |
|---|---|---|---|---|---|---|---|---|---|---|---|---|---|---|---|---|---|---|---|
| 1 | 647.8 | 799.5 | 864.2 | 899.6 | 921.8 | 937.1 | 948.2 | 956.7 | 963.3 | 968.6 | 976.7 | 984.9 | 993.1 | 997.2 | 1001 | 1006 | 1010 | 1014 | 1018 |
| 2 | 38.51 | 39.00 | 39.17 | 39.25 | 39.30 | 39.33 | 39.36 | 39.37 | 39.39 | 39.40 | 39.41 | 39.43 | 39.45 | 39.46 | 39.46 | 39.47 | 39.48 | 39.49 | 39.50 |
| 3 | 17.44 | 16.04 | 15.44 | 15.10 | 14.88 | 14.73 | 14.62 | 14.54 | 14.47 | 14.42 | 14.34 | 14.25 | 14.17 | 14.12 | 14.08 | 14.04 | 13.99 | 13.95 | 13.90 |
| 4 | 12.22 | 10.65 | 9.98 | 9.60 | 9.36 | 9.20 | 9.07 | 8.98 | 8.90 | 8.84 | 8.75 | 8.66 | 8.56 | 8.51 | 8.46 | 8.41 | 8.36 | 8.31 | 8.26 |
| 5 | 10.01 | 8.43 | 7.76 | 7.39 | 7.15 | 6.98 | 6.85 | 6.76 | 6.68 | 6.62 | 6.52 | 6.43 | 6.33 | 6.28 | 6.23 | 6.18 | 6.12 | 6.07 | 6.02 |
| 6 | 8.81 | 7.26 | 6.60 | 6.23 | 5.99 | 5.82 | 5.70 | 5.60 | 5.52 | 5.46 | 5.37 | 5.27 | 5.17 | 5.12 | 5.07 | 5.01 | 4.96 | 4.90 | 4.85 |
| 7 | 8.07 | 6.54 | 5.89 | 5.52 | 5.29 | 5.12 | 4.99 | 4.90 | 4.82 | 4.76 | 4.67 | 4.57 | 4.47 | 4.42 | 4.36 | 4.31 | 4.25 | 4.20 | 4.14 |
| 8 | 7.57 | 6.06 | 5.42 | 5.05 | 4.82 | 4.65 | 4.53 | 4.43 | 4.36 | 4.30 | 4.20 | 4.10 | 4.00 | 3.95 | 3.89 | 3.84 | 3.78 | 3.73 | 3.67 |
| 9 | 7.21 | 5.71 | 5.08 | 4.72 | 4.48 | 4.32 | 4.20 | 4.10 | 4.03 | 3.96 | 3.87 | 3.77 | 3.67 | 3.61 | 3.56 | 3.51 | 3.45 | 3.39 | 3.33 |
| 10 | 6.94 | 5.46 | 4.83 | 4.47 | 4.24 | 4.07 | 3.95 | 3.85 | 3.78 | 3.72 | 3.62 | 3.52 | 3.42 | 3.37 | 3.31 | 3.26 | 3.20 | 3.14 | 3.08 |
| 11 | 6.72 | 5.26 | 4.63 | 4.28 | 4.04 | 3.88 | 3.76 | 3.66 | 3.59 | 3.53 | 3.43 | 3.33 | 3.23 | 3.17 | 3.12 | 3.06 | 3.00 | 2.94 | 2.88 |
| 12 | 6.55 | 5.10 | 4.47 | 4.12 | 3.89 | 3.73 | 3.61 | 3.51 | 3.44 | 3.37 | 3.28 | 3.18 | 3.07 | 3.02 | 2.96 | 2.91 | 2.85 | 2.79 | 2.72 |
| 13 | 6.41 | 4.97 | 4.35 | 4.00 | 3.77 | 3.60 | 3.48 | 3.39 | 3.31 | 3.25 | 3.15 | 3.05 | 2.95 | 2.89 | 2.84 | 2.78 | 2.72 | 2.66 | 2.60 |
| 14 | 6.30 | 4.86 | 4.24 | 3.89 | 3.66 | 3.50 | 3.38 | 3.29 | 3.21 | 3.15 | 3.05 | 2.95 | 2.84 | 2.79 | 2.73 | 2.67 | 2.61 | 2.55 | 2.49 |
| 15 | 6.20 | 4.77 | 4.15 | 3.80 | 3.58 | 3.41 | 3.29 | 3.20 | 3.12 | 3.06 | 2.96 | 2.86 | 2.76 | 2.70 | 2.64 | 2.59 | 2.52 | 2.46 | 2.40 |
| 16 | 6.12 | 4.69 | 4.08 | 3.73 | 3.50 | 3.34 | 3.22 | 3.12 | 3.05 | 2.99 | 2.89 | 2.79 | 2.68 | 2.63 | 2.57 | 2.51 | 2.45 | 2.38 | 2.32 |
| 17 | 6.04 | 4.62 | 4.01 | 3.66 | 3.44 | 3.28 | 3.16 | 3.06 | 2.98 | 2.92 | 2.82 | 2.72 | 2.62 | 2.56 | 2.50 | 2.44 | 2.38 | 2.32 | 2.25 |
| 18 | 5.98 | 4.56 | 3.95 | 3.61 | 3.38 | 3.22 | 3.10 | 3.01 | 2.93 | 2.87 | 2.77 | 2.67 | 2.56 | 2.50 | 2.44 | 2.38 | 2.32 | 2.26 | 2.19 |
| 19 | 5.92 | 4.51 | 3.90 | 3.56 | 3.33 | 3.17 | 3.05 | 2.96 | 2.88 | 2.82 | 2.72 | 2.62 | 2.51 | 2.45 | 2.39 | 2.33 | 2.27 | 2.20 | 2.13 |

Degrees of freedom for the numerator ($v_1$)

*(Continued)*

**Table C.4** (Continued)

$$F_{0.025, \nu_1, \nu_2}$$

Degrees of freedom for the numerator ($\nu_1$)

| $\nu_2$ \ $\nu_1$ | 1 | 2 | 3 | 4 | 5 | 6 | 7 | 8 | 9 | 10 | 12 | 15 | 20 | 24 | 30 | 40 | 60 | 120 | ∞ |
|---|---|---|---|---|---|---|---|---|---|---|---|---|---|---|---|---|---|---|---|
| 20 | 5.87 | 4.46 | 3.86 | 3.51 | 3.29 | 3.13 | 3.01 | 2.91 | 2.84 | 2.77 | 2.68 | 2.57 | 2.46 | 2.41 | 2.35 | 2.29 | 2.22 | 2.16 | 2.09 |
| 21 | 5.83 | 4.42 | 3.82 | 3.48 | 3.25 | 3.09 | 2.97 | 2.87 | 2.80 | 2.73 | 2.64 | 2.53 | 2.42 | 2.37 | 2.31 | 2.25 | 2.18 | 2.11 | 2.04 |
| 22 | 5.79 | 4.38 | 3.78 | 3.44 | 3.22 | 3.05 | 2.93 | 2.84 | 2.76 | 2.70 | 2.60 | 2.50 | 2.39 | 2.33 | 2.27 | 2.21 | 2.14 | 2.08 | 2.00 |
| 23 | 5.75 | 4.35 | 3.75 | 3.41 | 3.18 | 3.02 | 2.90 | 2.81 | 2.73 | 2.67 | 2.57 | 2.47 | 2.36 | 2.30 | 2.24 | 2.18 | 2.11 | 2.04 | 1.97 |
| 24 | 5.72 | 4.32 | 3.72 | 3.38 | 3.15 | 2.99 | 2.87 | 2.78 | 2.70 | 2.64 | 2.54 | 2.44 | 2.33 | 2.27 | 2.21 | 2.15 | 2.08 | 2.01 | 1.94 |
| 25 | 5.69 | 4.29 | 3.69 | 3.35 | 3.13 | 2.97 | 2.85 | 2.75 | 2.68 | 2.61 | 2.51 | 2.41 | 2.30 | 2.24 | 2.18 | 2.12 | 2.05 | 1.98 | 1.91 |
| 26 | 5.66 | 4.27 | 3.67 | 3.33 | 3.10 | 2.94 | 2.82 | 2.73 | 2.65 | 2.59 | 2.49 | 2.39 | 2.28 | 2.22 | 2.16 | 2.09 | 2.03 | 1.95 | 1.88 |
| 27 | 5.63 | 4.24 | 3.65 | 3.31 | 3.08 | 2.92 | 2.80 | 2.71 | 2.63 | 2.57 | 2.47 | 2.36 | 2.25 | 2.19 | 2.13 | 2.07 | 2.00 | 1.93 | 1.85 |
| 28 | 5.61 | 4.22 | 3.63 | 3.29 | 3.06 | 2.90 | 2.78 | 2.69 | 2.61 | 2.55 | 2.45 | 2.34 | 2.23 | 2.17 | 2.11 | 2.05 | 1.98 | 1.91 | 1.83 |
| 29 | 5.59 | 4.20 | 3.61 | 3.27 | 3.04 | 2.88 | 2.76 | 2.67 | 2.59 | 2.53 | 2.43 | 2.32 | 2.21 | 2.15 | 2.09 | 2.03 | 1.96 | 1.89 | 1.81 |
| 30 | 5.57 | 4.18 | 3.59 | 3.25 | 3.03 | 2.87 | 2.75 | 2.65 | 2.57 | 2.51 | 2.41 | 2.31 | 2.20 | 2.14 | 2.07 | 2.01 | 1.94 | 1.87 | 1.79 |
| 40 | 5.42 | 4.05 | 3.46 | 3.13 | 2.90 | 2.74 | 2.62 | 2.53 | 2.45 | 2.39 | 2.29 | 2.18 | 2.07 | 2.01 | 1.94 | 1.88 | 1.80 | 1.72 | 1.64 |
| 60 | 5.29 | 3.93 | 3.34 | 3.01 | 2.79 | 2.63 | 2.51 | 2.41 | 2.33 | 2.27 | 2.17 | 2.06 | 1.94 | 1.88 | 1.82 | 1.74 | 1.67 | 1.58 | 1.48 |
| 120 | 5.15 | 3.80 | 3.23 | 2.89 | 2.67 | 2.52 | 2.39 | 2.30 | 2.22 | 2.16 | 2.05 | 1.94 | 1.82 | 1.76 | 1.69 | 1.61 | 1.53 | 1.43 | 1.31 |
| ∞ | 5.02 | 3.69 | 3.12 | 2.79 | 2.57 | 2.41 | 2.29 | 2.19 | 2.11 | 2.05 | 1.94 | 1.83 | 1.71 | 1.64 | 1.57 | 1.48 | 1.39 | 1.27 | 1.00 |

**Table C.4** (Continued)

$F_{0.01, \nu_1, \nu_2}$

| $\nu_2$ \ $\nu_1$ | 1 | 2 | 3 | 4 | 5 | 6 | 7 | 8 | 9 | 10 | 12 | 15 | 20 | 24 | 30 | 40 | 60 | 120 | ∞ |
|---|---|---|---|---|---|---|---|---|---|---|---|---|---|---|---|---|---|---|---|
| | | | | | | | Degrees of freedom for the numerator ($\nu_1$) | | | | | | | | | | | | |
| 1 | 4052 | 4999.5 | 5403 | 5625 | 5764 | 5859 | 5928 | 5982 | 6022 | 6056 | 6106 | 6157 | 6209 | 6235 | 6261 | 6287 | 6313 | 6339 | 6366 |
| 2 | 98.50 | 99.00 | 99.17 | 99.25 | 99.30 | 99.33 | 99.36 | 99.37 | 99.39 | 99.40 | 99.42 | 99.43 | 99.45 | 99.46 | 99.47 | 99.47 | 99.48 | 99.49 | 99.50 |
| 3 | 34.12 | 30.82 | 29.46 | 28.71 | 28.24 | 27.91 | 27.67 | 27.49 | 27.35 | 27.23 | 27.05 | 26.87 | 26.69 | 26.60 | 26.50 | 26.41 | 26.32 | 26.22 | 26.13 |
| 4 | 21.20 | 18.00 | 16.69 | 15.98 | 15.52 | 15.21 | 14.98 | 14.80 | 14.66 | 14.55 | 14.37 | 14.20 | 14.02 | 13.93 | 13.84 | 13.75 | 13.65 | 13.56 | 13.46 |
| 5 | 16.26 | 13.27 | 12.06 | 11.39 | 10.97 | 10.67 | 10.46 | 10.29 | 10.16 | 10.05 | 9.89 | 9.72 | 9.55 | 9.47 | 9.38 | 9.29 | 9.20 | 9.11 | 9.02 |
| 6 | 13.75 | 10.92 | 9.78 | 9.15 | 8.75 | 8.47 | 8.26 | 8.10 | 7.98 | 7.87 | 7.72 | 7.56 | 7.40 | 7.31 | 7.23 | 7.14 | 7.06 | 6.97 | 6.88 |
| 7 | 12.25 | 9.55 | 8.45 | 7.85 | 7.46 | 7.19 | 6.99 | 6.84 | 6.72 | 6.62 | 6.47 | 6.31 | 6.16 | 6.07 | 5.99 | 5.91 | 5.82 | 5.74 | 5.65 |
| 8 | 11.26 | 8.65 | 7.59 | 7.01 | 6.63 | 6.37 | 6.18 | 6.03 | 5.91 | 5.81 | 5.67 | 5.52 | 5.36 | 5.28 | 5.20 | 5.12 | 5.03 | 4.95 | 4.86 |
| 9 | 10.56 | 8.02 | 6.99 | 6.42 | 6.06 | 5.80 | 5.61 | 5.47 | 5.35 | 5.26 | 5.11 | 4.96 | 4.81 | 4.73 | 4.65 | 4.57 | 4.48 | 4.40 | 4.31 |
| 10 | 10.04 | 7.56 | 6.55 | 5.99 | 5.64 | 5.39 | 5.20 | 5.06 | 4.94 | 4.85 | 4.71 | 4.56 | 4.41 | 4.33 | 4.25 | 4.17 | 4.08 | 4.00 | 3.91 |
| 11 | 9.65 | 7.21 | 6.22 | 5.67 | 5.32 | 5.07 | 4.89 | 4.74 | 4.63 | 4.54 | 4.40 | 4.25 | 4.10 | 4.02 | 3.94 | 3.86 | 3.78 | 3.69 | 3.60 |
| 12 | 9.33 | 6.93 | 5.95 | 5.41 | 5.06 | 4.82 | 4.64 | 4.50 | 4.39 | 4.30 | 4.16 | 4.01 | 3.86 | 3.78 | 3.70 | 3.62 | 3.54 | 3.45 | 3.36 |
| 13 | 9.07 | 6.70 | 5.74 | 5.21 | 4.86 | 4.62 | 4.44 | 4.30 | 4.19 | 4.10 | 3.96 | 3.82 | 3.66 | 3.59 | 3.51 | 3.43 | 3.34 | 3.25 | 3.17 |
| 14 | 8.86 | 6.51 | 5.56 | 5.04 | 4.69 | 4.46 | 4.28 | 4.14 | 4.03 | 3.94 | 3.80 | 3.66 | 3.51 | 3.43 | 3.34 | 3.27 | 3.18 | 3.09 | 3.00 |
| 15 | 8.68 | 6.36 | 5.42 | 4.89 | 4.56 | 4.32 | 4.14 | 4.00 | 3.89 | 3.80 | 3.67 | 3.52 | 3.37 | 3.29 | 3.21 | 3.13 | 3.05 | 2.96 | 2.87 |
| 16 | 8.53 | 6.23 | 5.29 | 4.77 | 4.44 | 4.20 | 4.03 | 3.89 | 3.78 | 3.69 | 3.55 | 3.41 | 3.26 | 3.18 | 3.10 | 3.02 | 2.93 | 2.84 | 2.75 |
| 17 | 8.40 | 6.11 | 5.18 | 4.67 | 4.34 | 4.10 | 3.93 | 3.79 | 3.68 | 3.59 | 3.46 | 3.31 | 3.16 | 3.08 | 3.00 | 2.92 | 2.83 | 2.75 | 2.65 |
| 18 | 8.29 | 6.01 | 5.09 | 4.58 | 4.25 | 4.01 | 3.84 | 3.71 | 3.60 | 3.51 | 3.37 | 3.23 | 3.08 | 3.00 | 2.92 | 2.84 | 2.75 | 2.66 | 2.57 |
| 19 | 8.18 | 5.93 | 5.01 | 4.50 | 4.17 | 3.94 | 3.77 | 3.63 | 3.52 | 3.43 | 3.30 | 3.15 | 3.00 | 2.92 | 2.84 | 2.76 | 2.67 | 2.58 | 2.49 |

(Continued)

Table C.4 (Continued)

$$F_{0.01, v_1, v_2}$$

| | | | | | | Degrees of freedom for the numerator ($v_1$) | | | | | | | | | | | | |
|---|---|---|---|---|---|---|---|---|---|---|---|---|---|---|---|---|---|---|
| $v_2$ \ $v_1$ | 1 | 2 | 3 | 4 | 5 | 6 | 7 | 8 | 9 | 10 | 12 | 15 | 20 | 24 | 30 | 40 | 60 | 120 | ∞ |
| 20 | 8.10 | 5.85 | 4.94 | 4.43 | 4.10 | 3.87 | 3.70 | 3.56 | 3.46 | 3.37 | 3.23 | 3.09 | 2.94 | 2.86 | 2.78 | 2.69 | 2.61 | 2.52 | 2.42 |
| 21 | 8.02 | 5.78 | 4.87 | 4.37 | 4.04 | 3.81 | 3.64 | 3.51 | 3.40 | 3.31 | 3.17 | 3.03 | 2.88 | 2.80 | 2.72 | 2.64 | 2.55 | 2.46 | 2.36 |
| 22 | 7.95 | 5.72 | 4.82 | 4.31 | 3.99 | 3.76 | 3.59 | 3.45 | 3.35 | 3.26 | 3.12 | 2.98 | 2.83 | 2.75 | 2.67 | 2.58 | 2.50 | 2.40 | 2.31 |
| 23 | 7.88 | 5.66 | 4.76 | 4.26 | 3.94 | 3.71 | 3.54 | 3.41 | 3.30 | 3.21 | 3.07 | 2.93 | 2.78 | 2.70 | 2.62 | 2.54 | 2.45 | 2.35 | 2.26 |
| 24 | 7.82 | 5.61 | 4.72 | 4.22 | 3.90 | 3.67 | 3.50 | 3.36 | 3.26 | 3.17 | 3.03 | 2.89 | 2.74 | 2.66 | 2.58 | 2.49 | 2.40 | 2.31 | 2.21 |
| 25 | 7.77 | 5.57 | 4.68 | 4.18 | 3.85 | 3.63 | 3.46 | 3.32 | 3.22 | 3.13 | 2.99 | 2.85 | 2.70 | 2.62 | 2.54 | 2.45 | 2.36 | 2.27 | 2.17 |
| 26 | 7.72 | 5.53 | 4.64 | 4.14 | 3.82 | 3.59 | 3.42 | 3.29 | 3.18 | 3.09 | 2.96 | 2.81 | 2.66 | 2.58 | 2.50 | 2.42 | 2.33 | 2.23 | 2.13 |
| 27 | 7.68 | 5.49 | 4.60 | 4.11 | 3.78 | 3.56 | 3.39 | 3.26 | 3.15 | 3.06 | 2.93 | 2.78 | 2.63 | 2.55 | 2.47 | 2.38 | 2.29 | 2.20 | 2.10 |
| 28 | 7.64 | 5.45 | 4.57 | 4.07 | 3.75 | 3.53 | 3.36 | 3.23 | 3.12 | 3.03 | 2.90 | 2.75 | 2.60 | 2.52 | 2.44 | 2.35 | 2.26 | 2.17 | 2.06 |
| 29 | 7.60 | 5.42 | 5.45 | 4.04 | 3.73 | 3.50 | 3.33 | 3.20 | 3.09 | 3.00 | 2.87 | 2.73 | 2.57 | 2.49 | 2.41 | 2.33 | 2.23 | 2.14 | 2.03 |
| 30 | 7.56 | 5.39 | 4.51 | 4.02 | 3.70 | 3.47 | 3.30 | 3.17 | 3.07 | 2.98 | 2.84 | 2.70 | 2.55 | 2.47 | 2.39 | 2.30 | 2.21 | 2.11 | 2.01 |
| 40 | 7.31 | 5.18 | 4.31 | 3.83 | 3.51 | 3.29 | 3.12 | 2.99 | 2.89 | 2.80 | 2.66 | 2.52 | 2.37 | 2.29 | 2.20 | 2.11 | 2.02 | 1.92 | 1.80 |
| 60 | 7.08 | 4.98 | 4.13 | 3.65 | 3.34 | 3.12 | 2.95 | 2.82 | 2.72 | 2.63 | 2.50 | 2.35 | 2.20 | 2.12 | 2.03 | 1.94 | 1.84 | 1.73 | 1.60 |
| 120 | 6.85 | 4.79 | 3.95 | 3.48 | 3.17 | 2.96 | 2.79 | 2.66 | 2.56 | 2.47 | 2.34 | 2.19 | 2.03 | 1.95 | 1.89 | 1.76 | 1.66 | 1.53 | 1.38 |
| ∞ | 6.63 | 4.61 | 3.78 | 3.32 | 3.02 | 2.80 | 2.64 | 2.51 | 2.41 | 2.32 | 2.18 | 2.04 | 1.88 | 1.79 | 1.70 | 1.59 | 1.47 | 1.32 | 1.00 |

**Table C.5** Glossary of terms

**For use with industry survey and second academic survey (ASME/NSF product realization process survey).**

1. **Knowledge of the product realization process:** a framework for development and delivery of new products into the market in an integrative manner. Examples include stage-Gate, Spiral Models, etc.

2. **Bench marking:** the systematic process for evaluating the product, services, and work processes of organizations that are recognized as representing "best practices" for the purpose of organizational improvement .

3. **Concurrent engineering:** an approach to new product development where the product and all of its associated processes, such as manufacturing, distribution and service, are developed in parallel.

4. **Corporate vision and product fit:** a business portfolio management process used by the most innovative companies to check that new product ideas fit with the current corporate vision and strategy, but which provides for dynamically shifting corporate strategy to accommodate a truly new but supportable direction.

5. **Business functions/marketing, legal, finance, etc.:** working knowledge of contemporary business practices in marketing, legal, finance, purchasing and accounting.

6. **Industrial design:** the methodologies for creating and developing product concepts and specifications that optimize the function, value and appearance of products and systems for the mutual benefit of both user and manufacturer.

7. **Project management tools:** use of manual and computer-based tools such as planning and scheduling which develop the time frame of a project, including tasks to be completed, resources required, and checkpoints for accessing progress.

8. **Budgeting:** the process of anticipating project cash flow; how much money will be spent and when.

9. **Project risk analysis:** the process of determining possible risk to a project schedule and goals; and the likelihood of such an occurrence.

10. **Design reviews:** the scheduled-in checkpoints for assessing the design progress towards meeting product requirements and budget.

11. **Information processing:** acquisition, organization, management, control, communication, and utilization of information in support of PRP decision-making.

12. **Communication:** the ability to clearly and logically communicate ideas, information, and data orally and in written form to others in a way that engages the intended audience and addresses different learning styles.

13. **Sketching/drawing:** the ability to clearly illustrate ideas and design by freehand sketching.

14. **Leadership:** the ability to influence a group to work toward and reach a goal or objective, frequently within a specific time frame and within a given budget.

15. **Conflict management:** the ability to work with others in a way that helps to turn opposition of interest or ideas into a constructive, productive experience.

16. **Professional ethics:** the ability to conform to standards of conduct determined by one's profession, in alignment with team and corporate standards.

17. **Teams/teamwork:** the ability to work with diverse, multi-discipline team members to successfully reach a goal or objective.

18. **Competitive analysis:** the process of analyzing the business environment to collect relevant competitive intelligence and then integrating this information into new product development and delivery strategies.

19. **Creative thinking:** the process of generating ideas, that frequently emphasizes: (a) making and expressing meaningful new connections, (b) thinking of many new and unusual possibilities, and (c) extending and elaborating on alternatives.

*(Continued)*

**Table C.5** (Continued)

20. **Tools for "customer-centered" design:** (QFD, KJ, voice of customer, conjoint analysis, concept engineering, Taguchi method, etc.): using disciplined approaches to planning, communicating, evaluating, and documenting customer requirements and then translating them into design activities.

21. **Solid modeling/rapid prototyping systems:** using systems that can create a physical prototype directly from a CAD representation.

22. **Systems perspective:** the up-front identification of system components and their interactions for the purpose of optimizing the performance of the system as a whole.

23. **Design for assembly:** making the product easier to assemble, thereby reducing cycle time during production.

24. **Design for commonality-platform:** designing a "building block" that will be common to more than one product.

25. **Design for cost:** meeting customer requirements while minimizing cost of all aspects of the product, including production, assembly, distribution and maintenance.

26. **Design for disassembly:** designing new products so they can be taken apart for processes such as recycling or disposal.

27. **Design for environment:** designing environmentally conscious new products.

28. **Design for ergonomics:** designs that take into account human physiology and human factors in conjunction with how the product will be used or operated.

29. **Design for manufacture:** designed to maximize ease of manufacture by simplifying the design through part-count reduction, developing modular designs, minimizing part variation, designing a part to be multi-functional, etc.

30. **Design for performance:** designed to perform to product requirements under a wide variety of manufacturing and user operating conditions .

31. **Design for reliability:** designing the product so it "works the first time, every time" for the life of the product (decreasing cycle failure).

32. **Design for safety:** design so that manufacture of and the use or abuse of the product minimizes product liability.

33. **Design for service/repair:** considering during the design cycle how the product components and sub-assemblies will be easily accessed for service, repair and maintenance.

34. **CAD systems:** the computer-aided "drafting boards" that allow a user to define a new product by (a) creating images and (b) assigning geometric, mass, kinematics, material other properties to the product.

35. **Geometric tolerancing:** agreed-upon convention of symbols and terms used on engineering drawings to connote geometric characteristics and other dimensional requirements.

36. **Finite elements analysis:** mathematical model for analyzing the precise location where physical processes like stress and heat transfer occur.

37. **Design of experiments:** methodology for obtaining valid information during experimentation or testing.

38. **Value engineering:** A systematic approach to evaluating design alternatives that seeks to eliminate unnecessary features and functions and to achieve required functions at the lowest possible cost while optimizing manufacturability, quality, and delivery.

39. **Mechatronics (mechanisms and controls):** the precise actuation and control of mechanical devices through electronics.

**Table C.5** (Continued)

40. **Process improvement tools:** a set of tools and techniques, usually used by teams, that help to graphically organize, display and analyze data to determine root causes and continually improve work processes.

41. **Statistical process control:** statistical techniques used to monitor, control and improve process performance over time by studying variation and its source.

42. **Design standards (e.g., UL, ASME):** standards which describe how a required performance can be achieved by prescribing the physical or dimensional characteristics of a product or system and its manufacture or fabrication.

43. **Testing standards (e.g., ASTM):** the determination, by technical means, of the properties, performance, or elements of materials, products, services, systems, or environment, which may involve application of established scientific principles and procedures.

44. **Process standards (e.g., ISO 9000):** knowledge of and conformance to domestic and international business process quality standards.

45. **Product testing:** the understanding of test procedures to satisfy product specifications and customer requirements.

46. **Physical testing:** methodologies which test for the types of stress – including tensile, compressive, shearing, thermal, shock, vibration, fatigue, etc.

47. **Test equipment:** instrumentation, sensors and test tools, used for product performance testing.

48. **Application of statistics:** methodology for effectively designing test and analyzing test data using statistical techniques that are founded on probability theory.

49. **Reliability:** a sub-set of statistical engineering methodology that predicts performance of a product over its intended life cycle and understanding of the effects of various failure modes on system performance.

50. **Materials planning – inventory:** managing the handling and movement of materials in manufacturing.

51. **Total quality management:** a systematic, organization-wide approach to quality that stresses continually improving all processes that deliver products and services, with the major outcome of "delighting" the customer.

52. **Manufacturing processing:** processes that are used to create or further refine "work pieces" such as molding and casting, machining, extruding, stamping, forming, bonding, welding, coating, plating, painting, fabrication and assembly.

53. **Manufacturing floor/workcell layout:** setting up a dedicated area where electronic and/or mechanical component assembly – processes are connected together to produce a finish product.

54. **Robotics and automated assembly:** technology that allows assembly machines to accommodate more types of products, with quicker changeover, by incorporating the use of programmable robots or programmable automation devices.

55. **Computer integrated manufacturing:** total integration of automated design and manufacturing processes.

56. **Electro-mechanical packaging:** technology for putting together electronic and mechanical components into a "packaged form" so that they perform a required function.

**Table C.5** For use with industry follow-up survey (ASME/NSF product realization process survey).

1. **Teams/teamwork:** the ability to work with diverse, multi-discipline team members to successfully reach a goal or objective.

2. **Communication:** the ability to clearly and logically communicate ideas, information, and data orally and in written form to others in a way that engages the intended audience and addresses different learning styles.

3. **Design for manufacture:** designed to maximize ease of manufacture by simplifying the design through part-count reduction, developing modular designs, minimizing part variation, designing a part to be multi-functional, etc.

4. **CAD systems:** the computer-aided "drafting boards" that allow a user to define a new product by (a) creating images and (b) assigning geometric, mass, kinematics, material and other properties to the product.

5. **Professional ethics:** the ability to conform to standards of conduct determined by one's profession, in alignment with team and corporate standards.

6. **Creative thinking:** the process of generating ideas, that frequently emphasizes: (a) making and expressing meaningful new connections, (b) thinking of many new and unusual possibilities, and (c) extending and elaborating on alternatives.

7. **Design for performance:** designed to perform to product requirements under a wide variety of manufacturing and user operating conditions.

8. **Design for reliability:** designing the product so it "works the first time, every time" for the life of the product (decreasing cycle failure).

9. **Design for safety:** design so that manufacture of and the use or abuse of the product minimizes product liability.

10. **Concurrent engineering:** an approach to new product development where the product and all of its associated processes, such as manufacturing, distribution and service, are developed in parallel.

11. **Sketching/drawing:** the ability to clearly illustrate ideas and design by freehand sketching.

12. **Design for cost:** meeting customer requirements while minimizing cost of all aspects of the product, including production, assembly, distribution and maintenance.

13. **Application of statistics:** methodology of effectively designing test and analyzing test data using statistical techniques that are founded on probability theory.

14. **Reliability:** a sub-set of statistical engineering methodology that predicts performance of a product over its intended life cycle and understanding of the effects of various failure modes on system performance.

15. **Geometric tolerancing:** agreed-upon convention of symbols and terms used on engineering drawings to connote geometric characteristics and other dimensional requirements.

16. **Value engineering:** A systematic approach to evaluating design alternatives that seeks to eliminate unnecessary features and functions and to achieve required functions at the lowest possible cost while optimizing manufacturability, quality and delivery.

17. **Design reviews:** the scheduled-in checkpoints for assessing the design progress towards meeting product requirements and budget.

18. **Manufacturing processing:** processes that are used to create or further refine "work pieces" such as molding and casting, machining, extruding, stamping, forming, bonding, welding, coating, plating, painting, fabrication and assembly.

19. **Systems perspective:** the up-front identification of system components and their interactions for the purpose of optimizing the performance of the system as a whole.

20. **Design for assembly:** making the product easier to assemble, thereby reducing cycle time during production.

**Table C.6** Vulnerable zone distances for rates of release and level of concern screening – rural, F atmospheric stability, low wind speed (3.4 miles per hour); distances are given in miles for quantities of release up to 500 pounds/minute.

| Rate of Release (#/min) | QR 0.0001 | 0.0004 | 0.0007 | 0.001 | 0.002 | 0.0035 | 0.005 | 0.0075 | 0.01 | 0.02 | 0.035 | 0.05 | 0.075 | 0.1 | 0.25 | 0.5 | 0.75 | 1.0 | 2.0 | 5.0 | 10.0 |
|---|---|---|---|---|---|---|---|---|---|---|---|---|---|---|---|---|---|---|---|---|---|
| 1 | 9.0 | 2.5 | 1.7 | 1.3 | 0.9 | 0.6 | 0.5 | 0.4 | 0.3 | 0.2 | 0.2 | 0.1 | 0.1 | 0.1 | 0.1 | ** | ** | ** | ** | ** | ** |
| 2 | * | 4.5 | 2.8 | 2.1 | 1.3 | 0.9 | 0.8 | 0.6 | 0.5 | 0.3 | 0.3 | 0.2 | 0.2 | 0.1 | 0.1 | 0.1 | 0.1 | ** | ** | ** | ** |
| 3 | * | 6.7 | 3.9 | 2.9 | 1.7 | 1.2 | 1.0 | 0.8 | 0.6 | 0.4 | 0.3 | 0.3 | 0.2 | 0.2 | 0.1 | 0.1 | 0.1 | 0.1 | ** | ** | ** |
| 4 | * | 9.0 | 5.1 | 3.7 | 2.1 | 1.5 | 1.2 | 1.9 | 0.8 | 0.5 | 0.4 | 0.3 | 0.2 | 0.2 | 0.1 | 0.1 | 0.1 | 0.1 | ** | ** | ** |
| 5 | * | * | 6.3 | 4.5 | 2.5 | 1.7 | 1.3 | 1.0 | 0.9 | 0.6 | 0.4 | 0.3 | 0.3 | 0.2 | 0.1 | 0.1 | 0.1 | 0.1 | 0.1 | ** | ** |
| 8 | * | * | * | 7.1 | 3.7 | 2.4 | 1.8 | 1.4 | 1.2 | 0.8 | 0.5 | 0.4 | 0.4 | 0.3 | 0.2 | 0.1 | 0.1 | 0.1 | 0.1 | ** | ** |
| 10 | * | * | * | 9.0 | 4.5 | 2.8 | 2.1 | 1.6 | 1.3 | 0.9 | 0.6 | 0.5 | 0.4 | 0.3 | 0.2 | 0.1 | 0.1 | 0.1 | 0.1 | ** | ** |
| 15 | * | * | * | * | 6.7 | 3.9 | 2.9 | 2.1 | 1.7 | 1.1 | 0.8 | 0.6 | 0.5 | 0.4 | 0.3 | 0.2 | 0.1 | 0.1 | 0.1 | 0.1 | ** |
| 20 | * | * | * | * | 9.0 | 5.1 | 3.7 | 2.7 | 2.1 | 1.3 | 0.9 | 0.8 | 0.6 | 0.5 | 0.3 | 0.2 | 0.2 | 0.1 | 0.1 | 0.1 | ** |
| 25 | * | * | * | * | * | 6.3 | 4.5 | 3.2 | 2.5 | 1.5 | 1.1 | 0.9 | 0.7 | 0.6 | 0.3 | 0.2 | 0.2 | 0.2 | 0.1 | 0.1 | ** |
| 30 | * | * | * | * | * | 7.6 | 5.3 | 3.7 | 2.9 | 1.7 | 1.2 | 1.0 | 0.8 | 0.6 | 0.4 | 0.3 | 0.2 | 0.2 | 0.1 | 0.1 | 0.1 |
| 35 | * | * | * | * | * | 9.0 | 6.2 | 4.2 | 3.3 | 2.0 | 1.3 | 1.1 | 0.8 | 0.7 | 0.4 | 0.3 | 0.2 | 0.2 | 0.1 | 0.1 | 0.1 |
| 40 | * | * | * | * | * | * | 7.1 | 4.8 | 3.7 | 2.1 | 1.5 | 1.2 | 0.9 | 0.8 | 0.4 | 0.3 | 0.2 | 0.2 | 0.1 | 0.1 | 0.1 |
| 45 | * | * | * | * | * | * | 8.0 | 5.3 | 4.1 | 2.3 | 1.6 | 1.2 | 1.0 | 0.8 | 0.5 | 0.3 | 0.3 | 0.2 | 0.2 | 0.1 | 0.1 |
| 50 | * | * | * | * | * | * | 9.0 | 5.9 | 4.5 | 2.5 | 1.7 | 1.3 | 1.0 | 0.9 | 0.5 | 0.3 | 0.3 | 0.2 | 0.2 | 0.1 | 0.1 |
| 60 | * | * | * | * | * | * | * | 7.1 | 5.3 | 2.9 | 1.9 | 1.5 | 1.2 | 1.0 | 0.6 | 0.4 | 0.3 | 0.3 | 0.2 | 0.1 | 0.1 |
| 70 | * | * | * | * | * | * | * | 8.4 | 6.2 | 3.3 | 2.1 | 1.7 | 1.3 | 1.1 | 0.6 | 0.4 | 0.3 | 0.3 | 0.2 | 0.1 | 0.1 |
| 80 | * | * | * | * | * | * | * | 9.7 | 7.1 | 3.7 | 2.4 | 1.8 | 1.4 | 1.2 | 0.7 | 0.4 | 0.4 | 0.3 | 0.2 | 0.1 | 0.1 |
| 90 | * | * | * | * | * | * | * | * | 8.0 | 4.1 | 2.6 | 2.0 | 1.5 | 1.2 | 0.7 | 0.5 | 0.4 | 0.3 | 0.2 | 0.1 | 0.1 |

**Table C.6**  (Continued)

| | | | | | | | | 9.0 | 4.5 | 2.8 | 2.1 | 1.6 | 1.3 | 0.8 | 0.5 | 0.4 | 0.3 | 0.2 | 0.1 |
|---|---|---|---|---|---|---|---|---|---|---|---|---|---|---|---|---|---|---|---|
| 100 | * | * | * | * | * | * | * | 9.0 | 4.5 | 2.8 | 2.1 | 1.6 | 1.3 | 0.8 | 0.5 | 0.4 | 0.3 | 0.2 | 0.1 |
| 120 | * | * | * | * | * | * | * | * | 5.3 | 3.3 | 2.5 | 1.8 | 1.5 | 0.8 | 0.6 | 0.4 | 0.4 | 0.3 | 0.2 |
| 140 | * | * | * | * | * | * | * | * | 6.2 | 3.7 | 2.8 | 2.0 | 1.7 | 0.9 | 0.6 | 0.5 | 0.4 | 0.3 | 0.2 |
| 160 | * | * | * | * | * | * | * | * | 7.1 | 4.2 | 3.1 | 2.3 | 1.8 | 1.0 | 0.7 | 0.5 | 0.4 | 0.3 | 0.2 |
| 180 | * | * | * | * | * | * | * | * | 8.0 | 4.6 | 3.4 | 2.5 | 2.0 | 1.1 | 0.7 | 0.6 | 0.5 | 0.3 | 0.2 |
| 200 | * | * | * | * | * | * | * | * | 9.0 | 5.1 | 3.7 | 2.7 | 2.1 | 1.2 | 0.8 | 0.6 | 0.5 | 0.3 | 0.2 |
| 250 | * | * | * | * | * | * | * | * | * | 6.3 | 4.5 | 3.2 | 2.5 | 1.3 | 0.9 | 0.7 | 0.6 | 0.4 | 0.2 |
| 300 | * | * | * | * | * | * | * | * | * | 7.6 | 5.3 | 3.7 | 2.9 | 1.5 | 1.0 | 0.8 | 0.6 | 0.4 | 0.3 |
| 350 | * | * | * | * | * | * | * | * | * | 9.0 | 6.2 | 4.2 | 3.3 | 1.7 | 1.1 | 0.8 | 0.7 | 0.5 | 0.3 |
| 400 | * | * | * | * | * | * | * | * | * | * | 7.1 | 4.8 | 3.7 | 1.8 | 1.2 | 0.9 | 0.8 | 0.5 | 0.3 |
| 450 | * | * | * | * | * | * | * | * | * | * | 8.0 | 5.3 | 4.1 | 2.0 | 1.2 | 1.0 | 0.8 | 0.5 | 0.3 |
| 500 | * | * | * | * | * | * | * | * | * | * | 9.0 | 5.9 | 4.5 | 2.1 | 1.3 | 1.0 | 0.9 | 0.6 | 0.3 |

To find distance: Find nearest LOC across top. Use the lower LOC value for in-between numbers. This is a conservative approach. Find nearest QR on left column. Read across and down to find distance in miles. Multiply miles by 1.6 to get kilometers.

Notes: * No distance estimated because method is not valid for distances greater than 10 miles.

** No distance estimated because method is not valid for distances greater than 0.1 miles.

# Appendix D

**Table D.1** Composition of weathered gasoline.

| Component Number | Chemical Formula | $M_{w,i}$ (g) | Initial Mass Fraction | Initial Mole Fraction |
|---|---|---|---|---|
| 1. Propane | $C_3H_8$ | 44.1 | 0.0000 | 0.0000 |
| 2. Isobutane | $C_4H_{10}$ | 58.1 | 0.0000 | 0.0000 |
| 3. *n*-Butane | $C_4H_{10}$ | 58.1 | 0.0000 | 0.0000 |
| 4. *trans*-2-Butene | $C_4H_8$ | 56.1 | 0.0000 | 0.0000 |
| 5. *cis*-2-Butene | $C_4H_8$ | 56.1 | 0.0000 | 0.0000 |
| 6. 3-Methyl-1-butene | $C_5H_{10}$ | 70.1 | 0.0000 | 0.0000 |
| 7. Isopentane | $C_5H_{12}$ | 72.2 | 0.0200 | 0.0269 |
| 8. 1-Pentane | $C_5H_{10}$ | 70.1 | 0.0000 | 0.0000 |
| 9. 2-Methyl-1-butene | $C_5H_{10}$ | 70.1 | 0.0000 | 0.0000 |
| 10. 2-Methyl-1,3-butadiene | $C_5H_8$ | 68.1 | 0.0000 | 0.0000 |
| 11. *n*-Pentane | $C_5H_{12}$ | 72.2 | 0.0114 | 0.0169 |
| 12. *trans*-2-Pentene | $C_5H_{10}$ | 70.1 | 0.0000 | 0.0000 |
| 13. 2-Methyl-2-butene | $C_5H_{10}$ | 70.1 | 0.0000 | 0.0000 |
| 14. 3-Methyl-1,2-butadiene | $C_5H_8$ | 68.1 | 0.0000 | 0.0000 |
| 15. 3,3-Dimethyl-1-butene | $C_6H_{12}$ | 84.2 | 0.0000 | 0.0000 |
| 16. Cyclopentane | $C_5H_{10}$ | 70.1 | 0.0000 | 0.0000 |
| 17. 3-Methyl-1-pentene | $C_6H_{12}$ | 84.2 | 0.0000 | 0.0000 |
| 18. 2,3-Dimethylbutane | $C_6H_{14}$ | 86.2 | 0.0600 | 0.0744 |
| 19. 2-Methylpentane | $C_6H_{14}$ | 86.2 | 0.0000 | 0.0000 |
| 20. 3-Methylpentane | $C_6H_{14}$ | 86.2 | 0.0000 | 0.0000 |
| 21. *n*-Hexane | $C_6H_{14}$ | 86.2 | 0.0370 | 0.0459 |
| 22. Methylcyclopentane | $C_6H_{12}$ | 84.2 | 0.0000 | 0.0000 |
| 23. 2,2-Dimethylpetane | $C_7H_{16}$ | 100.2 | 0.0000 | 0.0000 |
| 24. Benzene | $C_6H_6$ | 78.1 | 0.0100 | 0.0137 |
| 25. Cyclohexane | $C_6H_{12}$ | 84.2 | 0.0000 | 0.0000 |

**Table D.1** (Continued)

| Component Number | Chemical Formula | $M_{w,i}$ (g) | Initial Mass Fraction | Initial Mole Fraction |
|---|---|---|---|---|
| 26. 2,3-Dimethylpentane | $C_7H_{16}$ | 100.2 | 0.1020 | 0.1088 |
| 27. 3-Methylhexane | $C_7H_{16}$ | 100.2 | 0.0000 | 0.0000 |
| 28. 3-Ethylpentane | $C_7H_{16}$ | 100.2 | 0.0000 | 0.0000 |
| 29. 2,2,4-Trimethylpentane | $C_8H_{18}$ | 114.2 | 0.0000 | 0.0000 |
| 30. *n*-Heptane | $C_7H_{16}$ | 100.2 | 0.0800 | 0.0853 |
| 31. Methylcyclohexane | $C_7H_{14}$ | 98.2 | 0.0000 | 0.0000 |
| 32. 2,2-Dimethylhexane | $C_7H_{18}$ | 114.2 | 0.0000 | 0.0000 |
| 33. Toluene | $C_7H_8$ | 92.1 | 0.1048 | 0.1216 |
| 34. 2,3,4-Trimethylpentane | $C_8H_{18}$ | 114.2 | 0.0000 | 0.0000 |
| 35. 2-Methylheptane | $C_8H_{18}$ | 114.2 | 0.0500 | 0.0468 |
| 36. 3-Methylheptane | $C_8H_{18}$ | 114.2 | 0.0000 | 0.0000 |
| 37. *n*-Octane | $C_8H_{18}$ | 114.2 | 0.0500 | 0.0468 |
| 38. 2,4,4-Trimethylhexane | $C_9H_{20}$ | 128.3 | 0.0000 | 0.0000 |
| 39. 2,2-Dimethylheptane | $C_9H_{20}$ | 128.3 | 0.0000 | 0.0000 |
| 40. *p*-Xylene | $C_8H_{10}$ | 106.2 | 0.1239 | 0.1247 |
| 41. *m*-Xylene | $C_9H_{10}$ | 106.2 | 0.0000 | 0.0000 |
| 42. 3,3,4-Trimethylhexane | $C_9H_{20}$ | 128.3 | 0.0250 | 0.0208 |
| 43. *o*-Xylene | $C_8H_{10}$ | 106.2 | 0.0000 | 0.0000 |
| 44. 2,2,4-Trimethylheptane | $C_{10}H_{22}$ | 142.3 | 0.0000 | 0.0000 |
| 45. 3,3,5-Trimethylheptane | $C_{10}H_{22}$ | 142.3 | 0.0250 | 0.0188 |
| 46. *n*-Propylbenzene | $C_9H_{12}$ | 120.2 | 0.0829 | 0.0737 |
| 47. 2,3,4-Trimethylheptane | $C_{10}H_{22}$ | 142.3 | 0.0000 | 0.0000 |
| 48. 1,3,5-Trimethylbenzene | $C_9H_{12}$ | 120.2 | 0.0250 | 0.0222 |
| 49. 1,2,4-Trimethylbenzene | $C_9H_{12}$ | 120.2 | 0.0250 | 0.0222 |
| 50. Methylpropylbenzene | $C_{10}H_{14}$ | 134.2 | 0.0373 | 0.0297 |
| 51. Dimethylbenzene | $C_{10}H_{14}$ | 134.2 | 0.0400 | 0.0319 |
| 52. 1,2,4,5-Tetramethylbenzene | $C_{10}H_{14}$ | 134.2 | 0.0400 | 0.0319 |
| 53. 1,2,3,4-Tetramethylbenzene | $C_{10}H_{14}$ | 134.2 | 0.0000 | 0.0000 |
| 54. 1,2,4-Trimethyl-5-ethylbenzene | $C_{11}H_{16}$ | 148.2 | 0.0000 | 0.0000 |
| 55. *n*-Dodecane | $C_{12}H_{26}$ | 170.3 | 0.0288 | 0.0181 |
| 56. Napthalene | $C_{10}H_8$ | 128.2 | 0.0100 | 0.0083 |
| 57. *n*-Hexylbenzene | $C_{12}H_{20}$ | 162.3 | 0.0119 | 0.0078 |
| 58. Methylnaphthalene | $C_{11}H_{10}$ | 142.2 | 0.0000 | 0.0000 |
| Total | | | 1.0000 | 1.0000 |

Used by permission from Johnson, P.C., Stanley, C.C., Kemblowski, M.W., and Colthart, J.D., "A practical approach to the design, operation, and a monitoring of in situ soil venting systems," *Ground Water Monitoring and Remediation*, 10(2), 159–178.

**Table D.2** Physical properties of regular gasoline constituents.

| Component Number | $P_i^v$ (20°C, atm) | $T_B$ (1 atm, °C) | $S_i$ (20°C, mg/l) | $k_{ow}$ |
|---|---|---|---|---|
| 1. Propane | 8.500 | −42 | 62 | 73 |
| 2. Isobutane | 2.930 | −12 | 49 | 537 |
| 3. *n*-Butane | 2.110 | −1 | 61 | 946 |
| 4. *Trans*-2-Butene | 1.970 | 1 | 430 | 204 |
| 5. *cis*-2-Butene | 1.790 | 4 | 430 | 204 |
| 6. 3-Methyl-1-butene | 0.960 | 21 | 130 | 708 |
| 7. Isopentene | 0.780 | 28 | 48 | 1,862 |
| 8. 1-Pentene | 0.700 | 30 | 148 | 710 |
| 9. 2-Methyl-1-butene | 0.670 | 31 | 155 | 525 |
| 10. 2-Methyl-1,3-butadiene | 0.650 | 34 | 642 | 323 |
| 11. *n*-Pentane | 0.570 | 36 | 40 | 2,511 |
| 12. *trans*-2-Pentene | 0.530 | 36 | 203 | 708 |
| 13. 2-Methyl-2-butene | 0.510 | 38 | 155 | 525 |
| 14. 3-Methyl-1,2-butadiene | 0.460 | 41 | 1,230 | 148 |
| 15. 3,3-Dimethyl-1-butane | 0.470 | 41 | 23 | 1,350 |
| 16. Cyclopentene | 0.350 | 50 | 158 | 871 |
| 17. 3-Methyl-1-pentene | 0.290 | 54 | 56 | 1,820 |
| 18. 2,3-Dimethylbutane | 0.260 | 57 | 20 | 4,786 |
| 19. 2-Methylpentane | 0.210 | 60 | 14 | 6,457 |
| 20. 3-Methylpentane | 0.200 | 64 | 13 | 6,457 |
| 21. *n*-Hexane | 0.160 | 69 | 13 | 8,710 |
| 22. Methylcyclopentane | 0.150 | 72 | 42 | 2,239 |
| 23. 2,2-Dimethylpentane | 0.110 | 79 | 4.4 | 16,600 |
| 24. Benzene | 0.100 | 80 | 1,780 | 135 |
| 25. Cyclohexane | 0.100 | 81 | 55 | 3,236 |
| 26. 2,3-Dimethylpentane | 0.072 | 90 | 5.3 | 16,600 |
| 27. 3-Methylhexane | 0.064 | 92 | 3.2 | 22,400 |
| 28. 3-Ethylpentane | 0.060 | 94 | 3.2 | 22,400 |
| 29. 2,2,4-Trimethylpentane | 0.051 | 99 | 2.2 | 42,660 |
| 30. *n*-Heptane | 0.046 | 98 | 3 | 30,000 |

**Table D.2** (Continued)

| Component Number | $P_i^v$ (20°C, atm) | $T_B$ (1 atm, °C) | $S_i$ (20°C, mg/l) | $k_{ow}$ |
|---|---|---|---|---|
| 31. Methylcyclohexane | 0.048 | 101 | 14 | 11,220 |
| 32. 2,2-Dimethylhexane | 0.035 | 107 | 1.5 | 57,544 |
| 33. Toluene | 0.029 | 111 | 515 | 490 |
| 34. 2,3,4-Trimethylpentane | 0.028 | 114 | 1.8 | 42,658 |
| 35. 2-Methylheptane | 0.021 | 116 | 0.9 | 77,625 |
| 36. 3-Methylheptane | 0.020 | 115 | 0.8 | 77,625 |
| 37. *n*-Octane | 0.014 | 126 | 0.7 | 104,700 |
| 38. 2,4,4-Trimethylhexane | 0.013 | 131 | 1.4 | 147,911 |
| 39. 2,2-Dimethylheptane | 0.011 | 133 | 0.3 | 199,526 |
| 40. *p*-Xylene | 0.0086 | 138 | 198 | 1,413 |
| 41. *m*-Xylene | 0.0080 | 139 | 162 | 1,585 |
| 42. 3,3,4-Trimethylhexane | 0.0073 | 140 | 1.4 | 147,911 |
| 43. *o*-Xylene | 0.0066 | 144 | 175 | 589 |
| 44. 2,2,4-Trimethylheptane | 0.0053 | 149 | 0.8 | 389,000 |
| 45. 3,5-Trimethylheptane | 0.0037 | 156 | 0.8 | 389,000 |
| 46. *n*-Propylbenzene | 0.0033 | 159 | 60 | 4,786 |
| 47. 2,3,4-Trimethylheptane | 0.0031 | 160 | 0.8 | 389,000 |
| 48. 1,3,5-Trimethylbenzene | 0.0024 | 165 | 73 | 12,883 |
| 49. 1,2,4-Trimethylbenzene | 0.0019 | 169 | 57 | 12,883 |
| 50. Methylpropylbenzene | 0.0010 | 182 | 6.8 | 33,884 |
| 51. Dimethylbenzene | 0.0007 | 190 | 21 | 44,668 |
| 52. 1,2,4,5-Tetramethylbenzene | 0.00046 | 196 | 3.5 | 12,883 |
| 53. 1,2,3,4-Tetramethylbenzene | 0.00033 | 205 | 21 | 12,883 |
| 54. 1,2,4-Trimethyl-5-ethylbenzene | 0.00029 | 210 | 7 | 204,000 |
| 55. *n*-Dodecane | 0.0004 | 216 | 0.004 | 1,537 |
| 56. Napthalene | 0.00014 | 218 | 33 | 1,738 |
| 57. *n*-Hexylbenzene | 0.00010 | 226 | 1.3 | 309,000 |
| 58. Methylnaphthalene | 0.000054 | 241 | 27 | 7,943 |

Used by permission from Johnson, P.C., Stanley, C.C., Kemblowski, M.W., and Colthart, J.D., "A practical approach to the design, operation, and a monitoring of in situ soil venting systems," *Ground Water Monitoring and Remediation*, 10(2), 159–178.

# Index